Transportation Management with SAP® TM

SAP PRESS

SAP PRESS is a joint initiative of SAP and Galileo Press. The know-how offered by SAP specialists combined with the expertise of the Galileo Press publishing house offers the reader expert books in the field. SAP PRESS features first-hand information and expert advice, and provides useful skills for professional decision-making.

SAP PRESS offers a variety of books on technical and business-related topics for the SAP user. For further information, please visit our website: *www.sap-press.com*.

Carter, Bauer, Lange, Dalm, and Persich
SAP Extended Warehouse Management: Processes, Functionality, and Configuration
2010, 847 pp., hardcover
ISBN 978-1-59229-304-9

Martin Murray
Discover Logistics with SAP: SAP ERP and SAP SCM (2nd Edition)
2014, 412 pp., paperback
ISBN 978-1-59229-926-3

Othmar Gau
Transportation Management with SAP LES
2008, 574 pp., hardcover
ISBN 978-1-59229-169-4

Jawad Akhtar
Production Planning and Control with SAP ERP
2013, 1033 pp., hardcover
ISBN 978-1-59229-868-6

Bernd Lauterbach, Dominik Metzger, Stefan Sauer,
Jens Kappauf, Jens Gottlieb, and Christopher Sürie

Transportation Management with SAP® TM

Galileo Press

Bonn • Boston

Galileo Press is named after the Italian physicist, mathematician, and philosopher Galileo Galilei (1564—1642). He is known as one of the founders of modern science and an advocate of our contemporary, heliocentric worldview. His words *Eppur si muove* (And yet it moves) have become legendary. The Galileo Press logo depicts Jupiter orbited by the four Galilean moons, which were discovered by Galileo in 1610.

Editor Emily Nicholls
Acquisitions Editor Katy Spencer
Copyeditor Ruth Saavedra
Cover Design Graham Geary
Photo Credit iStockphoto.com/2312011/© GBlakeley
Layout Design Vera Brauner
Production Graham Geary
Typesetting SatzPro, Krefeld (Germany)
Printed and bound in the United States of America, on paper from sustainable sources

ISBN 978-1-59229-904-1

1st edition 2014

Library of Congress Cataloging-in-Publication Data
Lauterbach, Bernd.
Transportation management with SAP TM / Bernd Lauterbach, Stefan Sauer,
Dominik Metzger, Jens Kappauf, Christopher Sürie, and Jens Gottlieb.
pages cm
Includes bibliographical references and index.
ISBN-13: 978-1-59229-904-1
ISBN-10: 1-59229-904-0
ISBN-13: 978-1-59229-906-5 (e-book : alk. paper)
ISBN-13: 978-1-59229-905-8 (e-book : alk. paper) 1. Materials handling--Data processing.
2. Shipment of goods--Data processing. 3. SAP transportation management. I. Title.
TS180.6.L38 2013
658.7'810285--dc23
2013037970

Contents at a Glance

Dear Reader,

Just as I was preparing to write this Dear Reader note, our editorial intern arrived in the office and recounted her transportation management headache from the night before. During her shift at a national bakery chain, the managerial team learned that the freight order scheduled for 6 a.m. delivery would be delayed. As the day wore on, the new ETA became 3 p.m. (right before the dinner rush) and then 5 p.m. (during the dinner service); the truck finally arrived at 9:30 p.m.

But the inconvenience of the truck's fifteen-hour delay exceeded the challenge of unloading it during peak dining hours. During service, the bakery ran out of four different cheeses and five different kinds of bread, which tangibly lowered customer satisfaction and heightened service team stress. The yogurt order, which had been damaged en route, became another casualty of the botched delivery. It wasn't until well after midnight that the team was able to deal with the freight to completion.

Transportation Management with SAP TM offers practical examples like this one to illustrate the best way to achieve smooth transportation execution—that is, to move your goods from point A to point B without becoming fodder for a water-cooler horror story. Its expert authors walk you through the steps to simplify, enhance, and monitor those crucial processes, giving you the tools to avoid snags with large-scale implications and tips for navigating possible obstacles at every stage.

We encourage you to visit our website at *www.sap-press.com* and share your feedback about *Transportation Management with SAP TM*. Because your comments and suggestions are the most useful tools to help us improve our books, we're very eager to hear how this book impacted your transportation processes and helped keep deliveries arriving on time.

Emily Nicholls
Editor, SAP PRESS

Galileo Press
Boston, MA

emily.nicholls@galileo-press.com
www.sap-press.com

Contents

Foreword

The transportation of goods is an age-old topic that has affected and connected people from different continents and regions since time immemorial. Today, modern man and his economy are wholly reliant on efficient transportation logistics. Even simple items that he requires on a daily basis are often procured from other continents.

Because the global energy situation is worsening and energy procurement is becoming increasingly cost-intensive—along with the fact that the number of services associated with transportation logistics is expected to rise rather than to fall—it is now more important than ever to organize and execute transportation logistics processes in an efficient, cost-sensitive, and energy-conscious manner.

SAP Transportation Management (SAP TM) is an important building block of well-organized and energy-efficient transportation logistics that are linked to business partners. As a result of its modern architecture, which is based on the Business Object Processing Framework, SAP TM provides consignors and logistics service providers with the ideal platform for a modern transportation processing structure. SAP TM is a comprehensive system offering a plenitude of options and functionality to model, configure, and implement transportation-relevant processes and operations.

This book was written in 2013, so its documented functionality and most of the illustrations and screenshots are based on the current release of SAP TM (version 9.0), which was available for unrestricted shipment. Where appropriate, it offers a sneak preview of latest SAP TM 9.1 functionality, launched in the final days of 2013.

To document its broad functionality and to support project teams with their implementation, SAP provides in-depth system documentation, solution manager content, release notes, and installation guides at a very granular level. These accompanying documents are publically available and include scenario descriptions and detailed explanations on how to deploy and configure SAP TM. In this book, we try to avoid repeating this information, but instead recommend resources for learning more.

Structure of the book

This book first provides the necessary background information on transportation and logistics in general and on the SAP Business Suite and SAP TM in particular. By introducing the individual modules and functional building blocks of SAP TM, the goal of this book is to provide a "big-picture" perspective of SAP TM and its components: how they work together and how this transportation system maps business requirements into software functionality.

To this end, we have divided this book into the following chapters, beginning with an introduction:

- Chapter 1: SAP Transportation Management Foundations
- Chapter 2: Solution Architecture and Technological Concepts
- Chapter 3: Master Data
- Chapter 4: Transportation Requirements and Order Management
- Chapter 5: Transportation Capacity Management
- Chapter 6: Transportation Planning
- Chapter 7: Carrier Selection and Subcontracting
- Chapter 8: Transportation Execution and Monitoring
- Chapter 9: Transportation Compliance
- Chapter 10: Transportation Charge Management
- Chapter 11: Charge Settlement
- Chapter 12: Integration with Other Components
- Chapter 13: Transportation Process for Shippers
- Chapter 14: Transportation Process for Less than Container Load in Ocean Freight
- Chapter 15: Transportation Process for Air Freight-Forwarding
- Chapter 16: Implementation Best Practices
- Chapter 17: Summary and Outlook

Marginal notes

Throughout this book, we use marginal text to help you locate topics of interest as you scan the text. Marginal notes appear beside paragraphs or sections with related information.

Transportation Management with SAP TM is not a book with the intention to document all necessary configuration steps of SAP TM. It can certainly not answer all evolving questions, but we hope to give you the right tool to ask the right questions and to better understand the essential functionality and issues involved.

Target audience

This book has been written for the following audiences:

- This book is targeted to everyone looking for a lucid, informed introduction to transportation management with SAP TM. Thus, each chapter describes in detail a specific functional area or business process, providing an overview of the underlying functionality and how it is used in practical business use. In this regard, we address SAP beginners and employees in departments where SAP TM is to be implemented, as well as students who wish to obtain an impression of the core processes in transportation management and their mapping in SAP software.
- We also write for ambitious users of SAP Business Suite who, in addition to relevant logistical processes, want a look at process integration and the up- and downstream functions, as well as their mapping in SAP TM.
- Last, we turn to management staff and IT decision-makers who are considering the implementation of SAP TM and wish to obtain an overview of functionality and building blocks.

Acknowledgments

Numerous colleagues and friends contributed to the successful completion of this book. It is therefore founded on their knowledge; they answered questions, provided tips, and were valuable discussion partners, and every one of them deserves a big thank you.

In particular, we would now like to extend our thanks to the following colleagues (in alphabetical order): Ananth Bhat, Steffen Brueckner, Kathrin Dyckerhoff, Tobias Eisel, Andreas Esau, Thomas Engelmann, Dragos Florescu, Stefan Förster, Torsten Geiger, Hanna Gradzka, Sabine Hamlescher, Martin Hentschel, Patrik Hornig, Holger Hüne, Dr. Hans-Jörg Kersten, Dr. Christian Knirsch, Kiran KS, Eric Osterroth, Jan Rumig, Stefan Sahm, Kai Seela, Peter Schilbe, Michael Sinibaldi, Akiko Sudo, Markus Urbanek, Peter Wadewitz, and Marcus Zahn.

Sincere thanks are also due to Emily Nicholls at SAP PRESS, who supported this book project all of the way, from concept to completion. Her effective and great collaboration was a valuable contribution to the realization of this book.

Above all, we would like to express our special thanks to our wives, partners, and families:

- Yumi Kawahara with Kai and Yuki
- Anna Camilla Robinson and Renate Metzger
- Isabella Mayer
- Susanne Kappauf with Leni and Anni
- Pia Penth
- Martina Bunte-Sürie with Vicky and Charlotte

For innumerable evenings and weekends, our wives and partners had to do without their husbands and partners, and our children without their fathers. They tolerated this with patience and still gave us the necessary support and confidence to finish this book.

Sincerely,
Bernd Lauterbach, **Dominik Metzger**, **Stefan Sauer**, **Jens Kappauf**, **Jens Gottlieb**, and **Christopher Sürie**

Transportation logistics usually represents the single most important element in logistics costs. Thus, the logistician needs to understand both logistics core operations and how the functionality of transportation management software helps them to stay competitive and keep their promises.

Introduction

The most imperative challenge facing many managers in recent years has been to regain lost market share and secure new competitive advantages. The impetus for this trend continues to be the ubiquitous tendency toward globalization and the ensuing intensification of international competition. *Customer orientation*, *lean management*, and *reengineering* are the buzzwords that characterize these efforts.

Many companies need to reorganize their value-added processes, whereby special attention should be paid to the interfaces between the increasingly important sales and procurement markets. Within this context, there is hardly a corporate function that has grown as much in significance in recent years as logistics. Treated until just a few years ago as an operational aid and an object of isolated rationalization efforts, logistics—especially in the age of supply chain management—is now considered an essential element of strategic corporate leadership.

Supply chain management

Over time, the profession of supply chain management has evolved to meet the changing needs of the global supply chain. According to the Council of Supply Chain Management Professionals (CSCMP):

> *Supply chain management encompasses the planning and management of all activities involved in sourcing and procurement...and all logistics management activities. Importantly, it also includes coordination and collaboration with channel partners, which can be suppliers, intermediaries, third-party service providers, and customers. In essence, supply chain*

management integrates supply and demand management within and across companies.

Therefore, supply chain management spans all movement and storage of raw materials, work-in-process inventory, and finished goods from point of origin to point of consumption. It can be considered as the operational component of logistics, including quantification, procurement, inventory management, transportation and fleet management, and data collection and reporting.

Further Reading

This book focuses on the software features and functions SAP Transportation Management (SAP TM) provides to support the configuration and implementation of integrated supply chain models. These models promote the collaboration and seamless linkages between modes of transport, transportation processes, and people responsible for managing transportation within the supply chain network.

Legal regulations, documentation guidelines and associated laws governing goods movement customs and restrictions, and internationally valid contracts are very complex. In addition to international and regional regulations, industry-specific documents also vary according to the mode of transport used.

To help you learn more about supply chain management, commercial and legal requirements, regulations, contractual relationships, global standards, and regional compliance rules, we have provided a list of references, links, and further reading at the end of this book.

Transportation management

Supply chain management includes logistics activities plus coordination and collaboration of staff, levels, and functions. In this context, *transportation* refers to the movement of products from one location to another as they make their way from the beginning of a supply chain to the customer's hands. So any supply chain's success is closely linked to the appropriate use of transportation. In this context, we have to differentiate between internal and external transport.

Internal and external transportation

Internal transport within an operation usually takes place from one production line to another or between different departments in a warehouse. In contrast to internal transportation, *external transport* usually involves a shipment from the supplier to the customer, between various factories, or between a company's warehouses. Therefore, external transport is based on a contractual relationship between business partners and consists of the cargo, the means of transport, and the transportation process.

In this introduction we consider transportation from the perspectives of external transportation—that is, from the viewpoint of both logistics service providers and carriers and from the viewpoint of consignors. We start with a brief introduction to transportation and logistics, introduce the protagonists and areas of responsibility, and give you a short overview of the different transportation modes and processes.

Transportation and Logistics

Transportation management is part of the supply chain process. It plans, implements, and controls the efficient, effective forward and reverse flow and storage of goods, services, and related information between the point of origin and the point of consumption to meet customers' requirements.

The role of transportation

Transportation execution plays a key role in every supply chain because products are rarely produced and consumed in the same location. Together with an increasing trend to outsource services and relocate production facilities, transportation services have become increasingly commoditized but still offer great optimization potential. In today's business environment, which is dominated by globalization, increased price competitiveness, and more demanding operational and financial performance, the pressure on supply chain management is increasing. Within the context of overall networks and logistics processes, the operation of transportation determines the efficiency of moving products and plays a crucial part in the manipulation of logistics to meet customer expectations. One example is the *next-day shipment* option, which has almost become a de facto standard and benchmark, not only for express couriers.

While the progress in techniques and management principles improves the movement of cargo, delivery speed, service quality, operation costs, use of facilities, and energy efficiency, the stakes are high and the challenges enormous.

The value of transportation

Whether we talk about internal or external processes, logistics is about creating value for customers, suppliers, and the firm's stakeholders. Value in logistics is primarily expressed in terms of *time* and *place*. That is, products and services have no value unless they are in the possession of the customers at the right time and at the right place.

Transportation refers to the movement of products or cargo from one location to another. From an execution perspective, transportation and inventory maintenance are the primary cost-absorbing logistics activities. Between one and two thirds of an enterprise's logistics costs are spent on transportation. Therefore, the operational significance of logistics for many companies still lies in its rationalization potential. In general, a reduction of logistics costs should improve corporate success by achieving a competitive advantage.

Transportation and inventory

Transportation is essential because no modern firm can operate without providing for the movement of its raw materials or its finished products. In the past, transportation may have been an isolated business process, but alongside warehouse management, it is now increasingly considered within the context of overall networks and logistics processes.

Inventories are also essential to logistics management because it is usually not possible or practical to provide instant production or ensure delivery times to customers. Inventories serve as buffers between supply and demand so that needed product availability can be maintained for customers while providing flexibility for production and logistics in seeking efficient methods for product manufacturing and distribution, as shown in Figure 1.

Commoditization of transportation services

In recent years, the transportation industry has become increasingly commoditized. Both a relatively easy time-to-market for new providers and increasing market globalization and liberalization have made it possible for new competitors to continuously enter the market, which has lowered prices for relatively simple transportation services. At the same time, there is an increase in the demand for transportation services as a result of globalization and the relocation of production to low-cost countries. Transportation resources—in other words, the means of transport—are becoming increasingly scarce, while the mode of transport is becoming more and more significant. Transportation management is never an isolated process.

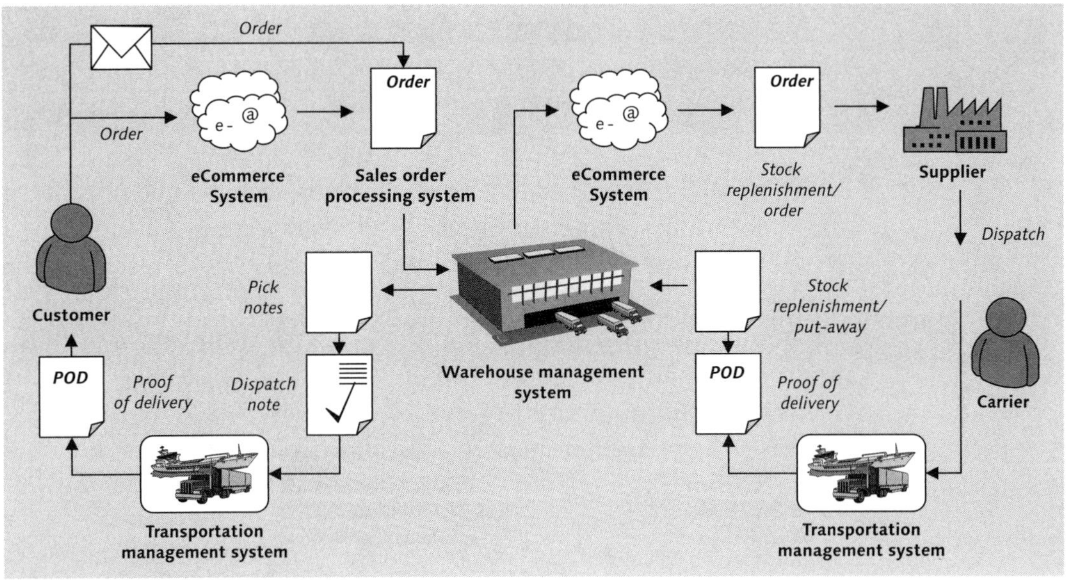

Figure 1 Supply Chain Execution

Protagonists and Roles

Many manufacturers and retailers use software solutions to manage their supply chain and reduce inventory and storage costs, with the possibility of quick delivery to the client. In this context, any successful supply chain is not only linked to the use of adequate transportation modes or transportation network design, but relies on the relationship and seamless collaboration of all parties involved. In general, there are two key players in any transportation that takes place within the supply chain:

- In most cases, the *consignor* or *shipper* is the party that requires the movement of the product between two points in the supply chain. In special situations, the *consignee* can act as initiating party.
- The *logistics service provider* (LSP) is the party that organizes transports of products or cargo.

Note

Throughout this book, we use the terms *logistics* and *supply chain* interchangeably, as well as *consignor* and *shipper*.

When making transportation-related decisions, factors to be considered vary depending on whether one takes the perspective of a carrier or a shipper. A carrier typically makes investment decisions regarding the available transportation infrastructure, while making operating decisions to maximize the return from these assets. A shipper, in contrast, uses transportation or outsourced services to minimize the total cost while providing an appropriate level of responsiveness to his customers. Figure 2 shows the business relationship between those involved in transportation.

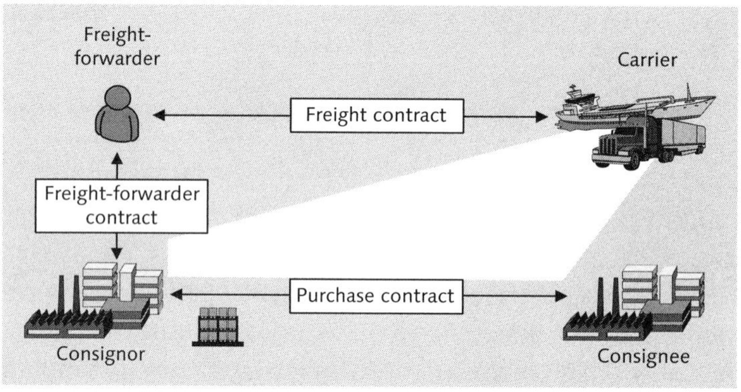

Figure 2 Contractual Relationships

Contractual relationships

A sales or purchasing order documents the sale of goods transaction between the shipper and consignee. It generally also contains attributes that regulate the transportation of goods. These attributes are usually the *incoterms* between the parties to the contract. These terms regulate the transfer of risk and the allocation of the goods to be conveyed. Incoterms are issued by the International Chamber of Commerce (ICC) and are published on its website.

In SAP systems, the order document is represented by the sales order on the consignor side and the purchase order on the consignee side.

Depending on how the incoterms are structured, the consignor or the consignee (or both, if necessary) commissions the transportation of goods.

Subcontracting

The special feature in this contractual relationship is *subcontracting*—contracting out transportation services or handing such services over to subcontractors. If goods are lost or damaged, the consignee or consignor cannot contact the carrier directly because they generally do not have a direct contractual relationship with the carrier. In this case, the contact person is the freight-forwarder who entered into a contract with the carrier. The carrier acts only on behalf of the 3PL—which is short for the third-party logistics personnel that physically transports the goods to the final consignee. In SAP TM, the contractual relationships can be modeled with a freight agreement (please refer to Chapter 10).

Let's briefly introduce these two roles and the factors that affect transportation decisions for different members of the supply chain.

Shippers

Shippers are the major players in a transportation network. They generate the transportation demand for LSPs, carriers (who supply the transportation services for moving the demand), and the network itself (typically composed of multi-modal services and terminals). The interactions of these players and their individual behavior, expectations, and often conflicting requirements affect their transportation decisions.

Shipper decisions

Transportation decisions made by shippers not only include the design of the transportation network and the choice of the adequate means of transport, but also consider the goal of minimizing the total cost of fulfilling a customer order while achieving the promised or even guaranteed responsiveness. Therefore, their decisions are usually based on a trade-off between different cost elements and their customers' expectations, as well as the margins generated from different products and customers.

SAP TM for shippers

Shippers from manufacturing and trading industries usually have a high demand for transportation services. In many cases, their requirements call for a solution that not only documents the transportation needs but executes its process steps such as centralized consolidation for multiple business units, optimized planning, carrier selection, and detailed freight rating. Please refer to Chapter 13 to learn more about the transportation process for shippers.

SAP TM offers good support for many industry-specific aspects, especially in consumer products, retail, chemicals, mills, and the high-tech industries. Future releases of SAP TM will enhance coverage of additional industries.

Logistics Service Providers

To survive in the face of the competitive pressures that arise from the increasing globalization of corporations, organizations today need to perform their activities in the most effective and cost-efficient manner possible. The availability of transportation capacity is critical to efficient and effective logistical operations. When this capacity becomes constrained, it is difficult for shippers to move products to customers in a cost-effective and timely manner. Thus, shippers seek greater reliability at a lower total cost.

Logistics outsourcing

The Council of Supply Chain Management Professionals defines logistics as

> *the process of planning, implementing, and controlling procedures for the efficient and effective transportation and storage of goods including services, and related information from the point of origin to the point of consumption for the purpose of conforming to customer requirements. This definition includes inbound, outbound, internal, and external movements.*

Logistics outsourcing thus refers to the procurement of any of these activities from an LSP. In general, the scope of a logistics outsourcing relationship can be limited to a particular set of tasks such as *transportation* and *storage*.

Role of the LSP

The role of LSPs in transportation management often goes beyond transporting things from one point to another. Today, successful logistics also requires comprehensive services, with a high level of process visibility and flexibility. SAP TM helps LSPs manage both the flow of goods and relationships between partners. Although LSPs' main business is transportation processing, we would like to distinguish between *freight-forwarders* and *carriers*.

Freight-forwarders

Like *freight brokers* and freight-forwarders, 3PLs organize and arrange the transportation of goods, whereas carriers physically transport goods. A freight-forwarder is a 3PL who acts on behalf of importers, exporters, or other companies or people to organize the safe, efficient, and cost-effective transportation of goods. They match shippers' freight with the optimal carrier and arrange the best means of transport, taking into account the type of goods and customer delivery requirements.

SAP TM for freight-forwarders

As just mentioned, freight-forwarders offer logistics and transportation services to their customers. In many cases, freight-forwarders do not execute transportation services themselves, but instead subcontract these services to carriers. The organization of transportation is their core business, and they need a powerful and flexible solution that can cover all aspects of transportation including quote and contract management; order management, execution, and subcontracting; capacity and booking management; price, cost, and profitability calculation; and customer and vendor settlement.

SAP TM combines these features with the ability to cover all modes of transport and intermodal transportation from door to door. In addition to transportation services, freight-forwarders offer a variety of value-added services such as trade compliance management and warehousing.

Contract logistics

In addition to operating as organizers of transportation or storage, 3PL enterprises increasingly also provide packaging, order management, customs clearance, and other value-added services. However, the main business of freight-forwarders is to organize the stream of goods between their customers and their customers' customers (also known as consignees).

SAP TM for contract logistics

Contract LSPs (also 3PLs and 4PLs) manage the complete logistics business for their customers. This means that they deal not only with transportation, but also with warehousing, value-added services, and often even with order management and invoicing in the name of their customers. Transactional coverage of contract logistics processes usually requires using SAP ERP as a sales order management system with integration into SAP Extended Warehouse Management (EWM) and SAP TM. It is therefore a kind of amalgamation of a shipper and a freight-forwarder's implementation.

Carrier

Many 3PLs have their own fleet and operate on certain routes or for certain customers or segments. At the same time, they organize and render transportation services. Such enterprises are known as *forwarders acting as carriers*. In air and sea cargo, carriers are generally multinational corporate groups such as large shipping companies and air cargo subsidiaries of airlines. The overland transportation market is shaped by medium-sized and micro enterprises that operate small fleets.

A carrier's goal is to make investment decisions and set operating policies that maximize the return on its assets. A carrier such as an airline, container shipping line, railroad, or trucking company must account for the following costs when investing in assets or setting pricing and operating policies:

- Fixed vehicle associated with operating resources
- Fixed operating cost related to infrastructure
- Trip-related costs, independent of the transported quantity

SAP TM for carriers

From a high-level view, carriers have similar business requirements as freight-forwarders. They, too, sell transportation services to their customers, but usually they also have the obligation to execute a part of the end-to-end transport chain with their own assets and transportation resources. Carriers are heavily dependent on their own network capacity and operating efficiency. SAP TM supports carriers' business with an integration of asset and equipment management and resource planning capabilities. Many carriers' specific processes (e.g., from the railway industry of container shipping lines) can be supported with SAP TM.

All enterprises operating as carriers have one thing in common: They try to generate a profit by making optimal use of the operating resources they deploy. By using these resources optimally, the goal is to minimize the high fixed costs associated with operating resources such as trucks, ships, and aircraft and the variable costs that arise as a result of deploying them.

Modes of Transport

During both prosperous and challenging economic times, most warehouse and logistics managers come under relentless pressure to achieve cost savings while improving supply chain performance. With the ever-increasing demands of customers, suppliers, and end users, coupled with the additional pressure of constant legislative changes, selecting the right mode of transportation can be crucial to meeting these challenges, decreasing transportation costs, and staying competitive.

Transportation costs

Supply chains use a combination of different modes of transportation; all of these differ in performance and characteristics. In this section, we consider the various transportation processes from the perspective of the modes of transport available, as well as in terms of cost and pricing structure.

Figure 3 shows the principle of the transport costs with different transport modes. Air freight is generally much more expensive than both types of land transport shown, but the storage cost might be lower. Thus, in terms of total cost, for the supply chain logistics operator, air freight might be the most reasonable transport mode for a particular transport purpose (e.g., transport of manufactured goods with high value and small volume or particularly fresh products with relatively high value per unit).

Load and shipment

In the following chapters, we repeatedly discuss shipments and loads. In this book, these terms are defined as follows:

- A *shipment* is a delivery from a point of departure to a point of destination. A *shipment* can take place over several legs.
- A *load* consolidates shipments from several consignors or shipments to several consignees. In some cases, a load may consist of only one shipment. In this case, the load is a *direct load*.

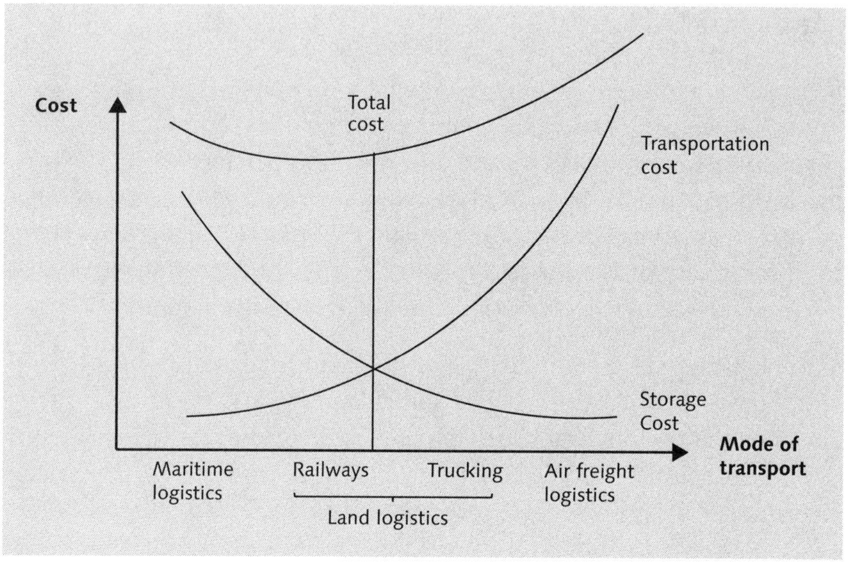

Figure 3 Transport Modes and Costs

Maritime Logistics

Water transport, by its nature, is limited to certain areas and is ideally suited for carrying very large loads at low cost. The maritime industry plays an important role in international freight and is organized either by the shipping companies themselves, by their agents, or by freight-forwarders acting as non-vessel operating common carriers (NVOCCs).

Cost considerations

Water transport can provide a cheap and high capacity-carrying conveyance for consumers, giving it a vital position in the transportation of particular goods, such as crude oil, grains, or large quantities of containers. Its disadvantage is that it needs a longer transport time and that its schedule is affected by the weather. To save costs and enhance competitiveness, current maritime logistics firms tend to use large-scale vessels, alliances with other shipping lines, and cooperative operation techniques. Moreover, current maritime customers care about service quality more than delivery price. Thus, it is necessary to build new logistics concepts to increase service satisfaction (e.g., real-time information, accurate time windows, and cargo tracking systems). The operation of the maritime transport industry can be divided into three main types:

Maritime shipping types

- In *container liner shipping*, the business is based on the movement of standardized containers on vessels that move regularly on voyages calling a sequence of ports. The unit of measurement in container cargo is the *TEU* (short for twenty-foot-equivalent-unit), which designates the space consumed by a 20-foot container.
- The characteristics of *tramp shipping* are individual transport routes and schedules with irregular transport pricing. It usually is used to move particular goods, such as dry bulk cargo and crude oil.
- *Industry shipping* defines special movements including heavy load project cargo (e.g., for construction of a refinery or a plant relocation), vessel charters, and oil and gas equipment movements.

Containerized cargo, FCL, and LCL

Containerized cargo is utilized to move a large percentage of world's general cargo. Via ocean transport, cargo is handled in two different ways:

- From a container liner perspective, cargo is always kept inside a container. The liner simply looks at full containers of different types, such as standard containers, refrigerated containers (reefers), or flat rack containers for large cargo items (called *out-of-gauge*).
- From a freight-forwarding perspective, the service provider distinguishes between *full container loads* (FCLs), where a customer orders a complete container for shipment, and *less than container loads* (LCLs) where the shipper's cargo is consolidated with cargo of other shippers in a container. In the LCL case, the forwarder pays for the full container and sells parts of its capacity to his customers to achieve maximum profitability. Chapter 14 provides more details about the transportation process of LCL ocean freight.

FCL and LCL combinations

For the mixed case of FCL/LCL, a shipper sends a shipment as a full container, and a freight-forwarder breaks it down into partial shipments if, for example, the load has to go to several consignees. For LCL/FCL, the shipment is delivered as a partial shipment and then consolidated by the freight-forwarder into a full container.

Air Freight Logistics

Air freight logistics is very important for different industries and service organizations to complete the supply chain functions and to keep their promises. The advantages it offers include quick delivery, reduced risk of damage, improved security, and flexibility, especially for time-critical parts and components. The disadvantage is higher freight prices. Therefore, air freight is typically selected when the value per unit is relatively high and the speed of delivery is important. In most cases, this transportation chain is *multi-modal* because air cargo transportation always requires pre-carriage to the airport and usually on-carriage transportation from the destination airport to the freight-forwarder or customer.

Schedule and charter

In air cargo, a distinction is made between scheduled flights and charter flights. Within scheduled flights, a further distinction is made between passenger planes that can also transport cargo (PAX) and cargo-only aircrafts (CAOs). Generally, cargo is loaded onto these planes to better utilize their capacity. Cargo and plane choice plays a role in both general planning and dispatching because particularly large and bulky goods that are not suitable for lower-deck storage in passenger planes must be transported on cargo planes. The difference between PAX and CAO is also important when transporting hazardous goods; some hazardous goods can only be transported on cargo planes.

IATA

In air cargo, numerous special rates and surcharges apply to standard cargo. These are defined by the International Air Transport Association (IATA), which is the umbrella organization for airlines. Its tasks include the standardization of air cargo processes, settlement and clearing between airlines and agencies (freight-forwarders and travel agents), the standardization of airport codes, and the production of anonymous statistics.

Agency business

The agency business is another special feature of air cargo. Airlines use freight-forwarders or agents to accept and process freight orders on their behalf. They pay an agency fee, which is an additional percentage charged on top of the freight rate. Agents are then responsible for processing the shipment, delivering it to the airport, and documenting it.

Please refer to Chapter 15 to learn more about air freight transportation.

Land Logistics

Land logistics is a very important link in logistics activities. It extends the delivery services for air and maritime transport from airports and seaports. The most positive characteristic of land logistics is the high accessibility level in land areas. The main transport modes of land logistics are railways and trucks.

Railways

Railway transport has advantages such as high carrying capacity, lower influence of weather conditions, and lower energy consumption. Disadvantages include the high cost of essential facilities, difficult and expensive maintenance, low flexibility to react to urgent demands, and time consumption in organizing railway carriages.

Rail carriers incur a high fixed cost in terms of rails, locomotives, railcars, and yards. There is also a significant trip-related labor and fuel cost that is relatively independent of the number of cars, but remains constant for the distance traveled and the time taken. Rail transport is priced to encourage large shipments over long distances. Prices display economies of scale in the quantity shipped and the distance traveled. The price structure and the heavy load capability make rail an ideal mode for carrying large, heavy, and high-density products over long distances—meaning that it's a good fit for low-value shipments that are not very time-sensitive.

Road transportation

In most countries, trucks are the dominant mode of land-based freight transportation. Road freight transport advantages include lower investment costs and high accessibility, mobility, and availability. Its disadvantages are low capacity, lower safety, and slow speed.

As is the case with other modes of transport, it may make sense in road transportation to bundle certain shipment sizes together (depending on size) for parts of the shipment leg to maximize fleet utilization and to better distribute the costs. Therefore, the trucking industry can be divided into two major segments: *less than truck load* (LTL), and *full truck load* (FTL).

LTL operations

LTL operations charge based on quantity and travel distance, typically priced to encourage shipments in small lots. LTL shipments take longer than TL shipments because other loads need to be picked up and dropped off. A key to reducing LTL costs is the degree of consolidation that carriers can achieve for the loads carried. LTL carriers use consolidation centers where trucks bring in small loads originating from one geographical area and leave with many small loads destined for the same geographical area. The goal is to minimize costs through consolidation without hurting delivery time and reliability. The key issue—and the challenge for each transportation system—is the assignment of loads to trucks and scheduling and routing of pickup and delivery. To learn more about freight scheduling, please refer to Chapter 6.

FTL operations

FTL operations charge by the full truck, independent of the quantity shipped. Rates vary with the distance traveled. Therefore, FTL operations have relatively low fixed costs, and owning a few trucks is often sufficient to enter the business. As a result, the market is very fragmented and is regionally dominated by multiple players. Because idle time and travel distance between successive loads add to cost in the truck load industry, carriers try to schedule shipments to meet service requirements while minimizing their trucks' idle and empty travel time.

FTL pricing displays economies of scale with respect to the distance traveled and the type and size of the trailer. FTL shipping is suited for transportation between manufacturing facilities and warehouses, between suppliers and manufacturers, or simply to cover the last mile of multimodal transportation.

Multimodal Transportation

Multimodal transport refers to a combination of at least two modes of transport. This results in an integrated transport chain that takes advantage of the strength of each alternative. The main characteristics of multi-modal transportation are *transshipment* terminals that allow efficient cargo handling between short-distance and long-distance traffic as well as application of standardized and reusable loading units.

But you can organize combined freight transport in different ways. In general, trucks cover short distances between the loading or unloading locations and the transshipment point. Long-distance haulage is often done by trucks as FTL or groupage system services in national or continental transportation. However, in international or transcontinental movements, long-distance haulage is mostly conducted by other means of transport such as train, vessel, or plane. Depending on the shipment attribute, it may make sense to move the main carriage from one means of transport to another. There are several ways of doing this.

Combined container transport

In combined container transport, standardized loading units are transshipped along different means of transport, applying various combinations of land, water, and air transportation. Trailer shipment refers to a combination of rail and road haulage. In this context, the *rolling road* usually describes the carriage of whole trucks (including both tractor and trailer) on low floor trains. The second alternative that contributes to additional cost savings is forwarding trailers without the tractor. However, this option requires a second tractor at the place of arrival. *Swap body transport* is basically similar to container transport, where loading units are handled by overhead cranes at a transshipment location. *Roll-on roll-off* (RoRo) traffic refers to the carriage of freight vehicles on ships over a certain distance.

Flexibility of multi-modal shipments

These multi-modal modes of transport combine the flexibility of trucks with the economies of scale of long-distance forwarding means of transport. However, additional handling processes are cost- and time-consuming. Ideally, the benefits of utilizing different means of transport outweigh the expenses. This mainly depends on the distance to be covered, the efficiency of the transshipment locations, and the goods to be conveyed.

Transportation Processes

If you consider value chains from start to finish, you'll notice that they actually no longer concern chains, but networks. Likewise, processes no longer run sequentially; instead, some processes are run in parallel, are increasingly subcontracted to third parties, and are divided into sub-processes. The number of participants in business processes is on the rise, and the coordination and reconciliation efforts are much greater than they were for horizontal value chains processed within an enterprise.

Network design

The design of a transportation network impacts the performance of a supply chain by establishing the infrastructure within which operational transportation (scheduling and routing) decisions are made (please refer to Chapter 3, Section 3.2). A well-designed transportation network allows a supply chain to achieve the desired degree of responsiveness at a low cost, managing both shipments and loads; the delivery process describes how the transportation network has been modeled. It can be direct, together with other deliveries, or it can concern a partial load. We now briefly introduce the typical transportation options and outline the main differences between the various modes of transport.

Direct Shipments

In a direct shipping network, full loads travel directly from the consignor to the consignee, the routing of each shipment is specified, and the supply chain manager only needs to decide on the quantity to ship and the mode of transportation to use. This decision typically involves a trade-off between transportation and inventory costs.

Cost considerations and trade-offs

The major advantage of a direct shipment transportation network is the elimination of intermediate warehouses and its simplicity of operation and coordination. Shipment decisions are completely local, and the decision made for one shipment does not necessarily influence others. At the same time, the transportation time from consignor to consignee is short because each shipment goes direct. They are either large shipments that fully utilize the means of transport, or goods that urgently need to be transported. This process is justified if the quantities requested by the consignee (e.g., replenishment lot sizes) are close to a transportation load and the available space is used optimally. For smaller quantities, however, direct transportation tends to be too costly. Costs are dependent on the type of carrier and the mode of transport (e.g., whether an FTL or LTL is used).

Milk runs and direct shipment

A *milk run* is a route in which a means of transport (typically a truck) either delivers products from a single consignor to multiple consignees or goes from multiple consignors to a single consignee. In direct shipping with milk runs, a supplier delivers directly to multiple consignees on a truck, or a truck picks up deliveries from many shippers destined for the same consignee. This method can be used to avoid LTL shipping by consolidating shipments to multiple consignees on a single truck, resulting in better utilization of the truck and lower costs.

Partial Loads

The 3PL can also optimize costs by combining loads—in other words, by transporting shipments from several consignors together but calculating and settling each shipment as a single shipment, thus increasing profit. If 3PLs act as carriers, they must be aware of the costs associated with providing their service so that they can profit from their enterprise.

With this option, consignors do not send shipments directly to consignees. For a partial load, a means of transport is used to deliver loads from several consignors to one consignee or loads from one consignor to several consignees. This process is usually divided by geographical region or mode of transport, making use of local distribution centers.

Distribution centers

Consignors send their shipments to the distribution center, from which it is forwarded to the consignee. In this context, the distribution center is an extra layer between shipper and consignee and is typically used to store inventory or serve as a transshipment location for multi-modal transports.

Inventory hubs

Inventory hubs allow shippers to consolidate shipments and to reduce supply chain costs when consignors are located far away from the consignees and transportation costs are high. From a carrier's perspective, a distribution center enables a supply chain to achieve economies of scale for inbound transportation to a point close to the final destination because each consignor sends a large shipment to the hub containing product for all consignees in the region covered by the hub. In addition, especially to replenish retail stores, the hub can hold inventory and send products to the retail stores in smaller replenishment lots. If the replenishment lots for the stores served by the hub are large enough to achieve economies of scale, the hub doesn't need to hold inventory. In this case, especially for products with a large, predictable demand, the hubs typically *use cross-docking* to reduce inventory and to move products into the supply chain faster.

Collective load

Small shipments are transshipped via a hub. Hubs are the central point of a star-shaped geographical area of short-haul routes, also described as a *hub-and-spoke system*. Shipments are consolidated at a central location and then assigned long-haul routes.

Linking short- and long-haul transportation

Collecting the shipments is described as *short-haul transportation*, and the connection between two hubs is described as *long-haul transportation*. In the sending hub, the incoming shipments are deconsolidated from short-haul transportation and scheduled for the corresponding destination routes. The destination route is also simply known as a route, and it has a corresponding goods issue area in the transshipment depot. The same applies to short-haul transportation, which also has corresponding goods issue areas in the transshipment depot. In the receiving hub, short-haul transportation is also known as rolling-out or roll-loading. Here, the freight list known as the *cartage note* confirms that the consignee has accepted the goods.

Overnight freight

The long-haul leg of the transportation process is often executed overnight and is therefore known as overnight freight, especially in the case of 24-hour transportation service where the customer has been promised transit times of less than 24 hours. During the day, shipments are collected until a cutoff time and consolidated in a hub. At night, the shipments are transported from hub to hub and then delivered by the next morning or early afternoon.

Line haul and ad hoc transportation

In collective loads, a distinction is made between line haul and ad hoc transportation. Similar to rail transportation, *line haul* or deadline transportation takes place at fixed times, regardless of whether it is being utilized. Ad hoc transportation takes place only if certain utilization limits have been reached. Therefore, it is generally not possible to guarantee transit times. Instead, transit periods are specified. Line haul exists only on the long-haul leg of the transportation process. In short-haul transportation, the goods are collected according to the orders.

Transportation currently constitutes a core process in virtually all enterprises. In this regard, standardization and integration with related processes are key factors in creating transparency and efficiency. In this introduction, we briefly introduced the core concepts and elements of transportation and logistics to lay the foundation for more advanced concepts.

Chapter 1 gives you an overview on the foundations of SAP Transportation Management as part of the SAP Business Suite. You will learn how SAP NetWeaver provides the infrastructure to run business processes and how SAP TM is used and integrated as part of the SAP Business Suite to handle offering, selling, assigning, planning, subcontracting, steering, and settling transportation services.

In this chapter, we present the foundations of SAP Transportation Management. First we introduce you to the basic principles of SAP NetWeaver and the SAP Business Suite, and then explain how SAP TM functionally covers the various aspects of transportation management.

1 SAP Transportation Management Foundations

SAP Transportation Management (TM) is part of the SAP Business Suite and consists of business processes that concern the offering, selling, assignment, planning, subcontracting, steering, and settlement of transportation services. To familiarize you with the core concepts of the system and the role of SAP TM as part of the SAP Supply Chain Management (SCM) platform, we divide this chapter into the following sections:

- Section 1.1 explains how SAP systems work and how SAP NetWeaver enables the business processes executed in the SAP Business Suite.
- Section 1.2 gives an overview of how SAP TM fits into the SAP Business Suite and covers the various components that interact with SAP Transportation Management.
- Section 1.3 introduces important SAP TM principles and business objects, and explains on a high level how SAP TM works before giving you a preview of the content of the following chapters.

1.1 SAP Business Suite and SAP NetWeaver

What is the SAP Business Suite?

The SAP Business Suite is a comprehensive and fully integrated family of business software applications that allows large and small organizations to do transactional planning, execution, and documentation of end-to-end business processes. Safe and flawless process execution makes it possible to achieve cost savings and ensure a seamless audit trail. The built-in configurability and flexibility of the SAP Business Suite enables easy adoption of business processes and development of new business portfolios.

Built on the SAP NetWeaver platform, the SAP Business Suite applications support the best practices of various industries. The integrated components for financials, controlling, production, procurement, marketing, sales, service, supply chain management, and risk and compliance management interact in a powerful way to facilitate transportation management processes in the manufacturing and shipper-focused industries (e.g., consumer products, retail, mill, and chemicals) as well as in the logistics services industry (e.g., freight-forwarding and carriers). Logistics service providers, or LSPs, are especially dependent on the robustness and completeness of the transportation management-related applications because the software is used to manage their core business.

The whole application suite and its components are based on a configurable interaction of business objects, which can be used in the context of the necessary business processes to represent the logical flow of work tasks and data required to manage the business successfully.

Business Objects

The term *business object* is used in many places and contexts throughout this book. Business objects such as sales orders, invoices, business partners, and equipment are representations of real-world business documents or processes in an IT system. They encapsulate and structure data belonging to a business-related context and provide a set of methods to create, maintain, and delete the object and to exchange the data via interfaces. No activities ever breach the integrity of the business object.

1.1.1 Layers and Components of an SAP System

Different ways to define the system layers

To look at a system with as complex architecture as the SAP Business Suite, we can look at the layers of the software from different perspectives:

- Technical software layers
- Technical hardware layers
- Application component layers

Software layer view

First, let's consider the two essential software layers of an SAP system, which are used to perform the vital steps of all business processes:

- **Basis or SAP NetWeaver layer**
  SAP NetWeaver is the application platform that provides the application server, system kernel, and database required to run any of the SAP Business Suite applications. You can find more details on SAP NetWeaver in Section 1.1.2.
- **Application layer**
  The SAP Business Suite provides the majority of the SAP business applications. You can find a detailed overview in Section 1.2. The components used to run transportation processes are all part of the SAP Business Suite.

Figure 1.1 gives an overview of the software layers of an SAP system.

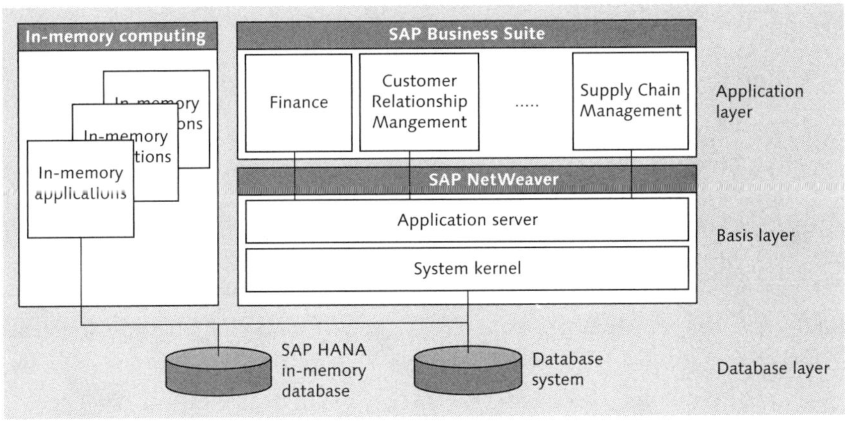

Figure 1.1 Software Layers of an SAP System

Hardware layer view

The second view is presented by the *server layers*, which show the system from a hardware-heavy, technical perspective:

- **Database layer**
  The database server is the component that hosts the database of the SAP system. This can be either the SAP-provided in-memory database SAP HANA, SAP MaxDB, or a database from a third-party vendor such as IBM's DB2. There is usually a single logical database server in the DB layer of an SAP system instance.
- **Application and integration layer**
  Application servers are hardware components on which application programs such as SAP TM transactions are executed. Integration servers provide messaging and integration capabilities to connect business processes running on different application servers. By scaling the application servers in a system (i.e., the number of physical servers, number of processors/cores per server, size of memory per server), an SAP system can be adjusted to host small to extra-large communities of users working in parallel.
- **Internet access layer and mobile access layer**
  Internet servers provide web access to application programs. This layer is especially useful if application programs need to be provided to large communities of Internet users (e.g., in web portals). Mobile access servers such as Sybase Unwired Platform offer access to application programs through mobile device frontends.
- **Presentation layer**
  The presentation layer provides end user access to application programs via various SAP access technologies such as SAP GUI and SAP NetWeaver Business Client (NWBC) or other desktop or mobile user interfaces.

Technically, it is possible for all layers to run on a single machine. However, in real-world implementations, there is a dedicated database server, one or more application servers, and a presentation layer workplace (usually a local desktop PC or laptop) per end user. This setup is called three-tier architecture and is the most commonly used model. To adjust the system performance with varying user numbers, you can easily scale the system by engaging additional application servers to host the additional user requests.

1.1.2 SAP NetWeaver Application Platform

Functional view of SAP NetWeaver

SAP NetWeaver is a complete application server that allows you to run SAP applications and partner products. It offers a variety of functional blocks that support the application running on top of it:

- The *application server* has an ABAP and Java stack that serve as an abstraction layer to technically run the applications on different operating and database systems. It provides the runtime environment for all Basis and application processes. SAP TM makes comprehensive use of the SAP NetWeaver ABAP stack.
- *Process integration* (SAP NetWeaver PI) is part of the Basis applications and supports interconnectivity between SAP applications in an in-house environment and external systems or business partners. It is accompanied by business process management that allows you to define communication patterns for complex business processes. SAP TM coordinates internal and external communication via SAP NetWeaver PI using enterprise services.
- *Information integration* supports master data management to coordinate master data distribution and quality. Knowledge management provides full text search and retrieval of information from unstructured data. Business intelligence functionality allows you to capture, retrieve, and derive key performance indicators from information from all SAP Business Suite components and external sources.
- *User integration* allows users to access the system through various channels such as portals, mobile devices, and the local client software of web browsers. It also includes collaboration scenarios.

Technical overview of SAP NetWeaver

If we climb down one staircase into the basement of SAP NetWeaver, we can look at the SAP system foundation from a more technical point of view.

Sitting on one of the multiple possible operating systems (OSs), SAP NetWeaver is equipped with an operating system-dependent kernel, which shields the application layers above from the specific behavior and requirements of the hardware and operating system. This architecture, therefore, allows unified operation of the application and harmonized access to system and hardware functions such as database and printer access. SAP NetWeaver can run on top of following operating systems and hardware platforms:

- Linux (RHEL, SLES, RedFlag Linux, zLinux) on IA32, PPC, X86_64, IBM zSeries
- AIX
- HP-UX on PA-Risc and IA64
- Solaris on SPARC and X86_64
- Windows NT, IA32, IA64, and X86_64
- IBM OS/400 and z/OS (OS/390)

Supported database systems include SAP HANA, SAP MaxDB, Microsoft SQL Server, IBM Informix Dynamic Server, IBM DB2, and Oracle. Figure 1.2 shows an overview on the technical components of SAP NetWeaver.

ABAP kernel as stable execution basis

The ABAP and Java runtime systems of SAP NetWeaver are the foundation of most applications that are executed in a homogenous and stable environment. ABAP is very important to SAP because most application functions are implemented in this proprietary programming language. Having started as a COBOL-like reporting language, ABAP has evolved into a powerful object-oriented programming tool whose language constructs provide excellent support to build business applications. Embedded into the ABAP runtime environment, processes executed by a user can be shielded very well from those of other users so that problems in one executed program do not impact other running processes (which is sometimes a disadvantage in Java Application Servers).

Access to SAP Program Source Code

ABAP is not considered flashy, but SAP did develop the language to be a best-fitting, reliable, and fast platform for business applications. SAP views it as so important and fundamental that the complete source code for the business applications is visible and delivered with the system. With a little programming knowledge, you can start the ABAP Workbench (Transaction SE80) and browse around in the coding (e.g., in SAP TM) to find out how it works. The code availability also allows you to enhance and reuse program coding by implementing enhancement spots, for example.

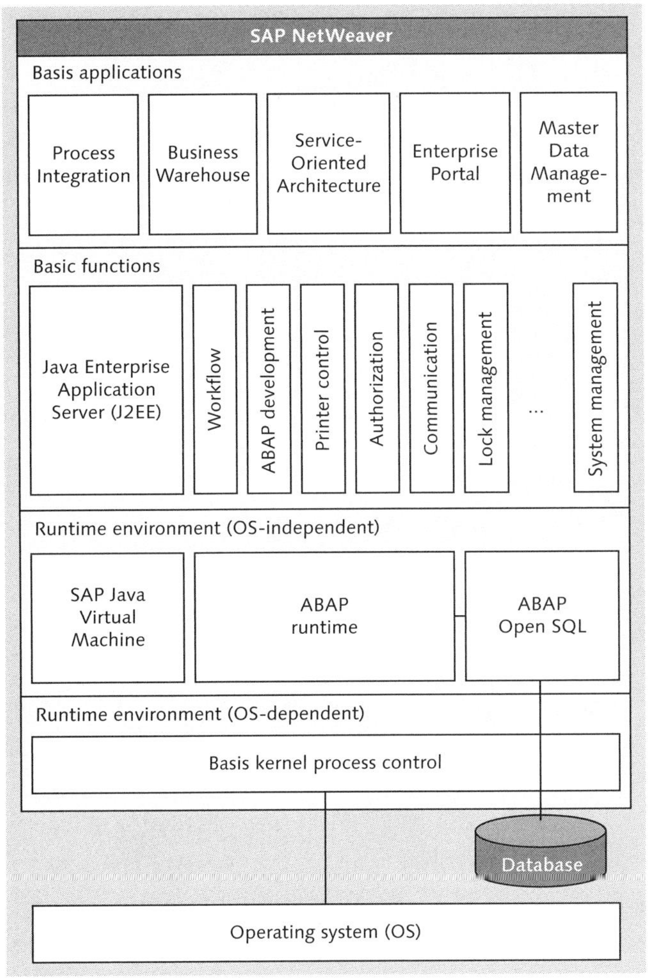

Figure 1.2 Technical Components of SAP NetWeaver

SAP's Java platform

Compared to ABAP applications, decoupling Standard Java J2EE applications is weaker, which could be a big disadvantage for business software because an application running out of control could easily bring down the whole Java Application Server, immediately affecting all other users and processes on it. SAP's investment into its own Java Virtual Machine resulted in a Java runtime environment that provides the same stability and decoupling as the ABAP Application Server (AS). The SAP Java AS is fully certified according to Java EE guidelines.

1.1.3 SAP NetWeaver Functions that Support Business Applications

The SAP NetWeaver ABAP runtime environment offers a plethora of system functions to all applications running on application servers. These services deal with harmonized access to the application environment (e.g., printing or communication) or with security support. At an overview level, the following are the most important basic functionalities:

Managing transactions and processes

- **Transaction management**
  Transaction management implemented in SAP NetWeaver and the well-designed applications guarantee consistency of related business data. If, for example, an invoice for an order is created, then the transaction management ensures that not only is the invoice created correctly, but the invoicing status and data in the related order are updated within the same transaction. You either get everything consistently done or, in case of partial inconsistency, all stored data is rolled back to its state when the transaction was started.
- **Technical process control (work processes and load balancing)**
  By means of process control, the system balances the work load of the currently running processes and transactions so that each user gets an allocation of processing time to finalize his transaction. Newly started processes are automatically allocated to the least-loaded application server in the system environment.

Managing security

- **Locking server**
  The locking server ensures that only one user or transaction at a time can alter a single business object (e.g., user A updates order 123 by adding a new item even as user B deletes order 123). For the first user accessing the object, the locking server ensures that any subsequent users only get display access to the object until the locks are released (usually when user A clicks the SAVE button). In SAP TM, an extended concept is implemented that works with *optimistic locks,* which allow multiple users to work in change mode either on separate parts of the same object or to only lock after updates have been committed. Additionally, locking can be implemented on sub-object level, meaning that multiple users can concurrently work on one object, each having their own work set of object data.

- **User management and authorization**
  To manage access rights to applications, you can define user roles with a comprehensive list of authorizations. The user roles can be assigned to named users who can log on to the system. After logon, the authorizations linked to the roles assigned to the user control each activity that a user can carry out with a transaction or object. A call center service agent may, for example, create new orders and create invoices, but he may not be allowed to correct disputed invoices. All these settings can be very individual for each company, so the definition and setup is usually part of the implementation phase of the SAP software.

Output control and workflows

- **Output management for printing and other channels**
  SAP NetWeaver allows you to manage all printers centrally in the system. In addition to technical integration of printing devices, printing queues and the integration of printing activities into the application layer are provided. In newer installations, Adobe Document Server is often used to actively manage the document output. SAP also offers the capability to use the Adobe Forms Designer and interactive PDF forms.

  In the context of output management, SAP NetWeaver offers options for sending created documents by fax or as an email attachment instead of printing them. The output management services allow you to automatically create an archived version of the document in a document management system.

- **Communication and service architecture**
  Communication management in SAP NetWeaver supports internal communication processes (*application to application*, or A2A) as well as external processes (*business to business*, or B2B). For all applications, there are plenty of enterprise services (web services with WSDL definitions) defined in the *Enterprise Service Repository* (ESR), which allow you to connect SAP services to other systems via SAP NetWeaver PI. The enterprise services allow you to execute activities on business objects. You can find detailed information at the Enterprise Services Workplace at *http://esworkplace.sap.com*.

- **Workflow management**
  SAP NetWeaver includes a complete workflow management system that can be integrated with all applications, the office functionality of the SAP Business Suite, and the business process management.
- **Development workbench**
  SAP NetWeaver contains a complete development environment for development coordination (packaging of objects), all dictionary objects (database tables, structures, data elements, etc.), ABAP objects (programs, function modules, classes, etc.), user interfaces (Web Dynpro, Business Server Pages, etc.), enterprise services, and other objects such as authorizations, transactions, and message classes. The development workbench allows you to test all the objects instantaneously and has very strong drill-down capabilities such as from call to method and further to referenced structures and data elements. Figure 1.3 shows an overview of the development workbench.

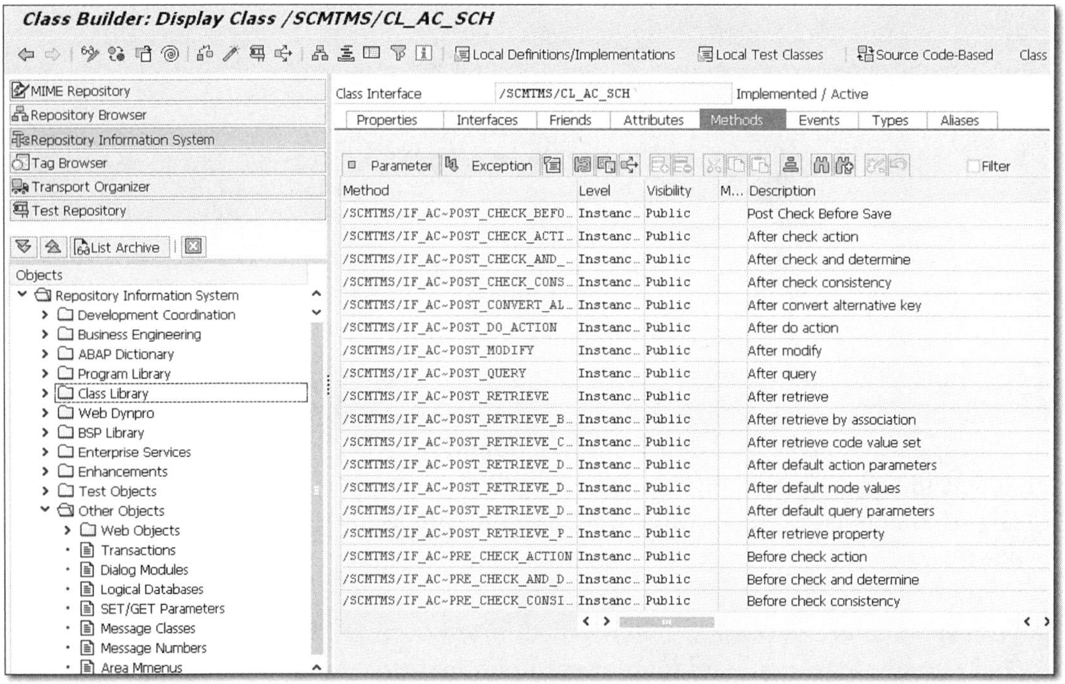

Figure 1.3 ABAP Development Workbench: Development Objects and Details of an SAP TM Class Interface

- **Change and transport organizer**
  Change and transport organizer is a tool for managing centralized and decentralized ABAP Workbench development projects and Customizing projects (system and process configuration). Development objects and customizing settings can be clustered in transport requests to achieve a structured release of consistent development and configuration content and allow export from a development system and import into a test or production system.

1.1.4 Important Business Concepts in SAP Systems

Three essential concepts realized in SAP systems make adoption of business requirements more easy and flexible. These concepts affect SAP NetWeaver as well as the business applications and components built on top:

- Client concept
- Organizational units
- Customizing and Implementation Guide

The SAP client concept

The *client* of an SAP system is a concept involving a logical separation of data and work areas within one technical system installation. Using the client, multiple independent organizations can work in a single system in completely separated work spaces with data being shielded from access by unauthorized users. Figure 1.4 offers an overview of the technical and logical organizational layers of an SAP system.

From a technical perspective, the client is the first primary key field of most of the database tables used in SAP NetWeaver and the SAP Business Suite. If a user logs on to an SAP system, he has to explicitly log on to a client—that is, enter the client number (000-999). The client has to exist and the user ID needs to have already been created in the client. Changing clients is not possible without logging off and on again. If a user works in a client, then access to data in another client is not possible because the Basis layer blocks access automatically. (There are very few exceptions for maintenance programs that work across clients and some basic settings that are client-independent.) A process requiring access to or communication between several clients in a system can be set up using SAP NetWeaver PI as an integration tool. You should create no more than 150 clients in one system to limit the organizational overhead of system maintenance.

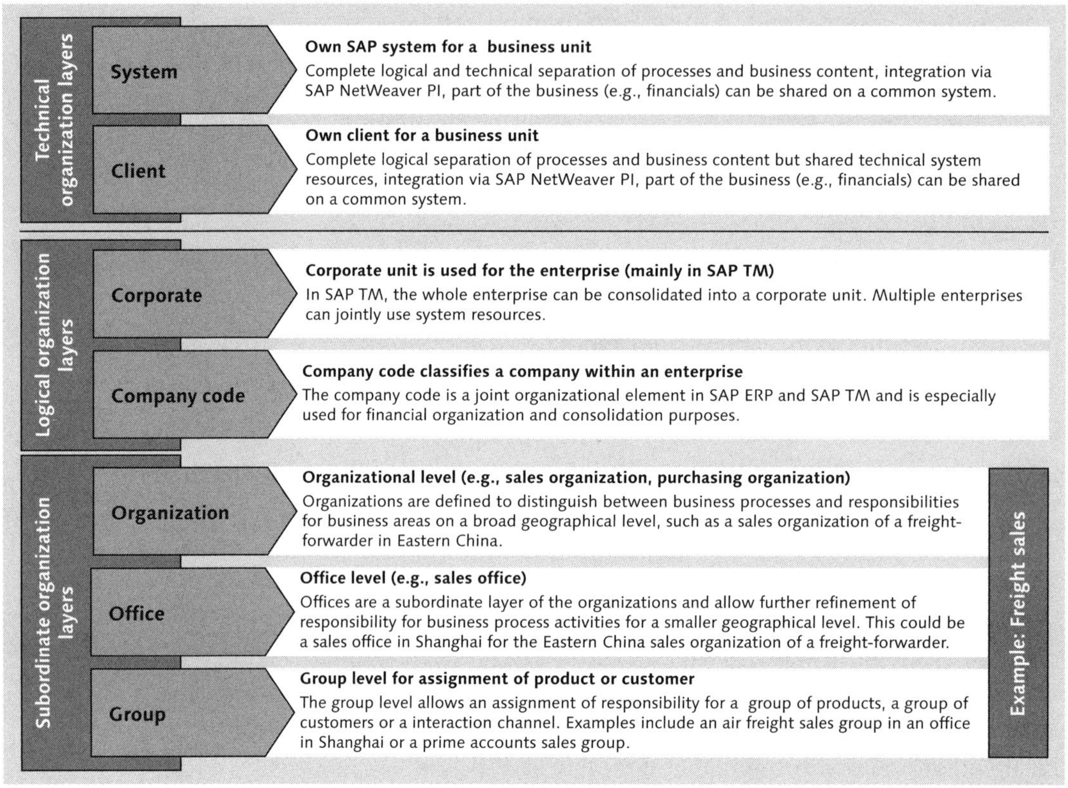

Figure 1.4 Organization Levels Provided in an SAP System

Organizational unit in a company

The two lower-level tiers in Figure 1.4 (logical organization layers and subordinate organization layers) show the organizational layers of an SAP Business Suite system. The logical layer can be used to implement an enterprise's processes with multiple companies. The subordinate organization layer allows you to assign users in the same company to work in different areas of responsibility (e.g., purchase organization in United States and sales organization in China). We give a more detailed explanation of organizational structure in Chapter 3, Section 3.1.

Customizing and Implementation Guide

The Implementation Guide (IMG) is a tool to organize a project implementation—that is, to set up business processes in a project or at a customer's site. You can create multiple projects, each holding Customizing tasks for specific business areas. Customizing tasks allow you to set up the configuration in the SAP system client. The configuration controls which process can be executed in what particular way, which consistency checks are executed, and which data determinations are done. Figure 1.5 shows part of the reference IMG and a Customizing task in an SAP TM system.

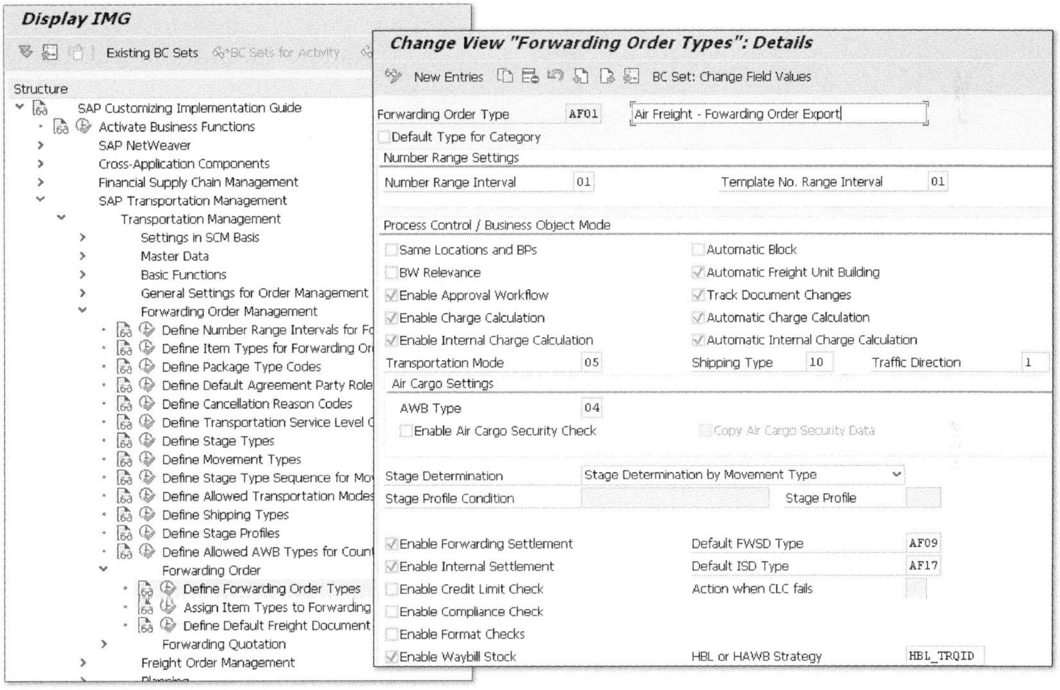

Figure 1.5 Customizing IMG and Configuration of a Forwarding Order Type for Air Freight in an SAP TM System

The following are some examples of typical Customizing tasks:

- Defining the countries where you do business and which currencies you use
- Defining order types and their functional details
- Defining price calculation rules and document printing rules

The Implementation Guide can also work on an overall task assignment as the *Reference IMG*, giving you access to all Customizing tasks in the whole system. The structure of the Implementation Guide reflects the different business areas to be configured. There are always joint SAP NetWeaver-related tasks such as country and communication settings. In addition—and depending on the installed components—you will find application-specific tasks. In an SAP ERP system instance, there is a setting for procurement, finance, sales, logistics, production, and so on. In an SAP TM instance, you'll find configurations for master data, forwarding order management, planning, and others.

Customizing allows you to define very detailed settings; in fact, in SAP TM there are more than 300 different configuration tasks. You can get a system up and running with minimum settings, but if you need it, a high level of flexibility is available.

1.1.5 SAP NetWeaver Application Components

SAP NetWeaver is home to some components that can be classified as Basis components used by a variety of application systems:

- SAP NetWeaver PI and enterprise service-oriented architecture (ESOA)
- SAP BusinessObjects Business Intelligence (BI)
- SAP NetWeaver Master Data Management (MDM)
- SAP Enterprise Portal (EP)

SAP NetWeaver PI and SOA

SAP NetWeaver PI handles interconnectivity within a system landscape and connectivity to business and communication partners outside of the enterprise. SAP NetWeaver PI can work in both A2A and B2B scenarios. Beyond the SAP-based integration, SAP NetWeaver PI can also integrate to other messaging platforms and integration backbones such as Microsoft BizTalk and IBM WebSphere. Using specialized adapters, you can achieve integration with EDI networks and standards (EDIFACT, ANSI X.12, ODETTE, VDA, etc.).

SAP NetWeaver PI comes with a complete landscape definition tool and various monitoring and maintenance tools. The System Landscape

Directory (SLD) allows you to define the complete system landscape from an integration perspective. In addition, business process management allows you to define and execute complex business processes including workflows that require multiple communication steps between business partners, and it coordinates the correct status and activities.

> **SAP TM and SAP NetWeaver PI**
>
> SAP NetWeaver PI is used for SAP TM. All A2A connections between SAP ERP and SAP TM are by default routed and managed by SAP NetWeaver PI.

Business intelligence

SAP NetWeaver BW is an enterprise-wide data warehouse that allows you to collect data from all components of the SAP Business Suite. The data is organized and provided in InfoProviders, which enable access through various queries to extract key performance indicators relevant to your business. In combination with the SAP BusinessObjects tools, connection with many non-SAP data sources is possible. In addition, graphical dashboards can be created easily. We explain SAP BusinessObjects BI, especially in the context of SAP TM use cases, in Chapter 12, Section 12.2.

A new level of business intelligence has been introduced through the in-memory component SAP HANA. In contrast to conventional BI installations, when you use an SAP HANA installation, all data is permanently kept in memory, where you can access it very quickly. With SAP HANA, Big Data isn't scary anymore.

SAP NetWeaver Master Data Management

SAP NetWeaver Master Data Management (MDM) offers centralized management of master data within the system landscape. Master data entities such as business partners only need to be maintained once and are distributed consistently to all connected and related systems to ensure high data quality. Consider a few typical master data errors that you should avoid:

- Misspelling: *Gorge Miller* instead of *George Miller*
- Missing data: *Palo Alto* without proper zip code

- Wrong format: Phone numbers with wrong or without proper separators, such as *1234567* instead of *123-4567*
- Wrong codes: *CAN* instead of *CDN* for the Canadian dollar
- Duplication: Duplicate master data records, such as *G. Smith* and *George Smith* for the same customer

SAP NetWeaver MDM offers some tools to detect duplicate master data entries or to correct misspelled entries. SAP BusinessObjects tools such as SAP Data Services, SAP Data Quality Management, and SAP Data Integrator support many daily and one-time maintenance tasks.

SAP NetWeaver Enterprise Portal

SAP NetWeaver Portal provides user-centric access to all relevant information that a user in a specific role needs. It integrates the user interfaces of the SAP Business Suite applications that are relevant to the user as well as business analytics, business-related content from web pages, office document access, and third-party applications. Authorization is simplified by a single sign-on mechanism, which requires that the user only log on one time in order to work with all connected systems and services. The portal role assigned to the user controls his authorization for data access.

1.2 SAP TM as Part of the SAP Business Suite

The SAP Business Suite is a comprehensive and fully integrated family of business software applications for large, medium, and small enterprises. It allows companies to plan, execute, and document their end-to-end business processes on a standardized software platform.

History of the SAP Business Suite

The SAP Business Suite was built as a successor of SAP R/3, an enterprise resource planning system that was initially launched in 1992. After several successful releases, it became apparent that a single monolithic system like R/3 has certain disadvantages. Therefore, SAP decided in 2000 to advance R/3 in multiple steps into SAP ERP 6.0, which became more powerful and flexible from a release and component strategy. On top, it also developed specialized and well-integrated components like SAP Advanced Planner and Optimizer (APO) and SAP Customer Relationship

Management (CRM). Today, the SAP Business Suite is a comprehensive set of well-integrated components with a yearly release of *enhancement packs* (EhPs), which allow customers to upgrade with less effort compared to the full releases of earlier times.

SAP Business Suite components and SAP TM

Let's explore the components and subcomponents of the SAP Business Suite that are relevant to SAP TM:

- Of course, SAP TM itself
- SAP ERP with Financial Accounting and Controlling (FI-CO), Logistics General (LO), Sales and Distribution (SD), Logistics Execution (LE), Materials Management (MM), and Plant Maintenance (PM)
- SAP Event Management
- SAP Environment, Health, and Safety (EHS) Management as an integral part of the SCM Basis layer
- SAP Global Trade Services (GTS) as part of the SAP Governance, Risk, and Compliance (GRC) solution
- SAP Extended Warehouse Management (EWM)
- SAP Customer Relationship Management (CRM)

We should mention a few additional SAP Business Suite components since they are of great importance for some businesses, even if they are not yet integrated with SAP TM out of the box or just have been integrated based on a custom development project:

- SAP Billing and Revenue Innovation Management (BRIM), which comprises Convergent Invoicing, Collection and Cash Management, Receivables Management, Dispute Management, and Financial Customer Care (see Chapter 11, Section 11.3)
- SAP Internet Pricing and Configurator (IPC), which has been integrated with SAP TM contract and order management on a project basis to allow flexible, rule-based selling of logistic services

Figure 1.6 gives an overview of the SAP Business Suite components used for transportation management and some of their functional areas.

SAP ERP
- Partner master data
- Sales and distribution
- Invoicing
- Convergent invoicing
- Invoice verification
- Finance and controlling
- Credit limit check
- Tax calculation
- Equipment maintenance

SAP TM
- Transportation master data
- Transportation order management
- Sales order integration
- Dispatching and planning
- Optimization
- Subcontracting and tendering
- Cargo management
- Strategic freight management
- Equipment management
- Capacity management
- Product catalog
- Spot quotations
- Tariff and rate definition
- Charge calculation
- Draft customer/supplier invoices
- Document management
- Output management
- Order workflow control

SAP BusinessObjects BI/ SAP HANA
- Analytics
- Business Context
- Reporting
- Dashboards

SAP GTS/EHS Management
- Import
- Export
- Trade compliance
- Security
- Hazardous goods

SAP Event Management
- Tracking and tracing
- Visibility
- Alerting
- Execution and event performance
- Demurrage processing

SAP EWM
- Inbound handling
- Deconsolidation
- Consolidation
- Outbound handling
- Storage
- Handling steps
- Warehouse planning

SAP CRM
- Interaction center
- Customer 360° View
- Marketing
- Opportunity management
- Contract and sales workflow
- Financial customer care
- Incident management

Figure 1.6 SAP TM-Related Component Ecosystem of the SAP Business Suite

1.2.1 SAP TM

When it comes to transportation, cargo, and freight, SAP TM is the main component of the business suite. It delivers the foundation for all mainstream processes and activities. As outlined in the introduction chapter, transportation management functionality can be used in various business situations:

SAP TM in different industries

- **Transportation and logistics for shippers**
  Shippers from manufacturing and trading industries plan, optimize, and organize their own shipments using SAP TM.
- **Transportation management for freight-forwarders**
  Freight-forwarders use SAP TM to handle their freight services on the buying and selling side and to broker between supply and demand. They try to optimize the supply chain and their own profitability.

Figure 1.6 gives you an overview of how SAP TM components are used to achieve forwarders' goals.

- **Transportation management for carriers**
  Carriers have similar business requirements as freight-forwarders, but additionally they focus on their own assets and equipment. SAP TM supports carriers' business by integrating asset and equipment management and resource planning capabilities.
- **Transportation management for contract logistics providers**
  Contract LSPs (also 3PLs and 4PLs) manage the complete logistics business for their customers, including order management and invoicing in their customers' names. As a sales order management system with integration with SAP EWM and SAP TM, SAP ERP supports this business.

We provide a more in-depth functional overview of SAP TM in Section 1.3.

1.2.2 SAP ERP

The central component of the SAP Business Suite is still SAP ERP, which incorporates a variety of components that are directly or indirectly related to logistics and transportation processes.

Sales and Distribution (SD)

Sales and Distribution (SD) is an SAP ERP component that encompasses a range of functions dealing with pricing and selling tangible goods and services. It is used a lot in manufacturing and trading industries to manage the complete sales process of goods to customers. The following are typical tasks handled in SD:

- Management of quotations and sales orders
- Conducting availability checks for goods and materials managed in sales orders
- Creation of scheduling agreements
- Credit limit and risk management in association with sales orders
- Sales price determination and invoice creation (billing document)
- Trade compliance management for shippers

In SD the sales order is typically the source of transportation demand, which is transmitted to an SAP TM system either at the shipper's premises (transportation department) or at an LSP. Based on sales order data, an *order-based transportation requirement* (OTR) represents the demand in SAP TM.

In combination with SAP TM, you can collectively do transportation planning for sales orders, allowing you to consolidate one or multiple orders in one shipment or split an order into multiple shipments. You also can jointly plan orders from different systems or clients, which is not possible using the SAP ERP transportation component (LE-TRA). Transportation scheduling can be done synchronously in SAP TM for a sales order created in SAP ERP if no global available-to-promise (ATP) or product selection is required. Tendering, subcontracting, and execution in SAP TM are possible based on SAP ERP orders prior to delivery creation. You can find further details on SAP ERP order integration with SAP TM in Chapter 4, Section 4.1.

SD is also used in combination with SAP TM to do customer invoicing. Since SAP TM is capable of creating draft or pro forma invoices, but does not have an invoice settlement functionality, the SAP ERP functions for billing (SD-BIL) and integration into financials (FI-CO) are used. You can find details about this integration in Chapter 11.

Logistics Execution (LE)

Logistics Execution comprises deliveries, shipping, warehousing, and shipment management in SAP ERP. The transportation demand generated by these processes is already well scheduled and detailed. Logistics Execution is typically used to handle the following tasks:

- Centralized and decentralized warehouse management with task and resource management and yard management
- Delivery preparation and documentation
- Goods issue processing
- Transport organization and documentation (SAP LE-TRA)
- Direct store delivery management
- Handling unit management
- Returns management for deliveries and orders

The shipping process of Logistics Execution (LE-SHP) mainly deals with getting goods and materials picked from a storage location, packed and consolidated into one or multiple physical units called *handling units* (e.g., boxes and pallets), and assigned to a logical unit called a *delivery*, which is prepared to be shipped to a single consignee.

One or multiple deliveries may then be consolidated into *shipments* (LE-TRA), which are transportation objects used to document a direct move or a single leg of a longer transportation move of goods from a shipper to a consignee. Transportation management capabilities concentrate on documenting the shipment details, doing a simple, rule-based route determination, preparing shipment paperwork, setting execution status, executing simple tendering to subcontractors, and finally calculating costs and settlements with the subcontractors. If required, transportation optimization for multiple deliveries can be achieved in combination with SAP APO Transportation Planning and Vehicle Scheduling (TPVS) or a third-party optimization tool.

Why Does SAP Offer Multiple Transportation Solutions?

The fact that you can find multiple transportation solutions in the SAP Business Suite has historical and specialization reasons. The shipment component in SAP ERP Logistics Execution was created as an integral part of a shipper's outbound or inbound process and has been intensively used by thousands of SAP ERP customers for more than 15 years. Although it offers comprehensive functionality, it has some limitations that make it too inflexible to address general transportation needs or to be the foundation of a complete transportation suite and platform:

- Dependency of shipped goods on material master data, which you need to have in order to create deliveries (an LSP usually does not have this master data of shipped goods)
- No joint inbound and outbound moves can be handled (either outbound delivery or inbound pickup shipments)
- No planning of shipments for demands coming from multiple SAP ERP systems or clients (i.e., enterprises using multiple ERP instances or clients cannot use LE-TRA as centralized transportation software)
- No order management to support selling and billing of transportation services (i.e., LE-TRA is not easily usable for LSPs)

> Specifically, SAP Transportation and Distribution (SAP TD) has been developed for SAP Oil & Gas to arrange downstream transportation (delivery to gas stations), covering all the specific needs of shipping fossil fuels, including pipeline-based transportation. Upstream processes and their corresponding immanent trading requirements are handled using the SAP Traders and Schedulers Workbench (TSW).

Like the SD component, Logistics Execution is a source of transportation demand that is communicated to SAP TM. This is reflected by SAP ERP delivery data used to create delivery-based transportation requests (DTRs) in SAP TM, as well as SAP ERP shipments transmitted to feed SAP TM freight orders for a tendering process. You will find more detailed explanations in Section 1.3 and Chapter 4, Section 4.1.

Materials Management (MM)

The focus of Materials Management (MM) is procurement and inventory management of tangible goods and materials for sale or for use in production processes. The following are typical processes handled with MM:

- Processing of purchasing requests and purchasing orders
- Inventory management of materials and products, including material evaluation for accounting and material price management
- Invoice verification for invoices related to goods and service delivery
- Self-billing process support
- Stock-taking and stock correction
- Management of material master data
- Handling of inbound deliveries including goods receipt processes
- Supplier returns management

MM processes for purchasing, inbound delivery handling, and returns management provide a purchase order and inbound delivery-based integration to SAP TM. This is similar to the SD and LE integration. OTRs and DTRs are used in SAP TM to reflect demand (for details, see Chapter 4, Section 4.1).

As described, SAP TM only creates draft invoices. For settling supplier invoices, MM provides invoice verification and self-billing capabilities, which also integrate into the financial components (FI-CO). You can find details about SAP TM supplier draft invoice integration in Chapter 11.

Plant Maintenance (PM)

Plant Maintenance (PM) is used to manage master data for equipment and to plan, organize, and monitor maintenance of all equipment that is required to sustain a company's operations. In many cases, this is production machinery and equipment. For transportation, this includes active and inactive transportation equipment such as trucks, trailers, containers, and railcars. SAP TM provides integration for transportation equipment. SAP PM is responsible for managing the master data and maintenance plan of the equipment, whereas SAP TM is responsible for its active planning and utilization within the transportation processes. When equipment needs maintenance, it has to be removed from transportation plans. Otherwise, if maintenance is approaching and the equipment is not at the right location, you need to organize transportation to the repair shop. This interaction is covered by the integration between SAP PM and SAP TM.

SAP ERP-SAP TM integration points

Recall that there are various integration points between SAP ERP and SAP TM. Figure 1.7 gives an overview of the logical process integration steps.

SAP ERP industry solutions: Retail

SAP for Retail offers a large set of retail-specific functionalities such as point-of-sales integration or management and planning of regular routes. This solution portfolio reuses the main objects of SD and LE and therefore offers the same integration concepts to SAP TM that the standard sales, shipping, and procurement functionalities offer. However, since retail also requires some specific functionalities on the transportation side, we give you some insight on this in Chapter 13.

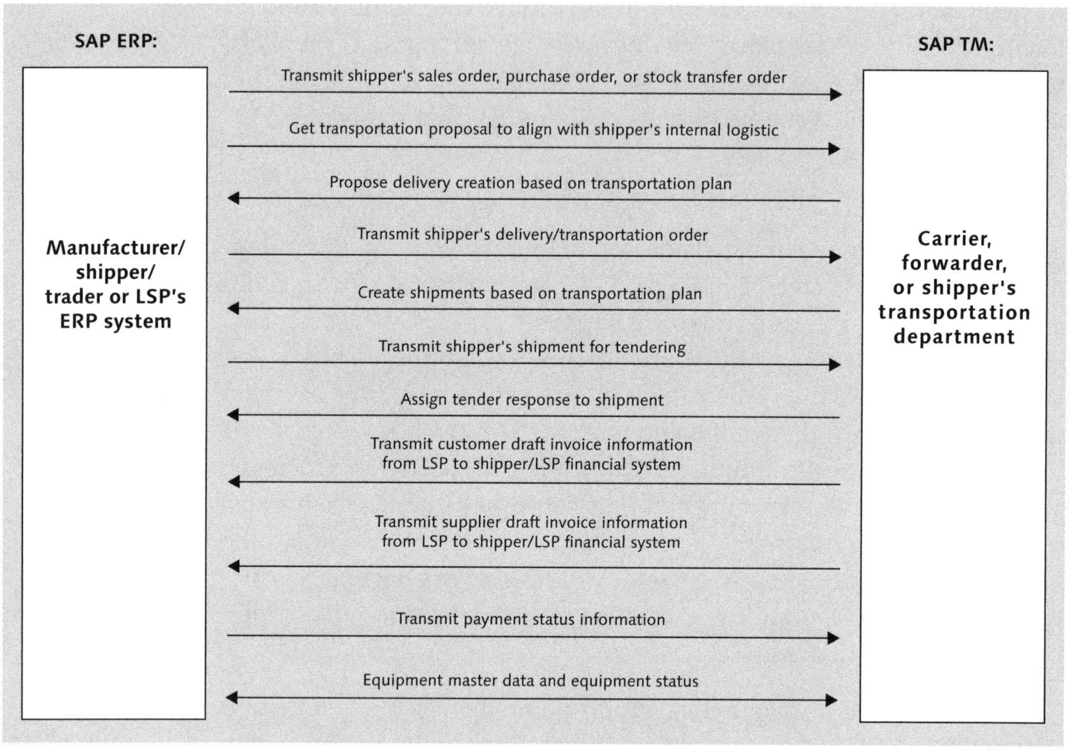

Figure 1.7 Logical Integration Points between SAP ERP and SAP TM

1.2.3 SAP Event Management

Tracking transportation processes

SAP Event Management is a versatile tracking and tracing tool that connects to multiple SAP Business Suite components to provide visibility of business processes and object status. Furthermore, SAP Event Management allows automated process control for intentional reactions to events and status messages or responses for unexpected events or missing milestone reporting.

SAP Event Management provides process tracking and control information that matches the processes in SAP TM for shipment status tracking, consolidation status tracking, equipment tracking, and operational instruction tracking. Bidirectional integration between SAP TM and SAP Event Management means that SAP TM sends process and milestone information to SAP Event Management, which then tracks the process execution and compliance based on information and reporting received from inside or outside the company. Then SAP Event Management posts information back to SAP TM to update the transportation processes. You can find details on SAP Event Management and its integration with SAP TM in Chapter 8, Section 8.2.

1.2.4 SAP Environment, Health, and Safety Management

Dangerous goods management

SAP Environment, Health, and Safety (EHS) Management allows you to manage dangerous goods compliance in connection with logistics processes running in SAP ERP, SAP TM, or SAP EWM. It consists of a rules engine and documentation and phrase management. The rules engine allows you to load predefined content corresponding to the various dangerous goods regulations for different situations, modes of transport, and countries (such as ADR and IMDG regulations). Documentation and phrase management support the creation and output of legally required dangerous goods documentation and safety data sheets for dangerous goods transports. We explain SAP EHS Management and its use in connection with SAP TM in more detail in Chapter 9, Section 9.2.

1.2.5 SAP Global Trade Services

Trade compliance

SAP Global Trade Services (GTS) is part of SAP Governance, Risk, and Compliance (GRC) Management. It covers a wide range of trade compliance checks and processes that must be handled in transportation. Due to increasing legal regulation and security checks, transportation has gotten more and more complex from the view of exporting from and importing into countries. Because trade compliance is mostly handled under the responsibility of local governments or groups of countries, the processes and requirements for proper export and import declarations are highly diverse.

SAP GTS supports SAP TM in the order management area, where functionality for general trade compliance and blacklist screening are provided. Furthermore, the shipment process is supported with export and import compliance. You can find more details on SAP GTS in Chapter 9, Section 9.1.

1.2.6 SAP Extended Warehouse Management

Warehousing

Warehouse management is a very common functionality for both shippers and LSPs. Many shippers operate a warehouse to store production supplies and finished goods to be shipped to their customers. In the logistics service industry, warehouses can be utilized in various ways:

- A contract logistics warehouse to keep the inventory for a business partner for whom the LSP takes over the logistics processes
- A hub warehouse to cross-dock cargo between different vehicles or means of transport without intermediate storage
- A consolidation warehouse (e.g., container freight station) where cargo is collected and then consolidated into a transport unit, which is then moved to a destination (see Chapter 14 and Chapter 15 for the corresponding consolidation processes)

SAP EWM functional overview

SAP Extended Warehouse Management (EWM) is a standalone warehouse management system with a rich spectrum of warehouse and material handling functionality. Besides warehouse inventory management, there is plenty of support for the following:

- Yard management
- Unloading and goods receipt handling
- Quality management and dangerous goods handling
- Consolidation of goods into handling and transport units
- Deconsolidation of transport and handing units
- Picking and put-away
- Shift and workforce planning
- Value-added services and kitting
- Provisioning and loading

SAP EWM is fully integrated into the logistics processes of SAP ERP, and so is automatically integrated to SAP TM. In addition, you can also connect SAP EWM directly to SAP TM; this integration has been implemented in various customer projects and is now also available as a standard feature. In Chapter 12, Section 12.2, you can find more details on the integration of SAP EWM, SAP ERP, and SAP TM.

1.2.7 SAP Advanced Planner and Optimizer

Optimization for ERP logistics

SAP Advanced Planner and Optimizer (APO) has been released to provide SAP ERP logistics processes with optimization capabilities. A major goal has been to enable planning that spans multiple SAP ERP implementations in a company. Because large manufacturing companies often run multiple SAP ERP systems (e.g., for each country or each business unit), a central planning and optimizing engine was required to consolidate the data of the SAP ERP logistics transactions into a joint plan, whose result subsequently was broken down into the local SAP ERP system's scope and communicated back to update the local planning. SAP APO supports various demand/supply-focused planning situations, such as the following:

- Supply network planning
- Demand planning
- Global available-to-promise (ATP)
- Supplier collaboration
- Back order processing

Transportation planning with SAP APO-TPVS

For transportation planning, the SAP APO-TPVS component is integrated with SAP ERP sales orders, deliveries, and shipments. It supports optimized transportation planning of orders and deliveries and creates shipments in SAP ERP. The TPVS optimizer engine is capable of solving complex multi-modal planning problems within large transportation networks in a shipper environment. The optimizer engine became the foundation of the SAP TM optimizer component and has been extended to also support LSP processes.

Are SAP APO-TPVS and SAP TM Complementary?

One question that often arises in conjunction with SAP APO and SAP TM is how far the two components can be used together when it comes to logistics. As of summer 2013, there is no full integration of a supply plan created in SAP APO and a transportation plan created in SAP TM. Looking at more strategic planning functions such as supply network planning and demand planning, there is an integration option via SAP ERP so that a shipper can optimize supply processes on a high level with SAP ERP and SAP APO and subsequently use the results as input for SAP TM-based transportation planning.

For detailed planning integration (e.g., using global ATP sourcing decisions during transportation planning to achieve a best decision concerning material supply and transportation costs), loose integration via SAP ERP currently limits the optimization potential.

Since SAP APO does not play a significant role in SAP TM processes, we do not provide any further insight into its functionality in this book.

1.2.8 SAP Customer Relationship Management

LSPs need to manage customer relations

Staying close to customers and providing them with excellent service that meets their requirements has become one of the main differentiators in the transportation and logistics service industry. For freight-forwarders and carriers whose core business is logistics services, sales price increases need to have a tangible upside. Therefore, customer intimacy—the art of staying close to customers—has become an attractive driver of rising market share.

An LSP should be informed about the requirements of his customers. Accurately timed and targeted marketing campaigns for his logistic services connect his logistical capabilities with the sensed customer demand. Precise transformation of opportunities into tailor-made contracts is one more key to success—for both the customers' business and the LSP's profitability. Furthermore, good contracts with well-defined service-level agreements and operating procedures can guide and control the operation to achieve legal compliance and business profitability. In this way, a modern LSP closely and proactively interacts in customer service with external and internal partners and stays on top of customer-relevant incidents in marketing, sales, logistics execution, compliance, and financials.

SAP Customer Relationship Management (CRM) can be the platform that bundles all the tools needed for powerful customer service. Especially with the new SAP CRM on SAP HANA and 360Customer, you can achieve proactive customer information and 360-degree insight into all aspects of the customer relationship. Central access to all functions is provided by the Interaction Center (IC) Web Client, which also offers telephony integration to automatically determine the correct master data for the calling party upon recognition of the caller's telephone number. The retrieved master data set can then be used throughout the customer contact, be it for marketing or sales activities, complaint handling, incident management, or financial customer care such as management of disputes. Figure 1.8 shows the five pillars of customer interaction, which it makes sense to handle via SAP CRM in the logistics service business.

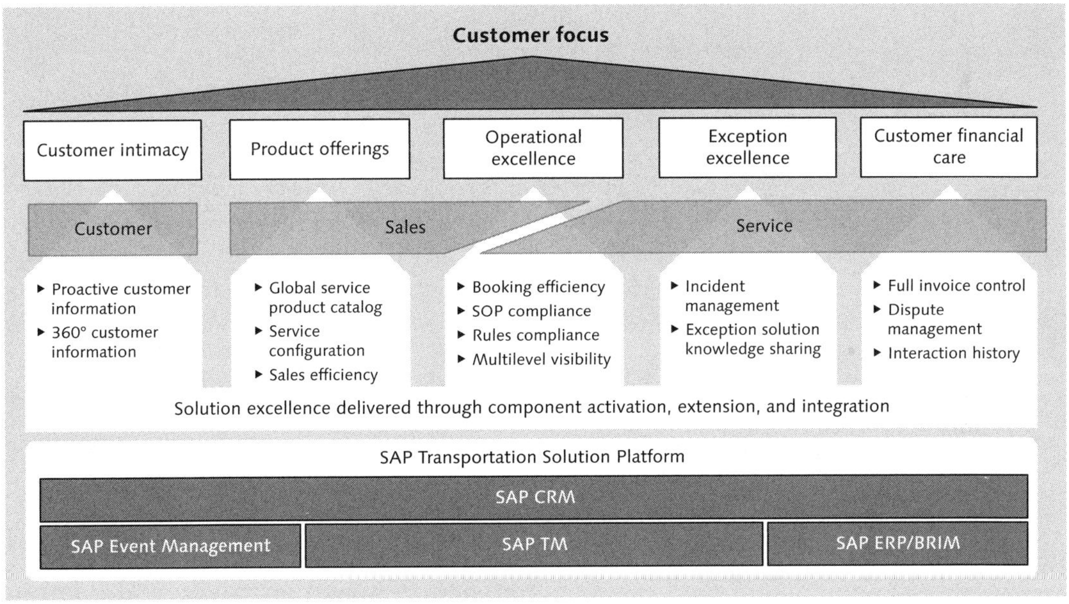

Figure 1.8 Customer Relationship Functions for LSPs

Not all of the functions listed in the figure are currently fully integrated with SAP TM. Product offerings and operational efficiency have the largest footprint when it comes to standard SAP TM and SAP CRM support. Once the customer has been identified, and the corresponding master and transactional data has been loaded, you can use SAP CRM to create quotes, contracts, and forwarding orders in SAP TM, making use of the defined global product catalog and its services, existing contracts, tariffs and rates, the assigned standard operating procedures (SOPs), and the defined compliance rules.

However, we have seen customer project implementations at worldwide freight-forwarding companies that far exceed current standard integration because all SAP TM objects in the customer system are embedded into corresponding SAP CRM objects, allowing efficient use of SAP CRM activity management to control marketing and sales workflows. In addition, SAP TM and SAP Event Management have been integrated with SAP CRM Incident Management to allow real-time notification to customer service about all issues coming up in the logistics network, thus enabling immediate incident resolution—if required—in collaboration with the customer.

One more extremely helpful tool is Financial Customer Care (FCC), which makes all SAP Billing and Revenue Innovation Management functions involving customer invoices, customer payments, disputes, collections, and cashiering available to customer service representatives. You can find details on the standard integration of SAP CRM and SAP TM in Chapter 4 and Chapter 10.

1.2.9 SAP Billing and Revenue Innovation Management

SAP Billing and Revenue Innovation Management (BRIM) has been consolidated out of multiple components that were previously promoted separately:

- Convergent Charging (CC) is a high-speed price calculation engine that allows event-based pricing (formerly known as Highdeal software).
- Convergent Invoicing (CI) is an extremely flexible billing solution that offers billing control by multiple customer profiles (contract accounts). Billing can be driven by events, and grouping of billing items into final invoices is rule based.
- Financial Contract Accounting (FI-CA) is a sub-ledger accounting system that offers very flexible accounting implementations and the ability to integrate into accounts receivable (FI-AR) as a general ledger.

- Financial Customer Care (FCC) allows you to dispute invoices on the invoice item level (e.g., if a customer complains about 1 of 100 invoice items, he can pay 99 items and leave only the one item open). It also helps you to manage your receivables and collections and do cashiering (in logistics businesses, many orders still have to be prepaid in cash before the cargo is moved). FCC runs as part of SAP ERP (like CI and FI-CA) but is virtually integrated into SAP CRM, where customer service has direct access to all relevant receivables-related functions out of the interaction center.

SAP BRIM has been integrated with SAP TM based on a custom development project and is available as an add-on to SAP TM (as Repeatable Custom Solution). We explain BRIM in more detail in Chapter 11, Section 11.3.

1.3 Functional Overview of SAP TM

By now you know that SAP TM is the main component used to administer any kind of transportation processes, either shipper- or LSP-focused.

History of SAP TM

As a functional component, SAP TM already has a long history within SAP logistics systems and components. Shipper-focused solutions have been provided to large, worldwide operating customers since 1987, when the mainframe system R/2 was released with a module called RT, which allowed the organization of multi-modal transports. Its successor on the well-established ERP system R/3 was the Transportation component of Logistics Execution (LE-TRA), which was started in 1993 and released with R/3 3.0 in 1995. Functionally enhanced in the following releases, it grew into a comprehensive solution for shippers to manage their transportation requirements including cost calculation and settlement. In 2000 it was enriched by the TPVS module of SAP APO, which brought transportation network optimization to SAP ERP.

Although still widely used in a large community of manufacturing and trading companies, LE-TRA has been overtaken in functional richness and integration capabilities by SAP TM. Figure 1.9 displays the history of SAP's transportation software.

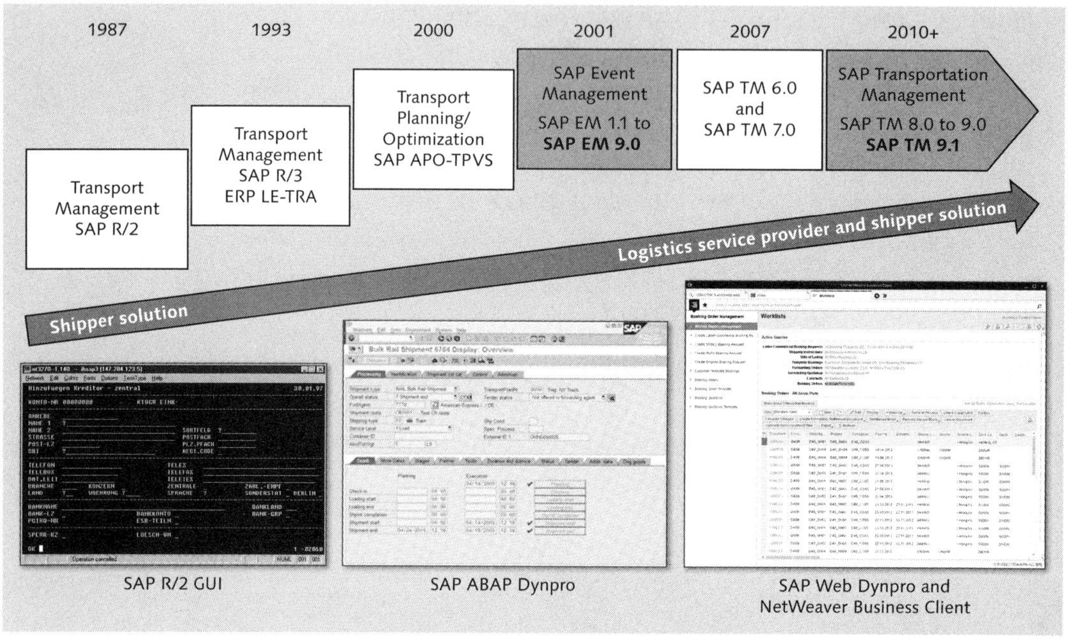

Figure 1.9 History of Transportation Components

After a less-than-successful start in 2007, SAP TM was redesigned and relaunched in 2010 as SAP TM 8.0, which mainly focused on shipper processes but also allowed forwarding by road transportation. In SAP TM 8.1, functionality for ocean freight-forwarding (FCL and LCL) was added, opening the door into the LSP market. This was extended with TM 9.0 into air freight-forwarding. Meanwhile, SAP TM is going into its fourth release; it provides a very comprehensive transportation management platform with SAP TM 9.1, which has brought enrichments for rail and truck carriers and shippers. Based on SAP TM 9.0, multiple enhancements have been made by SAP custom development to support railways or container shipping lines. For more information on container shipping lines, see Chapter 16.

SAP TM handles comprehensive scenarios

The organizational capabilities and functional abundance of SAP TM enable its use for a variety of shipper- and LSP-related transportation and logistics processes, such as direct, multistop and multistage, full truck load (FTL), less than truckload (LTL), and ocean carriage door-to-door or port-to-port for full container load (FCL) or less than container load (LCL). Consolidated air freight scenarios, rail carriage, and intermodal processes are supported.

Figure 1.10 shows an example of a door-to-door air freight scenario from Bremen, Germany to Bangkok, Thailand that uses local processes such as pickup in Bremen and delivery in Bangkok and consolidated long-haul truck moves for pre-carriage as well as capacity reservation and schedule-based movements such as the air freight main leg from Frankfurt to Bangkok. Different flight options have to be evaluated to find the most cost-efficient way and to guarantee compliance with customer service-level agreements (SLAs).

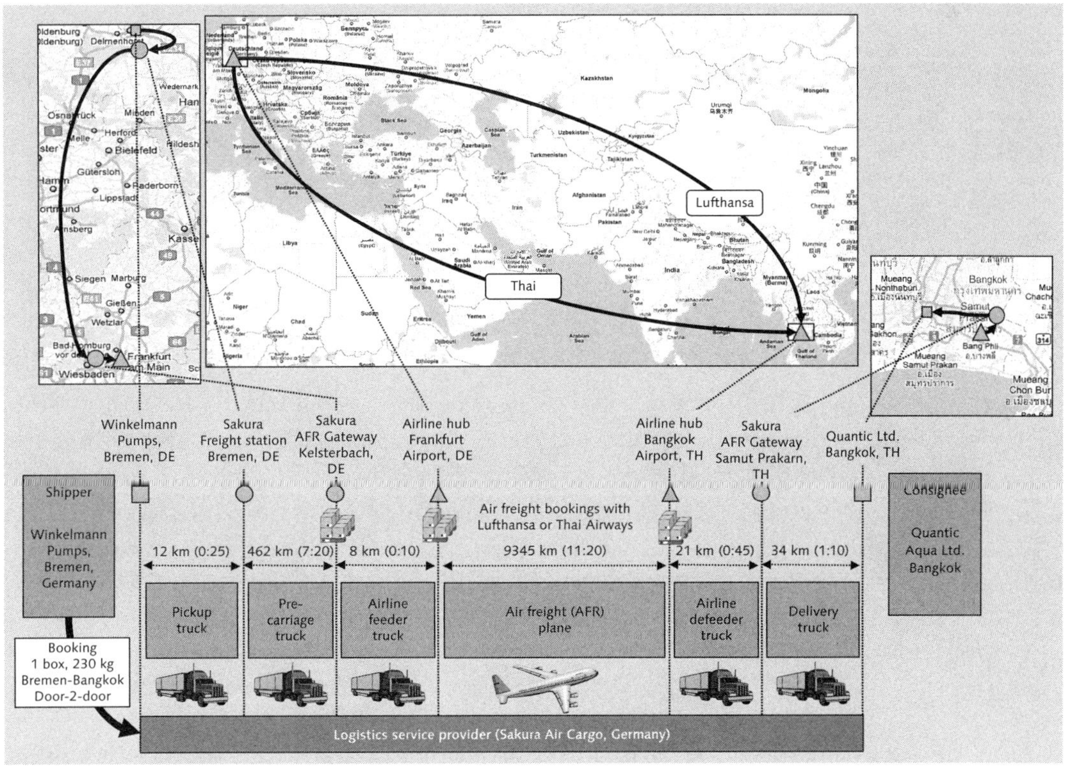

Figure 1.10 Example of Door-to-Door Air Freight Scenario

Role for logistics scenarios

Looking at an example like the air freight scenario, you can imagine that the geographical and mode-specific aspects imply that the operational and transactional execution have to be handled by different people in different roles. At an LSP, customer service representatives, local station operators, documentation teams, and long-haul dispatchers are actively involved with various tasks. This "handshake" is organized in SAP's transportation solution, where, with appropriate authorization, each participating person or group gets access to role-specific functionality.

Figure 1.11 breaks the process down into example transactional tasks required for an air freight transport with the involvement of six groups and the corresponding task assignments.

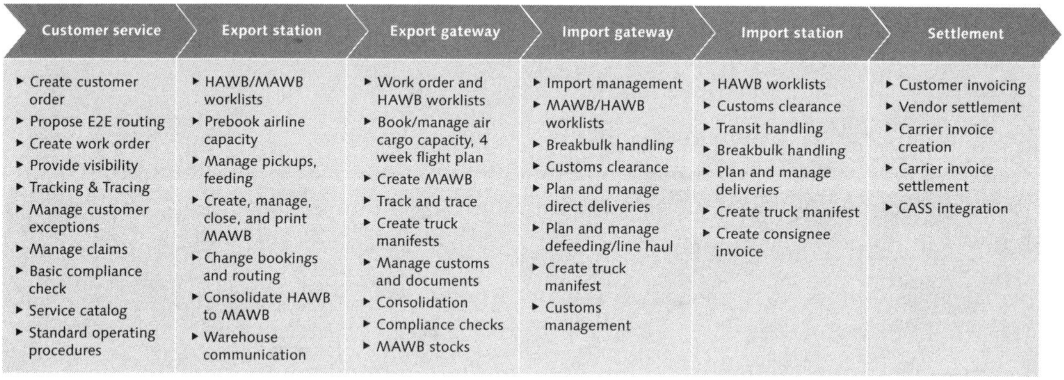

Figure 1.11 Air Freight Process Tasks in Different Organizations

Integration between SAP TM instances

SAP TM provides powerful workflow capabilities that easily allow assigning and controlling the flow of information and tasks within and between areas of responsibility in a company. Beyond that, SAP TM can integrate processes between several SAP TM instances (e.g., from a shipper using SAP TM as his local transportation planning system in combination with SAP ERP, to an LSP managing his customer orders with SAP TM).

Figure 1.12 shows an example of this type of situation. The shipper, the LSP, and the carrier are all running their own SAP TM instances. For the shipper, SAP TM is linked to his SAP ERP logistics system to arrange transportation. For the LSP and the carrier, SAP TM is used to handle their core business—that is, selling, arranging, buying, and executing freight and cargo moves for their customers.

Cross-system process flow

In Figure 1.12, the process flow across the systems is as follows:

1. The shipper receives an order for goods from the customer and starts a sales and delivery process in his SAP ERP system.
2. To arrange transportation, he uses SAP TM to plan shipments based on sales orders and deliveries.
3. The planned transportation orders for subcontracting are now communicated as door-to-door service orders to the LSP, who receives them as customer orders for transportation services.
4. The LSP uses planning tools in his own SAP TM instance to create transportation orders that he might subcontract to one or more carriers (such as a local truck fleet) or execute himself if he can.
5. In subcontracting, the carrier receives the transportation orders from the LSP as customer orders in his SAP TM system and executes them.
6. As the middleman, the LSP now needs to settle customer receivables and vendor payables.
7. The customer receives an invoice from the LSP for the services ordered.
8. The carrier receives a payment for the services executed for the LSP.

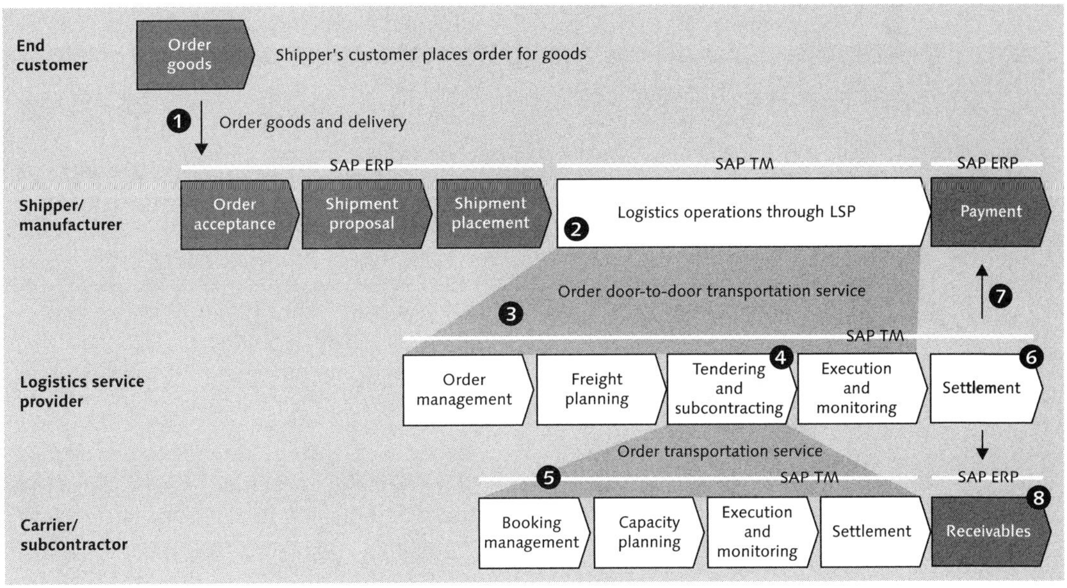

Figure 1.12 SAP TM Instances of Different Usage Levels Can Be Integrated

1.3.1 Mode of Operation of SAP TM

SAP TM business objects

Let's take a closer look at the overall mode of operation of SAP TM without going into too much detail. First, we introduce the main business objects of SAP TM. Table 1.1 lists their naming conventions, abbreviations, and definitions.

Business Object	Abbreviation	Definition/Description
Order-based transportation requirement	OTR	An order object representing the transportation demand from a sales, purchasing, or stock transfer order in an SAP ERP system.
Delivery-based transportation requirement	DTR	An order object representing the transportation demand from an inbound or outbound delivery in an SAP ERP system.
Forwarding quotation	FWQ	An offer from a carrier or LSP to an ordering party for the transportation of goods, which contains information about the price and other conditions related to the transportation services.
Forwarding order	FWO	An order from an ordering party to a carrier or LSP to transport goods from a shipper to a consignee in accordance with agreed terms and conditions.
Freight unit	FU	A set of goods that are transported together across the entire transportation chain. A freight unit can include constraints for transportation planning.
Resource		A machine, means of transportation, or other asset with a limited capacity that fulfills a particular function in the supply chain (e.g., truck, container, crane).
Schedule		A sequence of stops with related recurring departure and arrival times that is valid for a specified time period. Cargo associated with the schedule may move along the sequence or any part of it.

Table 1.1 Main Business Objects of SAP TM

Business Object	Abbreviation	Definition/Description
Freight booking	FB	An order providing transportation capacity whose execution is planned by a carrier, for example, a ship owner. The freight booking contains the plan for the logistical processing (e.g., fixed departure times of the ship).
Freight order	FO	An order whose execution is planned by a carrier or the shipper. The order contains the plan for the logistical processing (e.g., when and onto which vehicle freight units are to be loaded and planned departure times for the vehicle), and execution data.
Forwarding agreement	FWA	A long-term contract that represents the contractual relationship with a customer to whom you are selling transportation services.
Freight agreement	FA	A long-term contract that represents the contractual relationship with a carrier from whom you are buying transportation services.
Forwarding settlement document	FWSD	A document that is sent to SAP ERP to request the creation of an invoice for logistic services to be sent to a customer.
Freight settlement document	FSD	A document that is sent to SAP ERP requesting the verification of an invoice for logistic services received from a supplier or carrier.

Table 1.1 Main Business Objects of SAP TM (Cont.)

The business objects described in Table 1.1 are used in combination with various processing tools and engines to run transportation management operations. Figure 1.13 shows a typical process flow and the interaction of objects, engines, and some master data.

Requirements Management

The following descriptions and numbering refer to Figure 1.13. The Requirements Management component comprises all functions involving SAP TM integration with the SAP ERP logistics processes. It is frequently used in shipper transportation management, where a company needs to manage transportation for their own distribution and procurement tasks, but may also be applied in a tight collaboration in contract logistics.

OTRs are created from SAP ERP sales, purchasing, or stock transfer orders **A1**. By means of the Transportation Planning component (see description later in this section), freight units and freight orders are created **A2**, which are used to create delivery proposals **A3** that are sent back to SAP ERP to create deliveries according to the transportation plan. The deliveries may be altered in SAP ERP and, once released, are transferred to SAP TM, resulting in the creation of DTRs that consume the planning result based on the previously used OTRs **A4**.

In this way, the initial planning result is reused and adjusted without starting from scratch. Based on the freight orders or freight bookings of the planning result (which are actually the same objects as the ones described later in this section under subcontracting and tendering), you can subsequently create shipments in SAP ERP **A5**, which allows execution within the SAP ERP logistics context.

Order Management

Order Management is the main component for LSPs, where they start the sales processes for their customers. Transportation demand from customers is received as a customer order that creates a forwarding order in SAP TM **B1**. Even if the name of the object is closely related to freight-forwarding, the same object is also used for carriers to handle customer orders (such as *booking* for air and ocean cargo or a *waybill* for rail cargo). In the context of forwarding orders, a variety of processing and check functionality can be utilized:

- Transportation proposals to get routing options
- Credit limit checks for the order against SAP ERP Financial Supply Chain Management (FSCM)
- Blacklist and denied party screening against SAP GTS compliance lists
- Dangerous goods checks against mode- and country-specific rules defined in SAP EHS Management

A forwarding order can be based on a previously created forwarding quotation, which is a spot quote for a transportation service given to a customer **B2**.

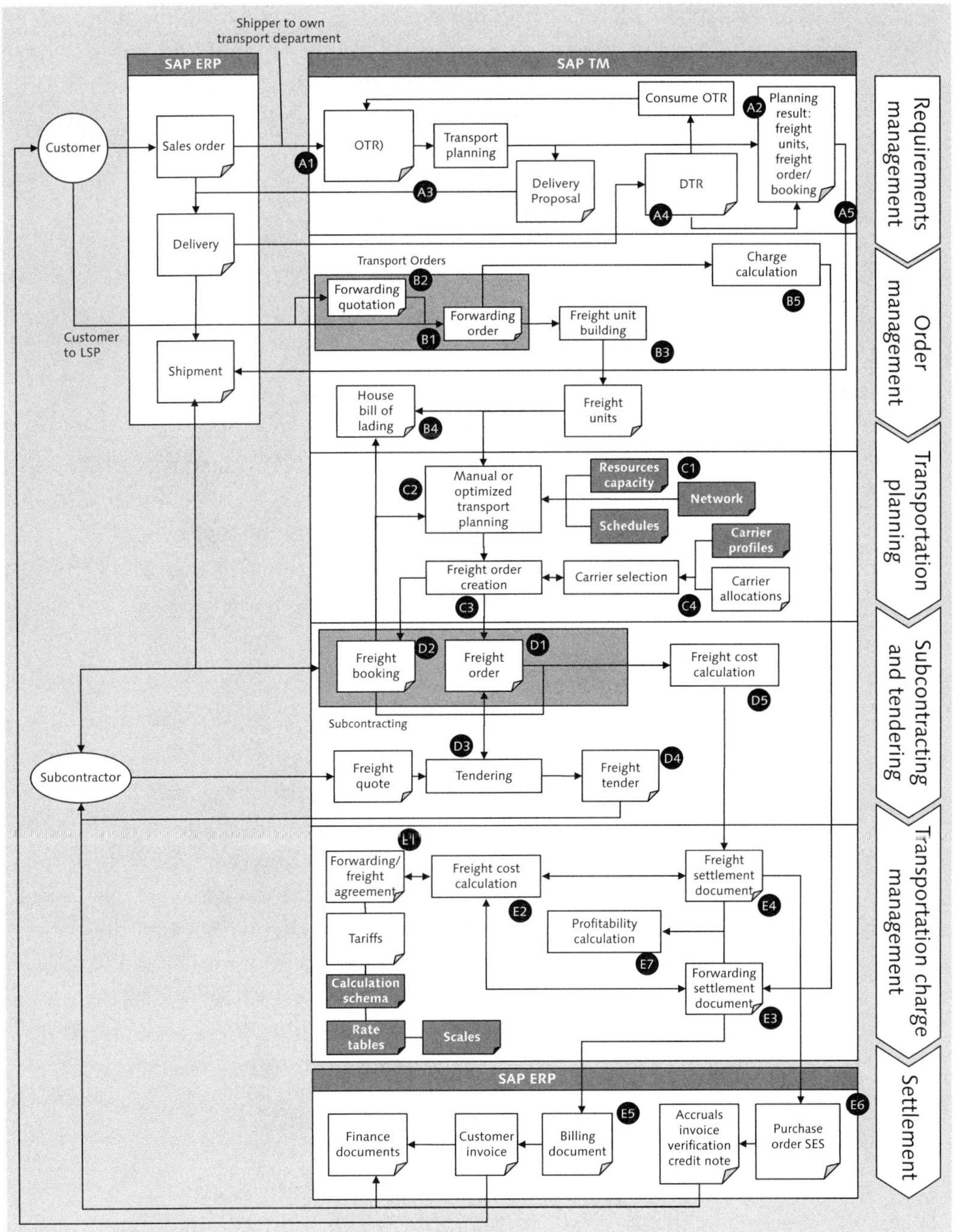

Figure 1.13 Typical Usage of SAP TM Business Objects

Forwarding orders are used to create freight units from the items of the order B3. Freight unit building is a planning service used to break large orders down into pieces that can be physically handled separately (e.g., on pallets or in truckloads) and that also can go different ways based on logistical or regulatory considerations. At a later stage, house bills of lading are generated based on the freight units and assigned freight bookings B4 to comply with legal documentation regulations.

Forwarding orders integrate with SAP TM Transportation Charge Management, from where they use services such as charge calculation to determine the price components of a customer order B5.

Transportation Planning

Transportation Planning provides manual and optimizer-supported planning capabilities to SAP TM and allows multi-modal end-to-end planning with consideration of real and virtual transportation costs. It can be used to create a release-ready transportation plan and to determine a list of transportation proposals for possible routings of an order. Transportation Planning involves owning and making use of a variety of network and resource master data (which are used to define the network), resources with their capacity, and schedules for recurring multistop connections C1.

Operational planning is done in the transportation cockpit as a user interface that allows manually performing activities like assigning freight units to a means of transport and running a parameterized optimization over the whole or a selected subset of the transportation demand. For this activity, the freight units, network, capacity, resources, and schedules—as well as existing and new freight orders and freight bookings—are taken into account C2. Based on the planning result, freight orders are created or freight units are assigned to freight bookings C3. During or after this step, carrier selection can be executed, which allows the assignment of one or multiple carriers to a freight order according to carrier profiles, allocation rules, and selection rules C4.

Subcontracting and Tendering

The Subcontracting component provides order objects representing the relation to service vendors or carriers. Freight orders can be created manually or as a result of a planning run D1. They provide an individually planned definition of a consolidated or unconsolidated freight move to be executed with a means of transport (e.g., a milk run in a city or an FTL move across a country). Freight bookings are used to represent a capacity allocation prebooked on a particular means of transport, which is often run on a schedule and operated by a carrier within his network D2. Examples are air freight capacity reservations for a master airway bill (MAWB) or booked container capacity on a container vessel. For freight bookings, SAP TM offers comprehensive capacity management that allows you to plan and allocate the capacity required to execute transportation services as a forwarder or carrier (e.g., maintenance of master flight plans and four-week flight plans).

Freight bookings usually have an assigned carrier (airline, ocean liner, or railway). For freight orders, individual tendering can be done based on a list of preferred vendors D3. Tendering can be executed as peer-to-peer, broadcast, or open. The partners may be integrated either via B2B messaging or through access by a vendor portal. Freight tender and freight quote objects make it possible to keep track of the process and provide the decision basis for carrier selection D4. The same way Order Management is integrated into SAP TM Transportation Charge Management, Subcontracting uses this component to determine the costs of a move including apportionment to the single freight units of a consolidated shipment D5.

Execution

The freight order and freight booking business objects used in Subcontracting are also the foundation of the execution process and cargo management. SAP TM tools support a variety of processes for these two objects:

- Loading and unloading of consolidated transportation units and loose cargo
- Discrepancy handling in hubs and stations
- Consolidation of cargo to and deconsolidation from transportation units (e.g., containers and pallets)
- Status management for cargo and shipments
- Creation of legal, regulatory, and operational documentation such as manifests and bills of lading

Transportation Charge Management

When it comes to cost, revenue, and profitability in transportation, SAP TM Transportation Charge Management is the component of choice. It integrates into order and subcontracting processes and provides structured contract, tariff, and rate data for all calculations.

In SAP TM, agreements are the main contract objects that can be used as forwarding agreements to represent a customer contract as well as freight agreements to embody a supplier contract E1. From a structural perspective, these contract types are similar because they contain tariffs referring to calculation schemas that hold a list of charge elements that are applicable for charge calculation either as prices for a customer or costs for a supplier. Each charge element typifies a certain kind of rate, fee, or surcharge (e.g., basic sea freight or port congestion surcharge). To efficiently negotiate a supplier contract or close a customer contract, SAP TM comes with a component called Strategic Freight Management, which allows the creation of contract quotes to suppliers and has analytical tools to analyze and compare bids. From a selected bid, a freight agreement can be assigned to the corresponding supplier.

The freight cost calculation engine analyzes the charge elements of a calculation sheet and calculates the correct amounts in the correct currencies and with the appropriate exchange rates E1. The calculation results are stored in the charge substructures of the forwarding order (customer invoice) or the freight order and freight booking (supplier invoice).

Settlement

From the forwarding order, freight order, and freight booking business objects, you can create draft invoices for settlement integration to SAP ERP. To create customer invoices, one or more forwarding settlement documents are created from the forwarding order and then are sent to SAP ERP to create billing documents (E3 and E5). For supplier invoice settlement, freight settlement documents are created from the freight order or freight booking; then they are transferred to SAP ERP to create accruals and later to create a purchase order and service entry sheet (SES) for invoice verification or self-billing (E4 and E6). Based on the charges calculated on the cost and revenue side, SAP TM produces a profitability calculation for either standard or real costs E7.

1.3.2 Overview of the Following Chapters

So let's walk through an overview of what you will find in the following chapters.

- **Chapter 2: Solution Architecture and Technological Concepts**
  The second chapter gives an overview of the technical architecture of SAP TM and how integration is done. We explain the business object modeling basics (BOPF, business object modeling, BAdIs) and describe the foundational tools of SAP TM, which are referenced often in later chapters. These tools include the Business Rules Framework (rules engine, BRF+), Post-Processing Framework (printing/output), and personal object worklists (POWL) as a central means to provide workflow capabilities. The third section describes technical integration by services and change handling within SAP TM (Change Controller).

- **Chapter 3: Master Data**
  This chapter explains the general master data of SAP ERP and SAP TM (e.g., business partners) and transportation-specific master data such as networks and resources. It also gives an overview of CIF integration used for master data transmission between SAP ERP and SAP TM.

- **Chapter 4: Transportation Requirements and Order Management**
  This chapter explains how to create and manage transportation orders in SAP TM. It uses shipper scenarios and forwarder/carrier scenarios to explain the capabilities of the customer order objects and quotations and the provision of their functionality for customer service and sales.

- **Chapter 5: Transportation Capacity Management**
  For shippers and LSPs, management of freight capacities and schedules is an important aspect of being able to move cargo. In this chapter, the capacity management process is described, including allocations, freight bookings, and their interplay with schedules.

- **Chapter 6: Transportation Planning**
  Transportation planning deals with the activities involved in the assignment of cargo items to vehicles or reserved capacities on trucks, trains, planes, or vessels. This chapter describes freight units as the

basis for planning and transportation units to model truck and trailer scenarios, and gives an overview of the interactive and optimized planning capabilities of SAP TM. It introduces and explains configuration and utilization of the transportation cockpit, the optimizer, transportation proposals, and configuration profiles.

- **Chapter 7: Carrier Selection and Subcontracting**
  This chapter explains freight orders and their use as subcontracting documents. The freight orders can be subcontracted to carriers or other service providers. Relevant carriers are determined through carrier selection. Subsequently, a tendering process can be executed to determine the best available price, conditions, and availability of the selected carriers. Carrier selection is a part of the optimization process that allows you to propose and select one or multiple carriers for subcontracting.
- **Chapter 8: Transportation Execution and Monitoring**
  Execution and monitoring deal with handling freight and providing visibility of shipments. In Chapter 8 we explain the different options to manage the cargo status of freight (i.e., freight document functions, discrepancy handling, loading status, and paperwork). In addition, we describe aspects of export and import handling in international supply chains and the corresponding setup in SAP TM. SAP Event Management is SAP's solution to provide visibility and tracking and tracing functionality to support the SAP TM processes. This chapter highlights the features of SAP EM and the processes that allow out-of-the-box integration with SAP TM.
- **Chapter 9: Transportation Compliance**
  This chapter deals with compliance issues arising from trade regulations and hazardous cargo. It explains the integration and functionality of SAP GTS, which supports various functions from blacklist screening to export and import compliance handling. Various tasks of handling, checking, and documenting hazardous cargo are done using SAP EHS Management. Like SAP GTS, SAP EHS Management is integrated with the SAP TM processes.
- **Chapter 10: Transportation Charge Management**
  This chapter explains the setup of agreements (contracts) and the definition of tariffs and rates. Calculation of charges within forwarding

and freight orders is described in detail. We also build a link to the global service product catalog because the service product is used as a core element of agreements. Charge calculation master data with service products, forwarding and freight agreements (contracts), calculation sheets, rate tables, scales, maintenance functions, upload and download, and contract determination are described.

Once a contract is determined on which the charges can be calculated, the individual characteristics of forwarding and freight orders are evaluated to execute the calculation and generate a list of charge items to be invoiced or paid in the charge settlement process.

- **Chapter 11: Charge Settlement**
  Charge settlement deals with the invoicing of all costs and revenues that are relevant to transportation. Forwarding and freight settlement documents are created to initiate payments or transfer. The chapter explains invoice creation and invoice verification in SAP ERP, which is based on the settlement documents. There are special means to distribute costs and to conduct internal settlement between LSP organizations. Additionally, flexible settlement with SAP ERP convergent invoicing can be achieved based on the SAP TM forwarding settlement documents.
- **Chapter 12: Integration with Other Components**
  This chapter describes the integration between SAP TM and SAP EWM, analytics (SAP BusinessObjects BI), and mobile applications.
- **Chapter 13, Chapter 14, and Chapter 15: Process Descriptions**
  Three process description chapters provide insight into the details of a shipper's outbound process, an LSP ocean LCL process, and a consolidated air freight process.
- **Chapter 16: Scenario Implementation Best Practices**
  This chapter highlights specific topics to keep in mind when you implement an SAP TM process at a customer site. It also gives an overview of how to "flavor" SAP TM processes and enhance standard functionality to cover areas that are not fully supported by the standard SAP TM, using liner shipping as an example.
- **Chapter 17: Summary and Outlook**
  This chapter concludes the coverage of SAP TM by providing a synopsis of the challenges the industry faces and the future direction of SAP TM.

When you get started with SAP Transportation Management, you'll quickly notice two important technological concepts that are different from the traditional SAP ERP technology. The user interface now uses current web technology, which enables the user to customize the UI easily, and data is now modeled in an object-oriented way.

2 Solution Architecture and Technological Concepts

As you read this book, especially in Chapter 4 through Chapter 11, you'll notice that we rarely use transaction codes and ABAP programs. There's a reason for this: SAP Transportation Management (TM) does not use the SAP GUI interface; it uses web user interfaces instead.

So what does this mean? SAP TM's web-based user interface (UI) can be used without the user having to install any frontend software on his computer. This is especially helpful for users who don't use the application frequently or only execute one business transaction in the system. In the shipping process described later in this book, you can also see that it is very easy for external business partners to access the system and collaborate in transportation-related business processes.

This chapter will give consultants and technical experts insight about how to customize the UI in order to adjust the terminology and screen layout to the customer's needs.

Like other SAP applications developed after the year 2000, SAP TM has moved away from the traditional framework of retrieving both master data and transactional data directly from database tables. The new framework, called the Business Objects Processing Framework (BOPF), encompasses both data storage and data processing. It is the framework of choice not only in SAP TM, but also in other SAP applications such as

SAP Business ByDesign and SAP HANA. We delve deeper into the contents of data storage and data processing in this chapter.

For readers who have already dealt with the traditional technological concepts of SAP applications, the BOPF may seem cumbersome at first glance. However, many developers agree that this new framework simplifies the way of designing a process in a program. In this chapter, you learn that the link between different pieces of information is much closer than with the traditional framework; getting information and understanding how information is related is more tangible because the different pieces of information now follow a hierarchical structure that can also be illustrated by technical drawings.

Although this chapter is not designed to make you a development expert who knows all the tricks and terms of the BOPF, we do want to give you a basic understanding of how you can interact with the BOPF so you understand where to find data and where to get started if you are looking for the root cause of a problem.

Don't Get Confused about Names!

Even though we will be talking about business objects in this chapter, it is not related to SAP BusinessObjects products such as SAP BusinessObjects Business Intelligence (BI) or SAP NetWeaver Business Warehouse (BW). However, we *do* talk about integrating SAP TM into the analytics applications (and thereby SAP BusinessObjects) in Chapter 12, Section 12.2.

For this chapter, the term *business objects* only refers to the Business Objects Processing Framework used in SAP TM.

As we will discuss in Section 2.3, SAP TM uses various tools and frameworks that allow a consultant or even the user to customize and personalize the application without having to consult a programmer. These features come in very handy because they reduce the number of modifications.

The reduction of modifications is a big advantage when it comes to support and troubleshooting. The coding itself is still standard, and customizing of the system is done in a different layer. If customizing has a negative effect on system behavior, you can undo it; removing the customizing layer quickly resets the functionality to standard. No programming is involved in this process.

In this chapter, we give you a glimpse at the most important tools and frameworks used in SAP TM and explain how to use them in the SAP TM context. Remember, of course, that each of the tools described in this chapter has more functionality than what is described here; we focus on the most common use cases in SAP TM. Many other resources are available that give a detailed focus on all of the functionalities of these tools and frameworks.

2.1 Technological Foundation of SAP TM

If you've ever dealt with software architecture, you probably have heard the term *service-oriented architecture* (SOA) countless times. SOA is not only an architecture for exchanging data between different business partners or systems, but also a new way of data modeling within a system itself.

Business Object Processing Framework (BOPF)

The BOPF is the central technological foundation of SAP TM and incorporates the idea of modeling data in an SOA-compliant way. As a consultant or application expert, you will not find a way around eventually dealing with this framework.

The BOPF models the storage of data in an object-oriented way, but also merges the processing of data into the same framework. Therefore, data storage and data processing are closely linked.

When you access Transaction /BOBF/CONF_UI, you will find an overview of all business objects used in SAP TM. Figure 2.1 shows how these business objects are grouped into different types. From a technological point of view, these objects are all alike, but from a business point of view, this grouping makes sense.

Let's look at a few of these types. *Business process objects* store transactional data such as customers' transportation requests or transportation orders. As the name suggests, you can find master data stored in *master data objects*, such as vehicle resources or locations. Especially when you are doing transportation planning in SAP TM, you will enter various profiles and settings that are used for optimizer planning, carrier selection, and so on. This data is stored in *metadata objects*.

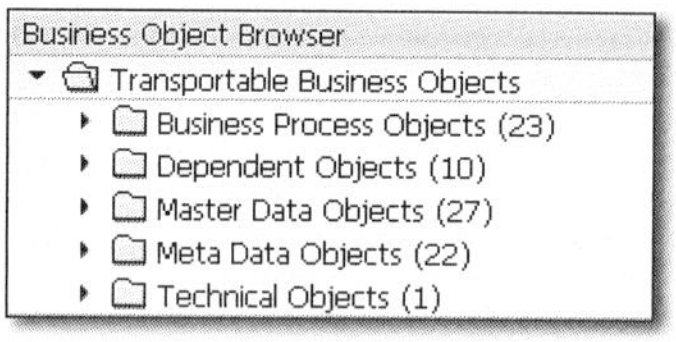

Figure 2.1 Business Object Types

2.1.1 Storing Data in the BOPF

Data storage in the BOPF

Because the BOPF stores data in an object-oriented way, each document or master data item is stored as an object instance in the database. Throughout this chapter we use a *freight order* (the order of a transportation service) as an example to illustrate how data is stored.

The freight order has some unique header data such as the order number and the carrier assigned. The special characteristic of this information is that it can occur only once per freight order—with good reason. A freight order with two numbers does not make sense.

But the freight order also has information that can have a different cardinality in any order. If you look at the items in the order, you will see that some freight orders carry only one item, while others have multiple items. All items, however, are always linked to unique information, such as the order number. We could say that the item information is assigned to the header information.

Business object node structure

The BOPF reflects the information structure in its business object structure. When you access Transaction /BOBF/CONF_UI, take a look at the freight order's technical setup by double-clicking on the business process object /SCMTMS/TOR. A business object consists of different nodes, each with its own purpose. All business objects have one characteristic in common: the superordinate *root node*, which contains the header information we just introduced. The subordinate nodes contain information that is assigned to the header information, such as the items. Subordinate nodes can occur multiple times within one instance of a business object.

Note

When we look at the technical setup of a freight order, we're talking about an *instance* of a business object. The business object itself is the definition of the structure. An instance uses this structure and is filled with data.

As you see in Figure 2.2, the item information (ITEMS OF THE TRANSPORTATION ORDER) is stored in node ITEM_TR, while other information (optional for a freight order) is stored in other nodes.

Information storage in a node

All information is stored in the nodes in a structured way. If you double-click on a business object node, you can "zoom in" and take a detailed look at the structure definition of the node.

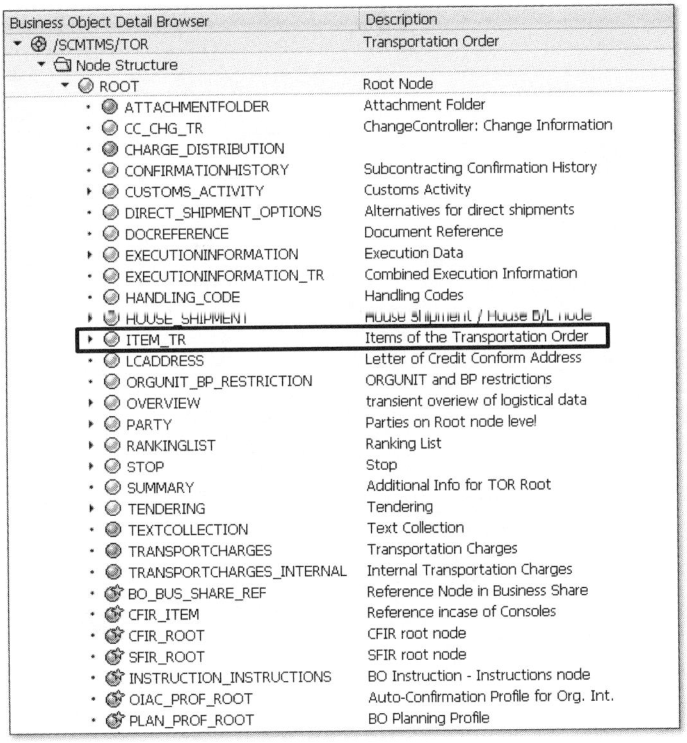

Figure 2.2 Overview of Business Object Nodes

As you can see on the right side of Figure 2.3, a simple *Data Dictionary* (DDIC) structure is assigned to each node. This structure is used as a *data structure*. The data structure contains all fields that can be filled with information such as the freight order number, carrier, and so on. Double-clicking on the data structure takes you to the DDIC (Transaction SE11) so you can see the structure.

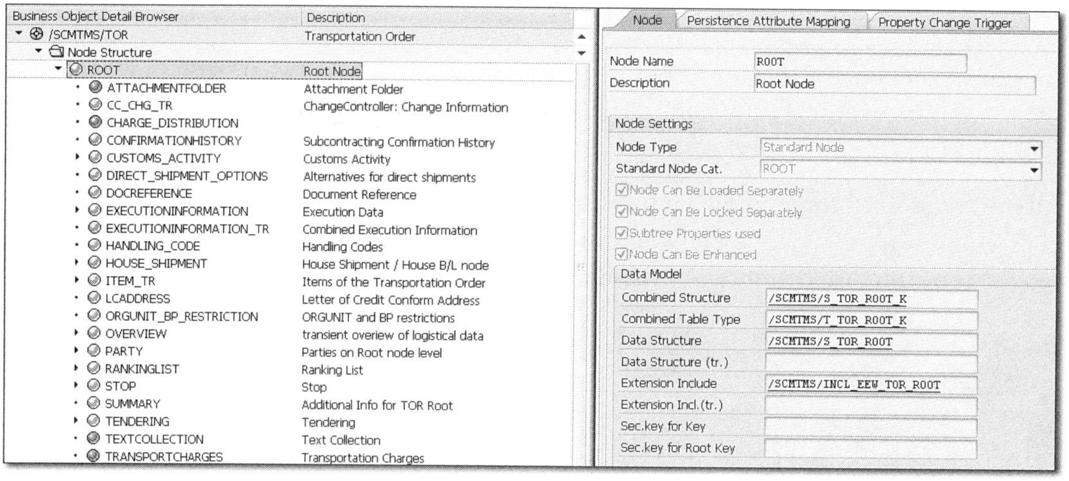

Figure 2.3 Node Information

Data structures and database keys

Figure 2.3 also shows that a combined structure (shown in Figure 2.4) is assigned in addition to a data structure. The combined structure looks exactly like the data structure except that it has three additional fields at the start of the structure:

- Each node instance can be identified in the database with a unique, 32-digit hexadecimal *key*, such as `005056AC01921ED1BEE25DC2FC88401C`.
- In order not to lose the link to its direct superordinate node, each node carries the *parent key* of its superordinate node.
- The *root key* field contains the key of the root node instance that this node instance is part of.

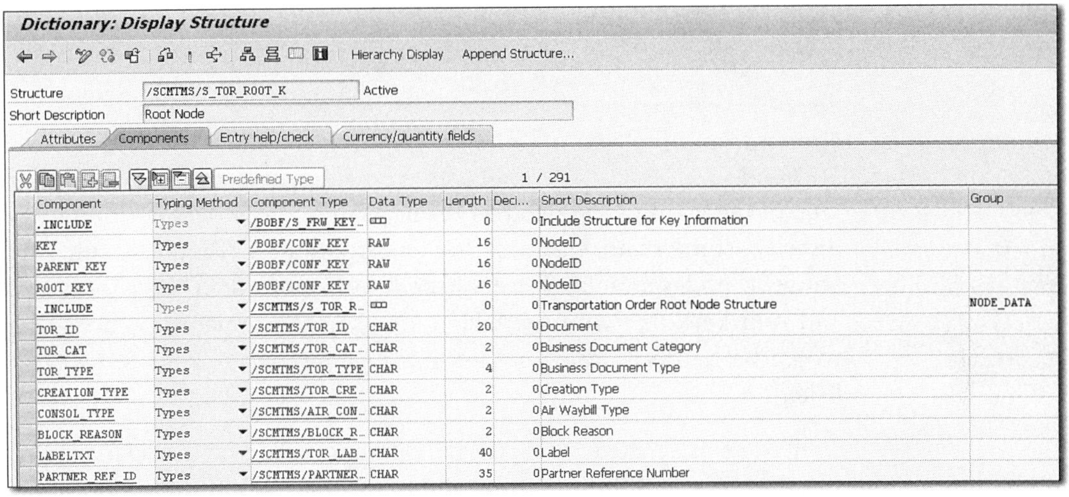

Component	Typing Method	Component Type	Data Type	Length	Deci...	Short Description	Group
.INCLUDE	Types	/BOBF/S_FRW_KEY...		0	0	Include Structure for Key Information	
KEY	Types	/BOBF/CONF_KEY	RAW	16	0	NodeID	
PARENT_KEY	Types	/BOBF/CONF_KEY	RAW	16	0	NodeID	
ROOT_KEY	Types	/BOBF/CONF_KEY	RAW	16	0	NodeID	
.INCLUDE	Types	/SCMTMS/S_TOR_R...		0	0	Transportation Order Root Node Structure	NODE_DATA
TOR_ID	Types	/SCMTMS/TOR_ID	CHAR	20	0	Document	
TOR_CAT	Types	/SCMTMS/TOR_CAT...	CHAR	2	0	Business Document Category	
TOR_TYPE	Types	/SCMTMS/TOR_TYPE	CHAR	4	0	Business Document Type	
CREATION_TYPE	Types	/SCMTMS/TOR_CRE...	CHAR	2	0	Creation Type	
CONSOL_TYPE	Types	/SCMTMS/AIR_CON...	CHAR	2	0	Air Waybill Type	
BLOCK_REASON	Types	/SCMTMS/BLOCK_R...	CHAR	2	0	Block Reason	
LABELTXT	Types	/SCMTMS/TOR_LAB...	CHAR	40	0	Label	
PARTNER_REF_ID	Types	/SCMTMS/PARTNER...	CHAR	35	0	Partner Reference Number	

Figure 2.4 Combined Structure in Transaction SE11

The "Family" in Object-Oriented Modeling

In object-oriented modeling, "family" terms are used instead of the lengthy terms *superordinate* or *subordinate.* (IT people tend to find short terms if abbreviations do not do the trick.) Therefore the terms *parent node* and *child node* are more often used here.

Database tables

We have talked about *how* data is stored, but we have not yet talked about *where* data is stored. And even though we're introducing new ways to model data in nodes and objects, at the end of the day we are back in the world of database tables.

The BOPF only introduces new means for data modeling; data storage is done with database table technology that you already know about from SAP products such as SAP ERP. However, the database tables contain the 32-digit keys of the node instances with which we can easily find node instances on the database tables. The details of a node like the one in Figure 2.5 give you information about where data is stored. The structure of the database table is defined with a DDIC table type that can also be found in the node information. The table type is the *combined table type* that you can also see in Figure 2.3 just above the entry for the data structure. It usually has the combined structure type as line type.

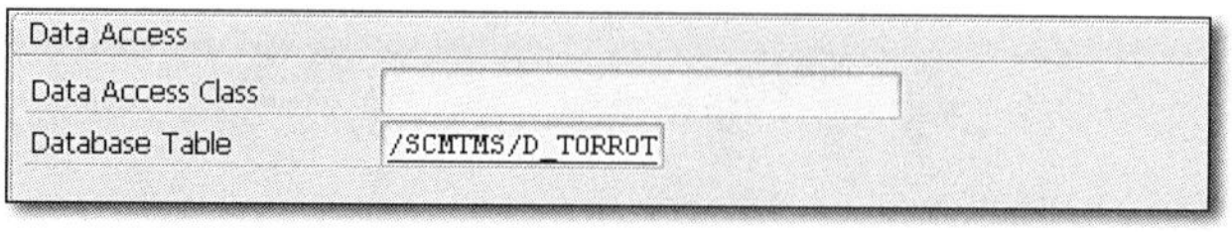

Figure 2.5 Database Table Assigned to a Business Object Node

Node associations

One of the goals of the BOPF is for users and developers not to need to directly connect to database tables any more. Therefore the link between nodes should be established with more than the parent keys and root keys that we have talked about in the previous section—that would be too technical. Another, more general way needs to be established.

Nodes are assigned with a special element, called an association, provided by the BOPF. An *association* is a logical link between two node instances. In Figure 2.6 there are associations from the root node to other nodes (/SCMTMS/TOR • NODE ELEMENTS • ROOT • ASSOCIATIONS). When you are retrieving data from a business object instance, follow the path from node to node using the association as a kind of bridge to get required information. For example, if we have only the number of a special freight order and want to know what items are on it, we start at the root node and then use the association between the root node and the item node to get to the information in the item node. Once we are on the item node instance, we can look for the field in the DDIC structure assigned to the node.

To take a closer look at an association's setup, you can use Transaction /BOBF/CONF_UI to access the business object, but instead of opening the node structure, open the node elements, and expand any node. There you will find a folder called ASSOCIATIONS, as shown in Figure 2.6.

Recall that a freight order can contain several items and therefore several instances of the item node. The association defines whether several nodes can be assigned or only one. As you can see in the CARDINALITY field on the right side of Figure 2.7, several item nodes can be assigned to the root node because the cardinality of the association is 1:0...n.

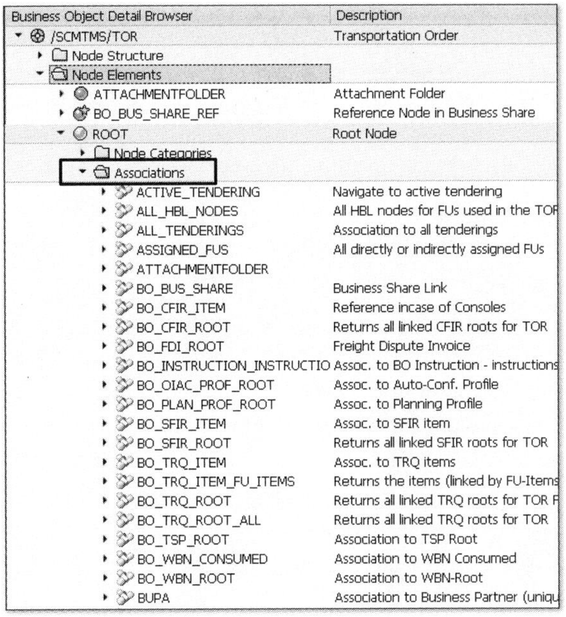

Figure 2.6 Associations as One Example of Node Elements

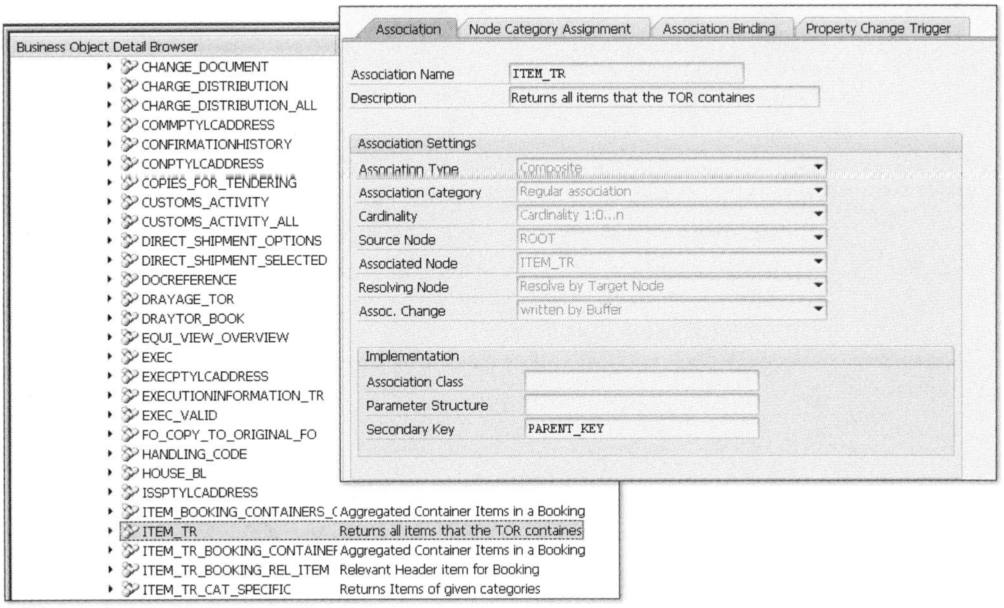

Figure 2.7 Association Details

Keep It Nice and Tidy

When you browse through the associations, you will find some associations from the root node to another node with the cardinality 1:1. You might be wondering why, in this case, another node is needed at all and why the information from the associated node is not put into the root node.

From a technical point of view, there's really no reason to do this. But from a logical point of view, there *is* a good reason: to keep the model nice and tidy.

Let's consider an example from real life. Your kitchen has a drawer for cutlery, a drawer for pans, and a drawer herbs and spices. (Hopefully you have more than three drawers in your kitchen, but for this example, three is enough.) Now, instead of using separate drawers, you *could* put everything onto a big shelf—but you would probably have a problem finding anything easily and quickly.

The same applies to the business object model. The more fields a node contains, the bigger the database table will be. A business object node can, in this example, be compared with a drawer; the fields can be compared with the items in your kitchen.

The performance of database accesses depends on the number of fields in one line. Therefore the number of fields in one line should be kept slim. For this reason there are some node associations with the cardinality 1:1; the information of the associated node is stored in a different database to improve database reading performance.

Displaying data stored on database

Let's return the goal of preventing users and developers from having to connect to databases. This also applies to consultants and technical experts. When working with SAP TM, you should no longer use Transaction SE16 to display data in the background. Instead, stick to the business object model and display data in the object-oriented way.

Displaying object node instances

To do this, you can use Transaction /BOBF/TEST_UI. Enter the business object in the corresponding field and then select a query, as shown in Figure 2.8.

Query

A *query* is a predefined search for business object node instances. Developers use queries to find node instances that carry certain information. Though we don't delve deeper into queries in this chapter, you should know that you can use queries in Transaction /BOBF/TEST_UI to find node instances.

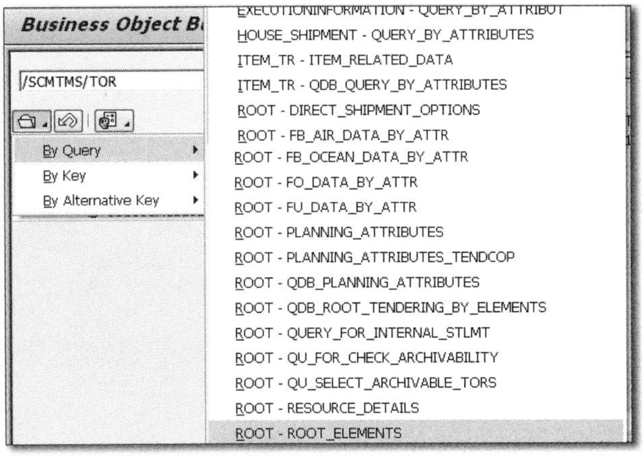

Figure 2.8 Query on Root Node

The node instances are displayed in a table view. In the example where you want to get information about an item in a freight order and you only know the number, you can use a query to display the root node first. Then you can use the association to the items to get to the item information, as shown in Figure 2.9.

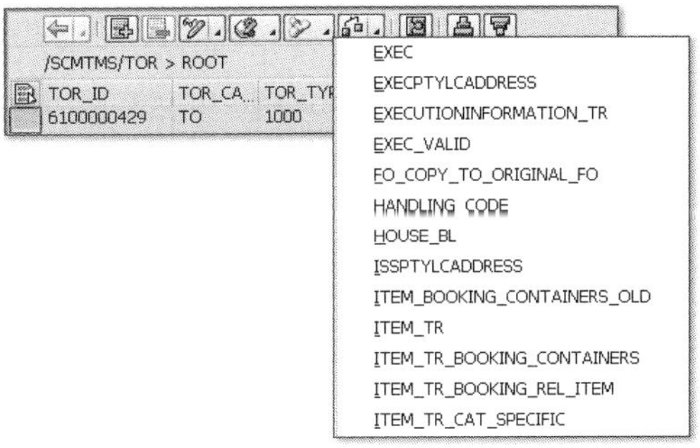

Figure 2.9 Execute Association in Transaction /BOBF/TEST_UI

2.1.2 Data Processing with the BOPF

Built-in methods

As already mentioned, the BOPF is not only a way of modeling the storage of data but also handles data processing. If you used Transaction /BOBF/CONF_UI earlier in Section 2.1.1 and browsed through the different node elements, you probably noticed that many node elements are assigned to a node:

- Node categories
- Associations
- Determinations
- Validations
- Actions
- Queries
- Alternative keys
- Status variables
- Status derivators
- Status schemas
- Attribute value sets

Don't worry—we won't go through all of the node elements in this chapter. However, it is worth taking the time to look at the most important node elements. Associations help to establish a link between two business object nodes and therefore mainly support the consistent modeling of data storage, but the node elements we want to deal with now are used for the data processing part of the BOPF methodology.

The three node elements responsible for built-in data processing are determinations, actions, and validations. You can browse through these elements the same way you browsed through the associations earlier.

Determinations

If you know a little bit about how SAP ERP was coded, you know that even when performing the most elementary changes to data, you need to establish the links to all follow-up activities in the coding. If a user wanted to add some custom logic to the follow-up activities, then the coding needed to be enhanced. This meant that to add custom logic, a developer had to know exactly where in the code certain things happen. With the BOPF, this problem has been solved with determinations. The framework itself calls *determinations* after one of the create data, read data, update data, or delete data (CRUD) methods is called. To define when a certain determination should run, all you have to do is select a checkbox.

Let's take a look at an example: the determination for number drawing for our freight order. As you can imagine, the logic of number drawing needs to be executed only upon creation of the freight order.

When you look at the details of the determination DET_DRAW_NUMBER shown in Figure 2.10 (or when you double-click on the determination and then go to the REQUEST, READ & WRITE NODES tab), the following is defined: that a number is only drawn when the root node instance is created.

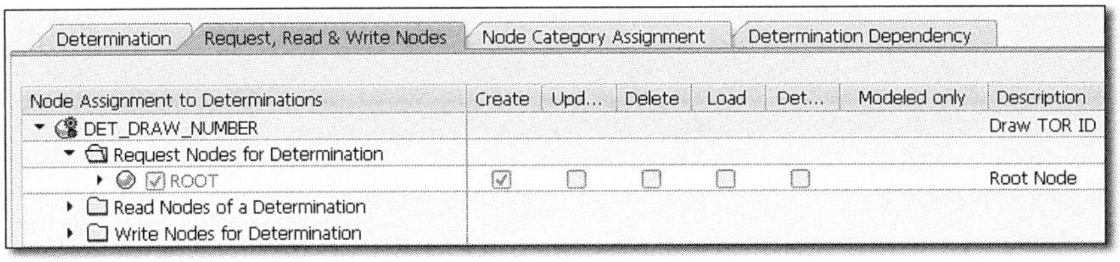

Figure 2.10 Triggering a Determination

Now that we know *when* the determination is called, we can take a look at *how* it is called and how the coding to be processed is found.

When you go back to the DETERMINATION tab, which is shown in Figure 2.11, you will find that a class is assigned to the determination. This class is called whenever the determination has to be executed.

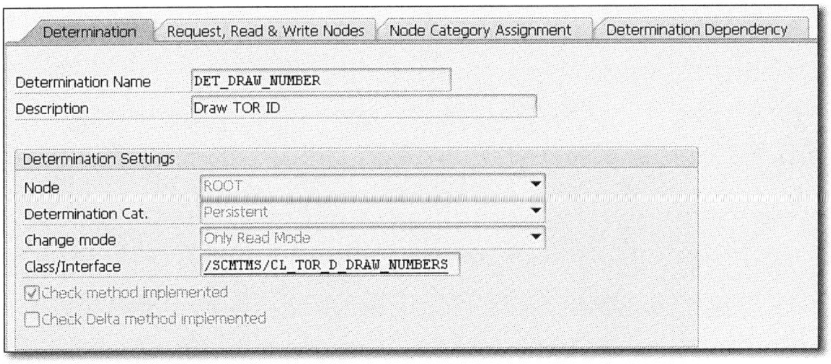

Figure 2.11 Determination Details

Naming Conventions

Whenever you find a class with the name /SCMTMS/CL_TOR_D_*, you can assume that it is a class designed for a determination of the business object TOR.

The naming convention is that after the namespace /SCMTMS/ and the usual identifier CL, there is an abbreviation of the business object that is being dealt with and one letter to determine whether the class is for a determination (D), action (A), validation (V), or query (Q).

Each class assigned to node elements uses an interface that provides three methods, as you can see on Figure 2.12. When a determination is executed, the framework calls the EXECUTE method of the interface. Further methods can be added to the class, but they will not be considered by the framework; instead, they need to be called by the EXECUTE method.

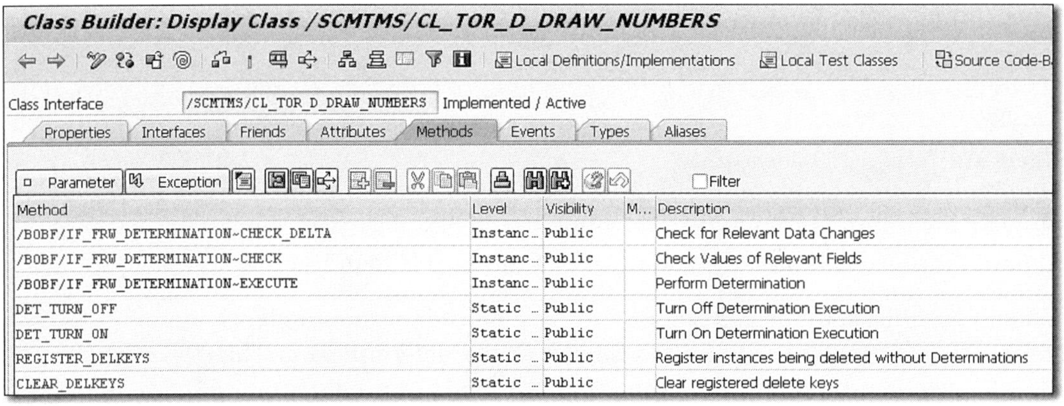

Figure 2.12 Class of a Determination

Actions

Though determinations are always called by the framework, they do not influence the business process in greater terms. Since the course of a business process cannot always be foreseen and is often influenced by people, program logic also needs to react to user interaction.

Processing a business object node according to a user's input is done by actions. An *action* is program logic that is externally called, usually by a button on the UI.

Say, for example, that we have put together a full truckload of various cargo items. We want to check whether the combination of goods on the truck is feasible or if some goods may not be transported together. Therefore, we want to execute an incompatibility check on the freight order, so we use the corresponding button on the freight order UI.

Note

This chapter only discusses what happens to the data. We cover how to set up an incompatibility check later in the book.

The UI action is linked to an action of the business object node, and the corresponding program logic is processed. The BOPF uses exactly the same approach as with determinations. Each action, like the action of the incompatibility check in Figure 2.13, has a class assigned to it that contains an interface with an EXECUTE method that is called by the framework.

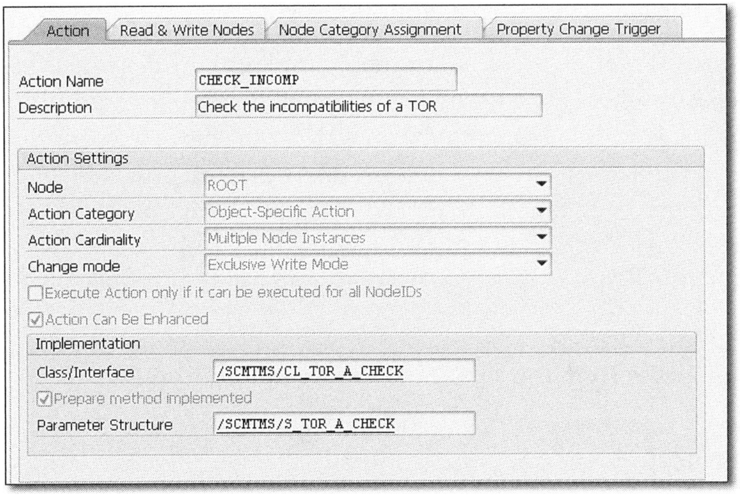

Figure 2.13 Action Details

Validations

As with determinations and actions, *validations* are also a piece of coding that is called and executed by the BOPF. The BOPF offers two different types of validations: the action validation and the consistency validation.

Action validations are associated to an action of the business object node and are called *before* the action is called. The action validation checks whether the action may be executed. For example, if we call the action that sends a transportation order to a carrier, an action validation checks beforehand whether a carrier is already assigned to the transportation order. If not, the action validation fails and the action will not be executed.

Consistency validations, on the other hand, are called after a change has been made to a business object node. Consistency validations, therefore, are *not* directly linked to an action, but monitor the change of a node instance, just like the determinations. However, while determinations may change some data on the node instance, the validations do not change anything on the node instance. Instead, they validate the consistency and create warning or error messages to notify the user.

For example, when you create a freight order, the save action is called. During this action, the determination for number drawing, is called since one of the CRUD methods was executed. The validation of number drawing is called after the save action is finished.

As mentioned earlier, without built-in program logic, the developer has to know exactly where the custom enhancement must be inserted. With the BOPF, this is often no longer necessary. Now the developer can simply add program logic using guided procedures that are provided when a business object is enhanced. Standard and custom logic are then displayed next to each other, making it look like one final product in the end (which, in fact, it is).

2.1.3 Business Add-ins

BAdIs

Rest assured, supporters of the "good old" SAP techniques! There are still some relics from the SAP applications we all know, like the Business Add-ins (BAdIs).

BAdIs are enhancements to the standard program logic without any modifications. SAP provides *enhancement spots*, which are specific places in ABAP coding where the customer can insert special program logic and alter data that needs to be processed.

A very popular example of an enhancement in SAP TM is the BAdI for optimizer pre-processing. Before handing over data to the optimizer, the user can add or change information that was previously gathered by the SAP TM application.

To create this BAdI, you need to go to Transaction SE18, as shown in Figure 2.14, and enter the name of the BAdI or enhancement spot. The SAP TM-specific BAdIs start with `/SCMTMS/`, following the naming convention. When you do an F4 search on the BAdIs, you will see that SAP TM offers over 100 BAdIs.

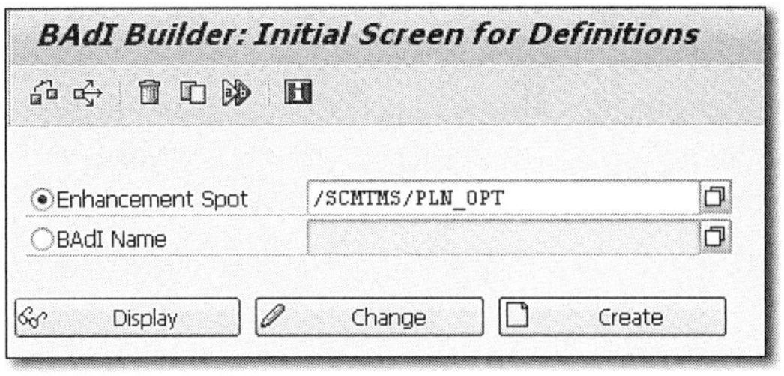

Figure 2.14 Input of an Enhancement Spot or BAdI

In an enhancement spot, there is usually a sample implementation to help the developer get started with the custom enhancements.

You can forward-navigate to the enhancement implementation (like the one shown in Figure 2.15) and to the enhancement's implementing class. The implementing class uses an interface that provides the developer with an information structure that is passed on from the standard coding and, after execution of the custom logic, is passed back to the standard program flow again.

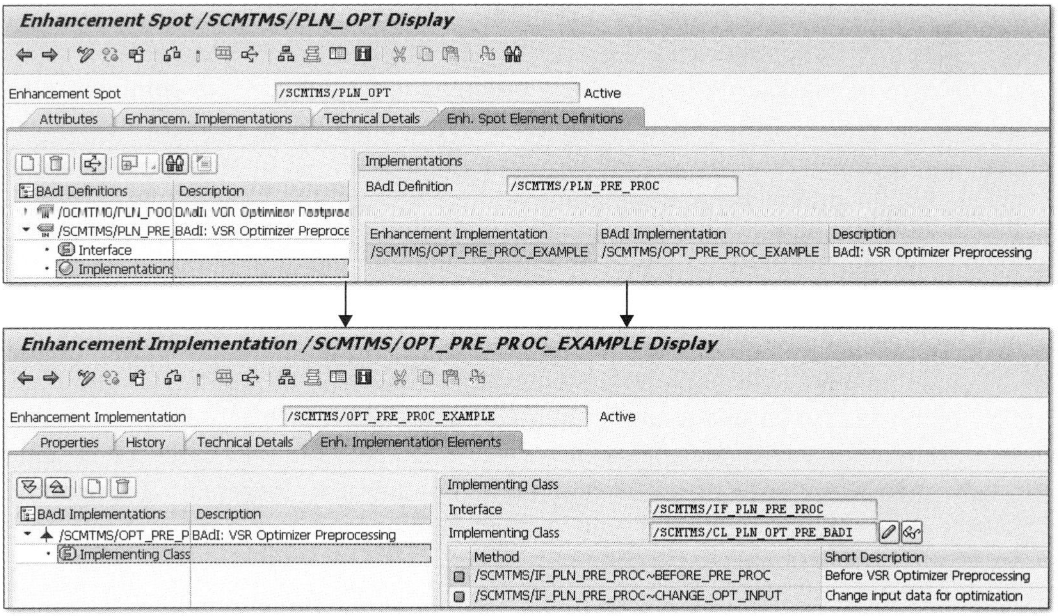

Figure 2.15 Enhancement Implementations

> **Note**
>
> Because this book aims to give you insight about what you can do with SAP TM as a technical expert, we won't delve deeper into how to implement BAdIs or any ABAP coding at this stage.

2.2 User Interface Technologies

Web Dynpro ABAP technology

In addition to the new way of storing and processing data, SAP TM also uses a new user interface technology. Where former SAP applications relied on SAP GUI user interfaces, the new applications use web UI technology.

The shift to web UI technology is accomplished by implementing user interfaces with Web Dynpro for ABAP—that is, using ABAP coding designed to be transformed into markup language that can be rendered by browsers.

SAP TM cannot exist *completely* without SAP GUI transactions, but the border between the usage of SAP GUI and web UI technology is clearly marked. While technical experts and consultants can continue to use SAP GUI for Customizing and system monitoring, business users rely on the SAP NetWeaver Business Client (NWBC) to perform all business transactions.

The SAP NWBC is structured differently from SAP GUI. Transaction codes are now obsolete, and you use a menu structure akin to web pages to navigate to your business transaction, as shown in Figure 2.16.

Not only is SAP NWBC capable of rendering Web Dynpro ABAP pages, but it can also display SAP GUI transactions such as the resource master data maintenance, which is part of the SCM Basis and therefore remains an SAP GUI transaction. Instead of using two different UI applications while configuring and testing SAP TM, you only have to use one.

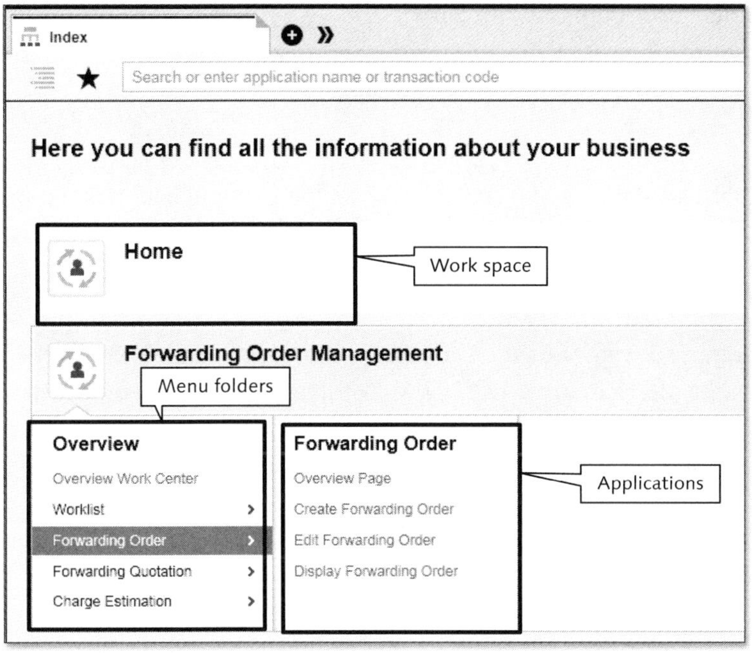

Figure 2.16 Menu Structure in SAP NWBC

There are different ways of opening SAP NWBC and each one has a specific purpose:

- **Zero-Footprint Client**

  System administrators use Zero-Footprint Client

  The Zero-Footprint Client is used by users such as Basis experts and system administrators who perform most of their activities in SAP GUI.

  Start the Zero-Footprint Client via Transaction NWBC, which can be called in SAP GUI. Calling this transaction opens your standard browser, displaying a list of roles that are assigned to the user. When you choose a role, the browser displays the menu defined in the role.

  Remember to confirm that your browser is supported by SAP.

  Browser-based NWBC for external users

  As you learn in Chapter 7, business partners such as carriers can log on to your SAP TM system and respond to tenders or requests for quotations. When doing so, they use a hyperlink that takes them to the SAP TM system in their standard browser. The advantage is that carriers don't have to install additional software to access your system with their user name because the standard browser is used with the zero-footprint approach.

Note

Even though the SAP NWBC user interfaces are displayed in a browser, you should only use the menu functionality within the UI and refrain from using browser functionalities such as the BACK button or browser favorites.

Business users use the installed NWBC

- **Client-installed SAP NWBC**
  The preferred way of using SAP TM is the client-installed SAP NWBC because it offers many advantages. Compared to the Zero-Footprint Client, the user does not have to track what roles are assigned to him (information the business user should not be concerned with, anyway). Additionally, you can decide how many sessions you want to open. If you want to open a new transaction in a new window, simply click on a menu entry in SAP NWBC while pressing [Shift] to open a new window. With SAP NWBC 4.0, tabs were introduced. Just like with current browsers, different applications can now be opened in the same window in different tabs.

Unlimited Modes

SAP GUI is restricted to six sessions opened in parallel, but there is no such restriction in SAP NWBC. Be aware, though, that opening several sessions in parallel consumes more resources on the application server.

Furthermore, with many sessions open, it's likely that you are locking data in a session that you may not need any more. For that reason, SAP NWBC has an automatic timeout functionality that releases lock entries after a certain time of inactivity. If this happens, you get an error message and you have to return to your transaction or refresh the page by pressing [F5].

Integration of TM roles to the SAP Portal

- **Portal**
  Due to the web UI capability of SAP TM business transactions, you can also integrate the SAP TM user interfaces into the SAP NetWeaver Portal if one user is using different SAP applications together in one business role.

The Importance of the Browser

Whether you are using the zero-footprint approach or the client-installed SAP NWBC, the standard browser installed on your computer is always important.

You may not notice that the standard browser's rendering functionality is in place, even when you are using the installed version of SAP NWBC, but the browser takes over the rendering activities when using SAP TM.

Keep your browser version up to date to ensure optimum rendering performance.

2.2.1 Floorplan Manager and Floorplan Manager BOPF Integration

After having talked about the BOPF, the first major technological pillar of SAP TM in Section 2.1, we will now talk about the UI technology, which is the second major technological pillar of SAP TM. The UI technology is supported by the Floorplan Manager (FPM) framework. The FPM is a tool that helps the developer build user interfaces with different building blocks.

Each user interface consists of a so-called FPM application that defines one use case in the SAP TM system. The forwarding order, for example, is one FPM application. As you see in Chapter 4, several FPM application configurations are designed for the forwarding order, each for different use cases (e.g., ocean forwarding order, air forwarding order, and so on).

Assignment of FPM components to application

Apart from the FPM application, the developer can now use different *UI building blocks* (UIBBs) to build a UI. The FPM framework provides predefined components called *generic UI building blocks* (GUIBBs) for this. For SAP TM, we use the following components:

- **Overview page**
  The overview page defines the general layout of the screen. It provides a global toolbar on top of the screen and can embed several other components.
- **Form**
  The form GUIBB gives you the opportunity to assemble several text

and display fields of a flat structure (such as the root node) on the screen.

- **List**
  The list GUIBB displays the content of a table. For example, the table in the EXECUTION tab of the forwarding order is made of a list GUIBB.
- **Tree**
  Hierarchical relations can be shown in a tree GUIBB. Very common examples of tree GUIBBs are the items in the forwarding order.
- **Tab**
  As you have seen in most UIs, in SAP TM the information in a document is divided into different tabs. These tabs are also built on a provided GUIBB.
- **FBI view**
  The FBI view defines the link between the UIBB and the data in the BOPF. We talk about the FBI view later in this section. The FBI view is also a GUIBB in FPM.

Each FPM application must have an overview page to provide the global toolbar. On the overview page, you can assign several UIBBs that can be built based on the listed GUIBBs. This assignment is depicted in Figure 2.17.

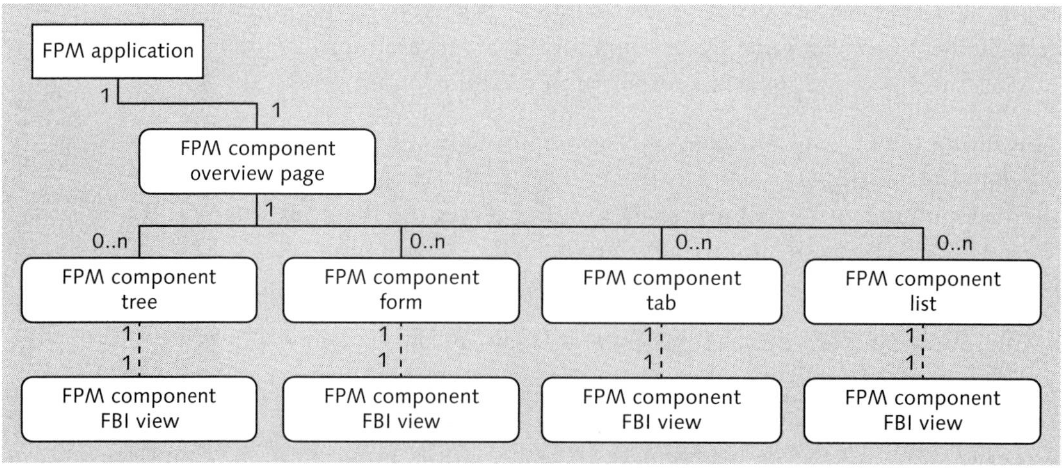

Figure 2.17 Assignment Hierarchy of Components in the FPM

As you can see by the shape of the boxes, the overview page and the assigned UIBBS are all FPM components, only of different types. The type of FPM component (meaning GUIBB) is described in the second line of the corresponding boxes.

Feeder classes

So far we have only defined the general look of the UI, but we haven't done anything about the link between the data in the system and how it can be put onto the screen.

This link is established in the FPM framework using so-called *feeder classes*, which provide the UIBBs with both data and a field catalog. The *field catalog* provides metadata about the data (e.g., which columns are available in a list). In the SAP TM system, the feeder classes are also where the *Floorplan Manager BOPF Integration* (FBI) framework comes into play.

Floorplan Manager BOPF integration

The FBI framework provides generic feeder classes that developers can reuse and assign to the UIBBs they have chosen on the overview page. FBI provides one feeder class per GUIBB; this is necessary because every GUIBB requires its own feeder class that delivers information specific to and specifically structured for the GUIBB.

Apart from the feeder class, the FBI also provides an FBI-specific GUIBB, which is the FBI view. The FBI view establishes the link between the BOPF node and the UIBB. Each UIBB has one FBI view. You can see this assignment in the component configuration or component Customizing. As you can see in Figure 2.18, especially in the lowest screen, the FBI view contains the business object and the business object node of the data that should be displayed in this UIBB. Furthermore, the FBI view contains a UI structure that consists of all fields that should be available on the UI.

Node Structure versus UI Structure

Why can't we use the actual node structure as the structure providing fields to the UI? Because the concept of the UI structure differentiates between data stored on the database and data to be read by a user.

For example, the creation date of a document is stored as a timestamp in one field of the node structure, but this timestamp is not readable to any user. Therefore, the UI structure contains three fields for the creation date: date, time, and time zone.

Figure 2.18 FBI View

At runtime, the FBI view fills the fields of the UI structure and passes it to the feeder class that passes the data on to the FPM-based UI. All communication between the BOPF and FPM is therefore channeled through the FBI-specific feeder class and the FBI view.

2.2.2 Defining User-Specific Roles and Menus

Displaying roles in NWBC

Recall that SAP NWBC displays menu structures that were defined in the user's role. To make the menu structure transparent to the user, SAP has predefined some roles according to the tasks different users may have in their daily business.

Each role is assigned a menu that is then displayed in SAP NWBC. If a user is assigned numerous roles, all of the menus are displayed in SAP NWBC.

To alter the menu of a role, go to Transaction PFCG and copy a provided role into your namespace. Access your newly created role and go to the MENU tab as shown in Figure 2.19, which has a folder structure like the SAP Easy Access menu in SAP GUI. The first folder always represents a workspace that is displayed in the workspace area of SAP NWBC. The subfolders and applications are displayed in the menu area of SAP NWBC.

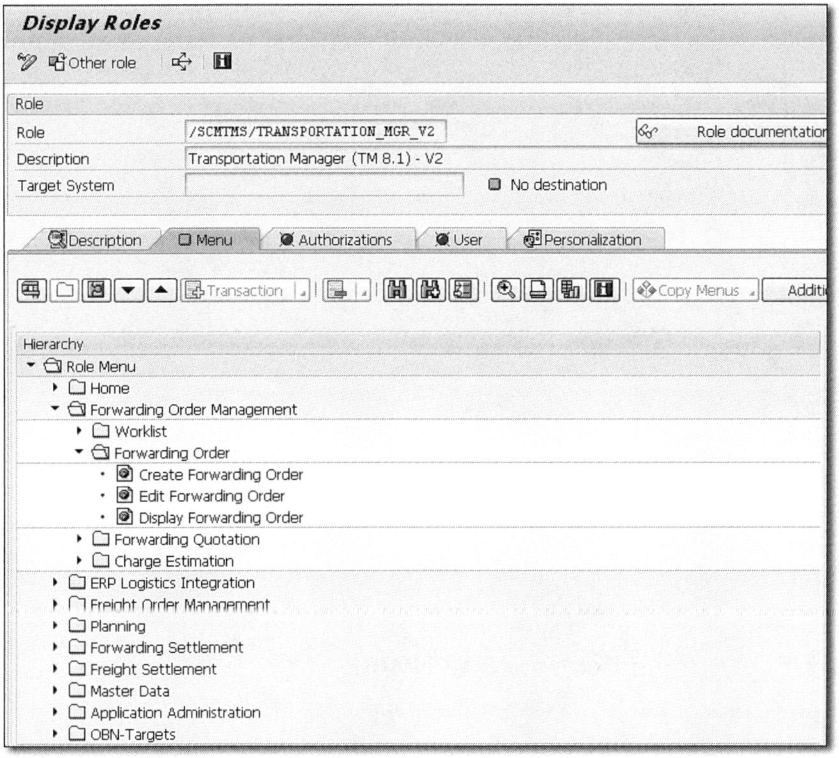

Figure 2.19 Menu Structure in PFCG

Drag and drop transactions from folder to folder. Right-click on the application to display its details, where you can also add parameters, as shown in Figure 2.20.

For example, if you want to add one menu entry where an explicit freight order number is displayed, you need to find the root key of the freight order. Once you have the root key, you can enter it as the KEY parameter. To navigate directly to your freight order, add the SKIP_INITIAL_SCREEN parameter and set the value to X, and your freight order is displayed straight away.

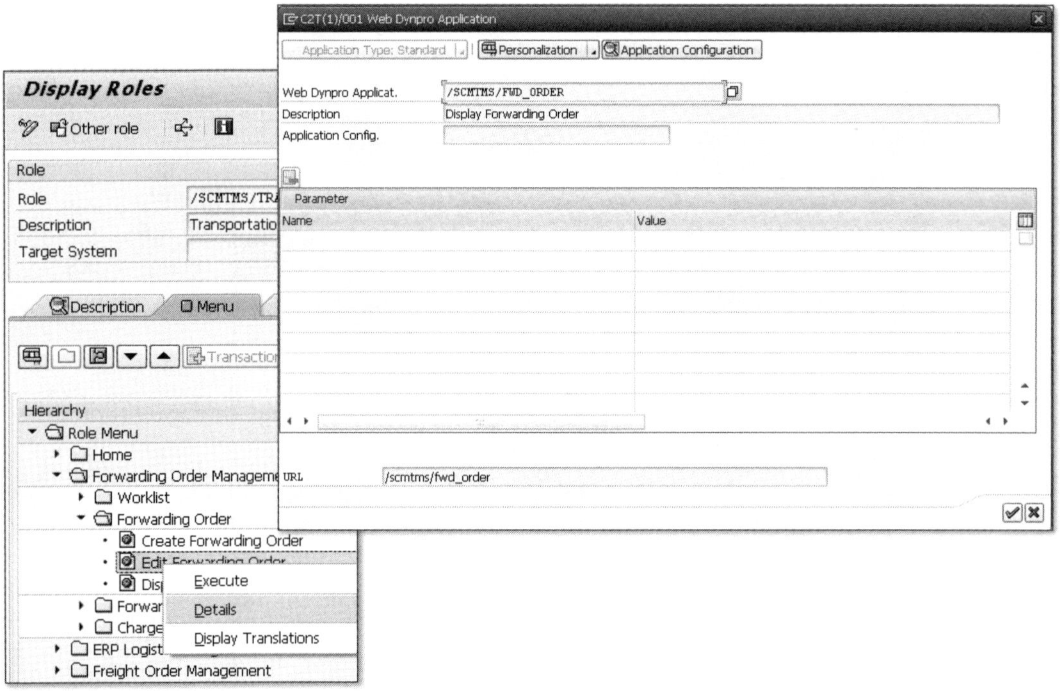

Figure 2.20 Application Details

2.2.3 Filtering Roles for SAP NWBC

As mentioned before, SAP NWBC displays all the menus that the roles assigned to the user contain. In a customer environment, you will quickly see that there are numerous roles, and therefore numerous menus assigned to one user. Many of the menus are not necessary for using SAP TM in a business process context, so you might want to filter which menus are supposed to be displayed in SAP NWBC and which ones aren't. To do this, use the *Service Interface Component Framework* (SICF) and its nodes. Go to Transaction SICF and create a new node in the path displayed in Figure 2.21: */default_host/sap/bc/nwbc*.

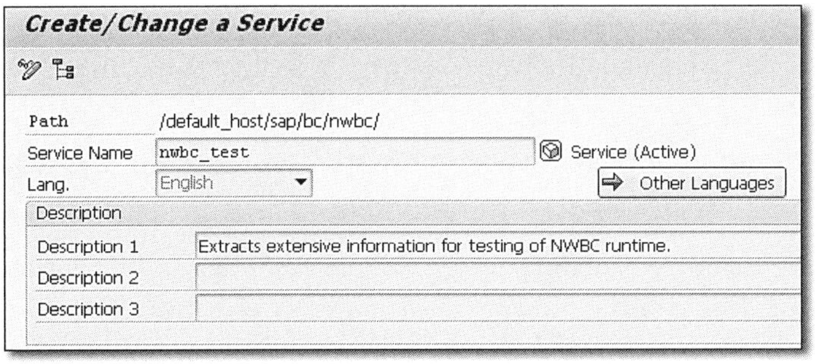

Figure 2.21 Service Node Definition

After creating this node, you can assign roles that you want to display in SAP NWBC to this node. This node represents a so-called *NWBC cockpit*. In Transaction PFCG, you can perform this assignment by declaring the newly created node to be COCKPIT, as shown in Figure 2.22.

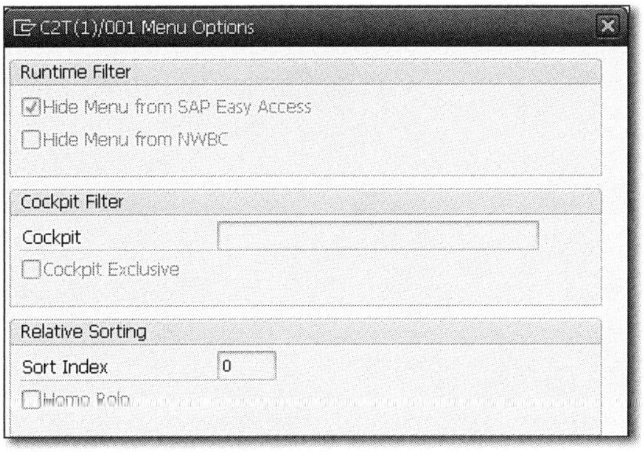

Figure 2.22 Menu Options

Last, add the name of the node to the URL defined in the NWBC logon tab. Simply append the node name to the URL, separated by a /.

SAP NWBC Logon Screen

As in SAP GUI, you need to define some details in order to access the system. In SAP NWBC, the main information is the URL. You can find the URL by starting the Zero-Footprint Client in SAP GUI. The URL shown in the address bar of the browser is the one you need to define in the SAP NWBC logon screen.

2.2.4 Customizing Screen Areas

Compared to SAP GUI, one big advantage of SAP NWBC is the ability to customize screens. As a technical expert or consultant, you can now perform changes to the UI in Customizing mode without touching any lines of code or making modifications to the system.

When browsing through a business transaction, you might come across some fields that you either don't need or want to rename. With SAP NWBC and Web Dynpro ABAP, you can change the UI in Customizing mode. From the business transaction, right-click on the screen area you want to customize, and choose TECHNICAL HELP, as shown in Figure 2.23. A new pop-up opens, and you can navigate to the UI Customizing by clicking on the link of the component Customizing of the current view.

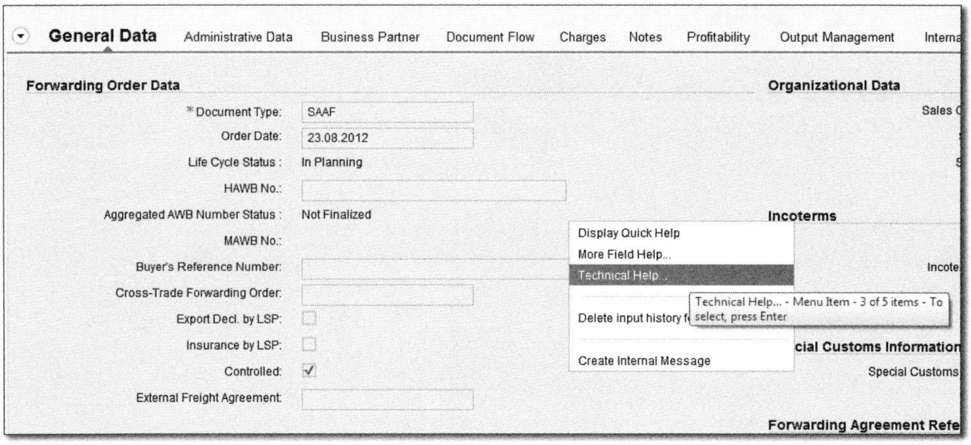

Figure 2.23 Opening Technical Help in SAP NWBC

In the UI Customizing, you can add and remove fields, change labels, and change the way fields are displayed. Your changes can be seen directly in the preview screen of the Customizing application.

2.3 Frameworks and Tools Used throughout SAP TM

Business process execution in SAP TM is not only based on new SAP TM-specific developments, but also uses various tools that enhance SAP TM functionality and support business process execution:

- BRF+
- Incompatibilities
- Post-Processing Framework
- Document creation and adaption
- Optimizer server
- SAP Visual Business
- Process Controller Framework

Please note that some of these tools are third-party tools that require a license, independent of the SAP license for SAP TM. If you've worked with other SAP applications before, you've probably come across some of these tools already.

In general, the purpose of using these tools is to enable the user to customize the system in more detail with no modifications or coding.

2.3.1 Condition Framework Using BRF+

The most prominent framework you will come across when using SAP TM is the *Business Rules Framework plus* (BRF+), which provides conditions to help the automatic decision making in SAP TM business process execution.

Conditions are tools that determine an input value based on master data or transactional data and derive an output value from the determined input value. Their use is widely spread throughout the SAP TM application; you will find possible action areas for them in all process areas.

Examples of use

The following are the most common functions:

- Document type determination
- Determination of organizational units
- Determination of the applicant rate tables in charge management
- Incompatibilities
- Determination of loading/unloading time of a freight unit
- Determination of delivery time windows for a freight unit

Take the determination of the organizational units of a freight order as an example. In this case, the condition is processed upon creation of the freight order.

The condition type used for this process step is `/SCMTMS/TOR_ORGUNIT`. In Customizing you can see this condition type by following the IMG path SAP Transportation Management • Transportation Management • Basic Functions • Conditions • Define Condition Types. The definition of the condition type includes three important details that you might need for your condition creation. These are boxed in Figure 2.24.

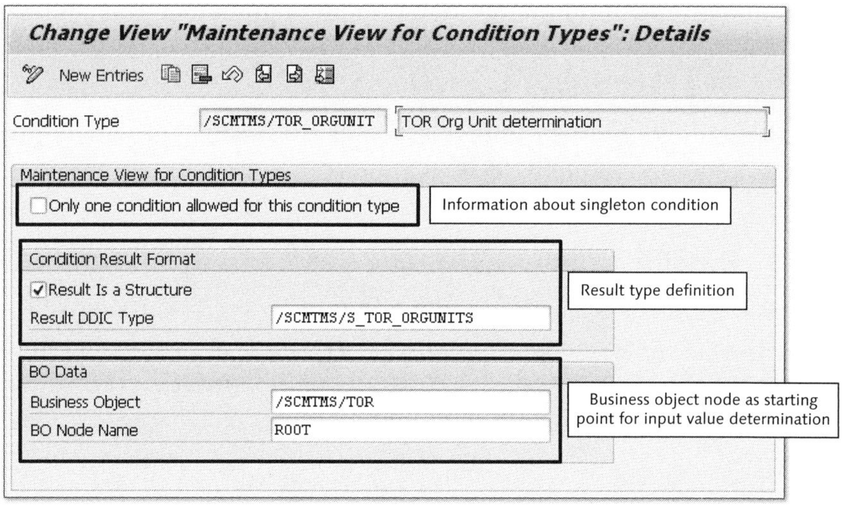

Figure 2.24 Condition Type Definition

Singleton conditions

The first checkbox indicates whether more than one of this condition type is allowed in the system. Some condition types, called *singleton conditions*, may only have one condition defined in a system. Note that their point of call cannot be influenced by customizing. For example, there is one condition type that determines the document type of the document created by a transferred SAP ERP sales order. In Customizing, there is no means of determining which one of this condition type should be called.

If there can be more than one condition of a condition type, as in our example, then there is a field where this condition has to be entered in order to be processed.

Condition result type

The second crucial piece of information is the DDIC type of the result. In Customizing, you can enter a data element or structure that is used as the output of the condition. If a structure is used, the corresponding flag in customizing must be set.

BO Data condition type

If you're building more complex infrastructures, then the last piece of information in Figure 2.24 is the most important. In the BO DATA section you can see where the condition starts searching for an input value. The business object and business object node is displayed here. We come back to this information later in this chapter.

Creating a condition

You will create a lot of conditions to automate business decisions when you're setting up your SAP TM system. In our example, we want to determine the planning and execution organization based on the source and destination location of the freight order.

Our formula is pretty easy: If the source location of the freight order is Istanbul and the destination location of the freight order is Berlin, then the planning and execution organization is Turkey. If the freight order goes the other way around, the planning and execution organization is Germany.

We start creating our condition by following the menu path APPLICATION ADMINISTRATION • GENERAL SETTINGS • CONDITIONS • CREATE CONDITION.

We choose the corresponding condition type (which in our case is /SCMTMS/TOR_ORGUNIT) and specify that we want to create a condition based on a *decision table*, which is also the example in Figure 2.25. This is the most commonly used type of condition. With the decision table, you can map a combination of input data to explicit output data, as you will see later.

In some cases, you might want to use a condition with direct business object access. This condition determines the input data (which, in this case, is also the output data). Direct business object access is commonly used when defining incompatibilities, which we deal with later in this chapter.

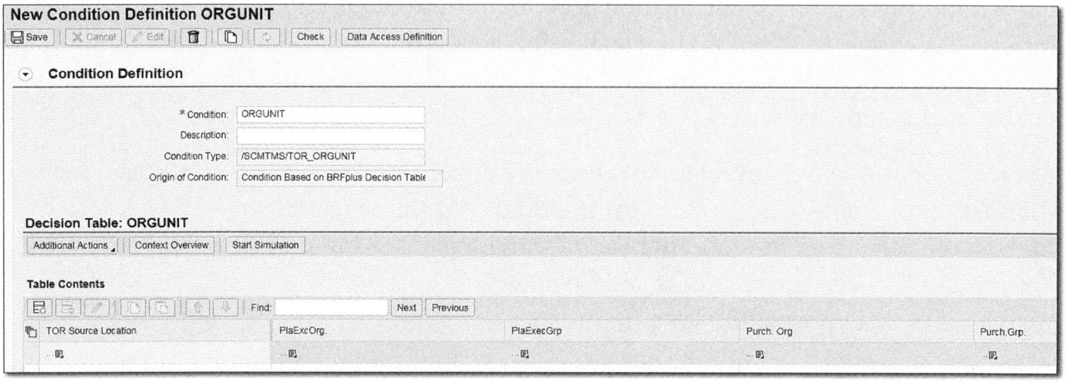

Figure 2.25 Condition Definition

When we define the condition, we see a decision table. Input columns have a blue background, and output columns have a green background. You can see that, by default, the source location of the freight order is taken as an input value. You can change the input data by clicking on the DATA ACCESS DEFINITION button at the top of the screen.

In Figure 2.26, you can see that there is one predefined *data access definition*: the source location of the freight order. You can add more input values by adding a line to the table. With F4 help, you can see what data access definitions are offered for this condition type. If you know that your input value can be found on the root node of the business object /SCMTMS/TOR, which represents the freight order, you can leave the DATA ACCESS DEFINITION field empty and enter the business object, business object node, and field directly in the corresponding fields.

Please note that you can only enter the business object and the node that were defined as the starting point in Customizing.

For example, we want to add the destination location of the freight order as a second input value. Because this data access definition is not predefined and the information is not stored on the root node of the TOR business object, we need to build our own data access definition.

Data crawler

Since the information we want to retrieve is not stored on the root node, we need to use a *data crawler* to use associations to navigate to the node where our field is. To do so, we need to define a data crawler profile via the Customizing menu path SAP TRANSPORTATION MANAGEMENT • TRANSPORTATION MANAGEMENT • BASIC FUNCTIONS • CONDITIONS • DEFINE DATA-CRAWLER PROFILE.

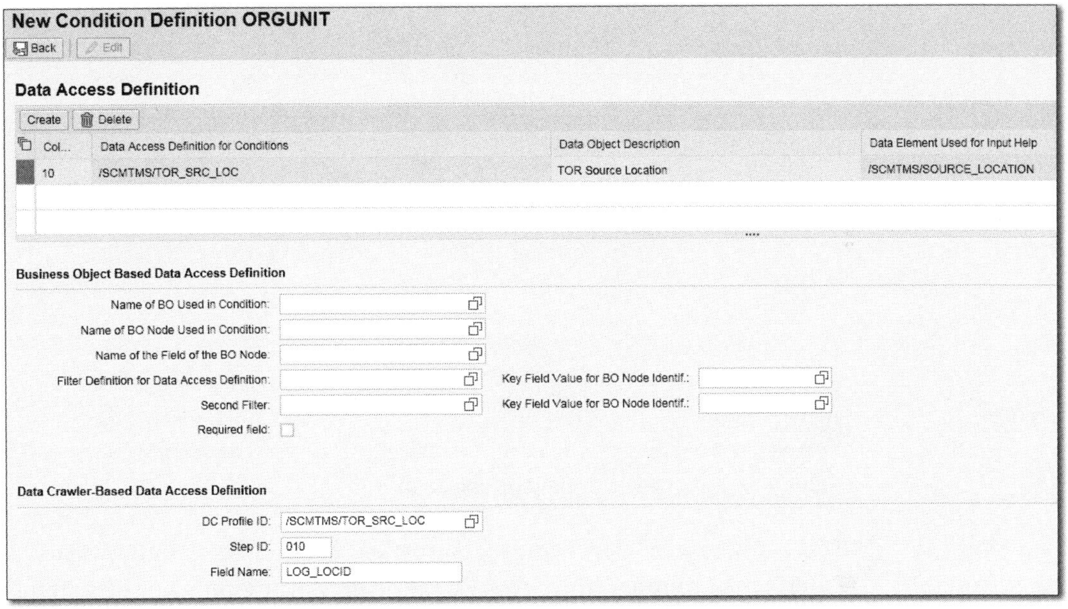

Figure 2.26 Data Access Definitions of a Condition

Now we want to create a new data crawler profile. After assigning a meaningful name and description (in our case, DESTINATION_LOCATION) we have to define where the data crawler should start "crawling" the data on the business object. The data crawler has to start at the starting point of the condition; in our case, this is the business object /SCMTMS/TOR and the node ROOT. In the upper part of Figure 2.27, you can see how this starting point is defined.

Now we need to define where we want to go. To do this, we define a step for the data crawler, which is done using the PATH STEPS in the dialog structure as shown on the left side of Figure 2.27. The lower part of this figure then shows how the path step is maintained.

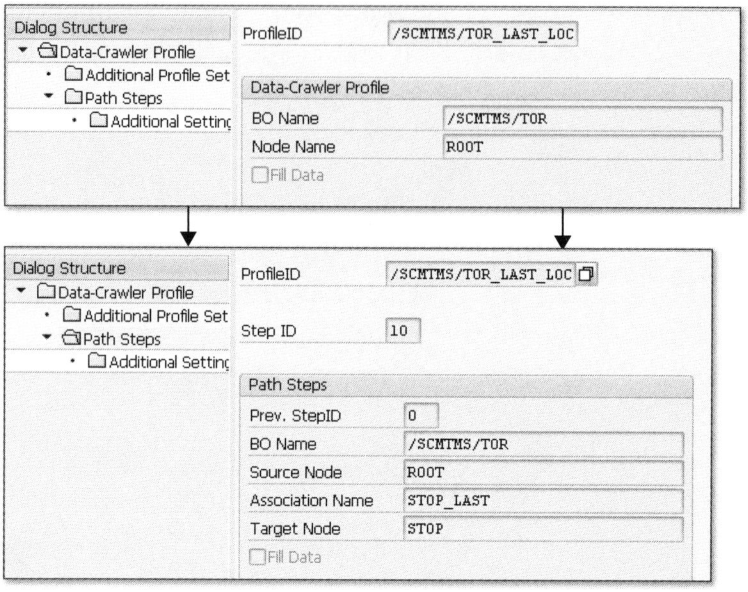

Figure 2.27 Data Crawler Definition and Steps

The data crawler step represents our use of an association. When we enter a step number and press [Enter], the starting point of this step is automatically updated. The only thing we have to maintain is the association we want to use. Fortunately, the business object /SCMTMS/TOR offers an association that not only returns data of the node STOP (where all stops of the freight order are stored), but also offers an association that filters out all stops except the destination location.

Therefore, we use the association STOP_LAST. Since we've now reached the point where we want to be with our data crawler to retrieve data, we select the Fill Data checkbox shown at the bottom of Figure 2.27.

Notice that in the lower area of Figure 2.26 that you can enter the data crawler profile directly on the screen where you define your data access definitions for your conditions. But to enable reuse of the data crawler, we want to assign the data crawler profile to a new global data access definition.

To do so, we create a new data access definition in Customizing, following the menu path SAP TRANSPORTATION MANAGEMENT • TRANSPORTATION MANAGEMENT • BASIC FUNCTIONS • CONDITIONS • DATA ACCESS DEFINITION. Here we create a new data access definition and assign the data crawler profile, define the step in which we want to retrieve data, and enter the field we want to read. The field is the component name of the DDIC structure assigned to the node that we have navigated to with our data crawler.

To help your users enter the right input data, enter the data type of the field in DATA ELEMENT FOR F4 HELPS. If you do not know the data type of your input, you can look it up in the DDIC structure of the business object node.

Since condition types can start at different business object nodes, not all data access definitions created in Customizing may be used in a condition of a certain type. Therefore, data access definitions are assigned to condition types in Customizing. When you follow the Customizing path SAP TRANSPORTATION MANAGEMENT • TRANSPORTATION MANAGEMENT • BASIC FUNCTIONS • CONDITIONS • ASSIGN CONDITION TYPE TO DATA ACCESS DEFINITION, you can maintain a corresponding entry to assign the data access definition we just created to the condition type /SCMTMS/TOR_ORGUNIT. If you set the DFLT DAD field to DATA ACCESS DEFNINITION IS DEFAULT FOR CONDITION TYPE (X), your data access definition will be displayed in every new condition of this type.

In our new condition, we can now enter our data access definition in the corresponding field. The information concerning data crawler profile and input help is automatically entered.

Maintaining a decision table

If you click on BACK, you will go back to the decision table and maintain values by clicking on the link. We can now enter the data in the table, as shown in Figure 2.28.

When maintaining more complex decision tables, note that the system always reads decision tables from top to bottom. If a corresponding combination of input values is found, the system stops looking through the table and continues the business process with the output value(s) found.

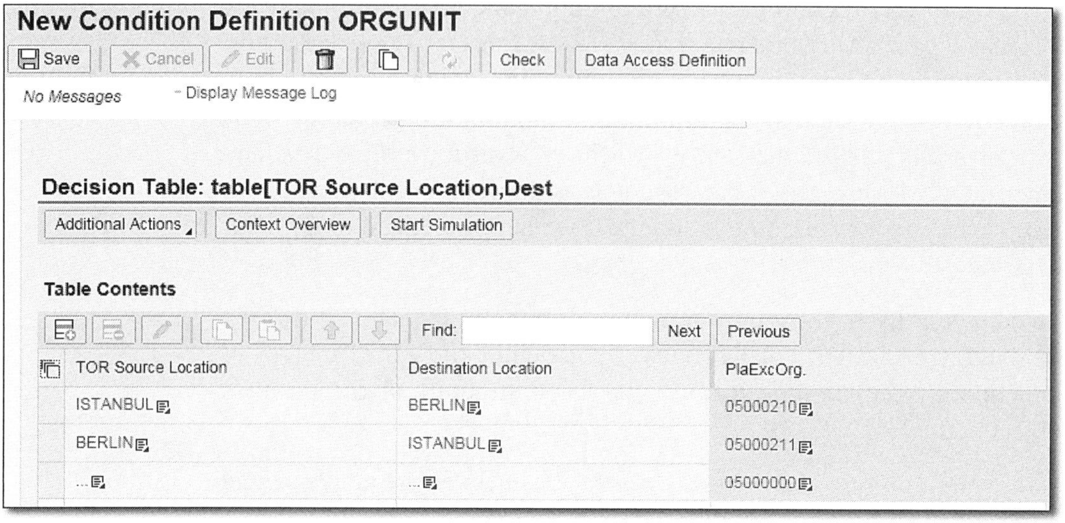

Figure 2.28 Decision Table

In our example, we now have a third planning and execution organization: Europe. This organizational unit is responsible for all other freight orders except for the ones between Istanbul and Berlin. We therefore enter a third line in the decision table, leaving the input value empty (which acts as a wildcard), and enter the planning and execution organization. In this case, it is very important to enter this line as the last line in the decision table because if this line is the first, all freight orders will get the planning and execution organization Europe.

Organizational Units in Decision Tables

Organizational units have an ID and an organizational unit number. In decision tables, only the number will be displayed. Therefore, you do not see the IDs Germany, Turkey, and Europe that we have used in our example in Figure 2.28, but only the organizational unit numbers.

After saving the condition, you can now assign your condition to a freight order type, as shown on Figure 2.29. The planning and execution organization will always be determined using your condition when you create a freight order of this type.

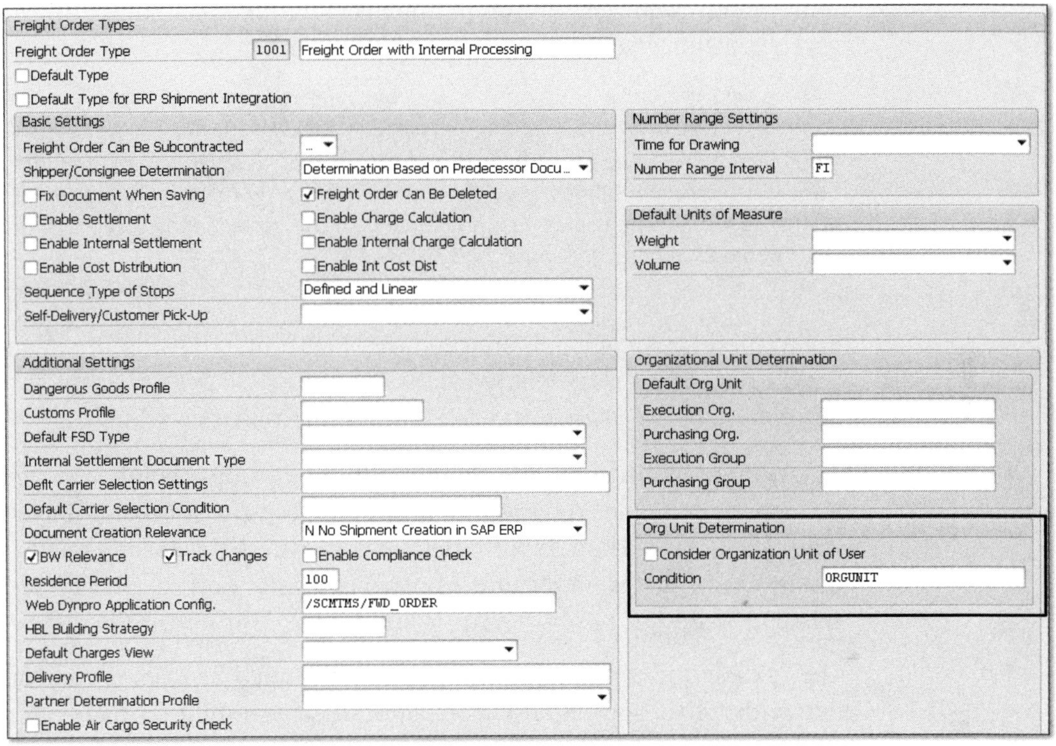

Figure 2.29 Organizational Unit Determination in Freight Order Customizing

> **Transporting Conditions**
>
> If the system is set up so that Customizing can be transported when entering a condition in edit mode, you will be asked to record your changes to the condition in a software Customizing transport request.
>
> Conditions can be transported to other systems. Be aware that this makes sense in some circumstances but not all; in our example, using organizational unit numbers does not make sense because the organizational unit numbers will be different in another system. However, conditions can also be created in systems where normal Customizing in Transaction SPRO is not possible.

2.3.2 Incompatibilities

Incompatibilities are used to implement loading constraints and routing constraints. Loading constraints identify goods as incompatible with other goods, vehicles, or compartments. Routing constraints identify goods or vehicles as incompatible with locations.

Note

This section about incompatibilities aims to give you an overview of the functionality incompatibilities provide and how they work. We take a more detailed look at the use of incompatibilities in the corresponding sections of this book, especially in Chapter 6, Section 6.3.4.

You can create incompatibilities in SAP NWBC by following the menu path Application Administration • Planning • General Settings • Incompatibility Definitions • Create Incompatibility.

Incompatibility area and incompatibility type

An incompatibility has two main settings: incompatibility area and incompatibility type. You can find both at the top-right corner of Figure 2.30. The Incompatibility Area field defines where this incompatibility can be used—that is, in the entire optimizer run or only in carrier selection and so on.

The Incompatibility Type field defines what should be made incompatible with what. You define incompatibilities by setting up two conditions. Depending on the incompatibility type you have chosen, the condition type of both conditions is predefined.

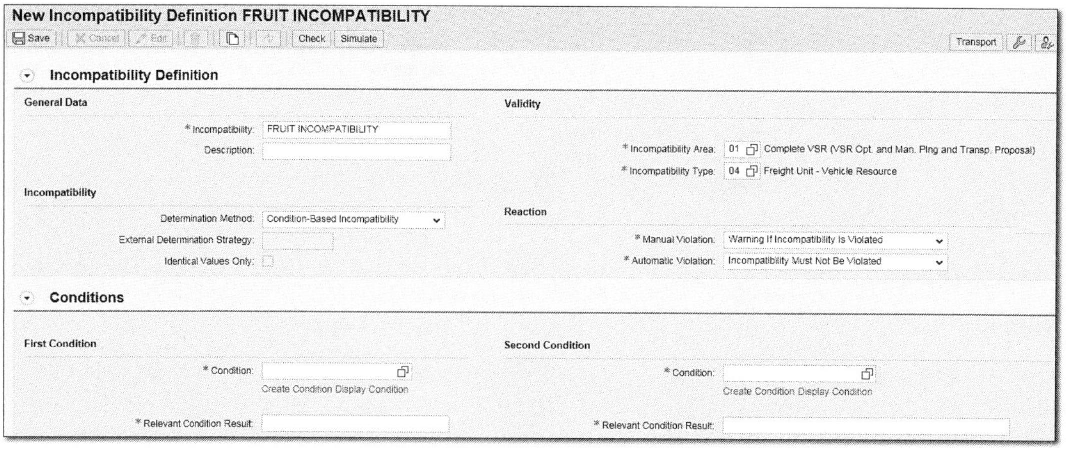

Figure 2.30 Incompatibility Definition

For example, say we want to define that a certain truck cannot carry fruit and vegetables, so we choose the incompatibility type FREIGHT UNIT – VEHICLE RESOURCE. The condition types of the two conditions being compared are now predefined. By clicking on CREATE CONDITION, you can create a new condition from the incompatibility definition screen.

On the left side of Figure 2.30 we create the first condition: FRUIT INCOMPATIBILITY. Since the freight unit itself does not carry any unique information that identifies it as fruit or vegetable, we need to fine-tune the condition.

We know that the products APPLE, PEAR, and ORANGE are fruits. Therefore, we define the decision table in such a way that if the input (the data access definition looks at the product ID of the freight unit's product item) is one of these, then the output is FRUIT, as illustrated in Figure 2.31.

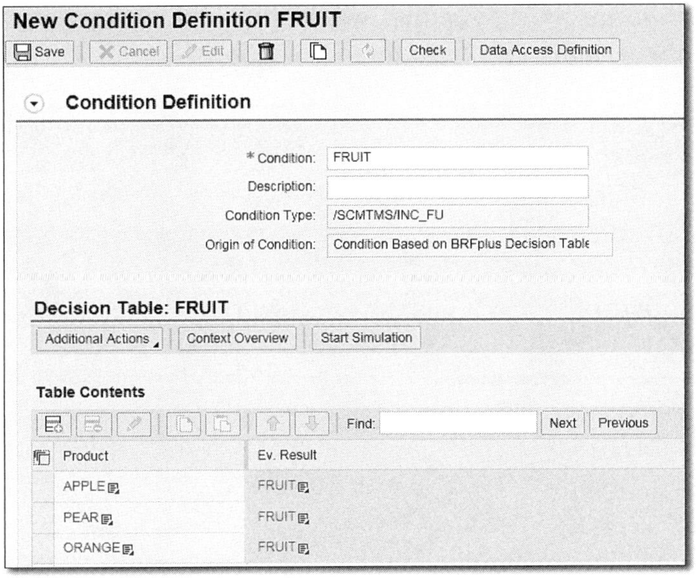

Figure 2.31 Grouping Different Input Options to One Value for Incompatibility

The first condition is automatically entered into the corresponding field in the incompatibility definition screen when you save it. We now enter the value FRUIT in the CONDITION RESULT field.

Now we need to create the second condition: the condition determining the resource name. In this condition, we can use the direct business object access condition because we only need the name of the truck. Set the condition access type to direct business object access and define a corresponding data access definition.

In the incompatibility definition screen, we now set the condition result of the second condition to TRUCK1. Now the truck is incompatible with all freight units that are defined as FRUIT in the first condition.

Grouping Products for Transportation Purposes

This example doesn't have any grouping information on the transportation quality of the freight unit. In general, we recommend storing that information in the product's *transportation group*.

Incompatibility settings

Once we have saved our incompatibility definition, we can group it together with others in incompatibility settings, which are shown in Figure 2.32. In this incompatibility setting, we can define whether we want equal behavior in manual and optimizer planning.

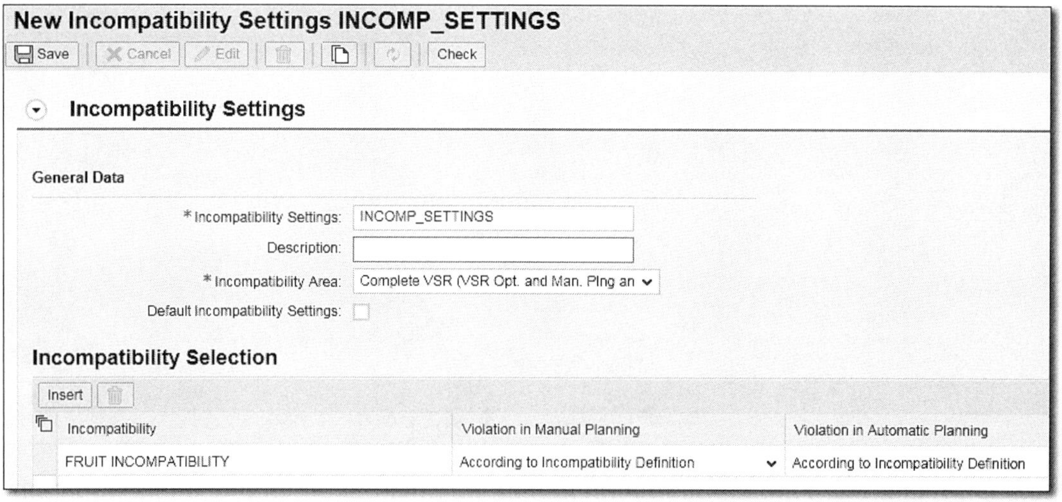

Figure 2.32 Incompatibility Settings

We delve deeper into the use of the different incompatibility types when we talk about planning with SAP TM later on.

2.3.3 Post-Processing Framework

The Post-Processing Framework (PPF) is a tool in SAP TM used to execute program logic that is considered as a follow-up action to a certain business process step.

The PPF is used for the following tasks, or *actions* (to name a few):

- Document printing
- Sending messages such as e-mail, fax, or EDI
- Workflow triggers

All actions are defined in the PPF with a *schedule condition* that defines whether the action needs to be executed and a *processing time* that defines when to execute the action.

In general, it is important to note that PPF actions can only process information that is already saved to the database. This prevents you from accidentally sending preliminary transportation orders to a carrier, for example.

The transaction for configuring the PPF is not accessible from SAP NWBC. Instead, you must enter Transaction SPPFCADM in the SAP GUI. On the initial screen of this transaction, you can see various PPF applications. The one that is important for SAP TM is /SCMTMS/TRANSPORTATION. Select this application and choose DEFINE ACTIONS PROFILES AND ACTIONS.

An action profile like the one shown in Figure 2.33 bundles all actions concerning a specific business process area (e.g., the action profile /SCMTMS/TOR bundles all actions important for the freight order).

Let's briefly consider the most important settings of the PPF.

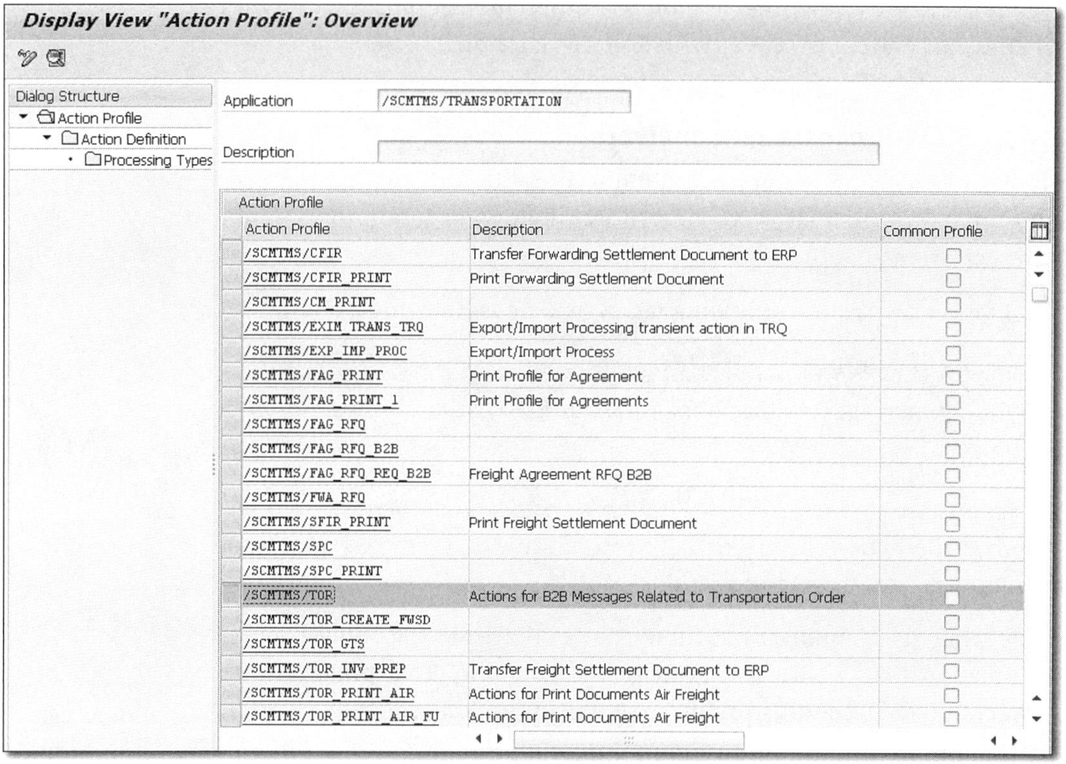

Figure 2.33 Action Profiles Containing Action Definitions

In Figure 2.34, you can see in the ACTION SETTINGS area where we can define the processing time of the action, and whether the action should be scheduled automatically or by a batch job.

The processing type of the action can be one of the following:

- Trigger alert
- Method call
- Workflow
- Smart Forms actions
- External communication

The type you choose determines the required information for the detailed processing.

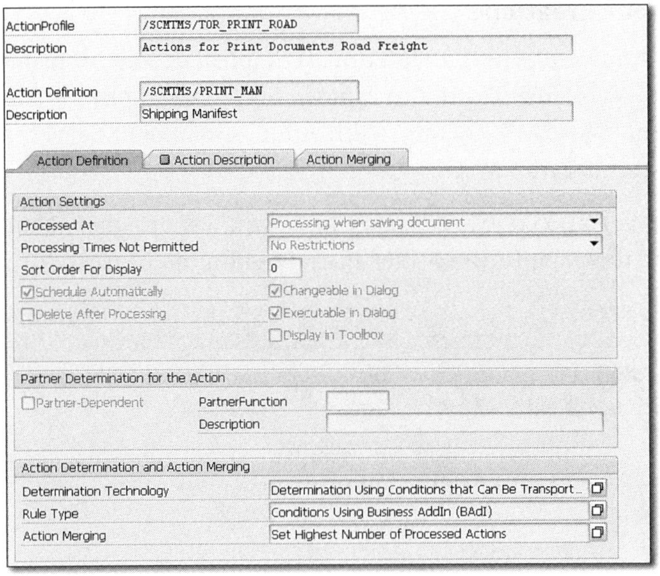

Figure 2.34 Action Definition

Schedule conditions define whether an action should be processed at all. The most common way of defining a schedule condition is to implement the BAdI EVAL_STARTCOND_PPF.

Schedule conditions for PPF actions

You can also define start and scheduling conditions in Transaction SPPFCADM by choosing the application /SCMTMS/TRANSPORTATION and selecting CONDITION CONFIGURATION (TRANSPORTABLE CONDITIONS). Figure 2.35 shows an exemplary start condition.

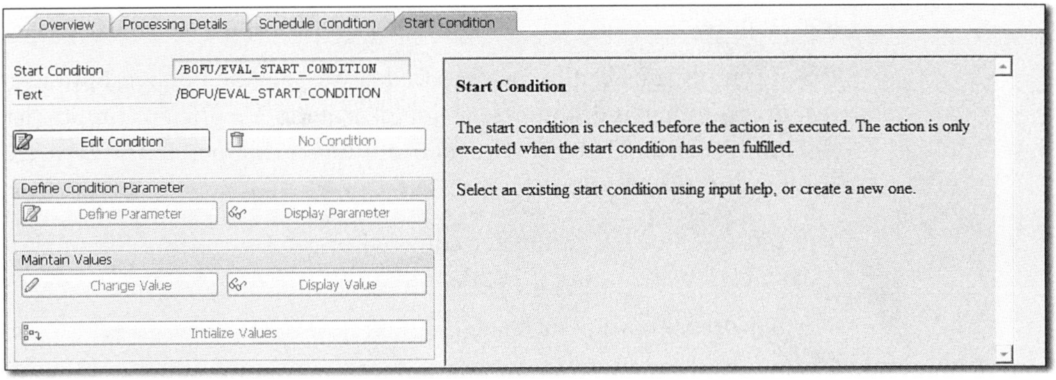

Figure 2.35 Start Condition of an Action

2.3.4 Document Creation

The previous section mentions that you can accomplish document printing with PPF. However, SAP TM is also capable of designing print forms to a certain extent. SAP TM integrates with Adobe Document Server, a tool that can be run within SAP GUI.

Start Transaction SFP and choose the print form you want to alter. Then add new fields to the print form or alter the layout on the LAYOUT tab.

Adobe Document Server

Please note that Adobe Document Server requires its own license and its own plug-in installed to use it within SAP GUI.

We will not go deeper into altering print forms in this chapter. For more information, consult resources that cover the Adobe Document Server.

2.3.5 Optimizer

SAP has already used optimization programs in various supply chain management (SCM) applications, such as SAP APO-TPVS, a predecessor of SAP TM. The optimizer is a C++-based application designed to solve transportation problems in a mathematical way. (Bear in mind that transportation problems might not be solved completely but can only be handled with a best approach. This applies to the optimizer, too.)

The optimizer can be run on hardware different from the application servers running SAP TM. In fact, we recommend this, because the system requirements of SAP TM and the optimizer differ.

Optimization runs can be made parallel. To provide enough optimizer resources, either the optimizer should provide a sufficient number of slots or you should connect several optimizer engines to the SAP TM application, as shown in Figure 2.36.

You can monitor the optimizer connections in Transaction RCC_CUST. As you can see in Figure 2.36, the optimizer provides six applications:

- TVSR for optimizer routing and scheduling
- TVSS for scheduling of existing freight orders

- TSPS for carrier selection
- TVRG for transportation proposals
- TSFM for strategic freight management
- TVSO for load optimization

RCCF: Destinations for Engines

Dest. ID	Appl.	Short Text	Status	M...	P...	Comm. Type	Communication Connection
TSFM01	TSFM	Strategic Freight Management Optimi...	Active	10	1	RFC	OPTSERVER_TSFM01
TSPS01	TSPS	Transportation Service Provider	Active	10	1	RFC	OPTSERVER_TSPS01
TVRG01	TVRG	Transportation Proposal	Active	10	1	RFC	OPTSERVER_TVRG01
TVSO01	TVSO	Load Optimization	Active	10	1	RFC	OPTSERVER_TVSO01
TVSR01	TVSR	Vehicle Scheduling and Routing engine	Active	10	1	RFC	OPTSERVER_TVSR01
TVSS01	TVSS	Vehicle Scheduling engine	Active	10	1	RFC	OPTSERVER_TVSS01

Figure 2.36 Optimizer Connections

We take a detailed look at the use of all these applications later in the book. For now it is only important to know that these six applications are only executable files on the optimizer server. No master data or transactional data is stored on the optimizer engine except for log files.

If you want to update your optimizer program, simply download the newest version of the executable files from the SAP Service Marketplace (*www.service.sap.com*) and replace the existing files on the server with the new ones.

2.3.6 SAP Visual Business

In SAP TM, you can display your transportation network and your transportation planning using a map. Furthermore, SAP Visual Business is also used to illustrate the load planning in the Transportation Cockpit or in the freight order. The application offers three options where a map can be used:

- Transportation cockpit
- Freight order
- Transportation network cockpit

The map in SAP TM is displayed using SAP Visual Business, which helps with visualizing the data in SAP systems. It is also used, for example, to visualize events in SAP Event Management or for the 3D view of vehicles in load planning (described in Chapter 6, Section 6.3.3 and Section 6.3.5).

As of SAP TM 9.1, use the application version SAP Visual Business 2.1.

Note

SAP Visual Business only provides the technological means to display a map in the SAP TM application and offers some user interaction possibilities. However, SAP is not a map provider, which means that an external map provider needs to be called from the SAP TM system.

In most cases, the map is called from the SAP TM application with the integration of a GIS provider; we explain this further in Chapter 3, Section 3.2.9.

SAP Visual Business settings in SAP TM

In the SAP TM system, you can run SAP Visual Business without worrying about most settings. But make sure that the business function `FND_VISUAL_BUSINESS` is activated and check whether you have activated the map functionality in Customizing by following the IMG path SAP Transportation Management • Transportation Management • Basic Functions • Geographical Map • Define Settings for Geographical Map.

Installing SAP Visual Business and SAP Active Component Framework

You also need to do some setup activities on your client computer in order to display the map. First install SAP Visual Business on your client. You can find the installation package in the SAP Service Marketplace.

In addition to the SAP Visual Business application, you need to install the SAP Active Component Framework (ACF). It, too, is available in the SAP Service Marketplace.

Whitelist certificate

After you've installed the two applications, install an ACF certificate on your client. The certificate is generated by the SAP TM system via Transaction WDR_ACF_GEN_CERT.

Once the certificate is generated, you can install it by following the IMG path SAP Transportation Management • Transportation Management • Basic Functions • Geographical Map • Visual Business • ACF • Install ACF Whitelist Certificate or by using Transaction WDR_ACF_WLIST if you can't access Customizing. In this transaction, you need to select the previously generated whitelist certificate and choose INSTALL CERTIFICATE.

Note

It is important to note that the installation of the certificate is only for the client computer on which you have executed these steps. If you want to work from a different computer on the same SAP TM system, you need to repeat these steps. Furthermore, if you use computers with multiple user accounts on the operating system, please note that the certificate is installed in the user-specific folders in the operating system. A new user on the same computer has to install the certificate again. However, the certificate is valid for all SAP users on one computer.

Stay Connected for the Map

Technically, SAP Visual Business sends requests to the map provider, which returns pictures that are put into the SAP Visual Business application. This means that whenever you want to display the map in SAP Transportation Management, you have to have a running Internet connection on your computer.

2.3.7 Process Controller Framework

Another tool in SAP TM that simplifies enhancements of the standard logic is the Process Controller Framework.

In many functional areas of SAP TM, strategies provided by the Process Controller Framework are performed. A *strategy* is an order of methods being performed one after the other. Since all methods included in a strategy contain the same signature, it is very easy for developers to add their own methods to a strategy.

For example, you can define the strategy for performing manual planning in the planning profile. As you can see in Figure 2.37, there are several strategies to choose from. If you compare these strategies in Customizing (follow the IMG path SAP TRANSPORTATION MANAGEMENT • SCM BASIS • PROCESS CONTROLLER • ASSIGN METHODS TO A STRATEGY), you can see that they all have common methods, but some strategies also offer additional methods.

Method assignment to Strategy

Strategy	Strategy Description	Method	Method Description	Sequence
VSRI_1STEP	Interactive planning and Carrier Selection	VSRI_PRE		10
VSRI_1STEP	Interactive planning and Carrier Selection	VSRI		20
VSRI_1STEP	Interactive planning and Carrier Selection	VSRI_POST		30
VSRI_1STEP	Interactive planning and Carrier Selection	VSRI_TSPS	Interactive Planning + TSPS	40
VSRI_CHK	Interactive Planning Strategy + check	VSRI_PRE		10
VSRI_CHK	Interactive Planning Strategy + check	VSRI		20
VSRI_CHK	Interactive Planning Strategy + check	VSRI_POST		30
VSRI_CHK	Interactive Planning Strategy + check	VSRI_CHECK		40
VSRI_DEF	Default Interactive Planning Strategy	VSRI_PRE		10
VSRI_DEF	Default Interactive Planning Strategy	VSRI		20
VSRI_DEF	Default Interactive Planning Strategy	VSRI_POST		30
VSRI_SCH	Interactive planning and scheduling	VSRI_PRE		10
VSRI_SCH	Interactive planning and scheduling	VSRI		20
VSRI_SCH	Interactive planning and scheduling	VSRI_POST		30
VSRI_SCH	Interactive planning and scheduling	VSRI_SCHED		40

Figure 2.37 Method Assignment to Strategies

If you want to enhance the logic of manual planning, add your own methods to the strategy or create a completely new strategy.

Services define which strategies may be used in which context. In SAP TM, a service is strictly linked to a use case. A strategy containing several methods can then be assigned to a service. This establishes the hierarchy shown in Figure 2.38.

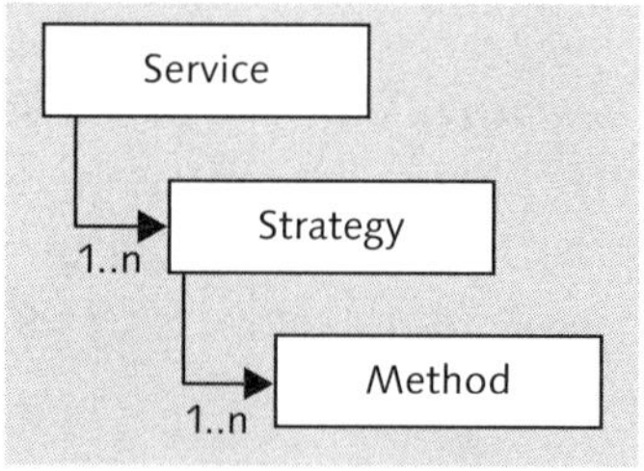

Figure 2.38 Process Controller Hierarchy

2.4 Integration Technology with SAP TM

So far in this chapter we have talked about the technologies that enable business logic within SAP TM. In Chapter 1, Section 1.2, we saw that

SAP TM can be seen as part of an SAP Business Suite and SAP SCM. Therefore, technologies that enable business logic within SAP TM are not sufficient without integration with other SAP applications.

Figure 2.39 shows that SAP TM uses various integration technologies to communicate with other applications. We take a look at these integration technologies in this section.

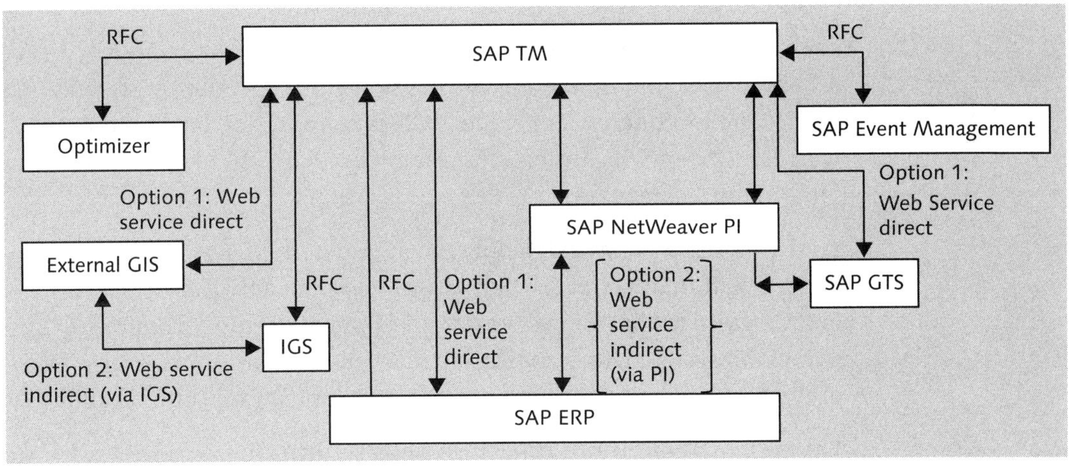

Figure 2.39 Overview of SAP TM Integration Technology

2.4.1 SOA-based Integration with SAP NetWeaver PI

SAP NetWeaver PI as middleware

When you look at the early stage of SAP R/3, you can see that the entire business logic was meant to run on one system. All the data gathered in business transactions was supposed to be stored on a central database from which it could be retrieved by different application modules.

Nowadays, with the big increase in data volume and the high level of interconnectivity of business partners (both internally and externally), system landscapes have become heterogeneous. Not only data is spread throughout different systems; business logic is also run in several applications.

To orchestrate both the supply of data and the business logic for the entire system landscape, you need to establish a central instance that is aware of all systems in the system landscape and knows where data can be provided. In the SAP system landscape, this central instance is SAP NetWeaver Process Integration (PI).

SAP NetWeaver PI provides a platform that allows different interfaces to communicate with each other. The communication is established using a uniform technology: web services. SAP NetWeaver PI not only acts as a directory for systems and applications in a landscape, but also supports and monitors communication among systems.

Service-oriented architecture (SOA)

The concept of service-oriented architecture (SOA) was established once it became clear that the entire business logic of an enterprise could not be executed in one system alone. SOA is supposed to enable interaction between systems regardless of their respective system architectures or programming languages.

Web services are essential helpers of SOA that implement the concept of the interaction of systems. Web services contain a data structure, usually in XML format, that shows what data can be passed on from or to a system in the system landscape. In the respective applications, *proxies* interpret the web service and execute business logic in the application.

The system applications offer their web services to a service broker (in our case, SAP NetWeaver PI). The service broker acts as a web service directory; you could also say the service broker is like a dating agency for business systems. All systems in the system landscape can now access the directory of web services in SAP NetWeaver PI and download the web services descriptions into their applications; there, they can start implementing business logic that either fills the XML data structure in order to send data to other systems (this system is the *service consumer*) or implements logic to interpret the XML to use the data (this system is the *service provider*). After it executes the business logic, the service provider can respond with another XML file to the service consumer.

We could look at more theory to understand the integration of systems with SOA and web services, but it might be best to examine it using a concrete example. Let's take a look at the solution architecture for the integration of an SAP ERP sales order into SAP TM.

Architecture of Sales Order Integration

Let's assume that the system landscape we have is an SAP TM 9.1 application that wants to interact with SAP ERP. To integrate Materials Management (MM) and Sales and Delivery (SD) orders, you need to have at least SAP ERP 6.04 SP9 in place. (If you don't, take a look at Section 2.4.2, which describes system integration without predefined content in both application systems.)

The first step is to register both SAP TM and SAP ERP in the system landscape directory of SAP NetWeaver PI so that it is aware of both systems and their application purposes. (The content of SAP NetWeaver PI is constantly updated, which means it knows what an SAP TM application is for.)

Once the systems are registered in the System Landscape Directory (SLD), the web services can be published to the respective systems. A web service consists of two service interfaces—one interface for each application. The service interface can be loaded into the application system.

Let's look at the service interface on the SAP TM side. The service interface for SAP ERP order integration is called `IntracompanyTransportationRequestRequest_In`, and you can find it in the SAP TM system using Transaction SPROXY. If you pay close attention, you will see that once you have confirmed the transaction code, there is a notification in the message area that the Enterprise Services Repository (ESR) is started. Because the ESR is situated in the SAP NetWeaver PI system, the information we see in Transaction SPROXY is actually loaded from the connected SAP NetWeaver PI system.

Figure 2.40 shows the numerous service interfaces available for SAP TM 9.1. The service interfaces enable not only communication between SAP systems within a company (application-to-application, or A2A, communication), but also the communication between SAP systems and legacy systems or systems of external business partners (business-to-business, or B2B, communication).

There are a few important things to consider about how the web service works when you are selecting the service interface for integrating SAP ERP orders into SAP TM.

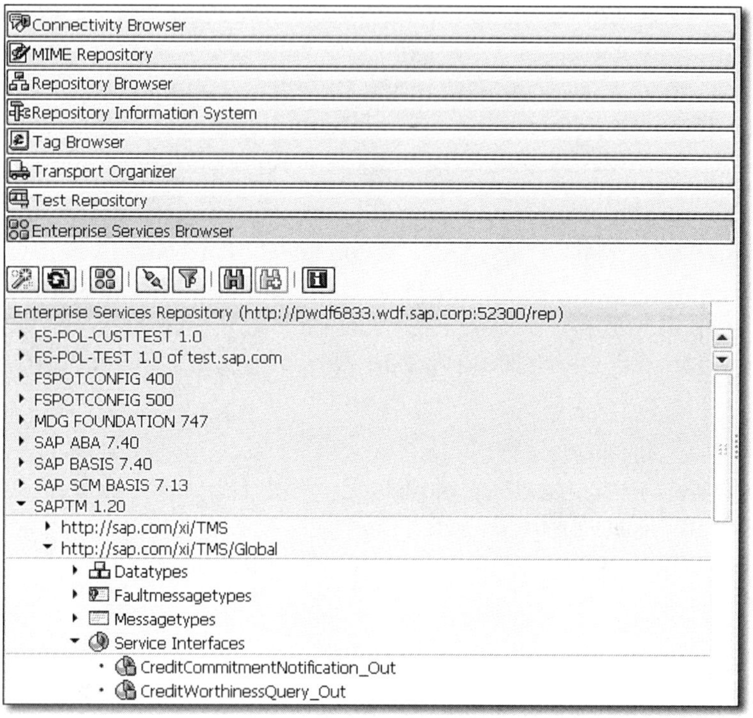

Figure 2.40 Service Interfaces in Transaction SPROXY

On the detailed view of the service interface, you can see a number of different tabs. Get familiar with the data structure that is passed to the SAP TM system. If you go to the External View tab and expand one node after the other, you will see that the XML message's data structure consists of two parts: the message header and the transportation request itself. The message header contains metadata about the message, especially stating from which system this message was sent and which system was supposed to be the recipient.

The important information for us is in the transportation request, which indicates what data can be passed to the SAP TM system. Notice that the data structure is somewhat similar to the data structure of the business object /SCMTMS/TRQ of which an instance will be created with this web service.

Global data types

When data is exchanged between two systems, both systems need to be able to interpret the data that is passed. So if the service consumer sends a date to the service provider, the service provider needs to know that the field will be no longer than eight digits. This cannot be taken for granted, because the SAP ERP system has its own data types that are not communicated to SAP TM. Therefore, global data types are used.

Global data types are defined in SAP NetWeaver PI. Here we can create a global data type for a delivery date. Though global data types are based on core data types such as string, integer, and so on, you can also encounter aggregated data types, meaning a structure containing several global data types. These global data types within an aggregated data type are again based on core data types.

The entire XML message format consists of fields using global data types. Once the service interface is loaded into the application system, the proxy generation will perform the next necessary steps.

Proxy

So far we have only looked at the data structure of the service interface's message. But remember that the SAP TM system cannot yet work with the data types contained in the message, so we need to perform a proxy generation. A proxy is an application system-owned interpreter of the web service's service interface; it converts the global data types of the message into data types the SAP TM system can work with. Let's take a look at what the proxy generation achieved with the city name of the shipper party.

The city name was based on a global data type (upper box), but as you can see in Figure 2.41, the proxy has generated its own data element in the SAP TM system (lower box).

The global data type defined that the city name should be based on the core data type string and should not be longer than 40 characters. If you take a look at the generated data type in Transaction SE11, you can see that this setup has been taken over in the new data element in SAP TM.

If aggregated global data types are used, the proxy will generate a structure type.

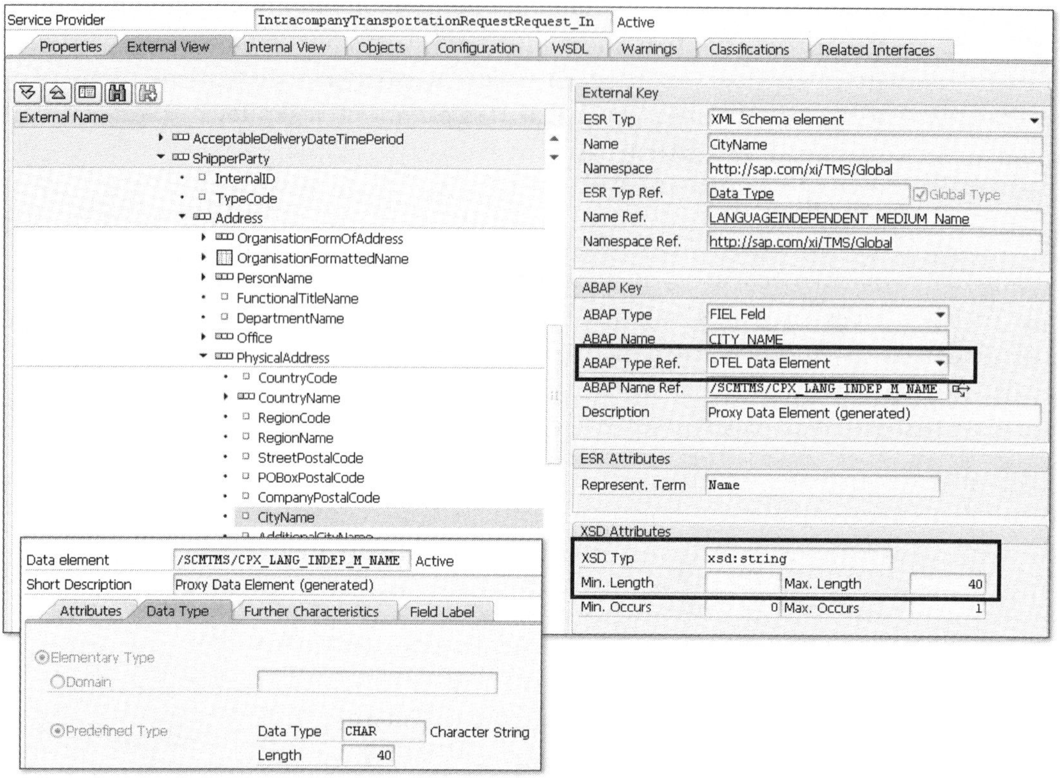

Figure 2.41 Data Element Generated by Proxy

Proxy provider class

So far we have only looked at the data structure of the service interface in SAP TM. But the OTR or DTR will not be created unless some business logic is executed. As we have said, a web service consists of a common data structure and business logic on the service consumer's side as well as on the service provider's side. In a communication between the service consumer and the service provider, neither knows exactly what is happening on the other side of the communication channel. The business logic of the communication partner is supposed to be a black box. This especially makes sense if communication via web services has been established between two systems with different programming languages.

The proxy not only generates the data elements in the system; it also creates a *provider class* (Figure 2.42). Use of the provider classes is comparable to the use of the classes assigned to actions, validations, and determinations that we discussed earlier. The provider class consists of a constructor and an executing method that is called once a message of this service interface is received in the system.

When you display the service interface in Transaction SPROXY, go to the PROPERTIES tab shown in Figure 2.42. Here you will find the provider class that was also generated by the proxy (lower part of the upper screen). When delving deeper into the implementation of this class (the middle screen shows methods provided by the class), you will see that, in our example, the instance of the business object TRQ will be created. The lower screen shows an excerpt from the actual coding implemented in the method.

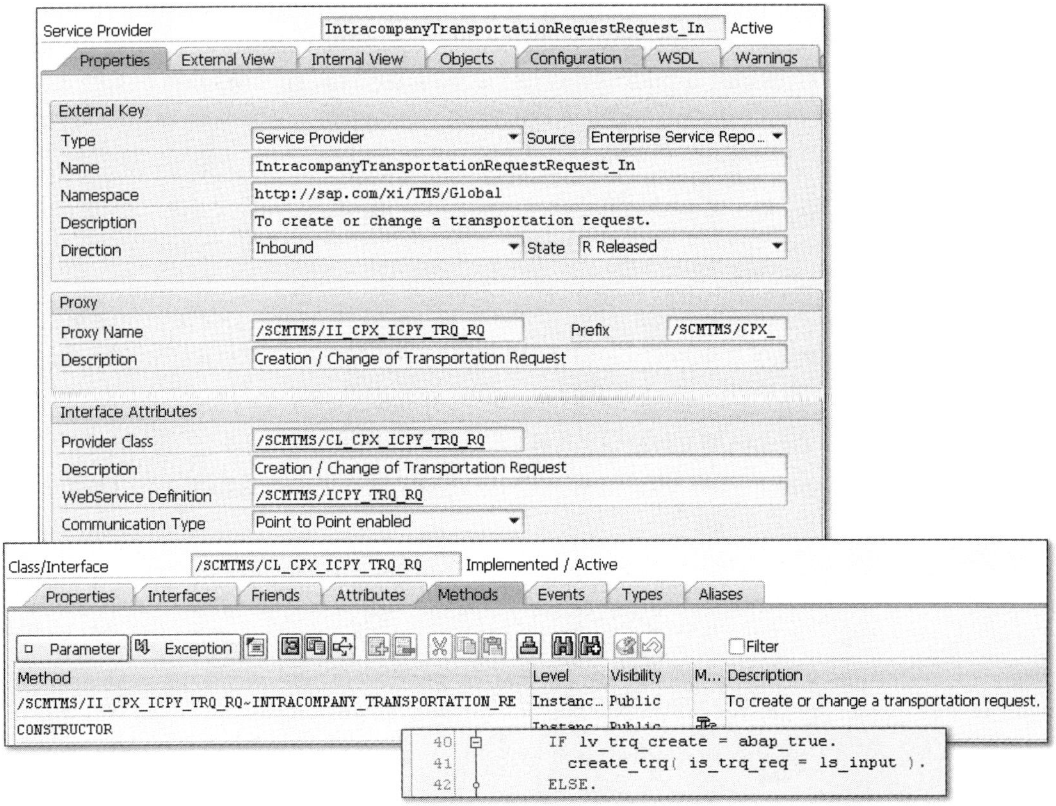

Figure 2.42 Proxy Provider Class

This Is Not the Entire Story about SOA

Communication between two SAP systems using SOA is a very complex topic for which many resources are available. This section is supposed to give you a quick glimpse at how the integration of transactional data between SAP ERP and SAP TM works. For more details about how to set up an SLD and how to create web services, another resource may be more helpful.

Recall from Figure 2.39 that you have two options for transferring an SAP ERP order from SAP ERP to SAP TM: either using SAP NetWeaver PI as middleware or using a direct web service-based communication channel. Both options have advantages and disadvantages, and the choice is yours.

Web service-based communication between SAP ERP and SAP TM was introduced with SAP TM 8.0. The web services for the standard integration are delivered with the SAP TM-specific content for SAP NetWeaver PI. In both application systems, you need to activate some business functions in order to utilize the service interfaces. We describe these in further detail in Chapter 4.

We have now looked at the service interface for receiving information from an SAP ERP document to create a corresponding document in SAP TM. The same settings can be observed in the SAP ERP system, where the service interfaces for the service consumer's side are present. You can get a detailed look at the SAP ERP service interfaces in Transaction SPROXY in the SAP ERP system.

As outlined in Table 2.1, SAP TM 9.1 offers several web services for A2A communication. These web services can be clustered into different functional areas of SAP TM.

Service	Service Consumer	Service Provider
Transportation Requirements (OTRs and DTRs)		
Create transportation requirements	SAP ERP	SAP TM
Cancel transportation requirements	SAP ERP	SAP TM
Sales order scheduling	SAP ERP	SAP TM

Table 2.1 Overview of A2A Services in SAP TM

Service	Service Consumer	Service Provider
Freight Settlement		
Trigger invoicing of SAP ERP shipment	SAP TM	SAP ERP
Cancel invoicing of SAP ERP shipment	SAP TM	SAP ERP
Transfer freight settlement document to SAP ERP	SAP TM	SAP ERP
Cancel freight settlement document	SAP TM	SAP ERP
Delivery Proposals		
Propose inbound delivery creation	SAP TM	SAP ERP
Propose outbound delivery creation	SAP TM	SAP ERP
Shipment Integration		
Create shipment based on freight order	SAP TM	SAP ERP
Cancel shipment based on freight order	SAP TM	SAP ERP
Create freight order based on shipment	SAP ERP	SAP TM
Cancel freight order based on shipment	SAP ERP	SAP TM

Table 2.1 Overview of A2A Services in SAP TM (Cont.)

Update of Transferred Documents

Please note that in most cases the web services that create documents in the other system are also used for updates. The proxy provider class evaluates whether a document has already been created and updates the corresponding document. Therefore, no duplicate documents are created in the service provider's system.

2.4.2 Integration with SAP NetWeaver PI without Predefined Content

If you take a closer look at the capabilities of SAP NetWeaver PI, you will see that this middleware application is used not only as a service broker and system landscape directory, but also for supporting message mapping in B2B communication. B2B communication takes place when two companies exchange data with their respective systems.

B2B communication between SAP TM systems

Let's walk through another example. We are a carrier, and our customer (a fashion company) wants to order transportation services. The fashion manufacturer uses an SAP TM system and has implemented various shipper scenarios (which you will see in Chapter 13). Once the fashion manufacturer has finished transportation planning, an order for transportation execution will be placed. To do this, the shipper sends out a request for quotation to us, the carrier. As a carrier, we are also using an SAP TM system. As you will see in more detail in the next chapters, the document flow will be the following: the shipper has created a freight order and now sends us a request for quotation. Because we are a carrier and the shipper's freight order is a transportation request, we will then create a forwarding quotation or forwarding order for the shipper's freight order (depending on the exact process).

This B2B communication can be established using SAP NetWeaver PI. Once the shipper's request for quotation has arrived in our system landscape, SAP NetWeaver PI will determine which inbound service interface needs to be triggered to which recipient based on the sender's system information and the message type. If the data structures of the request for quotation and the forwarding order's service interface do not match, SAP NetWeaver PI can also perform a mapping of fields. The service interface on our SAP TM side will then automatically create a forwarding quotation or forwarding order.

Table 2.2 lists all available B2B service interfaces that SAP TM provides.

Using SAP TM and Other Systems

In this example, we assumed that both business partners use SAP TM applications. Unfortunately, this cannot be taken for granted.

Do note, though, that the mapping of messages in SAP NetWeaver PI can be much more sophisticated than in this example. SAP NetWeaver PI can also map incoming IDocs to service interfaces.

Service	Service Consumer	Service Provider
Transportation Requirements (Forwarding Order and Forwarding Quotation)		
Create transportation requirements	External	SAP TM
Cancel transportation requirements	External	SAP TM
Confirm transportation requirements	SAP TM	External
Tendering		
Send transportation order	SAP TM	External
Cancel transportation order	SAP TM	External
Send request for quotation	SAP TM	External
Receive request for quotation response	External	SAP TM
Freight Booking		
Send freight booking	SAP TM	External
Cancel freight booking	SAP TM	External
Confirm freight booking	External	SAP TM
Send transportation waybill	SAP TM	External

Table 2.2 Overview of B2B Services in SAP TM

Synchronous or Asynchronous?

Web services can be performed synchronously or asynchronously.

All A2A and B2B web services in SAP TM are performed asynchronously. This makes sense because in most cases a user has to validate information that was entered using a service interface. For example, if a freight booking was sent to a carrier, the carrier needs to evaluate whether the requested space is still available. In the case of a synchronous communication, the shipper would not be able to work until the carrier has responded at least with a technical confirmation of the message receipt.

There is one exception: sales order scheduling. Since the information determined by SAP TM is crucial for continuing sales order creation, this information is passed on synchronously.

Order integration with SAP ERP 6.04 and lower

Recall from Section 2.4.1 that the predefined SOA-based communication between SAP ERP and SAP TM for order integration only works if an SAP ERP 6.04 SP9 system or higher is in place. Older versions of SAP ERP do not offer the business functions and service interfaces needed to use the predefined content for communication as explained in this chapter.

To integrate SAP ERP without predefined SAP NetWeaver PI content, you have to establish IDoc communication out of SAP ERP that is sent to SAP NetWeaver PI, where the IDoc needs to be mapped to the predefined service interface for OTR and DTR creation. Please note that in this case shipment integration and delivery proposals also need to be configured this way. Furthermore, direct communication between SAP ERP and SAP TM that bypasses SAP NetWeaver PI is no longer possible.

2.4.3 Monitoring Integration Messages with Service Interfaces

As you have seen, communication with SAP TM is usually established using service interfaces with or without SAP NetWeaver PI as middleware. In implementation projects, you might encounter situations where the communication setup was finished, but the incoming messages do not yet achieve the desired result. To troubleshoot, we recommend monitoring the incoming and outgoing messages in SAP TM to see which fields were filled incorrectly or what error messages occurred during processing of the message.

The monitoring of service messages is not specific to SAP TM but is common to many SAP applications. However, it is worth taking a look at it now because it rounds out the topic of communication between different systems.

With Transaction SXMB_MONI, you can display all messages that have arrived in your SAP TM application or have left the SAP TM system. After you enter the transaction, select MONITOR FOR PROCESSED XML MESSAGES and then define further selection criteria, such as the name of the interface or the timeframe.

The first column of the list of messages shows the status of the message and whether it was processed correctly. Consider a few common examples:

- Checkered flag: The message was processed successfully.
- Green flag: The message has been recorded, but the queue needs to be started manually.
- White flash on red button: An error has occurred.
- Green arrow: The message was not processed successfully but can be restarted in Transaction /SAPPO/PPO2.

By double-clicking on the STATUS DETAILS field, you can view the XML message that was sent or received. If an error has occurred, you can see the error message on the right side. If you want to view the content of the message, navigate to XML MESSAGE • INBOUND MESSAGE • PAYLOAD • MAINDOCUMENT, and the message content will be displayed in the lower window on the right side of Figure 2.43. This is Window 2.

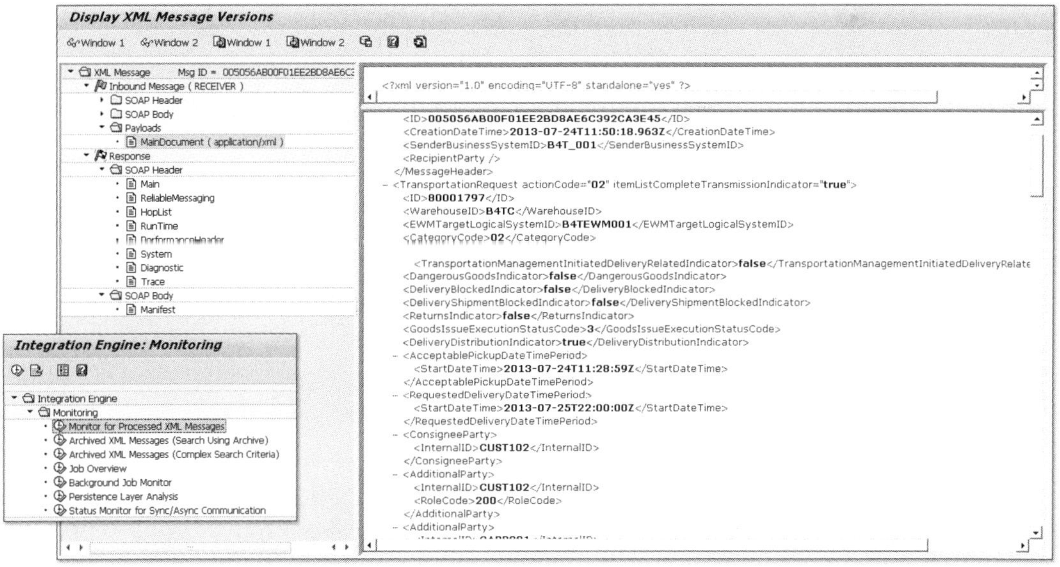

Figure 2.43 Monitoring the XML Message

Sometimes the message is not processed correctly because a field was not filled or was filled incorrectly. In this case, if you want to test whether this is the only issue with the message, you don't need to resend data from SAP ERP; instead, simulate the message entry. When displaying the message content as explained, download this XML file to your computer by clicking on the DOWNLOAD WINDOW 2 button on top of the message display screen (see Figure 2.43).

After that, go to Transaction SPROXY and enter the service interface you used. Notice the TEST button in the menu bar like in Transaction SE80. If you enter the test mode, you will be able to upload your XML file again, manually alter data in the XML editor, and then process the service interface using your manipulated message. Please keep in mind that you need to trigger `COMMIT WORK` manually in order to write data on the database. This can be done in the general menu of this transaction by following the menu path EXTRAS • TRIGGER COMMIT WORK (shown in Figure 2.44).

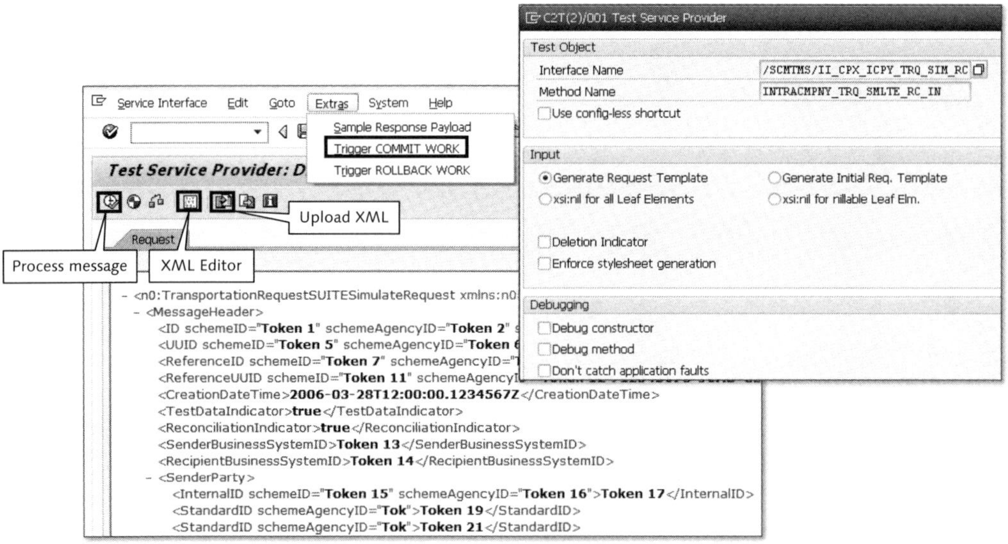

Figure 2.44 Simulating Entry of a Service Interface Instance

2.4.4 RFC-based Communication

Figure 2.39 shows that SAP ERP communicates with SAP TM not only using web services via SAP NetWeaver PI, but also using *remote function calls* (RFCs).

RFC-based communication is used to integrate master data from SAP ERP to SAP TM. Since we are not dealing with individual transactional data here, but instead with mass data, RFC communication is used instead of web services because web services are usually designed for communication concerning transactional data.

Integrating master data from SAP ERP to SAP TM is done using standard technology that is also used with other SAP SCM applications: the Core Interface Framework (CIF). You will learn more about the setup of the CIF in Chapter 3.

Master data integration with SAP ERP is established using RFC communication, as is the integration of SAP TM and SAP Event Management. SAP Event Management can be installed on the same system as SAP TM. However, you need to use Transaction SM59 to define an RFC connection to the SAP Event Management system, even if it is installed on the same system as SAP TM. We cover the setup of the SAP Event Management integration in Chapter 8, Section 8.2.

2.5 Conclusion

This chapter introduced you to the technological foundation of SAP TM, showing the two major technological pillars of SAP TM: the Business Objects Processing Framework, which takes over the tasks of data modeling, storage, and processing, and the user interface technology built on the FPM framework. We also covered how the data stored in the BOPF is transferred to the user interface technology using FBI.

While talking about UI technologies, we explored how the SAP NetWeaver Business Client interacts with the user and how different authorization role menus are put into the NWBC-based user interface. SAP TM screens can be customized without any coding and modifications.

Next we looked at other tools and frameworks used in SAP TM such as the BRF+, the Post-Processing Framework, and the Process Controller Framework. These tools facilitate customizing and adapting SAP TM to the business use case.

The final part of this chapter covered the technical integration of SAP TM with other business applications. In general, we have differentiated between an A2A communication using predefined content on both application systems and B2B communication, where only predefined content in SAP NetWeaver PI is available for SAP TM.

This technical information about SAP TM is meant to build a foundation for the SAP TM business processes described in Chapter 3 through Chapter 11.

Master data serves as a cornerstone for any business process. All central business objects and procedures, such as order management, planning, subcontracting, and shipment costing, are based on master data. In addition, it maps both the internal organization and partner business relationships, and is essential for specifying the transportation network and resource utilization.

3 Master Data

In the previous chapter, we gave you an overview of the solution architecture and technological concepts of SAP Transportation Management (TM). Before we take a closer look at transportation requirements, planning, and execution, we dedicate this chapter to explaining the mandatory and optional master data, while making you familiar with the most important terminology and configurations.

The master data foundation

Master data is an integral part of any planning system and is the cornerstone of any business process. All central business objects and processes that are key to transportation management are based on logistical master data. This data, apart from general master data such as organizational structures and business partners, typically includes the transportation network and resources, which together describe how transportation orders can be executed. Therefore, whether you work with a standalone SAP TM system or run SAP TM integrated into an SAP ERP system, master data is important because it supports both planning and execution activities.

The master data imperative

In contrast to SAP ERP, one of the key features of SAP TM is that it allows logistical processes to be executed independently of master data relating to business partners and transported goods. For third-party logistics providers (3PLs) who mainly provide transportation services, it is essential that the transaction data in SAP TM can be created without existing master data.

Therefore, planning and execution can be performed in SAP TM with only a minimal master data set created and maintained exclusively in SAP TM. However, when the focus is on organizational and financial processing, it is critical to have a complete master data record in an appropriate structure. In those cases, SAP TM seamlessly integrates with SAP ERP. In this chapter we explain both logistics master data representing the transportation network and general master data and resources used to execute the transportation of goods. All other master data and transportation mode-specific settings are explained in the relevant context.

General master data

To support end-to-end transportation scenarios for shippers and LSPs, and to ensure consistency with the execution process, SAP TM can require a set of general master data, for example, to model your organizational structure and avoid the recurring maintenance of addresses or product attributes. We take a look at this general master data in Section 3.1. By seamlessly integrating with SAP ERP and avoiding master data maintenance, SAP TM can use and reuse existing customer and product master data from SAP ERP.

But if master data is not mandatory, and if SAP TM is being used in a typical 3PL scenario, the system can be operated without some of the general master data, such as products and business partners. In all other cases, master data can either be created in SAP TM or integrated from SAP ERP.

Necessity of Master Data in SAP TM

In contrast to SAP ERP, with SAP TM you can get away with using only a minimum set of master data. Transactions can even be created without existing master data such as products, locations, and business partners. Optionally, to avoid frequent address entries and for ease of maintenance, you can also create the necessary master data in SAP TM or reuse existing data from SAP ERP.

Integration

When SAP TM is integrated with SAP ERP, the SAP ERP system is frequently the leading master data system. In this case, products and business partners are maintained in the SAP ERP system, where they are maintained as materials and customers, before being transferred to SAP TM via the Core Interface Framework (CIF). We will describe this integration and the transfer of master data in Section 3.1.5.

Logistics master data

Several master data elements are needed to support proper transportation planning and execution. In Section 3.2 we explain how these elements are created and combined to outline a transportation network.

Resources

For transportation execution and the capacity needed to perform the transportation activity, SAP TM uses resource master data such as vehicles and trailers. To reflect availability and operating times, and to handle transportation orders, you can set up calendar and handling resources. The configuration and maintenance of resources is explained in Section 3.3.

Let's begin by exploring the use of general master data in SAP TM.

3.1 General Master Data

All corporate departments use general master data. Since it maps both the internal organization and the partner and business relationships, general master data serves as a cornerstone for any business process. Clearly defined master data is of central importance for well-regulated financial processing in particular. The delineation of these departments and the maintenance of their master data is analogous to organizational structures and corporate responsibilities (see Figure 3.1).

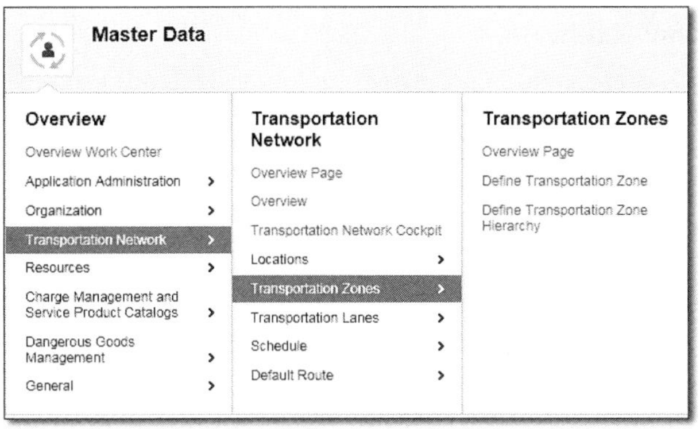

Figure 3.1 Master Data Maintenance in SAP TM

In this section, starting with the definition of the organizational structure, we present a thorough overview of the most important general master data, its significance for transportation processes, and, if applicable, its integration and distribution.

3.1.1 Organizational Structure

The individual elements of an organizational structure are used to map an enterprise in an SAP system. These organizational structures determine the operational framework in which all sequences and functions of logistics and financial processes occur. They also reflect the legal and organizational structure of a company and form the basis for data organization in the SAP Business Suite by enabling a variety of perspectives with regard to the master data, depending on the functional business area.

Organizational models

Organizational master data allows you to create organizational models, which are the legal, geographic, or organizational boundaries for the organizations that take part in the transportation process. By representing the structure of a company, and being used to determine responsibilities, the organizational units build a framework in which all relevant business processes occur.

Figure 3.2 shows an example of what this structure can look like when both SAP ERP and SAP TM are used. We go into further detail about this organizational structure in the upcoming subsections.

Internal Organizational Structure

A transportation service provider like Sunshine Logistic shown in Figure 3.2 usually has an organizational structure that can be set up as a combination of organizational elements. The simplest structure may consist of a single employee who is responsible for various tasks. However, in a larger enterprise, or in the case of a logistics service provider (LSP), the structure may be divided into various organizational areas. Figure 3.3 shows the structure, cardinality, and relationships between the organizational elements in SAP ERP and SAP TM.

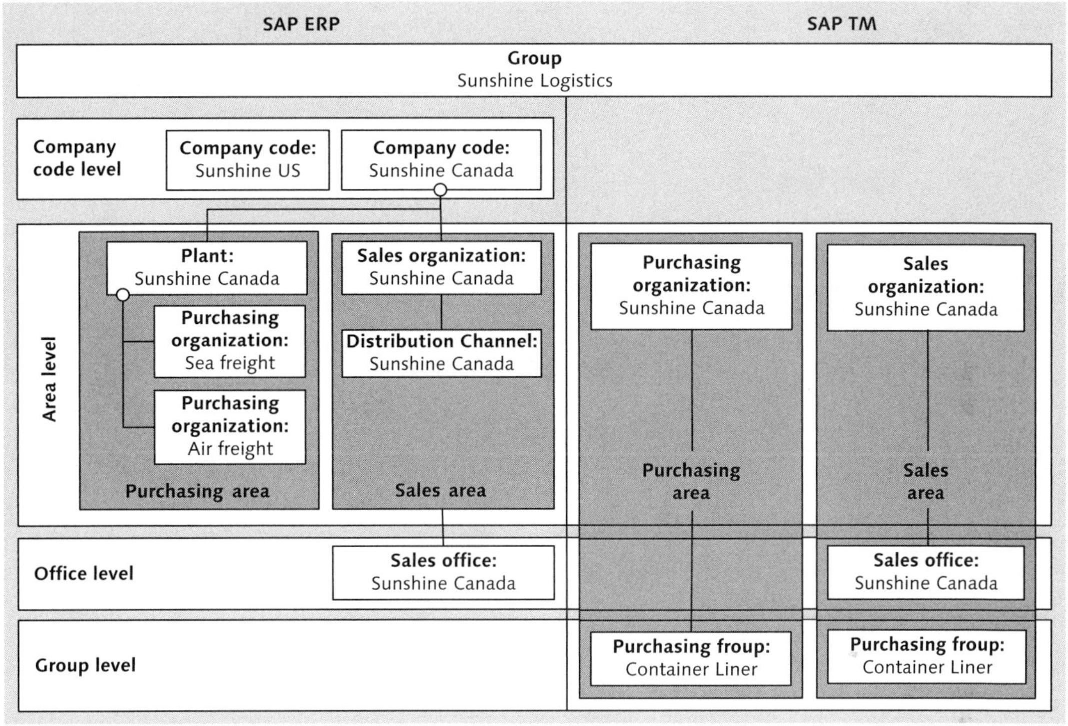

Figure 3.2 Organizational Structures in SAP ERP and SAP TM

In SAP TM, organizational units can be modeled independently from SAP ERP. They are categorized in three basic types:

- Sales (sales organization)
- Purchasing (purchasing organization)
- Planning and execution (transportation planning point)

These organizations can be further divided into groups and offices; employees acting on different roles are typically assigned to organizational groups.

Company and corporate organization

The optional corporate organization usually serves as the highest node and entry point in the organizational structure, whereas in SAP ERP the company organization corresponds to the company code. Like the company code, the corporate organization also defines the local currency and usually represents the legal entity of the company.

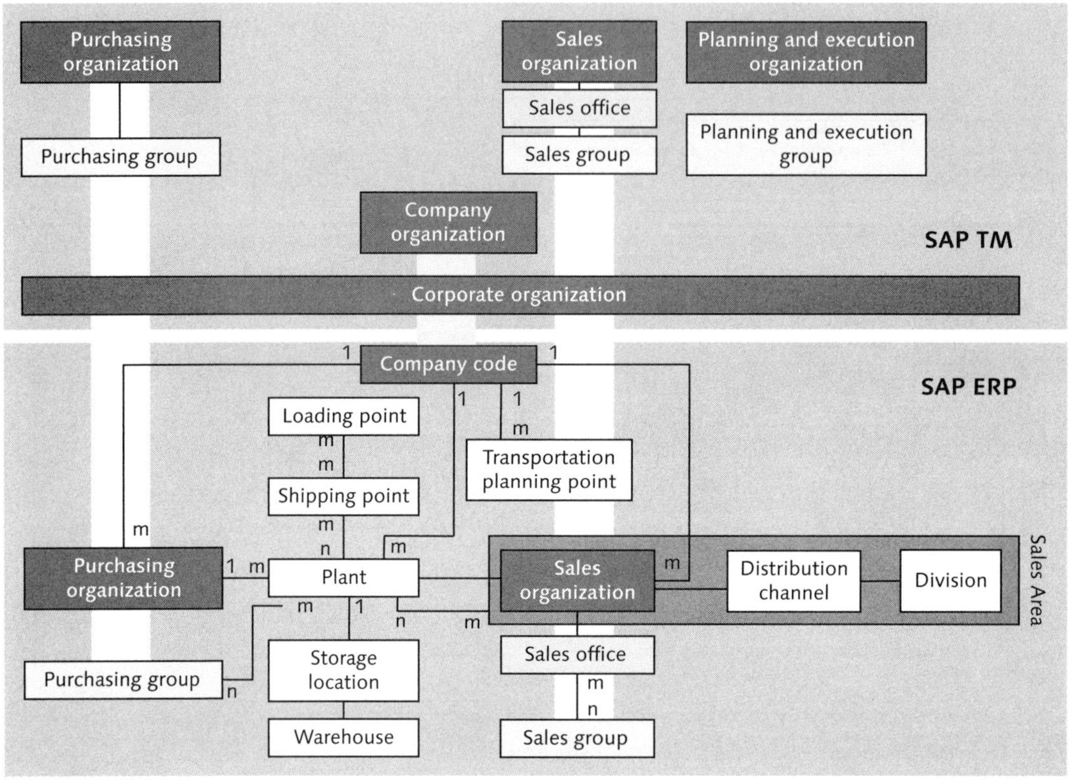

Figure 3.3 Mapping Organizational Structures in SAP ERP and SAP TM

Sales organization, office, group

Let's begin by looking at the sales organization, which organizes and structures the sale of logistics services and executes these services. It can consist of a hierarchy of suborganizations (e.g., for each country), sales offices (East Coast, West Coast, etc.), and sales groups (air freight, sea freight, etc.). You can also assign information to distribution channels (industrial customers, large customers, etc.) and divisions (container line haul, door-to-door, etc.).

In SAP ERP the distribution channel normally describes the channel through which a transportation order has been received, such as by key account or direct sales. In a transportation context, though, the distribution channel may also be used to distinguish between less than container load (LCL), full container load (FCL), less than truck load (LTL), and so on, to name a few.

A division is an organizational unit in SAP ERP that reflects the responsibility for sales and profits from a saleable service. In a transportation scenario, the division can be used to distinguish between air, sea, and land freight, for example.

In SAP TM, the sales organization has a hand in the following kinds of transactions:

- Forwarding quotations
- Forwarding orders
- Customer contracts for the sale of freight-handling services
- Forwarding settlements

Sales Organizations in SAP ERP and SAP TM

If a consignor uses SAP TM in conjunction with SAP ERP SD, he must take into account the differences in semantics between the sales organization in the SD order and in the SAP TM forwarding order.

The SD order is used for the sale of goods, whereas the SAP TM forwarding order is used for the internal or external sale of transportation services. Therefore, the consignor should implement different sales organizations for each scenario in a consignor-specific system.

In simple scenarios, one general sales organization in SAP TM is sufficient.

Purchasing organization and group

The next basic type of organizational unit is the purchasing organization, which is used to organize an enterprise according to purchasing demands. The actual procurement of materials and services always takes place in relation to a purchasing organization. You need purchasing organizations and purchasing groups for subcontracting. The purchasing organization arranges and executes all purchasing transactions relating to the logistics services provided by carriers and freight-forwarders.

The purchasing organization can be divided into several suborganizations (e.g., for each country, each shipment type, etc.). Several purchasing groups (for each shipment type, region, etc.) can be assigned to each purchasing organization.

The following types of transactions are generally associated with the purchasing organization in SAP TM:

- Freight orders
- Service orders
- Freight bookings
- Freight settlements

Planning and execution organization and group

For transportation planning and execution, which is the final organizational unit type, you have to define planning and execution groups. These represent the different organizational units that are responsible for land, sea, and air freight in different geographical regions. In this context, the planning and execution organization organizes the dispatching of the accepted shipment orders and the planning of the loads that are to be shipped; executes planning; and either executes the activities required or oversees their execution if they are outsourced.

The following activities are associated with the planning and execution organization in SAP TM:

- Trailer documents and resources
- Freight orders

Creating the Organizational Structure

In SAP TM, the master data for the organizational units of the transportation service provider is handled as a set of classifying characteristics in various business objects. All of this data (e.g., the sales organization in order documents) is mandatory.

Because a direct relationship with financial grouping objects (e.g., company codes, accounts, internal orders, etc.) is deliberately *not* established in SAP TM, the organizational data is transferred to SAP ERP or the connected external billing system for settlement and is used there for financial assignment.

Organizational integration

If SAP TM is used with SAP ERP as an external billing system, you can set up the same organizational structures in both systems to enable a meaningful assignment of the sales and purchasing processes to the subsequent settlement processes. An example of this kind of structure is provided in Figure 3.3, which compares organizations in SAP TM and SAP ERP.

In this context, the company organization corresponds to the SAP ERP company code and is used by transportation charge management functionality for invoicing and charging. When we create organizational units and staff assignments, we can distinguish between two kinds:

- Organization units
- Positions

An organizational unit is an object that is used to map the corporate structure of a company in an organizational model in an SAP system. This is done with various organizational elements and attributes. When you create the organizational element, you assign two things to it:

Organizational unit functions

- The *organizational unit function* describes the purpose of an organizational unit. Functions include purchasing, sales, planning and execution, company, corporate, and forwarding house.

Organizational unit roles

- The *organizational unit role* defines the organizational element's level within the organizational hierarchy. The following roles can be selected: organization, office, and group.

The hierarchical relationship of organizational elements, based on the roles and functions, is fixed. For organizational units with a sales function, you can only assign organizational elements with a group unit rule to the organizational elements with an office unit role. The element with the office unit role can only be assigned to an organizational unit with an organization unit role. For organizational units with a purchasing or planning and execution unit function, you can only assign organizational units with a group unit role to organizational units with an organization unit role. You can also assign employees to the organizational unit, which you need to do in order to be able to use the workflow capabilities of SAP TM.

Creating an organization and staffing

To create organizational structures in SAP TM, follow the SAP NetWeaver Business Client (NWBC) menu path MASTER DATA ORGANIZATION • CREATE ORGANIZATION AND STAFFING. To change these structures, follow the menu path MASTER DATA • GENERAL MASTER DATA • CHANGE ORGANIZATION AND STAFFING. Alternatively, you can use Transaction PPOME (see Figure 3.4).

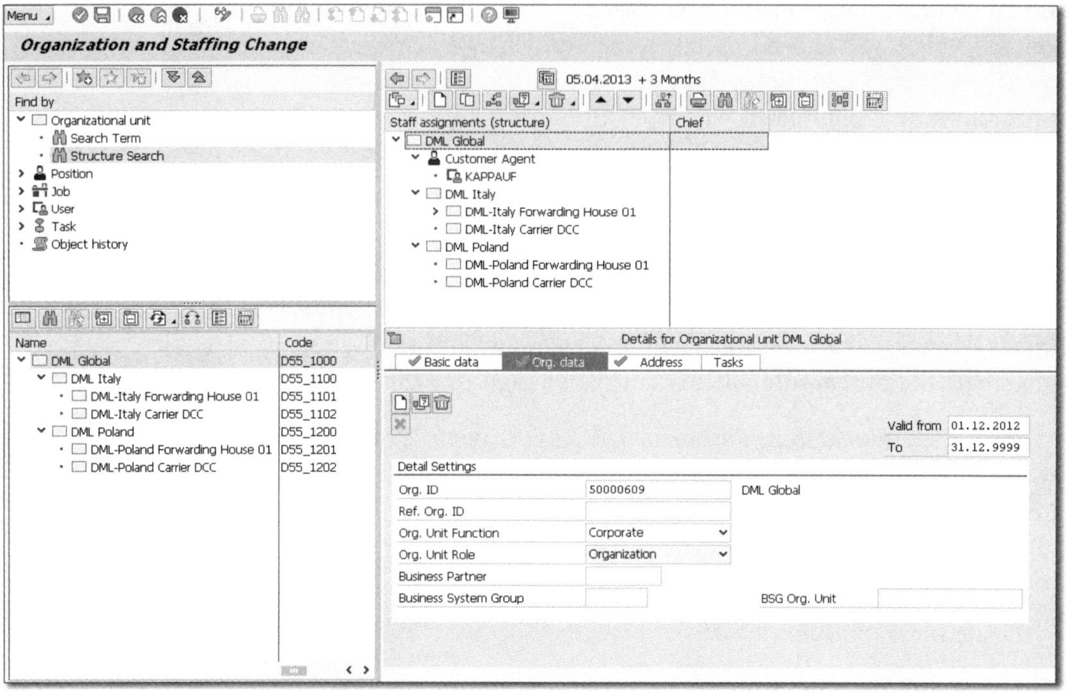

Figure 3.4 Organizational Hierarchy of an LSP

> **Note**
>
> When an organizational unit is created, the system automatically creates a business partner in the background. This business partner is used in transactional documents to represent the assigned organizational unit. Therefore, you have to configure the internal number assignment in the IMG for SAP TM by following the menu path Cross-Application Components • SAP Business Partner • Business Partner • Basic Settings • Number Ranges and Groupings. Please also refer to Figure 3.18 later in this chapter.

Positions

The system reuses the human relations functionality of the organizational plan to create and maintain organizational units, while employees can be assigned to positions. The *position* is an organizational unit that is mainly used to cascade and represent a hierarchy for workflow processing. SAP TM can use SAP Business Workflow if a process requires the involvement of a large number of agents in a specific sequence and if users should respond to errors and exceptions or if they are supposed to approve a business document (e.g., to check whether a forwarding order meets specific commercial conditions).

Employee assignment

In this context, you can define the organization's hierarchical structure and assign employees with specific tasks to individual organizational elements. Consider the following employees:

- The tendering manager, who is responsible for freight tendering in a purchasing organization
- The sales manager, who is responsible for credit limit checks in a sales organization and is the recipient of the relevant workflow tasks
- A purchasing agent, who is responsible for invoice verification
- Transportation planners, truck drivers, loading clerks, and warehouse personnel, who are assigned to an organizational element that is categorized as planning and execution

Note

For more information about SAP Business Workflow and how it can be used and deployed, we recommend *Practical Workflow for SAP* by Ginger Gatling et al. (SAP PRESS, 2009).

Creating and Merging the Organizational Hierarchy

To integrate SAP TM with SAP ERP or SAP Supply Chain Management (SCM), SAP TM provides a report to upload and merge the definitions of an organizational hierarchy from SAP ERP or SAP SCM (shown in Figure 3.5). You can find this report in the IMG for SAP TM by following the menu path Transportation Management • Master Data • Organizational Management • Organizational Model • Create and Merge Organizational Hierarchy or by launching Transaction /SCMTMS/ORG_INT.

With this report you can upload and merge organizational structures from SAP ERP or SAP SCM. It reads the existing structure in the source system, displays the organizational hierarchy, and then creates and saves the corresponding elements in the SAP TM organizational model.

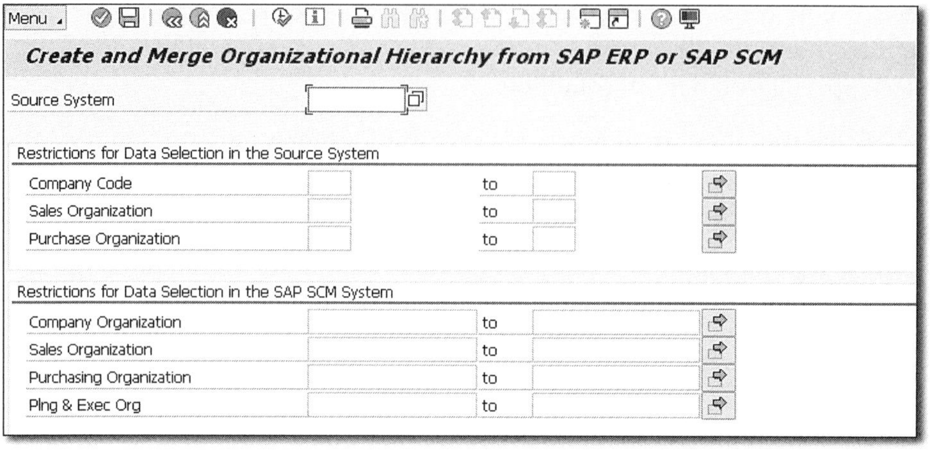

Figure 3.5 Report to Create and Merge the Organizational Hierarchy

Uploading from SAP ERP/SAP SCM

Based on the selection and restricting the mapping-relevant elements, you can select the organizational units from the source system. To upload and integrate hierarchies from SAP ERP, you must first define the following organizational elements:

- Sales organizations, including sales offices and sales groups
- Purchasing organizations, including purchasing groups

To upload definitions of organizational hierarchies from an SAP SCM system into SAP TM, you must first define organizational units with the following roles and functions in your SAP SCM system:

- Sales organizations, including sales offices and sales groups
- Purchasing organizations, including purchasing groups
- Execution and planning organizations, including execution and planning groups

Merging into SAP TM

To refine the selection of existing definitions of organizational hierarchies from the source system, you can enter the relevant company code, sales organization, and purchasing organization. To transfer the selected elements to the transportation system, you can either drag the selection to the SAP TM organizational hierarchy or use the buttons. You can remove the transferred elements before the organizational hierarchy has been saved. Once the data has been saved, though, you have to make adjustments in the organizational model itself.

3.1.2 Product

Product master data does not have a definitive set of semantics in transportation logistics. Instead, it may differ radically depending on the role and perspective of the user. By definition, product master data classifies, identifies, and characterizes materials, articles, and services that are purchased, sold, manufactured, or provided as a service and that remain essentially unchanged over a long period of time.

Materials in SAP ERP

From the shipper's point of view, the material master data defined in SAP ERP includes the deliverable, producible, and sellable goods that a transport demand can produce in a logistics process. The materials can be maintained in SAP ERP with their attributes and various quantities and then be allocated to organizations. In addition, you can define various types of transport materials and equipment in the material master (such as pallets, pallet cages, and cardboard boxes), which can also represent transport demand through their use in packaging one or more other materials.

In addition to the obligatory definition of material number and description, you also have to define the base unit of measure (e.g., count, box, or kilogram). Via the base unit, you can define additional quantity units with the conversion factors. The indication of the gross and net weights and volume is especially important for logistics processing because these values are taken into account for the capacity calculation of combined shipments. Volume refers to the volume occupied by a material during transport, *not* the net contents of a unit of material. In the material master, there is a sales view in which you can define the delivering plant and transportation group as transport-relevant attributes.

Transportation group

The transportation group is a categorization criterion that allows you to categorize materials that have the same defined processing conditions. Examples of values in the transportation group include palletized goods, refrigerated goods, and dairy products. If the material is classified as dangerous goods, you need to create a dangerous goods master record for transportation processing. SAP Environment, Health, and Safety (EHS) Management lets you save the necessary identifications and definitions for the various norms and carriers. Here, you can store dangerous goods classes and codes, material characteristics, rules for loading materials together, paper print definitions, and other details for dangerous goods definition. Each material classified as a dangerous good requires its own, separate dangerous goods master record.

Products in SAP TM

Product master data in SAP SCM, which is also used by SAP TM, can be divided into two basic types:

- **Product master data that maps precisely defined materials sold, purchased, or transported in connection with a logistics agreement**
  This is generally used in shipper processing of transports (the exception being contract logistics) and is similar to the material master definition of the SAP ERP system. It is employed in the traditional shipper solution and is also supported by SAP TM.
- **Product master data that constitutes a classification or grouping of various materials or represents a service**
  This is the product view of an LSP, in which the situation regarding the product master is considerably more diverse. You can only effectively use this type of product master when employing SAP TM as a service provider solution.

Consignor's perspective

From a consignor's perspective, the product master represents all of the materials and articles distributed by that consignor. The individual articles are described in precise detail, with information including a unique identification key, unique characteristics, and possible quantity specifications. In the SAP system, this data is usually defined centrally in the SAP ERP material master and is then distributed from there to other systems (such as SAP TM) with either some or all of its characteristics. Figure 3.6 shows a product master data record of a finished product that was originally defined as a material in SAP ERP and then transferred to SAP TM. Note that we discuss transferring master data from SAP ERP to SAP TM in Section 3.1.5.

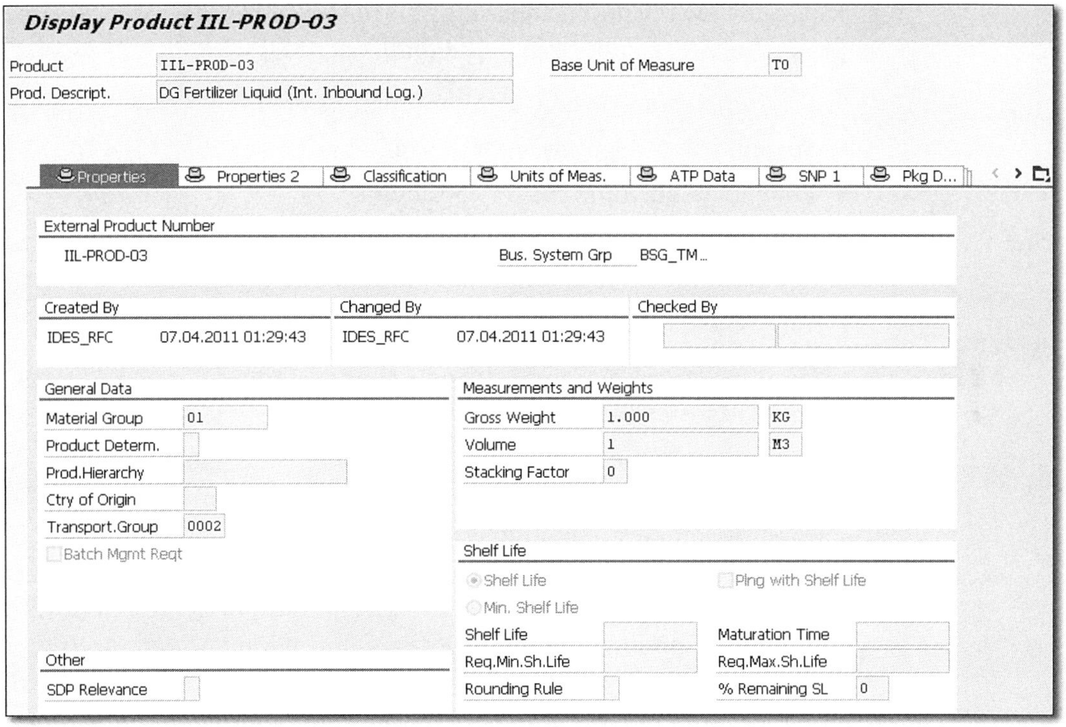

Figure 3.6 Product Master in SAP SCM

The LSP's perspective

Transportation services offered by an LSP are often commissioned with reference to standard material types or material groups as product master records. Such grouping can be done in the necessary granularity (with three to eight digits), using elements such as the commodity or harmonized system (HS) code, UN hazardous materials number, or other standards. The material group can be used to define generally valid characteristics for all shipments with reference to a particular material group (e.g., freight group or description). Other data (weight) can only be depicted in a general way and must be individually entered in the transport request. In transport processes in which full loads are frequently requested and transported (e.g., in container line operations or railway operations with full railcars), the product master records are usually defined based on transport equipment. The content of the transport equipment is often only roughly specified and not precisely known at the time of the initial order. Therefore, the range of ways in which

LSPs can view the product master is much more diverse. In this case, the product master can be used in the following ways:

- Precisely defined products in third-party logistics
- Standardized freight codes and material groups
- Roughly defined product categories
- Categories of transport equipment in which products only represent the outer packaging of the materials being transported
- Service products

Product Maintenance

Precisely defined products

The SAP TM system usually contains complete product master data records for 3PL contract partners (if an SAP ERP system is not used). An LSP that is contracted to provide warehouse and distribution services to a customer usually has a precise definition of the customer's product master, which can then be used to process transportation logistics. In this case, the product master records are created in SAP TM directly or are transferred there from the customer system, and they contain all essential characteristics required to provide the transportation service.

Freight codes and material groups

Standardized or custom freight codes or material groups (e.g., statistical goods numbers) are used to ensure appropriate grouping and classification of products. Standardized freight codes or material groups are often used as product master records if an LSP focuses mainly on the provision of transportation services not bound by contract.

To provide the granularity required (which is three to eight digits), these codes may be based on commodity codes or HS codes, UN hazardous material numbers, or other standards. General characteristics that apply to all loads can also be defined with reference to the material group (e.g., freight group or description).

Other details (e.g., weight) can only be represented in a generalized way and must be entered individually in the transportation order. Standardized freight codes and material groups are frequently used in rail logistics, for example, where they are used directly to calculate freight charges.

Figure 3.7 shows the definition of a product master record based on the customs tariff number (six digits). The weight of the load is entered in the transportation order in this case (base unit of measure kg). The product master permits scales based on intervals of 1 kg.

Roughly defined product categories

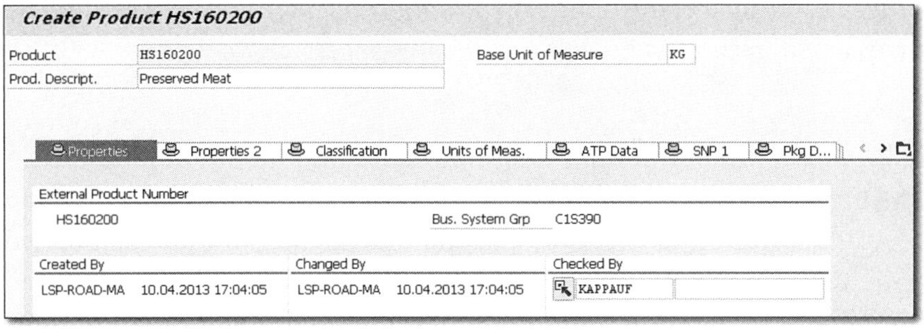

Figure 3.7 Product Master for an HS Code

Broad categories of products are defined to represent a significant simplification of the actual situation. As with standardized material groups, custom grouping criteria can be used to divide products into product categories. The relevant products can be created the same way in the product master. This type of rough classification may be sufficient for processes that do not involve customs processing or hazardous materials

Categories of transport equipment

In this case, products merely represent the outer packaging of the actual material transported. In business processes where full loads are generally commissioned and transported (e.g., container line haul, rail transportation with full railcars), it is useful to define product master records based on transport equipment. These cases often involve the transportation of large numbers of the same or similar containers or railcars, the contents of which only need to be defined in general terms and, in many instances, cannot even be specified when the order is initially created. However, the transport equipment must be defined (e.g., a 20-foot refrigerated container or a 67-foot flatcar). Only the required number of transport equipment products is defined as the load, and more precise details about the goods to be transported are added later.

Service products

In SAP TM, an LSP's product represents the services operated by that provider and therefore does not refer to the material goods that are transported (as in the case of express service providers). As of SAP TM

9.0, you can create service product catalogs in master data. We describe service products in Chapter 10, Section 10.1.4.

No product master representation

When SAP TM is operated without product-related master data, all goods to be transported are only entered as text in the transportation order. All load-specific and transportation-relevant details are directly maintained in the order itself.

3.1.3 Dangerous Goods

The handling of dangerous materials is regulated by numerous laws and regulations. Essential master data must be managed and maintained with the most up-to-date security regulations. Goods receiving and goods issue processing, warehouse operations, labeling, and printouts must be adapted to meet the requirements for handling dangerous materials and goods.

Dangerous goods

The dangerous goods functions in SAP TM enable you to ensure the safe transportation of dangerous goods in compliance with international regulations.

If a product has been classified as dangerous, you need to create a dangerous goods master record to check this product during transportation processing. SAP TM then checks the transportation business documents for the relevant regulations and ensures that all necessary information is included and parameterized in the business document. When you perform dangerous goods checks for your business documents, the system bases the checks on the relevant dangerous goods data. In a non-shipper scenario, if an LSP is supposed to handle dangerous goods, he usually does not have dangerous goods master data in his system.

Dangerous goods for LSPs

As of SAP TM 9.0, LSPs can directly maintain dangerous goods-related data in the forwarding order if a dangerous goods profile has been assigned to the corresponding forwarding order type. If no dangerous goods profile is maintained within a forwarding order type, it is not possible to transport dangerous goods. In all other cases, the document-based dangerous goods records can be created automatically when SAP TM creates a forwarding order based on data sent by the business partner (e.g., the shipper). We provide more information about how to maintain the forwarding order type in Chapter 4, Section 4.2.1.

Content Loader

As of SAP TM 9.0, you can use a content loader to import SAP EHS Management regulatory content, downloaded from the SAP Service Marketplace, directly into the dangerous goods master data tables in SAP TM. This functionality is especially useful for LSPs who do not have the dangerous goods master data in their system and usually receive data from their shippers, either in written or electronic form. It helps them to minimize the effort required to maintain and enter relevant data in a forwarding order and can be used as a template for document-based dangerous goods records. The content itself comprises dangerous goods regulations, the necessary texts such as substance names and phrases, and the required Customizing data.

EH&S Services

In a shipper scenario, SAP EHS Management lets you save the necessary identifications and definitions of the various norms and carriers. It can be integrated with SAP TM and stores dangerous goods classes and codes, material characteristics, rules for loading materials together, paper print definitions, and other details for dangerous goods definition. SAP SCM Basis also offers EH&S Services, and, for some IMG activities, it is possible to exchange the configuration between SAP EHS Management and EH&S Services in SAP SCM.

Scope of this section

SAP EHS Management contains several components to meet legal regulations and achieve compliance in the area of product safety and dangerous goods management. Its comprehensive configuration goes far beyond the scope of this book.

In this section we describe both the master data and classification of the dangerous goods master record, as well as the relevant settings and parameters to maintain phrases and text outputs. This master data is typically provided by the shipper, while document-based dangerous goods checks are based on the data in the relevant business documents and the basic configuration of EH&S Services in SAP TM. This configuration is part of transportation compliance and is therefore explained in Chapter 9.

Dangerous Goods Master

When you check dangerous goods, you are checking the product you want to transport against the rules you have defined in Customizing. This can be a check of certain aspects of the dangerous goods master data or the combination of the goods you want to transport. The check reflects the national and international regulations regarding the transportation of dangerous goods. The regulations depend on the transportation mode and the countries crossed while transporting the goods, and the check is triggered by product master attributes.

Product master attributes

Figure 3.8 shows a product and the relevant settings to define whether hazardous substance data exists for this product and how to perform a dangerous goods check. These attributes can be found in the product master in the Storage tab. Here you can specify whether the product is a hazardous substance or environmentally relevant. In this case, further EHS checks and reporting will be executed.

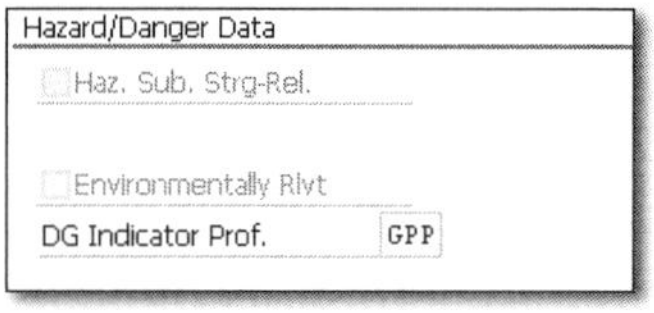

Figure 3.8 Dangerous Goods Indicator in Product Master

The actual execution of the check, as well as the output of dangerous goods documents, is controlled by a *dangerous goods indicator profile*. The profile itself contains a combination of indicators that are maintained in the dangerous goods check settings of the EH&S Services; in Figure 3.8, this is GPP (dangerous goods, relevant for document output and checks). You can find these settings by following the IMG menu path SAP Transportation Management • SCM Basis • EH&S Services • Dangerous Goods Management • Dangerous Goods Checks and Dangerous Goods Documents • Common Settings • Specify Indicator Profiles for Product Master. These are the only settings in the product master.

Executing compliance checks

The system can perform checks automatically; in addition, you can trigger the checks manually. Due to different risks in transporting dangerous goods, the system carries out checks in different ways and at different process steps to ensure safety during transportation. For example, the system can perform checks during freight unit building and vehicle scheduling and routing optimization. This ensures the compliance of the resulting freight unit and freight orders. In addition, you can perform checks on all relevant business documents. If there are dangerous goods errors, you can correct them and perform the check again.

Dangerous goods master

In SAP TM the dangerous goods master contains the data required to perform dangerous goods checks and generate dangerous goods documents and papers according to the applicable dangerous goods regulations and laws. To create a dangerous goods master record, you assign a dangerous goods regulation to an existing product and add other data.

Customizing Settings Needed to Activate Dangerous Goods

As a prerequisite to maintain dangerous goods parameters, the dangerous goods master pages have to be activated and the necessary screens and sequence have to be assigned in Customizing. If this hasn't been done yet, follow the instructions in SAP Note 1541010.

Once you have created a dangerous goods master record, it becomes available for dangerous goods checks and for creating dangerous goods documents. To create a new dangerous goods master record or display or edit an existing one, select the SAP NWBC menu path MASTER DATA • DANGEROUS GOODS MANAGEMENT • DANGEROUS GOODS MASTER. Alternatively, you can use Transaction /SEHS/DGD02 to change an existing record.

For example, Figure 3.9 shows the dangerous goods master of the liquid fertilizer from Figure 3.6.

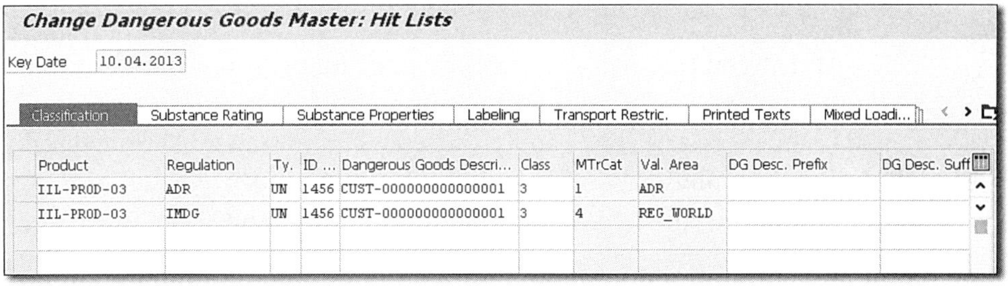

Product	Regulation	Ty.	ID ...	Dangerous Goods Descri...	Class	MTrCat	Val. Area	DG Desc. Prefix	DG Desc. Suff
IIL-PROD-03	ADR	UN	1456	CUST-000000000000001	3	1	ADR		
IIL-PROD-03	IMDG	UN	1456	CUST-000000000000001	3	4	REG_WORLD		

Figure 3.9 Dangerous Goods Master

To meet national and international regulations for this product, this master record has been created with reference to specific regulations, a validity area, and a mode of transport category.

Classification data

Legal regulations are relevant for the transportation of dangerous goods. You classify the master data by assigning a validity area and mode of transport category to the dangerous goods regulation. The classification assigns classes and codes and therefore specifies how the transportation is restricted or should be executed. The validity areas for dangerous goods data records are defined according to the applicable dangerous goods regulation and specify the countries, regions, jurisdictions, or organizational units in which the parameters of the dangerous goods master are valid. To carry out dangerous goods checks and generate dangerous goods papers, you must ensure that the validity areas for the dangerous goods master records do not overlap. In addition, the mode of transport category specifies the type of transport to which the rule applies—for example, road, rail, inland waterway, sea, air (cargo and passenger), or pipeline.

Figure 3.10 shows the underlying structure and relationship of some of the most important parameters of the dangerous goods master record and the "handshake" with the EHS configuration. (To learn more about EH&S Service configuration and compliance checks, please refer to Chapter 9.)

Substance rating, properties, restrictions

In addition to the classification data, the dangerous goods master also contains information about the dangerous substances themselves (see Figure 3.10). To help you be compliant with legal regulations, dangerous products are accompanied by detailed information about how they are supposed to be filled and transported, their storage conditions, and what to do in the event of an accident.

In SAP TM you can maintain not only first aid measures, but also details about how and under what circumstances the products are transported and which restrictions apply. These circumstances are typically described by physical and chemical conditions or the characteristics of the product itself, such as temperature, density, and flash and boiling points. Transportation restrictions are mainly concerned with quantity limitations and segregation, as in the case of mixed loading.

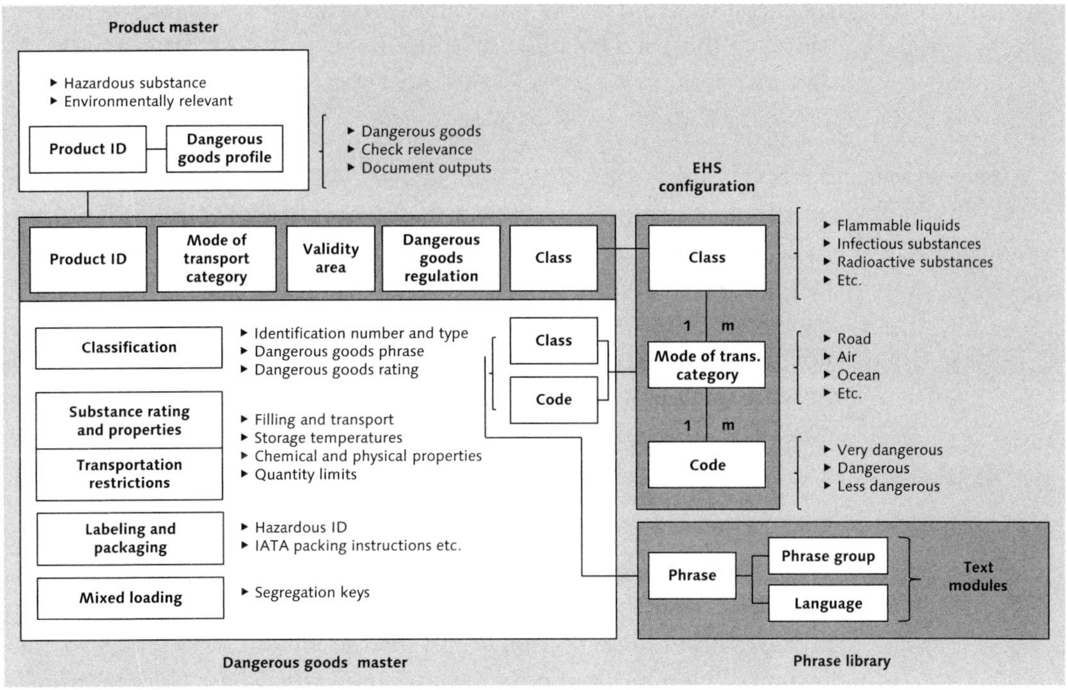

Figure 3.10 Structure of Dangerous Goods Parameters

The classification of the products and substances that present a danger during transport is based on classes and codes, which are assigned to the dangerous goods regulations specified in the dangerous goods master (Figure 3.11).

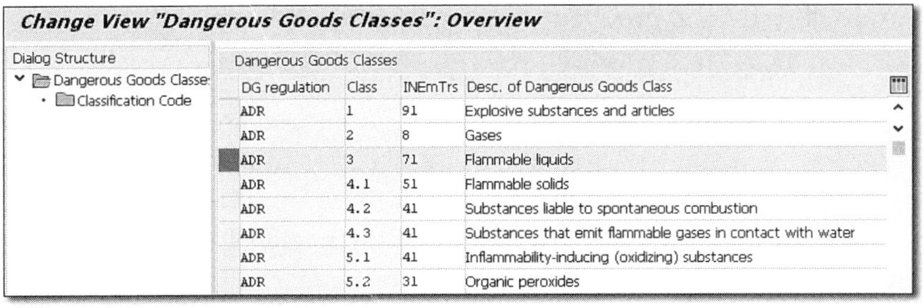

DG regulation	Class	INEmTrs	Desc. of Dangerous Goods Class
ADR	1	91	Explosive substances and articles
ADR	2	8	Gases
ADR	3	71	Flammable liquids
ADR	4.1	51	Flammable solids
ADR	4.2	41	Substances liable to spontaneous combustion
ADR	4.3	41	Substances that emit flammable gases in contact with water
ADR	5.1	41	Inflammability-inducing (oxidizing) substances
ADR	5.2	31	Organic peroxides

Figure 3.11 Assignment of Classification Codes to Classes

The assignment of classification codes to dangerous goods classes and regulations is part of the SAP SCM Basis configuration in the SAP TM

IMG. You can find the relevant EHS settings for the dangerous goods master in the IMG by following the menu path SAP Transportation Management • SCM Basis • EH&S Services • Dangerous Goods Management • Dangerous Goods Master.

Labeling and documentation

Labels on tanks, trucks, and containers indicate the type of risk of the hazardous or dangerous substance and clearly identify the dangerous goods. For example, toxic substances carry a danger label with a black skull and crossbones on a square white background. The declaration, usually warning placards and alphanumeric keys in the top half of the label, follows internally agreed-upon symbols and code systems. It provides information about required extinguishing media, required personal protection measures, and possible reactions of the substance.

Phrase Management

Whether the physical execution of a transport occurs by sea, land, or air, it usually crosses borders and is handled by people speaking different languages. Therefore, warnings and instructions regarding how to handle dangerous goods can be language-dependent. These texts are usually printed on specifications and first aid measures; they are typically used for document creation, dangerous goods texts on dangerous goods documents, and reporting.

The system offers a central tool to manage these texts. The language-dependent text modules are called *phrases* and are part of the master data for dangerous goods handling in SAP TM. Phrases are managed in phrase libraries and grouped together in phrase groups.

Phrase libraries, groups, codes

The *phrase library* defines the phrase assignment and origin. Using the import functionality provided by SAP EHS Management, you can upload purchased or company-specific phrase libraries to SAP TM and merge them with your existing phrase library. You can also update active phrases after importing a new version of the passive library by creating phrase references from the passive phrases to phrases in the active phrase library.

For each library, a *phrase group* is used to classify phrases. Each phrase belongs to a single phrase group and might have different assigned phrase codes. These codes are optional language-dependent abbreviations for individual phrases.

Figure 3.12 shows the phrase CUST-000...1. This phrase code is assigned to the dangerous product shown in Figure 3.9. It belongs to phrase group DG-TEXT and is used for to generate dangerous goods documents that will be send via EDI. No phrase code is maintained for the dangerous goods text in English (EN).

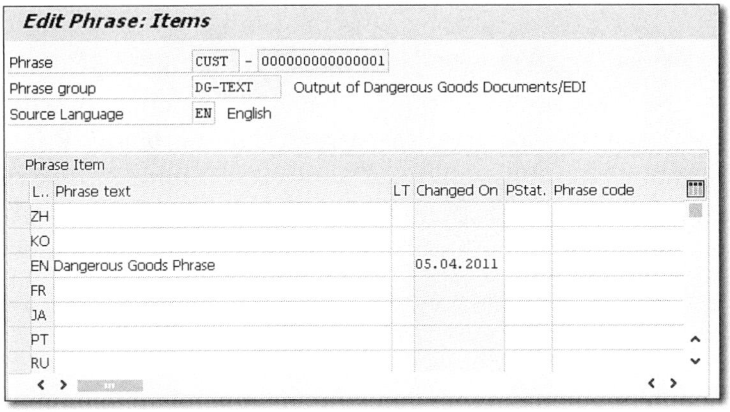

Figure 3.12 Editing Dangerous Goods Phrases

3.1.4 Business Partner

For transportation execution, a business partner is typically a person or organization in which a company has a business interest. Whether it is a single entity or a group of business partners, this master data object is used for a variety of business transactions.

Customers and vendors

In general, *business partners* are all legal entities or individuals with whom a company maintains business contacts. SAP systems usually differentiate between *customers* and *vendors*. From an accounting point of view, all customers with whom a company is in contact are *debtors*. Suppliers (or vendors) who provide deliveries or services are called *creditors*. A business partner can be a debtor and a creditor at the same time and therefore have different business partner roles.

Role concept in SAP

Business partners, as well as the roles they assume for your company, are managed centrally. According to this *role concept*, the business partner is defined as a general business partner first, and then his business partner roles are assigned. Each role might contain specific data that is relevant for the role. This way, there is no need to store redundant data, since the general business partner data is independent of a business partner's function- or application-specific extensions. Therefore, when a business partner is first created in an SAP system, the business partner role general business partner is automatically assigned and populated with general data such as name, search terms, and so on.

Business partner in SAP TM

In SAP TM business partners are all organizations, enterprises, and individuals with a fixed or loose working or order-based relationship with a consignor or LSP. This relationship may be defined by long-term contracts that are negotiated between the parties involved or by ad hoc activities (such as a quotations and orders). In this context we are talking about business partner master data for customers or vendors. You can, for example, create customer master records for the following business partners:

- Ordering parties
- Shippers
- Consignees
- Bill-to parties

The following are some examples of the partners that can be created as vendors:

- Freight-forwarders
- Carriers
- Transshipment locations

Integration with SAP ERP

As mentioned at the beginning of this chapter, SAP TM allows logistics processes to be executed largely without the existence of business partner master data. However, a business partner master record is virtually indispensable when it comes to settlement and billing. If you use SAP ERP for settlement, you should initially create your business partners as customers and vendors in SAP ERP if possible and then transfer them to other systems. (To learn more about the transfer of master data from SAP ERP to SAP TM, please refer to Section 3.1.5.)

Creating a Specific Business Partner for New Customers

You can create a specific business partner as a new customer, use this business partner in the transportation order (shipment request), and assign customer-specific data in the order. This option allows you to take orders from new customers. The new customer partner can then easily be replaced later, once the new business partner is centrally created and distributed.

Defining Business Partners

As mentioned before, business partners can be centrally maintained and defined using the SAP role concept. In SAP TM you can find the relevant maintenance transactions in SAP NWBC by following the menu path MASTER DATA • GENERAL • DEFINE BUSINESS PARTNER or by launching Transaction BP. Then assign additional roles to each business partner created. Figure 3.13 shows the definition of a customer (ordering party for transportation services) created in the roles of business partner (general), financial services business partner, and bill-to party. You can maintain the following information for all general business partners using the provided tabs:

- ADDRESS
  The business partner's main address
- ADDRESS OVERVIEW
  Additional addresses with a note on usage (e.g., mail address, delivery address)
- IDENTIFICATION
  Additional ID numbers to identify the business partner for communication (e.g., the IATA agent code of an air freight service provider, Standard Carrier Alpha Code, or commercial register number)
- CONTROL
  Business hours and tax classification
- PAYMENT TRANSACTIONS
  Details for payment transactions, including bank details and payment card details
- STATUS
  Status information and lock flags

Additional texts from the ADDITIONAL TEXTS tab (e.g., addresses or signatures) can be used for printing addresses or signatures on documents such as air waybills. To support printing in several languages, define the language of the texts and enter additional texts for one text type in different languages for one business partner instance.

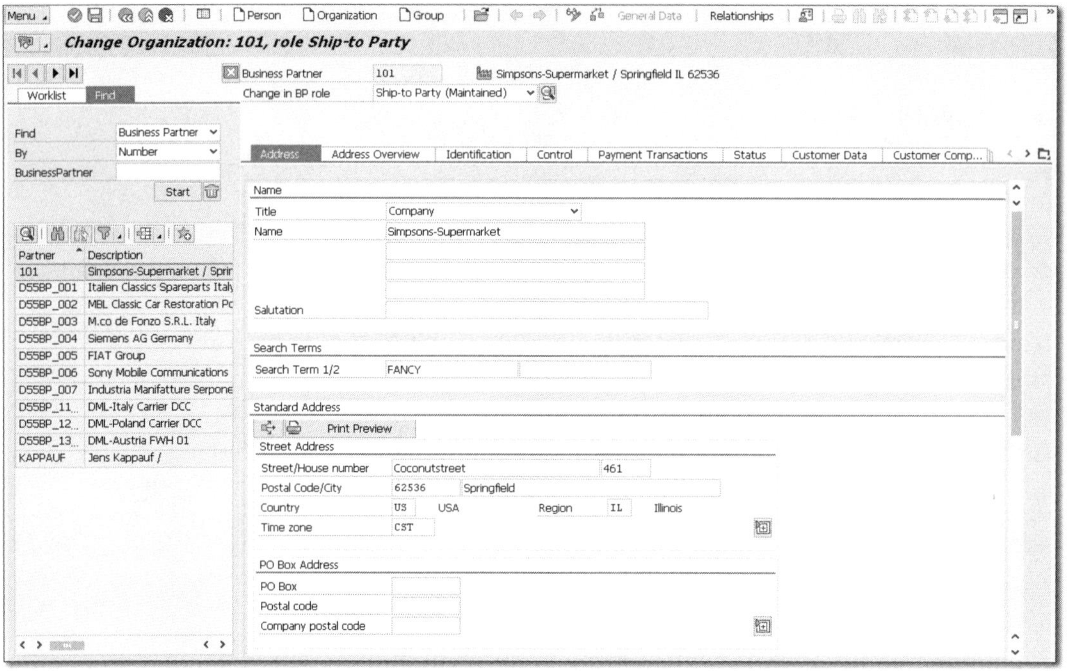

Figure 3.13 Business Partner Master Record for a Customer

Role-specific data

Depending on the business partner role, additional information or special tab pages are available for SAP TM. For business partners with the ship-to or sold-to party role, you can also maintain the location instances (CUSTOMER DATA) to which these business partners are assigned. Along with all assigned locations and their descriptions you can also assign the organizational units representing the customers own corporate structure (CUSTOMER COMPANY ORG. DATA). In addition, the organizational unit role is used to represent an organizational unit of a carrier or shipper organization (e.g., a specific sales organization), and can also be used as a carrier business partner or even as an ordering party for intracompany transportation.

Business Partner for Vendors and Carriers

For the vendors of an LSP such as other freight-forwarders or carriers, you can create a business partner in the role of carrier, as shown in Figure 3.14. In addition to the general business partner attributes, the vendor role offers the possibility to maintain service level codes, descriptions, and carrier codes (VENDOR DATA), such as Standard Carrier Alpha Codes (SCAC) for land and sea transportation, and airline codes for air cargo operations.

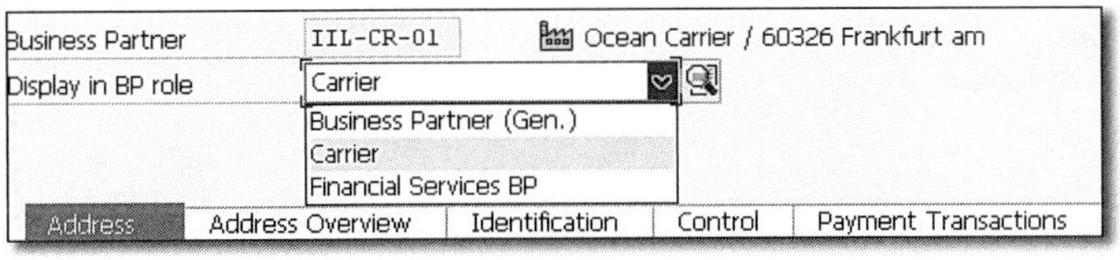

Figure 3.14 Business Partner Master Record with a Carrier Role

Carrier profiles

Carriers are business partners with the carrier role. To enable you to make effective use of the carrier or freight-forwarder in the planning, dispatching, tendering, and subcontracting areas of SAP TM, and to store information that is used in a similar constellation for multiple carriers, you can create a *carrier profile*. These profiles can be maintained in SAP NWBC via MASTER DATA • GENERAL • DEFINE CARRIER PROFILE or Transaction /SCMTMS/TSPP. You can only maintain one profile for each carrier.

Profile parameters

The carrier profile is used to define carrier- and transportation lane-specific parameters and attributes to define the service provider's range of responsibilities, transportation capabilities, and service level. These parameters typically include the following:

- Routes operated in the transportation network, as shown in Figure 3.15
- Freight codes, product freight groups, and transportation groups
- Transport equipment used or available (please also refer to Section 3.3.3)
- Fixed and dimension-based transportation costs for transportation optimization

Transportation proposals

These profiles are used during transportation planning to come up with a transportation proposal and then to determine the carrier. In this context, carrier determination can also take into account internal costs, as well as transportation lane-specific parameters such as restrictions or the general availability of necessary equipment. Internal costs can be defined per kilometer or mile and for a specific transportation lane and are also maintained in the carrier profile.

We'll go into more detail about transportation proposals and carrier selection in Chapter 6 and Chapter 7, respectively.

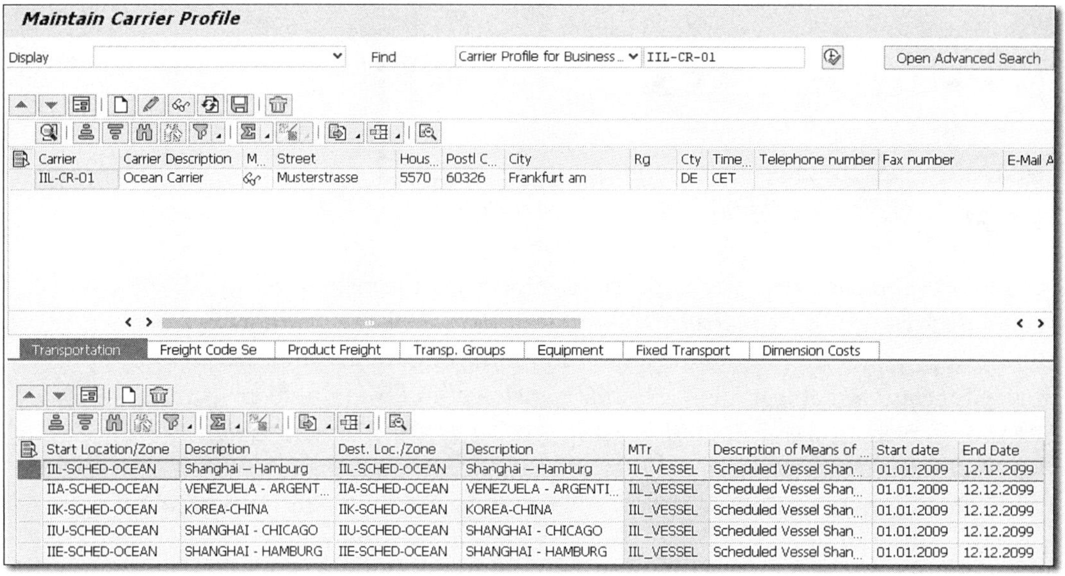

Figure 3.15 Carrier Profile for an Ocean Carrier

Employees and Internal Organizational Units

When you define the organization of an enterprise, business partners are automatically created for the individual organizational units. These are created with the organizational unit role and can be used directly in SAP TM to map business transactions within the enterprise.

Employees It is also possible to define a business partner's employees as business partners themselves if they occupy a dedicated role in your business partner's enterprise (e.g., the carrier's dispatcher who is personally responsible for your enterprise). These employees are defined in the role of employee (see Figure 3.16). Because of the option of creating hierarchies and relationships between business partners, you can then assign the employee as a subordinate business partner of the carrier and assign a relevant function description to the employee to clarify his role. This definition is required, for example, if you issue invitations to tender to carriers (see Chapter 7) and want to allow the business partner's employee to view and respond to these directly in SAP TM using the Internet collaboration portal.

Figure 3.16 shows the definition of a business partner employee and the relationship between this employee and the main business partner (carrier) from the previous example.

Drivers

For transportation execution, a driver is a person who can operate vehicles (see also Section 3.3.1) and performs transportation-related tasks. As of SAP TM 9.0, drivers are no longer maintained as resources but are defined as business partners with the role driver.

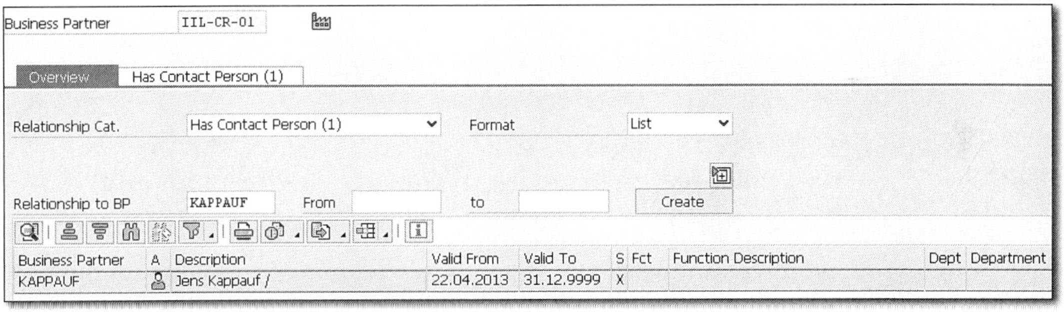

Figure 3.16 Business Partner and Relationship Definition for Employees

Figure 3.17 illustrates that this role not only provides additional parameters such as qualifications, but also allows you to specify the driver's availability and create shift sequences and absences, similar to the resource availability mentioned in Section 3.3.2. Although the system does not check the consistency of required or offered qualifications (e.g., to handle dangerous goods or to operate a specific vehicle), these qualifications held by the driver typically include licenses or certain permissions. Drivers are assigned to freight orders as a business partner on the BUSINESS PARTNER tab. The configuration of qualifications can be found under RESOURCES in SAP TM via the IMG menu path

SAP Transportation Management • Transportation Management • Master Data • Resources • Maintain Settings for Qualifications.

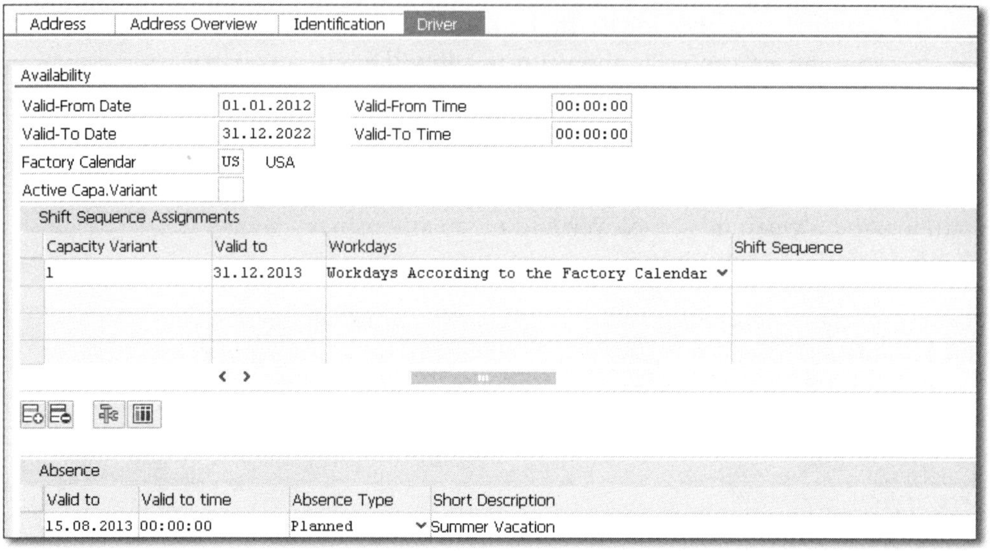

Figure 3.17 Business Partner with a Driver Role

3.1.5 Transferring Master Data between SAP ERP and SAP TM

To guarantee seamless integration of various SAP systems and the business transactions processed with them, certain master and transaction data are distributed among the systems. When SAP TM has been implemented as a standalone system, transportation demand and master data can be exclusively created in the transportation system. You need to synchronize master data whenever an SAP TM instance is implemented that will be integrated with an SAP ERP system.

Configuring the Integration

SAP TM is part of SAP SCM. Technically, it is based on the SAP SCM Basis component, in which the SAP SCM master data is made available. You can use the proven technology of the CIF to transfer the master data between SAP ERP and SAP SCM. The CIF is also used to transfer master data between SAP ERP and SAP TM.

APO Core Interface Framework

The APO CIF is a real-time interface for the integration of SAP ERP with SAP SCM. The CIF enables the provision of initial data and the supply of SAP SCM-based systems with data alterations.

The connection between the two systems is achieved via a remote function call (RFC) connection. The special feature of this type of communication between the systems lies in the asynchronous processing of the data transfer. This means that the data from the sending (SAP ERP) system is first buffered and then transferred, or that it is transferred and subsequently buffered by the receiving (SAP SCM) system, then processed.

Queued RFC

Outbound and inbound processing is performed in sequence in an outbound or inbound queue. In the event of an error caused by a failed network connection, for example, this queue saves all transfers and enables seamless continued processing after the error has been located and eliminated. The queue is a type of waiting line that enables real-time exchange and processing of information, making SCM-based planning in real time possible. This type of RFC invocation is known as a *queued remote function call*, or qRFC. Figure 3.18 shows the master data entities that need to be transferred so that you can use SAP TM.

After the initial transfer, the CIF allows you to execute the subsequent distribution (e.g., creation or change of a master data record) automatically in the background.

Configuration in SAP ERP

As far as the technical integration between SAP ERP and SAP TM is concerned, we assume that the two systems have already been connected and all relevant settings have been made. Although the following section describes the most relevant settings of the interface configuration, we would also like to highlight the detailed, publicly available SAP Implementation Guides.

SAP Documentation for CIF Integration

You can find this information at *http://Service.sap.com/support*. Follow the menu path Release & Upgrade Info • Installations & Upgrade Documentation. At SAP Help (*http://help.sap.com*) follow the menu path SAP Business Suite • SAP Supply Chain Management. The CIF integration and the relevant settings can be found in the documentation for SCM Basis under Integration Via Core Interface (CIF).

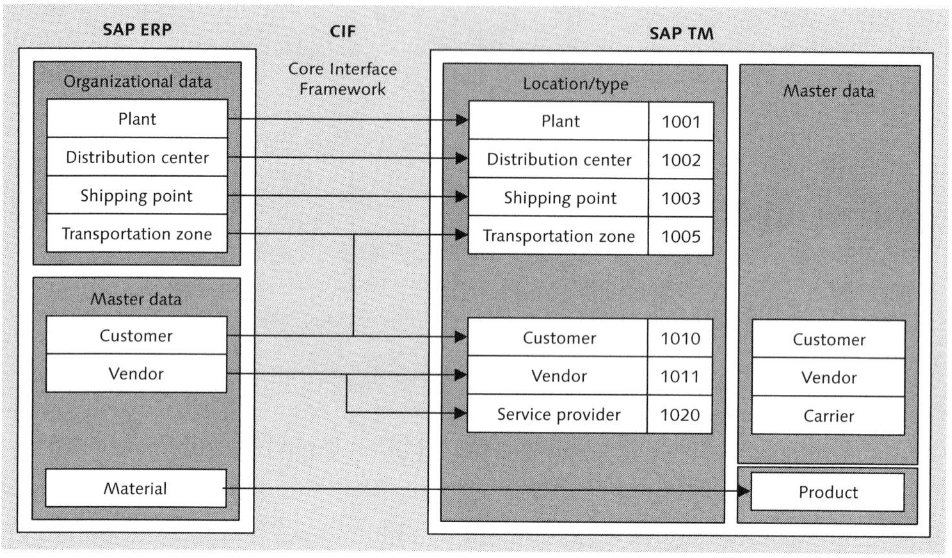

Figure 3.18 Transferring Master Data between SAP ERP and SAP TM

You must complete the technical preparations listed in Table 3.1 in SAP ERP before you can use the CIF to exchange data with SAP TM.

Activity	Transaction	Description
Set up an RFC connection to SAP TM	SM59	Defines the connection between SAP ERP and SAP TM.
Define a version of SAP TM	NDV2	The basic version of SAP TM must be defined for the correct CIF function to be used: SAP_APO, 51.
Define inbound queues	CFC1	Defines the operation mode (T) and queue type (I).
Activate change pointers	BD61	General activation of the master data transfer after changes
Select the relevant change pointers for the CIF master data transfer	BD50	Select the following message types: CIFCUS, CIFMAT, CIFMTMRPA, CIFSRC, CIFVEN, CIF_SUBC.
Master data transfer settings	CFC9	Set all master data transfers to 2 (BTE transfer, immediately).

Table 3.1 Technical Steps in SAP ERP to Prepare CIF Master Data Transfer

Activity	Transaction	Description
Inbound and outbound queue scheduler	SMQR, SMQS	Register all queues with the name or destination CF*.
Register queue display	SMQE	Register the CIFQEV02 program for all queues.
Activate CIF logging	CFC2	Set the level of detail for the logging function.

Table 3.1 Technical Steps in SAP ERP to Prepare CIF Master Data Transfer (Cont.)

Configuration in SAP TM

You must complete the technical preparations listed in Table 3.2 in SAP TM before data can be exchanged with SAP ERP using the CIF.

Activity	Transaction	Description
Set up an RFC connection to SAP ERP	SM59	Defines the connection between SAP ERP and SAP TM.
Define a business system group	/SAPAPO/C1	A business system group represents a unit within the enterprise, which is defined on the basis of legal, business-related, administrative, or geographic factors.
Assign logical systems to the business system group	/SAPAPO/C2	As a rule, SAP ERP and SAP TM should belong to the same business system group.
Inbound and outbound queue scheduler	SMQR, SMQS	Register all queues with the name or destination CF*.
Activate CIF application statistics	ASACT	Activate the application statistics for SAP functions.

Table 3.2 Technical Steps in SAP TM to Prepare CIF Master Data Transfer

Application-Specific Configuration

To enable a successful transfer of master data, the first thing you need to do is assign the SAP ERP vendor account groups to the location types that are used in SAP TM.

In the SAP ERP Customizing settings, follow the menu path INTEGRATION WITH OTHER SAP COMPONENTS • ADVANCED PLANNING AND OPTIMIZATION • APPLICATION-SPECIFIC SETTINGS AND ENHANCEMENTS • SETTINGS AND ENHANCEMENTS FOR SHIPMENTS • ASSIGN VENDOR ACCOUNT GROUP TO APO LOCATION TYPE. Maintain the assignment as shown in Figure 3.19.

Assign Vendor Account Group to APO Location Type

Account group	Location type
0001	Vendor
0005	TSP

Figure 3.19 Assigning a Vendor Account Group to a Location

Second, the transportation group definitions in SAP ERP and SAP TM must be aligned. In SAP ERP, you'll find the transportation groups in Customizing under LOGISTICS EXECUTION • BASICS • ROUTES • ROUTE DETERMINATION • DEFINE TRANSPORTATION GROUPS. In SAP TM you must maintain the corresponding transportation groups in Customizing under SAP TRANSPORTATION MANAGEMENT • SCM BASIS • MASTER DATA • PRODUCT • MAINTAIN TRANSPORTATION GROUP.

Integration Model and Selection Model

In general, integration models contain the necessary parameters to indicate which master data in the SAP ERP system is to be selected and transferred. The model is activated after its generation. This means that data to be transferred to SAP SCM is selected via the integration model. The data involved is generally either material-related or material-independent objects. Material-related objects include, in particular, materials and plants, contracts, delivery plans, and procurement information records necessary for procurement, as well as customer orders and planned independent requirements as the basis for requirement determination in SAP SCM. Material-independent objects include shipping points and customer and vendor master data. Customers and vendors are mapped as locations in SAP SCM, similarly to plants in SAP ERP.

One-Way Transfer

The transfer of master data is only performed in one direction: from SAP ERP systems to the SAP SCM system. Changes to a master record are not transferred back to the SAP ERP system.

Creating the integration model

Before you can execute an initial master data transfer, you must create and activate the integration and selection model. The integration model specifies which data objects are selected from SAP ERP and then transferred to SAP TM.

Call Transaction CFM1 to create the integration and selection model. Create a new model here and enter the logical system of SAP TM. You can create separate integration models, divided on the basis of master data, or one general integration model for all master data to be transferred. To create the integration model, first mark the object types (e.g., customer masters) to be selected on the CREATE INTEGRATION MODEL selection screen. Next, select specific selection criteria that further restrict the object types you have already selected. The required settings are shown in Figure 3.20. To generate the model, select EXECUTE and then SAVE. The system then displays the number of data records in each category that are relevant for the transfer.

Automatically Creating Business Partners

When creating the integration model, you must specify whether you want to create a business partner, a business partner and a location, or only a location in SAP TM. Business partners are necessary for the commercial part of the business, whereas locations are necessary for planning and executing transportation. This means that whenever cargo needs to be picked up or delivered, a location is necessary.

To create a business partner automatically, ensure that the mode of transfer for the business partner is set to the value 2, as shown in Figure 3.21. This ensures that both an SAP TM location and a business partner are created for customers and vendors defined in the SAP ERP system.

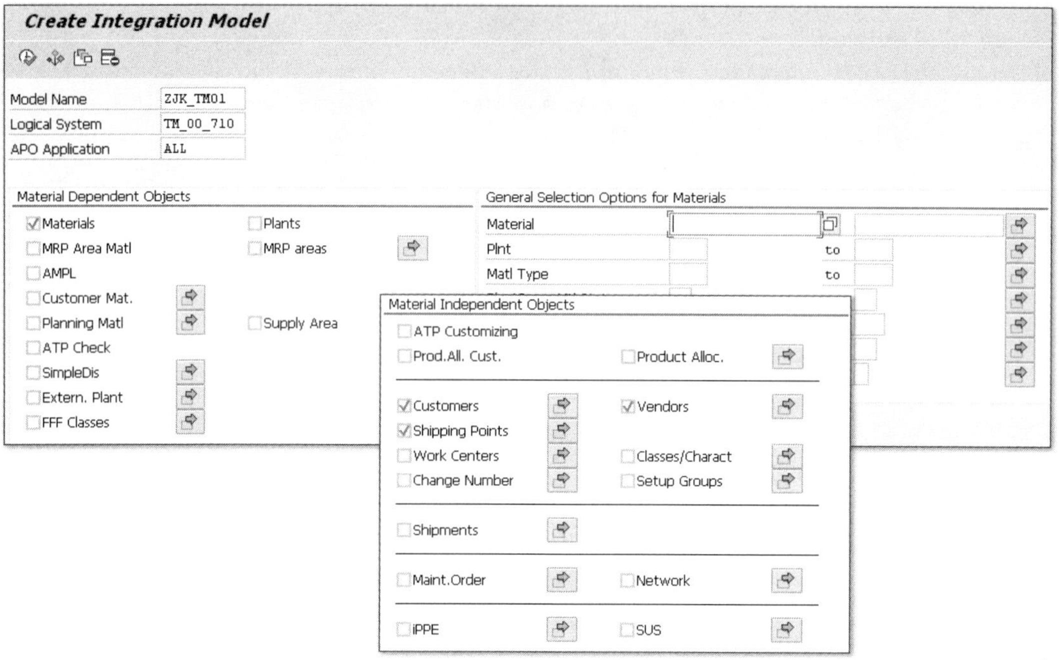

Figure 3.20 General Integration Model to Transfer All Relevant Master Data

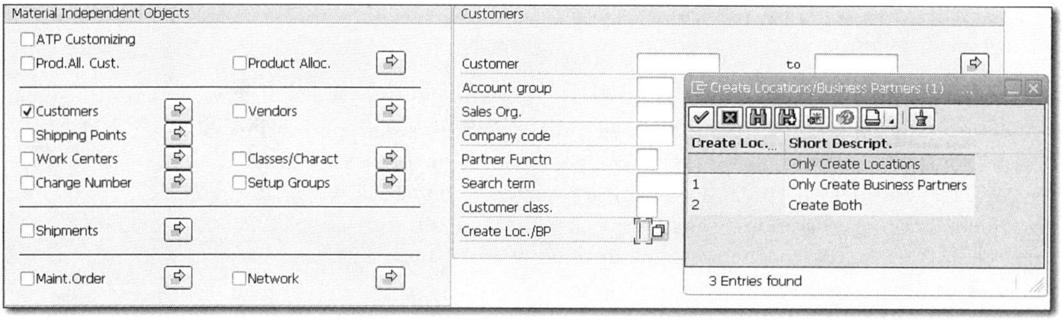

Figure 3.21 Automatically Creating Business Partners

Activating the integration model

Activating an integration model causes an initial transfer of master data from SAP ERP to SAP TM. The integration models to be activated are compared with the integration models that are already active. In this way, the transfer restricts itself to differences; in other words, the system transfers only data for filter objects that are not contained in any active integration models.

Starting the data transfer

You can activate the integration model once it has been generated and saved. Use Transaction CFM2 to do this, and enter the model to be activated in the selection criteria. After the selection, you can activate the new version of the integration model. To do this, select the latest version and click on ACTIVE/INACTIVE. The system then activates the integration model, and the master data is transferred to SAP TM based on the selection criteria for the model. The ACTIVATE OR DEACTIVATE INTEGRATION MODEL screen is shown in Figure 3.22.

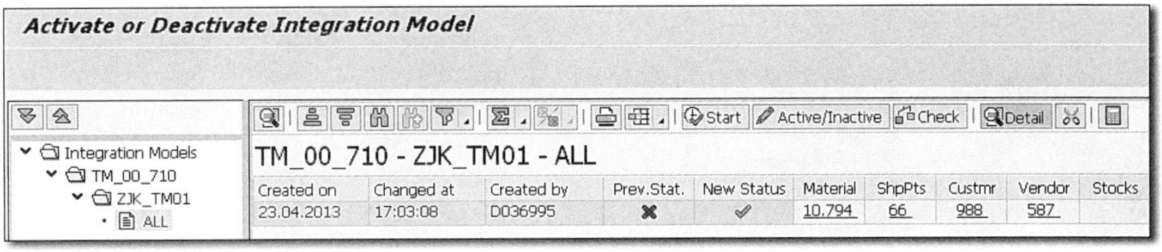

Figure 3.22 Activating the Integration Model

Schedule background job

The data transfer between SAP ERP and SAP TM is executed automatically based on CIF configuration. Master data changes in SAP ERP are tracked and recorded by change pointers. Activating a generated integration model checks and analyzes these change pointers, determines the changed master data, and finally transfers it to the target system.

The generation and activation of the integration models themselves can also be automated and configured as a background job using SAP ERP Transaction SM36. The execution of this job can be scheduled to occur periodically, typically every hour. Once the previously used and now deactivated integration models have been deleted, the execution steps of this job strictly follow the manual steps just described. Generating and activating a new integration model is based on the selected parameters, while the data selection itself is done via report variants for the programs in Table 3.3.

Step	Program	Transaction	Description
1	RIMODDEL	CFM7	Deletes the deactivated integration models of the previously executed background job.
2	RIMODGEN	CFM1	Generates a new version of the integration model, taking account of new master data. For example, if a new material master record has been created, and this new master data is within the selection range of the integration model, the system automatically creates a new inactive version of the integration model.
3	RIMODAC2	CFM3	Deactivates the old version of the integration model and activates a new version including all data that will be transferred to the target system. The report compares the old version of the integration model with the new runtime version and subsequently triggers the delta-upload of changed data. The next periodic processing of the background job (step 1) will then delete the old (inactive) version of the integration model.

Table 3.3 Background Job Execution Steps

Prerequisites

The report variants of these programs are a prerequisite for maintaining the execution steps and scheduling the automatic data transfer as a background job. To create these variants, use Transaction CFM1 first to generate the integration model, and save your selection as a variant of report RIMODGEN. For the first step (see Table 3.3), call Transaction CFM7 and select the inactive variant you created in the previous step. Make sure to select only the inactive version (the SELECT INACTIVE IMs ONLY checkbox) and save the settings as a new variant for the report RIMODDEL. Step 3 should activate the integration model. Execute Transaction CFM3 and select the generated integration model; after, save your selection as a report variant (RIMODGEN). Once the background job has been scheduled, the system will automatically transfer master data between SAP ERP and SAP TM.

This section introduced the general master data, the main characteristics of the organizational structure, the business partners and products, and the integration between SAP ERP and SAP TM. Before we explain the resources, let's continue with the logistics master data that's used to define and specify the transportation network.

3.2 Transportation Network

The transportation network defines direct reachability between your locations and transshipment locations, which together define how freight can be transported between your locations. Direct reachability between locations A and B means that B can be directly reached from A by a transportation option such as a vehicle resource or a schedule. Transshipment locations allow reloading from one transportation option to another. The network definition is essential for automatic planning, which determines the best path through the network and assigns the most suitable carriers for given transportation demands and for charge calculation considering freight and forwarding agreements that contain rates defined on trade lanes as geographical basis.

The following sections introduce locations and all other the concepts and tools that enable you to model your transportation network from reachability, transshipment locations, and business partner relation perspectives.

- Section 3.2.1 introduces locations, which form the nodes in the network at which goods are loaded or unloaded. Locations represent the most basic network concept since they define the source and destination of any transportation.
- Section 3.2.2 describes transportation zones, which allow you to group locations and thereby define reachability and transshipment options in an aggregated fashion.
- Transportation lanes are explained in Section 3.2.3. They define reachability between locations and zones on a means of transport and carrier level.

- Section 3.2.4 introduces schedules, which express recurring reachability at fixed dates and times along a predefined location sequence for a given mode or means of transport, as described in Section 3.3.1.
- Section 3.2.5 explains how to define transshipment locations, which are required for intermodal transportation or any scenario involving consolidation and deconsolidation.
- Section 3.2.6 describes default routes, which define paths from source to destination through a sequence of transshipment locations. They implicitly define transshipment locations to be used from source to destination and they can also predefine schedules or carriers to be used for the stages from source to destination.
- Section 3.2.7 introduces trade lanes as an additional perspective on the network, which is used to define business relationships to your customers (forwarding agreements) and carriers (allocations, business shares, freight agreements).
- Section 3.2.8 describes the transportation network cockpit, which is a powerful tool to visualize all objects of the transportation network on a map, to search for specific objects, and to maintain objects on the map. Its text-based cousin, the transportation network overview POWL, is presented, too.
- Finally, Section 3.2.9 describes the integration with geographical information systems, which is essential to determine the geographical coordinates of locations, and distances and durations between locations, as well as providing the geographical map data for a graphical visualization of a map.

3.2.1 Locations

A *location* represents a logical and/or physical location where goods are delivered, picked up, or transshipped, or where trucks and trailers get coupled or uncoupled. To define a location, select the menu path MASTER DATA • TRANSPORTATION NETWORK • DEFINE LOCATION. Enter the name and location type, which you can select from a list of more than 15 standard location types. These include, for example, production plant (1001), distribution center (1002), customer (1010), airport (1110), and container freight station (1130).

You can maintain additional data in the following tabs (see Figure 3.23):

- In the General tab, you can maintain identifiers such as the UN/LOCODE and the IATA code, which are commonly used in ocean and air freight businesses, respectively; the geographical data specifying longitude, latitude, and altitude; and the precision level that was used to determine the geographical coordinates. In addition to the time zone and the assigned business partner, the tab also contains the priority, which can be used to define location-dependent nondelivery costs or earliness and lateness costs to be considered in automatic planning.
- The Address tab contains the default address, PO Box address, and other contact details. The geographical coordinates are determined automatically as soon as you maintain the country code (see Section 3.2.9 on integration with GIS). By following the menu path Application Administration • Master Data • Transportation Network • Location: Mass Geo-Coding, you can determine geographical coordinates for many locations in one step.
- Alternative identifiers for the location can be maintained in a dedicated tab. You can predefine the possible types of alternative location identifiers by selecting the IMG menu path SAP Transportation Management • Transportation Management • Master Data • Transportation Network • Location • Configuration for Alternative Location Identifiers.
- The TM tab allows you to maintain the minimum and maximum goods wait times, which are described in more detail in Section 3.2.5. In addition, you can define the trailer-handling capability that determines whether a trailer swap or recoupling is allowed. You can also specify air cargo security information; classify the location as known, account, or unknown shipper; and set the known shipper code and its expiration date.
- You can assign calendar and handling resources in the Resources tab. These are described in more detail below.
- The Additional tab contains supplementary information about the locations. You can determine which fields are displayed in this tab by following the IMG menu path SAP Transportation Management •

Transportation Management • Master Data • General Settings • Maintain Freely-Definable Attributes.

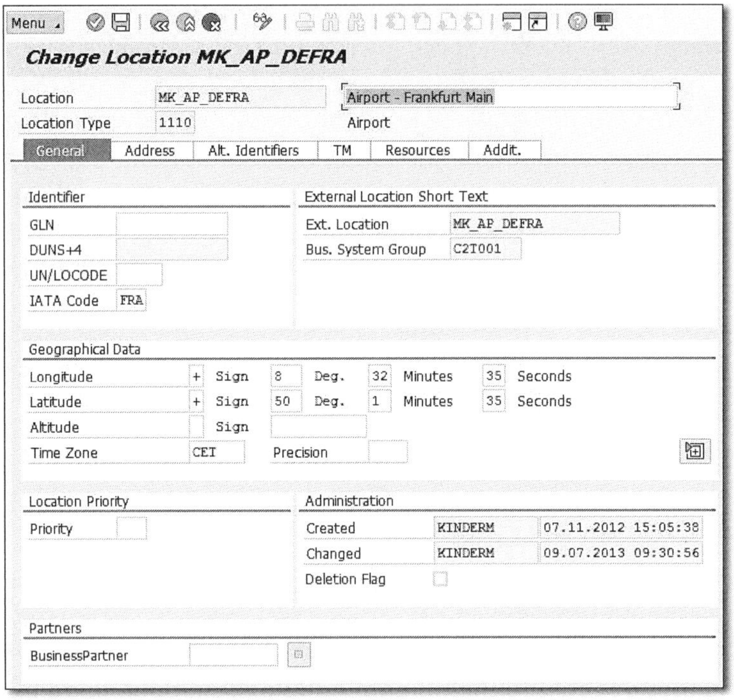

Figure 3.23 Maintaining a Location

Assigning operating times/handling resources

You can assign handling or calendar resources (operating times) for inbound and outbound transportation in the Resources tab. A calendar resource defines during which time intervals loading and unloading can take place at the location. In addition to this, a handling resource can restrict the number of loading and unloading activities that can be handled in parallel at the location. Thus, both calendar resources and handling resources affect the scheduling of unloading and loading activities at the location. You also have the option of not assigning a calendar or handling resource. In this case, loading activities at the location are not subject to any time restrictions. See Section 3.3.2 for an introduction of handling resources and calendar resources.

Figure 3.24 shows the maintenance of the inbound and outbound resources for a location in the Resources tab. By entering the values for consumption, you can define how much of the capacity offered by the relevant handling resource is consumed by a loading or unloading activity. You can define the same handling resource for inbound and outbound activities.

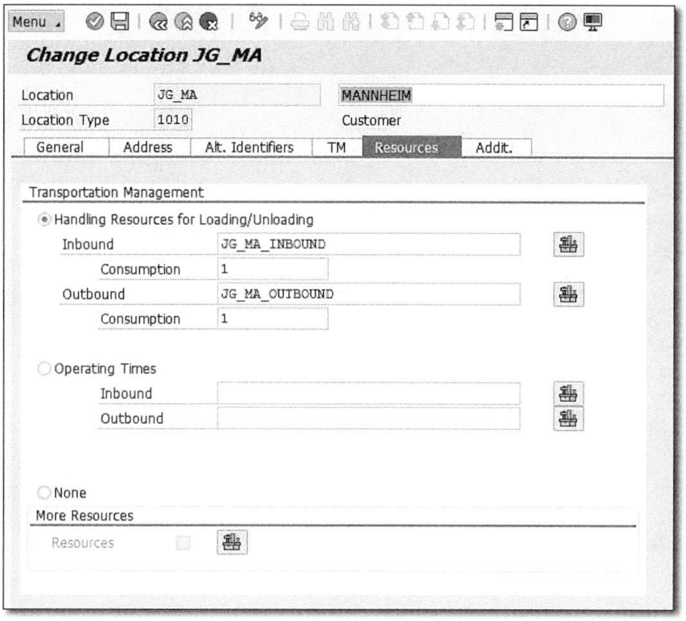

Figure 3.24 Defining Inbound and Outbound Resources for a Location

Handling resources/calendar resources

You can also click on the button next to the MORE RESOURCES field to define handling resources or calendar resources that are dependent on the means of transport. This enables the use of several handling resources at the same location (e.g., different loading ramps for truck and rail transportation). If a vehicle type is not used in the vehicle-type-dependent settings for a location, the general inbound and outbound resources for the vehicle type apply.

In the example shown in Figure 3.25, unloading activities at location JG_MA are scheduled in accordance with handling resource JG_MA_INBOUND_BIG if means of transport HH_BIG is used, whereas unloading activities are scheduled on the basis of calendar resource JG_MA_CAL_INBOUND if means of transport HH_SMALL is used. Loading activities are scheduled according to handling resource JG_

MA_OUTBOUND_BIG for means of transport HH_BIG and according to calendar resource JG_MA_CAL_OUTBOUND for means of transport HH_SMALL. The VALID column allows you to determine whether to use the calendar resource (value 1), the handling resource (value 2), or neither resource (value 0) for scheduling for each means of transport.

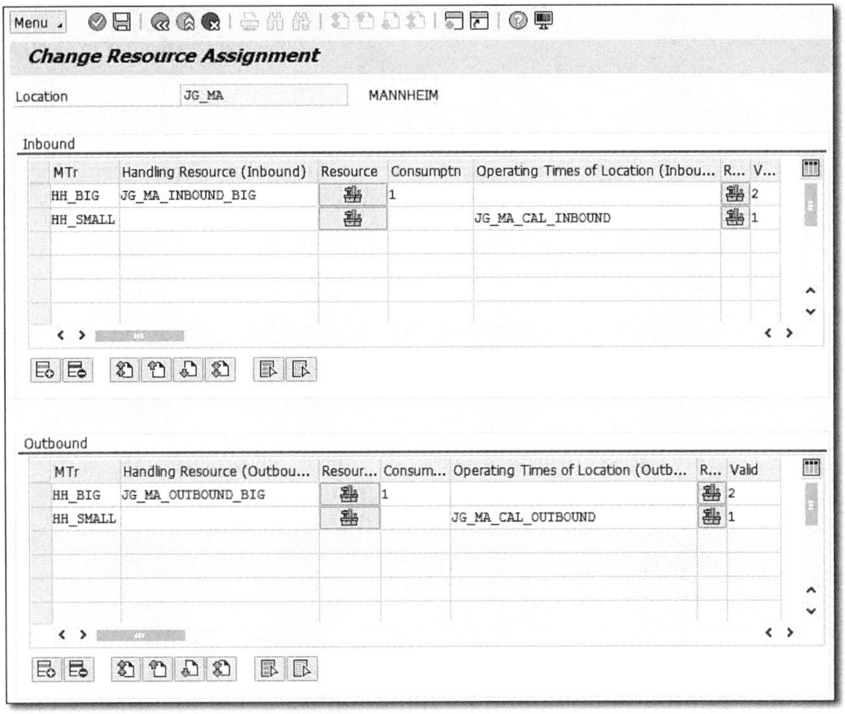

Figure 3.25 Defining Means of Transport-Dependent Inbound and Outbound Resources for a Location

Using a Resource for Inbound and Outbound Transport or for Several Locations

A calendar resource or handling resource can be used for inbound and outbound transportation at a single location or at several locations simultaneously. This means that you only need to maintain representative templates for opening times in calendar resources as a one-time activity, and you can then reuse these same templates for many different locations that have the same opening times.

Transferring locations from an ERP system

Locations can be transferred from an enterprise resource planning (ERP) system to SAP TM. If you set up a connection from SAP ERP to SAP TM, CIF functionality is used for this transfer.

One-time locations

You can store any locations that you do not want to store permanently in the system as master data as one-time locations. These are used whenever it is necessary to enter location data, such as address details, but a reference to the master data record is not possible or desirable. A one-time location is defined by the name of the organization, the address or communication data, or a combination of these details. One-time locations are typically used when you create forwarding orders from new ordering parties, and the locations of these parties are not defined in the master data.

3.2.2 Transportation Zones

A transportation network usually contains a large number of locations. Defining relationships such as transportation lanes or trade lanes between pairs of locations is possible, but it is a time-consuming and error-prone task. A definition on the location-to-location trade lane level for agreements, allocations, and business shares is not required in many transportation businesses. Instead, these need to be maintained on a more aggregated level.

A *transportation zone* is a group of locations; it allows the transportation network to be represented and maintained in an aggregated form. Relationships such as transportation lanes or trade lanes can be defined on the zone level, which is a much more compact, manageable, and less error-prone solution.

Defining transportation zones

To define a transportation zone, select the menu path MASTER DATA • TRANSPORTATION NETWORK • DEFINE TRANSPORTATION ZONE and click on the button for creating a transportation zone. When you enter the name and description of the transportation zone, the screen for maintaining the transportation zone appears. You can also search for existing zones and edit one of them (see Figure 3.26).

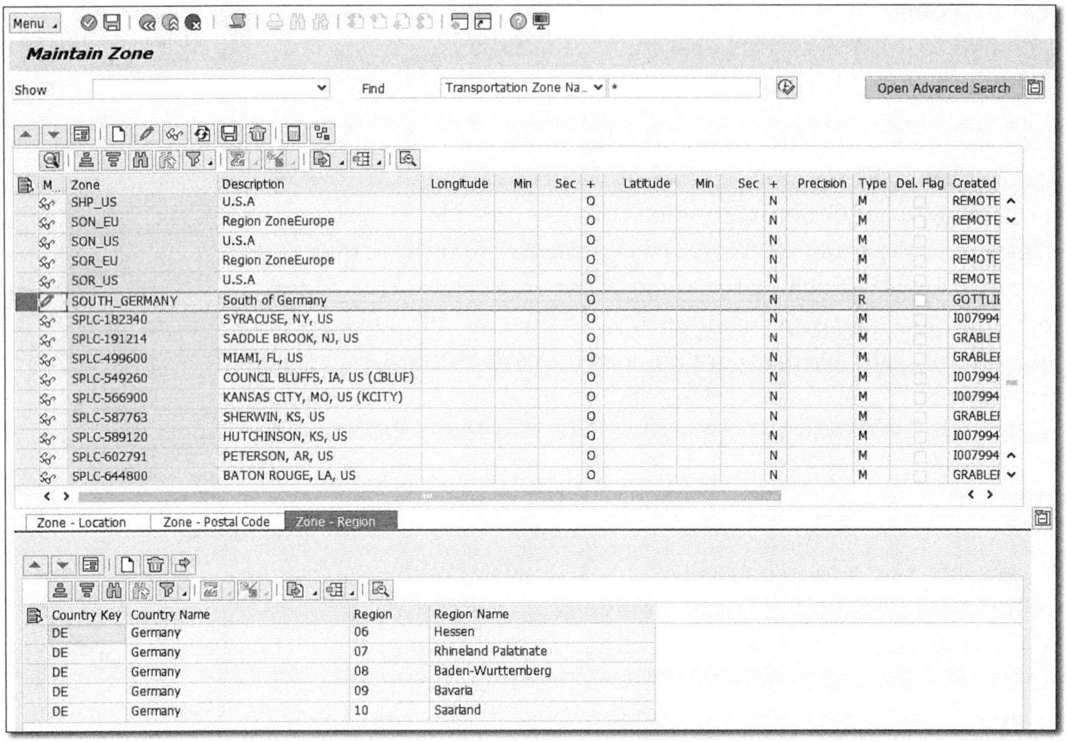

Figure 3.26 Maintaining a Transportation Zone

Explicit/implicit location assignment

In the Zone – Location tab, you can explicitly assign locations to the zone and exclude locations from the zone. The Zone – Postal Code tab allows you to assign a set of valid postal code ranges of a specific country to the zone, and in the Zone – Region tab, you can include a set of countries and regions. The last two tabs result in an implicit assignment of those locations to the zone, which match the postal code ranges, the countries, or the regions defined for the zone. This implicit assignment is very useful because of its robustness, since new locations entering the transportation business are automatically included in the respective transportation zones. If you want to know the locations included in a zone, just select the zone and press the Display included locations button.

Zone types

The zone type is determined automatically and indicates if the zone contains only direct location assignments (direct zone), only postal code assignments (postal code zone), only region and country assignments (region zone), or a combination of these (mixed zone).

Geographical coordinates

You can define the geographical coordinates of a zone, which are used to display the zone in the transportation network cockpit. Alternatively, you can use the CALCULATE COORDINATES button or determine the geographical coordinates for many zones in a single step by selecting the menu path APPLICATION ADMINISTRATION • MASTER DATA • TRANSPORTATION NETWORK • CALCULATE TRANSPORTATION ZONE COORDINATES.

Zones for countries and regions

If you need many zones—say, one per country or one per region—you can select the menu path APPLICATION ADMINISTRATION • MASTER DATA • TRANSPORTATION NETWORK • CREATE ZONES FOR COUNTRIES AND REGIONS to create all the required zones in a single step by choosing the countries and regions relevant for your business.

Transportation zone hierarchy

The transportation network can be structured to a greater degree using the transportation zone hierarchy, in which transportation zones can be assigned to other, higher-level transportation zones. All locations within a transportation zone automatically also belong to the higher-level transportation zone. The transportation zone hierarchy is used when determining the distance and duration between locations and has a far-reaching impact on many business processes, such as charge calculation, transshipment scenarios, *vehicle scheduling and routing* (VSR) optimization, and carrier selection.

Maintaining the transportation zone hierarchy

To maintain the transportation zone hierarchy, follow the menu path MASTER DATA • TRANSPORTATION NETWORK • DEFINE TRANSPORTATION ZONES • DEFINE TRANSPORTATION ZONE HIERARCHY. As shown in Figure 3.27, use the name RELH_ZONE for the transportation zone hierarchy and the name RELHS_ZONE for the hierarchy structure, and select CREATE or CHANGE. The RELHS_ZONE hierarchy is delivered as standard. The customer-specific transportation zone hierarchy must be maintained in the RELH_ZONE hierarchy.

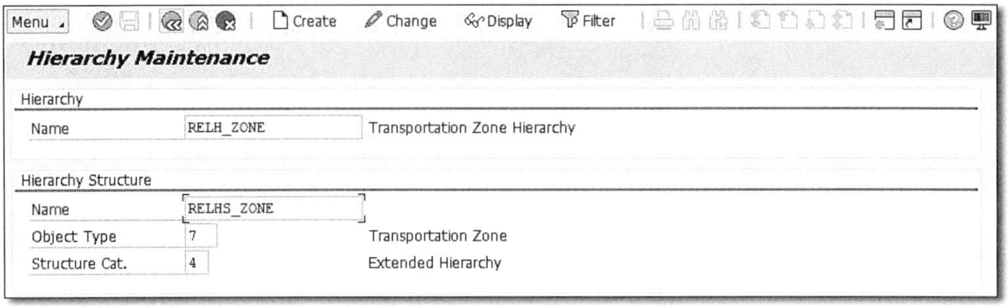

Figure 3.27 Choosing the Transportation Zone Hierarchy

Maintenance of the transportation zone hierarchy is shown in Figure 3.28.

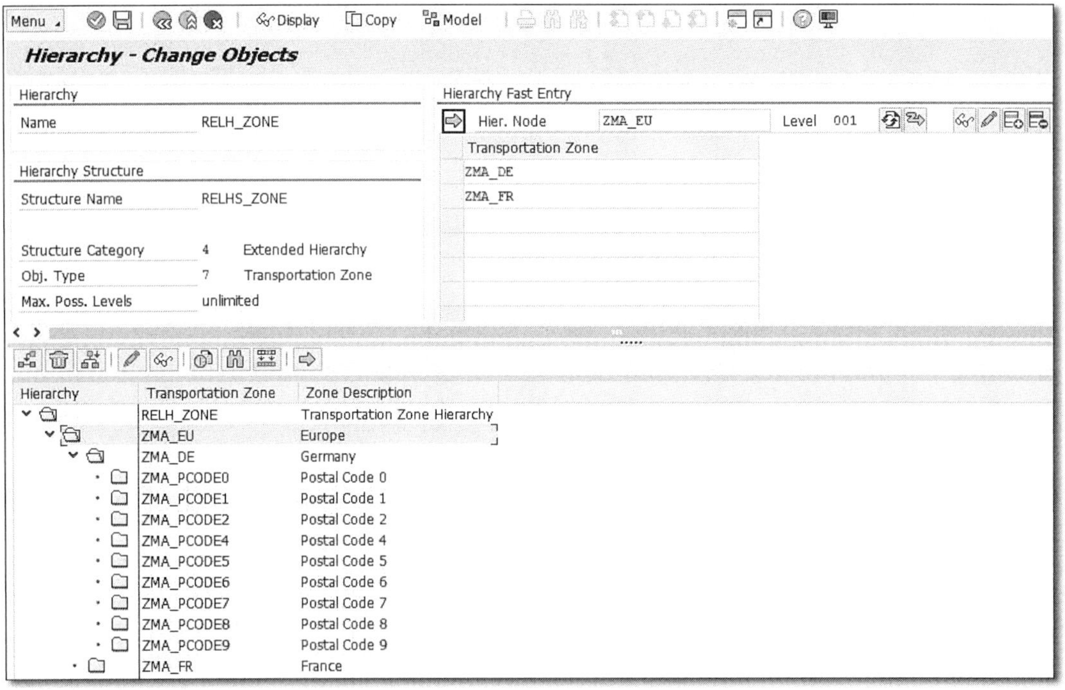

Figure 3.28 Maintaining the Transportation Zone Hierarchy

The maintenance screen is structured around an overview of the existing transportation zone hierarchy displayed as a tree structure and the HIERARCHY FAST ENTRY area, where you can enter new hierarchy nodes. The

RELH_ZONE hierarchy element is the root node in the hierarchy, which means that all other transportation zones are directly or indirectly subordinate to this element. You can double-click to select a parent node in the tree display, enter one or more transportation zones in the HIERARCHY FAST ENTRY screen area, and click on the COPY button to add these to the hierarchy. To delete transportation zones from the hierarchy, select a hierarchy element in the tree display and click on the DELETE OBJECTS FROM HIERARCHY button. This deletes all *subordinate* elements from the hierarchy. To remove all objects from the hierarchy, delete the root (note, however, that the RELH_ZONE element is retained to allow you to add new elements to the hierarchy later).

Careful Modeling of Transportation Zones and Zone Hierarchy

Use the transportation zones and transportation zone hierarchies to model your transportation network as compactly as possible, and avoid redundancies when defining transportation zones and in the transportation zone hierarchy.

Remember that the same location can be contained in several transportation zones and that transportation zones can overlap in this way. When you are creating a new zone, the system informs you by a message if it overlaps with an existing zone.

Two transportation zones are considered to be overlapping if they have at least one location in common and to be equivalent if they contain the same set of locations. Equivalent transportation zones are always overlapping, but overlapping transportation zones are not necessarily equivalent.

We recommend that you avoid defining equivalent transportation zones because a very large number of implicit combination options may arise for automatic planning (transportation proposals and VSR optimization), both during automatic determination of transportation lanes, distance, and duration (see Section 3.2.3) between locations, and when determining transshipment locations. This may lead to long runtimes in certain cases.

If you have maintained equivalent transportation zones, you should delete all but one of these. Also check overlapping transportation zones to determine whether the overlaps are useful from a business perspective or whether they could be eliminated.

Keep your transportation zone hierarchy as flat as possible.

3.2.3 Transportation Lanes

Transportation lanes describe the reachability of locations within the transportation network. They are defined by three elements:

- A start, which may be a location or a transportation zone
- A destination, which also may be a location or a transportation zone
- A means of transport

A valid transportation lane indicates that the destination can be directly reached from the source by resources of the means of transport. When you maintain the start or destination, a transportation zone always represents all of the locations it contains. If a transportation lane is defined between two transportation zones, this means that the means of transport operates between all locations in the start transportation zone and all locations in the destination transportation zone. A transportation lane from a (start) location to a (destination) transportation zone indicates that the means of transport operates between that location and all locations in the transportation zone. An intrazone lane (a transportation lane where the start and destination transportation zones are the same) indicates that the means of transport operates between any two locations within this zone. A location transportation lane (a transportation lane where the start and destination location are the same) has no influence on reachability between locations but can be used for the initialization of allocations.

Compact Maintenance of Transportation Lanes

Check whether the reachability of locations in your transportation network can be modeled by transportation lanes with transportation zones as the start and/or destination. In most cases, fewer transportation lanes are required if you define them using transportation zones rather than using locations directly.

To maintain a transportation lane, follow the menu path Master Data • Transportation Network • Define Transportation Lane. You can maintain transportation lanes in the following tabs:

- Transportation Lane
  To create or change individual transportation lanes, enter the start and destination and select Create or Change.

- Intra-Zone Lane/Location Lane
  Enter a location or transportation zone. You can then create or change a transportation lane from the location to the same location (location transportation lane) or from the transportation zone to the same transportation zone (intrazone lane).
- Mass Maintenance (Create)
  Use an existing transportation lane as a template for generating new transportation lanes. You can overwrite existing transportation lanes, leave them unchanged, or enhance them with additional information (see Figure 3.29). You can also specify whether the duration and distance are to be copied from the template or recalculated.
- Mass Maintenance (Display/Change)
  You can define selection criteria for transportation lanes (start, destination, start location type, destination type, etc.) and then display or change the selected transportation lanes based on these criteria.

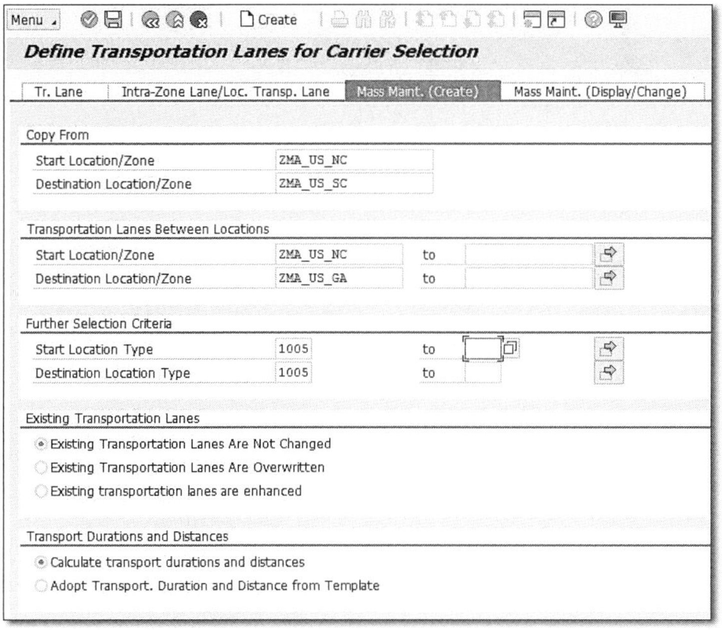

Figure 3.29 Mass Creation of Transportation Lanes

The screen for maintaining an individual transportation lane and the mass maintenance screen have similar structures. The mass maintenance screen is shown in Figure 3.29. Here you can create several transportation lanes at the same time, assign both a means of transport and a carrier to each, and maintain all of the transportation lane parameters.

The left side of Figure 3.30 shows the transportation lanes.

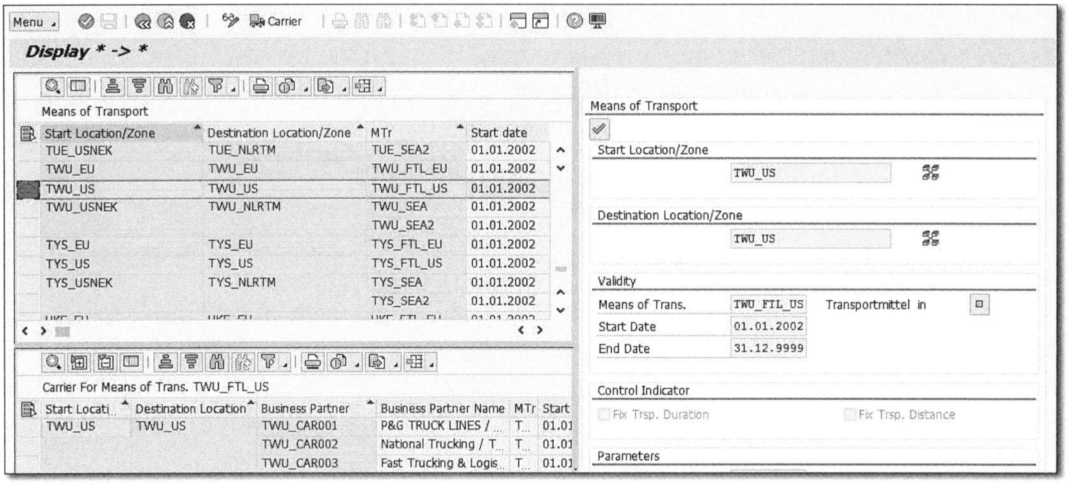

Figure 3.30 Displaying Transportation Lanes

The means of transport for the transportation lanes are displayed in the MEANS OF TRANSPORT screen area in Figure 3.30, with one row representing each means of transport in a transportation lane. You can double-click on a row to select the corresponding means of transport for a transportation lane and maintain the following parameters:

- The validity dates for the transportation lane.
- The distance between the start and destination and the amount of time it will take the means of transport to travel that distance (i.e., the duration). You can generate a proposal using the button provided.
- Two control indicators that indicate whether the specified duration and distance are to be overwritten by an automatic distance and duration determination.
- The PRECISION field indicates whether the distance was calculated automatically based on the straight-line distance (value: 0000) or with geographical information system (GIS) precision (0100) or whether it was entered manually (1000).

- The quantity, distance, and minimum costs, of which the last is only relevant for destination-based distance costs (the use of these costs is controlled in the cost profile, which is discussed in Chapter 6, Section 6.4.3).

Additional parameters allow you to control the carrier selection (see also Chapter 7, Section 7.1.4):

- You can specify whether business shares are to be taken into account in carrier selection, which tolerances apply when an excess or shortfall occurs, and which penalty costs are to be used in cases that fall short of or exceed the business share.
- You can define a strategy for carrier selection by selecting the RELEVANT FOR CARRIER SELECTION field and specifying whether costs and/or priorities are to be used, which costs are to be used (internal costs, costs from SAP ERP or from SAP TM Charge Calculation), and whether or which continuous moves are permitted.
- You can decide whether the planning period or the minimum and maximum capacities defined for the carriers are to be used for the initialization of allocations (see Chapter 5, Section 5.2).
- You can also define several means of transport for a transportation lane. To do this, click on the CREATE button for a new entry in the MEANS OF TRANSPORT screen area.

Carrier selection

For each means of transport, you have the option of assigning carriers in the CARRIER area (in the same way you assign a means of transport) and maintaining additional details for each carrier that is relevant for carrier selection:

- Internal costs
- Priority
- Arrival and departure windows for continuous moves
- Maximum distance for continuous moves
- Discounts for continuous moves
- Desired business share for the carrier
- Minimum and maximum capacity for initializing the allocation of the carrier (see Chapter 5, Section 5.2)

You can control the effects of the parameters maintained for the transportation lane on carrier selection in the carrier selection settings (see Chapter 7, Section 7.1.4). Note that direct shipment option determination (see Chapter 6, Section 6.1.2) uses carrier selection to choose the best option, and thus the carriers on the transportation lane represent the only alternatives considered.

Transportation lane, distance, and duration determination

The automatic transportation lane, distance, and duration determination selects the appropriate transportation lane for a means of transport and a specified start and destination location, and determines the corresponding distance and duration between the start and destination. This automatic determination is used, for example, by automated planning procedures and VSR optimization or transportation proposals; it enables the integration of GIS tools to spare users the laborious task of manually maintaining distances and durations between locations (see Section 3.2.9).

Zone hierarchy/ means of transport hierarchy

The selection of a transportation lane for a start and destination location is necessary because the hierarchies of transportation zones (see Section 3.2.2) and means of transport (see Section 3.3.1) may give rise to several transportation lanes, which subsume the means of transport, start, and destination location.

You can make settings in Customizing to specify the sequence in which the three hierarchies (start, destination, and means of transport) are to be taken into account when determining the transportation lane. To do this, select the IMG path SAP Transportation Management • Transportation Management • Master Data • Transportation Network • General Settings for Transportation Network Determination. The Consider hierarchical relationships between means of transport first and Consider source hierarchy first parameters allow you to select one of the following four access sequences:

Possible access sequences of hierarchies

- 1. Means of transport, 2. Start, 3. Destination
- 1. Means of transport, 2. Destination, 3. Start
- 1. Start, 2. Destination, 3. Means of transport
- 1. Destination, 2. Start, 3. Means of transport

The access sequence determines which hierarchy is taken into account first when a transportation lane is determined.

Let's consider an example that is based on a scenario with a means of transport T1, a higher-level means of transport T2, locations A and B, zone ZA (which contains location A), and zone ZB (contains location B).

When a transportation lane from A to B is requested, the transportation lane determination function goes through the possible combinations in the sequence listed below, following the access sequence 1. Means of transport, 2. Destination, 3. Start:

1. Means of transport T1, Start A, Destination B
2. Means of transport T2, Start A, Destination B
3. Means of transport T1, Start A, Destination ZB
4. Means of transport T2, Start A, Destination ZB
5. Means of transport T1, Start ZA, Destination B
6. Means of transport T2, Start ZA, Destination B
7. Means of transport T1, Start ZA, Destination ZB
8. Means of transport T2, Start ZA, Destination ZB

Most specific lane first

The first combination in this sequence for which a transportation lane exists returns the transportation lane as a result. The underlying principle here is to select more specific transportation lanes first, which allows you to maintain "general" transportation lanes and to refine exceptions with "more specific" transportation lanes.

Determining distance and duration

The automatic distance and duration determination determines a start location S, a destination location De, a distance Di(S,De), and a duration Du(S,De) for a given means of transport T. Depending on the configuration, the distance from the transportation lane can be used, or, alternatively, it can be calculated on the basis of a straight-line distance or with an external GIS tool, which takes account of the existing road network and so on.

When a specific request is submitted for (T,S,De), the system first checks whether a transportation lane exists for (T,S,De). If it finds one, the distance and duration values maintained for the transportation lane (T,S,De) are returned directly as the result for Di(S,De) and Du(S,De).

If a transportation lane (T,S,De) does not exist, a transportation lane (T′,S′,De′) is determined in accordance with the configured access sequence. This transportation lane (T′,S′,De′) is a "superior" (higher-level) transportation lane in at least one of the hierarchies (means of transport, start, or destination). If the GIS quality parameter is set for means for transport T′ (for information about maintaining a means of transport, see Section 3.3.1), the GIS tool calculates the distance Didyn(S′,De′). If GIS quality is not selected for T′, then Didyn(S′,De′) is calculated as the product of the straight-line distance between S′ and De′ and the distance factor for means of transport T′.

The distance Didyn(S,De) is then determined the same way for means of transport T on the basis of the GIS quality parameter of T. The requested distance Di(S,De) is then calculated as follows: Di(S,De) = Didyn(S,De) × Di(S′,De′)/Didyn(S′,De′). The relationship between the distance Di(S′,De′) maintained for the transportation lane (T′,S′,De′) and the result of the dynamic distance calculation Didyn(S′,De') is thus also used to determine the specific distance requested. All distance calculations are based on the geographical coordinates of S and De (or S′ and De′).

If the distance of transportation lane (T,S,De) or (T′,S′,De′) is not maintained, then Di(S,De) = Didyn(S,De) and Di(S′,De′) = Didyn(S′,De′), respectively.

The duration calculation for Du(S,De) is essentially the same as the calculation of Di(S,De). Note, however, that the three speeds of the means of transport are required to call the GIS tool. If the three speeds are not maintained, the GIS tool is not called, and the duration is calculated using the straight-line distance and the average speed of the means of transport. If two relevant means of transport have identical entries for the three speeds, the GIS tool would only be called once per source and destination pair, and the determined distance and duration would be used for both means of transport.

If the duration of transportation lane (T,S,De) or (T′,S′,De′) is not maintained, then Du(S,De) = Dudyn(S,De) and Du(S′,De′) = Dudyn(S′,De′), respectively.

3.2.4 Schedules

Schedule, departure rules, departures

A schedule represents a recurring transportation that follows a predefined location sequence. It can be used to model regular ship, air, road, or rail transportation and is valid for a specific period of time. The schedule contains a set of departure rules defining the pattern of days on which the transportation is possible, the times of departure and arrival among the location sequence, and cutoff and availability times needed for transshipment scenarios. A departure represents one instance of transportation along the whole location sequence, with all arrival times and departure times being determined by the departure time at the first location of the sequence. The departures of a schedule can be generated based on a departure rule of the schedule or be maintained manually. A departure defines that goods can be transported along the location sequence, or a subsequence of it, respecting the pre-determined departure and arrival times for the locations. Thus, a schedule defines reachability in the transportation network, according to a predefined location sequence and given the departure and arrival times of the stops.

Example departure rule/departure

Figure 3.31 shows an example of a carrier flight schedule with location sequence Munich, Chicago, and Los Angeles and two departure rules, which can be used to generate the departures (called *flights* for the air mode of transport), as shown in Figure 3.32.

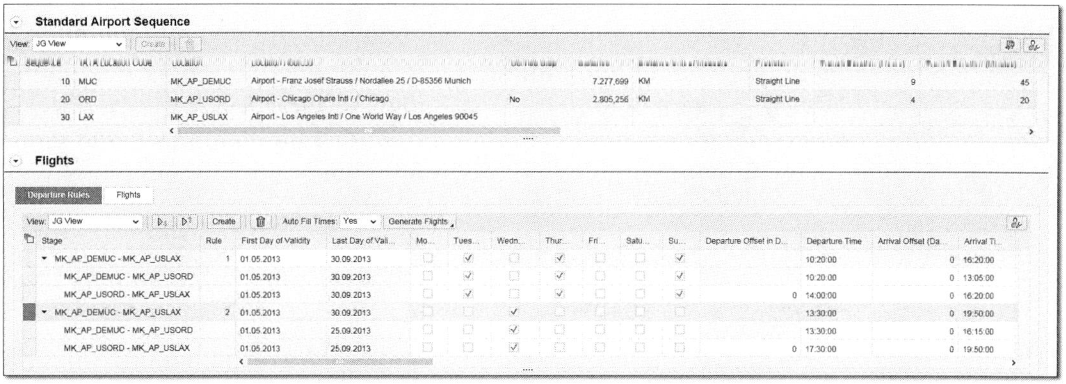

Figure 3.31 Carrier Flight Schedule with Location Sequence and Two Departure Rules

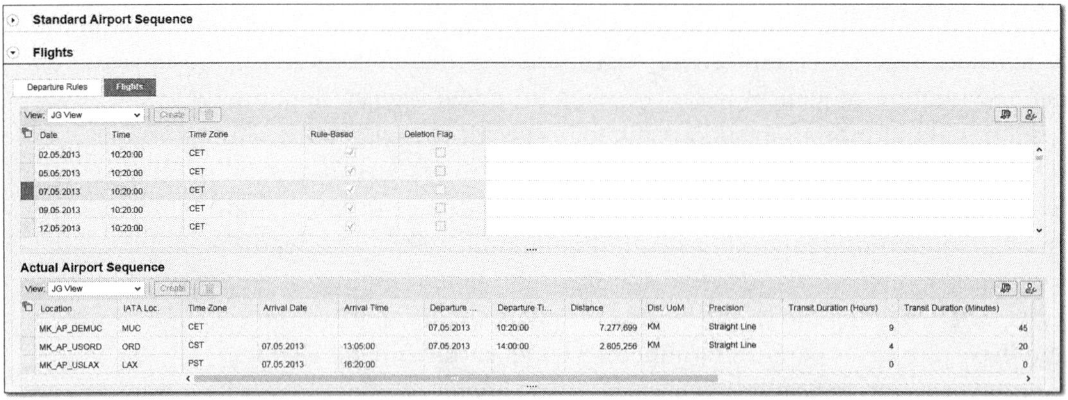

Figure 3.32 Carrier Flight Schedule with Departures (Flights)

Schedule-based freight documents

Schedules can be consumed by creating a freight document based on a schedule's departure, either by manual (ad hoc) creation (see Chapter 5, Section 5.4.1 and Chapter 7, Section 7.2.1), capacity management (see Section 5.1.2), or transportation planning (see Chapter 6, Section 6.3.5, Section 6.3.6, and Section 6.4). The schedule-based freight document inherits schedule data from the location sequence; the relevant times for departure, arrival, cutoff, and availability; and any available capacities. Capacity management allows the systematic creation of freight documents for a set of schedules and all their departures in a predefined time period (e.g., for the next six months). Transportation planning chooses the best departures for a given set of freight units, either by creating new freight documents for the chosen schedule departures or by consuming already existing schedule-based freight documents, which may stem from capacity management or a previous transportation planning step.

Schedule types

Each schedule has a type chosen from the types maintained in the IMG menu path SAP Transportation Management • Transportation Management • Master Data • Transportation Network • Schedule • Define Schedule Types.

The schedule type defines how schedules of this type can be used. You can define the following.

- Default Type
  One of all schedule types can be marked as default. If you create a new schedule and do not specify a schedule type, this default type is chosen.

- MODE OF TRANSPORT
  You can choose one of the modes of transport defined in the system (see Section 3.3.1), such as road, rail, sea, or air.
- GATEWAY
  This specifies whether the schedule has two gateways, one source and one destination, or no gateway at all. The corresponding schedules are called gateway schedules and carrier schedules, respectively, and their roles are described shortly.
- DIRECT
  For a gateway schedule, you can define whether the two gateways are directly connected or whether other locations such as ports or airports are used in between, which is the most common use case. The corresponding schedules are called direct gateway schedules and indirect gateway schedules, respectively.
- REFERENCE
  If you have chosen an indirect gateway schedule type, you can define whether it refers to a carrier schedule or is maintained without reference to any carrier schedule.
- DOCUMENT TYPE
  This specifies the freight document type that is used for creating freight documents out of the schedule.
- ALLOCATION TYPE
  If you maintain an allocation type, you can create allocations of this type out of the schedule (see Chapter 6, Section 6.1 for capacity management and the interplay of schedules and allocations). Otherwise, you cannot create allocations out of the schedule.
- TEMPLATE
  This defines the schedule to be used as a template only; that is, the freight documents can be changed manually regarding location sequence and departure and arrival times. If the schedule is not used as a template, the location sequence and times cannot be changed manually and the freight document keeps a reference to the schedule. Independent of this parameter, automatic planning never changes location sequence or departure or arrival times.

- CC STRATEGY
  The change controller strategy is called after the schedule has been changed. For example, you can define a change controller strategy for automatic propagation of the schedule changes into the already created freight documents. Note that the standard does not propagate any schedule changes into referencing freight documents, allocations, or schedules. However, the reference data status of a referencing instance is updated to allow the user to identify the need for manual adjustments according to the schedule changes.
- DELETION STRATEGY
  The deletion strategy is called after the schedule has been manually deleted by the user or automatically deleted per report. It allows you to insert your own logic to be processed after schedule deletion.
- OFFSET TIME TYPE
  You can choose between two alternatives for defining cutoff and availability times in the schedule. One the one hand, with the RELATIVE option, your cutoff and availability entries in the schedule are interpreted relatively. A cutoff of one day and 12 hours means that the cutoff time for a departure at 8 a.m. is mapped into a cutoff time of 8 p.m. two days before. Similarly, an availability of 20 hours for an arrival at 5 p.m. is mapped into an availability time of 1 p.m. on the next day.

  On the other hand, with the ABSOLUTE option, the time is interpreted absolutely and the date is determined relatively by a minimum offset. A cutoff is defined by an absolute time and an offset in days. For example, given a departure at 10:00 a.m. and a cutoff defined by an offset of one day and a time of 11:30 a.m. results in a cutoff time 11:30 a.m. two days before. For a departure at 11:45 a.m. and the same cutoff values, the cutoff time would be 11:30 a.m. one day before.
- ONE ORDER
  If this parameter is active, only one freight order can be created per schedule departure, and this freight order covers the complete location sequence of the schedule. Thus, all freight units to be transported by this departure are consolidated into the same freight order for the departure. The schedule gets locked during planning to ensure that another planning session cannot create another freight document.

Deactivating this parameter, you can create multiple freight orders for one departure, either by manual planning or by automatic planning.

- Use Capacities
  You can use this flag to maintain schedules with capacities.
- Use Transportation Costs
  This parameter allows you to define internal planning costs for the schedule, such as fixed costs for using a departure and quantity costs.

You can also define a customer-specific user interface configuration for the schedule.

Carrier schedules

A carrier schedule contains a location sequence along which goods can be transported by a carrier with the given mode of transport. Each intermediate location in the sequence can be used to load and unload goods, whereas the first and last location only allow loading and unloading, respectively. The following carrier schedule types are delivered as standard: ocean carrier schedule (type 1000), carrier flight schedule (1100), road carrier schedule (1200), and rail carrier schedule (1300). Ocean carrier schedules connect ports, carrier flight schedules connect airports, and the other two types connect locations via road and rail, respectively. The location sequence can model ocean carrier schedules with 20 or more ports as well as direct flights between two airports and multistop flights along a sequence of airports. Carrier schedules are mostly used to represent transportation capabilities offered by external carriers, such as ocean carriers and road carriers. However, it is not mandatory to assign a carrier. This makes it possible to model a schedule for your own fleet or for an air freight scenario in which an airline offers its schedule but is represented by regional subsidiaries, which all can receive air freight bookings for the schedule.

Gateway schedules

A gateway schedule allows transportation from source gateway (for consolidation) to destination gateway (for deconsolidation). Gateways usually serve as transshipment locations (see Section 3.2.5); for ocean transport of containers, they are commonly called container freight stations. The following gateway schedule types are delivered as standard: sailing schedule (type 2000), sailing schedule with reference (2100), master flight schedule (2500), master flight schedule with reference (2600), and road gateway schedule (2200). Indirect gateway schedules such as sailing schedules and master flight schedules connect the source gateway to the destination gateway via a source (air-) port and a destination (air-) port. Indirect gateway schedules are used to model scenarios in which the transportation between the (air-) ports is mainly organized by consolidation at the source gateway, and deconsolidation is organized at the destination gateway.

The gateway-to-gateway connection wraps the port-to-port connection details. This allows the users to focus on cutoff times at the source gateway and availability times at the destination gateway, which implicitly consider the cutoff times at the source port and the availability time at the destination port, respectively. The departure rules and departures between the (air-) ports can be defined implicitly by reference to an underlying ocean carrier schedule or carrier flight schedule, respectively, or explicitly without any reference to a carrier schedule.

Sailing schedule with reference

For an ocean carrier schedule with the port sequence PT_DEHAM (Hamburg), PT_USPNJ (Newark), and PT_USCHS (Charleston)—with these ports having container freight stations CFS_DEHAM, CFS_USPNJ, and CFS_USCHS, respectively—you can define three sailing schedules covering the following locations:

1. CFS_DEHAM → PT_DEHAM → PT_USPNJ → CFS_USPNJ
2. CFS_DEHAM → PT_DEHAM → PT_USCHS → CFS_USCHS
3. CFS_USPNJ → PT_USPNJ → PT_USCHS → CFS_USCHS

Whereas the first and third sailing schedules use a direct port connection out of the underlying ocean carrier schedule, the second sailing schedule connects an indirect port connection with port PT_USPNJ as the intermediate port between the ports PT_DEHAM and PT_USCHS.

Connection schedules

Gateway schedules that refer to two or more carrier schedules are called *connection gateway schedules* and can be used by transportation planning and capacity management like any other gateway schedule. A significant portion of air freight-forwarding worldwide is handled by connection flights, which combine multiple flights offered by one or multiple airlines. If certain connection flights are used frequently in your business, you can create a connection master flight schedule with departure rules that combine departure rules of multiple carrier flight schedules. Figure 3.33 gives an example of a gateway in Munich to an airport in Munich to an airport in Chicago to an airport in Denver to a gateway in Denver. The departure and arrival times for the airport-airport stages stem from two referenced carrier flight schedules.

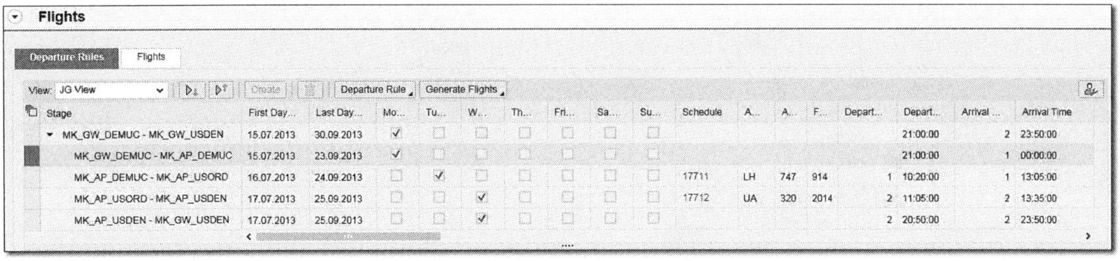

Figure 3.33 Departure Rule for a Connection Master Flight Schedule

In ocean freight-forwarding, the analogous concept of connection voyages is also used to combine multiple voyages offered by one or multiple ocean carriers.

Maintaining a carrier schedule

Create a new schedule by following the menu path Master Data • Transportation Network • Schedule • Create Schedule and choosing the schedule type. When creating a carrier schedule, you first define the General Data and the Standard Stop Sequence, as shown in Figure 3.34 for a road carrier schedule.

Maintain the validity and the stop sequence by iteratively adding the locations. Optionally, you can define the description, the means of transport, and the carrier, either explicitly or implicitly, by entering its standard carrier alpha code (SCAC) that is commonly used in the U.S. transportation industry. Additionally, you can maintain the following general data:

- Shipping Type
  You can choose one of the predefined shipping types such as FCL or LCL (see Chapter 4, Section 4.2.1 for an introduction). The schedule can be used for a freight unit at hand if the schedule and freight unit have the same shipping type or if one of the two shipping types is undefined.

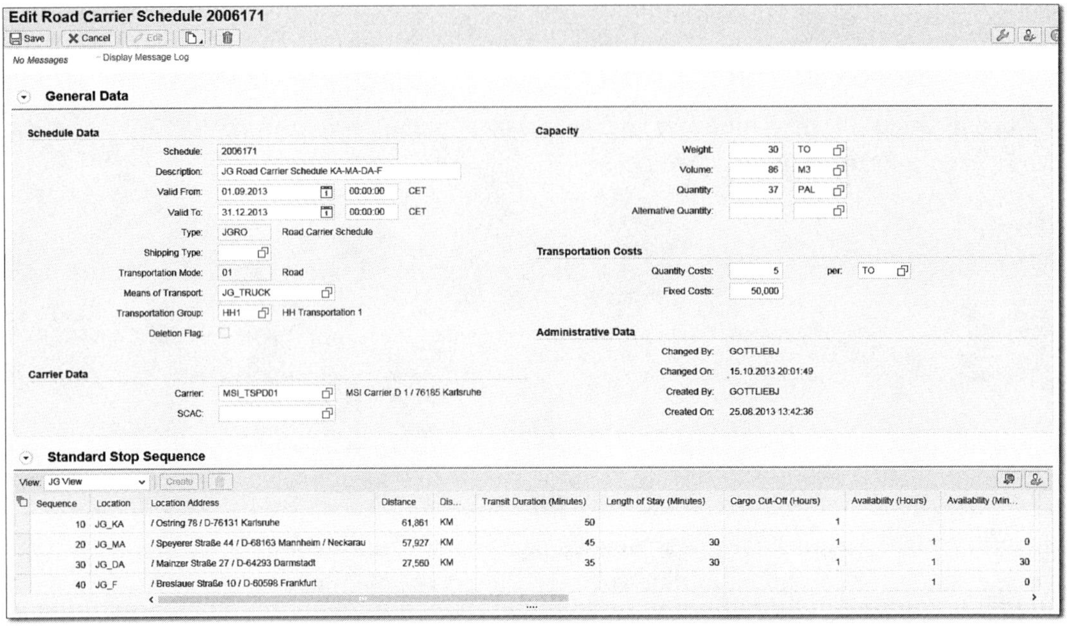

Figure 3.34 Road Carrier Schedule: General Data and Location Sequence

- Transportation Group
  Like the shipping type, the transportation group can be used to determine which freight units can be assigned to the schedule at hand.
- Capacity
  You can define weight, volume, and two additional capacity restrictions. These serve as templates for the capacities that are maintainable for departure rules, which themselves also serve as templates for the capacities maintained per departure generated for the departure rule. Note that the capacities can only be maintained if this functionality is activated in the schedule type.
- Transportation Costs
  You can define quantity costs and fixed costs, which are considered by automatic planning and apply per schedule departure. In the optimizer cost settings you can specify whether the quantity costs are multiplied by the distance traveled (see Chapter 6, Section 6.4.3). Like the capacity definition, these costs serve as templates for the departure rules that also serve as templates for the generated departures.

This functionality can only be used if it is activated in the schedule type.

- MODE-SPECIFIC FIELDS
  For ocean carrier schedules, you can define the loop to group related schedules along the same rotation. For carrier flight schedules you can define the airline code, the flight number, and the aircraft type code as well as whether the carrier is executing the flight itself, in order to identify code-shared flights.

The stop sequence serves as a template for the departure rules and allows you to maintain the following data:

- TRANSIT DURATION
  This represents the transit duration from a stop to its successor.
- DISTANCE
  The system automatically determines the distance between two consecutive stops based on the geographical coordinates of the locations. You can manually change the proposed distance.
- LENGTH OF STAY
  This represents the length of stay at an intermediate stop.
- CARGO CUT-OFF
  You can define the cutoff time that is considered by automatic planning. This value specifies when freight units must be delivered to the location at hand so that they can be transported by the schedule from this location. Depending on the schedule type, you can define it relatively or absolutely, as described above. In addition, it is also possible to define document cutoff and dangerous goods cutoff times.
- AVAILABILITY TIME
  You maintain the availability time considered by automatic planning, which defines when freight units delivered by the schedule to the location at hand can be picked up for further transportation departing from the location.

By clicking on the CREATE button in the DEPARTURE RULES tab, you can maintain multiple departure rules following the location sequence. Figure 3.35 shows an example of one departure rule defined for the road carrier schedule depicted in Figure 3.34. You can define the validity of the rule, which can be a subperiod of the schedule's validity; the pattern of days; the times of departure, arrival, cargo cutoff, document cutoff, dangerous goods cutoff, and availability; as well as the durations for transit and length of stay. With the AUTO FILL TIMES option, you can change one departure or arrival time and let the system propagate this to the other times. Defining a factor calendar allows you to suppress generation of departures on public holidays according to the calendar.

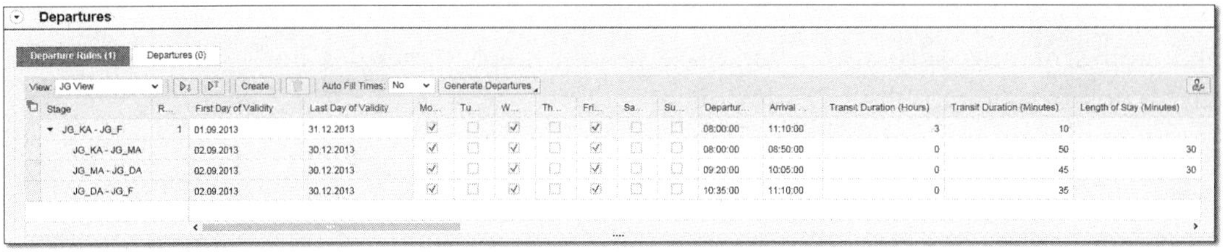

Figure 3.35 Road Carrier Schedule: Departure Rule

You can select multiple departure rules and generate departures by clicking on the corresponding button, either for the whole validity period or for an explicitly defined period. The departures are displayed in the DEPARTURES tab, as shown in Figure 3.36. All the times and durations are derived from the departure rule definitions. By clicking on the CREATE button, you can create a new departure. When you select a departure, all of these times and durations are displayed in the ACTUAL STOP SEQUENCE area. It is possible to manually change times and durations, even if they were generated based on a departure rule.

For ocean carrier schedules, you can maintain the vessel name and its International Maritime Organization (IMO) ship identification number on the departure rule and departure level. For carrier flight schedules, it is possible to mark intermediate stops as service stops (e.g., for refueling); service stops cannot be used for loading or unloading but appear in the air freight bookings created for the schedule, and to maintain the aircraft type code on the departure rule stage level. Moreover, for multistop flights you can define per stage whether the carrier is operating this stage itself or it is a code-share flight.

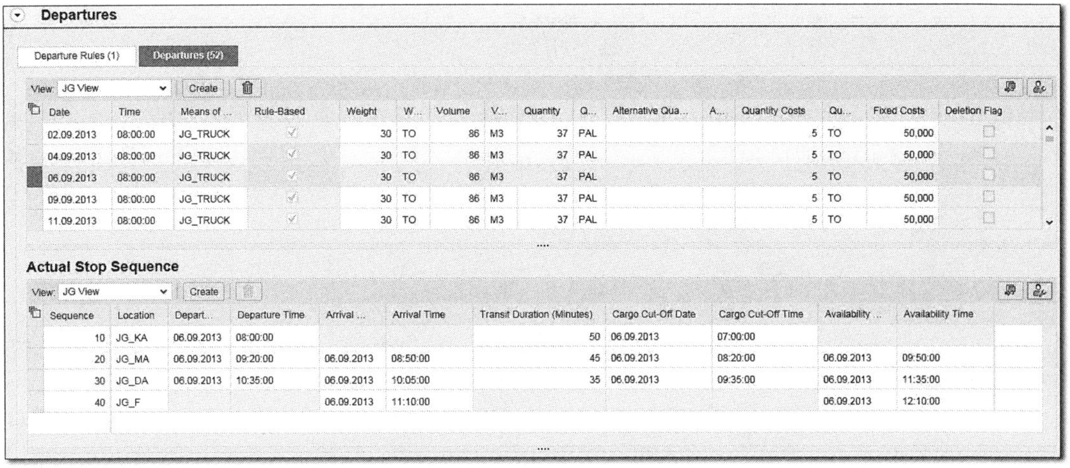

Figure 3.36 Road Carrier Schedule: Departures and Actual Stop Sequence

Maintaining a gateway schedule

Maintaining gateway schedules is similar to maintaining carrier schedules, but some important additional data has to be maintained. First, you maintain the source gateway and the destination gateway, with corresponding cutoff times for the source gateway, transit duration to source port, transit duration from the destination port, and availability time for the destination gateway, as shown in Figure 3.37 for a sailing schedule. Then you define the standard port sequence by entering the locations. By clicking on the SCHEDULE button and selecting the ASSIGN option, you can let each stage refer to an ocean carrier schedule, from which all times and durations are taken over in the sailing schedule. For a sailing schedule without reference, you have to manually maintain all this data in the sailing schedule itself.

You define the departure rules similarly to carrier schedules. The difference is that you can reference the referenced carrier schedules' departure rules by selecting the created departure rule's stage, clicking on the DEPARTURE RULE button, and selecting the ASSIGN option to select a departure rule of the referenced carrier schedule.

With connection gateway schedules, you just maintain more stages and link them to different schedules. Each departure rule stage refers to one departure rule of the referenced carrier schedule.

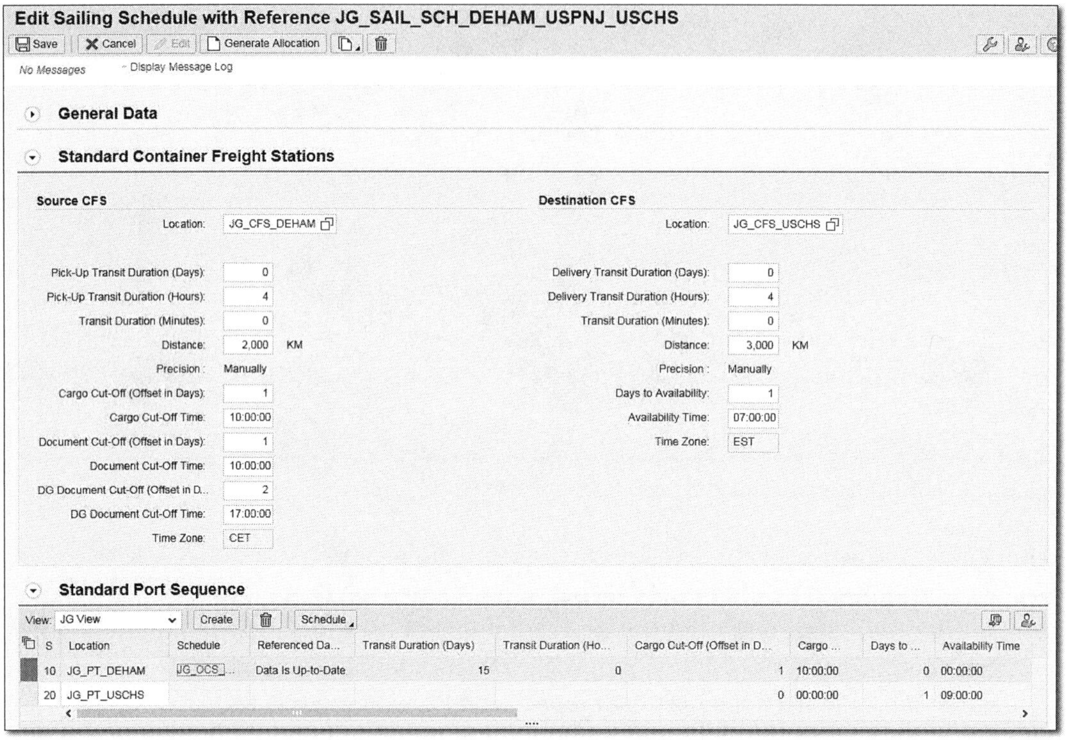

Figure 3.37 Sailing Schedule: Gateways (Container Freight Stations) and Location Sequence

The REFERENCED DATA STATUS field indicates whether the underlying carrier schedules have changed since the gateway schedule was created. This is quite useful if weekly updates of carrier schedules are uploaded into the system and you want to quickly check whether your gateway schedule is affected by these updates.

Uploading schedules

Several transportation businesses such as global shippers and global freight-forwarders make frequent use of the regular schedules of ocean carriers and airlines, which publish their schedules and update them on a regular basis. For any mode of transport, schedules can be uploaded by the function module /SCMTMS/BAPI_SCHEDULE_SAVEMULT, which offers create, update, and delete access to the schedules in the system. This approach requires mapping of the external data into the generic interface of the function module. Alternatively, you can run the report that's available via the menu path Application Administration • Master Data • Transportation Network • Schedule Upload, as shown in Figure 3.38, and use one of the standard-delivered upload strategies or build your own upload strategy.

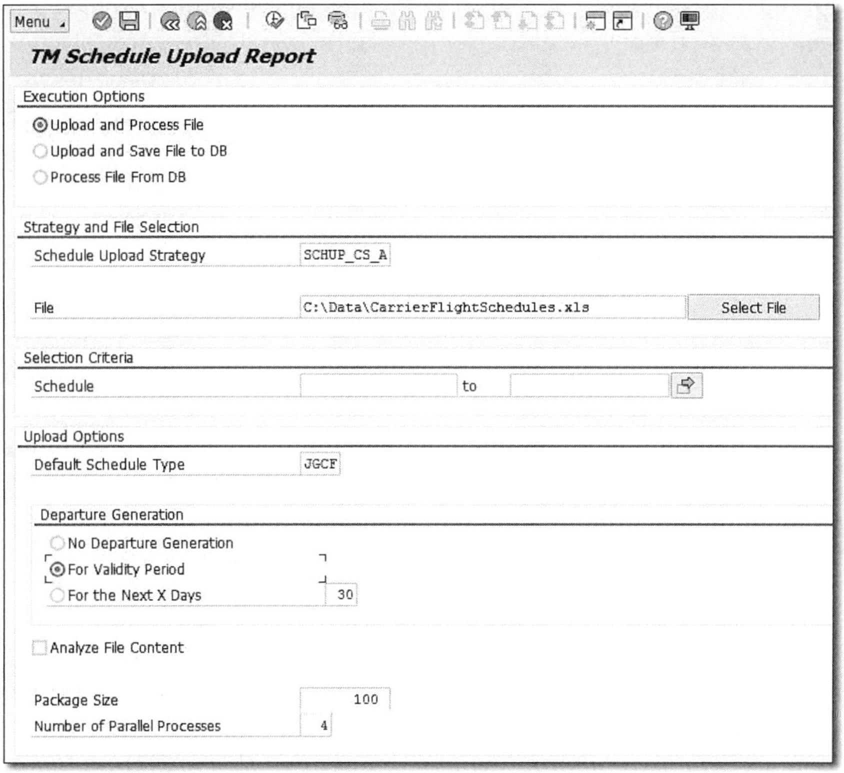

Figure 3.38 Schedule Upload Report

Standard upload strategies

For example, the carrier flight schedule upload strategy SCHUP_CS_A allows you to upload schedule data from *.xls* and *.csv* files with a specific column signature. These files are designed to represent flight schedules in a simple and compact way that is easy to understand and maintain. OAG, one of the leading data providers for flight schedules for most airlines worldwide, offers its data in the file format offered by the standard upload strategy SCHUP_CS_A, which allows a straightforward integration of flight schedules into SAP TM. See SAP Note 1857686 for more details on this efficient schedule integration, which frees the system integrator from building custom interfaces to all relevant airlines in the world.

Parameters in the upload report

The upload report offers direct processing of a file, which is recommended for small and medium files. If the file is very large or will be uploaded by background job, it can be uploaded into the database and then processed from the database. You can define an additional selection filter, such that only the matching schedules out of the files are uploaded. The schedule type can be predefined, and if it's not predefined, it has to be set in the file. You can trigger an automatic generation of departures, either for the whole validity period of the schedule or for a relative time period. The latter option is useful if you have departure rules that cover the whole next year but you are used to creating freight documents for only the next four weeks. You can extract additional data out of the schedule file via the ANALYZE FILE CONTENT option. In the carrier flight schedule scenario, this would return a list of the contained IATA location codes, airline codes, and aircraft codes; these are useful for verifying the completeness of your corresponding master data. The package size and number of parallel processes make it possible to adapt the schedule upload according to the expected file size, the available hardware, and the desired runtime of the upload.

Departure creation report

Whereas this report can automatically create departures, there are some scenarios in which you would want to upload schedules but don't create departures immediately. If you are handling a lot of schedule data, manually triggering the creation of departures may be a tedious task, so you can follow the menu path APPLICATION ADMINISTRATION • MASTER DATA • TRANSPORTATION NETWORK • CREATION OF SCHEDULE DEPARTURES and explicitly create departures for a set of schedules. You define a selection of the relevant schedules as well as the time period for which the departures will be generated. It is also possible to run this report in a simulation mode first to determine how many departures would be generated. As with the schedule upload report, you can also define package size and number of parallel processes.

Deleting schedules

If schedules have limited validity (e.g., half a year), and new schedules are regularly created in the system to reflect the corresponding transportation options (here, for the next half year), then the number of schedules in the system grows continuously. Usually, the outdated schedules are not used anymore, so we recommend that you remove them from the system to restrict the data volume and prevent users from wasting time with useless schedules.

Of course, you can manually delete schedules (e.g., by selecting them from a query result in the transportation network overview POWL; see Section 3.2.8). If you follow the menu path APPLICATION ADMINISTRATION • MASTER DATA • TRANSPORTATION NETWORK • SCHEDULE DELETION and run the report located there, you can define a selection of schedules, choose one of the following deletion options, and delete the matching schedules automatically.

- COMPLETE SCHEDULES
  The whole schedule is deleted, with all its departure rules and departures.
- DEPARTURES IN THE PAST
  This deletes only schedule departures in the past. The remaining departures and all departure rules are kept, even if they don't have a departure anymore.
- SCHEDULES WITHOUT DEPARTURES & IN THE PAST
  This acts like the previous option but also deletes schedules without any departure.
- CONSIDER DELETION FLAG
  This option deletes all schedules with an active deletion flag.

To avoid undesired deletion of schedules, you can first run a simulation of the found schedules by choosing the SHOW SCHEDULES/DEPARTURES option, as shown in Figure 3.39.

Deletion versus marked for deletion

In general, if a schedule is still referenced by freight documents, allocations, or other schedules, then it is only marked for deletion and so cannot be used for new references by those documents. A schedule is only deleted if there are no references to it. Thus, if you are running the schedule deletion report on a weekly basis, the deletion markers get set immediately, and the schedule is only deleted if the referencing documents have disappeared.

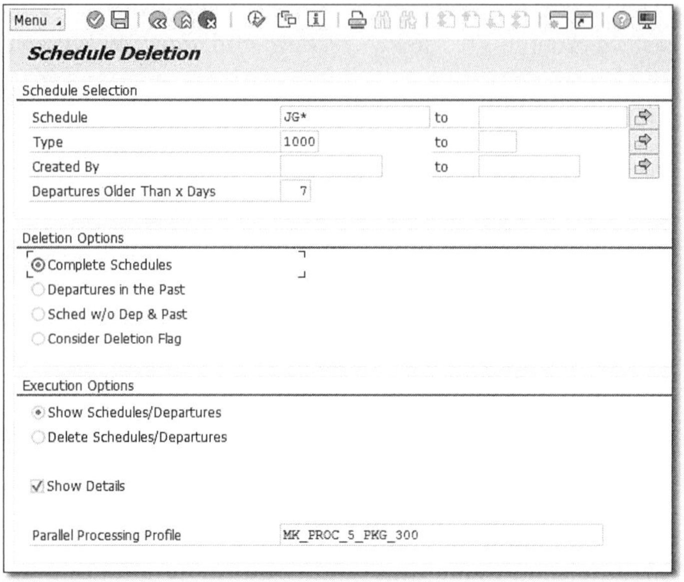

Figure 3.39 Schedule Deletion Report

Converting Old Schedules to New Schedules

SAP TM 8.1 and its predecessor releases contain a different schedule object with less functionality and a different maintenance screen.

With SAP TM 9.0 covering all the functionality of the old schedule by the (new) schedule as just described, the old object became obsolete and cannot be used anymore. However, customers who are upgrading from a previous release to SAP TM 9.0 can keep their old schedule instances in the system and transform them into instances of the new schedule by a report. See SAP Note 1682472 for more details about the conversion from old to new schedules.

3.2.5 Transshipment Locations

In global transportation networks, an individual transportation order is usually executed using multiple modes of transport in sequence. The mode of transport is changed at transshipment locations, where the goods are unloaded and reloaded.

Let's walk through the steps of an example in which goods are transported from Germany to the United States:

1. A truck transports the goods from the start destination in Germany to the port of Hamburg.
2. At the port, the goods are unloaded from the truck and reloaded onto a ship.
3. The ship carries the goods to the port of Newark.
4. The goods are unloaded and reloaded onto a truck.
5. The truck transports the goods to their destination.

In this scenario, the ports in Hamburg and Newark serve as transshipment locations. Whereas goods are unloaded and reloaded from and onto different modes of transport in this example (truck and ship), there are also transshipment scenarios where the mode of transport remains the same. This is the case, for example, with collection trips by truck, where the goods collected are unloaded and reloaded onto other trucks at a local depot, and these trucks then carry the goods along a long-haul route to another depot.

Explicit definition of transshipment locations

To define a location as a transshipment location, select the menu path Master Data • Transportation Network • Locations • Assign Transshipment Location. Define a set of locations or transportation zones and a transshipment location, and choose Create/Update Assignments to assign the transshipment location to the set of locations or transportation zones, as shown in Figure 3.40.

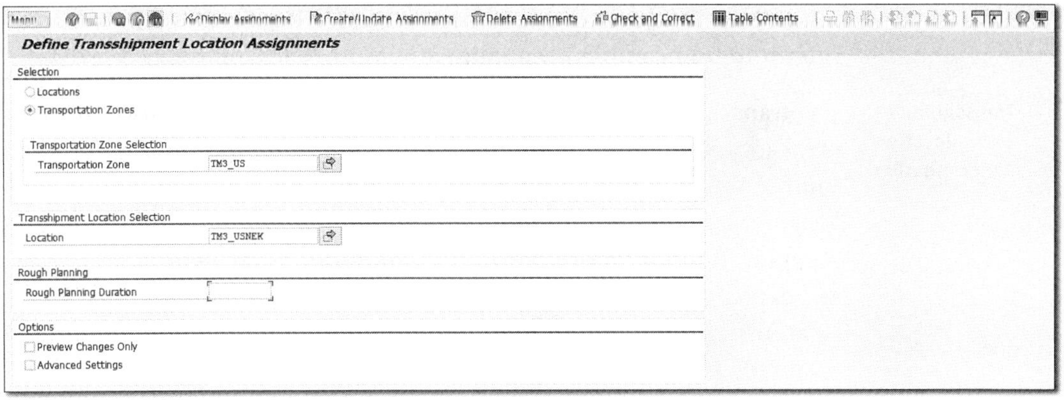

Figure 3.40 Defining Transshipment Location Assignments

You can also display already defined transshipment location assignments, as shown in Figure 3.41.

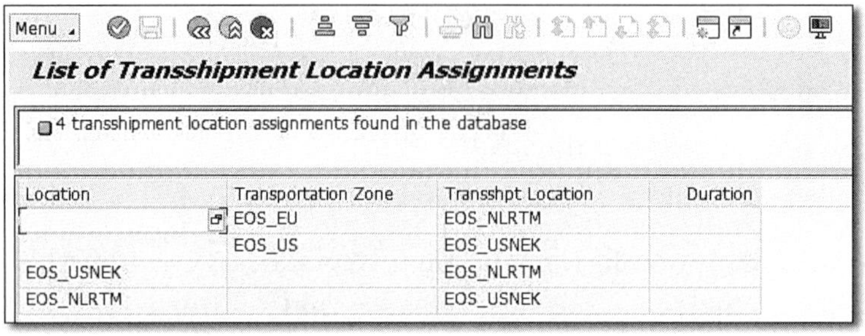

Menu

List of Transshipment Location Assignments

4 transshipment location assignments found in the database

Location	Transportation Zone	Transshpt Location	Duration
	EOS_EU	EOS_NLRTM	
	EOS_US	EOS_USNEK	
EOS_USNEK		EOS_NLRTM	
EOS_NLRTM		EOS_USNEK	

Figure 3.41 Transshipment Location Assignments

Rough planning

For each transshipment location assignment, you can also define a duration that is considered by rough planning (see Chapter 6, Section 6.4.2). If rough planning is activated and the duration is maintained, transportation from the transshipment location to the assigned locations and zones—and vice versa—is possible with the defined duration, without definition of a schedule or a transportation lane. This allows the modeling of intermodal scenarios in which, for example, pre-legs and subsequent legs are not planned in detail but should be planned in a rough way based on the maintained (rough) transportation duration instead. The rough planning concept represents a simple version of reachability for automatic planning, but no freight documents are created for the rough planned parts of the transportation. Therefore, we do not consider it to be a complete way for expressing reachability.

Transshipment location determination

The transshipment location determination finds the relevant transshipment locations for a given source and destination location. For a given source, it determines the transshipment locations that are explicitly assigned to the location and its zones. From the set T of these transshipment locations on, more transshipment locations are determined if they are explicitly assigned to one location of T. Similarly, transshipment locations are determined from the perspective of the destination location. This process determines the transshipment location network connecting the source and destination, and you can define a limit on the number of consecutive transshipment locations leading from source to destination (see Chapter 6, Section 6.4.2).

Explicit definition along default route

Note that default routes can also be used to define transshipment locations in an explicit fashion. However, the transshipment locations in a default route apply only if the transportation is planned along the default route. (Section 3.2.6 has more details on default routes.)

Implicit definition by automatic connection determination

In intermodal scenarios, many different schedules may touch one specific port. To allow the port to be used as a transshipment location for connecting two schedules arriving at and departing from the port, respectively, the port would have to be defined as a transshipment location for all other locations covered by the schedules. In a global ocean schedule network that connects hundreds of ports, this would be an enormous, error-prone maintenance task. In a global air freight network covering thousands of airports, the task would be even larger. Therefore, SAP TM offers automatic connection determination, which can be activated in the optimizer settings (see Chapter 6, Section 6.4.2). This means that any location covered by two schedules can serve as a transshipment location for all locations covered by the two schedules and therefore represents an implicit transshipment location definition. This significantly reduces the maintenance required for defining transshipment locations. If even the pre-legs and subsequent legs in a global network are served by schedules, no explicit transshipment location assignment is required anymore. Only if pre-legs and subsequent legs are *not* covered by schedules, then do the transshipment locations entering the schedule network have to be defined explicitly.

Transportation network = reachability network + transshipment network

Reachability and Transshipment Locations

There are three ways to define that location A can be reached from location B:

- Define a transportation lane from A to B, either on the location level or on appropriate zones.
- Define a schedule that goes from A to B, either directly or indirectly via other locations.
- Define a (rough) duration for a transshipment location assignment of A to B or vice versa, and use rough planning.

There are three ways to define B as a transshipment location for location A:

- Explicit assignment, as just described
- Explicit assignment along a default route, as described in Section 3.2.6
- Implicit assignment, as just described

> In scenarios that include transshipments (e.g., an intermodal scenario from Asia to Europe), you have to ensure that reachability and transshipment locations are defined appropriately. In scenarios without transshipments (such as a local distribution scenario within Bavaria, Germany), you have to ensure that the reachability within Bavaria is defined appropriately.
>
> If you want to transport goods from location A to location C, using location B as a transshipment location, you must implicitly or explicitly define B as a transshipment location for A or C and define reachability from A to B and from B to C, using any of these possibilities.
>
> From this perspective, the reachability network and the transshipment network together form the transportation network, where the reachability network is the set of all locations and their reachable relations, and the transshipment network is the set of all locations and their transshipment relations.

Constraints for transshipment duration

You can define a minimum goods wait time and a maximum goods wait time for a location; these are then taken into account in the scheduling of activities during the transshipment of a transportation order. The goods wait time of a transportation order is the length of time between the end of unloading and the start of reloading at the transshipment location.

For example, a minimum goods wait time of one day ensures that the goods delivered remain at the transshipment location for at least one day before they are transported further. With a maximum goods wait time of 72 hours, the goods delivered must not remain at the transshipment location for more than 72 hours before they are picked up.

3.2.6 Default Routes

Purpose

In complex transportation networks, there may be numerous possible paths from one source to one destination. For many businesses, only a few of these possible paths are reasonable; often, business experience or careful analysis of the network leads an organization to know the most desirable route for a given source and destination. Therefore, many transportation businesses are organized by default routes; these are static rules defining how goods are to be routed geographically through the global transportation network. For intermodal transports, default routes can predefine the sequence of transshipment locations, as mentioned in the last section. For truck and trailer scenarios involving dynamic recoupling, default routes can predefine the sequence of coupling and uncoupling locations for the trailer unit.

A default route defines a location sequence for a given source and destination, which can be locations or zones. For a transport from a source to a destination, the default route serves as a template guiding the transportation through the network.

Determining a matching default route

Given a source location and a destination location, the system determines matching default routes and chooses the most specific one. A default route matches a given source location S and destination location D if the default route's source is location S or a zone containing S and the default route's destination is location D or a zone containing D. If there are multiple matching default routes, the system chooses the most specific one; a direct location match is more specific than a zone match.

Source zone/destination zone

The ability to define default routes with a source zone and a destination zone significantly reduces the default route maintenance efforts. Let's consider an ocean scenario where goods from Germany are transported to the United States via ports in Hamburg and Newark. This could be expressed by maintaining many location-to-location default routes, each having one source location in Germany and one destination location in the United States. Assuming 20 source locations in Germany and 50 destination locations in the United States, this would result in 1,000 default routes. Defining one zone for Germany and one zone for the United States, you can define one default route from Germany to the United States, via the transshipment ports Hamburg and Newark. This zone-to-zone default route has the same effect as the 1,000 location-to-location default routes. Another advantage of the zone-to-zone default route is that you don't have to create new default routes if new source or destination locations show up in the source and destination zone.

Defining default routes

To define a default route, select the menu path Master Data • Transportation Network • Default Route • Create Default Route and enter the default route type, which you can select from the defined default route types. Besides maintaining the description and the validity period, you can restrict the applicability of a default route by selecting the DANGEROUS GOODS checkbox and setting the SHIPPING TYPE (see Figure 3.42). By maintaining the TRANSPORTATION STOPS list, you can define the following per stage: the mode of transport, the stage type, the schedule, and the carrier, which represent constraints for planning. For example, if a schedule is defined for a specific stage, planning will only create a transportation plan using the given schedule for this stage. Similarly, if a carrier is defined for a specific stage, it will be used by planning.

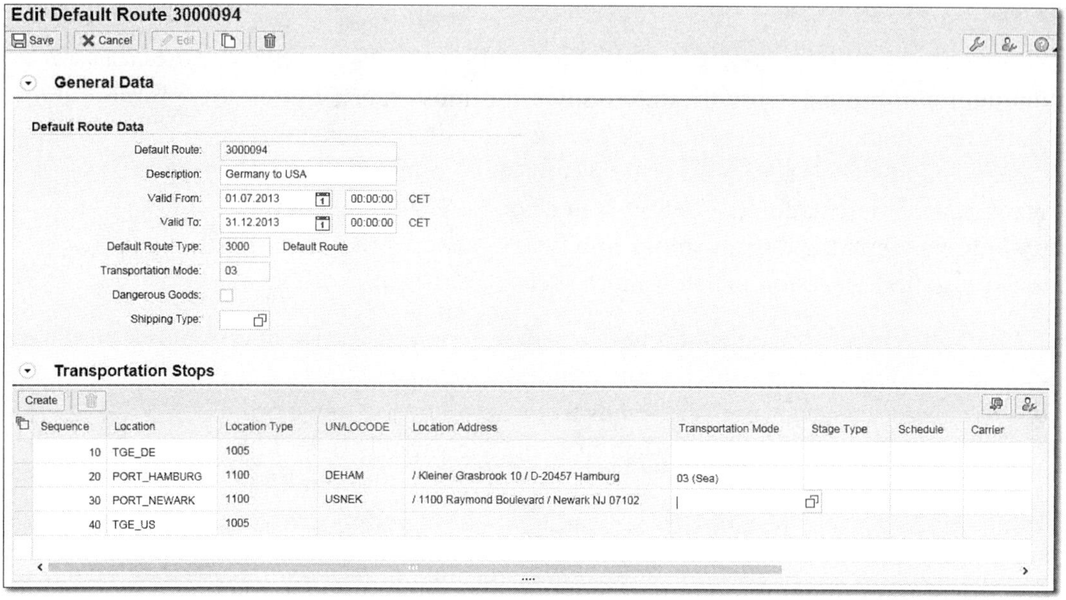

Figure 3.42 Default Route

Per stage, you can also define the responsible planning and execution organization and whether this organization has to explicitly check any freight unit assignments for freight documents created for this stage. Such definition of your organization interaction model allows you to structure responsibilities for different business units in your company (e.g., for ocean freight bookings on the main legs of your default routes; see Chapter 5, Section 5.4.1).

Default route types

Default route types can be maintained by selecting the IMG menu path SAP TRANSPORTATION MANAGEMENT • TRANSPORTATION MANAGEMENT • MASTER DATA • TRANSPORTATION NETWORK • DEFINE DEFAULT ROUTE

TYPES. For a default route type, you can define the mode of transport that is passed on to default routes created for this type. Similar to the shipping type, a default route can be applied to a forwarding order if both have the same mode of transport or one mode is undefined. You can also specify whether the default route is mandatory—that is, any freight unit with stages built according to a default route must use the mandatory default route. Either the freight unit stages were built by the corresponding freight unit building option, or they were created out of a forwarding order by the schedule selection functionality using default routes.

Use cases

Default routes can be used in various contexts for defining the stages of freight units, transportation units, and forwarding orders' actual routes, the latter including the selection of schedules. These use cases are described in the following sections.

Creating Freight Unit Stages

During freight unit building, freight unit stages can be automatically generated based on a default route. Given a freight unit at hand, the system determines a default route based on the freight unit's source location and destination location. Then the freight unit stages are created according to the default route's location sequence. Thus, the location sequence defines the transshipment locations to be used when moving goods from source to destination.

This functionality is useful if the transportation of goods is to be organized strictly in a rule-based fashion. For example, if goods moving from Germany to the East Coast of the United States will always be transported via the ports in Hamburg and Newark, this can be modeled by a default route with the source zone representing Germany as the first stop, the port in Hamburg as the second stop, the port in Newark as the third stop, and the destination zone representing the East Coast of the United States as the fourth stop.

As an alternative to applying default routes during freight unit building, the transportation cockpit allows you to apply default routes to selected freight unit stages, too (see Chapter 6, Section 6.3.5). Each selected freight unit stage is replaced by freight unit stages according to the most specific default route. Compared to the previous use case of strictly organizing transports according to default routes, this use case considers default routes as one option to guide the transportation through the network. However, the user can explicitly define any path through the network and use the default routes only on-demand (e.g., if he is not sure about a good path through the network). Note that the cockpit allows splitting and merging of freight unit stages, so the system always offers the planner the manual choice to change freight unit stages that were already created by default routes.

Creating Transportation Unit Stages

Freight unit building can also be used to create transportation units by specifying a transportation unit type as a document type in the freight unit building rule (FUBR). Like freight unit stages, transportation unit stages can automatically be generated based on a default route. Given a particular transportation unit, the system determines a default route based on the transportation unit's source and destination locations. Then transportation unit stages are created according to the default route's location sequence.

Different transportation unit types can cover different modes of transport, such as the trailer unit for a road and the railcar unit for a railway. For a trailer unit, the location sequence of the determined default route defines the locations where the trailer can be uncoupled from and coupled to a truck; thus, the corresponding trailer unit stages are to be assigned to road freight orders. For a railcar unit, the location sequence of the determined default route defines the locations where the railcar is uncoupled from and coupled to a train, and hence the corresponding railcar unit stages are to be assigned to rail freight orders.

In certain transportation scenarios, the trailers in the transportation network are moved back and forth along a predefined chain of coupling and uncoupling locations. This scenario can be modeled by a default route applied during freight unit building to create trailer unit stages. This rule-based approach simplifies manual planning a lot, allowing the planner to start with assigning trailer unit stages to road freight orders, and freeing the planner from manually defining trailer unit stages. As with freight units, the planner can also manually apply default routes to selected transportation unit stages in the transportation cockpit.

Creating Forwarding Order Stages and Selecting Schedules

For a forwarding order, the system can determine the available schedules per forwarding order stage, allowing the user to choose from the alternative schedules. After a schedule is chosen, its data (e.g., departure date and time and arrival date and time) is taken into the stages of the forwarding order's actual route.

The schedule determination also considers default routes. If a matching default route is found, new stages are proposed for the forwarding order according to the location sequence of the default route. If the default route considers more specific information for a stage—such as a carrier or a schedule to be used—then only schedules for this carrier or only the predefined schedule are offered to the user for the stage.

When the user has chosen one alternative, the proposed forwarding order stages (in the actual route) are created according to the default route, similar to the creation of freight unit stages and transportation unit stages by a default route. However, additional information such as carrier and schedule is also taken from the default route into the forwarding order stages.

The location sequence of the default route defines the transshipment locations to be used for the forwarding order. When creating freight unit(s) for the forwarding order, the freight unit stages are created according to the forwarding order stages, also considering the carriers and schedules that potentially were defined by the default route.

This functionality enables scenarios in which the actual routes of forwarding orders are predefined in a rule-based fashion. For example, such scenarios are relevant for ocean freight-forwarders in a less than container load (LCL) scenario, in which the transportation is organized by rules guiding freight through the network. In addition to defining the sequence of ports to be used for ocean transportation, even an ocean carrier and its ocean carrier schedule can be predefined per stage in a default route.

3.2.7 Trade Lanes

Relationships to customers and carriers

The relationships of your transportation business to your carriers and customers are usually structured in a geographical fashion. For example, you may have a forwarding agreement with a certain customer from Germany to North America, a freight agreement with an ocean carrier from Germany to the United States, and an allocation with another carrier (airline) from the airport in Frankfurt to the airport in Chicago, and you may have defined a business share between your two favorite road carriers within Germany to be 40% for the first carrier and 60% for the second carrier.

Purpose

Trade lanes specify the basic geographical relationships that structure all these business objects. A trade lane can have a mode of transport and a means of transport. The trade lane defines a direction for transportation, which is characterized by the orientation and a source and a destination (both being either a location or a transportation zone).

Defining trade lanes

You can define a trade lane by selecting the menu path Master Data • Transportation Network • Trade Lane • Create Trade Lane and entering a trade lane type. Then you can specify its description, mode of transport, means of transport, and the geographical definition consisting of the orientation, source, and destination (see Figure 3.43).

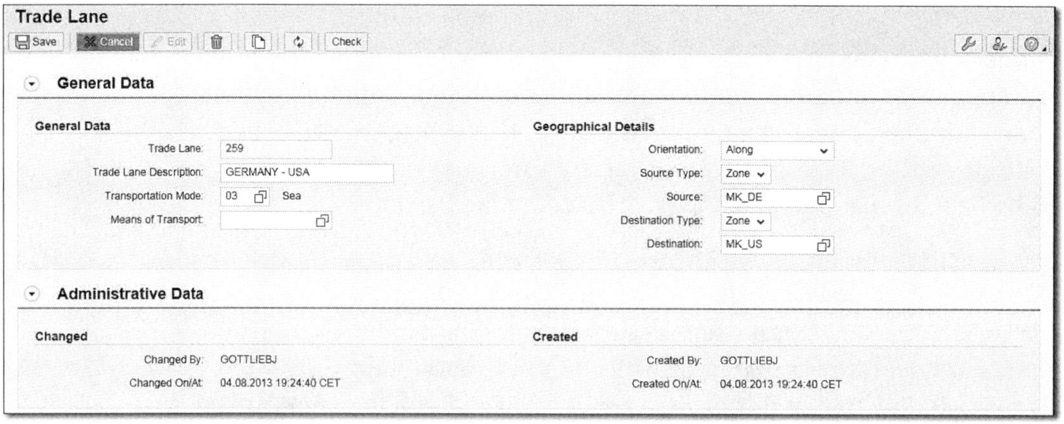

Figure 3.43 Trade Lane

Orientations

You can maintain the following orientations:

- **From**
  Covers transportation starting from a selected location or zone and reaching any other location.
- **To**
  Covers transportation starting from any location and reaching a selected location or zone.
- **Within**
  Covers transportation within a zone (i.e., starting from a location in the zone and reaching another location in the zone).
- **Along**
  Represents transportation starting from a selected location or zone and reaching another selected location or zone.
- **Inbound**
  Covers transportation within and to the selected zone (i.e., all transportation that ends in the zone and starts inside or outside the zone).
- **Outbound**
  Represents transportation within and from the selected zone (i.e., all transportation that starts in the zone and ends inside or outside the zone.

Trade lane type

The trade lane types can be defined via the IMG menu path SAP Transportation Management • Transportation Management • Master Data • Transportation Network • Trade Lane • Define Trade Lane Types. Here, you can specify the number range interval per trade lane type.

Transportation lane versus trade lane

Transportation Lane versus Trade Lane

Although both concepts characterize the direction of transportation between a source and a destination—both being locations or zones—they have different aims.

The transportation lane defines reachability—that is, which means of transport can be used to transport goods from a source to a destination and which carriers are available per means of transport. In addition, it contains many parameters that affect transportation planning and optimization and carrier selection. Thus, the transportation lane defines how transportation can be planned and executed in your network, so it has a significant impact on transportation processes.

The trade lane is just the geographical basis for multiple business objects that define the relationships to carriers and customers. It can be defined for a means of transport or a mode of transport or even without reference to a mode of transport. Its orientation concept is more generic than the direction in a transportation lane. The trade lane itself does not contain any control parameters, so it doesn't influence any transportation processes. However, in conjunction with objects such as forwarding agreements, freight agreements, allocations, and business shares, it is the key element to structure your business from a geographical perspective.

3.2.8 Transportation Network Cockpit and Overview POWL

Let's take a look at two complementary tools for providing transparency of the transportation network and searching and maintenance capabilities: the transportation network cockpit and the transportation network overview POWL.

Transportation network cockpit

The transportation network cockpit allows the display of the transportation network or parts of it on a geographical map, as well as the creation of master data objects such as locations and trade lanes on the map or starting the maintenance user interfaces described in the previous sections. Visualizing the network on a map is probably the best way to understand the network structure and to verify its correctness and completeness. Certain transportation network master data inconsistencies can be identified easily on the map, such as incorrect geo-coordinates for locations or missing geo-coordinates, which are frequently shown as a spot with coordinate (0,0) in the Atlantic Ocean, where there is definitely no island.

Search capabilities

By following the menu path Master Data • Transportation Network • Transportation Network Cockpit, you can define search criteria for locations, transportation zones, transportation lanes, trade lanes, schedules, and default routes. When you click on the Search button, the resulting objects are displayed on the map, as shown in Figure 3.44 for a set of schedules matching the schedule search criteria. You can iteratively add search results to the map and clear the map to start with a new search. You can focus on the map and the master data selection screen area, respectively, by hiding the other screen area.

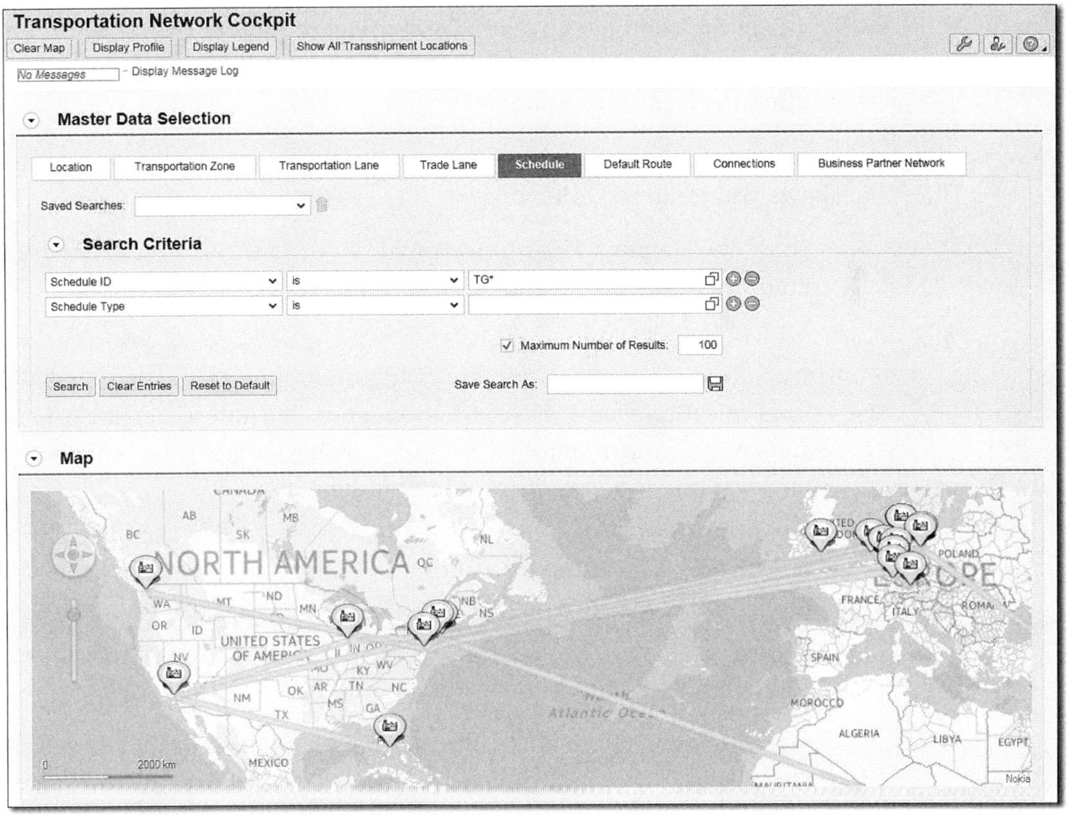

Figure 3.44 Transportation Network Cockpit

If you want to see the complete network offered to you by a carrier, select the BUSINESS PARTNER NETWORK tab to show all network master data for the chosen carrier. The set of all transshipment locations can be displayed by the corresponding button.

Searching for connections

Use the CONNECTIONS tab to search for paths from a source location to a destination location, which allows easily checking whether your network definition is complete from a reachability and transshipment location perspective. The found paths are displayed on the map and, in addition, if there is no direct reachability from source to destination, this is displayed as a red arc. If you expected connections to exist, but they are not shown, this helps identify whether a transshipment location or a schedule or a transportation lane is missing for the connection. Because the network definition is crucial for automatic planning, we recommend verifying the completeness and correctness of your network definition by the transportation network cockpit in case automatic planning does not find the transportation plan you expected.

More functionality

When you right-click on any object on the map, the context menu allows you to do the following:

- Show the details of the object, which opens the maintenance user interface of the object and allows editing.
- Hide the object.
- For a location, you can find assigned transshipment locations, related transportation zones, transportation lanes, schedules, default routes, and connections to another location; create a trade lane referring to the location; or move the location on the map.
- For a transportation zone, you can find assigned transshipment locations, related locations, transportation lanes, schedules, and default routes and create a trade lane referring to the transportation zone.
- Display arc-based objects such as transportation lanes, trade lanes, schedules, and default routes on the street level or as a straight line.

Using the context menu after right-clicking on the map itself, you can do the following:

- Search for addresses and display them on the map.
- Define whether arcs will be displayed as straight lines or on the street level.
- Remove highlighting from the results from the last search.
- Create a location at the current position on the map.
- Personalize the map to your needs. For example, specify the initial area shown on the map (default is the world) and choose among the map types (preconfigured maps from various GIS vendors).

Transportation network overview POWL

Whereas the transportation network cockpit represents the geographical view of the network, the transportation network overview POWL provides queries and worklists for all schedule types delivered in standard, default routes, and trade lanes.

As shown in Figure 3.45, you can select the menu path MASTER DATA • TRANSPORTATION NETWORK • OVERVIEW TRANSPORTATION NETWORK and choose the query for master flight schedules, for example. With standard POWL sorting, filtering, and personalization capabilities, you can quickly navigate through master flight schedules that are relevant for you. Many key characteristics are shown here, such as airline code, flight number, source and destination, number of intermediate stops (to quickly identify multistop flights), and an indicator for connection master flight schedules. The reference data status allows you to identify master flight schedules that refer to updated underlying carrier flight schedules and that may need manual adjustments (see Chapter 5, Section 5.1.3 for schedule capacity management). For any set of schedule instances selected in the result list, you can create freight documents and generate departures. You can also display, edit, copy, or delete a selected schedule instance or create a new one.

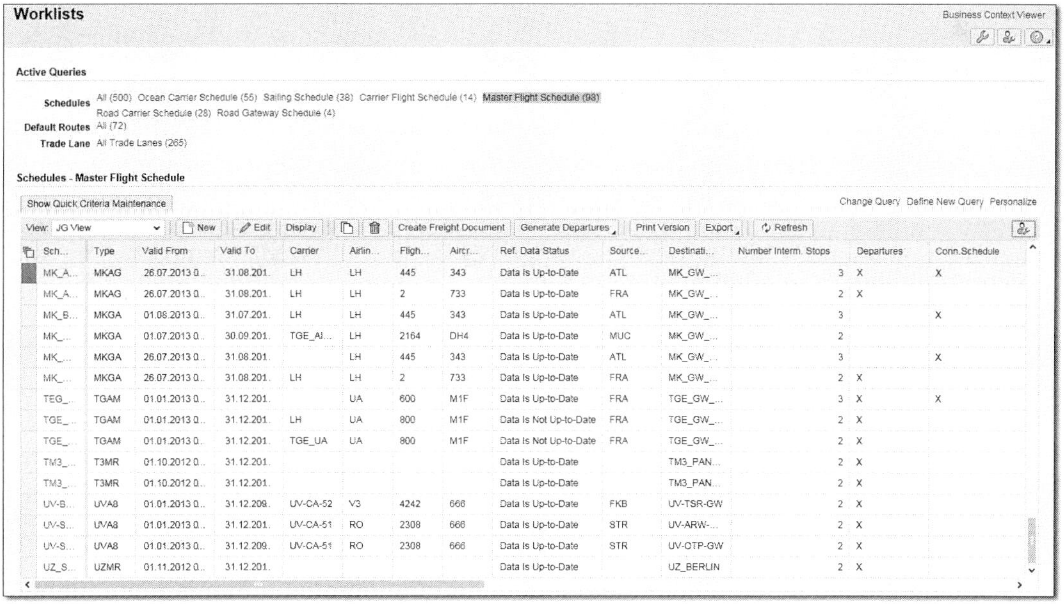

Figure 3.45 Transportation Network Overview POWL with Query for Master Flight Schedules

3.2.9 Integration of Geographical Information Systems

A *geographical information system* (GIS) captures and stores geographical data, enabling data analysis and visualization. In the context of transportation, it allows geo-coding, distance and duration determination, routing, and geographic map visualization; these are essential for informed decision making.

Basic Functionalities

Geo-coding

Given an address, the geo-coordinates representing longitude, latitude, and altitude are determined. Geo-coding is the prerequisite for any decision making based on geographical data. It is required to show locations at their right positions on the map and to enable distance and duration determination and routing.

Distance and duration determination

Given a source and a destination geo-coordinate, the distance and duration to reach the destination from the source are determined. Most GIS tools allow you to configure whether you want to determine the shortest or fastest path.

Distances are required to do optimizer-based planning, to minimize the total distance traveled when transporting freight, or to adhere to distance-based constraints for trucks. Distances are also required for charge calculation and settlement, where the charges are calculated based on the distance of the transportation.

Durations are also required to do optimizer-based planning, to minimize the total duration for a transport, or to consider constraints on the total duration of a transport.

The distance and duration determination is done by shortest-path algorithms in the GIS based on a graph of the transportation network, which usually contains the junctions of streets and highways as nodes and their connections as arcs with distances.

Routing

So given a source and destination geo-coordinate, the best route to the destination from the source is determined. Here, a route is represented by a sequence of geo-coordinates, starting with the source and ending with the destination. The routing is required to determine distance and duration, but in many cases only distance and duration are required and retrieved from a GIS tool. The routing information is only retrieved from a GIS tool if it will be visualized on a geographical map, like in the transportation network cockpit described in Section 3.2.8.

Routes, default routes, routing

Routing, Default Routes, and Vehicle Scheduling and Routing

The terms *route* and *routing* are used in various contexts.

In the GIS context discussed in this section, *routing* means determining a path through the street network, which is frequently called a route and specifies which streets, highways, and paths are to be used to travel from a source location to a destination location. The path connects the source and destination by very detailed intermediate physical positions (available only in the GIS), such as crossroads, highway exits, and so on, and thus it defines how a driver can execute a transportation from source to destination.

A *default route* is a statically defined path through a transportation network consisting of locations maintained in the SAP TM system. Usually, crossroads are not maintained as locations. Thus, a default route is defined on a higher level of aggregation.

The *vehicle scheduling and routing* (VSR) optimizer solves vehicle routing problems in the transportation network specified by locations in SAP TM. For a given fleet of vehicles and transportation orders, the optimizer decides which orders will be delivered by which vehicle. In addition, the optimizer determines the best possible route for each vehicle. Here, again, a route is a sequence of locations in the network. Like the default route, routes are defined on a higher level than the level of crossroads and so on.

While a default route is defined manually and statically, usually based on business experience, the routes obtained by the VSR optimizer are determined automatically and dynamically, based on optimization criteria. One of the criteria is the total distance traveled, which is based on pair-wise distances between the locations. Interestingly, these distances can be retrieved out of a connected GIS, which internally performs a routing task between two locations to determine the distance required by the VSR optimizer to make its routing decisions.

Geographical map visualization

The geographical map shows a set of transportation-relevant objects. These objects can be position-based (e.g., locations, zones, and resources) or relation-based (freight units, freight orders, trade lanes, and schedules). The relation-based objects can be shown as straight lines connecting two consecutive locations in an idealized way or based on the detailed routing information, which makes it possible to check the actual route along the street level. However, the higher detail level of an actual route comes with a longer response time since many more details have to be retrieved out of the connected GIS to draw the detailed segments on the street level.

Transparency on transportation network

Map visualization is extremely helpful to understand, analyze, and explain the transportation network structure, as explained in Section 3.2.8 for the transportation network cockpit. While a list of locations in a tabular representation is hard for anyone not familiar with the locations to interpret, a visualization of the locations on the map allows everybody to understand the relative geographical positions on the map and thereby also to assess neighborhood relations and distances between locations.

Interactive planning on map

When planning transports, a map of the freight units and planned freight orders enables the planner to identify geographical consolidation potentials, which are much harder to identify in tabular lists of freight units and freight orders. Moreover, the quality of an actual route of a freight order can easily be judged by checking it on a map. For example, good routes should not contain any crossing stages, which is hard to judge in a tabular stage list but is easily checked on a map. See Chapter 6, Section 6.3.5 for more details on map-based planning.

The routing can be visualized on the map, enabling the user to see which streets and highways are used to reach the destination from the source. This is helpful during execution of the transportation. For example, if information about traffic jams is also shown on the map, it enables the planner to judge whether there is a risk of delays and to trigger replanning if needed.

Integration and Technology

Open integration infrastructure

There is an existing market of GIS vendors; some offer global data, and others offer only regional data, such as for North America, Europe, or China. Some SAP TM users are familiar with certain GIS vendors when they start using SAP TM and want to continue using tools from these vendors. Some global companies even use multiple GIS tools (e.g., one per continent or region). For all these reasons, SAP TM does not contain its own GIS, but offers an open infrastructure into which any GIS tool or even multiple GIS tools can be easily integrated. This way, SAP TM users can choose which GIS tool(s) to integrate into SAP TM.

Integration technology

The integration of services for geo-coding, distance and duration determination, and routing requires ABAP knowledge. More information is contained in SAP Note 1685381, which contains a white paper that explains how to use and configure GIS services and a guide describing how to connect SAP TM to an external web service.

Direct connection via web services

It is state-of-the art to directly connect to a GIS tool by web services rather than using middleware between the GIS and SAP TM. Note that SAP TM still supports a standard integration via the GIS services contained in the SAP Internet Graphics Server (IGS) for compatibility reasons, but we don't recommend using it because it requires its own server, and its interface is tied to one vendor and uses only single distance calls; the result is acceptable performance only for small scenarios.

Map visualization by SAP Visual Business

The visualization of maps within SAP TM is covered by SAP Visual Business, which provides two-dimensional scenes on geographical maps and three-dimensional scenes such as the result of load planning for a truck, including the positions of loaded pallets in the truck. From a GIS perspective, SAP Visual Business can visualize any geographical map accessible by REST-based web services. See Chapter 2, Section 2.3.6 for more details on installing SAP Visual Business and *http://scn.sap.com/docs/DOC-43251* for configuring or changing a map provider for SAP Visual Business.

Running a GIS integration

While integrating a GIS provides a lot of value for many scenarios, this value has a cost, including GIS license fees and the effort to develop and configure the integration. We recommend that you build up and preserve know-how about the GIS tools you use because GIS products evolve, and you may want to integrate more GIS services later. Many GIS vendors offer hosted services and local installations. Especially for these installations, it is important to have local know-how to keep the system up and running.

GIS quality, stability, performance

We also recommend that you check the accuracy of the integrated GIS tool by explicitly checking known examples for geo-coding, distances, routing, and map visualization. It is also helpful to monitor the availability of the GIS integration because downtime may have a severe impact on transportation planning or charge calculation. From a performance perspective, we recommend using mass calls for distance and duration determination because determining a whole matrix in one step is an order of magnitude faster than determining a distance matrix by a set of individual calls. This is especially important when running optimization scenarios involving a large number of locations per optimizer run.

Not using GIS at all?

Of course, you can run SAP TM without GIS. In this case, geo-coordinates for locations are determined on a region and country level and can be maintained manually to get higher accuracy without a GIS. However, this may be a tedious, error-prone task. Without GIS, the distance determination can be done based on the straight-line distance, which is a reasonable approximation of the street-level distance for some scenarios. Alternatively, the distance can be maintained explicitly on the transportation lane level, which, again, is hard, error-prone work prone—especially if medium or large transportation networks are modeled. Nevertheless, all these options without GIS are also valid, so it's up to the user to judge the value of a GIS integration for the business, to decide whether to integrate a GIS tool, and to choose the right GIS vendor and option.

3.3 Resources

Resources play a central role in planning and execution in SAP TM and are required to physically transport goods within the transportation network.

Use of resources

Resources' attributes and configuration specify the available capacity and the hours in which they can be operated. Resource data is relevant to the planning of order dates, taking into account working times and the available capacities of the resources. In conjunction with booking, resources offer the capacity needed to perform transportation activities on freight units, such as transportation, loading, and unloading activities. Therefore, they might have different capacity dimensions, be able to contain shifts, and be assigned to a means of transport. In contrast to the general master data, resources must always be created in SAP TM.

SAP TM currently recognizes the following *resource types*:

- Vehicle resource
- Calendar resource
- Handling resource

Resource types

Goods are loaded onto and unloaded from vehicles and transportation unit resources and transported around the transportation network. Calendar resources are used to map the operating hours for a location, during which goods may be dropped off or picked up from this location. Handling resources map both operating hours and time-dependent capacities. These define the number of loading or unloading activities that can be executed simultaneously.

Before we explain the configuration of these resources (see Section 3.3.4), we briefly explain their main characteristics and relevance within the transportation process.

Drivers in SAP TM 9.0.

As of SAP TM 9.0, drivers are defined as business partners. Although they are often handled the same way as resources, we have decided to briefly explain them in the context of *employees* in Section 3.1.4. In addition, following the current menu structure in SAP TM, we briefly mention transportation units in Section 3.3.3.

3.3.1 Vehicles and Trailers

Vehicles are moving resources, including trucks, railcars, planes, and ships, which can transport goods between locations. All vehicles are assigned to a means of transport, which represents a class of vehicle and is used to define groups of vehicles.

Means of Transport

Because vehicles are moving resources, such as planes, trucks, and vessels, that can be used to transport goods between locations, all vehicle resources are assigned to a specific *means of transport*. This means of transport not only describes the characteristics of specific vehicle types, but also represents a class of vehicle, how the physical transportation between locations is executed, and the method of transportation.

The *transportation mode* is a code that is assigned to the transport method (means of transport) and used to indicate the actual mode of transportation and how the goods are transported. Therefore, a vehicle resource is an instance of a particular means of transport or a group of identical instances of a means of transport that can provide transportation services within the transportation network, making use of a specific transportation mode. To simplify planning, each means of transport is assigned to a single transportation mode, such as sea, air, or road.

Take, for example, a 48-foot dry container, a 60-foot rail boxcar, and a refrigerated truck. Each means of transport is assigned to a transportation mode and provides further characteristics that are relevant and are taken into account during the transportation planning process. Characteristics such as vehicle capacity, speed, cost structure, geographic availability, whether there is a company-owned fleet, temperature controls, and other parameters can be used to group vehicles in a means of transport. In that context, the vehicle resource is typically used to map the capacity and availability of vehicles, considering not only their dimensions, but also the combinations.

Therefore, we first explain the relationship between the means of transport and the transportation mode, the characteristics of vehicles and resources, their combinations and compartments, and then how they can be used to model a hierarchy of means of transport. In this context, the means of transport is assigned to vehicle resources in SAP TM master data. It is also assigned to a *transportation lane* (see also Section 3.2).

Transportation mode

Figure 3.46 shows the relationship between vehicle resources, means of transport, and transportation modes.

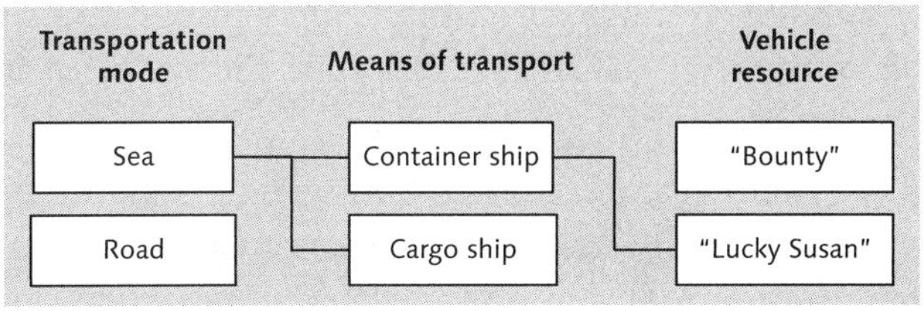

Figure 3.46 Relationships between Transportation Mode and Resources

We have seen that the transportation mode is a code defined to indicate the actual mode of transportation. To indicate how a product is to be transported, a *mode of transport category* and one of four *transportation mode categories* is assigned to each transportation mode: road, rail, sea, and air, as shown in the TModCat column in Figure 3.47.

To maintain the transportation mode in Customizing, select the IMG menu path SAP Transportation Management • Transportation Management • Transportation Lane • Maintain Transportation Mode.

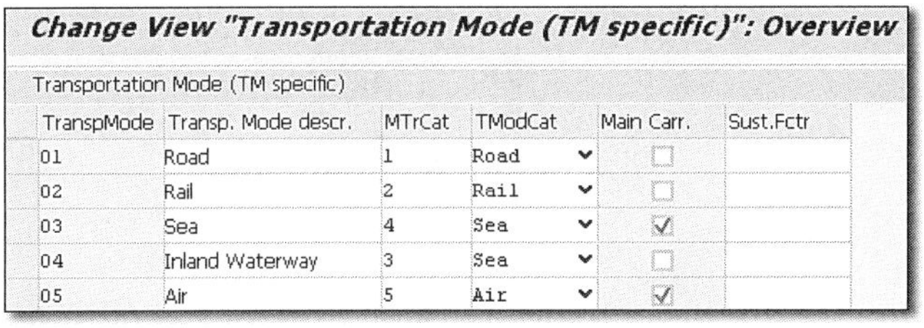

Change View "Transportation Mode (TM specific)": Overview

Transportation Mode (TM specific)

TranspMode	Transp. Mode descr.	MTrCat	TModCat	Main Carr.	Sust.Fctr
01	Road	1	Road	☐	
02	Rail	2	Rail	☐	
03	Sea	4	Sea	☑	
04	Inland Waterway	3	Sea	☐	
05	Air	5	Air	☑	

Figure 3.47 Configuration of the Transportation Mode in SAP TM

Transportation mode in SAP ERP

In SAP TM this code is assigned to the transportation method and should correspond to the transportation mode used in SAP ERP. To double-check the transportation mode in SAP ERP, select the IMG menu path Logistic Execution • Transportation • Basic Transportation Functions • Routes • Define Routes • Define Modes of Transport.

Means of transport configuration

To maintain a means of transport in Customizing, select the IMG menu path SAP Transportation Management • Transportation Management • Master Data • Resources • Define Means of Transport.

Figure 3.48 shows the maintenance screen where, in addition to the name, a description, and a standard code, the following information can be defined:

- The transportation mode is used to select the vehicle resources that are to execute the transportation. In this example, the resource is used for road transportation.
- The resource class is used to group resources that have the same characteristics or types (e.g., vehicle resources or handling resources). Please also refer to Section 3.3.4.
- The higher-level means of transport in the means of transport hierarchy, which may be undefined.
- An average speed and a low, medium, and high speed (which represent the average speeds when traveling in developed areas, on highways, and on motorways).
- A distance factor, which defines the ratio between the distance actually covered and the straight-line distance and whether a resource is traveling based on a fixed schedule (e.g., sailing or rail schedule).

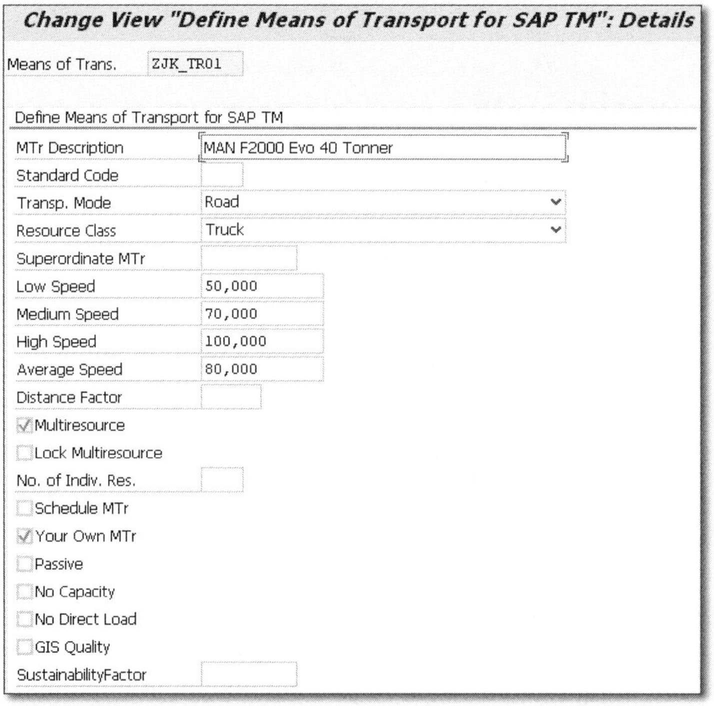

Figure 3.48 Means of Transport Configuration in SAP TM

General resource parameters

You can also specify that the means of transport is part of your own fleet (YOUR OWN MTR) or belongs to an external transportation service provider. In addition, a vehicle resource can represent a specific vehicle or be configured as a multiresource, both *active* and *passive*.

Active and passive vehicles

The means of transport not only determines whether a resource can be modeled with capacity constraints, but also whether, for example, a vehicle resource can move itself or has to rely on another resource to be pulled. We therefore distinguish between two kinds of resources.

- *Passive resources* such as trailers, barges without a tug, and railcars without an engine cannot move by themselves and therefore need to be coupled to an active resource (e.g., truck) that can move by itself and in this case tows the passive resource (trailer).
- *Active resources* are self-propelled, typically motorized vehicles such as trucks, ships, forklifts, and so on. An active vehicle can be driven independently of other vehicles, whereas a passive vehicle must be joined to an active vehicle in order to move. In this case, we refer to a vehicle resource combination, consisting of an active vehicle and one or more passive vehicles. Vehicle combinations are frequently used in overland transportation. In Europe, trucks often have just a single trailer. In Australia and North America, however, vehicle combinations often include several trailers. These combinations are usually known as *road trains*.

Multiresource flag

With the MULTIRESOURCE flag you can specify whether each resource has to be individually created or whether a single, active resource can be reused multiple times. This is useful if many resources (e.g., a 40-ton trucks) exist for a specific means of transport. Here you specify whether the resource is passive (that is, if it is incapable of moving independently) and whether it has transportation capacity on its own (please also refer to Figure 3.50).

GIS integration/ distance calculation

The GIS QUALITY option determines whether a GIS tool is used for the distance and duration determination. If this option is selected, the GIS tool calculates the distance and duration. The duration determination takes account of the low, medium, and high speeds specified. If a GIS tool is not used, the distance is calculated as a product of the straight-line distance and the distance factor, whereas the duration is based on the average speed entered.

Means of transport hierarchies

You can assign different means of transport to nodes in a means of transport hierarchy, in which specific means of transport are subordinate to general means of transport. This type of hierarchy is useful for a compact description of the transportation network because properties of the subordinate means of transport can be inherited from the higher-level means of transport. The example in Figure 3.49 shows a means of transport hierarchy with two independent means of transport (truck and ship), each with subordinate, specific means of transport assigned to them.

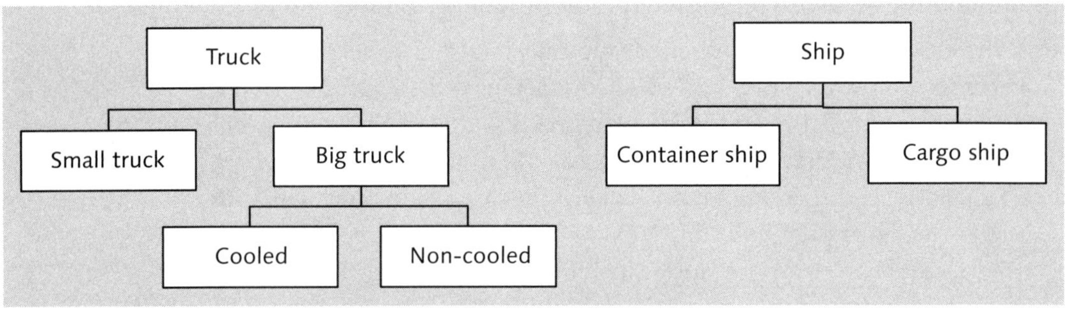

Figure 3.49 Example of a Means of Transport Hierarchy

Vehicle Combinations

Vehicle combinations are generally used to transport a larger quantity of goods with an active vehicle. In other words, the passive vehicles usually provide additional transportation capacity. In some scenarios, the trailers provide the only transportation capacity, and the active vehicles (the trucks) serve only to move the trailers. The coupling and uncoupling of vehicles is mapped in the system with coupling and uncoupling activities, which couple or uncouple a passive and an active vehicle. A means of transport combination is an instance of a particular means of transport or a group of identical instances of means of transport. They typically use a model-specific truck and trailer combination or an entire external fleet in one step.

Compartments

When mapping the loading space of a vehicle resource, you can use compartments to indicate a certain division of a truck or trailer into smaller units. Vehicles with compartments are often used to transport mixed goods. In retail, refrigerated and unrefrigerated goods are often transported in a single vehicle containing both cooled and uncooled compartments. Vehicles that deliver fuel to service stations have approximately five different compartments for storing different fuel types, like petrol and diesel. Using vehicles with compartments in these industries generally enables a significant reduction in transportation costs because, without these compartments, several vehicles would be required to deliver goods to individual locations.

Resource combinations

Figure 3.50 shows an example of a vehicle combination consisting of an active vehicle (truck) and a passive vehicle (trailer), both of which have three compartments. The arrows indicate how the quantity of goods loaded is dictated by the capacity restrictions at the different levels. The quantity of goods loaded into compartment C1 is limited by the capacity of C1, and the same applies to all the other compartments, C2 to C6. The quantity of goods loaded into compartments C1 to C3 is limited by the capacity of the active vehicle, whereas the quantity of goods loaded into compartments C4 to C6 is limited by the capacity of the trailer. Finally, the quantity of goods loaded into all compartments (C1 to C6) is limited by the capacity of the vehicle resource combination.

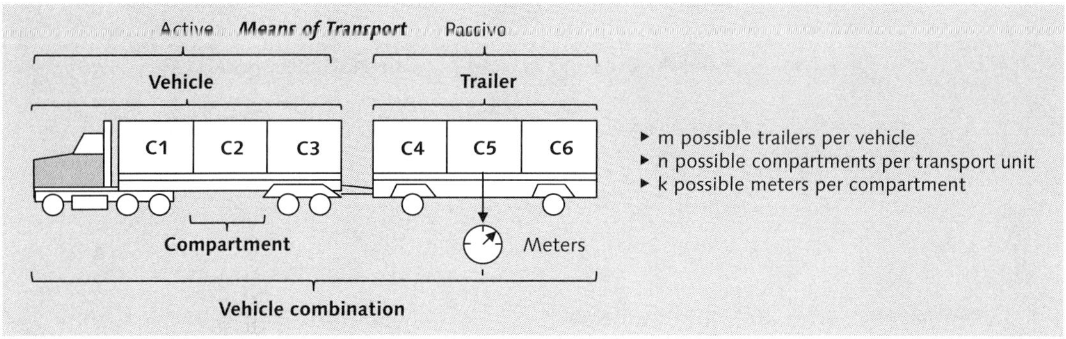

Figure 3.50 Vehicle Combination and Compartments

In this example, compartments indicate a division of the vehicle resource and break trucks and trailers into smaller units. They are typically used to improve transportation planning for oil and gas, chemical, retail, and consumer products. In addition to the definition of capacity constraints, compartments can also be used to assign dangerous goods to certain areas of a truck.

Combination attributes

For each means of transport combination, you can define a total capacity that represents the maximum load that can be transported (this does not include the weight of the active and passive vehicles themselves in the means of transport combination). You can also assign attributes to a means of transport combination to define incompatibilities. Therefore, vehicle combinations can be used to do the following:

- Model a specific truck and trailer combination
- Define the structure and number of objects within the combination
- Specify the capacity (maximum capacity and minimum capacity)
- Maintain additional, freely definable attributes
- Determine compatibilities/incompatibilities based on customizable attribute definitions
- Edit coupling/uncoupling durations per passive means of transport

Maintaining combinations

You can maintain means of transport combinations in the Customizing settings by selecting the IMG menu path SAP Transportation Management • Transportation Management • Master Data • Resources • Define Means-of-Transport Combinations (see Figure 3.51).

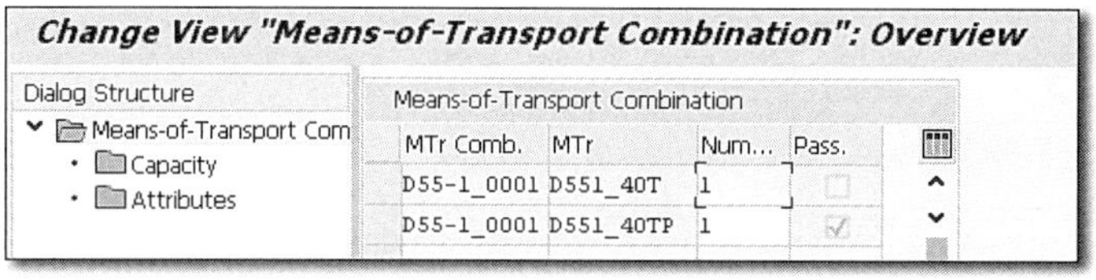

Figure 3.51 Maintaining Means of Transport Combinations

Combination options

You can specify how many different means of transport are included in each means of transport combination. A combination of a truck and one, two, or three trailers is used in land transportation. These means of transport combinations can be maintained as follows:

- Means of transport combination 1: one active means of transport (truck) and one passive means of transport (trailer)
- Means of transport combination 2: one active means of transport (truck) and two passive means of transport (trailers)
- Means of transport combination 3: one active means of transport (truck) and three passive means of transport (trailers)

Modeling Means of Transport Combinations

In vehicle scheduling and routing, a succession of alternative vehicle combinations are evaluated, beginning with the active vehicle and followed by each of the passive vehicles in sequence. It is therefore essential to avoid any gaps in the number of permitted vehicle combinations. There are no such gaps in this example because a combination is defined for each of the scenarios involving one, two, and three trailers.

However, if you had not defined the second means of transport combination, a gap would exist between the combination involving one trailer and the combination involving three trailers. In this case, automatic planning may not generate any transportation plans using three trailers. Means of transport combinations give rise to a very large number of possible combinations in automatic planning. Therefore, we recommend that you model your transportation scenarios with the fewest possible means of transport combinations and avoid "tricks" when it comes to modeling means of transport combinations.

Duration of coupling/uncoupling

An active vehicle is capable of moving in combination with different passive vehicles, as defined in the means of transport combination that is maintained. Note that an active vehicle may travel independently of any other vehicle for part of its journey before being coupled with and decoupled from various passive vehicles. These dynamic changes in the means of transport combinations moved by the active vehicle can be mapped in the system using coupling and uncoupling activities.

You can define a coupling duration and an uncoupling duration for a trailer. SAP TM schedules the uncoupling activity for a trailer when a different vehicle has to move it in the next stage. If the system has assigned an empty trailer to a stage, it schedules the coupling activity directly before the transportation activity and the uncoupling activity directly after the transportation activity.

As shown in Figure 3.52, you can define the duration of these coupling and uncoupling activities for each passive means of transport in the Customizing settings. To do this, select the IMG menu path SAP Transportation Management • Transportation Management • Master Data • Resources • Define Coupling/Uncoupling Duration.

Figure 3.52 Maintaining the Duration of Coupling and Uncoupling

Compartments and Capacity

You can use compartment types to define a fixed number of compartments for each resource. For example, you use the compartment type to define the capacity of the compartment, possible steps, and a compartment profile. In the compartment profile, you first specify the number and type of compartments and then assign this compartment profile to the means of transport (see Figure 3.53).

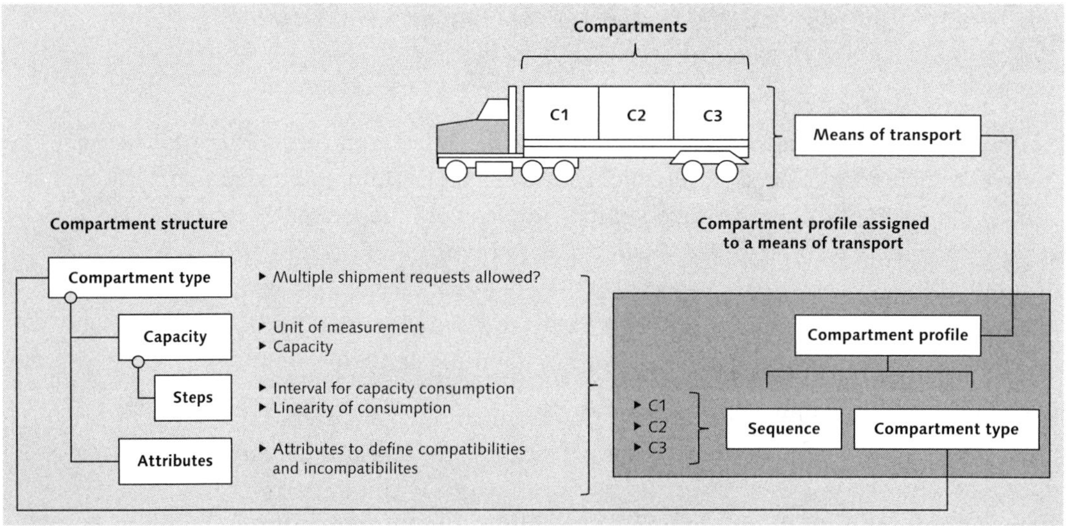

Figure 3.53 Compartment Types and Profile Structure

Maintaining compartments

To define compartment types in the Customizing settings, select the IMG menu path SAP Transportation Management • Transportation Management • Master Data • Resources • Define Compartment Type.

Figure 3.54 shows the maintenance screen. This compartment type specifies the capacity of the compartment, possible steps, and whether a compartment is allowed to contain freight units for one shipment request only. You can also define steps to avoid having the capacity consumption match the actual load.

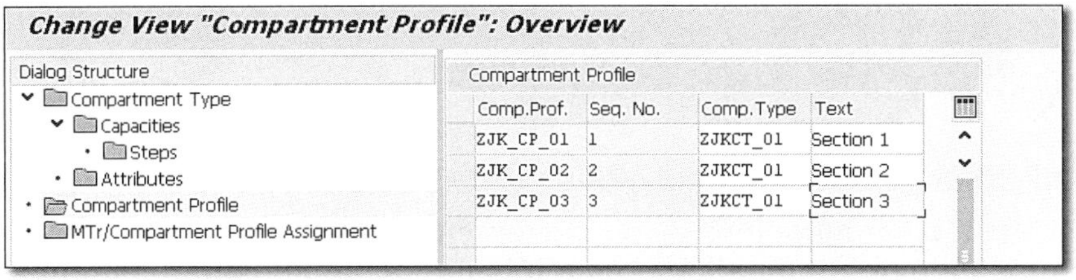

Figure 3.54 Maintaining Compartments

Compartment profiles

The compartment profiles contain the compartment type, together with the sequence representing the compartment structure. In the compartment profile, you enter the number and type of compartments that a means of transport can have. Next assign this compartment profile to a means of transport. You can also define attribute-based incompatibilities, such as whether a trailer can be coupled to a vehicle.

Fixed and flexible compartments

There is no attribute to specify whether the individual compartments are fixed or flexible. Instead, this is determined by the relationship between the total capacity of a vehicle resource and the total capacity of all compartments assigned to that resource. If a resource has *fixed compartments*, this means that the capacities of all of the assigned compartments add up to exactly the figure specified as the total capacity of the resource. If a resource has *flexible compartments*, then the capacities of all of the assigned compartments add up to a figure greater than that specified as the total capacity of the resource.

Modeling Compartments for Retail and Fuel Distribution

Let's look at the way means of transport and vehicle resources are frequently modeled for two industries:

- The retail industry uses two compartments, one with the attribute "cooled" and the other with the attribute "uncooled." Both compartments have a capacity that corresponds to the vehicle capacity.
- The fuel distribution industry uses five compartments with identical attributes. All have the same capacity, which equals one-fifth of the vehicle capacity.

The compartments in the retail scenario are flexible, meaning that the vehicle can be fully loaded with refrigerated goods, fully loaded with unrefrigerated goods, or loaded with a combination of the two. In the fuel distribution scenario, by contrast, the compartments are fixed. In other words, the total vehicle capacity can only be fully utilized by filling each compartment to its full capacity.

Compartment incompatibilities

You can also define incompatibilities for compartments, based on attributes. For more information, refer to the SAP TM system documentation, which provides a detailed explanation and several examples of how compartments are modeled.

Resource Capacity

For each compartment, you can define one capacity for each dimension (see Figure 3.53). Within a compartment, the capacity consumption is linear. By defining steps, you can model progressive capacity consumption. This is useful, for example, if the compartment has a door or if you want to use movable partitions. If you do not define steps, the capacity consumption matches the actual load; in other words, the capacity consumption is linear. This means that each load with a capacity that is greater than the capacity of the previous step and smaller than or equal to the capacity of this step consumes the complete capacity of this step. If you define the first step with a capacity of zero, the compartment does not consume any capacity if you leave it empty. Only if the capacity of the load is greater than zero does the system take into account the capacity of the next step, with capacity consumption based on this step.

The relationship between the total capacity of the vehicle resource and the total capacity of all assigned compartments determines whether the individual compartments are fixed or flexible. This is not an attribute of the compartment. A resource with fixed compartments means that the total capacity of all assigned compartments is the same as the total capacity of the resource. A resource with flexible compartments means that the total capacity of all assigned compartments is greater than the total capacity of the resource.

Capacity constraints

You can choose up to eight dimensions and units of measurement to describe the capacity. Note that the mass and volume are predefined by default. Planning can only take the vehicle capacity into account if the dimensions and units of measurement correspond to those that you have defined in an FUBR. In addition, the capacity of a resource may be time-dependent and is determined by a range of factors:

- If a downtime is currently defined for the resource, then only limited capacity is available for handling resources, whereas no capacity is available for other resources.
- If an active capacity variant is specified for the resource and if alternative data is defined for this variant in a capacity profile for the current date, then the system uses the capacity defined in the capacity profile.
- If an active capacity variant for which no capacity profile exists is specified in the resource, and an interval with a capacity that is currently valid is defined in this variant, then the system uses this capacity.
- If you have created a capacity profile in addition to the standard available capacity, then the system uses the capacity defined in the capacity profile.
- The system only uses the standard available capacity if it is unable to determine a currently valid capacity for the resource in any of the above checks.
- If a standard available capacity is not defined for the resource, then it has no capacity.

For more information, refer to the SAP TM system documentation, which provides detailed explanations of capacity, capacity variants, and capacity profiles.

Decreasing capacities

One special feature of capacity modeling is the modeling of decreasing capacities, which can be defined for (active and passive) vehicle resources and transportation unit resources. In some transportation scenarios, goods are transported for several customers simultaneously and are separated by partitions within the vehicle. These partitions reduce the total capacity of the vehicle because a partition itself consumes a certain amount of capacity. The decrease in the capacity of the vehicle or transportation unit resource can be defined on the basis of the number of stops. You can define decreasing capacities independently of compartments by selecting the SAP NWBC menu path Application Administration • Planning • General Settings • Decreasing Capacity Settings. The maintenance screen for decreasing capacities shown in Figure 3.55 allows you to define the decreases for ranges of stops for each means of transport. Each capacity decrease can be maintained as absolute or relative, and is based on the number of stops in the relevant stop range.

Means of Transp...	Load Unit	Stop Range: Start	Stop Range: End	Value Type	Capacity Decrease	Percentage
0001	TO	0002	0010	Single Value	15	Capacity Decrease Refers to Load Unit
0001	TO	0001	0002	Single Value	10	Capacity Decrease Refers to Load Unit
		0000	0000	Single Value	0,000	Capacity Decrease Refers to Load Unit

Figure 3.55 Maintaining Decreasing Capacities

3.3.2 Calendars and Handling Resources

In the previous section, we saw that resources not only play a central role in planning transportation activities, but also offer the capacities needed to physically transport goods within the transportation network. In addition to vehicles and trailers, handling and calendar resources mainly focus on working times and qualifications and contain the additional attributes and parameters needed to execute transportation activities. Now let's briefly highlight some of the main characteristics of these resources. The definition of resources and their main configuration is explained in Section 3.3.4.

Calendars

Calendar resources are assigned to locations and are used to map operating times. (For location-specific parameters, please also refer to Section 3.2.) Vehicle loading and unloading activities cannot take place at the assigned locations outside of these hours. To schedule the goods receipt and goods issue processing time and to define when a specific location is available for loading or unloading, assign calendar resources to a calendar and, optionally, to a shift. You can consider a multiple-shift operation when specifying operating times for a location and maintain downtimes, when a resource should not be available. Figure 3.56 provides an idealized representation of a calendar resource for an airport. In this example, the resource is assigned to airport location AIY and is used to model a planned downtime during the night.

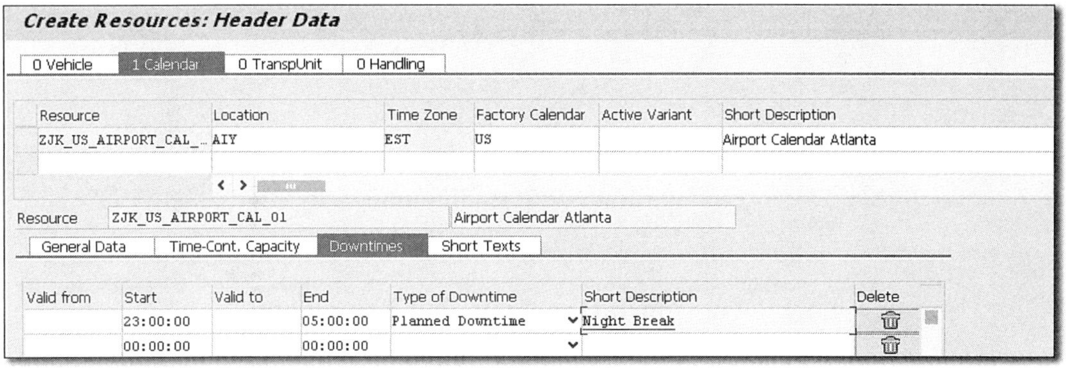

Figure 3.56 Calendar Resource to Schedule Airport Downtime

Handling Resources

Handling resources are used for handling transportation orders at a location, that is, for loading goods onto a vehicle (outbound) or unloading goods from a vehicle (inbound). For example, handling resources can be used to map loading ramps or doors, and they represent a generalization of calendar resources. In addition to operating times, they can also define restrictions regarding the maximum number of activities that can be executed simultaneously or the capacity the resource offers to allow the goods to be loaded onto or unloaded from vehicle resources. In addition to maintaining downtimes to specify the resource availability, you can also maintain the qualification requirements needed to use the handling resource (e.g., a special training) and additional equipment attached to the resource itself. Furthermore, you can maintain attributes to group handling resources into categories. Figure 3.57 shows a forklift being used at the same airport (AIY) from the previous example.

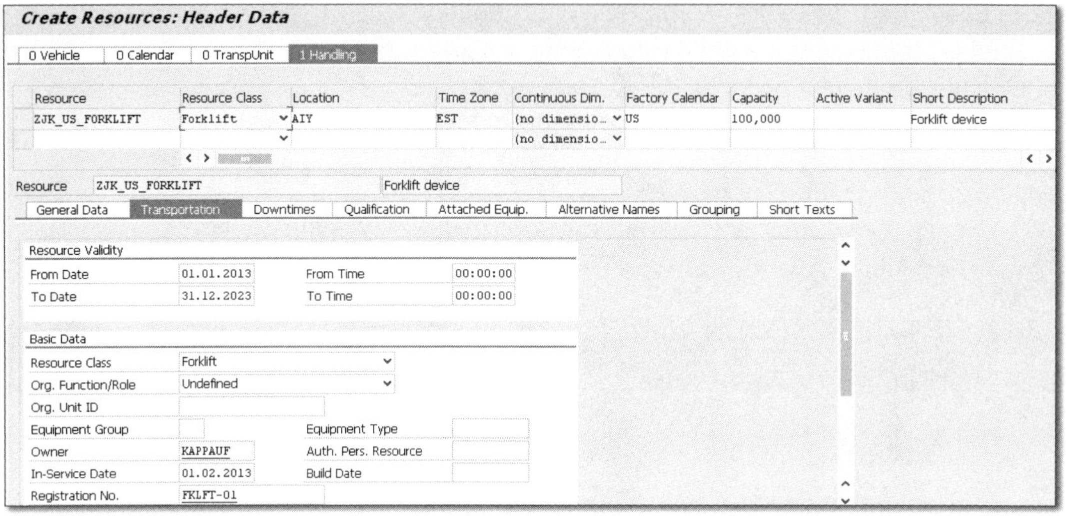

Figure 3.57 Handling Resource

3.3.3 Transportation Units and Equipment

As of SAP TM 9.0, you can create trailer units to model the logistical handling of transports with trailers. These trailer units, similar to transportation units in previous releases of SAP TM, might be required if a trailer is moved by several trucks. (To learn more about trailer units please refer to Chapter 6.)

Transportation Units

In previous releases of SAP TM, transportation unit resources represented the freight capacities that are assigned to vehicles so that they can be transported. They could have been assigned to both active and passive vehicles and enabled the modeling of containers. The essential difference between a passive vehicle and a transportation unit resource was that a passive vehicle was coupled to an active vehicle, and the vehicle resource combination then had more capacity than the active vehicle alone. By contrast, assigning a transportation unit resource did not increase capacity. It merely indicated that an active resource and a transportation unit resource are to be used for transportation. Like vehicle resources, transportation unit resources may have contained compartments.

Figure 3.58 shows aircraft ULD. In SAP TM 9.0 transportation unit resources can still be maintained, but they are currently not linked to any process functionality.

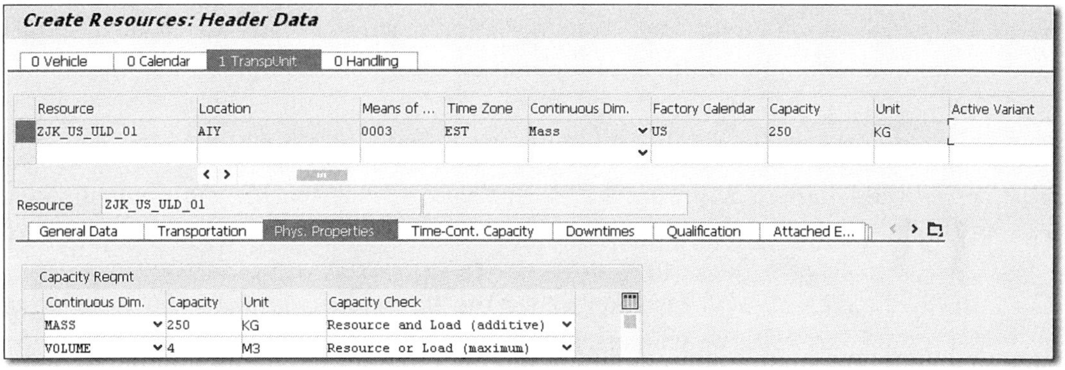

Figure 3.58 Transportation Unit Resource

Equipment Types

Similar to transportation units, physical equipment properties not only represent the freight capacities that are assigned to vehicles, but are also taken into account during the creation of freight units.

Freight unit creation

In this context, taking a container as an example, SAP TM consolidates the order items and displays the container as a freight unit. The freight unit (container) then inherits the physical properties of the assigned equipment type and equipment group. The process and functionality of the creation of freight units will be explained in Chapter 6, Section 6.1.3. The equipment group and equipment type can be defined in Customizing and are part of the resource master data configuration in SAP TM.

Carrier profiles/ forwarding quotations

In addition to freight unit building, equipment groups and types are also used in carrier profiles. In Section 3.1.4 we briefly mentioned that a carrier profile (see Figure 3.15) identifies the transportation capabilities of a carrier. Apart from the definition of freight code sets, transportation lanes, and carrier-specific parameters, it also contains equipment groups and equipment types (the kind of equipment that is owned by the carrier).

The physical equipment properties are also used for forwarding order or quotation items of the type container. Based on the item type configuration, the equipment group is entered automatically, and equipment types of this group are displayed for selection. Depending on the equipment group and type, the system automatically calculates the tare weight and capacity. This data, together with the physical property data of the equipment type, is relevant for the planner and is then available in the forwarding order and subsequent documents.

Equipment configuration

Equipment groups are transportation-mode-specific and are maintained in the Customizing settings of SAP TM via the IMG menu path SAP TRANSPORTATION MANAGEMENT • TRANSPORTATION MANAGEMENT • MASTER DATA • RESOURCES • DEFINE EQUIPMENT GROUPS AND EQUIPMENT TYPES. Here you specify the equipment groups and assign the corresponding equipment types. You can, for example, enter values for ocean-, land-, or air-specific equipment. This equipment represents different resource classes such as containers, unit load devices (ULDs), boxes, and trailers. Figure 3.59 illustrates different transportation modes (road, rail, and sea). For these transportation modes, and the equipment group CNS (standard container), multiple equipment (container) types have been assigned.

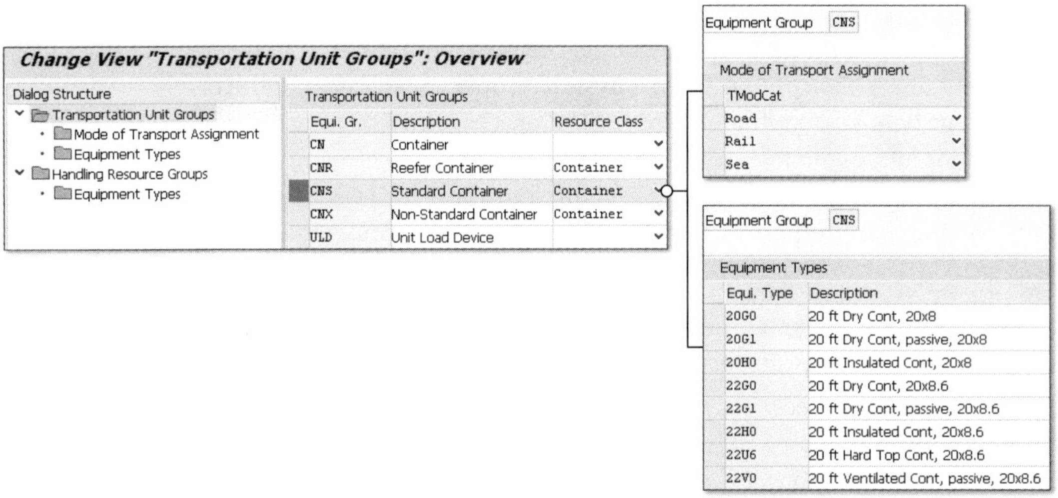

Figure 3.59 Equipment Configuration

3.3.4 Defining Resources

Resource worklist

Whether you enter the transaction from the SAP GUI or from SAP NWBC, the creation of a resource is exactly the same. To display an overview of existing resources of all types, you can use the POWL for resources, which you can find under MASTER DATA • RESOURCES. Simply click on the name of a resource to view its details.

Creating and changing resources

You can display, change, or create resources directly or from a template. From the initial screen enter a name and resource type, and then click on the CREATE RESOURCES button. This takes you to the main resource maintenance screen, where you can maintain all resource types with the exception of drivers. To create, change, or display resources in SAP NWBC, choose MASTER DATA • RESOURCES • DEFINE RESOURCE. When you create a new resource, you first need to specify a resource type and resource class.

Resource Classification

The SAP TM resource types are vehicle resource, calendar, transportation unit resource, and handling resource, as shown in Figure 3.60. You can also specify the resource class when you create a resource. Recall that because drivers are maintained as business partners, they cannot be selected here.

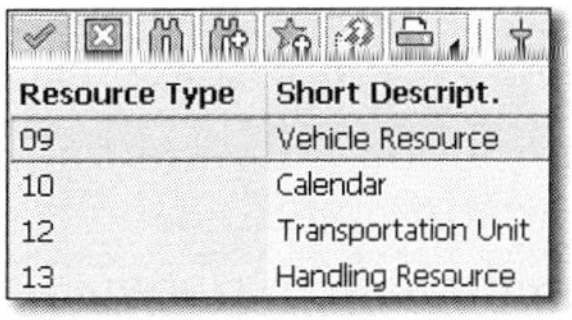

Resource Type	Short Descript.
09	Vehicle Resource
10	Calendar
12	Transportation Unit
13	Handling Resource

Figure 3.60 Resource Types in SAP TM

Grouping resource characteristics

In previous releases of SAP TM, resources were inferred from the means of transport (e.g., an *active* means of transport with a transportation mode *sea* was supposed to be a vessel). Since SAP TM 9.0, you can maintain *resource classes* to group resources with the same characteristics or type. You can use resource classes to specify the kind of vehicle or handling resources that can be used for your transportation activities at a specific location, such as forklifts or dollies. This enables you to see exactly what kinds of resources are available for transportation planning at a glance.

When you enter the means of transport in the resource master data, the system automatically determines the resource class based on the means of transport or equipment group. For example, if you have assigned the resource class truck to means of transport ZJK_TR01 and subsequently create a resource with means of transport ZJK_TR01 as in Figure 3.61, the system automatically determines that your resource is a truck.

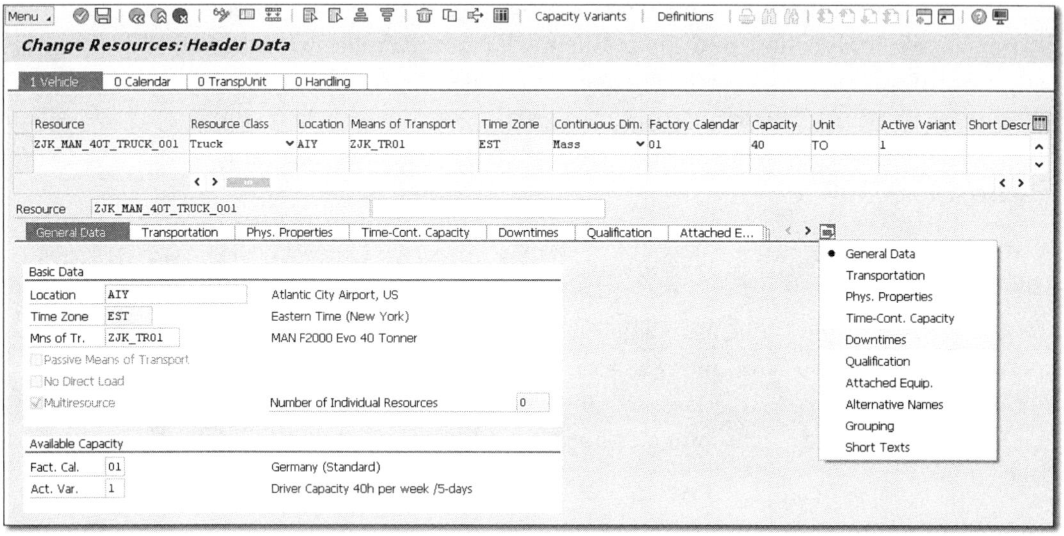

Figure 3.61 Maintaining a Vehicle Resource

Resource classes

You assign a resource type (vehicle, transportation unit, handling resource, or driver) to the resource class and enter a description, as shown in Figure 3.62. The resource class can then be assigned to a means of transport or equipment group (see Figure 3.59), thereby specifying the precise resource that is to be taken into consideration as soon as you enter the means of transport or equipment group. You can also specify the resource class when you create a resource, as described in the next section. You can find these settings by following the IMG menu path SAP TRANSPORTATION MANAGEMENT • TRANSPORTATION MANAGEMENT • MASTER DATA • RESOURCES • DEFINE RESOURCE CLASS.

Resource Class Definition

Res. Class	Res. Type	Description
1	Vehicle Resource	Truck
2	Vehicle Resource	Trailer
3	Vehicle Resource	Locomotive
4	Vehicle Resource	Rail Car
5	Vehicle Resource	Vessel
6	Vehicle Resource	Airplane
10	Transportation Unit Resource	Container
11	Transportation Unit Resource	Unit Load Device (ULD)
12	Transportation Unit Resource	Box
13	Transportation Unit Resource	Trailer
20	Handling Resource	Doors
21	Handling Resource	Dolly
22	Handling Resource	Forklift
30	Driver	Driver

Figure 3.62 Definition of Resource Classes

Resource Attributes

After you have specified the name, resource class, and resource type, you maintain the other resource attributes. The available attributes are dependent on the resource type and include general data parameters such as a location with corresponding time zone and a factory calendar for the new resource you want to create. Table 3.4 shows which attribute tabs are available for the standard resource types shown in Figure 3.60.

Attribute Tab/Resource Code	09	10	12	13
TRANSPORTATION	X		X	X
PHYSICAL PROPERTIES	X		X	
TIME-CONT. CAPACITY	X	X	X	
DOWNTIMES	X	X	X	X
QUALIFICATION	X		X	X
ATTACHED EQUIPMENT	X		X	X
ALTERNATIVE NAMES	X		X	X
GROUPING	X		X	X
SHORT TEXT	X	X	X	X

Table 3.4 Resource Attributes

General data

You can maintain a means of transport, time zone, dimension, factory calendar, and capacity (including a capacity unit) for the new vehicle you want to create. You also have the option of assigning a location to the vehicle. This is the location where the resource resides. The factory calendar is used in planning, as is the dimension and capacity. To create or change a factory calendar in Customizing, follow the IMG menu path SAP TRANSPORTATION MANAGEMENT • SCM BASIS • MASTER DATA • CALENDAR • MAINTAIN FACTORY CALENDAR.

In addition, you can enter the capacity requirements of a vehicle or transportation unit resource during transportation. The CONTINUOUS DIM. column shown in Figure 3.61 specifies in which dimensions (e.g., weight or volume) the resource is scheduled. The capacity defines the capacity and unit of measure of the resource. Together with the dimension, these settings determine the capacity and volume for building a freight unit.

Transportation

In the TRANSPORTATION tab, you can specify the resource validity and when the resource is available. You can also maintain additional basic data, such as the owner of the resource and its registration number, registration number country, and the in-service date of a vehicle, transportation unit, or handling resource.

Physical properties and capacity

In addition to the capacity maintained at the header level, you can specify additional capacities here; each is each described using a dimension (e.g., mass), an appropriate unit of measure (kilos or metric tons), and a capacity. These parameters, together with the location, the time zone, and the assigned factory calendar, are considered for transportation planning.

In releases prior to 9.0, a FINITE SCHEDULING flag was required for activating finite scheduling in addition to defining capacities. If you wanted automatic planning to take vehicle capacity into account, you selected the FINITE PLANNING field in the GENERAL DATA tab. When checked, this parameter specified that the existing capacity load on the resource be taken into account. For example, if the continuous dimension of the truck has been maintained as *mass*, and the capacity has been specified as 40 tons (as in Figure 3.61), the optimizer would fail if this capacity were exceeded while trying to build the freight units.

In the current release of SAP TM, you can choose up to eight dimensions and units of measurement to describe the resource capacity, while mass and volume are predefined by default. In this context, remember that transportation planning only takes the vehicle capacity into account if it corresponds to the dimensions and measurements defined in the corresponding FUBR. To learn more about freight unit creation, please refer to Chapter 6, Section 6.1.3.

Downtimes

The actual availability of a resource within the general availability depends on the factory calendar, downtimes (time segments in which resources are not available), and shift and break definitions. The DOWNTIMES tab allows you to define downtimes, each with a start and end time, a downtime type (either planned or unplanned), and a short description. The resource cannot be used during downtimes. You can define downtimes for all resource types and for the driver business partner. Downtimes override the availability of a resource and exclude vehicle resources from planning. For handling resources, downtimes can reduce the available capacity (but a remaining available capacity of 0 also excludes a handling resource from planning). A differentiation is made between scheduled downtime (e.g., for maintenance) and nonscheduled, unexpected downtime (due to a machine breakdown).

Downtimes differ from a regular, scheduled nonworking time based on the factory calendar or the shift sequence (e.g., break or holiday).

Qualifications

Here you can specify the qualifications required to use the resources to define qualification types and qualifications. Follow the IMG menu path SAP TRANSPORTATION MANAGEMENT • TRANSPORTATION MANAGEMENT • MASTER DATA • RESOURCES • MAINTAIN SETTINGS FOR QUALIFICATIONS. You use this function to specify the qualifications such as skills, capabilities, and permissions needed to operate vehicle and handling resources. You can rate qualifications with an efficiency factor for each resource, and you can use the incompatibilities to check the consistency of a required and offered qualification.

Attached equipment

You can assign attached equipment to the categories you have defined in the IMG under SAP TRANSPORTATION MANAGEMENT • TRANSPORTATION MANAGEMENT • MASTER DATA • RESOURCES • MAINTAIN SETTINGS FOR ATTACHED EQUIPMENT.

Alternative names/ short texts Here you can maintain alternative names, which can be used when selecting resources. You define categories of alternative names in the IMG under SAP Transportation Management • Transportation Management • Master Data • Resources • Specify Categories of Alternative Names.

Grouping You can create groups of vehicle resources, transportation unit resources, and handling resources based on attributes. To do this, select the IMG menu path SAP Transportation Management • Transportation Management • Master Data • Resources • Maintain Settings for Grouping Attributes.

In this chapter we covered both general master data and how it is integrated with SAP ERP, and SAP TM-specific master data like resources and the transportation network. These parameters and attributes are necessary to execute transportation requirements, which will be introduced in the next chapter.

Regardless of whether you're looking from the perspective of a shipper, an LSP, or a carrier, the transportation process starts with a request for transportation services. This chapter covers the different ways to start the process in SAP TM and how to record the requests in the system.

4 Transportation Requirements and Order Management

When we look at a very simplified, document-driven process flow in SAP TM, we see that not many documents are used to cover the transportation process. In fact, there are only three process steps (leaving out transportation charge management for the moment) that cover the transportation planning process, as you can see in Figure 4.1. In SAP TM, the starting point of the process is always a *transportation request* (TRQ), followed by freight unit and transportation order.

Figure 4.1 Transportation Planning Process Flow

SAP TM can be used to monitor and execute the transportation processes of many different industries, which indicates that transportation process approaches can vary significantly.

Transportation requests for shippers

These approaches require different ways of starting the transportation process in the system. For example, for a shipper, transportation is not the core business, so the main focus is laid not on the ordering and execution of transportation services, but on manufacturing, materials management, and so on. The shipper will most likely use an ERP system to cover these processes. Since the transportation process is not the core business, the leading system in the entire value-generating process should be the ERP system because it handles processes that produce the competitive advantage. The SAP TM system therefore *receives* information; it is not allowed to alter any information that is important for the manufacturing process.

To work within this scenario, SAP TM provides built-in ERP integration features from which you can integrate transactional data from SAP ERP to SAP TM to continue the process with regard to transportation services.

Transportation as core business

However, if we look at the transportation process from a logistics service provider (LSP) or carrier's perspective, the transportation process *is* the core process. Therefore, more emphasis needs to be laid on the information generated by the transportation process. Since the ERP functionality does not provide information that is detailed enough for the LSP and carrier's needs, the information needs to be generated in the SAP TM system, which is now the leading information (that is, process-driving) system.

So since the ERP integration does not deliver sufficient information for transportation as a core business, the transportation request needs to be created in SAP TM itself. So how can these different requirements be covered with only one document, as shown in Figure 4.1? The answer is that they cannot. Although the main process flow remains the same, we need to take a closer look at the transportation request process step. When zooming in on this step, we can see that several documents are necessary. However, it is important to remember that all documents and the respective process variants lead to the common next step: freight unit building (to be covered in Chapter 6).

Figure 4.2 shows the different transportation request documents. While the ERP order-based transportation requirement (OTR) and the ERP delivery-based transportation requirements (DTR) cater to the shipper's transportation process variant, the forwarding quotation and the forwarding order cover the LSP and carrier's requirements.

So let's delve deeper into both process variants and focus on how these documents support the start of the transportation process.

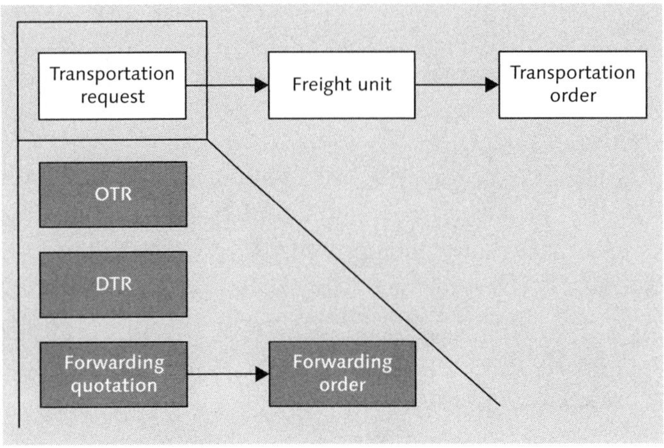

Figure 4.2 Transportation Request Documents

4.1 Integration of SAP ERP Documents

In Chapter 1, we introduced SAP ERP integration as a major component of SAP TM. In this section we take an in-depth look at how that SAP ERP integration can be established. First, it must be said that you do not need to be a Sales and Distribution (SD) component expert to establish the integration into SAP TM, but it *is* advantageous to know a little bit about the standard SD and Materials Management (MM) document flow.

SAP TM can integrate the following SD and MM documents:

- Sales orders
- Purchase orders
- Stock transfer orders
- Inbound deliveries
- Outbound deliveries

Prerequisites for SAP ERP integration

As always, you cannot immediately begin with integrating orders or deliveries from SAP ERP into SAP TM. First, you need to make sure the prerequisites for the SAP ERP integration are fulfilled. In general, there is only one major prerequisite: the master data. The master data needs to be integrated into SAP TM from the SAP ERP system using the Core

Interface Framework (CIF) that we introduced in Chapter 3. For a seamless SAP ERP integration, make sure the following master data is integrated:

- **Locations and business partners**
  Since SAP ERP does not have an entity called *location*, there is no 1:1 copy of the SAP ERP master data. Debtors and creditors as well as shipping points and plants can be integrated. Make sure you have set the integration model correctly—that is, for the debtors and creditors, both locations and business partners are created. For shipping points and plants, only create locations, and link those locations to a business partner representing your organizational unit.
- **Products**
  Again, we need to take a look at different vocabulary between SAP ERP and SAP TM. In the SAP SCM environment, we talk about "products." However, the master data copied from SAP ERP is referred to as the "material master data."
- **Organizational data**
  In the transportation process with SAP ERP integration, it is optional to assign a sales organization to a transportation request. However, if you want to represent the sales organization from the sales order in the SAP TM transportation request, you need to transfer the organizational model from SAP ERP to SAP TM.

Integration of Master Data with the Same Names

If you plan to link your SAP TM system to more than one SAP ERP system, make sure you do not run into trouble with duplicate names. Because the SAP ERP systems run independently, they use number range intervals that often overlap. When integrating this master data, for example, the SAP TM system will encounter a problem because it can only create one location with the same name.

The most common use case is the integration of shipping points and plants, which usually start with the same number.

Even if you are working with only one SAP ERP system, you will notice that shipping points and plants may have the same number. This also needs to be resolved in the SAP TM system upon master data creation.

SAP TM delivers some Business Add-in (BAdI) implementations that cover exactly this case. In Customizing, you can get to these BAdI implementations by following the IMG menu path SAP TRANSPORTATION MANAGEMENT • SCM BASIS • INTEGRATION • BADIS FOR SPECIFIC APPLICATIONS • LOCATION AND BUSINESS PARTNER • BADI: INBOUND PROCESSING FOR LOCATION. Here you will find three implementations ready to use.

- The implementation APOCF001_SYSDIF adds a suffix to the location name, identifying the logical system name of the SAP ERP system where this location was created.
- The implementation APOCF001_TYPEDIF adds a prefix telling you whether the SAP ERP counterpart is a customer, creditor, shipping point, and so on.
- The third implementation, APOCF001_TYPESYSDIF, adds both.

System landscape

The other prerequisite for the SAP ERP integration is the system landscape. In general, the SAP ERP integration uses web services for communication between SAP ERP, SAP NetWeaver PI, and SAP TM.

For the most part, the integration of transactional data between SAP ERP and SAP TM can be done with SAP NetWeaver PI as middleware or without, using direct communication. However, we recommend that you use SAP NetWeaver PI as middleware in order to have better monitoring capabilities.

Having briefly mentioned web services communication, we have to keep one thing in mind. While this chapter explains the technical integration with web services, this integration is only possible if you are using SAP ECC 6.04 SP9 or higher. For SAP ERP releases older (lower) than this, it is mandatory to establish the integration via SAP NetWeaver PI using IDocs.

For details about SAP ERP integration in terms of architecture, please refer to Chapter 2, Section 2.4.

Basis setup for SAP ERP integration

Before we get too far into talking about SAP TM, we should take some time to examine the general setup steps in SAP ERP; these are necessary to enable the integration into SAP TM. Some of the activities are standard Basis activities that are not specific to the SAP TM integration (such as creation of RFC users or RFC connections), so we'll only look at activities specific to the SAP TM integration here.

Some *business functions* need to be activated for the SAP TM integration. The following business functions are required; some of them are specifically designed for the SAP TM integration:

- LOG_TM_ORD_IN
- SD_01
- LOG_ESOA_OPS_2
- LOG_ESOA_OPS_3
- ESOA_OPS01

Again, some of these business functions are only available as of SAP ECC 6.04 SP09.

4.1.1 Sales Order Integration

The most commonly used type of integration between SAP ERP and SAP TM is *sales order integration*. With sales order integration, shippers use SAP TM to organize the transport of sold goods to the customer. Though Chapter 13 offers more details about the shipper's process, in this section we focus on how to create an OTR and what to do with it.

In the previous section, we started with the setup in SAP ERP that is standard for all other SAP ERP SD and MM documents. Now we take a look at the settings that are specific to the integration of sales orders.

Output determination

SAP ERP has a feature called *output determination*. Output determination in SD is used to control the output—meaning follow-up activities, messages, and documents—for sales, shipping, transportation, and billing. It is used to exchange information with external and internal systems that represent business partners when looking at it from a process perspective.

Output determination can automatically propose output for an SD document using the SAP ERP condition technique. Though the terminology is similar, *condition technique* in SAP ERP and *conditions* in SAP TM are two completely different things.

As you can see in Figure 4.3, the core of the SAP ERP output determination is the output determination procedure, which consists of steps that are supposed to be executed in the given order. Each step represents a condition of a defined condition type; this condition determines whether an output of the defined output type should be triggered.

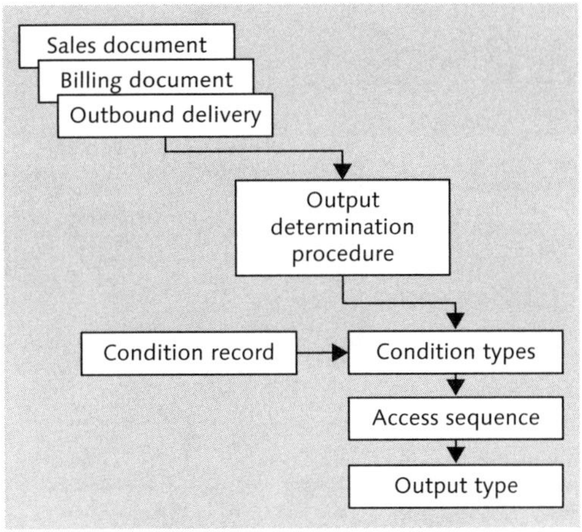

Figure 4.3 Output Determination in SD

The *condition type* is linked to the output type and can then contain condition records that define when an output should be triggered. In terms of structure, a *condition record* is a table (like the decision table in SAP TM conditions) that is used to find a value. However, in the SAP ERP condition technique, a condition type can contain several condition records of different structures (e.g., one condition record where the input data is sales organization and customer, and a second condition record where the input data is order type).

With the *access sequence*, you can define in which order the condition records should be processed. Since the condition type is linked to the output type, you do not need to define which output needs to be triggered. This is implicitly done in the output determination procedure, where you define a condition type.

So recall that the output type is linked to the condition type, meaning that once the condition type finds a suitable result, the output of this output type is triggered. In SAP ERP, several output types are available that represent print output, EDI messages, A2A messages, e-mail, and so on. Some of these are shown in Figure 4.4.

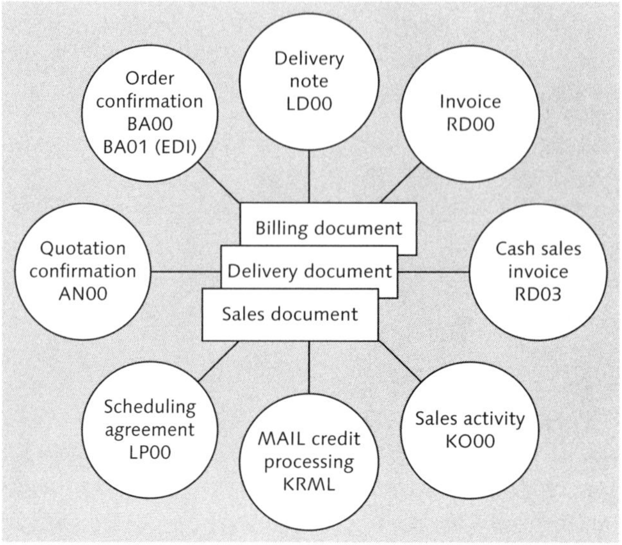

Figure 4.4 Output Types in SAP ERP SD

Output determination setup

For the integration of the sales order, you need to create a new output type (output type TRSO). To do this in your SAP ERP system, follow the IMG menu path Sales and Distribution • Basic Functions • Output Control • Output Determination • Output Determination using the Condition Technique • Maintain Output Determination for Sales Documents • Maintain Output Types.

To maintain the output type TRSO, you need to define general data, processing routines, and partner functions, as shown on Figure 4.5.

Once the output type is configured, you can add it to the output determination procedure by following the IMG menu path Sales and Distribution • Basic Functions • Output Control • Output Determination • Output Determination using the Condition Technique • Maintain Output Determination for Sales Documents • Maintain Output Determination Procedure. Simply add a new step with condition type TRSO and requirement 27 to procedure V10000.

Assignment of Output Determination Procedures

This chapter assumes use of the standard output determination procedure and the standard sales order type for integration.

However, it is clear that this might not always be the case. If you are using a different sales order type, you can assign an output determination procedure to the order type by following the IMG menu path SALES AND DISTRIBUTION • BASIC FUNCTIONS • OUTPUT CONTROL • OUTPUT DETERMINATION • OUTPUT DETERMINATION USING THE CONDITION TECHNIQUE • MAINTAIN OUTPUT DETERMINATION FOR SALES DOCUMENTS • ASSIGN OUTPUT DETERMINATION PROCEDURES.

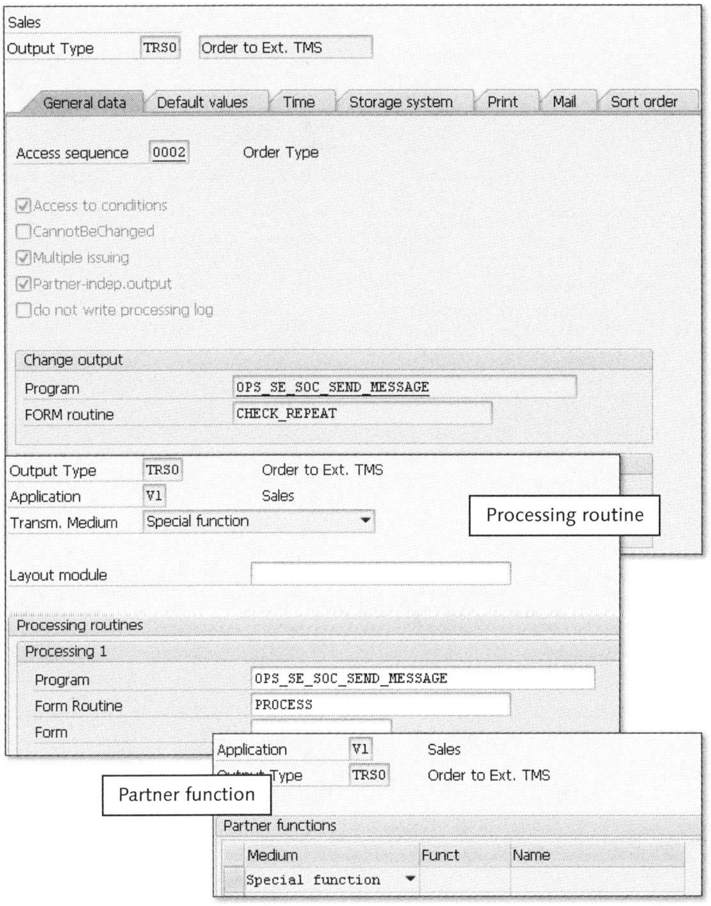

Figure 4.5 Output Type TRSO

Once the output determination is configured, we can activate the transfer of sales orders by following the IMG menu path INTEGRATION WITH

Activation of sales order transfer

Other mySAP.com Components • Transportation Management • Order Integration • Activate Transfer of Sales Documents in the ERP system.

As you can see in Figure 4.6, the activation of a sales document is always dependent on three things:

- Sales organization plus distribution channel and division
- Sales order type
- Shipping condition (if not relevant, you can leave this entry empty)

Sales Document Transfer

SOrg.	DChl	Dv	SaTy	SC	Ctrl Key	TM No.	Control Key Description
0001	01	01	FW01	01	0001	C001	Transfer Sales Order, O. Delivery; Order Scheduling inactive

Figure 4.6 Activation of Sales Documents

The activation of the sales document is done via a *control key*, which defines which document types should be transferred to SAP TM. Figure 4.7 shows the standard control keys.

Control Key Parameters

Ctrl Key	SO to TM	PO to TM	Outbd Del.	Inbd Del.	SO Sched.	Control Key Description
0001	☑	☐	☑	☐	☐	Transfer Sales Order, O. Delivery; Order Scheduling inactive
0002	☑	☐	☐	☐	☐	Transfer Sales Order; Order Scheduling inactive
0003	☑	☐	☑	☐	☑	Transfer Sales Order, O. Delivery; Order Scheduling active
0004	☑	☐	☐	☐	☑	Transfer Sales Order; Order Scheduling active
0005	☐	☐	☑	☐	☐	Transfer Outbound Delivery; Order Scheduling inactive
0006	☐	☑	☐	☑	☐	Transfer PO, I. Delivery; PO Scheduling inactive
0007	☐	☑	☐	☐	☐	Transfer PO; PO Scheduling inactive
0008	☐	☑	☐	☑	☐	Transfer PO, I. Delivery; PO Scheduling active
0009	☐	☑	☐	☐	☐	Transfer PO; PO Scheduling active
0010	☑	☐	☐	☑	☐	Transfer Inbound Delivery; Order Scheduling inactive
0011	☐	☑	☑	☐	☐	Transfer STO; O. Delivery; STO Scheduling inactive
0012	☐	☑	☑	☐	☐	Transfer STO; O. Delivery; STO Scheduling active
0013	☐	☑	☐	☐	☐	Transfer STO; STO Scheduling inactive
0014	☐	☑	☐	☐	☐	Transfer STO; STO Scheduling active
0015	☐	☑	☑	☐	☐	Transfer rSTO; O. Delivery; rSTO no Scheduling available
0016	☐	☑	☐	☐	☐	Transfer rSTO; rSTO no Scheduling available
0017	☐	☑	☑	☐	☐	Transfer rPO; O. Delivery; rPO no Scheduling available
0018	☐	☑	☐	☐	☐	Transfer rPO; rPO no Scheduling available
0019	☑	☐	☐	☑	☐	Transfer Returns Order, I. Delivery; no Scheduling available
0020	☑	☐	☐	☐	☐	Transfer Returns Order; no Scheduling available

Figure 4.7 Control Keys for SAP ERP Integration

In addition, you can see in Figure 4.6 that you can define an SAP TM number when activating the transfer of sales documents. However, you will notice when filling this field that no F4 help is available. The purpose of this field is to link one SAP ERP system to several SAP TM systems. If several SAP TM systems are linked to one SAP ERP system and the TM number is filled, SAP ERP decides to which SAP TM system the order should be sent, based on the entries made in the activation table explained with activating the transfer of sales orders. However, you can enter anything you like here. The "navigation" to which logical system the message is routed is done entirely in SAP NetWeaver PI, where the string sent from SAP ERP is interpreted and translated into a logical system. If you encounter such a situation, consult an SAP NetWeaver PI expert to ensure that the routing is performed correctly.

Error handling in SAP ERP

Once you have configured these SAP ERP settings (for more details of the configuration steps, please also consult the standard Solution Manager content for SAP TM), a message according to output type TRSO is always sent when a sales order has been created, updated, or deleted.

If you cannot see any document being created or updated in the SAP TM system, you can check on the SAP ERP side to see if an output was triggered. Access the sales order you have just saved and then go to the output overview by following the menu path EXTRAS • OUTPUT • HEADER.

If you see an output of type TRSO with a green traffic light, the output message was successfully created.

4.1.2 Integration of Materials Management Orders

The integration of MM orders is, in many regards, similar to the integration of SD documents. The most important customizing activity in SAP ERP for the transfer of MM documents is the specification of the control keys, as just described.

However, in contrast to SD documents, MM documents use the SAP ERP workflow technology for processing the output. Therefore, error handling as described for the sales order is not possible for purchase orders, stock transfer orders, or inbound deliveries.

4.1.3 Order-based Transportation Requirement

Let's shift our attention back to SAP TM. Once a sales order is created and the web services have successfully been executed, a document is created in the SAP TM system; this is the OTR, which represents the sales order data generated in SAP ERP and carries all information relevant for the transportation process.

The OTR is the corresponding SAP TM document that contains all transportation-relevant information from the SAP ERP orders. It can represent three things:

- Sales orders
- Purchase orders
- Stock transfer orders

For SAP TM, it doesn't matter whether the predecessor document was a sales order, stock transfer order, or purchase order. The OTR represents all of these SAP ERP document types.

OTR type determination

Remember from Chapter 2, Section 2.3.1, that there is a singleton condition that is used to define the OTR type of integration SAP ERP orders. You can create and define a condition of the condition type /SCMTMS/OTR_TYPE to differentiate OTRs by type. Among other information, you can also use the sales order type used in SAP ERP for the condition definition. There is a predefined data access definition that you can use for this. If the condition does not find a result, the default type will be used. The default type is defined in Customizing for OTR types with a corresponding flag.

Let's take a close look at the OTR document itself. You will notice that compared to other SAP TM documents, the amount of information is relatively low. This is because the only information stored on the OTR is transferred from SAP ERP and is crucial for the transportation process.

The most important difference from other documents in SAP TM is that there are no action buttons on the OTR, so the document is read-only, with no changes allowed. This is because in this scenario, the SAP ERP system is the information leading system, so any updates should come from SAP ERP.

On the GENERAL DATA tab, you will see the most important information, such as the OTR document type and the total weight and volume of the freight. There is also a field for a sales organization. Recall that you can integrate your sales organization from SAP ERP and put it on the OTR, but that it might not serve for the purposes of SAP TM.

Sales Organization in SAP ERP and SAP TM

When first looking at the empty SALES ORGANIZATION field on the OTR document, you might find it strange that the sales organization used in the SAP ERP sales order was not moved over to the OTR.

However, in most cases, this is correct. Since SAP TM covers the transportation process, the sales organization in this case is *not* the organizational unit selling the *products* anymore; the sales organization in the OTR is the organizational unit selling the *transportation services*. In most cases, these are different organizational units, if the shipper even sells the transportation of products at all.

On the bottom of the screen displaying the OTR document you can see the information area for the items. Since one OTR represents one sales order, all items of the sales order are displayed and listed here. Here you can also find the delivery date that was assigned to the items on the SAP ERP sales order. The delivery date in SAP ERP can be defined in many different ways, such as using a connected global available-to-promise (ATP) check, simple route determination in SAP ERP, or sales order scheduling using SAP TM. No matter which way you determine the delivery date in SAP ERP, it is a date the SAP TM system will work with.

Locations and transportation dates

On the LOCATIONS AND DATES/TIMES tab you can see the locations of the entire transportation. The source location is the shipping point or plant for which the sales order was created. The destination location is the location that was created from the customer transferred from SAP ERP via CIF. Note that the destination location is not for the sold-to party maintained in the sales order, but for the ship-to party.

You will notice that in many cases the source and/or destination locations are empty. This is because the source and destination location can vary from item to item, depending on the storage location or shipping point assigned to the items of the SAP ERP order. During freight unit building, several freight units are created for the OTR because the transportation locations of the items differ.

In Section 4.2 we describe how the transportation order clerk can manually predefine the routing of the transportation requirements by adding stages to the forwarding order. With OTRs, there is no such functionality. The transportation path is simply defined from shipping point to customer for an SAP ERP sales order, meaning that the OTR contains only one stage. (We describe one exception concerning incoterm locations later.)

After freight unit building, the freight unit stage can be split into several stages if the transportation process requires this. This process represents the difference between the *ordered route* and the *actual route*. As its name suggests, the ordered route is the routing that was ordered by the customer (for an integrated SAP ERP order, simply a direct transportation path). The actual route, on the other hand, is the route that was actually used. If the SAP ERP sales order was defined from a shipping point in Germany to a customer in the United States, the ordered route is simply from Germany to the United States. However, the actual route looks different—for example, one stage is from Germany to a port in the Netherlands, one stage is from the Dutch port to an American port, and one stage is from the American port to the customer.

So though the ordered route differs from the actual route, the actual route is not reflected on the OTR; the OTR only represents the transportation *requirement*, and therefore the ordered route.

OTR type Customizing

When looking at the Customizing of OTR types as depicted in Figure 4.8, you will see that, compared to other document types in SAP TM, there is not much to customize here. You can access the OTR type Customizing by following the IMG menu path SAP Transportation Management • Transportation Management • Integration • ERP Logistics Integration • Order-Based Transportation Requirement • Define Order-Based Transportation Requirement Types.

Rather than going into every detail of the type customizing, we only mention the most important topics here. Since you cannot do anything with the OTR itself, it is mandatory to create freight units in order to start planning. While we go into detail of freight unit building later in this book, it is still worth mentioning that we have some options in this area.

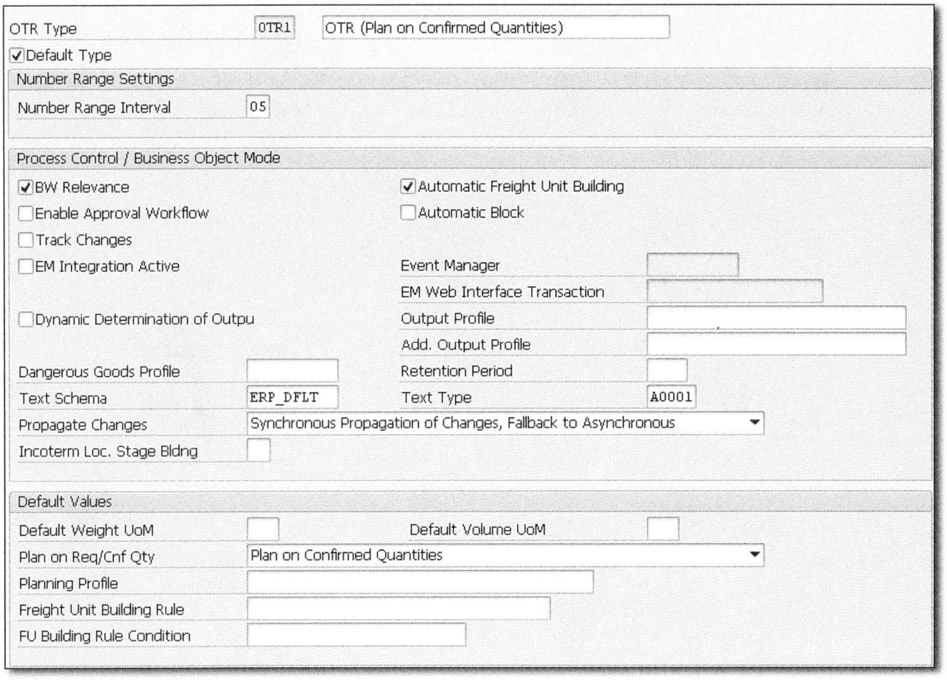

Figure 4.8 OTR Type Customizing

The Automatic Freight Unit Building checkbox defines whether a freight unit should be built automatically after the OTR is created. This makes sense in many use cases. If you consider consolidation of sales orders, however, you might not want to create freight units directly after creation, but instead only trigger freight unit building via a batch job. This way you can consolidate several OTRs to one freight unit.

Regardless of whether you want to create freight units automatically or using a batch job, you need to define how the freight units should be created. To do so, you define a *freight unit building rule* (FUBR) in Customizing. If the method of creating freight units depends on some values of the OTR, you might even want to use a condition to find the right FUBR.

The default units of measurement can be used to have consistent units of measurement in the SAP TM system. If the integrated material master contains unit of measurement conversions, the default units are used on the OTR's header information. The item information will keep the units of measurement from the original SAP ERP order.

Charge Calculation on OTRs

Recall that there are no buttons or actions a user can use on the OTR, which means that no charge calculation can be executed for the OTR.

The reason for this lies in the intended process. Since OTRs are used in the shipper's process, the SAP ERP system is the leading system—for potential surcharges for transportation services, too. Charge calculation in SAP TM is used to calculate selling rates for transportation services and is therefore used only on forwarding orders that are used in LSP or carrier's processes.

Therefore, the surcharges for transportation have to be imposed on the SAP ERP sales order already.

Incoterm locations

The General Data tab includes a UI section about the incoterms and incoterm location. The incoterm and the corresponding location are assigned to the order in SAP ERP, and that information is transferred to the SAP TM system.

In most releases of SAP TM, the incoterms and incoterm location are for information only, and the information does not have any direct impact on routing or charge calculation. However, as of SAP TM 9.1, the incoterm location defined in SAP ERP can have an impact on the routing of the freight unit in SAP TM.

In Figure 4.8 the Incoterm Loc. Stage Bldng field defines how the incoterm location of the SAP ERP order should be interpreted.

Depending on the Customizing entry you have chosen, the SAP TM system creates two stages. As shown in Figure 4.9, the first stage, represented by the bold line, leads from the source location to the incoterm location. Stage 2, shown as the dashed line, leads from the incoterm location to the destination location.

Depending on the scenario, the stage that is relevant for the shipper's planning can be stage 1 or stage 2. Remember, an OTR represents an

SAP ERP sales order as well as an SAP ERP purchase order and a stock transfer order. This means that OTRs can be used for both outbound transportation processes and inbound transportation processes.

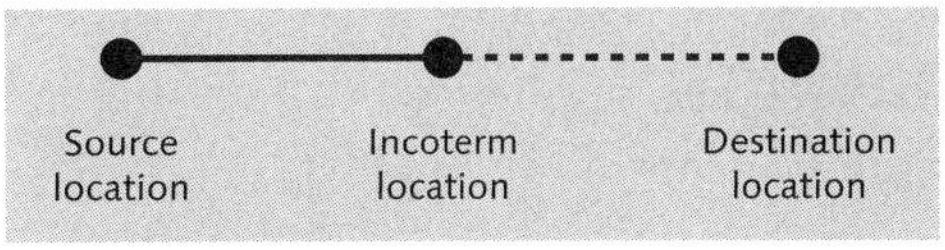

Figure 4.9 Stage Split on Incoterm Location

Therefore, the Customizing entries apply to both outbound and inbound processes. With this Customizing entry, you can define whether stage 1 or stage 2 is the stage the shipper using SAP TM is responsible for. The Customizing entry also defines whether the stage the shipper is not responsible for should be created at all. If it should be created, it will be created as a *statistical stage*, meaning it has no influence on the freight unit and therefore on the planning process steps.

If you have integrated an order that represents an internal transfer of goods, such as with a stock transfer order, and the shipper is nevertheless responsible for the entire transportation, you can also choose to create two relevant stages.

If no incoterm location is entered in the SAP ERP order, no stage splitting will be done in SAP TM.

Recall that there is no entity representing locations in SAP ERP. In SAP TM, locations are an essential part of the entire master data framework. The incoterm location entered in SAP ERP is therefore only a free text entry that can be entered when creating the order. However, to execute a stage splitting, a location master data must exist in SAP TM.

The incoterm location is still transferred to SAP TM as a free text. In SAP TM, the text is then mapped to a location master data. To do this, you need to assign a location master data to a text by following the SAP NWBC menu path Master Data • Transportation Network • Locations • Assign Location to Incoterm Location.

To recognize the incoterms in SAP TM, you need to maintain them in SAP SCM Basis Customizing via the IMG menu path SAP Transportation Management • SCM Basis • Master Data • Define Incoterms. In this Customizing activity, you can also define whether the incoterm requires an incoterm location.

Business partners/ party roles

In SAP ERP orders you can assign multiple customers and creditors to the order using different partner functions. In SAP TM you can see on the BUSINESS PARTNERS tab that the OTR also applies to the participation of different business partners in the transportation process.

However, you need to map the defined partner functions of the SAP ERP order to the *party roles* in SAP TM. You can define the party roles in Customizing by following the IMG menu path SAP TRANSPORTATION MANAGEMENT • TRANSPORTATION MANAGEMENT • MASTER DATA • BUSINESS PARTNERS • DEFINE PARTY ROLES. In a standard SAP TM system, the most common party roles are already predefined.

The mapping between the partner function and the party role defined in SAP TM is done using a BAdI. You can find a standard BAdI implementation for the mapping in your SAP TM system by following the IMG menu path SAP TRANSPORTATION MANAGEMENT • SCM BASIS • INTEGRATION • BADIS FOR SPECIFIC APPLICATIONS • BADI: MAPPING PARTNER FUNCTION INTO PARTY ROLE CODE. Please make sure you have previously integrated SAP ERP customers and creditors into SAP TM using the CIF integration.

Service levels

Service levels can have an impact on the way the transport is planned. If you are familiar with SD, you will know that in SAP ERP the *shipping conditions* are used to assign certain service-level agreements regarding the delivery of goods to a sales or purchase order.

If SAP TM is used for transportation planning, this information is crucial and therefore needs to be transferred to the OTR. On the GENERAL DATA tab you can see the SERVICE LEVEL field, which is the corresponding field in SAP TM. You can define service levels in SAP TM via the IMG menu path SAP TRANSPORTATION MANAGEMENT • SCM BASIS • MASTER DATA • BUSINESS PARTNER • DEFINE TRANSPORTATION SERVICE LEVEL CODES. The mapping between the ERP shipping conditions and the TM service level codes is purely done by name, so make sure that you use the same codes as in the SAP ERP system when defining service level codes in SAP TM.

In SAP TM, no program logic is directly attached to the service level codes, but the information is propagated into the freight unit, where you can use this information using conditions, incompatibilities, or other means to influence the planning. In addition, the charge calculation on the freight order can be influenced using the service level code that originated in the SAP ERP order.

Statuses

On the OTR you will find three important statuses: the planning status, the execution status, and the consumption status. The *planning status* shows whether freight units have been created (if so, the planning status is *In Planning*; if not, the planning status is *New*). If the freight units are not only created but also already planned on freight orders, then the planning status will change to *Planned*. If the freight unit is planned on freight orders, the *execution status* of the OTR shows whether the freight order is already executed, in execution, or not yet executed.

The *consumption status* of the OTR shows whether DTRs were already created and have consumed the freight units of the OTR. (We talk about freight unit consumption later in this chapter.) If the OTR is only partially consumed, this means that not all of the OTR's freight units have been moved to DTRs; some still remain linked to the OTR.

Text type mapping

If you want to propagate more information from the SAP ERP order to the OTR in SAP TM, you can do this in the NOTES tab on the OTR document. Because the OTR cannot be edited, the notes of the OTR need to come from the SAP ERP order.

In the SAP ERP order, you can assign information using *text types*. Map the SAP ERP text types to the SAP TM text types by following the IMG menu path SAP TRANSPORTATION MANAGEMENT • TRANSPORTATION MANAGEMENT • INTEGRATION • ERP LOGISTICS INTEGRATION • ORDER-BASED TRANSPORTATION REQUIREMENT • ASSIGN ERP TEXT TYPES TO TM TEXT TYPES FOR OTRS. Make sure you have previously defined SAP TM text types; many of these are predefined in a standard SAP TM system. If the mapping has been done, you can see the SAP ERP text information on the NOTES tab of the OTR document.

Defining SAP TM Text Types

You can define SAP TM text types via the IMG menu path Cross-Application Components • Processes and Tools for Enterprise Applications • Reusable Objects and Functions for BOPF Environment • Dependent Object Text Collection • Maintain Text Schema.

4.1.4 Integration of Deliveries

Recall that both SAP ERP orders and SAP ERP deliveries can be integrated into SAP TM. The integration setup on the SAP ERP side is very close to the integration setup of SAP ERP orders, so we will not go into details about the deliveries again. Please make sure you activate the transfer of the delivery documents in the SAP ERP IMG menu path Integration with Other mySAP.com Components • Transportation Management • Order Integration • Activate Transfer of Delivery Documents. Notice that this IMG activity looks very similar to what we have seen when activating the transfer of SAP ERP orders.

Also on the SAP TM side, the technical integration is almost the same as with SAP ERP orders. Another transportation request document is created, this one is the DTR. Technically, there is no difference between the OTR and the DTR; they are both instances of the business object /SCMTMS/TRQ and are only differentiated by different document names.

The similarity between the OTR and the DTR can also be seen when you compare the document type Customizing activities of both business objects. The DTR type Customizing contains exactly the same settings as the OTR, including settings concerning freight unit building, stage splits based on incoterm locations, and so on.

The DTR is the business document representation of the SAP ERP deliveries, as the OTR was for SAP ERP orders. When you are setting up the integration scenario between SAP ERP and SAP TM, you have different options for which documents you want to integrate and how the interaction of SAP ERP and SAP TM documents should work. The main question to answer is "What system should be the leading system for the transportation process?" The right answer to this question depends on what you want to achieve with the SAP ERP integration, as we will see in Section 4.1.5.

4.1.5 Integration Scenarios

If the SAP TM system is the dedicated system leading the transportation process, you would probably set up the integration scenario as shown in Figure 4.10.

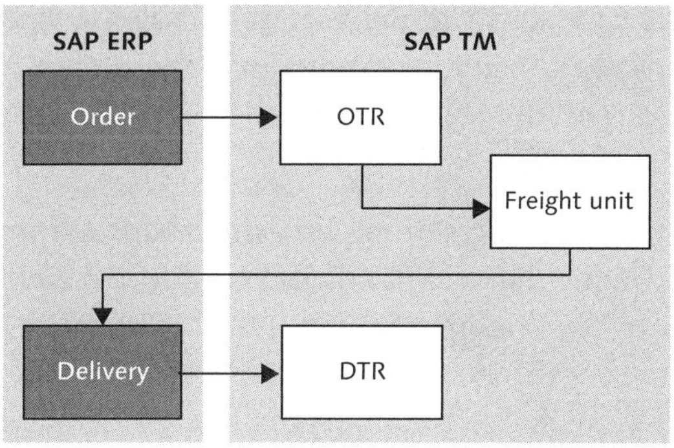

Figure 4.10 Integration Scenario with Delivery Proposals

The SAP ERP order is created in SAP ERP and then integrated into the SAP TM system. In the SAP TM system, the OTR document creates freight units automatically. Now the process continues in SAP TM. The freight units are planned on freight orders according to the delivery dates, service levels, incoterms, and so on.

The rationale for doing the planning in SAP TM is that SAP TM offers a more sophisticated means of transportation planning such as ocean and air schedules, connectivity to carriers, distance and duration information based on geographical information systems, and so on.

Once the planning of the freight unit is done, we know a more precise delivery date of the goods that were ordered in the SAP ERP order. This information can then be played back to the SAP ERP system, where delivery documents are created accordingly. This process step is called the *delivery proposal*.

Delivery proposal

Delivery proposals are used to propagate planning information from SAP TM to SAP ERP, where deliveries are then created. The process is called delivery proposal and not delivery creation because, in the end, SAP ERP decides how deliveries should be created. The information for the delivery proposal derives from the freight units.

Once you've finished planning the freight units in SAP NWBC, follow the menu path ERP Logistics Integration • Delivery Creation • Create Deliveries in ERP. When you enter a selection profile (preferably the same one you have used for planning), your freight units are displayed. Select the freight units you want to create delivery proposals for and choose Create Delivery Proposals. You can review the proposals at the bottom of the screen and then send the proposals to SAP ERP by clicking on Send Delivery Proposals to ERP.

The delivery proposals are created using settings and information from both freight units and OTRs. While the delivery dates are collected from the freight order that is now assigned to the freight units, the quantities are taken from the freight units themselves. SAP TM also tries to consolidate several freight units into one delivery proposal. If one OTR creates several freight units, which are then planned on the same freight order, SAP TM will consolidate these freight units to one delivery proposal. SAP TM also tries to consolidate several freight units of multiple OTRs if they are planned on the same freight order. However, the consolidation is only done if the SAP ERP order allows order consolidation. You can check this setting on the General Data tab of the OTR, where there is a flag telling you whether order consolidation is allowed.

Once the delivery proposals are sent to SAP ERP, delivery documents are created. SAP ERP might split one SAP TM-based delivery proposal by creating several deliveries, possibly due to additional split criteria in SAP ERP. However, SAP ERP does not change data such as dates or quantities sent from SAP TM and does not consolidate several delivery proposals into one delivery.

The delivery proposal in SAP TM also takes into account that SAP ERP consumes order schedule lines in chronological order. Therefore, SAP TM only proposes a delivery for a schedule line of an order item if all schedule lines of the same order item with earlier delivery dates already have an assigned delivery.

Delivery profile

In SAP TM you can also define how delivery proposals should be created. These settings are consolidated in a *delivery profile*. Though the delivery profile is optional, it makes sense to create a profile to reuse the same settings every time you want to create delivery proposals. You can define delivery profiles via the SAP NWBC menu path APPLICATION ADMINISTRATION • PLANNING • GENERAL SETTINGS • DELIVERY PROFILE. The settings you can define here concern how and if freight units can be consolidated into one delivery proposal and which freight units may not be consolidated.

One option of the delivery profile is to "fix" the planning result for freight units and freight orders by selecting the FIX PLANNING RESULTS checkbox shown in Figure 4.11. Fixing the planning means the planning results transferred to SAP ERP will not be changed in SAP TM any more. This is a good strategy because you might continue processing the transferred data in SAP ERP as well. You will learn more about fixing freight orders in Chapter 6. You can define how the delivery proposals should be created (e.g., one proposal per item, one proposal per OTR, and so on).

The last feature of delivery profiles that we want to mention is incompatibilities, which we introduced in Chapter 2. *Incompatibilities* in delivery profiles can be used to prevent certain freight units or items from being consolidated into one delivery proposal.

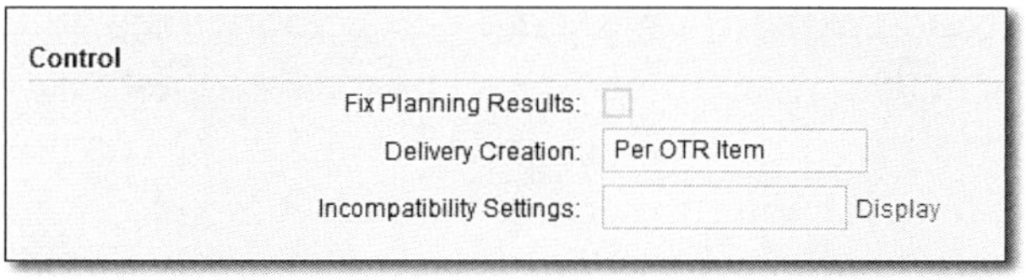

Figure 4.11 Delivery Profile

Delivery profiles are optional and especially unnecessary if you've already planned the freight units, because in this case, SAP TM considers the freight units consolidated on freight orders. Settings like incompatibilities are already taken into consideration during the planning process.

However, if you want to create delivery proposals *before* the freight units are planned, you can do this as well. In this case, the settings in the delivery profile are crucial to help SAP TM decide which freight units should be consolidated.

Delivery creation in background job

In a daily business, you will probably not manually create delivery proposals every time you have finished planning some freight units. Therefore, background report /SCMTMS/DLV_BATCH was created for you to run as a batch job. You need to define a selection profile selecting your freight units or freight orders; the background report does exactly what you can do interactively in SAP NWBC (skipping your review, of course).

After the delivery proposals have been sent to SAP ERP, deliveries are created. SAP TM gets information about the status of the delivery creation. The delivery type being used for the delivery creation is the one defined in the SD or MM setup. If you have activated the transfer to SAP TM of this delivery type as well, then not only will a short message about the delivery creation be transferred to SAP TM, but even more happens.

The delivery is integrated, and a DTR is created for the delivery, as already described in Section 4.1.4. You might be wondering what we can do with the delivery now, since the planning process has already happened. Recall that SAP ERP might create deliveries slightly differently from the SAP TM proposal; this information is then represented in the DTR.

OTR consumption

If you have looked at the Customizing of DTR types, you will see that you can also define whether freight units should be created automatically in the DTR type. In the process shown in Figure 4.10, the creation of freight units would be counterproductive because not only were freight units already created with the OTR, but they were already planned. The DTR will not, in this case, create new freight units. However, the freight units will be unassigned from the OTR and moved to the DTR. We call this process *OTR consumption* or *freight unit consumption*.

If you look into the document flow of the DTR now, you can see not only the SAP TM documents, but also the SAP ERP documents. What you will also notice is that the OTR is now a predecessor document of the DTR, and the DTR again is a predecessor document of the freight unit (even though the freight unit was created before the DTR).

Delivery split/update

There is only one difference in the DTR type Customizing: the setting concerning the delivery split/update types that can be assigned to the DTR type. A delivery split/update type determines how SAP TM handles updates or splits of freight units if SAP ERP deliveries are linked to these freight units. In general, we can differentiate between two scenarios of a delivery split or update: one triggered by SAP TM and the other triggered by SAP ERP or SAP Extended Warehouse Management (EWM). Please note that the delivery split or delivery update process only works for outbound deliveries, not for inbound deliveries.

A delivery split or update triggered by SAP TM is caused by a planning being done in the SAP TM system. This means that the freight unit was planned on a freight order, which caused an update of one of the following dates that are linked to dates in the SAP ERP delivery:

- Goods issue date
- Loading date
- Transportation start date
- Delivery date

Alternatively, the planner may have decided to split a freight unit that was integrated from an SAP ERP delivery and planned the two freight units on different freight orders. In this case, the original SAP ERP delivery has to be split. The new delivery is then integrated into SAP TM again, and the freight unit is reassigned to the new DTR.

With the delivery split/update type, you can define whether corresponding information should be sent to SAP ERP to make changes about the SAP ERP delivery. In addition, you can influence the planning behavior in SAP TM. You can find the Customizing activity for delivery split/update types via the IMG menu path SAP TRANSPORTATION MANAGEMENT • TRANSPORTATION MANAGEMENT • INTEGRATION • ERP LOGISTICS

Integration • Delivery-Based Transportation Requirement • Define Delivery Split/Update Types.

As you can see in Figure 4.12, you can first define whether an update or split should be sent to SAP ERP at all. You can also define if late planning changes are allowed. This means that if the delivery cannot be changed any more (e.g., because the goods movement status is already *Partially Completed* or *Completed*), then you can prevent planning changes of the corresponding freight unit in SAP TM by issuing an error message. Alternatively, you can warn the planner by issuing a warning message that allows late planning changes.

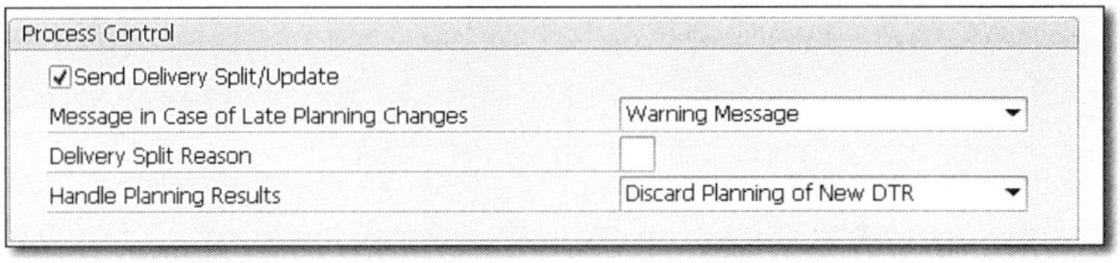

Figure 4.12 Delivery Split/Update

With the delivery split reason, you can propagate a split reason code from SAP TM to SAP ERP with which the SAP ERP delivery split is executed. Please note that the reason code defined in the SAP TM system needs to have the exact numeric value as in SAP ERP because the mapping is done purely by numeric values.

The last setting in Figure 4.12 concerns the second scenario—the delivery update triggered by SAP ERP or SAP EWM.

Delivery Update Triggered by SAP EWM

As of SAP TM 9.0, SAP TM can integrate with SAP EWM. However, the integration is always done via the SAP ERP document's delivery or shipment. Also, in this scenario, a delivery update triggered by SAP EWM only triggers the update of the delivery in SAP ERP. The update message to SAP TM is then done by SAP ERP.

Technically, there is no difference between the update triggered by SAP ERP and the update triggered by SAP EWM. Concerning SAP TM, the update message is, in both cases, only created after the SAP ERP delivery has been updated.

If the SAP ERP delivery was split, a new DTR is created, and the freight unit that was assigned to the original DTR is also split and reassigned accordingly. If the freight unit was already planned, you now need to decide what you want to do with the planning you have already performed. You can either discard the planning of the new DTR's freight unit, discard the planning of the original DTR's freight unit, or discard any planning that was performed with freight units connected to the updated delivery.

After you have defined the delivery split/update type, you can assign this type to the DTR type in Customizing.

In the integration scenario described previously, SAP TM was the leading system; it decided not only how the planning should be done, but also whether and how the items of the order were split or consolidated. In some cases, it makes sense to leave that decision in SAP ERP. Since OTR and DTR can work without each other in SAP TM, you can decide whether to leave out the integration of either one.

Figure 4.13 shows how the integration scenario works if SAP ERP is still the leading system for delivery creation and item order consolidation. In this case, the integration of the order is optional, represented by the dashed line on the figure. If an OTR is created, no freight units should be created for it. Only after the delivery is created for the order and the DTR is created in SAP TM should freight units be created.

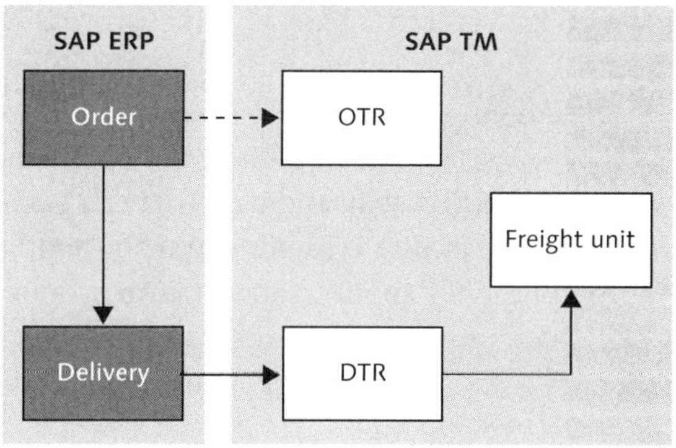

Figure 4.13 Integration Scenario without Delivery Proposals

Those freight units now better represent the splits and consolidations that have been done previously in SAP ERP. Planning can now start, though the planning results will not affect the delivery dates of the DTR anymore. Therefore, make sure the dates in the freight unit are considered as hard constraints during planning. (Refer to Chapter 6 for more information about soft and hard constraints in planning.)

4.1.6 Sales Order Scheduling

In addition to the creation of individual documents in SAP TM to process the transportation planning in a company, it is also possible to use SAP TM to schedule transports without creating any documents.

This functionality is called *sales order scheduling* and is, as the name suggests, only provided for SAP ERP sales orders. When creating an SAP ERP sales order, you have to provide some mandatory fields such as the sold-to and ship-to party, the material, the plant where the material is provided, and the delivery date requested by the customer.

At the top of the sales order screen you will find a new, SAP TM-specific button called DOCUMENT. When you click on this button, the sales order scheduling is executed. Sales order scheduling is supposed to define the exact dates for the sales order, meaning not only the delivery date but also the material availability date, loading date, and goods issue date.

Once the sales order scheduling is triggered, the sales order is transferred to SAP TM, where an OTR is created. Basically, the process is exactly like the actual creation of OTRs and freight units; the only difference is that this OTR and all subsequent documents will not be saved in the SAP TM system. The OTR type used for the sales order scheduling has to trigger freight unit building automatically and has to have a planning profile assigned. For the freight unit(s) created, a transportation proposal is started. You will find out more about transportation proposals in Chapter 6, Section 6.3.6. The first result of the transportation proposal is then transferred back to SAP ERP and the OTR, and freight units no longer exist in the SAP TM system.

Backward scheduling of transportation proposal

When the sales order is transferred to SAP TM, the requested delivery date entered on the item line is considered as the requested delivery start date in the freight unit. The transportation proposal therefore tries to take this date into consideration and determine the start date of the transport including the loading date. The loading date, transportation start date, and delivery date are then sent to the sales order in SAP ERP, where they will be considered in the following way:

- The delivery date will be the delivery date in the sales order
- The transportation start date will become the goods issue date in the sales order
- The loading date will be the loading date in the sales order
- The loading date will be the basis to calculate the material availability date in the sales order

Interaction with ATP check

Once the dates of the schedule lines in the SAP ERP sales order are defined, an available-to-promise (ATP) check can be triggered to find out whether the material can be provided in the plant requested on the calculated material availability date. If the ATP check fails to provide the material on the calculated date, it will provide the next possible date. With this new date from ATP, the sales order scheduling is triggered again, now with the newly determined material availability check as the basis for the transportation proposal, which is now doing a forward scheduling from the material availability date to determine a new delivery date.

Sales order scheduling vs. route-based scheduling

The advantage of the sales order scheduling interaction between SAP ERP and SAP TM compared to the route-based scheduling in SAP ERP is that it can take the more sophisticated SAP TM planning constraints in transportation planning into consideration. For example, if resources are not available or ocean schedules run infrequently, this can be taken into consideration in sales order scheduling but not in the route-based scheduling in SAP ERP.

Synchronous communication

Sales order scheduling is the only standard communication between SAP TM and SAP ERP that is done synchronously. This means that when you click on the DOCUMENT button in the SAP ERP sales order, you have to wait until the transportation proposal returns the results to the sales order. This is because the SAP TM system does not save any documents for this process, so there is no possibility of getting the documents in SAP TM other than right at the processing time.

4.2 Forwarding Orders and Forwarding Quotations

In Section 4.1 we talked extensively about the integration of transportation requests from SAP ERP. The integration of SAP ERP documents was introduced to SAP TM with release 8.0, which was meant to be the release focusing on the shipper's business.

As you have learned so far in this chapter, integrated transportation requests only serve the purpose of transferring the transportation-relevant information of SAP ERP orders to SAP TM to execute the transportation process there. OTRs and DTRs are not sophisticated enough to serve as the transportation request document for LSPs or carriers. In these businesses, more information is required than simply what goods need to be transported from where to where on what date. These businesses often add much more information to the transportation request when transportation service level codes, transportation charge management, and consolidation come into play. Simply stated, transportation is the core business of LPSs and carriers, which explains why more emphasis is put on the transportation information.

Like OTRs and DTRs, the forwarding orders and quotations generally act as the beginning of the operational process in transportation management. The issuing of a forwarding order creates a contract between two companies.

The forwarding order is often called the "typewriter" of SAP TM because the document needs to be created mostly manual. But this term underestimates the importance of the forwarding order for the transportation process. The forwarding order is the only document in which you can manually define the data that is used for the entire transportation process of planning, charge management, execution, and so on.

The central business object behind the forwarding order is the transportation request object, which we first discussed in the context of OTRs and DTRs. The forwarding order makes use of the same object as is used by OTRs and DTRs, which means that comparable data is written into the same database areas.

Creating forwarding orders

The forwarding order is the central order business object in SAP TM, and it helps you perform all of the important order-taking processing steps in order management. A forwarding order can be created in different ways:

- **Manual creation**
  This is the most common way of creating transportation requests in SAP TM. A customer relations manager of the LSP or carrier talks to the customer and enters all data manually in SAP NWBC.
- **Incoming EDI message**
  Many big LSPs and carrier companies have been using EDI communication for a long time. Therefore, it is only logical that they also prefer EDI communication for the creation of forwarding orders in their systems. SAP TM makes this functionality possible by providing B2B web services that automatically create a forwarding order. This web service integration can be compared to the integration of SAP ERP orders, but more information can be transferred, such as transportation stages, dates concerning the transportation, and so on.
- **Use of a template**
  In many cases, customers order the same type of transport on a regular basis. To avoid having to type in the same data every time, the order clerks of the LSP or carrier use templates. A new forwarding order can then be created as a copy from the template. We talk about templates later in this chapter.

Saving documents

A fundamental feature of the forwarding order is that the document can be saved at practically any time in the order-taking process. If you want to automatically create freight units from the forwarding order, a few prerequisites need to be met. However, the forwarding order can still be saved in any incomplete state. Once the forwarding order contains sufficient information for freight unit building, the freight units are created. This helps customers and LSPs to enter orders even if some information is missing. At LSPs or carriers it is also often the case that multiple employees are involved with the order-taking process. For instance, the customer's contact person enters all the customer's information, but the forwarding order is only complete once a transportation clerk has added data that is relevant for planning and execution.

POWL for forwarding order management

When you browse the SAP TM system in SAP NWBC, you will notice that like the SAP ERP integration, forwarding orders have a dedicated work area, called Forwarding Order Management. When you access this work area, you can see that the menu path WORKLISTS • OVERVIEW FORWARDING ORDER provides a POWL with all existing forwarding order documents and their related documents.

As you can see in Figure 4.14, SAP TM provides you with many preconfigured queries categorizing forwarding orders into the different transportation modes.

At the top of each POWL query are some buttons that you can use to perform actions on the forwarding order documents without opening the documents themselves. For process steps like collective invoicing and consolidated freight unit building (explained in detail in other chapters), this feature comes in very handy because the actions can be executed on several documents together.

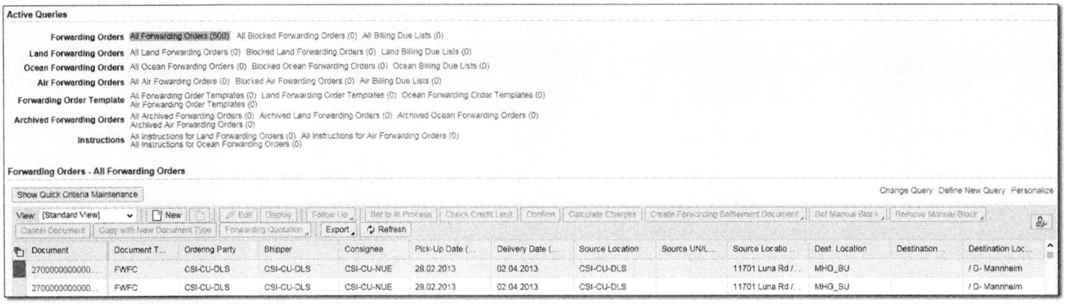

Figure 4.14 POWL for Forwarding Orders

By clicking on SHOW QUICK CRITERIA MAINTENANCE, you can set more filters on the predefined queries. A screen area opens where all possible filter values can be defined. The quick criteria are saved for the user so that every time you enter the POWL query again, your last entries will be remembered. If you want to avoid this, you can alternatively use the filter of the POWL table itself by clicking on a column's header.

If you have some selections you want to permanently display in the forwarding order POWL, you can use the CREATE NEW QUERY link on the top-right corner of the POWL table. In a guided procedure, you can enter your selection criteria, name the POWL query, and assign the query to one of the query categories (the lines on top of the table). Note that this newly created query is only available for your user and cannot be used by other users.

4.2.1 Forwarding Orders

Since the forwarding order is a very large and complex business object, this section is the longest section of the chapter. But you will not find pages explicitly dedicated to the Customizing of the forwarding order type. Instead, we go through the forwarding order document tab by tab and delve deeper into the functionality of the different fields. When Customizing influences the process flow of the fields, we mention this while covering the corresponding section of the forwarding order document. You can find the forwarding order type Customizing via the IMG menu path SAP TRANSPORTATION MANAGEMENT • TRANSPORTATION MANAGEMENT • FORWARDING ORDER MANAGEMENT • FORWARDING ORDER • MAINTAIN FORWARDING ORDER TYPES. If other Customizing activities are involved in the forwarding order, they are mentioned when they become relevant.

If you have worked with SAP TM 8.1, you will notice that SAP TM 9.1 has fewer links to create a forwarding order. In SAP TM 8.1, there were creation links for forwarding orders for different transportation modes—one link for a land forwarding order, one link for an air forwarding order, and so on. This differentiation has been removed, but the different transportation modes are still in place. You can predefine the transportation mode of a forwarding order in the forwarding order type. However, if you want to use one forwarding order type for several transportation modes, you can leave this setting empty. When you follow the SAP NWBC menu path FORWARDING ORDER MANAGEMENT • FORWARDING ORDER • CREATE FORWARDING ORDER, you will see that not only can you enter a forwarding order type, but you can also enter data such as the transportation mode, a template number, or a forwarding agreement to which the forwarding order is related. You will learn more about charge management for forwarding orders later in this chapter. If you haven't defined a transportation mode in the forwarding order document type, then you have to decide on a mode here. Please note that defining the transportation mode as air, for example, does not mean you cannot create transportation stages of other transportation modes such as road or rail. The transportation mode decision is only relevant for the naming of fields in the forwarding order and for what fields are displayed or hidden.

General Data

Once you have defined a document type, you are directed to the forwarding order document. If you haven't defined a template, you will see that the document is almost empty; only some fields are prepopulated. The General Data tab shown in Figure 4.15 displays the most important information that is globally relevant for the transportation contract between a customer and an LSP or a carrier.

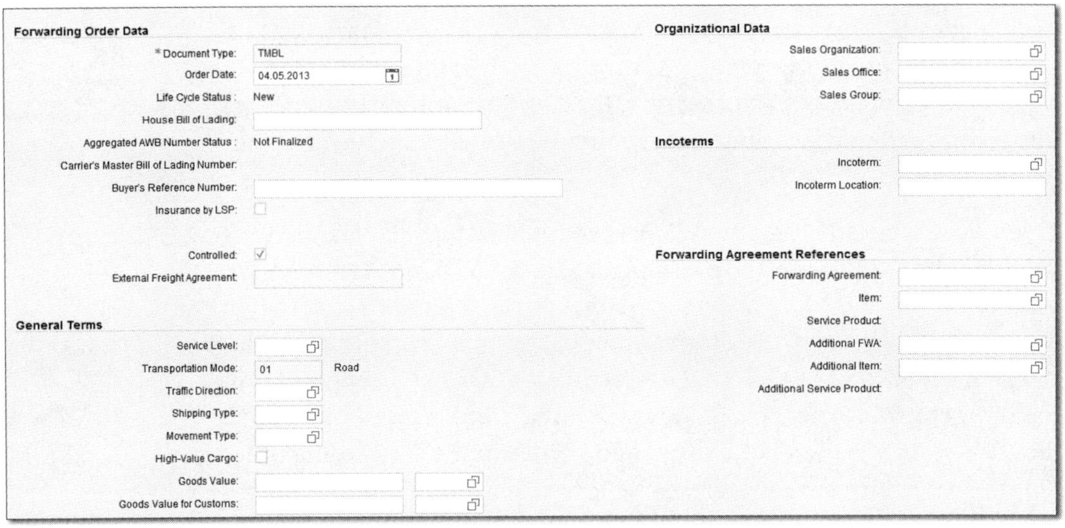

Figure 4.15 General Data of a Forwarding Order

The topmost field (apart from the document type) is Order Date. This field is prepopulated with the current date, but you can alter it as necessary. The order date is important not only for reference, but also for charge management. Rate tables that are used in charge management can have different validity dates. Depending on the settings you have made for charge calculation, you can calculate charges based on either the actual transportation date or the order date.

Organizational data and incoterms

In the top-right corner of the GENERAL DATA tab, you can find the fields for the organizational data—the sales organization, office, and group that is responsible for the forwarding order. In Chapter 3 we discussed the importance of organizational units in SAP TM, so we do not go into detail here again. However, it is important to note that entering a sales organization is one of the prerequisites of creating a freight unit. If no sales organization is entered, neither automatic freight unit building nor manual freight unit building will succeed.

Like with integrated SAP ERP orders, the incoterm is a very important piece of information about the transportation responsibilities. You can enter the same incoterms here that you already defined in Customizing. The incoterm location here is free text because it is agreed upon with the customer. Since you can define your stages in a forwarding order manually, no mapping or stage splitting are done based on the incoterm location entered.

However, the incoterm is considered in charge calculation and settlement. Depending on the incoterm you use, the stages are charged against different party roles of the forwarding order. You can assign a payer party role (that is, which party role should pay for which stage types) in Customizing by following the IMG menu path SAP TRANSPORTATION MANAGEMENT • TRANSPORTATION MANAGEMENT • FORWARDING ORDER MANAGEMENT • DEFINE DEFAULT AGREEMENT PARTY ROLES FOR STAGES. We look at stage types later in this section.

Uncontrolled transportation

The CONTROLLED checkbox in the FORWARDING ORDER DATA section on the GENERAL DATA tab (shown in Figure 4.15) defines whether the ordered transportation is a controlled or uncontrolled transport. With uncontrolled transports, the LSP organizes the entire transportation chain, including the main stage; however, the main stage is charged not from the carrier to the LSP, but directly to the ordering party of the entire transport. You can define an uncontrolled transport by deselecting the checkbox. If you deselect the checkbox, the EXTERNAL FREIGHT

Agreement field becomes editable. You can then enter the freight agreement that exists between the carrier and the ordering party.

When doing further planning of the freight units belonging to a forwarding order that have been declared uncontrolled transports, you can find the Controlled checkbox and the external freight agreement on the Terms and Conditions tab of the freight booking that covers the main stage of the freight units. Note that you can consolidate uncontrolled freight units on one freight booking only if they all have been assigned the same external freight agreement. The pre-carriage and the on-carriage of an uncontrolled transport is still charged from the carrier to the ordering party via the LSP. Figure 4.16 illustrates the difference between controlled and uncontrolled transports, which is a scenario that often takes place in sea and air transportation and is the main carriage process.

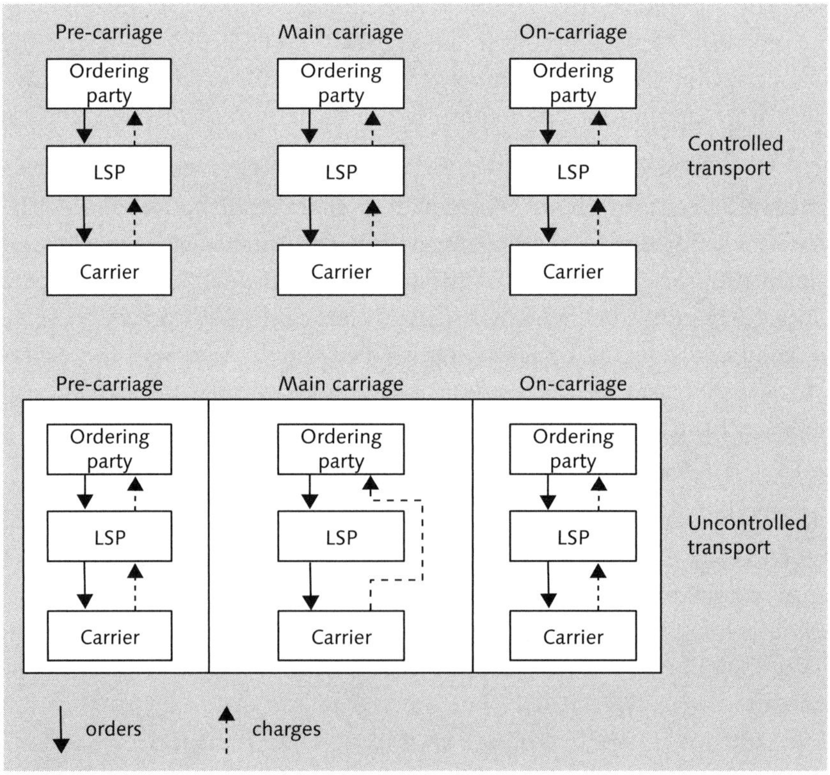

Figure 4.16 Uncontrolled and Controlled Transportation

We cover further details of the different transportation modes and their impact on the forwarding order later in this chapter.

Shipping types

One of the most important features of SAP TM 8.1 was the introduction of movement types and shipping types. These settings are crucial for the air and sea transportation processes; they influence the data entry and validation of the forwarding order. Shipping type and movement type are also prerequisites for the creation of freight units from a forwarding order.

The *shipping type* defines what kind of transportation the customer has ordered. In the transportation business, the differentiation is, roughly speaking, between a full container load (FCL) and a less than container load (LCL). In SAP TM these processes are shipping types. Considering that the terms FCL and LCL are mainly used in sea transportation, the shipping types also may depend on the transportation mode.

You can define shipping types in Customizing by following the IMG menu path SAP Transportation Management • Transportation Management • Forwarding Order Management • Define Shipping Types. In this Customizing activity, you define all the shipping types you need. The shipping type entered in the forwarding order document determines what items may be defined on the item area of the forwarding order. In Customizing you therefore need to define your shipping type, whether all cargo items need to be assigned to an equipment item (e.g., a unit load device in air, a container in sea, or a railcar in rail), or whether such equipment items are allowed at all.

Unit Load Device

When talking about packaging units in air transportation processes we will use the term *unit load device* (ULD) to represent special types of pallets or containers used in air transportation.

If you have entered the shipping type ULD that is depicted in Figure 4.17, you will get an error message in the forwarding order if not all of your items are assigned to equipment items. We take a look at the definition of items later in this chapter.

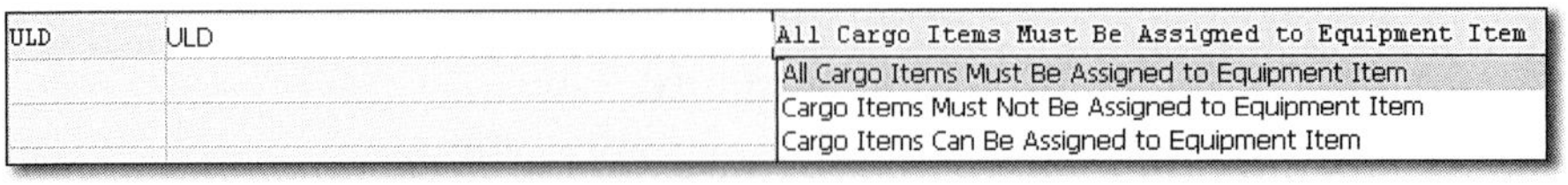

Figure 4.17 Shipping Type Definition

In this Customizing activity you can also define for which transportation modes the shipping type is valid. This covers the fact that the terminology is different in different transportation modes.

The shipping type can be entered manually on the forwarding order or be preset in the forwarding order document type Customizing.

Movement types

The second important feature introduced with SAP TM 8.1 is the *movement type*. The movement type in the forwarding order determines its routing. It is worth taking a detailed look at the impact of the movement type, since you have to deal with it when creating forwarding orders.

In general, movement types can be defined in Customizing in the IMG menu path SAP TRANSPORTATION MANAGEMENT • TRANSPORTATION MANAGEMENT • FORWARDING ORDER MANAGEMENT • DEFINE MOVEMENT TYPES. Here you can simply define your movement type and whether the location assigned to the shipper should automatically be taken as the source location of the forwarding order. The same setting can be done with the consignee and the destination location.

By now we have only defined a movement type code. The influence on routing is determined in the Customizing activity that you can find by following the IMG menu path SAP TRANSPORTATION MANAGEMENT • TRANSPORTATION MANAGEMENT • FORWARDING ORDER MANAGEMENT • DEFINE STAGE TYPE SEQUENCE FOR MOVEMENT TYPES.

Figure 4.18 shows the standard Customizing for the default movement type DD (door-to-door). As you can see, you can define a stage type sequence with sequence numbers. As with all sequence-related Customizing activities, the numbers do not have to exactly sequential, but the numerical ordering of the numbers. For the movement type, you define the stage types that are being used. We talk about stage types later in the section about the STAGES tab. For now we can state that we bring different stage types into a sequential order. In addition, we define whether a stage of a certain type may occur or has to occur. For the movement type in our example, only the main carriage (stage type 03) must occur. The STAGEPROP. checkbox defines whether some stages on the STAGES tab should already be created when the corresponding movement type is selected on the GENERAL DATA tab.

Define Stage Type Sequence for Movement Type

Mov. Type	Seq. No.	Stage Type	StgeTpeOcc	StageProp.	Det. Rule	Set. Rule
DD	1	01	Stage type can occur in any...	☑	Always Relevant for Planning	Not Relevant for Internal S...
DD	2	02	Stage type can occur in any...	☐	Always Relevant for Planning	Not Relevant for Internal S...
DD	3	03	Stage type must occur at le...	☑	Always Relevant for Planning	Not Relevant for Internal S...
DD	4	04	Stage type can occur in any...	☐	Not Pl.-Rel. if Ex. and Sal...	Not Relevant for Internal S...
DD	5	05	Stage type can occur in any...	☑	Not Pl.-Rel. if Ex. and Sal...	Ready for Execution

Figure 4.18 Stage Type Sequence for Movement Type

The next column defines whether the stage is relevant for planning or only statistical. On the STAGES tab you can manually assign a planning and execution organization to each stage. The decision about whether the stage is relevant for planning is based on the setting in this Customizing activity, which again considers the planning and execution organization. Some of the settings you can choose in this column compare the planning and execution organization with the sales organization of the forwarding order. If these two organizations belong to different companies or countries, the stage is not declared as relevant for planning. But you can also take the easier route here (if that applies to your business process) and simply state whether this stage is always or never relevant for planning.

The last column in Figure 4.18 (SET. RULE) specifies from what point in time an internal settlement of the stage can be created. You can choose different execution statuses here to define in what execution status the corresponding stage of the freight unit must be to start an internal settlement.

Shipper's and buyer's consolidation

Another transportation process that is often used in air and sea transportation is the *shipper's consolidation* in export transportation and the *buyer's consolidation* in import transportation. If your forwarding order has a corresponding transportation mode assigned to it, you can set this information on the GENERAL DATA tab of the forwarding order. The transportation for the LSP is considered like an FCL or ULD transportation because the shipper organizes the transportation to the port, but different house bills of lading need to be issued for each item of the forwarding order, even though only one freight unit was created.

House waybills

In the transportation business, the house bill of lading (HBL) number—not the forwarding order number—is the relevant number for identifying the entire transportation. You can find the corresponding field on the General Data tab of the forwarding order.

You can manually enter a waybill number or HBL number in this field. This is then propagated to the freight units that belong to this forwarding order.

However, the assignment of waybill numbers can also be triggered automatically using *waybill stocks*. To do this, activate the use of waybill stocks in the Customizing of the forwarding order document type by selecting the Enable Waybill Stock checkbox and assigning an HBL or house air waybill (HAWB) strategy. The strategy defines how the HBL should be created: per forwarding order, per container, per shipper/consignee combination, and so on.

To draw HBL numbers, you need to define waybill stock types in Customizing. Follow the IMG menu path SAP Transportation Management • Transportation Management • Master Data • Waybill Stock • Define Waybill Stock Types.

The waybill stock type defines how the HBL number is put together. Figure 4.19 shows how a waybill stock type is assigned to a transportation mode.

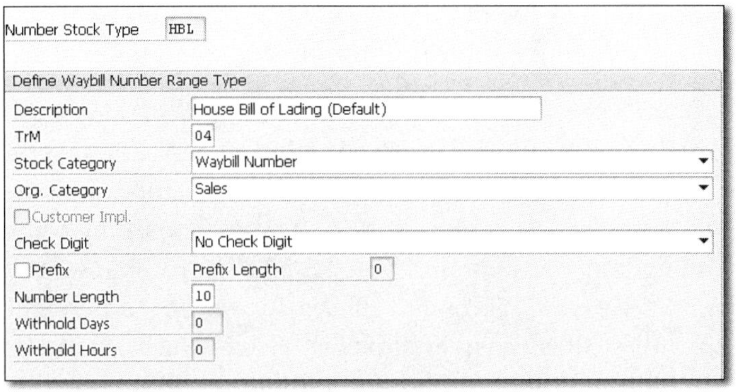

Figure 4.19 Waybill Stock Type Definition

You can further define whether the number resulting from this stock type should be used as a waybill number or a tracking number. On the bottom of this Customizing screen, you define how the number should be put together, whether a prefix will be assigned, the length of the number, and how a potential check digit should look.

You can also define how long the number should be withheld after being returned.

After defining a number stock type, we can use it to create a waybill stock. In SAP NWBC go to Master Data • General • Overview Waybill Stock to create a new waybill stock based on the number stock type. Here you can define a number range from which a number should be drawn. At the bottom of the screen, you can then define when this waybill stock should be taken into account by defining certain combinations of sales organizations and ordering parties. If the combination of ordering party and sales organization of the forwarding order matches one of the combinations defined in the waybill stock, a number is drawn from here.

If you choose to assign an HBL number automatically, the HBL field on the forwarding order is not editable. Instead you need to select HBL • Draw HBL Number from the action toolbar at the top of the forwarding order screen. This only works after the sales organization, ordering party, and cargo have been entered.

Note

You can find further details about waybill stocks and waybill numbers in Chapter 8, Section 8.1.

Goods value

On the General Data tab of the forwarding order, you can also enter the value of the goods that need to be transported. This information may become relevant for charge calculation or customs. Please note that this information is only a manual entry. The values are not taken from product master data.

Service level

We talked about the transportation service level codes when we discussed SAP ERP order integration. You can manually assign a service level code to the forwarding order. The service level might influence your planning and charge calculation if you have made corresponding entries.

If you want to define your own service level codes, do this via the IMG menu path SAP TRANSPORTATION MANAGEMENT • TRANSPORTATION MANAGEMENT • FORWARDING ORDER MANAGEMENT • DEFINE TRANSPORTATION SERVICE LEVEL CODES. If your service level code only applies to certain transportation modes, then you can define this here, too.

Items

At the bottom of the forwarding order document you will find the table of items that are ordered to transport in this forwarding order. You can define several items and create item hierarchies that represent how the packaging was done. Figure 4.20 shows a simple example of an item hierarchy. You can create an item hierarchy like this by choosing INSERT • CONTAINER to insert the first item. To create a subordinate item, select the superordinate item (in this case, the container) and again choose INSERT • PACKAGE. The new item will be created as a subordinate item that is displayed in a hierarchical way. Alternatively, you can enter the data of the items directly into the table and then create a hierarchy by dragging and dropping a subordinate item onto a superordinate item.

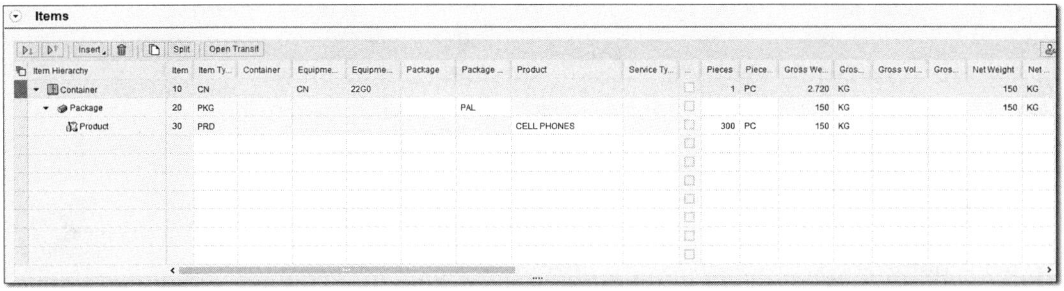

Figure 4.20 Item Area of the Forwarding Order

Item types When you enter an item, you are warned by an error message that entering an *item type* is mandatory. The item type encompasses several settings relevant for the item.

Figure 4.21 shows the information you can define in the Customizing of the item types. The ITEM CATEGORY field defines what kind of item this item type describes. It also impacts the way an item hierarchy can be built.

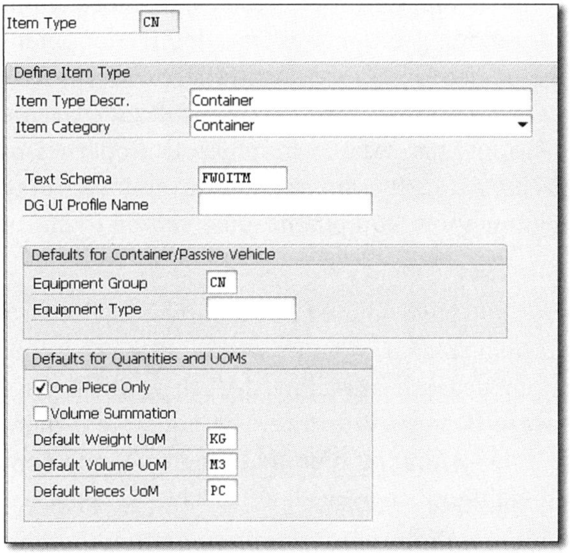

Figure 4.21 Item Type Definition

Item Hierarchy

SAP TM predefines what an item hierarchy can look like. Whether a subordinate-superordinate item relationship is allowed depends on the item category assigned to the item type. In general, SAP TM allows only the following item hierarchies:

- Container/trailer (may contain a package or product item)
- Package (may contain a package or product item)
- Product (must be the lowest item in the hierarchy)

You can assign text types that you have defined in Customizing (recall these from Section 4.1.3). The text is displayed on the NOTES tab of the forwarding order as well as on the freight unit representing this item.

If you want to define an item type that is relevant for dangerous goods processing, you can assign a dangerous goods UI profile to this item type. You will find further information about dangerous goods UI profiles and processing in Chapter 9.

If you have defined an item category as a container or passive vehicle, you can assign equipment groups and equipment types to the item type, and all relevant data from the equipment type is then automatically put on the item data in the forwarding order. You can define equipment types via the IMG menu path SAP Transportation Management • Transportation Management • Master Data • Resources • Define Equipment Groups and Equipment Types. The physical properties of the equipment type will be used as the tare weight in the forwarding order item. More information about equipment types can be found in Chapter 3, Section 3.3.

Recall that the shipping type you have defined on the forwarding order's general data defines what item categories you may use. If the shipping type defines that no equipment items are allowed, then you will get a corresponding error message if you try to insert an equipment item. However, if you have defined a shipping type that requires equipment items, you cannot create freight units before all items have been packed into equipment items.

In addition, you can limit the number of item types you can use on a forwarding order by assigning item types to the forwarding order type. You can do this in Customizing by following the IMG menu path SAP Transportation Management • Transportation Management • Forwarding Order Management • Forwarding Order • Assign Item Types to Forwarding Order Types. Once you have assigned some item types to your forwarding order type, no other item types can be selected in your forwarding order. If you have assigned several item types of an item category, you can define a default item type. When you insert a certain item category in the forwarding order, the default item type is automatically used. The default item type is especially important when using the B2B integration for automatically creating forwarding orders. Because the web service does not carry the information of the item type, this information has to be taken from Customizing. You can also assign instruction sets to the item types of a forwarding order.

Assignment of passive vehicles

If you define containers as equipment items in the forwarding order's items, you are working with anonymous entities, if you wish. Containers are not defined as master data such as vehicle resources. If you define a forwarding order for rail or road transportation, however, you can assign a passive vehicle from master data (a rail trailer or truck trailer) to the forwarding order and use it as an equipment item.

Quantities

When you define an item hierarchy, the weight of the products of the lowest items in the hierarchy are added up on the highest item level. On each level, the gross weight of the subordinate item is taken over as the net weight of the superordinate item and added with the tare weight to determine the gross weight. This again is taken over as the net weight of the next superordinate item, and so on.

If you have defined default units of measurement on the item level, the quantities of the lower item levels will be converted to the default unit of measurement. Please note that if you have chosen a product item from product master data, no quantities will be moved into the forwarding order item; the quantities need to be inserted manually.

In forwarding orders, other than OTRs and DTRs, you do not need product master data for your product items. Even though you can choose product master data for your product items, you can also insert free text into the corresponding field. This is especially handy for LSPs and carriers that transport different materials every day because they do not need to create master data for each material

Item split

If a customer orders the transportation of 300 cell phones, you can enter this in one item line defining a quantity of 300 pieces. However, your transportation process may require you to treat each cell phone individually to enable you to enter specific data for each phone or to simplify freight unit splitting. If so, you can still enter one item line and define pieces and weight for all 300 cell phones, select this line, and click on the SPLIT button at the top of the item table. The item line is automatically split into 300 individual lines. In addition, the gross weight entered is distributed equally among the new items.

Item details

When you select an item line, a new screen area appears below the item table. In this screen area you can enter several pieces of information specific to the individual item. This information is also propagated to the corresponding freight unit in freight unit building. The item details contain several tabs.

The first tab is PRODUCT DETAILS. You can enter item-specific details here. Some data, such as quantities, is moved from the item table. Other data you can only define here in this screen area. If you enter a goods value for an individual item, this information is also moved to the corresponding field on the GENERAL DATA tab of the forwarding order. The field is then not editable any more, but the system expects you to enter this information on every item as necessary. On the GENERAL DATA tab, the values of the individual items are then added up.

The LOCATIONS AND DATES/TIMES and BUSINESS PARTNER tabs are only visible in the item details if, in the forwarding order type Customizing, it was defined that locations, dates, and business partners cannot be globally defined in the forwarding order. The SAME LOCATIONS AND BPs flag in Customizing determines whether you are required to enter the transportation locations and business partners on the item level or whether it is sufficient to define them globally on the forwarding order.

If the locations and business partners are defined on the item level, this differentiation is also taken into consideration by freight unit building as a split criterion. The tabs in the ITEM DETAIL screen area are exactly the same as on the forwarding order, so we do not go into detail about defining business partners and locations here, but instead cover it later in this chapter.

You can add references to the item on the document REFERENCES tab. In many cases this tab needs to be filled manually because the document references might be system independent. You can reference a specific item of the referenced document. To differentiate different document types, you need to use business transaction document types that you can define in Customizing via the IMG menu path SAP TRANSPORTATION MANAGEMENT • TRANSPORTATION MANAGEMENT • GENERAL SETTINGS FOR ORDER MANAGEMENT • DEFINE BUSINESS TRANSACTION DOCUMENT TYPE CODES. As already mentioned, the document type codes can be defined system independently if you want to reference to a photo on a server, a phone call protocol, and so on. You can also reference an item from the references document. You can also define type codes for items following the IMG menu path SAP Transportation Management • Transportation Management • General Settings for Order Management • Define Business Transaction Document Item Type Codes. Document type codes and item type codes are not linked to each other. For implementation projects, we recommend that you use these type codes as often as the document references are used (i.e., charge calculation or routing decisions).

On the Customs tab you can define customs-relevant data. You will find more about customs declarations when we cover SAP Global Trade Services (GTS) integration in Chapter 9.

During freight unit building, the quantities of the forwarding order items will be moved into the freight unit. However, if in the execution process it is discovered that the actual quantities of the freight unit differ from the quantity declared in the forwarding order, the quantities will be changed in the freight unit. This results in a recorded discrepancy between the freight unit's item and the forwarding order's item. This discrepancy is displayed on the Discrepancies tab of the item details. The current status of the discrepancy handling is also recorded here.

Business Partners

Once you have defined all relevant data on the General Data tab and you have entered the items that need to be transported, you can now assign business partners to the transportation contract.

To sketch a very simple example first, only two business partners are assigned to the forwarding order: the shipper, where the goods are picked up, and the consignee, where the goods are delivered.

In the forwarding order, the business partners take over different responsibilities. These responsibilities are reflected by *party role codes* that you can see on the Business Partners tab. You can add party role codes in Customizing via the IMG menu path SAP Transportation Management • Transportation Management • Business Partners • Define Party Roles.

As you can see in Figure 4.22, the Business Partners tab includes a table where you can assign business partners to the party roles. The party roles Ordering Party, Shipper, and Consignee are mandatory in every forwarding order.

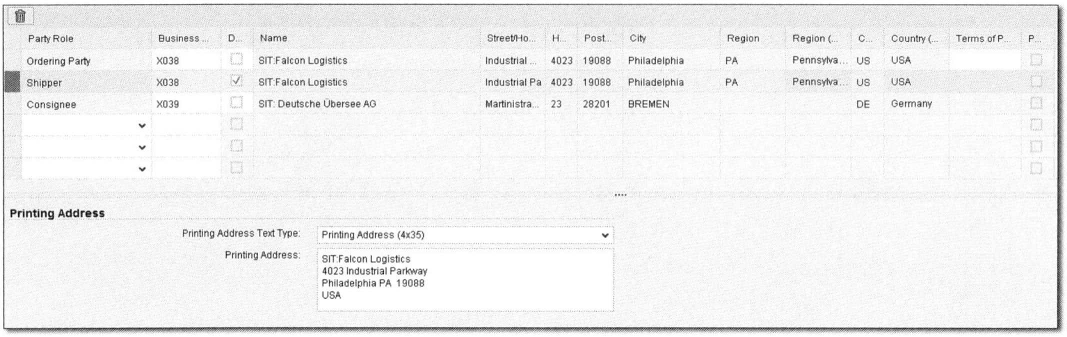

Figure 4.22 Business Partners in the Forwarding Order

The business partners you assign to the party roles need to be maintained as master data either using the CIF integration from SAP ERP or by manual creation in SAP TM. As you can see in the figure, the addresses of the business partners are taken from master data. However, if you want to use a different address than the one defined in the master data, you can do so by selecting the Deviating Address checkbox. If you select the row of the party role whose address you want to change in the forwarding order, you can change it in the Printing Address field below the table. This field is prepopulated with the address from master data. The address defined here is used for document printing.

Partner determination profile

Only three party roles are mandatory in a forwarding order unless you define differently. *Partner determination profiles* define which additional party roles are mandatory on a forwarding order. You define partner determination profiles in Customizing via the IMG menu path SAP Transportation Management • Transportation Management • Master Data • Business Partners • Define Partner Determination Profiles. In a partner determination profile, you select the party roles that you want to define as mandatory in the forwarding order. You can also define how business partners are copied from one party role to another. In the example in Figure 4.22, the ordering party and the shipper are the same business partner. Therefore, in the partner determination profile we could define that the business partner of the ordering party should be copied to the shipper, as shown in Figure 4.23.

Party Role	Sequence	Edit Level	Srce Type	Srce Role	Source Role Desc.	Not Mod.
6	2	Mandatory	Party Role	1	Ordering Party	☐

Figure 4.23 Partner Determination Profile

With the Not Modifiable column in the partner determination profile, you can also specify whether to allow printing address changes.

The partner determination profile can be assigned to the forwarding order type directly in Customizing. But depending on the transportation scenario, you might not be able to define the mandatory party roles with only one partner determination profile. Instead, the mandatory party roles can depend on the incoterms assigned to the forwarding order. Therefore, you can also define the partner determination profile depending on the incoterms and the forwarding order type. To do so, follow the Customizing menu path SAP Transportation Management • Transportation Management • Master Data • Business Partners • Assign Partner Determination Profiles Based on Incoterms.

Although entering business partners in the forwarding order is not really complicated, we cannot understate the importance of the entries. The assignment of a business partner to the ordering party is crucial for charge calculation, which we describe in Chapter 10.

Recall that business partners can be defined globally on the Business Partner tab of the forwarding order or individually for each item, depending on the Customizing of the forwarding order type. The definition is the same; it only has to be done on the item details.

Locations and Dates/Times

Some of the most important information for transportation planning is the definition of locations and times. On the Locations and Dates/Times tab, you can define the source and destination location of the entire transportation. Note that the routing is not defined here, but on the Stages tab. If you have defined that locations and business partners are not the same for every item, you can still define either a global source location or a global destination location and leave the other location empty on the global level to fill it in on the item level.

The location you choose on this tab needs to be maintained as location master data. If you have chosen a location master data, you will see that the corresponding fields in the address area of the screen are filled with that address.

Location codes

In the global transportation business, locations are always assigned a global identifier, which depends on the transportation mode. Depending on the transportation mode defined for the forwarding order, the global identifiers are displayed on the LOCATIONS AND DATES/TIMES tab. Because air transportation uses IATA codes and sea transportation uses UN/LOCODES, if you use an air forwarding order, you will see a field IATA CODE and if you use an ocean forwarding order, you will see a field UN/LOCODE. The codes are assigned in the location master data and prepopulated in the forwarding order when a location master data is selected.

Location/business partner link

If you have followed along with our example, you have defined business partners in the forwarding order. If you go to the LOCATIONS AND DATES/TIMES tab, you will see that locations are already entered. If the business partner assigned as the shipper has been assigned a location, this location is automatically taken as the source location of the forwarding order. The same applies to the consignee and the destination location. It also works the other way around, meaning that if you define locations first, the party roles shipper and consignee are filled automatically. However, if both party roles and locations are filled and you change one of them, no changes are made to the other.

One-time locations

If a customer calls and wants to get a transportation service from a location that you haven't used before, you do not have to define new master data for this location. Instead, you can just leave the LOCATION field in the forwarding order empty and type in the address directly, as shown on Figure 4.24. Entering the address creates a *one-time location*. The one-time location is no different from location master data—in fact, a new location master data item was created. The new location is represented by a number that you can use later to add more information in the location master data. Usually the number range for one-time locations starts with 1.

If you want to change this, follow the IMG menu path SAP TRANSPORTATION MANAGEMENT • TRANSPORTATION MANAGEMENT • MASTER DATA • TRANSPORTATION NETWORK • LOCATION • DEFINE NUMBER RANGE INTERVALS FOR ONE-TIME LOCATIONS.

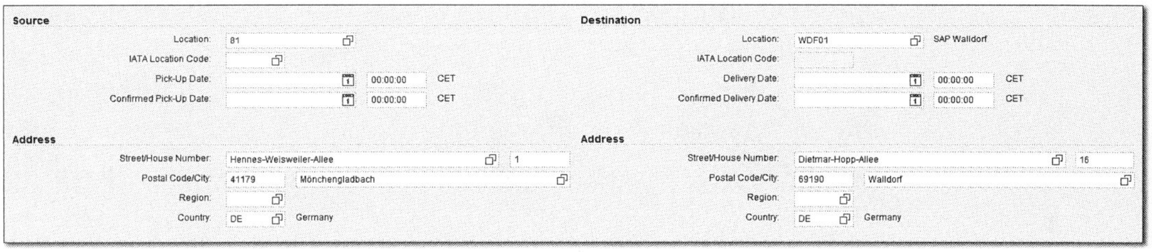

Figure 4.24 Locations with One-Time Locations

Take the Term "One-Time Location" Seriously

If you've started thinking that you no longer need to create locations in master data, be aware that one-time really means one-time. Every time you only enter an address in the forwarding order, a new one-time location is created, regardless of whether you have used the address before. This prevents freight unit consolidation because the consolidation requires same item locations. Furthermore, where-used frameworks are not useful anymore.

Therefore, we recommend that you use one-time locations in exceptional cases only. In other cases, take the time to create a location master first.

Pickup and delivery date

On the same tab as the locations, you can also define the ordered transportation dates for pickup and delivery. Please note that the dates entered on this tab are used for the entire transport, meaning the pickup date is the pickup date of the first stage, and the delivery date is the delivery date of the last stage of the transport. What you define here are the ordered dates; there is no validation of whether the time frame between pickup and delivery date is feasible (there is a validation of whether the delivery date is after the pickup date, in case of typing errors). Pickup and delivery dates are not both mandatory; one of the dates is sufficient for freight unit building.

The dates and times are defined in the forwarding order in the time zone that is used by the corresponding location. Transportation planning usually works not with a single time, but with a time frame for delivery or pickup. In freight unit building, the time defined in the forwarding order is rendered into a time frame if this is required. You will find more information about pickup and delivery time windows in Chapter 6.

Forwarding order confirmation

When you have defined the items of the forwarding order, you will probably come across some fields concerning confirmation. The LSP can confirm quantities and dates/times to the customer after order-taking. This process step can be automated with the AUTOMATIC CONFIRMATION flag in the forwarding order type Customizing. In addition, you can define on what data the confirmation should be done. This Customizing mainly concerns the dates to be confirmed. The confirmation can be done based on order data, meaning the ordered dates are simply confirmed. Another option in Customizing allows you to do planning first and then confirm to the ordering party the dates that result from planning. If you want to confirm manually, you need to enter dates in the corresponding fields of the confirmed dates.

If you don't want automatic confirmation, you can fill in the confirmation fields by clicking on the CONFIRM button in the action toolbar of the forwarding order document.

Stages

After creating source and destination locations and dates for the entire process, the LSP usually defines the exact routing, including any potential intermediate stops that are on the forwarding order. Note that in real life these activities are often done by different employees. The forwarding order can be saved in every state and passed on to the next team, so enriching the forwarding order with more data is not a problem.

The team responsible for routing the forwarding order uses the STAGES tab to define the route. On that tab you will find a table containing all stages that have been created for the forwarding order. In many cases, some stages are predefined here thanks to the movement type entered on the general data of the forwarding order. Recall that the movement type can be assigned mandatory and optional stage types. In Customizing you can decide whether some stages should be proposed on the forwarding order when a particular movement type is chosen.

Stage profiles

However, there is a second way of automatically assigning stages to the forwarding order. In Customizing of the forwarding order type, you can specify whether stage determination of the forwarding order should be done using the movement type or using a *stage profile*. Stage profiles can be defined in Customizing via the IMG menu path SAP TRANSPORTATION MANAGEMENT • TRANSPORTATION MANAGEMENT • FORWARDING ORDER MANAGEMENT • DEFINE STAGE PROFILES. This Customizing activity looks very similar to the Customizing activity of assigning stage types to the movement type, which was depicted in Figure 4.18. You can make the same settings here like in the movement type-related Customizing activity.

You might be wondering why there are two different ways of defining the same thing. With stage determination by movement type, you always get a fixed stage setup—the one you assigned to the movement type. With stage profiles, you can be more flexible.

Figure 4.25 shows an excerpt from Customizing of the forwarding order type. If you decide to use stage determination by stage profile, you can assign a stage profile directly to the forwarding order type. In this case, the stage setup will always be the same for the forwarding order, no matter which movement type was chosen. You can also determine the stage profile using a condition. This offers the flexibility to determine the stage profile based on any data from the forwarding order compared to a 1:1 assignment of a stage profile to a forwarding order type. We recommend that you fill in a stage profile even if you are using a condition. If the condition does not return a result, the assigned stage profile will be considered as a fallback solution.

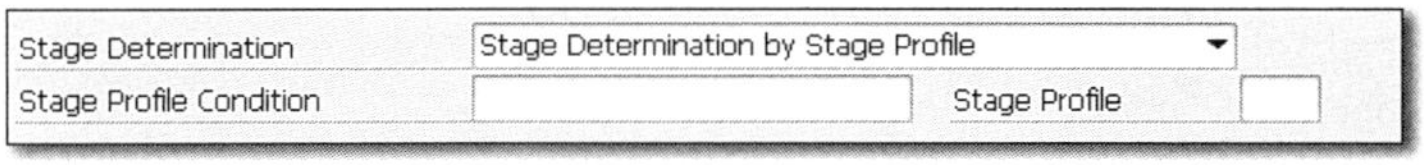

Figure 4.25 Stage Determination by Stage Profile

Stage Determination

Stage determination by movement type is independent of the forwarding order type. No matter which document type is used, the stage setup assigned to the movement type is always considered.

When using stage determination by stage profile, the stage determination depends more on the forwarding order type.

Stage types

We've talked a lot about how stage types are assigned to stage profiles or movement types. Now we should take a closer look at what stage types actually are.

Stage types define the characteristic of the transportation stage in the entire transportation. In Customizing you can create stage types by following the IMG menu path SAP TRANSPORTATION MANAGEMENT • TRANSPORTATION MANAGEMENT • FORWARDING ORDER MANAGEMENT • DEFINE STAGE TYPES.

In Customizing you simply assign a description to the stage type, as shown on the left in Figure 4.26. You also need to determine the stage category for the stage type. Stage categories are standard categories in the SAP TM system that cannot be enhanced. Like item categories, they cluster stage types into five groups. Several process steps in SAP TM use the stage category rather than the stage type. For example, in transportation charge management, you can decide for which stage categories the rate table should be taken into consideration. In addition, the stage category main carriage has several validations that we examine later.

Stage Type	Description	Stage Cat.	Ins. Set
00	Self delivery by customer	Customer Self-Delivery	
01	Pick-Up	Pre-Carriage	SET1
02	Pre-Carriage	Pre-Carriage	
03	Main Carriage	Main Carriage	
04	On-Carriage	On-Carriage	
05	Delivery	On-Carriage	
06	Self-Pickup by Consignee	Customer Pick-Up	

Figure 4.26 Stage Type Definition

Since instruction sets can be assigned to stage types, it might make sense to restrict stage types to certain transportation modes. You can do this with the Customizing activity found via the IMG menu path SAP TRANSPORTATION MANAGEMENT • TRANSPORTATION MANAGEMENT • FORWARDING ORDER MANAGEMENT • DEFINE ALLOWED TRANSPORTATION MODE FOR STAGE TYPE. Here you can decide which transportation modes are allowed for a certain stage type. You can also define a default transportation mode, which is used for the stage type once the stage type has been assigned to the forwarding order's routing.

Routing of forwarding order

Now that we have talked a lot about Customizing and the prerequisites for the forwarding order's routing, we return to the forwarding order document and start with a manual routing of the forwarding order.

As you can see in Figure 4.27, the table containing the stages differentiates between the ordered route and the actual route. As the names suggest, the *ordered route* is the routing that the customer orders or that is agreed upon with the customer. Therefore, the routing of the orders route is also considered by charge management.

Insert | Schedule | Freight Document | Actual Route | Set OI Status | Layout

Stage Description	Stage Type	Transport...	Source Location	UN/LOC...	City	Pick-Up Date	Pick-Up T...	Destination Location	UN/LOC...	City	Delivery Date
▾ Ordered Route							00:00:00				
▾ Complete Carriage		01 (Road)					00:00:00	WDF01		Walldorf	15.05.2013
▾ Item 10		01 (Road)	TM5_USNEK		Newark	08.05.2013	00:00:00	WDF01		Walldorf	15.05.2013
Stage 1		03 (Sea)	TM5_USNEK		Newark	08.05.2013	00:00:00	DEHAM	DEHAM	Hamburg	
Stage 2			DEHAM	DEHAM	Hamburg		00:00:00	WDF01		Walldorf	15.05.2013
Item 20		01 (Road)	DEHAM	DEHAM	Hamburg		00:00:00	WDF01		Walldorf	15.05.2013
▾ Actual Route							00:00:00				
Complete Carriage							00:00:00	WDF01		Walldorf	15.05.2013

Figure 4.27 Forwarding Order Routing

However, if the LSP wants to do a different routing (to optimize costs, for example), he can insert a different actual route. The *actual route* is moved to the freight unit so that transportation planning works with the stages and dates defined in the actual route.

If the movement type or the stage profile has proposed some stages already, those stages will be in the ordered route. The actual route will also contain those stages.

The source location and pickup date of the first stage is taken from the LOCATIONS AND DATES/TIMES tab, and so are the destination location and the delivery date of the last stage. If more than one stage is proposed, you need to enter the other locations and dates manually.

If you want to add stages, select a stage and insert another stage either before or after it. Alternatively, you can split the selected stage into two stages and enter new intermediate stops in the forwarding order.

Direct assignment of schedules

If you have defined schedules in your transportation network, you can directly assign a schedule instance (meaning a voyage or flight) to a stage in the forwarding order. To do so, click on the SCHEDULE button at the top of the stage table and then click on SELECT. A search help appears that looks for schedules that run between the locations of the stage. Once a schedule is selected, a booking for this schedule is created. The dates from the schedule are also propagated into the forwarding order's stage. The delivery date of a stage is always the earliest pickup date of the next stage unless you define a different pickup date. When you select the stage that you have chosen a schedule for, more information from the schedule (e.g., cutoff dates) is available below the stage table.

Direct creation of freight documents

If you do not want to assign schedules (e.g., because you are planning a road transport), you can create a freight document directly from the stage by choosing FREIGHT DOCUMENT • CREATE. The dates defined in the stage then serve as dates for the freight document. Alternatively, you can select an already created freight booking for your stage.

To do all of this, you need to define in Customizing what document types should be used for the freight documents and whether the features described above should be available on the forwarding order type. You can find all of these settings by following the IMG menu path SAP TRANSPORTATION MANAGEMENT • TRANSPORTATION MANAGEMENT • FORWARDING ORDER MANAGEMENT • FORWARDING ORDER • DEFINE DEFAULT FREIGHT DOCUMENT TYPES FOR STAGES.

You can choose a freight document type based on the following information:

- Forwarding order type
- Shipping type
- Stage type
- Transportation mode
- Sales organization

The freight document type can be either a freight order type or a freight booking type.

It is also possible to create one freight document for several consecutive stages, but make sure that all stages for which you want to create a common freight document have been assigned the same freight document type.

Transportation proposals

As you learned earlier in this book, SAP TM comes with an optimizer engine that can be used for routing and scheduling. So instead of manually creating stages with locations and dates, you can create a routing using the optimizer's capabilities. As a prerequisite, you need to assign a planning profile to your forwarding order type in Customizing; then select ACTUAL ROUTE • DEFINE ROUTE on the STAGES tab to start the optimizer.

The optimizer returns one or several results that you can choose from. If you accept one of the results, it is moved into the routing of the forwarding order. In Customizing of the forwarding order type, you can choose how the transportation proposal's result should be considered in the forwarding order. If only the routing is supposed to be copied into the forwarding order, the stages are filled accordingly with dates and locations. Alternatively, freight documents may already have been created from the transportation proposal, and they are then assigned to the forwarding order's stages.

In some cases, the actual routing differs from the ordered route. If the LSP has agreed with the customer that the actual route should be the basis for charge calculation, you can copy the actual route into the ordered route by choosing ACTUAL ROUTE • COPY ACTUAL ROUTE.

Transportation cockpit

If the employee taking the order is also responsible for the transportation planning, he can directly access the transportation cockpit from the forwarding order. Select FOLLOW-UP • START TRANSPORTATION COCKPIT in the action toolbar at the top of the document. The transportation cockpit is filled with the selection criteria that were assigned to the planning profile that is supposed to be used by the forwarding order (planning profiles can be assigned to forwarding order document types in Customizing).

Item view versus stage view

Since a forwarding order can contain items with different source and destination locations, the stages displayed on the STAGES tab sometimes need to be on the item level.

SAP TM offers an item view and a stage view for the stage table. You can switch between the stages by clicking on the LAYOUT button at the top of the stage table.

In our example, two items in the forwarding order have the same destination location but different source locations. We want to define a routing that sends item 1 from its source location to the source location of item 2, from where both items are transported together to their common destination location.

Figure 4.28 shows the item view before the insertion of the additional stage for item 10. After we split the stage of item 10, item 10 contains two stages, whereas item 20 still contains only one stage.

Item view before manual routing

Stage Description	Source Locati...	City	Destina...	City
▾ Ordered Route				
▾ Complete Carriage				
Item 10	WDF01	Walldorf	82	Mönchengladbach
Item 20	DEHAM	Hamburg	82	Mönchengladbach

Item view after manual routing

Stage Description	Source Locati...	City	Destina...	City
▾ Ordered Route				
▾ Complete Carriage				
▾ Item 10	WDF01	Walldorf	82	Mönchengladbach
Stage 1	WDF01	Walldorf	DEHAM	Hamburg
Stage 2	DEHAM	Hamburg	82	Mönchengladbach
Item 20	DEHAM	Hamburg	82	Mönchengladbach

Stage view after manual routing

Stage Description	Source Locati...	City	Destina...	City
▾ Ordered Route				
▾ Stage 1	WDF01	Walldorf	DEHAM	Hamburg
Item 10				
▾ Stage 2	DEHAM	Hamburg	82	Mönchengladbach
Item 10				
Item 20				

Figure 4.28 Item View and Stage View

When we switch to the stage view, we can clearly see which stages have to be planned separately for the two items and which stages can be planned together.

Validations

Some validations are performed when you do the routing of the forwarding order. Most validations concern whether all mandatory stage types have been used and whether the transportation modes assigned to the stage types are allowed.

Validations also check whether the stage type representing the main carriage uses the transportation mode for which the forwarding order was created.

Organizational interaction

The sales organization is responsible for entering the order, for example, but is not allowed to create freight bookings for specific transportation stages or send them to a carrier. So it can propose how to transport the goods (by specifying a route and a schedule and assigning a freight order or freight booking) and set the organization interaction status to *To Be Checked* in order to transfer the affected stages to the planning and execution. You can set the interaction status on the stage table by selecting SET OI STATUS • TO BE CHECKED. The planning and execution organization checks the proposal and the transportation stage details in the transportation cockpit. It can then confirm the data exactly as proposed in the forwarding order, change data such as the departure, and then confirm the proposal or reject it outright. The status for the stage in the forwarding order then changes to *Confirmed*, *Confirmed with Deviations*, or *Rejected*, respectively.

The prerequisite for this process step is that you have defined a planning and execution organization for the stage. You can directly assign an organizational unit in the stage table.

Charges and Internal Charges

The CHARGES and INTERNAL CHARGES tabs show the result of charge calculation. Even though charge calculation is a very important topic in forwarding order management, we do not go into great detail about charge management until Chapter 10. In general, these two tabs show the result of a performed charge calculation.

In Customizing of the forwarding order type, you can enable both internal and external charge calculation. Whether internal or external charge calculation is triggered depends on the combination of sales organization and ordering party in the forwarding order. The ordering party can also be an organizational unit from your own company.

In this Customizing activity you can also specify whether charge calculation should be triggered manually or automatically when the document is saved. If you want to trigger charge calculation manually, select CALCULATE CHARGES • CALCULATE INTERNAL CHARGES or CALCULATE CHARGES • CALCULATE CHARGES from the action toolbar at the top of the document.

Profitability

The PROFITABILITY tab is read-only but contains valuable information for the LSP business. This tab compares the expected revenue as determined by the charge calculation performed on the forwarding order and the expected costs that derive from the charge calculation on the freight documents related to the forwarding order.

On this tab you can differentiate between *planned profitability* and *expected profitability*. Though the two terms sound very similar, they are different because the data source for the profitability analysis is different. For planned profitability, the charges from the forwarding order's charge calculation are compared to the charge calculation that is done on the related freight documents. If cost distribution comes into play, only the costs distributed to this forwarding order are considered.

Cost Distribution

Often, several forwarding orders are consolidated on one freight document. However, when you perform a profitability analysis, the costs imposed on the freight document need to be distributed to the related forwarding order.

SAP TM offers cost distribution functionality, which we discuss in Chapter 11, Section 11.2.

Expected profitability, however, considers the data from the settlement documents, both the forwarding settlement document and the freight orders' or freight bookings' settlement documents.

Output Management

Like every business document in SAP TM, you can also trigger output for the forwarding order. We do not want to go into the details of setting up output here but only discuss how to assign output-related Customizing to the forwarding order and how to see the output on the forwarding order document.

In Customizing you assign output profiles to the forwarding order type. The output profile is defined in the Post-Processing Framework (PPF) (see Chapter 2, Section 2.3.3).

Once an output profile is assigned or dynamically determined, you can go to the OUTPUT MANAGEMENT tab on the forwarding order document. You will see an empty table. When you select GENERATE • ACTIONS INCLUDING CONDITION CHECKS, all output actions that meet the schedule conditions defined in the PPF are triggered.

The table is filled with the actions that meet the schedule conditions. You can see whether the actions have been processed and what kind of actions they are.

As you can see in Figure 4.29, we only have print actions in our example. This means the forwarding order document is supposed to be printed. When you select a line, more details about the action are listed below the table. When you select the DOCUMENT PREVIEW tab in the actions details view, you get a print preview of the document, filled with all the information that we filled in during this chapter or that was derived automatically.

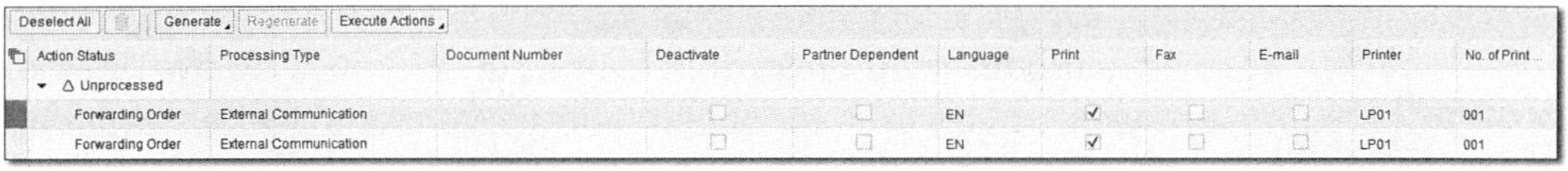

Figure 4.29 Output Management

Global Functions on the Forwarding Order

At this point we've systematically browsed through the forwarding order tab by tab. However, there are some functionalities of the forwarding order that cannot be directly assigned to tabs. We discuss these next.

Freight unit building

Recall from Figure 4.1 that the transportation process starts with a transportation requirement and then continues with planning using the freight unit document. Consequently, the forwarding order has to build freight units in order to continue the transportation process.

We discuss the freight unit building step in Chapter 6, Section 6.1, but for now we need to take a closer look at how to trigger it from the forwarding order either manually or automatically. You can define this in Customizing of the forwarding order type. If you choose to use automatic freight unit building, the freight unit will be created the first time the forwarding order is saved. If you make any planning-relevant changes to the forwarding order later, the freight unit is updated accordingly.

Prerequisites for Freight Unit Building

Whether you are using automatic or manual freight unit building, some fields in the forwarding order need to be filled in to create a freight unit:

- Sales Organization
- Source and Destination Location
- Items with Quantities
- Dates and Times
- Movement Type
- Shipping Type

In some cases, even more fields are required. You can check whether you have filled all required fields for freight unit building by clicking on the Check button in the action toolbar of the forwarding order.

You can also manually create freight units by selecting Follow-Up • Create freight units from the action toolbar. Manual freight unit building is often used if freight units might be built for several forwarding orders. In this case, you can select several forwarding orders from the POWL and then create freight units from the POWL. If dates, locations, sales organization, and so on are the same on several forwarding orders, the freight unit consolidates these forwarding orders into one freight unit.

To create freight units, you need to assign an FUBR to the forwarding order type in Customizing. If the way freight units should be built depends on data in the forwarding order, you can also assign a condition to determine the FUBR. Like with stage type profiles, you can specify an FUBR in addition to a condition in Customizing to have a fallback scenario in case the condition does not return any result.

Restricted processing

Sometimes, forwarding orders are only created for charge calculation and business administration reasons. In this circumstance, avoid passing these forwarding orders onto transportation planning; in Customizing of the forwarding order type, you can choose restricted processing of the forwarding order. This way, the forwarding order will always be blocked for planning and execution.

Credit limit check

SAP TM integrates with SAP Credit Management to provide the credit limit check feature. You can activate the credit limit check in Customizing of the forwarding order type. The activation of the credit limit check is only allowed if forwarding settlement is also allowed; this way, the credit limit check is performed when the forwarding order is created. The order data is passed on to SAP Credit Management, and the check is performed there.

In Customizing of the forwarding order type, you can decide what happens if the credit limit check fails. The negative check result can either be only informative and have no impact on the forwarding order, or can cause a planning and execution block until the credit limit check is successful.

A credit limit check is always performed again if any of the following information of the forwarding order has changed:

- Sales organization
- Credit limit check amount (usually the result of the charge calculation)
- Business partners
- Logistics data with influence on charge calculation

Canceling a forwarding order

If you need to cancel a forwarding order, you can do this either from the action provided in the POWL or directly from the document. If you cancel a forwarding order and freight units have already been created for this document, those freight units will be canceled and withdrawn from the freight documents on which they might have been planned.

Canceling Freight Units from Freight Documents

If freight units are canceled (e.g., because the related forwarding order was canceled), then the planning is withdrawn, which means the freight units will be taken off the freight order or the freight booking. If you want to notify the planner automatically about this change, you need to set up your freight documents accordingly by using a change controller strategy that handles this situation.

You can find more information on the setup of freight documents and change controller strategies in Chapter 7.

When you cancel the forwarding order by clicking on the CANCEL DOCUMENT button either on the document itself or on the POWL, you are asked to define a cancellation reason code. The code you choose can be used for analysis and is displayed on the GENERAL DATA tab of the forwarding order below LIFE CYCLE STATUS, which is changed to *Canceled*.

You can define cancellation reason codes in Customizing via the IMG menu path SAP TRANSPORTATION MANAGEMENT • TRANSPORTATION MANAGEMENT • FORWARDING ORDER MANAGEMENT • DEFINE CANCELLATION REASON CODES. It's not necessary to specify a cancellation reason code when canceling a forwarding order.

Page selector

As you have seen, the forwarding order contains a lot of information for business administration, transportation planning, charge calculation, and organizational interaction. In many LSP processes, these different process areas are usually performed by different areas within the company.

For the person who takes the order, it would be cumbersome to navigate through all of the tabs we have mentioned to enter relevant data. Therefore, he can use the page selector, which is located in the action toolbar in the top-right corner of the forwarding order document.

Recall from Chapter 2, Section 2.2.4 that you can customize the forwarding order screen according to your needs. However, it might be useful to switch between a fast order entry screen and the full-blown forwarding order document. SAP TM provides fast order entry screens for the air, land, and sea transportation modes because the required information depends on the transportation mode.

Figure 4.30 shows a fast order entry screen for land transportation. It displays only the data needed for order entry, including the business partners, sales organization, general terms, locations, dates, and items.

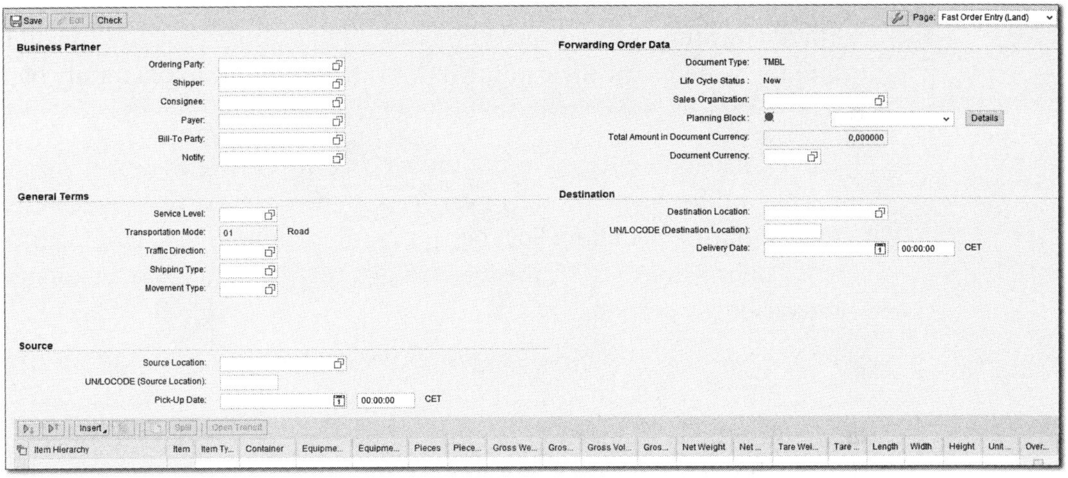

Figure 4.30 Fast Order Entry

If you need more information on the fast order entry screen, you can add more fields in screen Customizing.

Notice there is no action toolbar in Figure 4.30. The fast order entry only works for order entry. You need to switch back to the conventional forwarding order screen for processing the forwarding order.

EDI communication with the forwarding order

As already mentioned, the ordering party can send an order via EDI messages to make the LSP or carrier create a forwarding order. This process can be automated with SAP TM because the service interface `TransportationRequestRequest_In` works for all fields that might be relevant for creating a forwarding order. Make sure you have done the setup of the field and service mapping in SAP NetWeaver PI properly. The forwarding order will be created automatically if this service interface is triggered.

If you want to confirm data back to the ordering party, you can also use EDI communication with the service interface `TransportationRequest-Confirmation_Out`. For more information about the automatic creation of documents using service interfaces and communication with web services, return to Chapter 2, Section 2.4.

Summary

The forwarding order is the central document for entering information into the system. The forwarding order can be created automatically by a B2B service or manually by an order entry clerk.

The information entered in the forwarding order is relevant for charge calculation between the LSP or carrier and the order party, as well as for transportation planning. For transportation planning, the planning process can be influenced by the forwarding order by defining stages on the forwarding order.

After planning, execution, and charge calculation and settlement, the forwarding order can be used to do a profitability analysis to determine whether the customer's order for transportation services is profitable. Profitability analysis can also be used to decide up front whether the order should be accepted and confirmed.

However, there is a third way of creating forwarding orders. If a forwarding quotation was used for the interaction with the customer, all data from the forwarding quotation can be transferred to the forwarding order. This is described in detail in Section 4.2.3.

4.2.2 Charge Estimation

Customers often call the LSP to request a price for a transportation service. You can enter a forwarding order and perform charge calculation on the forwarding order, but entering a forwarding order takes a lot of time, and since the customer is on the phone, you might require a quicker way of calculating charges.

For this reason, you will find another menu entry in the work area forwarding order management in NWBC: charge estimation. The SAP NWBC menu path Forwarding Order Management • Estimate Charges •

CHARGE ESTIMATION takes you to a screen where you can quickly estimate the charges for the customer's order.

Note

To use the charge estimation, you need to have set up the Transportation Charge Management component. You can find more information on charge management in Chapter 10.

Notice that because the charge calculation itself usually works independently of the document type, no forwarding order type is necessary to start the charge estimation. The charge estimation screen in Figure 4.31 looks a little like the fast order entry of a forwarding order. On this screen you are asked to enter all data that is relevant for charge calculation. The following fields are mandatory:

- SALES ORGANIZATION
- SOURCE LOCATION
- DESTINATION LOCATION
- ORDERING PARTY

Other fields may not be mandatory for system validation but are often required to perform charge calculation (such as items).

If no pickup date is entered, the current system date is used as the pickup date for the charge calculation.

Hint

As described in Chapter 2, Section 2.2, SAP TM applications can be displayed in a browser using a hyperlink. Since this also applies to the charge estimation, think about providing the customer with the link to the charge estimation application. This allows the customer to use SAP TM's capability without contacting the LSP.

If you do this, make sure you thoroughly check your authorization setup so that the customer can only estimate charges for himself.

The charge estimation is read-only; that is, the estimated charges cannot be saved or turned into a forwarding order.

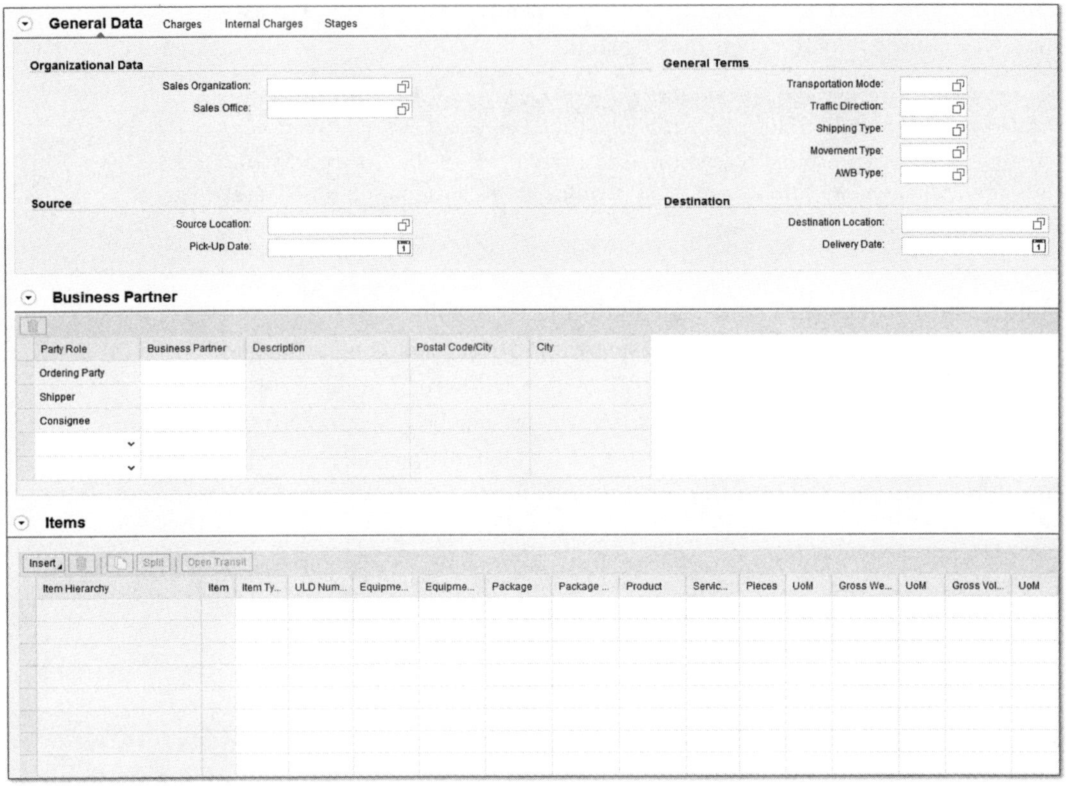

Figure 4.31 Charge Estimation Screen

4.2.3 Forwarding Quotations

We have spent a long time talking about the forwarding order, which is the document that represents the actual order or contract between an ordering party and an LSP or carrier.

Before you can create an order in the transportation process, you need to create a quotation, which is covered by SAP TM with the *forwarding quotation* document. This business document helps the ordering party send the data of a potential forwarding order with the quotation price. When the quotation is successful, the forwarding order can be created in relation to the quotation.

Creating a forwarding quotation

Notice that the forwarding quotation looks very similar to the forwarding order. In fact, you can do most of the things we talked about in Section 4.2.1 in the forwarding quotation. You can create forwarding quotations via the SAP NWBC menu path FORWARDING ORDER MANAGEMENT • FORWARDING QUOTATION • CREATE FORWARDING QUOTATION. Just like you did with the forwarding order, when creating a forwarding quotation, you need to define a forwarding quotation document type. You can specify a forwarding quotation document type in Customizing by following the IMG menu path SAP TRANSPORTATION MANAGEMENT • TRANSPORTATION MANAGEMENT • FORWARDING ORDER MANAGEMENT • FORWARDING QUOTATION • DEFINE FORWARDING QUOTATION TYPES.

When looking at the forwarding quotation itself and customizing its type, you can find many similarities between the forwarding quotation and the forwarding order. The following features and processes are handled exactly the same way as in the forwarding order:

- Item definition
- Defining items with different source or destination locations
- Item type assignment to document type

  You can find the corresponding Customizing activity via the IMG menu path SAP TRANSPORTATION MANAGEMENT • TRANSPORTATION MANAGEMENT • FORWARDING ORDER MANAGEMENT • FORWARDING QUOTATION • ASSIGN ITEM TYPES TO FORWARDING QUOTATION TYPES. Make sure that when creating a forwarding order out of a forwarding quotation, the same item types are assigned to the forwarding order type that is used for the forwarding order creation.
- Stage determination by either movement type or stage profile
- Automatic charge calculation when saving the document
- Partner determination by partner determination profile
- Transportation proposals
- Creation of one-time locations
- Canceling the forwarding quotation with a reason code

Not everything is the same, though. Figure 4.32 shows the most important differences between a forwarding quotation and a forwarding order. In the quotation you can specify a valid-to date, which is the deadline by which the quotation must be accepted or rejected.

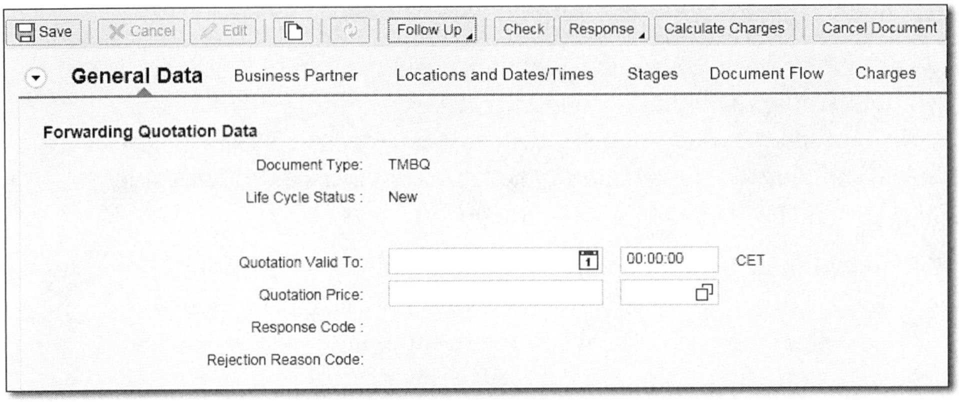

Figure 4.32 Forwarding Quotation Fields and Actions

Just like with charge calculation on the forwarding order, SAP TM can calculate the quotation price. This takes place, for example, if the customer calls the LSP or carrier and asks for a price. The price is calculated and can be submitted to the ordering party by selecting RESPONSE • SUBMIT from the action toolbar at the top of the screen.

Communication ordering party to LSP

The communication that takes place during the quotation process can vary from customer to customer. When communication occurs by phone, you can simply update the status of the quotation manually using the actions provided by the RESPONSE button. Below the quotation price, as depicted in Figure 4.32, you can see your response to the customer—whether you have accepted the quotation or rejected the quotation. When rejecting a quotation, you can specify a rejection reason; this can also be communicated to the ordering party. The forwarding quotation document is canceled once the quotation has been rejected.

The forwarding quotation also supports the communication between ordering party and LSP via EDI communication. In this case, the forwarding quotation can be created via the corresponding web service

TransportationRequestQuotationCreateRequest_In. This service interface provides all the necessary fields to create a forwarding quotation in SAP TM, just like manual creation of a forwarding quotation. The response of the carrier or LSP is then sent out to the ordering party with the service interface TransportationRequestQuotationConfirmation_Out.

Note

Even though the name of the service interface TransportationRequest-QuotationConfirmation_Out suggests that you can only use it to confirm or accept quotations, you can also use it to reject quotations.

Forwarding quotation tendering

A quotation can also be made in the course of a tendering process. Imagine that the ordering party also uses SAP TM and starts a tendering for a freight order in their system (more about triggering the tendering process can be found in Chapter 7, Section 7.3). A request for quotation is sent out to the LSP or carrier, and a forwarding quotation is created for the customer's request for quotation.

EDI Messaging from SAP TM to SAP TM

It is possible to start a tendering process and communication between an ordering party's SAP TM system and a carrier's SAP TM system. However, you need to make sure you have the correct SAP NetWeaver PI setup in place so that the outgoing B2B messages of the request for quotation are matched to the correct incoming B2B service interfaces.

If you want to use your forwarding quotation as part of the tendering process, you need to define this in Customizing of the forwarding quotation type. In the Customizing activity, change the setting QUOTATION MODE to WITH REQUEST FOR QUOTATION. The forwarding quotation document in SAP NWBC changes slightly. You get additional fields showing the response due date and a potential price limit.

As shown in the top-right corner of Figure 4.33, the response options are now restricted to ACCEPT and REJECT because you can only accept or reject a request for quotation and communicate a price.

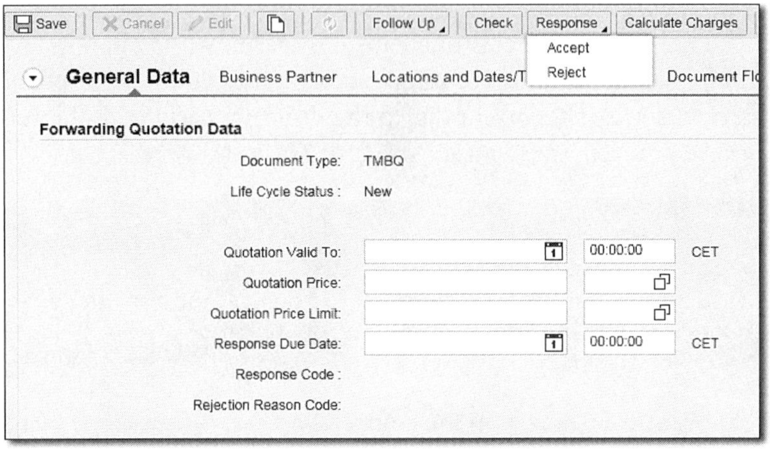

Figure 4.33 Forwarding Quotation with Request for Quotation

Processing a forwarding quotation

Whether you are using a forwarding quotation with or without request for quotation, the processing of a forwarding quotation is mostly the same. Processing a forwarding quotation is not much different than processing a forwarding order.

When the forwarding quotation has come in, you can do the routing of the quotation. This routing can be done manually on the STAGES tab. (Return to Section 4.2.1 for more information on how to do manual routing on this tab.) Alternatively, you can use a transportation proposal for the route determination of the forwarding quotation by specifying an FUBR in Customizing of the forwarding quotation type, similar to Customizing of the forwarding order type. No freight units will be built based on the forwarding quotation; the FUBR is only used for the simulation of the freight unit building and optimizer planning in the transportation proposal.

After you route the forwarding quotation, you can calculate charges for the document in exactly the same way as on the forwarding order. The prerequisite is again the correct combination of sales organization and ordering party and existing master data in SAP TM Transportation Charge Management. As before, the calculated price is displayed on the GENERAL DATA tab, as shown on the figures above. You can also manually overwrite the quotation price later. Only this value is communicated to the customer.

As with the forwarding order, you can view the detailed result of charge calculation on the CHARGES tab. Because the quotation does not serve as an actual order or contract, settlement is not possible based on the forwarding quotation document.

Forwarding order from forwarding quotation

If the customer has accepted the quoted price, you can now create an actual order for the quotation. But don't worry—you do not need to create a new forwarding order from scratch; you can create the forwarding order directly from the forwarding quotation by selecting FOLLOW UP • CREATE FORWARDING ORDER in the action toolbar of the forwarding quotation.

The forwarding order type that is used for creation must be defined in Customizing of the forwarding quotation type in the DEFAULT FWO TYPE field. Additionally, in Customizing of the forwarding quotation type you can specify how many forwarding orders may be created out of the forwarding order.

Additionally, a forwarding order can only be created out of a forwarding quotation if the quotation was already submitted.

All relevant data is copied from the quotation to the order. The most important data copying is probably the calculated charges. Since master data in SAP TM Transportation Charge Management might change during the quotation and order-taking process, you want to avoid calculating charges again on the forwarding order when a different result could occur. Therefore, the charge calculation results are copied to the forwarding order, and the status for charge calculation indicates that no further charge calculation is necessary. If you need to do a charge calculation on the forwarding order again, you can do so using the functionality covered in Section 4.2.1.

Assignment of forwarding quotation

The assignment of the forwarding quotation to the forwarding order is displayed in the document flow if you have created a forwarding order from a forwarding quotation. Later, the document flow of the freight unit and freight orders will document that the process has started with a forwarding quotation.

However, in some cases you create a forwarding order independently of the forwarding quotation even though a quotation exists for this workflow.

In this case, you can subsequently assign a forwarding quotation to a forwarding order.

On the forwarding order actions are available in the action toolbar that enable the retroactive assignment of the forwarding quotation to the forwarding order as shown in Figure 4.34. When you want to assign a forwarding quotation to the forwarding order, click on the ASSIGN FWQ button. A new pop-up appears in which you can enter the document number of the forwarding quotation. However, checks will be performed if the selected forwarding quotation aligns with the current forwarding order. The standard checks include whether the locations and dates in both documents match and whether the combination of ordering party and sales organization is the same.

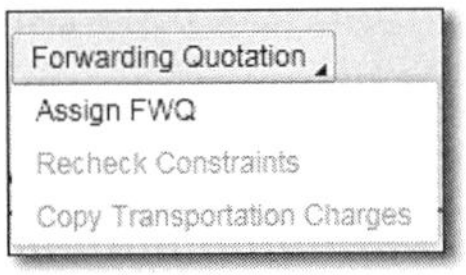

Figure 4.34 Assignment of Forwarding Quotation to Forwarding Order

You can enhance these checks with a BAdI found via the IMG menu path SAP TRANSPORTATION MANAGEMENT • TRANSPORTATION MANAGEMENT • BUSINESS ADD-INS (BADIS) FOR TRANSPORTATION MANAGEMENT • FORWARDING ORDER MANAGEMENT • ASSIGNMENT OF FORWARDING QUOTATION TO FORWARDING ORDER • BADI: EXTENSION OF CHECKS FOR ASSIGNMENT OF FORWARDING QUOTATION TO FWO.

When you are creating documents for an import/export process, you can also assign an import forwarding quotation to an export forwarding order. You will find more information on import/export processes in Chapter 14 and Chapter 15.

4.2.4 Creating Orders and Quotations with Reference

In the transportation business, LSPs and carriers often have a stock of regular customers who frequently order transportation services for the same route or with similar items. If you do not want to create new orders or quotations from scratch, you can use existing documents and create new ones from them.

Copying forwarding orders

One option is to copy an existing forwarding order (e.g., from a transportation service performed in the past) and update certain information, such as dates. When you copy the forwarding order, no link is established between the existing forwarding order and the copy. A forwarding order can also be copied with a new type, meaning the new forwarding order takes over most of the data from the existing forwarding order but is assigned a different forwarding order type. This method is used in import/export processes where an import forwarding order is created from the export forwarding order. (Again, Chapter 14 and Chapter 15 offer more information about import and export forwarding orders.) When you display a forwarding order, click on the COPY WITH NEW DOCUMENT TYPE button in the action toolbar to open a new tab in SAP NWBC. In this tab you can specify the new document type and, if necessary, a new transportation mode.

You can get to this screen shown in Figure 4.35 by following the SAP NWBC menu path FORWARDING ORDER MANAGEMENT • FORWARDING ORDER • CREATE FORWARDING ORDER.

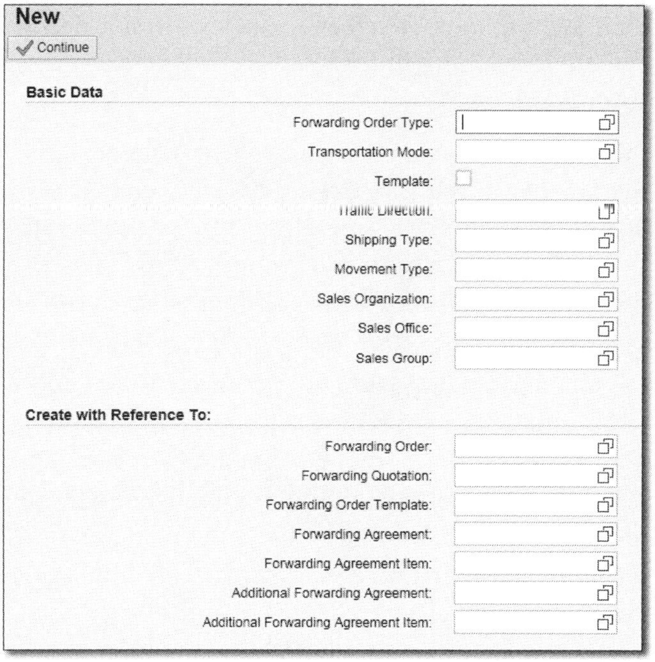

Figure 4.35 Initial Screen of Forwarding Order Creation

In the upper screen area, you define information that should be assigned to the new document you want to create. In the lower screen area, or the reference area, you assign references to existing documents. We revisit this screen in a few pages, so keep it in mind.

You can use this initial screen to create a forwarding order as a copy from an existing forwarding order. Enter the forwarding order type in the upper screen area and specify a forwarding order document number in the reference area. The new document acquires all the data from the reference forwarding order. However, as already mentioned, no reference is shown in the document flow.

Forwarding order templates

The risk with copying an existing forwarding order is that you have to be very careful with the data in the existing forwarding order. Discrepancies might result if you accidentally copy item quantities that the customer did not order.

To avoid this risk, you can create templates for forwarding orders. To create a template you follow the same menu path as when creating forwarding orders. On the initial screen, as shown on Figure 4.35, select the TEMPLATE checkbox in the upper screen area. You do not need to customize any separate forwarding order template types, but you can use the forwarding order types.

If you look at the template document, you will see that it closely resembles the forwarding order (which is no surprise, since it's a template for creating forwarding orders). Note that you cannot trigger charge calculation or freight unit building from forwarding order templates because these functionalities are reserved for the forwarding order document itself.

The template document is missing some tabs that appear on the actual forwarding order document: PROFITABILITY, ATTACHMENTS, INTERNAL CHARGES, OUTPUT MANAGEMENT, and HBL OR HAWB. These forwarding order tabs are concerned with the actual execution of the transportation service. Because the template is not meant to have anything to do with process steps that trigger transportation execution, freight unit building and charge calculation are disabled.

The biggest difference is that you cannot define any dates in the LOCATIONS AND DATES/TIMES tab. This is because templates should be timeless, meaning that in most cases it is the dates that differ from order to order.

Once you save the template, a document number is assigned. We recommend that you use a different number range for templates and actual forwarding orders to differentiate between the two document categories. You can assign the number range of both the forwarding order documents and the forwarding order templates in Customizing of the forwarding order type.

Now when you want to create a forwarding order from a template, you have different options. One option is to use the POWL that was shown in Figure 4.14 for forwarding orders. This POWL includes queries to find template documents. For example, you can search template documents for a certain combination of sales organization and ordering party or a specific routing. When you have found the right template, you can select it in the POWL and click on the CREATE FORWARDING ORDER FROM TEMPLATE button at the top of the POWL.

Alternatively, you can display the template by following the SAP NWBC menu path FORWARDING ORDER MANAGEMENT • FORWARDING ORDER • DISPLAY FORWARDING ORDER. If you know the document number of the template, you can insert it. The system automatically recognizes that you have chosen a template. If you don't know your document number, you can select the TEMPLATE checkbox on the initial screen. If this checkbox is selected, the F4 help only displays forwarding order template documents.

When displaying the template document, you can also click on the CREATE FORWARDING ORDER FROM TEMPLATE button from the action toolbar. If you want to copy the template with a new document type, as explained previously, the new document will also be a template document.

The third option is to use the initial screen of the forwarding order creation. In the reference area, you can specify a template document number. All data from the template is then copied into the forwarding order.

Like when you copy existing forwarding orders, no relationship between the template and the order is displayed.

Create forwarding order from quotation

Creating forwarding orders from the forwarding quotation is possible not only with the features described in Section 4.2.3. If you look at Figure 4.35 again, you will see that you can also define a forwarding quotation in the reference area.

Creating a forwarding order with *reference* to a forwarding quotation is the same process as creating a forwarding order from the forwarding quotation—meaning that the prerequisites for creating a forwarding order from a forwarding quotation must be met. In this case, the forwarding quotation is added to the forwarding order's document flow as a predecessor document.

Referencing forwarding agreements

In the reference area of Figure 4.35, you can assign forwarding agreements as a reference for the forwarding order that is being created. This referencing has a slightly different effect than the referencing of quotations, templates, and other orders.

No data is copied into the new forwarding order. The referencing here is used to avoid the system-based agreement determination, which is explained in detail in Chapter 10, Section 10.2. If you assign agreements as reference, these agreements will be considered when calculating charges on the forwarding order.

Templates for forwarding quotations

We've primarily concentrated on forwarding orders during our discussion of the template. But using templates is also possible for forwarding quotations; the following features are also applicable for quotations:

- Defining different number ranges for templates
- Creating forwarding quotation templates
- Reference area on the initial screen of the forwarding quotation creation
  - Referencing forwarding quotations
  - Referencing forwarding quotation templates
  - Referencing forwarding agreements
- Disabling charge calculation of forwarding quotation template

Note

It is usually not possible to copy forwarding quotations. There are no actions that allow you to copy quotations or quotation templates—neither with the same document type nor with a different document type.

Therefore, the only way to copy forwarding quotations is to use the reference in the reference area of the forwarding quotation creation screen.

4.2.5 Summary

Because the transportation requirement document marks the beginning of the transportation planning, charging, and execution process, it is a very important document for the whole process. The information entered in the transportation requirement is passed on to the next process steps.

The transportation requirement document is the only document in which information about the goods to be transported is entered into the system.

Consider Figure 4.36, which was shown at the beginning of this chapter in relation to which transportation requirement documents exist and how they are used. You should be able to explain the differences between the documents and give details about each of the documents depicted by the bottom-left boxes. No matter which of these documents are created, they all have the same technological basis: the transportation request business object. Document categories differentiate these documents.

Let's walk through it to review.

ERP integration

There are two ways of creating transportation requirement documents in SAP TM. The first option is the integration of SAP ERP order and delivery documents into SAP TM, which was explained in detail in Section 4.1. This option is used by the so-called shipper industry that uses SAP TM for the transportation of its own manufactured and sold or purchased materials.

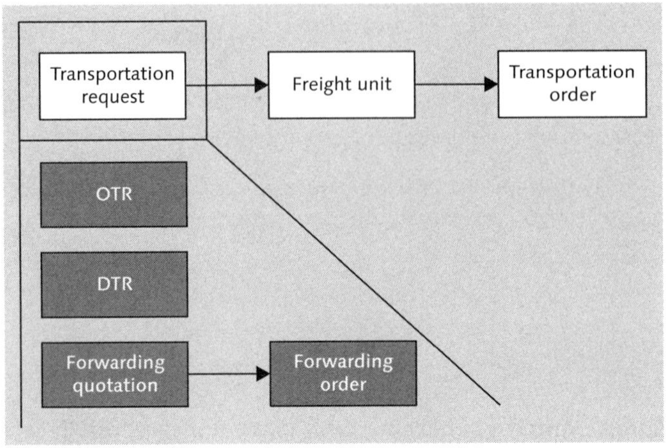

Figure 4.36 Recap of Transportation Requirement Documents

You can integrate the following SAP ERP orders:

- Sales orders
- Purchase orders
- Stock transfer orders

The SAP ERP order integration described in Section 4.1.1 and Section 4.1.2 works using the output determination procedure in SAP ERP. You need to do some configuration concerning the output management and the activation of the transfer of the orders from SAP ERP to SAP TM. A prerequisite for order integration is the integration of master data such as customers, creditors, and material that is used in the orders. The CIF is used to integrate master data.

If you want to integrate SAP ERP orders created in the MM component (purchase orders or stock transfer orders), use the output processing workflow.

Once the SAP ERP order has been transferred to SAP TM, an order-based transportation requirement (OTR) is created. For the OTR, a freight unit can be created to use in transportation planning. The OTR does not differentiate between SD and MM orders.

After the OTR's freight units have been planned, SAP TM can send a delivery proposal to SAP ERP. Deliveries can then be created based on the planning results in SAP TM, but the leading system that determines how deliveries are created is still SAP ERP.

The integration of created deliveries is configured like the integration of orders. After the integration of deliveries, a new document is created in SAP TM: the delivery-based transportation requirement (DTR). If a related OTR exists for the DTR, the DTR consumes the OTR's freight unit. In any case, you can configure different integration scenarios, depending on what SAP ERP documents should be integrated into SAP TM. These integration scenarios are described in Section 4.1.5.

Forwarding orders and quotations

The second option for creating transportation requirement documents in SAP TM is creating forwarding orders and forwarding quotations. These documents are used by LSPs and carriers who need more information about the requested transportation process than what is integrated from SAP ERP orders.

Forwarding orders can be created via EDI messages or manually. The forwarding order document combines various information:

- Information relevant for business administration
- Charge calculation-relevant data
- Data crucial for transportation planning

This information is displayed in the document in different tabs. The most important tabs of the forwarding order are explained in Section 4.2.1.

If customers inquire about the price of a transportation service based on existing agreements, no order has to be created. The charge estimation, covered in Section 4.2.2, calculates the charges of transportation services without creating any documents in SAP TM. This application could possibly also be provided to the customer directly if a customer requests charge estimations very often.

In a tendering process, the customer's enquiry will not necessarily result in the creation of an order. Therefore, the first step of the customer engagement is to create a forwarding quotation, as described in Section 4.2.3. With the help of the forwarding quotation, the customer's enquiry can be communicated to and registered by the LSP or carrier, who then communicates a price to the customer. This communication can also be done electronically using web services.

Once the customer agrees to the quoted price and terms, a forwarding order can be created from the forwarding quotation.

If forwarding orders or quotations are often created for the same routing or material, the person who take orders can make the task easier by creating template documents up front. When an order needs to be created, the data can be reused via templates and the employee can add data specific to the individual order. Templates are covered in Section 4.2.4.

Once the transportation requirement is completely entered and confirmed and freight units have been built, the transportation planning process can start. Let's continue in Chapter 5 with the creation of planning-relevant master data before going into the details of transportation planning in Chapter 6. In that chapter you will recognize much of the information that we have entered in either the SAP ERP order or delivery or the forwarding order.

Transportation capacity management allows you to plan consumption of your carriers' transportation capacities on a long-term, mid-term, and short-term basis. This helps you secure the capacities you need to run your transportation business and reduces transportation costs by early reservation and avoidance of ad hoc subcontracting.

5 Transportation Capacity Management

Transportation capacity management is the process of defining and using your carriers' transportation capacities on different time horizons and geographies. It starts with the long-term contractual part covered by freight agreements and associated capacities, which are represented by freight agreement allocations. Based on carrier schedules, mid-term capacity planning considers gateway schedules based on the carrier schedules and schedule-based allocations to plan the capacities for these gateway schedules. The short-term, operative part is the creation of freight documents according to the previously defined mid-term capacities. These operative freight documents are used in daily business, which assigns incoming new freight to the freight documents.

In this chapter, Section 5.1 presents the capacity management process, including the interplay of the business documents covering the strategic, tactical, and operative aspects; systematic creation of freight documents; and change management that enables you to react to changes in carrier schedules, which are the basis of your business.

Section 5.2 presents allocations, which are used to plan your consumption of your carriers' capacities on various geographical and time levels. The geographical levels range from location-to-location to zone-to-zone levels, and the time levels contain schedule departures as well as daily, monthly, and yearly perspectives. Allocations can be used to create freight documents with corresponding capacities. They are also used by automatic carrier selection to avoid exceeding the planned capacities.

Section 5.3 introduces business shares, which manage the distribution of freight to your carriers according to predefined target shares among the carriers for a specific trade lane. Like allocations, business shares are considered during automatic carrier selection.

Section 5.4 presents freight bookings, which are the freight documents that cover ocean freight and air freight transportation. Basically, these freight bookings can be used to reserve capacity from your carriers. Once the carrier has confirmed the booking, you can add freight to it. When you have completed your planning and want to execute the booking, you can send the content of the booking to your carrier, representing the legal document accompanying the execution of the transportation.

5.1 Capacity Management Process

Capacity management involves planning the transportation capacities your carriers offer to you. The capacity management process involves various business documents that represent the long-term contractual aspect, the mid-term capacity planning perspective, and the short-term view on capacities that can be used in daily business.

Let's begin in Section 5.1.1 with an overview of the capacity management process and the interplay of the involved business documents that reflect these perspectives. Section 5.1.2 describes how the (operative) freight documents can be systematically created based on the mid-term capacity planning. Section 5.1.3 presents the change management capabilities that help you to react to changes in your carriers' schedules.

5.1.1 Overview

Motivation

The capacity management process aims at defining and planning the transportation capacities that carriers offer to you and that you want to use for your transportation demands. We present the process from the perspective of a global air forwarding company, but the process or parts of it can also be used for ocean forwarding or shippers that systematically reserve and use their carriers' capacities for any mode of transport. Capacity management has the following main goals:

- To secure sufficient transportation capacity and avoid bottleneck situations in which you cannot transport what you need to transport
- To reduce transportation costs by using planned capacity to negotiate good long-term contracts and by avoiding expensive ad hoc bookings
- To ensure stable, reliable, and long-term relationships with your carriers

Contractual perspective

Capacity management involves the objects and the logical sequence of creating these objects as depicted in Figure 5.1. Negotiations with your carriers result in freight agreements, which specify the freight rates per trade lane and are explained in detail in Chapter 10, Section 10.1.1. You can define freight agreement allocations, which are assigned in the freight agreements and represent long-term capacity reservations per trade lane. The freight agreement and the assigned allocations resemble the contractual perspective on capacity management.

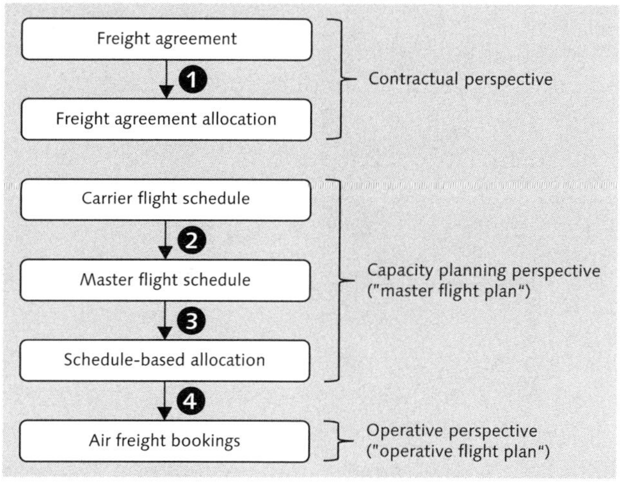

Figure 5.1 Overview of Capacity Management Process

Capacity planning perspective (master flight plan)

Carrier flight schedules can be uploaded automatically into the system as described in Chapter 3, Section 3.2.4; they represent the regular departures that you may want to use in your daily transportation business.

You can define master flight schedules; these connect a source gateway with a destination gateway and can refer to one carrier flight schedule or, in the case of a connection flight, to multiple carrier flight schedules offered by different airlines. In the master flight schedule, you can trigger creation of schedule-based allocations with buckets referencing the departures of the master flight schedule. This allows you to define capacities for the departures.

In the air freight business, it is common to define the schedule-based allocations for the next six to twelve months, and this plan is frequently called a *master flight plan*. In SAP TM, the master flight plan is represented by three objects: the carrier flight schedule to define the departures offered by the carrier; the master flight schedule to define cutoff times, source, and destination gateway; and the schedule-based allocation to define the capacities among the departures.

Operative perspective (operative flight plan)

When you reach the operative management of capacities, it is important to firmly reserve the planned capacities from the carriers, because otherwise the carrier may use them for other customers. These reservations are made by creating air freight bookings based on the master flight plan; for example, the departure date, time, and capacity are taken out of the master flight plan and put into the newly created booking. The air freight bookings are sent to the carrier and, once confirmed by the carrier, represent the operative flight plan that secures the short-term capacities. Usually, the air freight bookings are systematically created each week for the next six weeks. Whereas the master flight plan is used mainly internally and is usually not communicated to the carrier, the operative flight plan is the basis for daily business and is therefore aligned with your carrier.

Consuming the operative flight plan

Planning freight is done on the basis of the operational flight plan. New freight units can be assigned to the air freight bookings, iteratively consuming the capacity reserved by the bookings. If bookings were created only when the required freight units appear, there would be a high risk of not getting the carrier's confirmations or having to change the bookings' quantities each time new freight shows up.

Manual and automatic steps

You can manually perform the planning steps ❶, ❷, and ❸ in Figure 5.1. Step ❹ can also be done manually, but is usually performed automatically by the report to systematically create freight documents out of schedules, which we describe in Section 5.1.2.

Long-term, mid-term, and short-term perspectives

The capacity management process secures capacities on the long-term level with the freight agreement allocations and allows capacity planning on the mid-term level with the master flight plan, which is then used to secure capacities on the short-term level with the operative flight plan. This kind of hierarchical planning is common in business areas other than transportation (such as in supply chain management and production planning), where planning can take place on strategic, tactical, and operative levels.

Let's illustrate the capacity management process with typical examples for direct flights, multistop flights, and connection flights.

Example of direct flight

Figure 5.2 shows an example of a direct flight from Frankfurt (FRA) to New York (JFK). The carrier flight schedule with flight number LH-400 offers weekly departures, starting on Monday, July 1, 2013 at 9:00. The master flight schedule connects gateways in Frankfurt and New York and references the carrier flight schedule's departures. The schedule-based allocation references the master flight schedule's departures and assigns a capacity to each departure. The air freight bookings have been created based on the master flight schedule and the schedule-based allocation, with one booking per departure and taking the capacity of the allocation and the departure date from the master flight schedule.

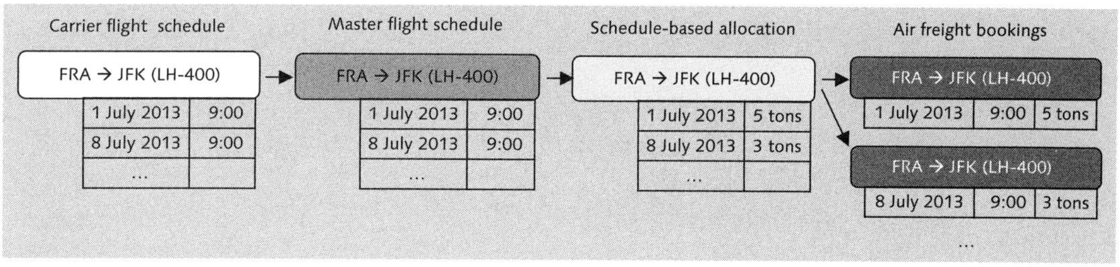

Figure 5.2 Example of Direct Flight

Example of multistop flight

Figure 5.3 shows an example of a multistop flight from Addis Ababa (ADD) to Kilimanjaro (JRO) and then to Mombasa (MBA). The carrier flight schedule with flight number LH-9664 offers weekly departures, starting on Monday, July 1, 2013 at 10:20 in ADD and 13:40 in JRO.

Assuming there are gateways in ADD, JRO, and MBA, you can create three master flight schedules: the first from ADD via intermediate airport JRO to MBA, the second from ADD to JRO, and the third from JRO to MBA. These master flight schedules reference the carrier flight schedule's departures and consume both stages, the first stage and the last stage, respectively, of the underlying carrier flight schedule. Each master flight schedule is referenced by one schedule-based allocation that assigns capacities to each departure.

The air freight bookings are created based on the master flight schedules and their schedule-based allocations. Although the three bookings on July 1 refer to the same physical flight from ADD to JRO to MBA, they are treated as independent bookings from the capacity perspective. The bookings from ADD to MBA, from ADD to JRO, and from JRO to MBA contain 2 tons, 1 ton, and 2 tons, respectively. Thus, the carrier receives two bookings that cover the first stage and have a joint capacity of 3 tons, and two bookings that cover the second stage and have a joint capacity of 4 tons.

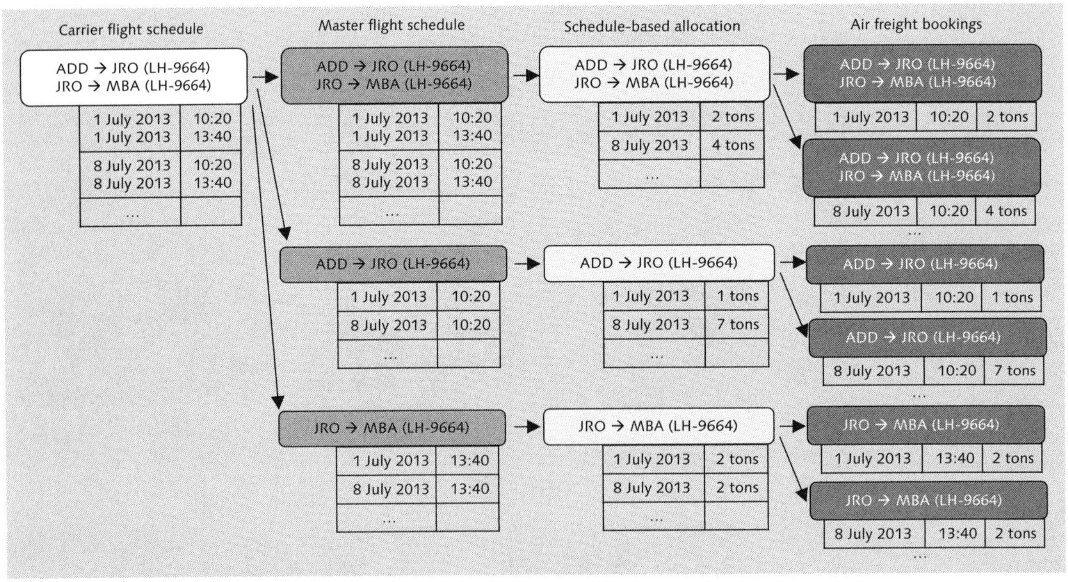

Figure 5.3 Example of Multistop Flight

Example of connection flight

Figure 5.4 shows an example of a connection flight from Frankfurt (FRA) to Caracas (CCS) to Bogota (BOG). There are two direct carrier flight schedules from FRA to CCS and from CCS to BOG, offered by two different carriers. Assuming there are gateways in FRA and BOG, you can create a master flight schedule connecting the departures of the first carrier flight schedule with the departures of the second carrier flight schedule. In the master flight schedule, you can define the carrier who will receive freight bookings created for the schedule. The (connection) master flight schedule is referenced by one schedule-based allocation to assign capacities to each departure. The air freight bookings are created based on the master flight schedule and their schedule-based allocations.

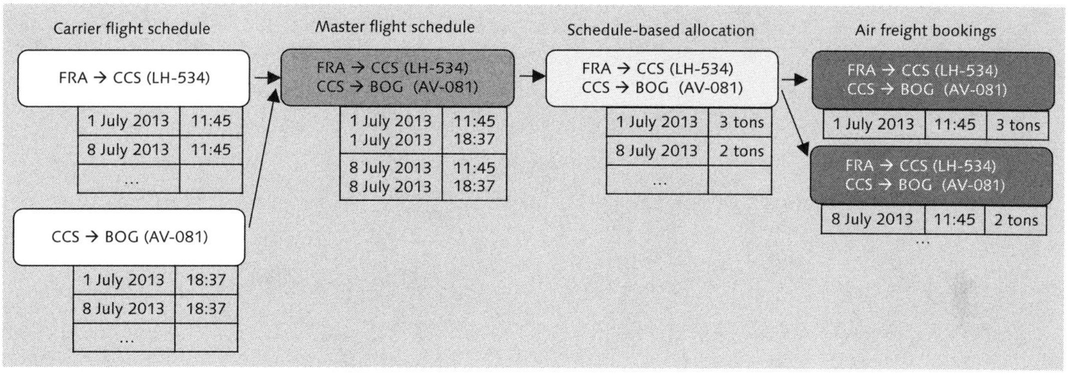

Figure 5.4 Example of Connection Flight

Allocation buckets and consumption

For a specific trade lane and carrier, the freight agreement allocation defines the planned capacity per time bucket. Allocations allow different bucket types, such as daily, weekly, monthly, quarterly, yearly, and schedule-based, which we explain in more detail in Section 5.2. Each allocation stores the available capacity per bucket and provides the already consumed capacity per bucket, which can be determined automatically.

You can define the trade lanes in the freight agreements and hence in the freight agreement allocations on the location-to-location level, but these are frequently defined on the zone-to-zone level to define rates and capacities on a more aggregated level. A transportation zone can represent a set of regions or countries, and you can easily create zones for all regions and countries in the world, as described in Chapter 3, Section 3.2.2.

For example, you may have a freight agreement allocation from Frankfurt to the United States and schedule-based allocations from Frankfurt (FRA) to New York (JFK) and from Frankfurt to Chicago (ORD). Then, air freight bookings from Frankfurt to New York would match both allocations and hence consume their capacities (as shown in Figure 5.5), with air freight bookings being identified by their master airway bill (MAWB) number.

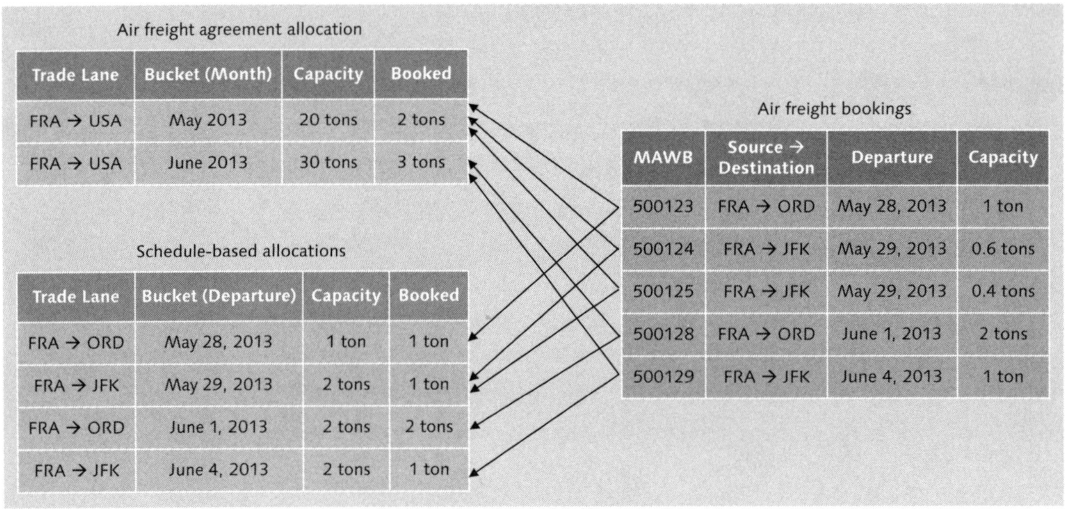

Figure 5.5 Air Freight Bookings Consuming Freight Agreement Allocation and Schedule-Based Allocation

Enabling the complete process

To model the whole capacity management process, you have to define the following:

- Freight agreement allocation type in the freight agreement type
- Schedule-based allocation type in the master flight schedule type that enables references

Using capacity management subprocesses

It is possible to use only a subprocess with a subset of the objects described above or variants of the process. For example, you can omit the master flight schedule and the schedule-based allocation and directly create an air freight booking with reference to a carrier flight schedule or, in case of a connection flight, multiple carrier flight schedules. In this case, you explicitly maintain the booking's capacity.

5.1.2 Systematically Creating Schedule-based Freight Documents

Report to create schedule-based freight documents

You can systematically create schedule-based freight documents by executing the report /SCMTMS/MP_SCHED_CREATE_TOR. In the PLANNING OPERATION tab, you can choose which freight documents will be created: air freight bookings, ocean freight bookings, road freight orders, or rail freight orders. For each selected freight document category, an individual tab allows you to define selection criteria, as shown in Figure 5.6, where you can see the schedule selection criteria to be used for creating air freight bookings.

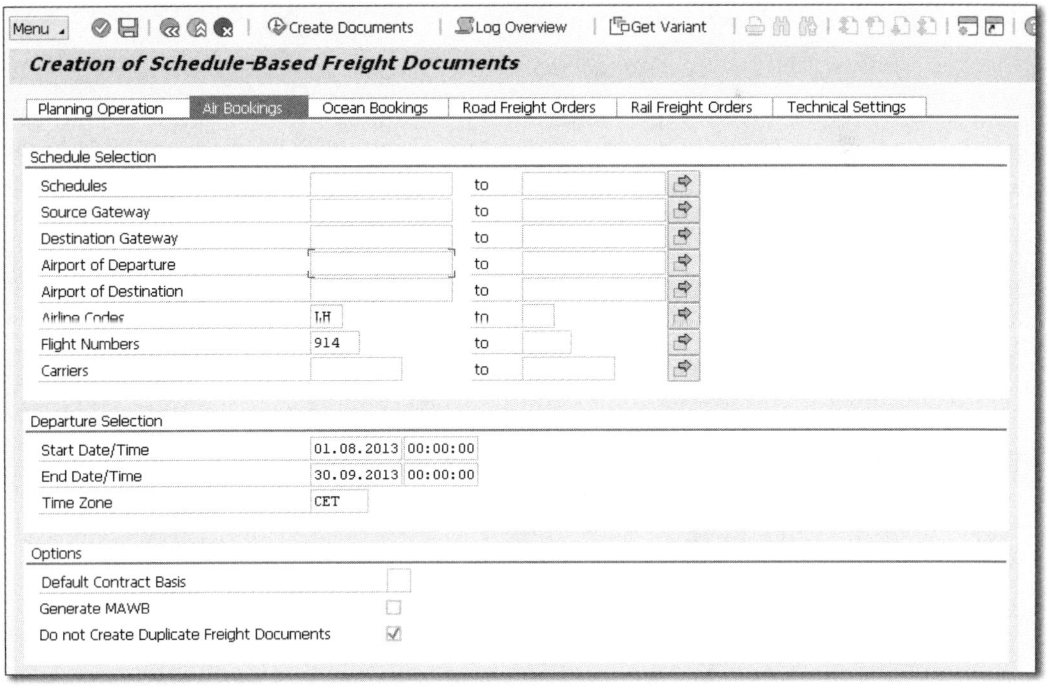

Figure 5.6 Schedule Selection Criteria of Report for Creating Schedule-Based Freight Documents

Departure selection, duplicate prevention

Besides defining the criteria, you also set the time period for which freight documents will be created based on the determined schedules' departures. If the report was run twice or with overlapping selection criteria, you may not want to create duplicate freight documents, so you should set the corresponding parameter accordingly. Some air freight-specific parameters are available, such as the default contract basis, which allows the creation of allotment, blocked space, charter, and ad hoc air freight bookings, and the trigger for MAWB creation.

Technical settings

The TECHNICAL SETTINGS tab allows you to define the following system behavior:

- Cancel processing if a selection error occurs.
- Cancel processing if a freight document creation error occurs.
- Display the selected departures of the determined schedules.
- The freight documents can be saved or just shown in simulation mode, without saving.
- The package size allows parallelization.
- A log is written as shown in Figure 5.7. You can define whether message details are added and how long the log files are available in the system.

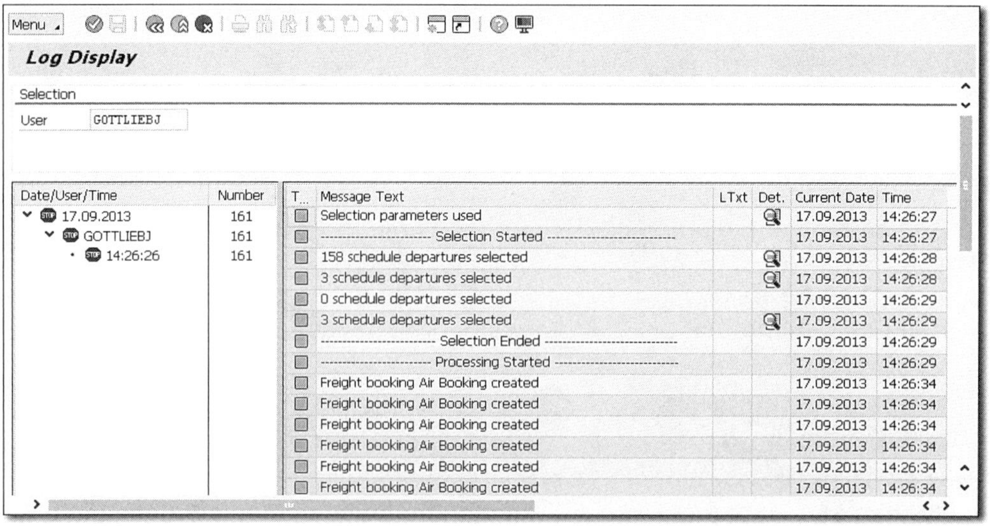

Figure 5.7 Log Information for Created Schedule-Based Freight Documents

5.1.3 Schedule Change Management

Carrier flight schedule changes

If you uploaded or manually created your carriers' schedules, defined schedule-based allocations, and created air freight bookings, then your operative flight plan is up-to-date. Frequently, the carrier changes his schedule by changing the departure date or time, omitting complete departures, or offering new departures. You can upload the changed schedule data as carrier flight schedules, and the system will adjust the previous versions accordingly. Alternatively, you can manually change the carrier flight schedule. This change has an impact on the subsequent business documents in the capacity management process.

Delay in first leg of connection flight

In the event of a delay of the first flight by 10 minutes in a connection master flight schedule that has a connection time of 3 hours, you can accept this delay because 2 hours and 50 minutes are still sufficient to bring the freight from the first to the second airplane. However, if the delay of the first flight is 3 hours, you cannot keep the connection and have to decide whether to take a later departure of the second flight, choose a different carrier flight schedule for the connection, or, in the worst case, give up this connection master flight schedule and create an alternative master flight schedule with a different connection airport.

Reference data status

For connection flights, in particular, a delay of the first flight may invalidate the complete connection. For this reason, there is no automatic propagation of the changes into the subsequent business documents. Instead, the master flight schedule, schedule-based allocation, and air freight booking have a reference data status that indicates whether the underlying carrier flight schedule has changed. This status field is shown in the corresponding POWL queries for these objects so that you can easily identify the affected objects and manually react to the changes. Within these objects, the reference data status is shown on the header level and on the detail level:

- In a master flight schedule, the stages of the departure rules indicate whether the underlying carrier flight schedule's departure rule has changed.

- In a schedule-based allocation, the departure buckets indicate whether the underlying departure has changed.
- In an air freight booking, the stages of the carrier routing indicate whether the underlying departure has changed.

Response to changes

The user interfaces of these three objects allow you to copy the most recent information from the underlying carrier flight schedule or to change the reference and take the data out of another carrier flight schedule. Thus, you can manually change the master flight schedule, schedule-based allocation, and air freight bookings so that they consider the changed carrier flight schedule. This is intended to be a manual step, since the reaction to the change may require interaction with the involved carriers (e.g., via phone or email). However, by implementing the appropriate change controller strategies, you can automate this change process according to your needs.

Changes for all modes of transport

Although explained in the context of air freight, the reference data status and the capabilities for manual reactions to carrier schedules' changes are available for gateway schedules, schedule-based allocations, and freight documents in general—that is, for all modes of transport.

5.2 Allocations

Allocation, bucket, consumption

An allocation represents the planned capacities for a carrier and a trade lane during a validity period. The capacities can be defined for multiple dimensions such as volume, weight, and 20-foot equivalent units (TEUs) and a sequence of time periods of the same granularity, which are frequently called (time) buckets. For each dimension and time period, the allocation captures the already consumed portion of the maintained capacity. All freight documents matching the carrier, trade lane, and validity period consume the corresponding bucket(s) of the allocation. As soon as a freight document is created, the matching allocations are determined asynchronously and updated according to the freight document's capacity. The consumed quantities are visible in the allocations and allow tracking of the capacities and their utilizations. Recall from Section 5.1.1 that one freight document can consume buckets of multiple allocations.

Figure 5.8 shows a schedule-based allocation for carrier LH, a trade lane from a gateway in Munich to a gateway in Denver, mode of transport air, and the validity from July 15 to September 30, 2013. The buckets represent departures of the underlying schedule 17713, which offers weekly departures starting on Monday, July 15 at 21:00. For each departure bucket, you can maintain the maximum gross weight and the maximum gross volume. The corresponding consumption is shown as well. In this example, the allocation was newly created, and no consumptions were recorded yet.

Carrier flight schedule based allocation

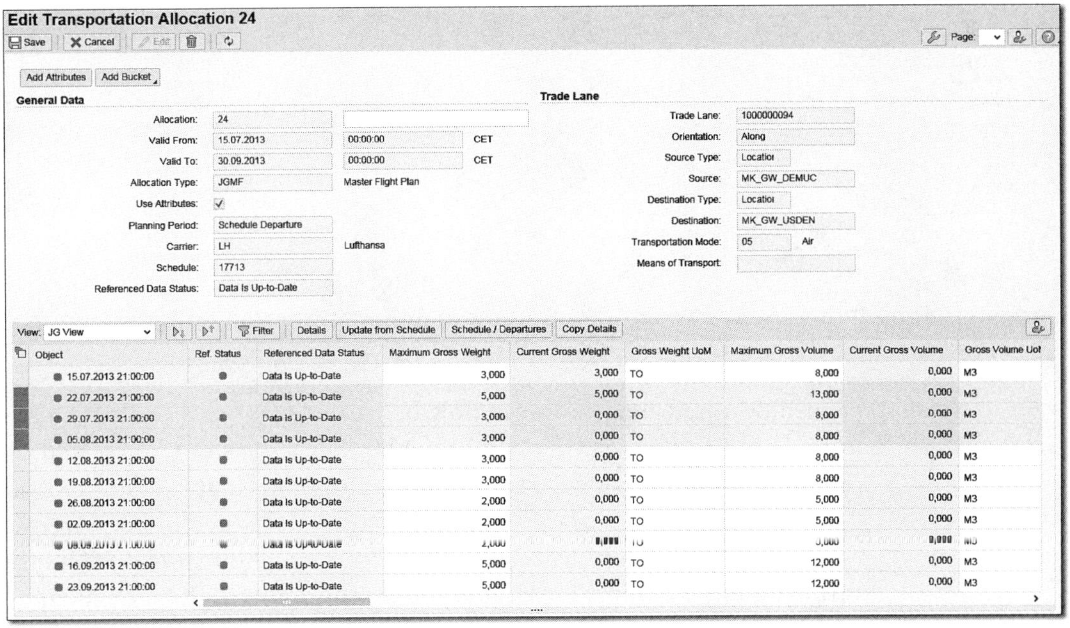

Figure 5.8 Schedule-Based Allocation for Air Freight

Figure 5.9 shows an allocation with monthly buckets covering transportation from China to Germany in 2014 by one ocean carrier. Here, the capacity is defined by the number of 20-foot containers per month, measured in TEU. Consumed weight and volume can also be tracked, although no capacity was defined.

Allocation with monthly buckets

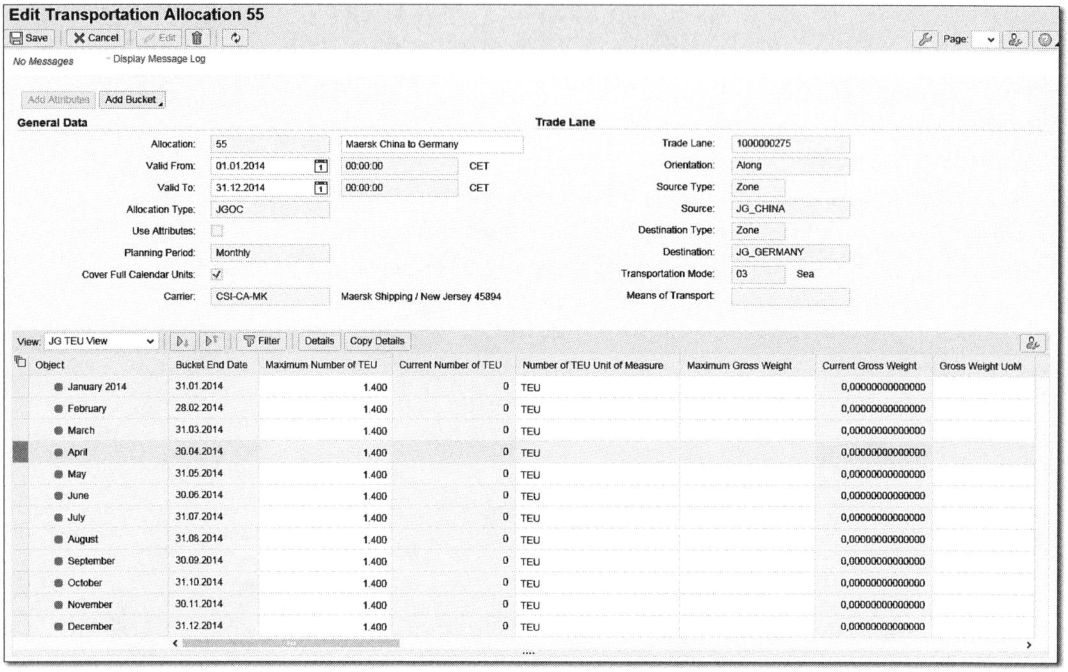

Figure 5.9 Monthly Allocation for Ocean Freight

Creating allocations

Schedule-based allocations can only be created from a schedule, as mentioned in Chapter 3, Section 3.2.4. Other allocations can be created by following the menu path PLANNING • ALLOCATION • CREATE ALLOCATION and specifying the allocation type. An allocation created out of a freight agreement is called *freight agreement allocation*, and its only difference from other allocations is that it is contained in the freight agreement.

Maintaining multiple allocations

Alternatively, you can also select PLANNING • WORKLIST • OVERVIEW PLANNING for the POWL and choose the query for allocations to maintain allocations. It is possible to select multiple allocations and maintain them in the same user interface, as shown in Figure 5.10. At the top, you can see general information about the allocations and can choose which of them will be shown in detail at the bottom, where the buckets are displayed.

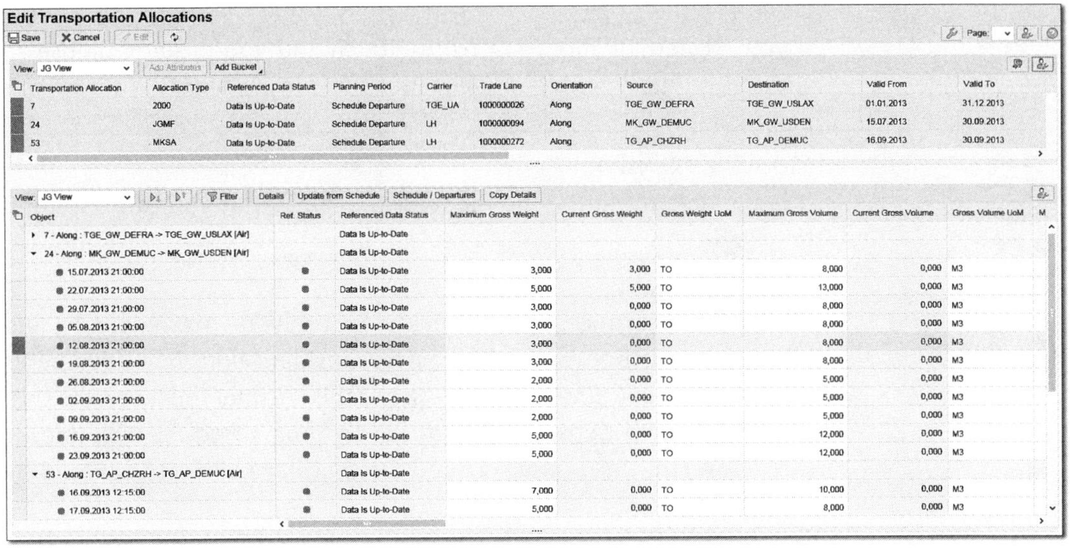

Figure 5.10 Maintaining Multiple Allocations in One User Interface

Filtering, comparing, updating

In the allocation maintenance user interface, you can filter allocations and buckets, for example, according to a start and end time, reference data status being not up-to-date, or their consumed quantity being above the maximum quantity. Whereas automatic carrier selection respects the allocations' capacities, it is possible to manually create freight documents that result in the capacity being exceeded. The user is informed about such a capacity violation by a warning message in the freight document and by the allocations' buckets showing higher consumption than capacity. Click on the SCHEDULES/DEPARTURES button to compare the current allocation's departure dates and times with those from the underlying schedule and decide which schedule data should be transferred into the allocation's buckets.

Allocation types

In Customizing, you can follow the menu path SAP TRANSPORTATION MANAGEMENT • TRANSPORTATION MANAGEMENT • PLANNING • GENERAL SETTINGS • DEFINE TRANSPORTATION ALLOCATION TYPES and define allocation types by specifying the following parameters:

- **Default Type**
  If you created an allocation from scratch and you did not choose an allocation type, the default allocation type is chosen; it is set by this parameter.
- **Mode of Transport**
  It is possible to define the mode of transport, but you can also omit this field.
- **Planning Period**
  You can choose among daily, weekly, monthly, quarterly, yearly, and schedule departure. Whereas the first considers all freight documents in the specified time period—also called (time) bucket—the last option refers to departures of an underlying schedule. This means that all freight documents created for that departure are covered by the bucket.
- **Schedule-Based Allocation**
  This specifies that the allocation depends on a schedule—that is, you can create the allocation only out of a schedule, for which the allocation type has been defined in the schedule type, as mentioned in Chapter 3, Section 3.2.4. The schedule determines the validity and trade lane of a schedule-based allocation.
- **Full Calendar Units**
  You can define whether the bucket fully covers a calendar unit or if it can start at any time but has a duration according to the planning period. If you use a daily planning period and don't use full calendar units, you can have a planning period from Monday 8:00 until Tuesday 8:00. If you use a monthly planning period and full calendar units, the bucket starts on 0:00 on the first day of the month and lasts until 0:00 on the first day of the next month.
- **Use Attributes**
  This allows you to create multiple buckets for the same period of time. The buckets consider different attribute combinations based on shipping type, contract basis, or handling code. Using handling codes for the upper deck and lower deck of an airplane, you can define two buckets with individual capacities, one for the upper deck and one for the lower deck. The concept of attributes is described shortly in more detail.

- Use Carrier Selection
  This defines whether allocations of this type are considered by carrier selection.
- Carrier Selection Unit of Measure
  If carrier selection is activated, you can specify the allocation's unit of measure that is considered for carrier selection. You may define allocations with volume, weight, and TEU quantities and choose carrier selection considering the TEU capacities.
- Update Quantity Automatically
  This defines whether a newly created allocation or bucket gets an automatic update of its consumed quantities.
- Bucket Overlapping
  A freight document may cover multiple buckets of the allocation. Using this parameter, you can define whether all covered buckets get consumed by the freight document or only the first covered bucket gets consumed. This parameter should not be changed if allocations already exist in your productive system because the buckets would contain data according to both consumption modes, which makes the quantities hard to interpret.
- Allocation BW Relevance
  This specifies whether the allocation type is relevant for analytics based on SAP NetWeaver BW.

Additionally, you can set the number range for the allocation type and assign your own UI configuration to the allocation type.

Converting old to new allocations

Converting Old Allocations to New Allocations

The allocation type concept was introduced with SAP TM 9.0. If you are using allocations from an earlier release of SAP TM and want to upgrade to release 9.0, you can use report /SCMTMS/TAL_CONVERSION to convert your old allocations into new allocations with an allocation type of your choice. You can either use the number range of the old allocations or define a new number range as defined by the allocation type.

Attributes and equipment

So it is possible to define multiple buckets for one time period, based on different attribute combinations. The prerequisite is that you have selected the Attributes option in the allocation type Customizing. Now you can maintain an allocation and introduce attribute nodes by clicking on the ADD ATTRIBUTES button, as shown in Figure 5.11, where two attribute nodes have already been created. The departure buckets are shown hierarchically below the attribute level. Each departure appears under each attribute node, which allows you to define the capacity for each attribute combination and departure.

Handling code

Standard attributes are the contract basis (which can be allotment, blocked space, charter, or ad hoc), the service level, and the shipping type. In addition, you can click on the DETAILS button to add one or multiple handling codes, which characterize the goods that can be transported and the required equipment, such as three unit load devices (as shown in Figure 5.11).

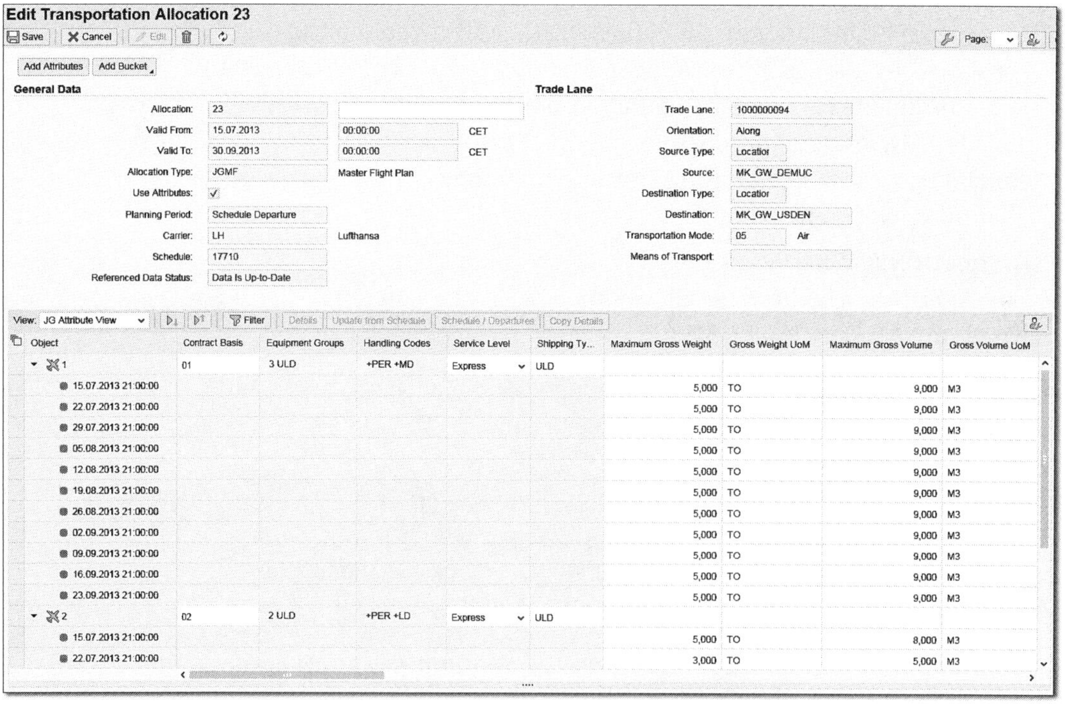

Figure 5.11 Schedule-Based Air Freight Allocation with Attributes

If you enter quantities in the attribute row, they are propagated into the departures' fields below. If you create freight documents for this allocation, one air freight booking will be generated per attribute and departure combination, and the additional information such as the handling codes and equipment is copied into the air freight bookings.

Handling Codes

Handling codes can classify goods and characterize how they are to be transported. You can define handling codes in Customizing, following the menu path SAP Transportation Management • Transportation Management • Basic Functions • General Settings • Define Handling Codes. For each handling code, you can add a description and specify whether it is to be used in external communication or whether it serves internal purposes only.

Handling code constraint mode

When adding handling codes in the attribute node of an allocation, you can choose among the following handling code constraint modes relevant for the freight documents created for the allocation:

- Handling Code Must Be Identical
  Only freight units with the same handling code can be assigned to the freight document. This constraint mode is shown in the allocation with the prefix +.
- Exclude Objects with This Handling Code
  Only freight units that do not have this handling code can be assigned to the freight document. This constraint mode is shown in the allocation with the prefix -.
- Not Relevant for Planning
  No constraints are imposed, and this handling code is used only for informational purposes. The handling code is directly shown in the allocation, without any prefix.

For example, defining the handling codes PER (perishables) and FRO (frozen goods) as external and C23 (special code 23) as internal, you can force the system to include perishables, to exclude frozen goods, and to mention special code 23 as a handling code for information purposes only. This combination would be displayed in the allocation as +PER – FRO C23.

Handling code constraint profile

If you want to use handling codes without allocations, you can select the menu path SAP Transportation Management • Transportation Management • Basic Functions • General Settings • Define Handling Code Constraint Profile in Customizing, capture a combination of handling codes and corresponding constraint modes in a handling code constraint profile, and assign it to a means of transport. When you create freight orders with that means of transport, the handling codes and their constraint modes are copied from the profile into the freight order.

Handling codes can be entered in air forwarding orders, are propagated into the corresponding freight units, can be defined for freight documents as just described, and are considered by automatic and manual planning, where the handling codes can be displayed in the transportation cockpit. If marked as external, they are contained in the air freight booking messages sent to the carriers.

5.3 Business Shares

Motivation

Though you may have your preferred carrier for a certain trade lane, you also collaborate with other carriers, perhaps to resolve bottleneck situations, peak demands, or other issues with your preferred carriers. Giving freight orders to other carriers only when you have severe problems may not be a good basis for a solid relationship.

To protect your relationship with the other carriers, you may decide to grant them a certain percentage of your transportation business each month. In this way, you can establish a stable relationship but still give most of your transportation business to your preferred carrier. In other scenarios, you may have an agreement with your two major carriers that each of them gets 50% of your transportation business in a certain region. This helps your relationship with the two carriers because they can rely on getting the agreed-upon amount of your business.

Business share, target shares

With SAP TM, you can define such target shares per carrier as a business share for a trade lane and means of transport or mode of transport on a daily, weekly, monthly, quarterly, or yearly basis. The business shares are used by carrier selection, as described in Chapter 7, Section 7.1.

Maintaining business shares

Via the menu path PLANNING • BUSINESS SHARE • CREATE BUSINESS SHARES, you can create business shares, as shown in Figure 5.12. In the TRADE LANES area, you can define the trade lanes for which you want to define business shares. For each trade lane, you can define the means of transport or mode of transport. In this example, the trade lane represents all transports within the transportation zone USA, and the business share is defined for means of transport 0001. In the TARGET SHARE area, you can insert the relevant carriers and assign a percentage for each. The BUSINESS

SHARE DETAILS area specifies the validity period of the business share, the business share period (which can be daily, weekly, monthly, quarterly, or yearly), and the unit of measure that is used to determine the percentages among the freight documents that match the trade lane and the (time) buckets. Like in the allocation type Customizing, you can also define whether the business shares refer to full calendar units.

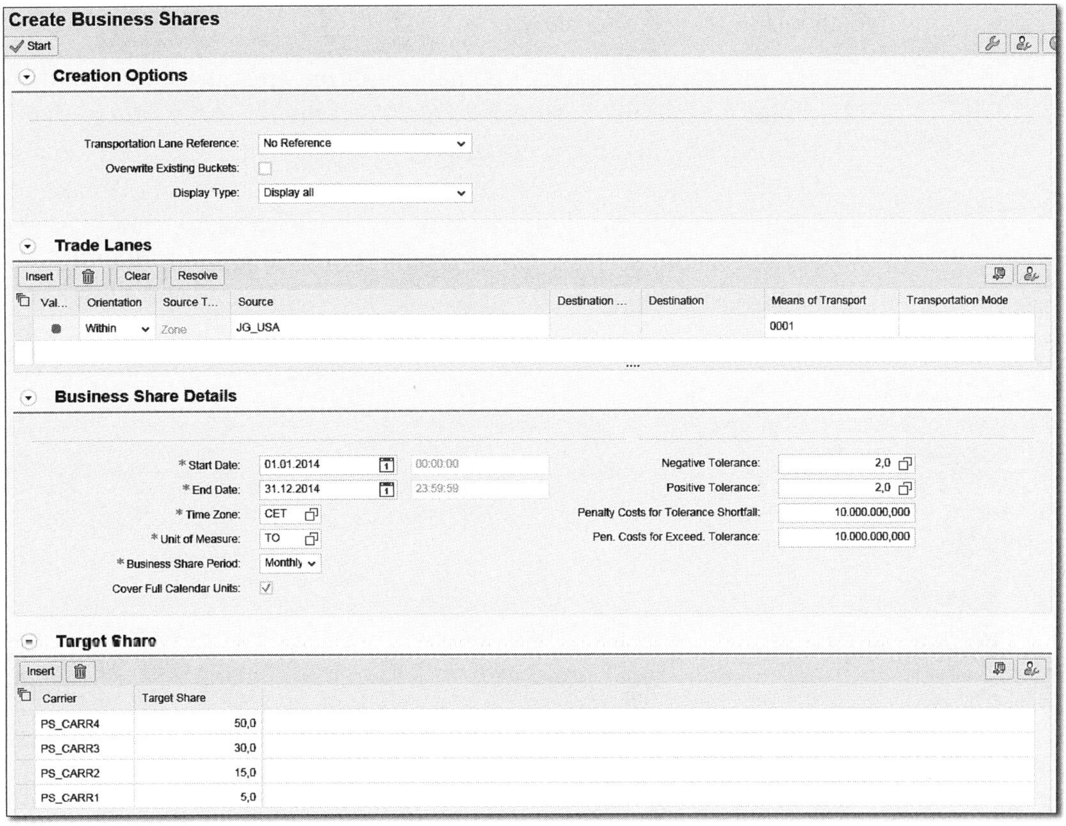

Figure 5.12 Creating Business Shares

Automatic carrier selection takes into consideration the negative and positive tolerances and the corresponding penalty costs for violating the tolerances; these are explained in detail in Chapter 7, Section 7.1.2.

The CREATION OPTIONS area specifies how the business share is created, with three options:

- Without reference to a transportation lane
- Only if a corresponding transportation lane exists
- Only if a corresponding transportation lane exists, with the data copied from the transportation lane

It is possible to define whether existing buckets are overwritten. The display type determines whether only newly created business shares or all business shares are shown.

Once you have maintained the data in all the areas, click on the START button to create the business shares. Then the created business shares are displayed and can be edited, as depicted in Figure 5.13. You can also edit business shares by selecting the menu path PLANNING • WORKLIST • OVERVIEW PLANNING and choosing the business share query in the POWL. You can select multiple business shares and edit them in one user interface, similar to the process for allocations described in Section 5.2.

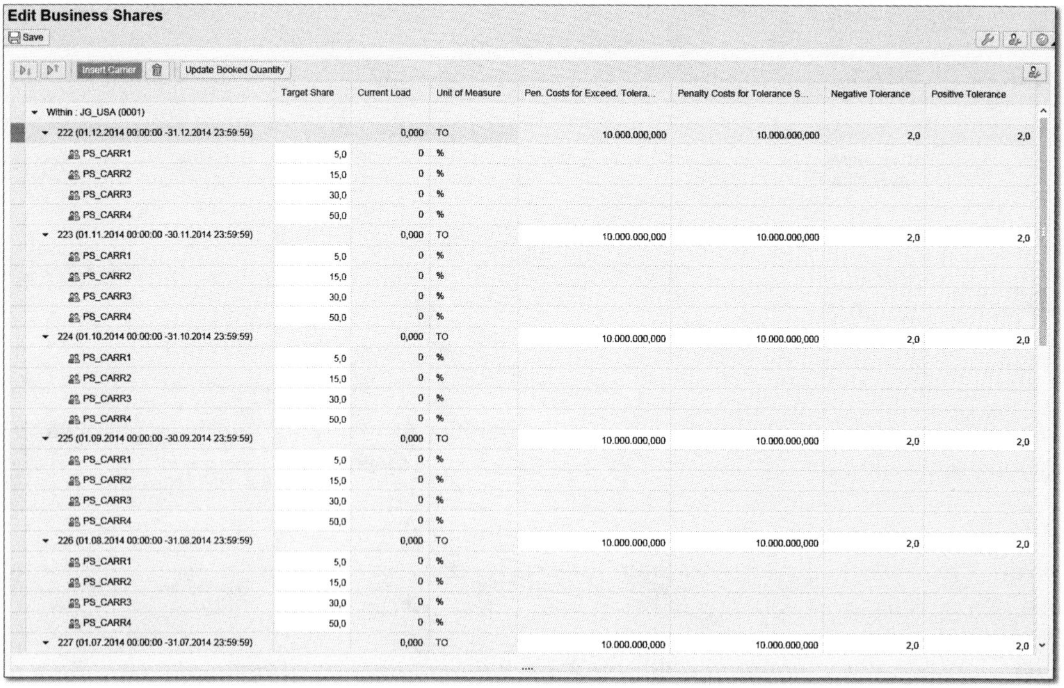

Figure 5.13 Maintaining Business Shares

The current values of the business shares are updated automatically when new freight documents are created, analogously to the allocations' consumption values. In the previous example, the business shares have just been created, so all carriers have a share of 0%. The shares are calculated per bucket.

You can also create business shares out of freight agreements, as mentioned in Chapter 10, Section 10.1.1.

5.4 Freight Bookings

Ocean/air freight bookings

Freight bookings are used to reserve freight space on a sea vessel or in an airplane. The corresponding mode-specific freight documents—called ocean freight bookings and air freight bookings, respectively—provide mode-specific information such as the vessel name or flight number on their user interfaces. The space reserved by freight bookings is consumed by assigning freight units to the freight bookings.

Legs of freight bookings

An ocean freight booking represents ocean transportation from a port of loading to a port of discharge, and an air freight booking represents air transportation from an airport of departure to an airport of destination. Freight bookings can cover a consolidation location before the source (air-) port and a deconsolidation location after the destination (air-) port, as illustrated in Figure 5.14. These consolidation and deconsolidation locations are called container freight stations in the ocean case and gateways for the air case. Since the term *gateway* is also used in ocean scenarios, we also use the term *gateway* in the general sense. Note that the main leg can consist of multiple stages in order to model connection flights or multistop voyages.

Bookings can be generated based on schedules, capture the results of planning (e.g., the assigned freight units), and are subcontracted to your carrier. As we cover in Chapter 7, you can directly define a carrier or use carrier selection to determine the carrier of a booking. You can send the freight booking to your carrier and explicitly set the response by the carrier, which can confirm the booking, confirm with deviations, or reject the booking.

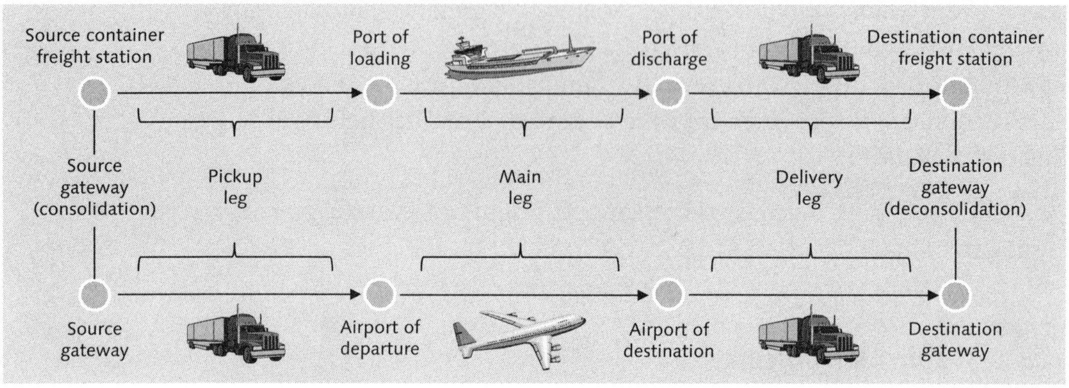

Figure 5.14 Structure of Locations and Stages of Ocean Freight Bookings and Air Freight Bookings

Charges can be calculated, freight settlement documents can be created, and costs can be distributed among the involved parties. Freight bookings also serve for transportation execution processes such as printing and tracking and tracing the progress of transportation.

So let's examine both kinds of freight bookings. Section 5.4.1 presents ocean freight bookings, reviewing the different functional areas in the user interface. Section 5.4.2 describes air freight bookings, focusing on how they differ from ocean freight bookings and additional air freight specifics.

5.4.1 Ocean Freight Bookings

Creating ocean freight bookings

There are many ways to create ocean freight bookings. You can follow the menu path Freight Order Management • Ocean • Create Ocean Freight Booking and manually create an ocean freight booking. You can also create ocean freight bookings from the corresponding POWL query, which is available via the menu path Freight Order Management • Ocean • Overview Ocean Freight Bookings. Alternatively, it is possible to copy ocean freight bookings, which means that the header and logistical data of the original booking but no freight unit assignments are copied.

You can also create ocean freight bookings by manual planning (see Chapter 6, Section 6.3), automatic planning (see Chapter 6, Section 6.4), or by capacity management, using the report to create schedule-based freight documents, as explained in Section 5.1.2. Manual and automatic planning can also change the freight unit assignments to freight bookings. It is also possible to create freight bookings from an ocean forwarding order for the stages of the actual route, as described in Chapter 4, Section 4.2.1.

User interface areas

Ocean freight bookings contain a lot of information that is structured on the user interface in multiple areas that are described in the following.

Business partners

In the BUSINESS PARTNER area, you can define the carrier, the executing carrier, the shipper, the consignee, and the communication party, as shown in Figure 5.15. The carrier confirms the booking and charges for it, although the executing carrier may be a different carrier.

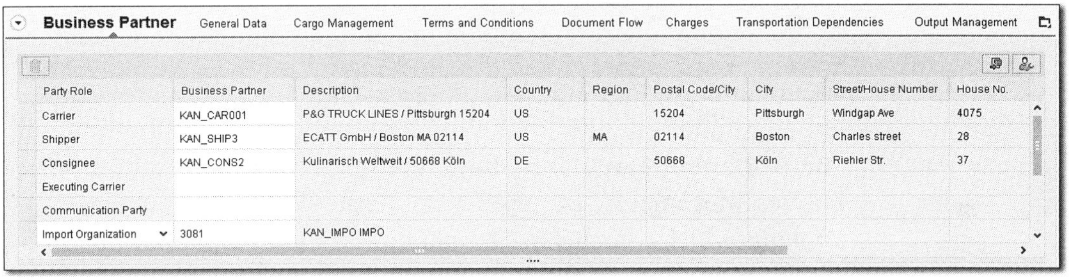

Party Role	Business Partner	Description	Country	Region	Postal Code/City	City	Street/House Number	House No.
Carrier	KAN_CAR001	P&G TRUCK LINES / Pittsburgh 15204	US		15204	Pittsburgh	Windgap Ave	4075
Shipper	KAN_SHIP3	ECATT GmbH / Boston MA 02114	US	MA	02114	Boston	Charles street	28
Consignee	KAN_CONS2	Kulinarisch Weltweit / 50668 Köln	DE		50668	Köln	Riehler Str.	37
Executing Carrier								
Communication Party								
Import Organization	3081	KAN_IMPO IMPO						

Figure 5.15 Business Partners

General data

The GENERAL DATA area in Figure 5.16 displays information about carriers, goods, capacity requirements, organizational data, and data about the voyage, vessel, and underlying schedule. It is possible to maintain a booking without reference to a schedule.

Terms and conditions

The TERMS AND CONDITIONS area specifies the incoterm, the incoterm location, and whether it is a controlled or uncontrolled transport. You can also define the freight term, which can be prepaid or collect and, together with the traffic direction (import or export), determines how freight settlement documents are created. Additionally, it is possible to define the movement type, the shipping type used for subcontracting, and whether consolidation at the source container freight station and destination container freight station is intended. If no consolidation is chosen, the corresponding container freight station is skipped.

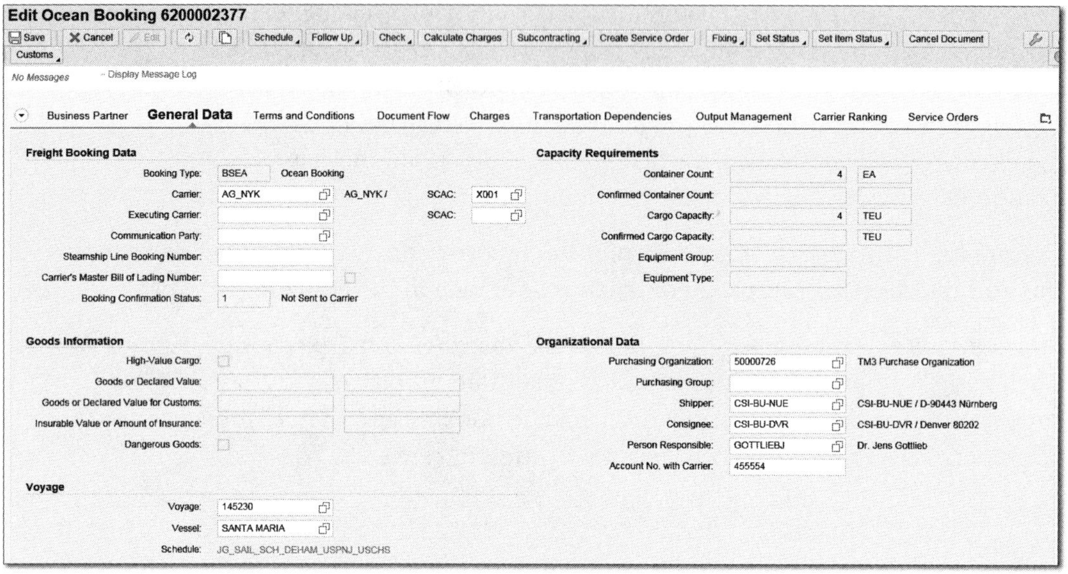

Figure 5.16 General Data of an Ocean Freight Booking

Document flow

The Document Flow area shows the relationships with the involved business documents, as depicted in Figure 5.17 for an ocean freight booking, which contains five freight units created for one original ocean forwarding order. By clicking on the hyperlinks, you can directly navigate to the involved documents.

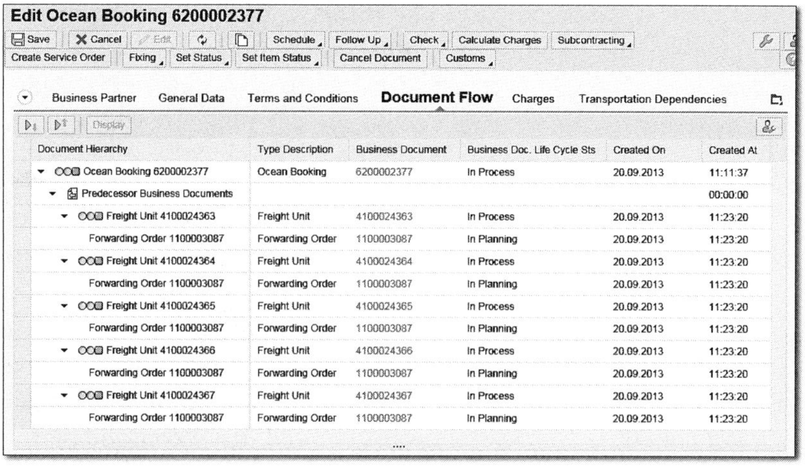

Figure 5.17 Document Flow

Transportation dependencies

The TRANSPORTATION DEPENDENCIES area is available only for ocean freight bookings and displays the dependencies to other freight documents, such as road or rail freight orders for the pre-leg or the subsequent leg. This is helpful to analyze time conflicts or assess the effects of potential delays in the transportation chain.

Cargo management

The CARGO MANAGEMENT area allows you to display and maintain information about the loaded cargo and its structure consisting of containers or unit load devices, freight units, packages, and products, as shown in Figure 5.18. You can insert new containers, packages, and products and assign freight units to containers by dragging and dropping them. It is also possible to insert new freight units selected according to their identifiers, forwarding orders, or arbitrary attributes. You can also distribute the items of an already assigned freight unit over several containers, unit load devices, or compartments of the ocean freight booking, air freight booking, or freight order, respectively. The freight unit's quantities can be split and distributed over multiple unit load devices.

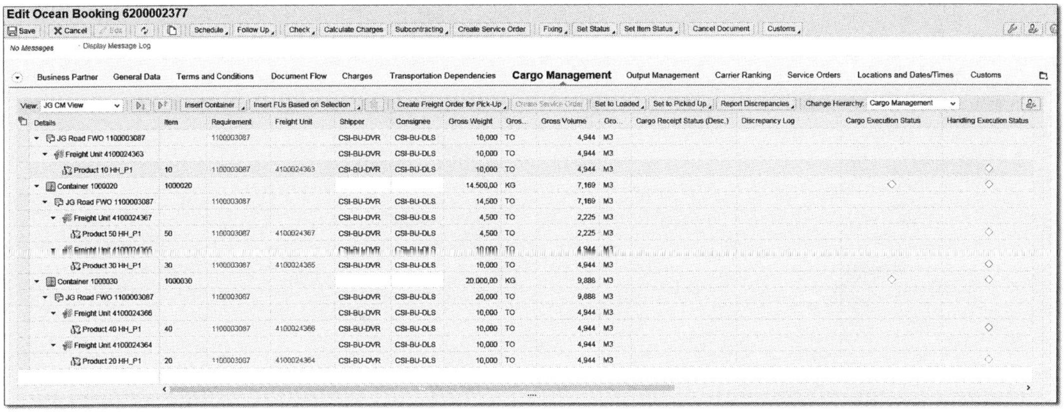

Figure 5.18 Cargo Management

Hierarchical view for cargo management

You can define your own hierarchical view to display the cargo structure according to your needs, as explained in Chapter 6, Section 6.3.4. Analogously to the transportation cockpit, you can dynamically switch between alternative hierarchical views.

Pickup and delivery freight orders

For ocean bookings, you can create pickup and delivery freight orders that cover the transport from source container freight station to port of loading and from port of discharge to destination container freight station, respectively. These freight orders cannot be planned and are for execution purposes only. Such freight orders can also be created for air freight bookings, covering the transport from source gateway to airport of departure and from airport of destination to destination gateway.

In cargo management, you can also set the pickup and loaded status and report discrepancies if the actually loaded goods differ from the planned and expected freight. This process is described in detail in Chapter 8, Section 8.1.2.

Locations, dates, and times The LOCATION AND DATES/TIMES area contains all relevant data about the source container freight station, port of loading, port of discharge, and destination container freight station and the corresponding departure date and time and cutoff dates and times for cargo, dangerous goods, and documents, as shown together with the stages in Figure 5.19.

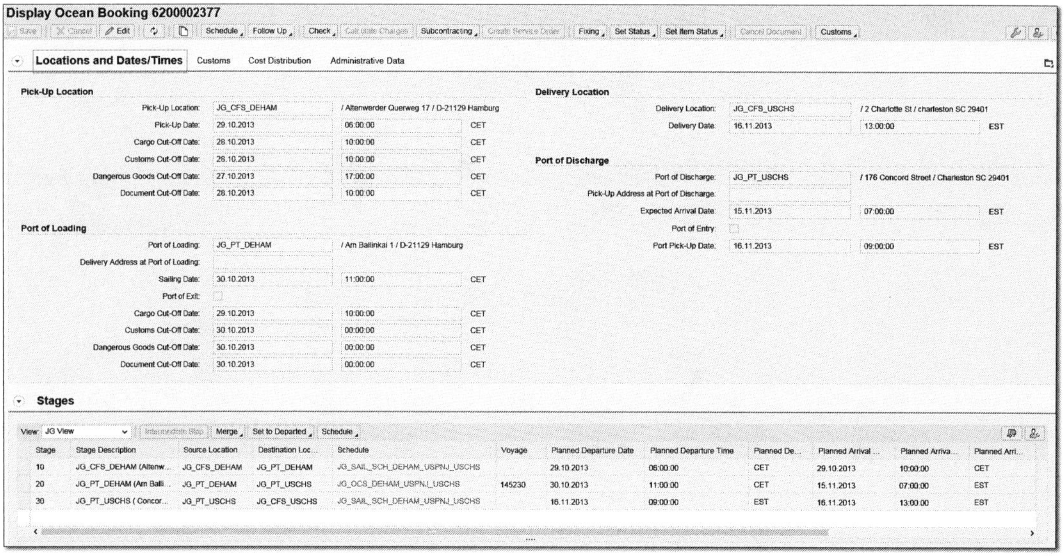

Figure 5.19 Locations, Dates and Times, and Stages

Stages The STAGES area shows the same information from the stage perspective, which is particularly helpful if the booking refers to schedules, for which the references are displayed, or if it contains more than three stages. More than three stages appear in ocean connection bookings referring to multiple underlying ocean carrier schedules or in multistop bookings, which we discuss later. By clicking on the SCHEDULE button, you can assign a schedule to the stage, unassign a schedule, or update the stage's data per the schedule's data, as mentioned in Section 5.1.3. If you manually create an ocean freight booking for a connection voyage without reference to a connection sailing schedule, you create multiple stages and assign a different ocean carrier schedule to each stage.

Capacity requirements

The CAPACITY REQUIREMENTS area lists the required and confirmed capacities and determines the corresponding cargo capacity, as shown in Figure 5.20. In this example, three equipment types refer to a 20-foot container, and the last equipment type refers to a 40-foot container, which is reflected in the corresponding cargo capacity that is measured in TEUs.

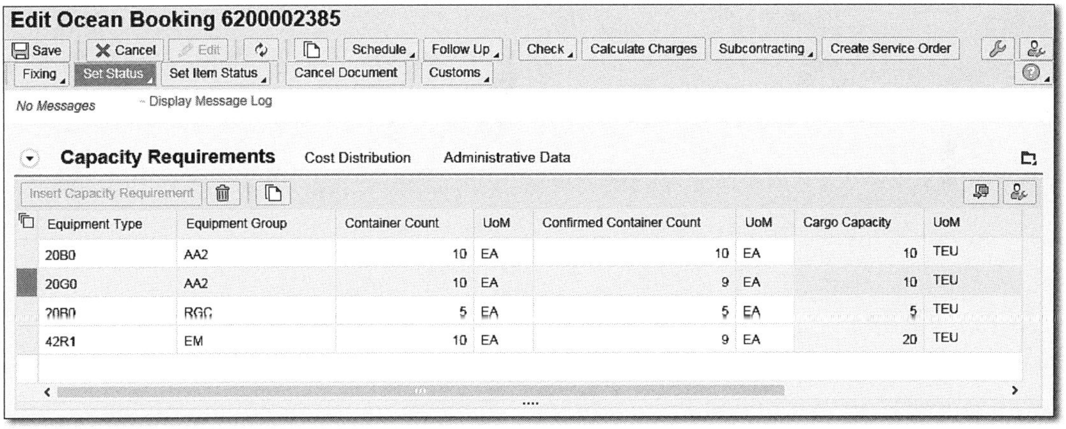

Figure 5.20 Capacity Requirements

Overview

The OVERVIEW area provides a quick overview of the stages and cargo in a hierarchical view, as depicted in Figure 5.21. You can expand the stages and see the cargo below, including the substructure of the involved containers. The planned start and end times for each stage are also shown. This area is useful if you want to see the most important information at a glance without having to gather the details that are spread over multiple areas.

Charges

The CHARGES area provides details about the determined charges for the freight booking. You can trigger charge calculation by clicking on the corresponding button in the toolbar of the freight booking. See Chapter 10 for more details on charge calculation.

Cost distribution

The COST DISTRIBUTION area shows details about cost distribution, which is relevant if multiple parties are involved and the charges of the freight booking should be distributed among these parties. More details on cost distribution are presented in Chapter 11, Section 11.2.

Customs

The CUSTOMS area displays information relevant for customs handling, such as the customs status, border crossing information, a list of customs activities, and item groups. Refer to Chapter 9, Section 9.1 for more details on customs handling.

Miscellaneous areas

The ADMINISTRATIVE DATA area displays when the freight booking has been created and last changed, and by whom. The CHANGE DOCUMENTS area shows, on a rather technical level, the changes that were made in the ocean freight booking, if that option was activated in the freight booking type. The ATTACHMENTS area allows you to store attachments such as documents and URLs in the freight document. References to other documents can be captured in the DOCUMENT REFERENCES area, and notes can be added in the NOTES area.

Freight booking types

You can define the booking types via the Customizing menu path SAP TRANSPORTATION MANAGEMENT • TRANSPORTATION MANAGEMENT • FREIGHT ORDER MANAGEMENT • FREIGHT BOOKING • DEFINE FREIGHT BOOKING TYPES.

Similarities to freight orders

The booking type contains standard sections such as BASIC SETTINGS; CHARGE CALCULATION AND SETTLEMENT DOCUMENT SETTINGS; NUMBER RANGE SETTINGS; CHANGE CONTROLLER SETTINGS; ADDITIONAL STRATEGIES; DEFAULT UNITS OF MEASURE for volume, weight, and quantity; DEFAULT MEANS OF TRANSPORT DETERMINATION; EVENT MANAGEMENT SETTINGS; OUTPUT OPTIONS; and ORGANIZATIONAL UNIT DETERMINATION. These are also contained in freight order types (described in Chapter 7, Section 7.2.1).

Differences from freight order types

Let's consider how freight bookings differ from freight orders:

- All booking types can be subcontracted, in contrast to freight order types, which can forbid subcontracting to cover transportation businesses fully relying on their own fleets. Though carrier selection and tendering are offered for freight orders, only carrier selection is possible for freight bookings. In most scenarios, the carrier is already known at the time of booking creation.
- Whereas freight orders allow star-shaped and unrelated stages, freight bookings only allow sequential stages.
- Freight orders cover self-delivery and self-pickup scenarios, which are not relevant for freight bookings.

Pickup and delivery freight orders

- You can create pickup and delivery freight orders for the stages from consolidation location to source (air-) port, and from destination (air-) port to deconsolidation location, respectively. The freight booking type can define the pickup freight order type and the delivery freight order type.

Multistop ocean freight bookings

- Freight booking with multiple ports of loading and ports of discharge

  Usually, an ocean freight booking transports goods along a port-to-port connection. In some ocean transportation businesses, it is common to have ocean freight bookings transporting goods along a port sequence. Freight can be loaded in all but the last port, and freight can be unloaded in all but the first port.

  Using this flag, you can enable ocean freight bookings with a port sequence that has more than two stops. Figure 5.22 shows an example with a port sequence of Hamburg, Newark, and Rio de Janeiro. Such ocean freight bookings can only be created manually, and this functionality is only offered for ocean scenarios.
- Service orders can be created from freight bookings, and you can specify the default service order type.
- You can choose between manually entering the goods' value on the header level or automatically aggregating it by the items' goods' values.

Defining main container items

- The container item source determines whether container items are taken from the forwarding order or are manually defined in the freight booking. In the first case, the item structure of the forwarding order is directly copied into the freight booking. In the second case, if you enter a number of containers, an equipment group, and an equipment type in the CAPACITY REQUIREMENTS area of the freight booking, corresponding container items are created automatically in the CARGO MANAGEMENT area. The assigned freight units become subitems of these container items.

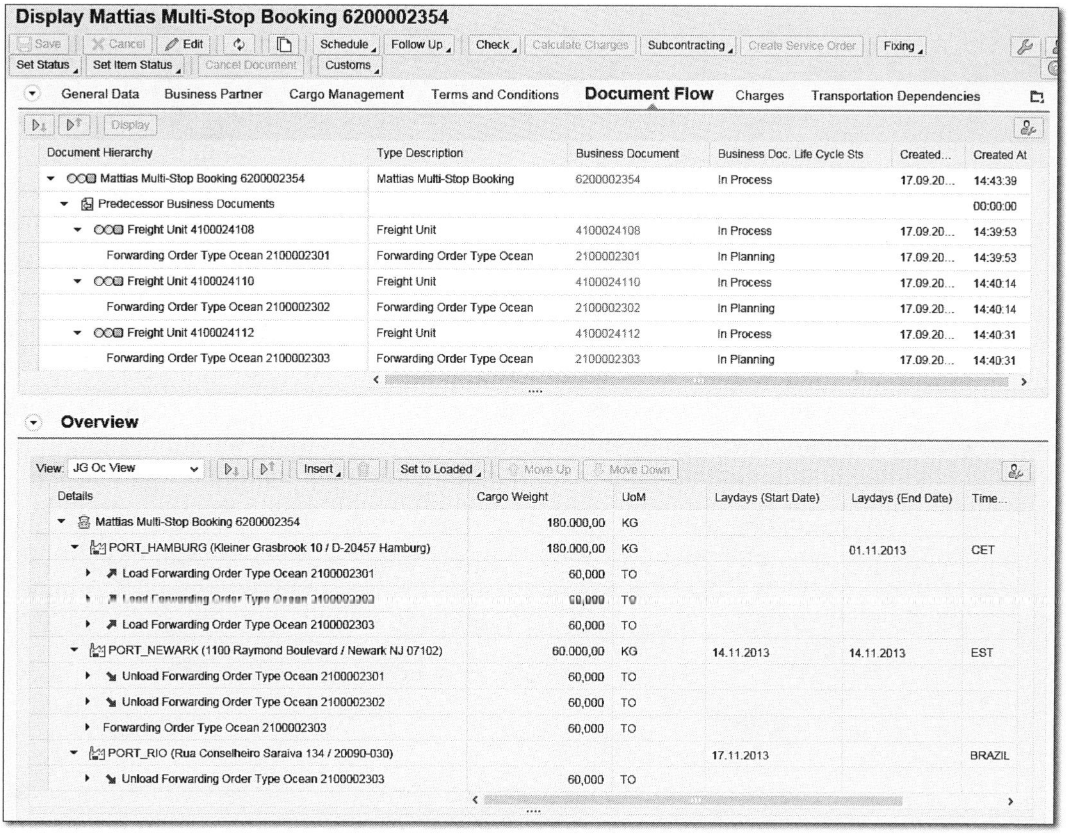

Figure 5.22 Example of a Multistop Ocean Freight Booking

- Freight bookings can be fixed when capacity planning is finished. Like for freight orders, you can also fix the document when it is created or choose not to fix it at all.

Omitting source, destination gateway

- SERVICE DEFINITION
  In addition to the default service level, service level condition, and traffic direction, you can also define the shipping type, the movement type, and whether a consolidation location and a deconsolidation location, respectively, are involved. Using the last two parameters regarding consolidation and deconsolidation locations, you can create port-to-port, port-to-gateway, gateway-to-port, and gateway-to-gateway ocean freight bookings and their air freight booking counterparts.

Organization interaction model

- PREDECESSOR DOCUMENT HANDLING
  The sales organization in a forwarding company can access freight bookings and assign forwarding orders and their freight units, but the final decision about the assignment is made in the planning and execution organization. The organization interaction status of the corresponding freight unit stages can request confirmation from the planning and execution organization where the capacity manager works. Once a forwarding order is assigned to the booking, the capacity manager has to check and confirm the assignment. He may accept or reject the assignment; in the second case, an alternative freight booking needs to be identified to ensure that the sold forwarding order can be executed. In the transportation cockpit, the fields for the organization interaction model are not visible in the standard lists, so you have to activate them in the corresponding views. Refer to Chapter 6, Section 6.1.2, which discusses the organization interaction model from the freight unit stage perspective.

  The ORGANIZATION INTERACTION parameter activates the organization interaction processing for freight bookings. If active, you have to define an auto-confirmation profile.

  The auto-confirmation profile allows you to automatically confirm freight unit stages, which means that they don't have to be checked manually by the planning and execution organization. Define these profiles using the Customizing menu path APPLICATION ADMINISTRATION • GENERAL SETTINGS • ORGANIZATION INTERACTION • AUTO-CONFIRMATION PROFILES • CREATE AUTO-CONFIRMATION PROFILE. For example, you can define that all assignments of quantities below 50 kg are confirmed automatically, which means that all assignments above 50 kg have to be confirmed manually by the planning and execution organization.

The UPDATE FROM PREDECESSOR parameter specifies whether the assignment of freight unit stages to a freight booking is processed asynchronously or synchronously. If you use the organizational interaction process, we recommend the asynchronous processing, which does not lock the freight booking and may therefore lead to exceeding the booking capacity. The assignment of the freight unit stage can then be confirmed automatically or manually.

- EXECUTION SETTINGS
  The same settings are offered as for freight orders. In addition, you can specify whether the carrier confirmation is a prerequisite for reaching the *Ready for Execution* status.
- CO-LOAD and AIR CARGO SECURITY CHECK
  These parameters and the corresponding authorization check are described in the context of air freight bookings next.

5.4.2 Air Freight Bookings

Similarity to ocean freight bookings

Air freight bookings have many functional similarities to ocean freight bookings. The following screen areas are identical to ocean freight bookings: CUSTOMS, COST DISTRIBUTION, SERVICE ORDERS, OUTPUT MANAGEMENT, DOCUMENT FLOW, ATTACHMENTS, NOTES, DOCUMENT REFERENCES, EXECUTION, and ADMINISTRATIVE DATA.

The BUSINESS PARTNER area is also identical to ocean freight bookings. Frequently, the carrier is different from the executing carrier, due to code-shared flights. For example, one carrier executes the flight under his flight code LH-577, and one or multiple other carriers offer it as their own flights (e.g., UA-344 and AC-349).

The LOCATIONS area is more compact than for ocean freight bookings and provides information about the airports of departure and destination and the expected departure and arrival dates and times. The cargo cutoff and availability times are also shown.

Booking area

The BOOKING area shown in Figure 5.23 contains a lot of general data such as the issuing carrier, the MAWB stock and the drawn number (see

Chapter 8, Section 8.1.1 for more details), the source and destination airport, and the expected departure and arrival dates and times. Source and destination gateway information and the corresponding cutoff dates and times are displayed, too. Organizational data such as the source organization and the destination organization can be maintained. The capacity requirements, unit load device information, and the handling codes with their constraint modes are all taken from the underlying master flight schedule, which is also indicated here. For more details on the use of handling codes, see Section 5.2.

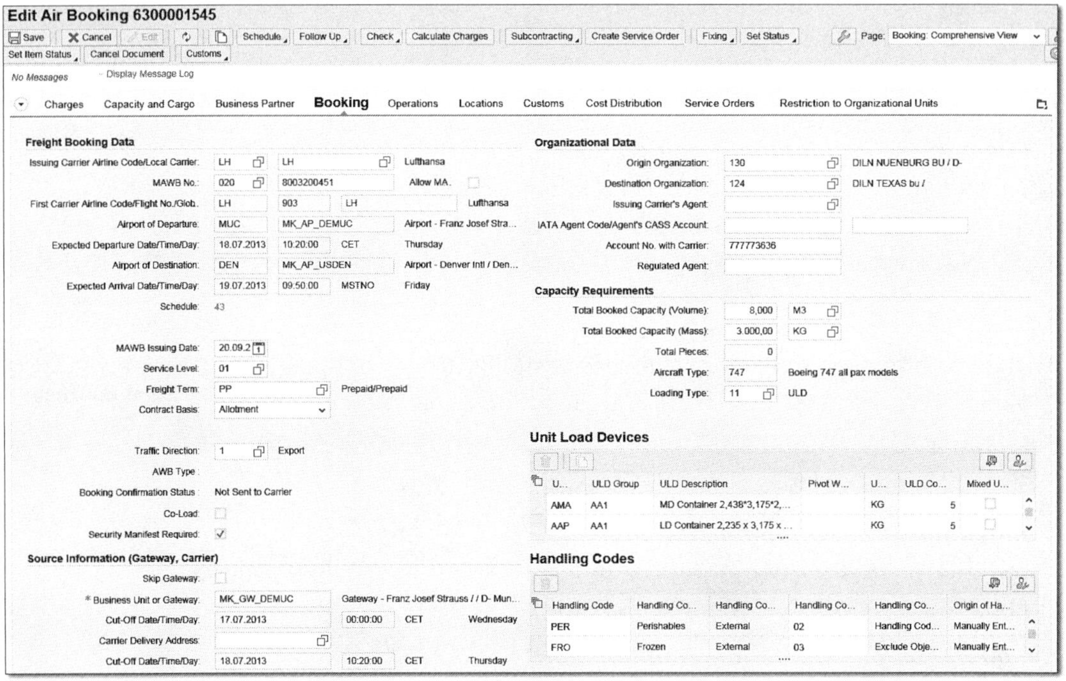

Figure 5.23 Booking Area of Air Freight Booking

Contract basis The contract basis is contained in the Booking area and can be selected from the values defined in Customizing. To do so, follow the menu path SAP Transportation Management • Transportation Management • Freight Order Management • Freight Booking • Define Contract Basis. The contract basis entries allotment, blocked space, charter, and ad hoc are delivered by default and reflect different levels of contractual commitment to the carrier of the air freight booking. The higher the commitment, the higher the cost of cancellation. For example, if you cancel an allotment, only low costs, if any, are incurred. If you cancel a blocked space booking, however, you usually have to pay the full freight amount to the carrier.

Carrier routing = stages

The BOOKING area also contains the CARRIER ROUTING section, shown in Figure 5.24, which provides the stages' information, similar to the STAGES area of ocean freight bookings. The example in the figure shows two stages of a connection flight from Munich (MUC) to Chicago (ORD) by Lufthansa and from Chicago to Denver (DEN) by United Airlines. For each stage, you see the referenced schedule and the reference data status, which is explained in Section 5.1.3.

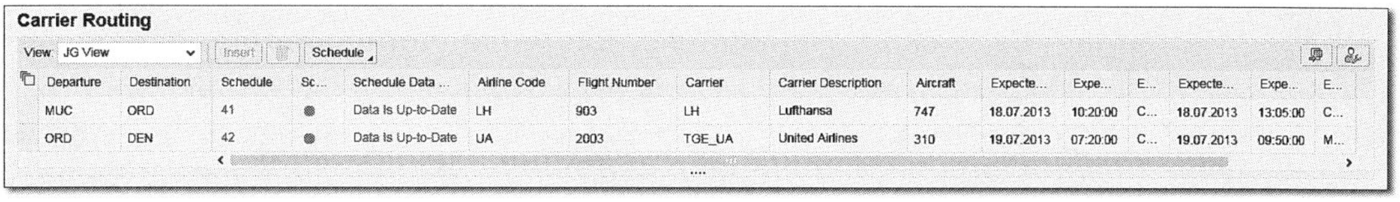

Figure 5.24 Carrier Routing

Charges and air waybill

The CHARGES area contains the same settlement content as ocean freight bookings but adds some air waybill-specific information. You can define whether the agreement rates from the standard rate tables or the TACT rates are shown on the printed air waybill. Similarly, you can specify whether settlement should use the standard rates or the TACT rates for the calculation of the charges relevant for settlement. It is also possible to activate printing of other charges in the house air waybill and the MAWB.

Capacity and cargo

The CAPACITY AND CARGO area contains cargo management like for ocean freight bookings and provides capacity information such as the booked volume and weight, the remaining capacity for volume and weight, and the utilization as a percentage. For air freight bookings, the cost efficiency is mainly determined by the density factor, which characterizes the ratio of volume to weight and is shown as well.

The OPERATIONS area contains general data such as the air waybill type and the issuing date of the MAWB, additional goods information about declared value, insurable value, and handling instructions.

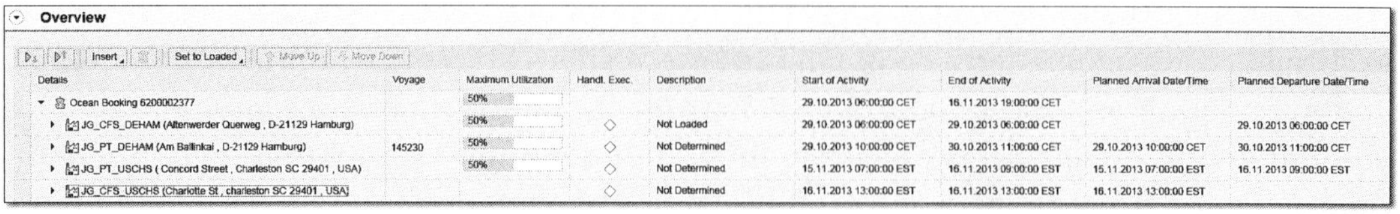

Figure 5.21 Overview of Ocean Freight Booking

Carrier ranking list

The CARRIER RANKING area shows the results of carrier selection, which is a ranked list of the available carriers. Automatic carrier selection chooses the best carrier according to the criteria you have specified. If, for some reason, you want an alternative carrier, you can check the other available carriers in the ranking list and select another carrier. See Chapter 7, Section 7.1 for more details on carrier selection.

Service orders

The SERVICE ORDERS area covers the service orders defined for the freight booking and its items. A service order can capture tasks such as customs clearance of the document or cleaning and fumigating of containers. It is possible to calculate charges for the covered services and to create settlement documents for the services. See Chapter 7, Section 7.2 for more details on service orders.

Status

The STATUSES area captures all kinds of status values, including lifecycle status, fixing status, archiving status, and more values for subcontracting, confirmation, invoicing, dispute cases, execution, customs, transmission to SAP ERP, consistency, and blocking information. The document status and the item status can be set by the corresponding buttons in the booking's toolbar. Refer to Chapter 7, Section 7.2.1 for a discussion of status values in the context of freight orders, which behave similarly to freight bookings, and to Chapter 8, Section 8.1.4 for a detailed description of execution-related status management.

Output management

The OUTPUT MANAGEMENT area allows you to print documents and send messages. See Chapter 2, Section 2.3.3 for more details on the underlying technology.

Execution

The EXECUTION area collects data about planned events, their expected date and time, and their actual date and time, which can be reported by SAP Event Management, as explained in Chapter 8, Section 8.2.

Published versus unpublished bookings

Usually, a special department is responsible for creating air freight bookings. The capacity managers in this department may already create bookings, but the bookings should become visible to planners and sales agents only after they have explicitly been published. To do so, the capacity manager can set the status to *Published*. The INTERMEDIATE PROCESSING parameter in the EXECUTION SETTINGS section allows you to define whether newly created bookings get the status unpublished or published. Unpublished bookings get a planning block status and therefore cannot be consumed or seen by other departments. Publishing removes this planning block status. A published booking can be unpublished.

Unrestricted versus restricted visibility

It is also possible to further restrict the visibility of the air freight bookings to certain organizational units within your company. The capacity manager can maintain several organizational units, together with their functions (sales, company, or forwarding house), in the RESTRICTION TO ORGANIZATIONAL UNITS area. With the SET TO PUBLISHED WITH RESTRICTIONS button, the capacity manager can publish the booking to the maintained organizational units. Authority checks for display and changing the air freight booking are also executed, according to the user's role and organization. Note that the organizational units can already be maintained in a master flight schedule and are then copied into the air freight bookings created for the master flight schedule.

Together, the publish concept, the restricted visibility concept, and the organization interaction model described in the Section 5.4.1 enable fine-grained access control for the air freight bookings in your company.

Air cargo security

Special security requirements arise for air cargo and are covered by air cargo security (ACS) checks and statuses. These can be activated in the forwarding order type Customizing with the ENABLE AIR CARGO SECURITY CHECK parameter, and in the booking type Customizing by the ACS CHECK and ACS AUTHORITY CHECK parameters in the ADDITIONAL SETTINGS section. The required air cargo security status is captured in forwarding orders and propagated to freight units. The available air cargo security status can be maintained in air freight bookings. It can also be defined in a master flight schedule and is copied into air freight bookings created from the master flight schedule.

The following standard ACS status values are available, ordered from highest to lowest security: *Secure for Passenger Aircraft* (SPX), *Secure for Cargo Aircraft* (SCO), and *Not Secure* (NSC). The air cargo security check determines whether the ACS status of a freight unit is compatible with the assigned air freight booking. Freight units with the status SPX can be assigned to air freight bookings with the status SPX or SCO, and freight units with the status SCO can be assigned to bookings with the status SCO. A freight unit with the status NSC gets the planning block status and hence cannot be assigned at all. In this case, the corresponding forwarding order has to be processed according to security guidelines until it becomes secure. Then the status can be changed and the document can be planned.

Automatic planning ensures that only compatible assignments are made. Depending on your user's authority, manual planning yields either a warning or an error message for an incompatible assignment.

You can maintain country-specific air cargo security status values in Customizing by following the menu path SAP TRANSPORTATION MANAGEMENT • TRANSPORTATION MANAGEMENT • BASIC FUNCTIONS • SECURITY • DEFINE AIR CARGO SECURITY STATUSES and assigning the country-specific values to the standard ACS status values (SPX, SCO, and NSC).

Known shipper

In some countries such as the United States, the air forwarder has to have known the shipper for a certain period of time, which can be defined in Customizing via the menu path SAP TRANSPORTATION MANAGEMENT • TRANSPORTATION MANAGEMENT • BASIC FUNCTIONS • SECURITY • DEFINE OFFSETS FOR CALCULATING KNOWN SHIPPER STATUS. If the forwarder knows the shipper for more than half a year, for example, the shipper's goods can be shipped via air freight. Otherwise, the forwarding order gets the status NSC, which means that the forwarder has to check the goods very carefully before transporting them by airplanes. The known shipper status can be maintained in the business partner or the corresponding location, and it is copied automatically into a newly created forwarding order if the COPY AIR CARGO SECURITY DATA parameter is active in the forwarding order type.

Co-load

In the co-load process, you transport goods on a flight and use the air waybill stock and contract of another forwarder, the consolidator. This scenario is relevant if you do not have a contract with a carrier for a certain destination or if you do not have enough freight for the destination. Co-loading is a purely manual process, in which you maintain the consolidator and the MAWB number received from the consolidator in your air freight booking. This process is enabled by the Co-load parameter in the air freight booking type.

This chapter has introduced the capacity management process and its building blocks to plan capacities from long-term, mid-term, and short-term perspectives. The next chapter explains transportation planning, which makes use of the planned capacities by assigning freight units to schedules and freight documents.

Planning transportation activities are a key component of any transportation management solution. A proper transportation plan can help to save money if it reasonably addresses the constraints present during its execution.

6 Transportation Planning

Transportation needs, transportation capacities

In previous chapters we introduced order-based transportation requirements (OTRs), delivery-based transportation requirements (DTRs), and forwarding orders as business documents that represent a transportation need in SAP Transportation Management (TM), while vehicle resources, schedules, and freight bookings were introduced as a means to represent transportation capacity. The key objective of transportation planning is to bring together transportation needs and transportation capacity in the most efficient manner.

The first step that can be attributed to planning upon the creation of a transportation need is the creation of freight units. Freight units represent transportation requirements for planning and are obtained from their predecessor business documents via freight unit building rules (FUBRs). Freight units represent transportable objects that are kept together from their source to their destination (e.g., pallets and containers).

In the planning process, freight units can be assigned to multiple freight orders in a transportation chain or consolidated into one freight order, as is done in a local delivery tour. Essentially, the planning step covers the assignment of freight units to freight orders; these freight orders are the result of planning.

Manual and automatic planning

This planning step can be performed manually or automatically. A purely manual planning step can assign freight units to a vehicle resource using drag and drop functionality in the transportation cockpit to create a freight order; it can make the same assignment using the optimizer called by a background job.

The main objective of planning is to support the user with reasonable guidance for manual planning as well as with powerful automation capabilities. The transition between both planning alternatives is smooth because automatically created plans can be adapted manually, and manual planning processes can make use of automation. For example, you can start the optimizer interactively in the transportation cockpit.

Selection profile, planning profile

The planning process is mainly configured by two profiles: the selection profile and the planning profile. The *selection profile* is responsible for the decision about *what* needs to be planned. It basically selects the freight unit (stages). The *planning profile* determines *how* to plan—that is, which transportation capacities are available for planning and which constraints (e.g., incompatibilities) need to be considered.

Transportation cockpit

The planning interface (transportation cockpit) is very flexible and configurable by the definition of various layouts so that it can process many structurally different planning scenarios (e.g., from planning a local road transport to planning overseas transportation chains).

This chapter is structured in the following way. In Section 6.1 we examine freight units and address the different properties of freight units defined by the freight unit type, their relationship to predecessor and successor business documents, and rules for their creation. Section 6.2 introduces transportation units, which can be used to represent the transportation plan for scenarios involving trucks and trailers, and locomotives and railcars. Section 6.3 deals with manual planning processes. This section encompasses planning process configuration including details on selection profiles, planning profiles, and transportation cockpit layouts—but also user-centric decision making based on transportation proposal functionality.

Section 6.4 deals with automated planning. In contrast to Section 6.3, the focus here is on processes and functionality that are not primarily interactive and user-driven. Background planning and detailed insight into the optimization capabilities are in the scope of this section.

6.1 Freight Units

Freight units

Freight units are an important element in the planning process because they provide the link between transportation requirements—for example, a forwarding order and the transportation document (i.e., the freight order). They can only be omitted in special circumstances if the transportation requirement exactly matches the to-be-created transportation document (called a *shortcut planning process*). This process is described in more detail in Section 6.1.3.

Let's start with the definition of a freight unit and then move into properties of the freight units defined in Customizing and the information stored in the freight unit. The last subsection of Section 6.1 deals with creating freight units.

6.1.1 Freight Unit Definition

The *freight unit* is a set of goods that is transported together through the entire transportation chain. The freight unit is the smallest unit that can be transported. This means that everything that is included in one freight unit stays together from its source to its ultimate destination—that is, it is always transported together.

Granularity

The granularity of freight units required for transportation planning depends a lot on your business scenario. For example, if you are an electronics manufacturer, a freight unit can be one of the following:

- One USB stick if you want to send this USB stick from your distribution center directly to the final consumer (e.g., as a parcel shipment)
- A container full of USB sticks if you want to replenish your distribution center in the United States from your factory in China

Note

The more freight units are formed, the more detailed and individually you can plan. However, this makes planning more complex and requires higher processing capacity and thus leads to longer runtimes. Therefore, we strongly recommend that you only define the granularity of freight units to the detail level required for your business scenario.

> Just imagine the system load and number of objects created if you had created freight units for each USB stick in a replenishment scenario.

Given the dependency of the "optimal" freight unit granularity on the business scenario, there is no general rule for how freight units should be created. However, the following uses cases can be distinguished:

- For general cargo, freight units may be created based on handling units (e.g., pallets) or a group thereof.
- For full container freight, a freight unit usually represents the container.
- For bulk product, a freight unit may represent a quantity that corresponds to the capacity available for the transportation of the product. For example, a forwarding order for 5,000 tons of fertilizer to be transported with railcars with a capacity of 50 tons each should yield 100 freight units of 50 tons each.

Relationship to business documents

Figure 6.1 shows how freight units relate to other objects in SAP TM. They are created based on transportation requirements (n:m relationship). One freight unit can have one or more transportation requirements as predecessor documents. This means that freight units can be used to consolidate transportation requirements from several forwarding orders considering the restriction that the freight unit will stay together on the complete transportation chain (i.e., predecessor documents must have the same origin, destination, dates, etc.). On the other hand, one transportation requirement can yield several freight orders. This is probably the more important case because there are many good reasons why one transportation requirement is split over several freight units:

- A sales order consists of several items that have different transportation characteristics (e.g., frozen pizza and fresh ravioli require different temperature conditions during transportation)
- A sales order with an item that has different schedule lines (e.g., transportation should be in weekly quantities and not together)
- A sales order with an item representing a large quantity of a bulk product (e.g., the full quantity needs to be split into quantities that fit with the capacity of transportation resources)

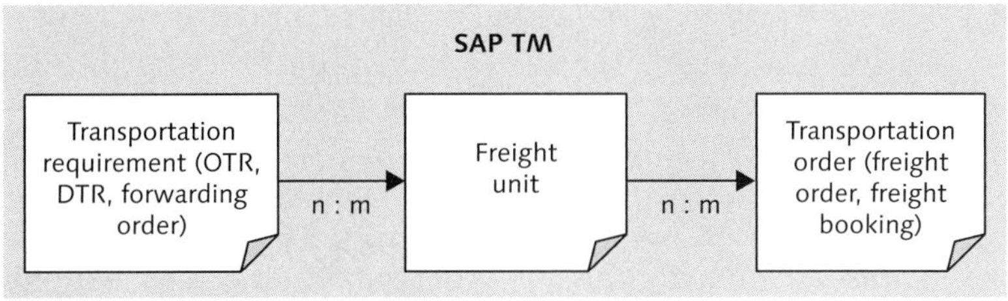

Figure 6.1 Relationship of Freight Units to Other Objects

In the planning process freight units are assigned to transportation orders (n:m relationship). Any type of assignment is allowed (1:1, 1:n, n:1, n:m) and depends on the business scenario:

- A 1:1 assignment can occur if a customer orders a full truck load directly from the plant to his warehouse.
- A 1:n assignment can occur for a container that is shipped across several stages. Then this container is assigned to one freight order (e.g., by truck) for the pre-carriage, to another freight order or freight booking (by sea) to represent the main carriage, and to a third freight order (by rail) to represent the on-carriage.
- An n:1 assignment typically occurs in distribution scenarios (e.g., if the freight order represents a truck transport that delivers the load to multiple customers [unloading locations]).
- An n:m assignment can occur as a result of any combination of these scenarios.

6.1.2 Properties of Freight Units

Whenever a freight unit is being created, it is created based on a certain freight unit type. The freight unit type defines the properties of a freight unit and can be defined in Customizing via the menu path Transportation Management • Planning • Freight Unit • Define Freight Unit Types.

Freight unit types

You should use different freight unit types based on your business requirements; that is, the electronics manufacturer may have different tracking and tracing requirements for freight units that represent final customer orders compared to freight units that represent stock replenishments for the distribution center. Thus, two different freight unit types can be used that are differently customized with respect to their execution tracking relevance and event management settings that govern integration with SAP Event Management.

In addition, you can influence the following properties of a freight unit in freight unit type Customizing:

- Number range settings
- Change controller settings
- Execution settings
- Direct shipment options
- Output options
- Organizational unit determination
- Additional settings (e.g., settings to influence the pickup and delivery window)

What? Where? When?

Because the freight unit represents the transportation demand in planning, it has to answer the following questions:

- *What* needs to be transported answers the question about the set of goods being transported. Relevant information is quantities and units of measure as well as characteristics (e.g., temperature conditions to be met during transport).
- *Where* answers the question about the source and destination location and potentially predefined transshipment locations defined as stages in the freight unit.
- *When* questions deal with the temporal aspect of transportation (e.g., when the freight unit should be picked up at the source and delivered at the destination location).

Pickup/delivery windows

To represent the dates and times of pickup and delivery in the freight unit, four time stamps are defined in the freight unit for both pickup and delivery. Figure 6.2 shows how these are obtained based on the requested dates and times defined in the forwarding order, forwarding quotation, OTR, or DTR.

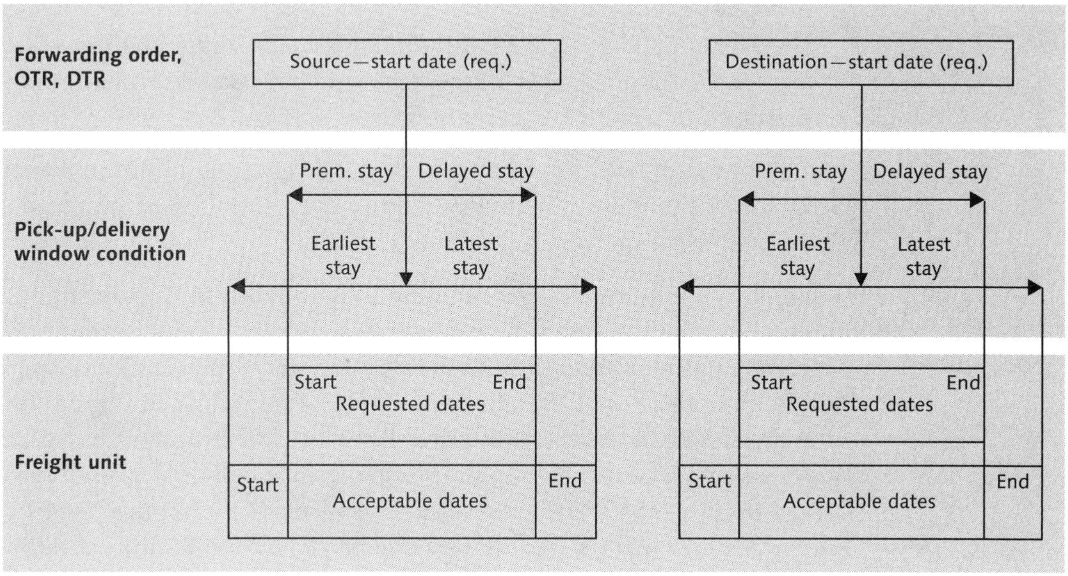

Figure 6.2 Schematic Description of Pickup and Delivery Windows

A condition with condition type /SCMTMS/TOR_TIMEWIND can be assigned to the freight unit type in freight unit type Customizing. Based on this condition, SAP TM calculates four time stamps for pickup and delivery, which will have the following interpretation later in automated planning:

- **Acceptable date—start**
  No pickup/delivery is allowed prior to this date (e.g., because the product has not been produced yet).
- **Requested date—start**
  This is the start of the desired pickup/delivery period. A pickup/delivery between the acceptable and requested start date is allowed, but can be penalized in planning as an earliness cost.
- **Requested date—end**
  This is the end of the desired pickup delivery period. A pickup/delivery within the requested start and end date does not incur any penalty costs in planning. This period can be used to represent the appointment time window agreed on between the supplier and/or customer at the source/destination location.

- **Acceptable date – end**
  No pickup/delivery is allowed after this date. Any pickup/delivery between the requested and acceptable end date is allowed but can be penalized in planning as a lateness cost.

Not all of these dates have to be defined. For example, if the customer accepts all deliveries no matter how early they are, then no acceptable start date for the delivery needs to be defined.

Using the concept of conditions in the freight unit type Customizing to define pickup and delivery time windows allows for a lot of flexibility in setting up business scenarios. If the requested or acceptable dates should not be the same for all freight units of the same freight unit type, the time window condition can be used to define different dates based on the destination location (customer) or goods included or any other relevant criteria of the freight unit. Thus, to increase the service level for important customers, these can be assigned a much tighter time window than less important customers.

Freight unit user interface

In the SAP TM user interface, freight units can be accessed from different origins: by using the display or edit freight units transactions by following the menu path PLANNING • FREIGHT UNITS, via the link in the document flow of other documents such as forwarding orders or freight orders, or via POWLs (e.g., PLANNING • WORKLISTS, query on freight units). Figure 6.3 shows the standard freight unit user interface.

The information contained in the freight unit is structured into different tabs. The following are the most relevant:

- GENERAL DATA
  The GENERAL DATA tab contains an overview of the freight unit:
  - SOURCE LOCATION (address and pickup window)
  - DESTINATION LOCATION (address and delivery window)
  - REQUIRED CAPACITY (quantities and unit of measure)
  - ORGANIZATIONAL DATA
  - FREIGHT UNIT TYPE
  - FREIGHT UNIT BUILDING RULE (see Section 6.1.3)

  Thus, the GENERAL DATA tab answers questions related to what, where, and when.

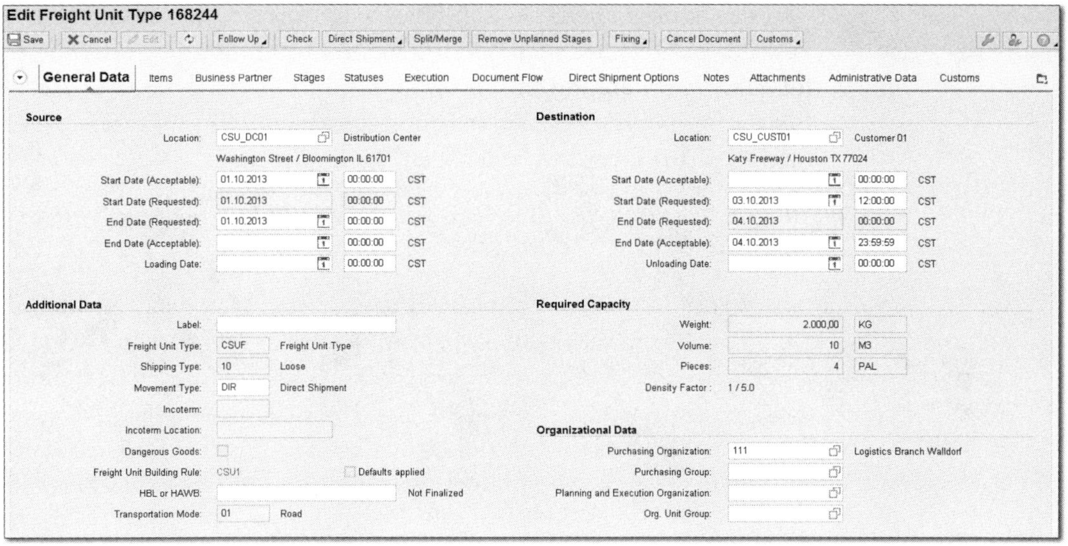

Figure 6.3 Freight Unit User Interface

- ITEMS

  The ITEMS tab displays the information related to the content of the freight unit. It shows you the hierarchy of items (container, package, and product) contained in the freight unit as well as the individual products with their quantities and units of measure.

- BUSINESS PARTNER

  On the BUSINESS PARTNER tab, you can find information about the relevant business partners (e.g., shipper and consignee).

Stages and transportation constraints

- STAGES

  The STAGES tab can contain one or more entries. In the simplest scenario, only one stage is present—the stage from the source of the freight unit to its destination. This implies no constraints for planning; that is, the freight unit can later be transported directly from its source to its destination or indirectly via transshipment locations—whatever is the most effective way for the business scenario.

  However, additional stages can be added to the freight unit either manually in the STAGES tab or by using the transportation proposal functionality (see Section 6.3.6 and Chapter 4, Section 4.2). If the transportation network allows sea transports from Europe to North

America to originate from Hamburg or Rotterdam, these options may be offered to a customer based on a transportation proposal in the forwarding order, and the customer's choice is represented as stages to or from Hamburg or Rotterdam in the freight unit. This stage information is considered to be a constraint in planning.

Note

The freight unit, which only exists once, appears in several virtual instances (stages) in planning. These instances can be planned independently from each other (e.g., pre-carriage, main carriage, and on-carriage). That is, different users can plan the individual stages at different times.

This is common business practice because the user responsible for U.S. domestic transport (e.g., on-carriage from port to customer) often is not familiar with domestic transport in Europe (pre-carriage from source to port) or ocean transport (main carriage from port to port). Furthermore, the individual stages are frequently not planned in the same sequence in which they occur, but rather the main carriage ocean is planned first based on the sailing calendar of the ocean vessel pre-carriage, and on-carriage is planned later.

Statuses

- STATUSES
  The freight unit has several statuses. Let's take a look at the most relevant:
  - The lifecycle status shows whether the freight unit is new, in process, completed, or canceled.
  - The fixing status determines whether the business document can be changed or not. Fixing only prevents the change of existing planning results.
  - Possible planning statuses are *Not Planned*, *Partially Planned*, and *Planned*, depending on whether none, some, or all stages of the freight unit have been planned.
  - The execution status of the freight unit is changed when any of the freight orders or bookings that the freight unit is assigned to are executed.
  - Freight units can be blocked separately for planning and execution. Use cases for planning blocks include when not all planning-relevant information is available or approvals are missing.

Use cases for execution blocks can be missing approvals or required prepayment.

- EXECUTION
  The EXECUTION tab provides the interface to SAP Event Management (see Chapter 8, Section 8.2). All events reported for a freight unit are displayed in this tab, including planned and actual dates and times. Expected events can be reported on the EXECUTION tab, and unexpected events can also be inserted there.
- DOCUMENT FLOW
  The document flow shows all related (predecessor and successor) documents for the freight unit.
- DIRECT SHIPMENT OPTIONS
  The direct shipment options can be generated automatically when the freight unit is created via a process controller strategy (default: DSO_DEF) based on the freight unit type Customizing settings or manually triggered from the freight unit user interface.

  Direct shipment options are generated for each carrier-service level combination based on a freight agreement. Thus, they represent "real" costs. They can be used if freight units are not consolidated during planning, but rather are assigned directly to a carrier. In planning using the optimizer, the most cost-effective price (direct shipment option) can be used as a reference cost to decide whether a consolidated solution for multiple freight units is more cost-efficient compared to the sum of direct shipment options for the individual freight units. If the direct shipment option is chosen, the freight unit needs to be converted to a freight order either manually or via a background report.
- NOTES
  Notes can be used to add texts to the freight unit.
- ATTACHMENTS
  Attachments can be any electronic documents (e.g., PDF files) or URLs. Files and URLs can be organized in a folder structure.

Direct shipment options

Organizational interaction status

Finally, an organizational interaction status can be maintained in the stages of the freight unit. The organizational interaction status relates to an internal business process between different organizational units (sales organization and planning and execution organization) of a logistics service provider (LSP). The sales organization may be allowed to create stages in a forwarding order, but the planning and execution organization is responsible for assigning the stages to schedules, freight bookings, or freight orders. The organizational interaction status that is maintained independently for each stage determines which organization is tasked with the next activity. Figure 6.4 shows how the different organizational units interact and how this is represented in the organizational interaction status of the freight unit stage.

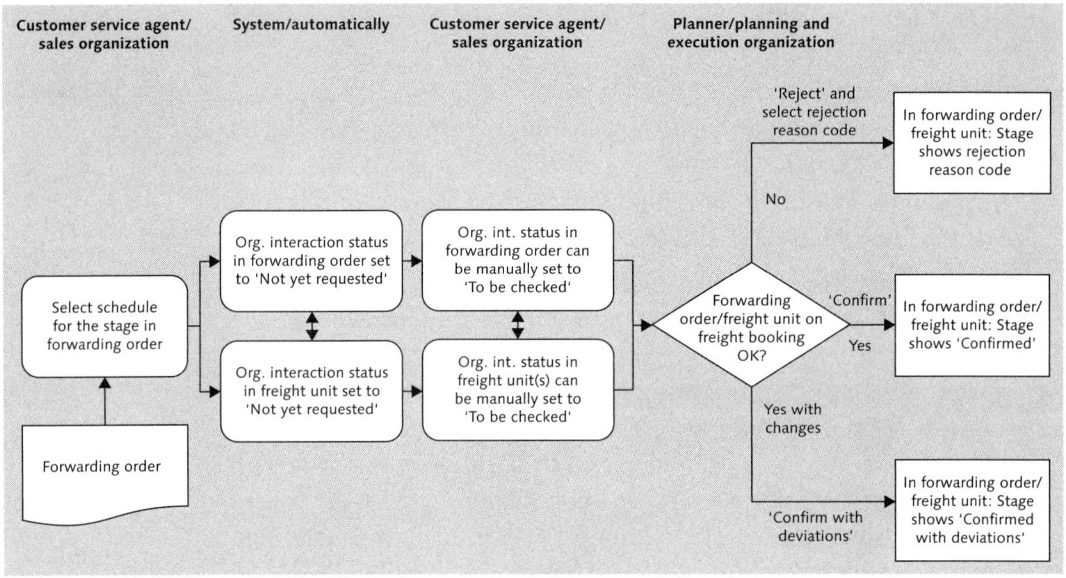

Figure 6.4 Organizational Interaction Process

6.1.3 Creating Freight Units

Automatic freight unit building

Figure 6.5 displays the triggers that start the creation of a freight unit from its predecessor documents (OTR, DTR, or forwarding order). Freight unit building is either triggered automatically or done manually. If freight units should not consolidate items from different predecessor documents, which is the most common case, then automatic freight unit building can be activated for the relevant document order type in Customizing:

- For forwarding orders: Transportation Management • Forwarding Order Management • Forwarding Order • Define Forwarding Order Types
- For OTRs: Transportation Management • Integration • ERP Logistics Integration • Order-based Transportation Requirement • Define Order-based Transportation Requirement Types
- For DTRs: Transportation Management • Integration • ERP Logistics Integration • Delivery-based Transportation Requirement • Define Delivery-based Transportation Requirement Types

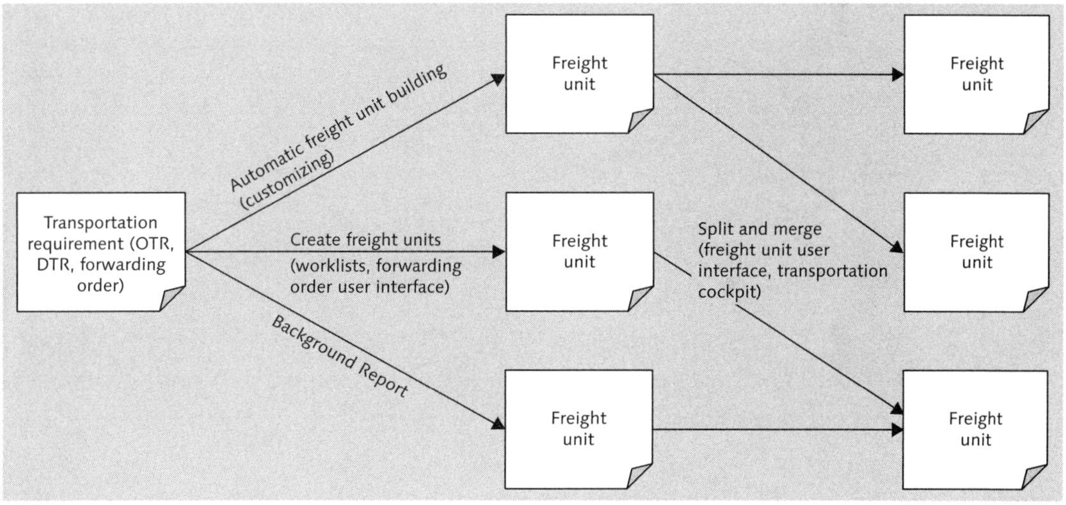

Figure 6.5 Trigger for Freight Unit Creation

In addition, freight unit creation can be triggered by a background report for forwarding order management preparation (/SCMTMS/TRQ_PREP_PLN). There are also manual options to trigger the creation of freight units either directly from the forwarding order user interface or via worklists for any of the possible predecessor documents (OTR, DTR, and forwarding order).

Freight unit changes

Once freight units have been created, you might have to change them. Changes from preceding business documents are propagated automatically to the freight unit, and change controller strategies assigned in the Customizing of the freight unit type (Transportation Management •

PLANNING • FREIGHT UNIT • DEFINE FREIGHT UNIT TYPES) govern the behavior of the reaction to a change. However, manual changes may be also needed (e.g., if a freight unit needs to be split into two parts because its full quantity cannot be assigned to a transportation resource due to a capacity limitation). For this purpose a split and merge transaction allows you to execute the required changes directly in the freight unit user interface or from the transportation cockpit.

Freight unit building rule

The creation of a freight unit is always based on an FUBR (shown in Figure 6.6). FUBRs are defined in APPLICATION ADMINISTRATION • PLANNING • GENERAL SETTINGS • FREIGHT UNIT BUILDING RULE.

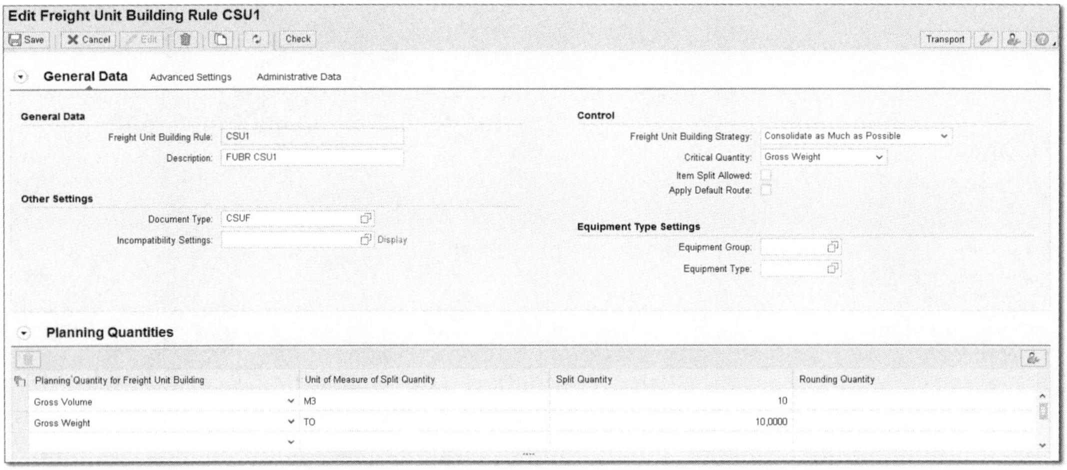

Figure 6.6 Freight Unit Building Rules

In the FUBR you can control the strategy that is used during freight unit creation. The CONSOLIDATE AS MUCH AS POSSIBLE option allows you to consolidate items from one or more different predecessor business documents into one freight unit. CONSOLIDATE PER REQUEST (COMPATIBLE PARTS) allows you to consolidate several items of the same predecessor business document, and CONSOLIDATE PER ITEM creates one freight unit per item of the predecessor business document.

Additionally, you can maintain whether the FUBR is allowed to split items into several freight units. This is generally allowed by all three strategies but may not be reasonable in certain planning scenarios.

The resultant business document type of the freight unit is another important setting in the FUBR. Although the name FUBR may indicate that the resultant business document is always a freight unit, this is not the case; in fact, the result of an FUBR can be either a freight unit, a trailer document, or a freight order. Which object is created is specified either in the DOCUMENT TYPE field or on the ADVANCED SETTINGS tab in the DOCUMENT TYPE DETERMINATION CONDITION field with a condition of type /SCMTMS/TOR_TYPE.

Creating a freight order directly from the FUBR is referenced as the *shortcut planning process* because in this case freight units as separate business documents are omitted—and additional planning steps are not required—because the freight order is created right away. A typical planning scenario that uses this feature is the zero-click scenario, which can be configured in the following way:

Zero-click scenario

1. The sales order is transferred to SAP TM, and the OTR determination condition (type /SCMTMS/OTR_TYPE) results in an OTR type with AUTOMATIC FREIGHT UNIT BUILDING activated.
2. The determined FUBR has a resultant business document type defined that creates a freight order.
3. A background job triggers carrier selection and tendering for the freight order.

In this process no user activity is required after saving the sales order in SAP ERP. A freight order is automatically created, tendered, and awarded to a carrier.

The FUBR also needs to consider incompatibilities, which we discuss in Section 6.4.4 in more detail. Assume that certain products are not allowed to be transported together; for example, ice cream and ketchup can't be shipped together because they have different temperature requirements. Thus, these two items of the OTR (and ultimately the sales order) need to be kept apart in freight unit building, although other items of the same sales order (e.g., chocolate ice cream and strawberry ice cream) can be consolidated into one freight unit. Incompatibilities can be used to express such a planning constraint, so incompatibility settings can be assigned in the FUBR.

Equipment

You can also use freight unit creation to consolidate items of a business document into a container and display the freight unit as a container. To be able to do this, you need to define EQUIPMENT GROUP and EQUIPMENT TYPE in the FUBR. In this scenario, SAP TM takes the physical properties defined in Customizing for the equipment into account (TRANSPORTATION MANAGEMENT • MASTER DATA • RESOURCE • DEFINE EQUIPMENT GROUPS AND EQUIPMENT TYPES; see also Chapter 3, Section 3.3).

Planning quantities

Last, you have to define planning quantities. Planning quantities are an integral part of the FUBR, because only planning quantities are copied into the freight unit from predecessor business documents. For each planning quantity, a split quantity and a rounding quantity can be defined. The split quantity defines the maximum value a freight unit can take in any of the planning quantities. If the gross weight in the OTR is 9 tons and the split value for gross weight is defined as 4 tons, three freight units with gross weight 4 tons, 4 tons, and 1 ton would be created. If the rounding quantity is defined as one piece, and the 9 tons from the previous example corresponded to six pieces of 1.5 tons each, then the result would be three freight units of 3 tons (two pieces) each.

Critical quantity

Critical Quantity

Finding the best possible assignment of items to freight units is a knapsack problem. Because of the combinatorial nature of this task, it is too computationally expensive to solve this kind of optimization problem during freight unit building.

Therefore, a heuristic is applied that can be influenced by maintaining the *critical quantity* in the FUBR. In this heuristic, all items are sorted in descending order based on their critical quantity and are assigned to freight units in this sequence. This heuristic provides the optimal result unless the items are very heterogeneous; that is, some items are very small but are heavy compared to large but light items.

Figure 6.7 provides an example in which the optimal and heuristic solution deviate from each other, assuming the critical quantity is gross weight and the split quantity for gross weight is 10 tons and for gross volume is 10 cubic meters.

OTR items			Heuristic assignment				Optimal assignment		
Item	Gross weight [to]	Gross volume [m3]	FU 1	FU 2	FU 3	FU 4	FU 1	FU 2	FU 3
10	8	2	X				X		
20	6	4		X				X	
30	4	6		X				X	
40	2	1	X						X
50	2	1			X				X
60	2	1			X				X
70	2	1			X				X
80	1	8				X	X		
Total	**27**	**24**	**10 / 3**	**10 / 10**	**6 / 3**	**1 / 8**	**9 / 10**	**10 / 10**	**8 / 4**

Figure 6.7 Assignment of Items to Freight Units

Freight unit building rule determination

Considering that FUBRs are so important, how does SAP TM determine which one to use? The first attempt is to read a condition of type /SCMTMS/FUBR from the predecessor business document type Customizing:

- For forwarding orders: Transportation Management • Forwarding Order Management • Forwarding Order • Define Forwarding Order Types
- For OTRs: Transportation Management • Integration • ERP Logistics Integration • Order-based Transportation Requirement • Define Order-based Transportation Requirement Types
- For DTRs: Transportation Management • Integration • ERP Logistics Integration • Delivery-based Transportation Requirement • Define Delivery-based Transportation Requirement Types

The next step is to determine the FUBR based on this condition. If there is no condition defined or the determination fails, the FUBR is determined directly from the predecessor business document type Customizing, and if nothing is maintained there, default settings are applied.

You can see which FUBR has been applied for any freight unit in the freight unit user interface on the GENERAL SETTINGS tab.

6.2 Transportation Units

Freight units represent the original transportation demands and abstracts from forwarding orders, OTRs, and DTRs. The transportation plans for vehicle resources and locomotive resources are represented by road freight orders and rail freight orders, respectively. The freight orders' stages define the path through the network.

Scenarios involving trailers and railcars can be modeled by transportation units, which are called trailer units and railcar units for the road and rail modes of transport, respectively. Like freight orders, transportation units have stages defining their path through the network.

We first present the general concept of transportation units, contrasting it with freight units and freight documents in Section 6.2.1. Then, Section 6.2.2 describes trailer units and relevant scenarios involving trucks and trailers. Section 6.2.3 discusses railcar units.

6.2.1 Transportation Units, Freight Units, and Freight Orders

Demand and capacity

While the freight unit represents a single transportation demand, the transportation unit can represent a set of transportation demands that may even have different source and destination locations. Suppose a trailer is moved from location A to B to C, delivering three freight units: the first from A to B, the second from A to C, and the third from B to C. The trailer itself cannot move, but it demands transportation by a vehicle. From this perspective, the transportation unit also represents a transportation demand. In contrast to freight units and transportation units, freight orders do not represent a demand, but they can satisfy the demand.

The road freight order represents a certain capacity, which can be used to load freight units. Similarly, freight units can be loaded into the trailer unit until its capacity is reached. Since trailers are a means to transport goods, they also serve as transportation capacity. In contrast to trailer units and freight documents, freight units do not *provide* capacity; they only *consume* capacity.

From these perspectives, the transportation unit unifies properties from freight units and freight documents. It represents transportation demand and capacity at the same time.

No moving capability

However, freight units and transportation units cannot move themselves. You always need a truck or a locomotive to move freight units and transportation units, which themselves also contain freight units. Freight orders represent the actual movements of active vehicle resources such as trucks and locomotives. Without freight orders, neither freight units nor transportation units can be transported. However, freight orders *can* represent transportation without transportation units and freight units; these represent empty moves, which may make sense, although you usually will attempt to avoid them.

Two assignment decision levels

Transporting a freight unit in a trailer unit requires coupling the trailer unit to the freight order. For truck and trailer scenarios, three object layers are relevant for two assignment decisions to be made: which freight units will be transported in which trailers, and which trailers will be moved by which vehicles?

Different quantities per stage

While the freight unit has the same volume and weight along all its stages, both the transportation unit and the freight order may have different (loaded) volumes and weights per stage.

Subcontracting only for freight orders

Freight orders can be subcontracted, but transportation units and freight units cannot be subcontracted. To be transported and subcontracted, they have to be assigned to freight orders.

Transportation unit is category

Roughly speaking, the transportation unit shares some similarities with freight units and others with freight orders, but it also differs from freight units and freight orders. Based on these criteria, the transportation unit is an object categorized between freight units and freight orders.

Transportation unit types

Each transportation unit has a specific type, which can be maintained in Customizing by following the menu path SAP TRANSPORTATION MANAGEMENT • TRANSPORTATION MANAGEMENT • PLANNING • TRANSPORTATION UNIT • DEFINE TRANSPORTATION UNIT TYPES. The similarity of transportation units to freight units and freight orders is directly reflected in the transportation unit type, so we won't describe it in detail; instead, refer to Section 6.1.2 for freight unit types and Chapter 7, Section 7.2.1 for freight order types. It is important to mention that you can assign item types to the transportation unit type, which is useful for multi-items for railcar units, as mentioned in Section 6.2.3.

6.2.2 Trailer Units

Trucks and trailers

Truck and trailer scenarios involve trailer resources being coupled and uncoupled to truck resources. Figure 6.8 shows different trucks and trailers. Box trucks have their own loading capacity, while tractors don't have their own capacity. Full trailers can be coupled to box trucks, and semi-trailers can be coupled to tractors. From a planning perspective, full trailers and semi-trailers are handled identically, so from now on, we only refer to *trailers.* We also do not differentiate between box trucks and tractors and simply use the term *trucks* for both. Only where required do we explicitly refer to tractors.

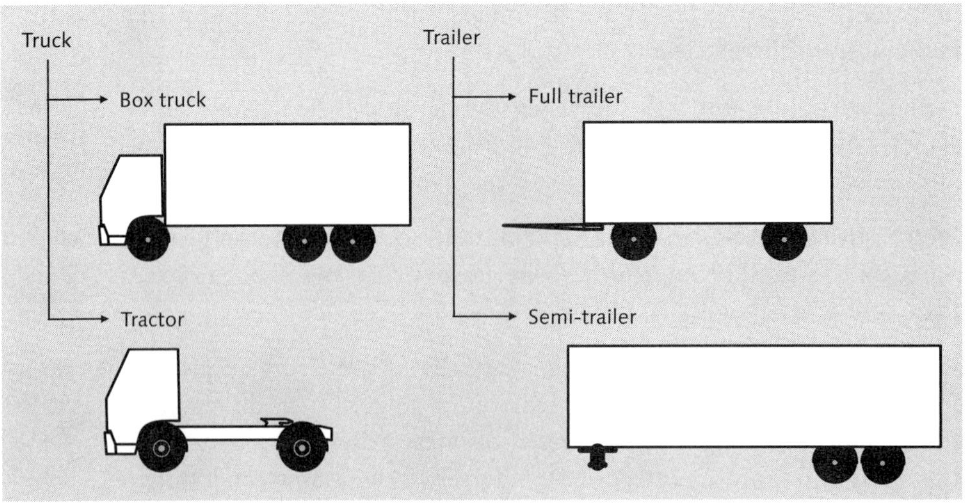

Figure 6.8 Truck and Trailer Examples

Trailer unit

Goods can be transported with trailers by loading the trailer, using a truck to move the trailer to its destination, and then unloading the trailer. Planning has to make the routing decisions for a trailer, and these decisions are represented in the trailer unit's stages. Since trailers cannot move on their own, each stage of a trailer unit must be assigned to one road freight order representing a vehicle that moves the trailer along the stage. A direct transport from source to destination would result in a trailer unit having one stage only.

Increasing planning complexity

There are multiple cases of increasing planning complexity:

- **Fixed assignment of trailer to truck**
  If a trailer resource is always attached to the same truck resource, the scenario could be modeled by a vehicle resource with two compartments—one for the truck itself and one for the trailer. This modeling does not require using trailer units to represent the trailer's plan because the road freight order for the truck always specifies the trailer's tour. The planning decision concerns whether the freight units are loaded into the truck or the trailer.
- **Flexible assignment of trailer to truck**
  All stages of a trailer unit are assigned to one truck, and different tours of a trailer resource are assigned to different trucks. Planning has to decide which freight units are transported in the trailer and which vehicle resource moves the trailer unit.
- **Dynamic recoupling**
  The trailer unit has at least two stages, and the stages are assigned to different vehicles. For example, the trailer unit is moved along locations A, B, and C; the movement from A to B is done by vehicle V1, and the movement from B to C is done by vehicle V2. Thus, the trailer is coupled to V1 at location A, moved to B, uncoupled from V1, coupled to V2, moved to C, and then uncoupled from V2.
- **Multiple trailers assigned to one vehicle**
  In some countries such as Australia, vehicles resources can carry two or three trailers. Such a vehicle combination is called a road train. Thus, each stage of a road freight order may be assigned to multiple trailer units.

- **Dynamic recoupling and multiple trailers assigned to one vehicle**
  This combination of the previous two cases has the highest planning complexity. Multiple trailer units can be moved by one road freight order, and one trailer unit can be moved by multiple road freight orders, one per trailer unit stage.

Two views on the same plan

The transportation plan is represented by trailer units and road freight orders, which resemble the perspectives of the trailer and of the truck. Due to the stage-wise assignments of trailer units to road freight orders, you cannot change the stages of the trailer unit without changing the stages of the corresponding road freight order, and vice versa. The planner has to think from the trailer perspective and from the truck perspective, which requires quickly changing the perspective or, ideally, having both perspectives at the same time—a function that is offered by the transportation cockpit, as described in Section 6.3.3.

Trailer swaps

Dynamic recoupling is frequently used in scenarios where the distance from source to destination is so large that the driver of the truck cannot make the trip in one shift. Trailer swaps are introduced to maximize utilization of the trailer fleet. Suppose one trailer has to be moved from Hamburg to Munich, and another trailer has to be moved from Munich to Hamburg. The first truck may take the first trailer from Hamburg to Kassel, which is located roughly half way between Hamburg and Munich. The second truck may take the second trailer from Munich to Kassel. In Kassel, both trailers are uncoupled and then coupled to the other truck. Then the first truck takes the second trailer from Kassel to Hamburg, and the second truck moves the first trailer from Kassel to Munich.

This plan allows each of the drivers to be back at home at the end of their shifts. The first trailer unit has stages from Hamburg to Kassel to Munich, and the second trailer unit has stages from Munich to Kassel to Hamburg. The first road freight order has stages from Hamburg to Kassel to Hamburg, and the second road freight order has stages from Munich to Kassel to Munich. Figure 6.9 shows the STAGES tab for the trailer unit from Hamburg to Munich, which includes the links to the corresponding road freight orders that are shown on the transportation cockpit's map in Figure 6.10.

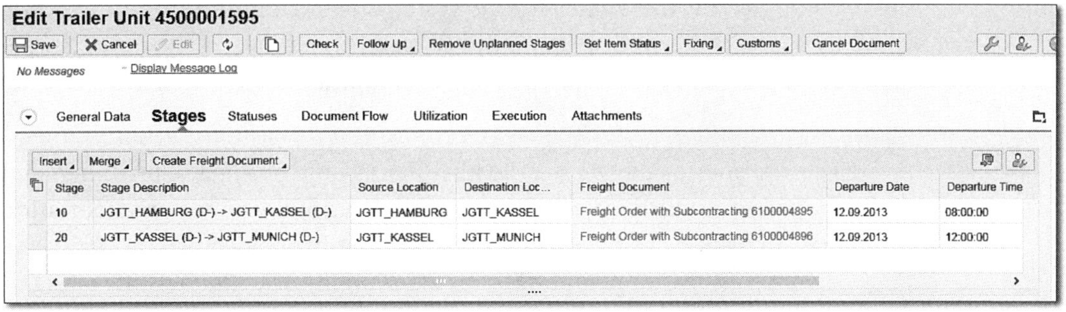

Figure 6.9 Trailer Unit Stages

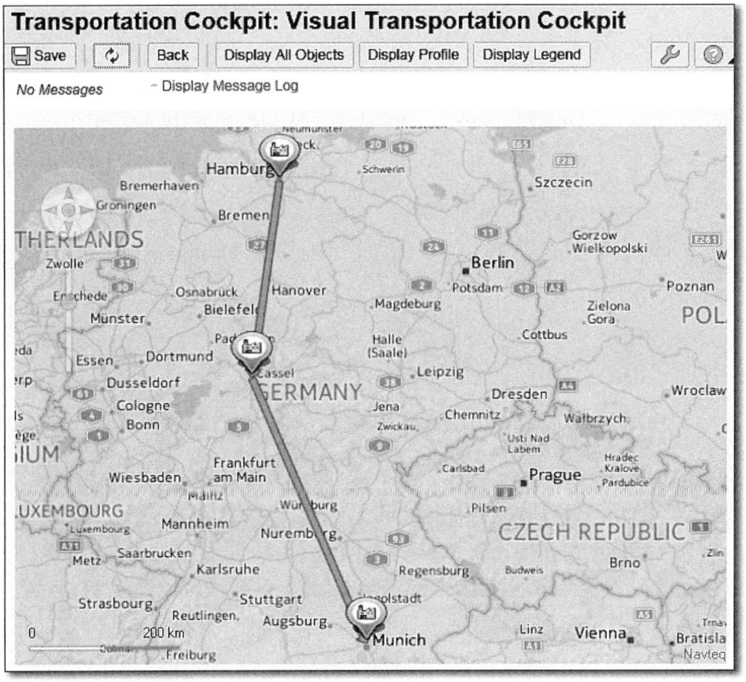

Figure 6.10 Mapped Trailer Freight Orders

In Australia, road trains may consist of tractors carrying two trailers in urban areas and three trailers elsewhere. Suppose that the trailers are systematically moved from A to B to C and then back from C to B to A. Two trailers can be carried between A and B, and three trailers between B and C. Here, the planning decision involves defining the road trains—in other words, which trailers are moved together by which tractor.

Using two default routes for the systematic movements from A to C and back, the trailer units' stages can be determined in a rule-based fashion, and then planning only has to decide about the tractor assignment to the trailer units' stages.

These scenarios can be handled by manual and automatic planning, and the transportation cockpit provides full visibility of the transportation plan by representing it from the trailer unit perspective and from the road freight order perspective, as mentioned in Section 6.3.3.

Creating trailer units

There are many ways to create trailer units:

- Manual planning in the transportation cockpit makes it possible to do the following:
  - Assign a freight unit to a trailer resource, which creates a trailer unit with the trailer resource and the freight unit assigned to it
  - Create a new trailer unit from scratch in the trailer unit hierarchy
  - Create a new trailer unit for a trailer resource in the trailer resource list
- Automatic planning
- Freight unit building, which creates trailer units if you define a trailer unit type as a document type
- Explicit creation via the menu path PLANNING • TRAILER UNIT • CREATE TRAILER UNIT

POWL query

You can follow the menu path PLANNING • WORKLIST • OVERVIEW PLANNING and choose the query for trailer units to get an overview of the trailer units in the system. From here, you can edit, create, and cancel trailer units.

6.2.3 Railcar Units

Rail transportation is done by trains, which may consist of one locomotive resource and a set of railcars. There are scenarios in which the whole train is subcontracted and freight units are directly assigned to rail freight orders. In other scenarios, planning is required on the individual railcar level, which is supported by railcar units that can play a role similar to trailer units in truck and trailer scenarios.

Manual planning supports the same functionality as for truck and trailer scenarios; that is, you can assign a locomotive to rail freight orders, create railcar units with assigned railcar resources and freight units, and assign a railcar stage to one rail freight order.

Manual planning

While trailer units and road freight orders have the same stages during their joint movement, railcar units and rail freight orders have only the stops in common, where the railcar is loaded, unloaded, coupled, or uncoupled. The main motivation for this modeling is the reduction of data volume because trains may have many more stops than a truck, and many more railcars are usually coupled to a rail freight order.

Stages of railcar units and rail freight orders

Some companies ship a lot of cargo by railcars. If 25 complete railcars full of a certain product are to be shipped, this would create an enormous number of railcar units. To reduce the number of documents, you can use the concept of multi-items to have one railcar unit that represents the load of 25 railcars. Via the menu path SAP Transportation Management • Transportation Management • Forwarding Order Management • Define Item Types for Forwarding Order Management in Customizing, you can define an item category as a passive vehicle resource and enable multi-items. The multi-item can represent multiple subitems without explicitly generating the subitems. Alternatively, the multiple subitems can be generated automatically by expansion. For the subitems, you have to specify an item type that can be a subitem of a multi-item.

Multi-items

6.3 Manual Planning

Manual planning concerns manually creating or changing the transportation plan. The transportation cockpit is the main user interface to display the transportation plan and to plan manually. You can configure how to use the transportation cockpit from several perspectives.

Section 6.3.1 introduces planning strategies that define the different steps performed in manual and automatic planning. You can use standard planning strategies and incorporate your own enhanced strategies, too.

Planning and selection profiles are described in Section 6.3.2, as well as additional settings, which together define the planning scenario and configure planning parameters.

Section 6.3.3 describes the need for configuring the transportation cockpit and explains how page layouts are used for configuration.

Hierarchical views and dual views represent special user interface elements that provide visibility of the substructures of relevant objects and enable efficient comparisons and replanning of freight documents and transportation units, respectively. These are described in Section 6.3.4.

Section 6.3.5 explains how to use the transportation cockpit by describing its entry options and all its functionalities (e.g., for manual planning, automatic planning, navigation through object lists and hierarchies, changing the optical appearance on the fly, etc.).

Section 6.3.6 presents the transportation proposal, which is a semi-automatic planning approach that combines automatic planning with a user interface that enables a manual decision about the determined alternative transportation proposals. We describe how to configure and use the user interface, as well as special parameters for automatic determination of transportation proposals.

Section 6.3.7 describes scheduling, which can be triggered out of the transportation cockpit and systematically determines start and end times for selected freight documents.

6.3.1 Planning Strategies

Standard planning strategies

A planning strategy defines the system behavior for different planning steps in manual planning and automatic planning. The system offers the following standard planning strategies:

- `VSRI_DEF` for manual planning
- `VSS_DEF` for scheduling
- `VSRI_SCH` for manual planning with subsequent scheduling
- `VSR_CHECK` for checking the plan
- `TSPS_DEF` for carrier selection optimization

- VSRI_1STEP for manual planning with subsequent carrier selection
- VSR_DEF for vehicle scheduling and routing optimization
- VSR_1STEP for one-step optimization (which calls vehicle scheduling and routing optimization and then carrier selection optimization) and for transportation proposals (carrier selection is called for each determined transportation proposal)
- ALP_DEF for load optimization

Defining your planning strategies

Each strategy consists of a sequence of methods. If you want to refine the standard behavior, you can build your own strategies, which can reuse the standard methods. Once defined, you can use your own strategies in the profiles to replace the default standard strategies. Refer to Chapter 2, Section 2.3.7 for more details on defining strategies and methods.

6.3.2 Profiles and Settings

Besides the configurability of the user interface, described in Section 6.3.3, many additional parameters define the planning scenario and how planning can be performed. The selection of objects relevant for planning is handled by selection profiles, and planning parameters are covered by planning profiles.

Selection Profiles and Selection Attributes

Selection profile

The selection profile defines which documents are considered for planning, and can be created via the menu path Application Administration • Planning • Selection Profiles • Create Selection Profile. The selection profile consists of three building blocks: time-related selection attributes, geographical selection attributes, and additional selection attributes, as shown in Figure 6.11. You can create the three attribute objects independently of each other and of the selection profile by following the menu path Application Administration • Planning • Selection Profile Attributes. The building block principle allows reuse of the same attribute definitions in different selection profiles. Selection according to the selection profile returns objects that meet the criteria of the time-related selection attributes, geographical selection attributes, and additional selection attributes. The selection profile allows you to define a limit for the number of selected objects.

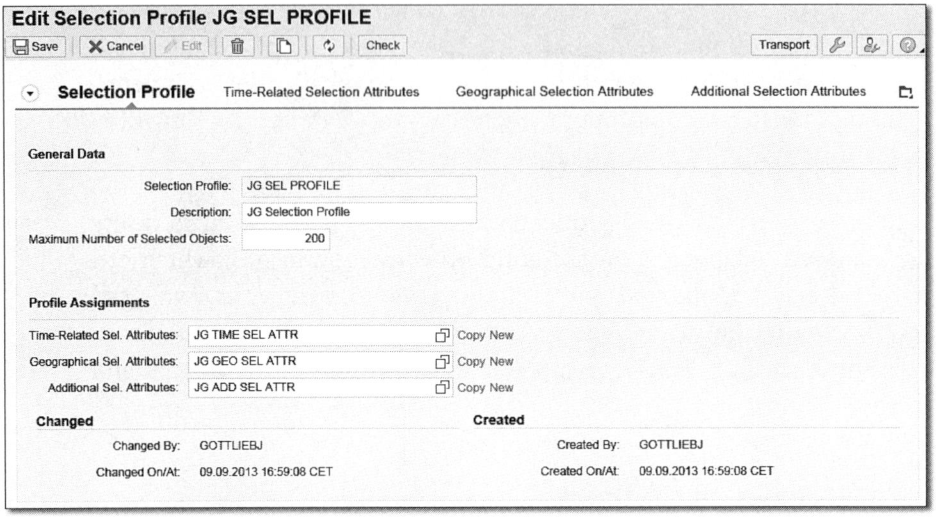

Figure 6.11 Selection Profile

Time-related selection attributes

The time-related selection attributes use two horizons—one for the pickup time windows and one for the delivery time windows, as shown in Figure 6.12. You can define the horizons either relatively or absolutely. Using the absolute option, you can explicitly maintain the pickup horizon and the delivery horizon. Using the relative option shown in Figure 6.12, you can define a range by specifying the number of days and an additional duration in hours and minutes. Additionally, you can define an offset in days, hours, and minutes and an offset direction that points into the past or into the future. The system takes the current time and adjusts it by the offset to determine the start of the horizon. Adding the range to this yields the end of the horizon. You can specify a factory calendar, which enables the system to consider nonworking days when determining the start of the relative horizons. If the determined start day is a nonworking day, the horizon's start is moved to the next working day, and the range remains the same. The relative horizon is useful if you always plan objects in advance (e.g., for the next 30 days).

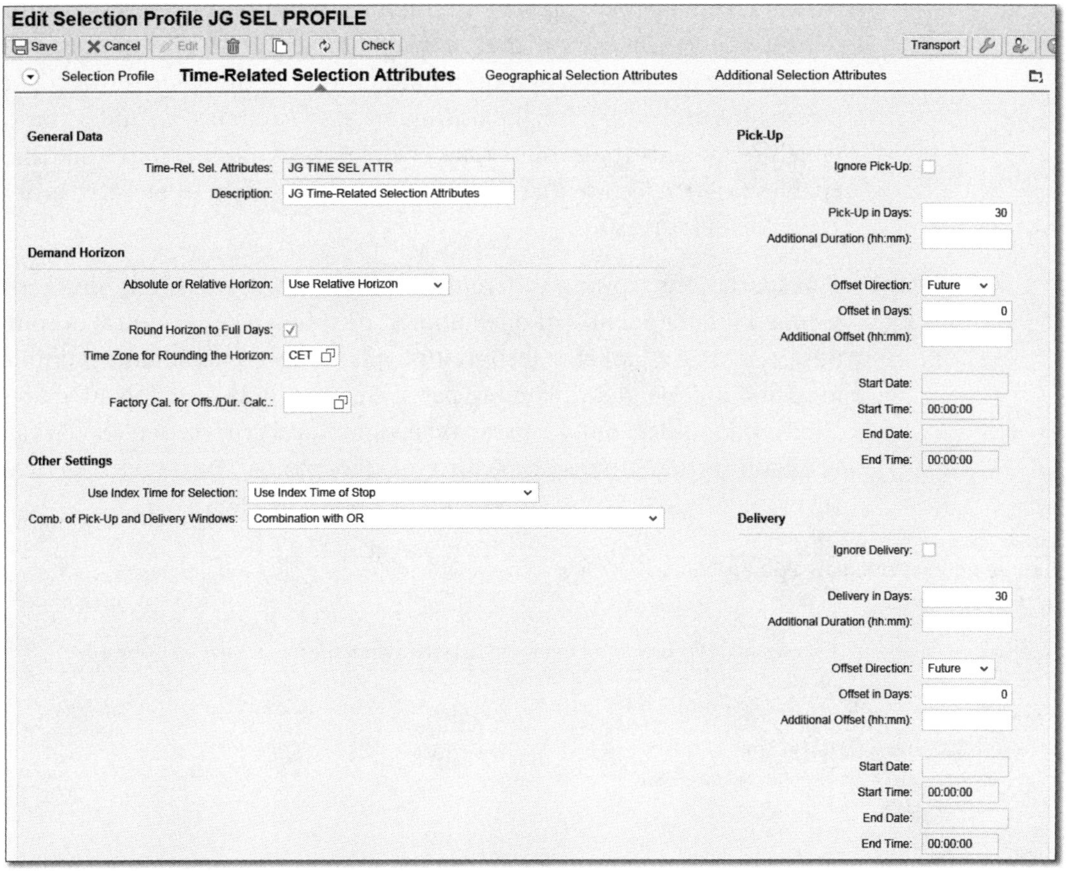

Figure 6.12 Selection Time Attributes

You can round the horizon to full days and define the time zone to be used for this rounding. Using the flags for IGNORE PICK-UP and IGNORE DELIVERY, you can focus selection on the dates of only pickup or delivery. If both are marked as relevant, you can define whether both pickup and delivery or only one of them must be within its horizon.

Alternatively, you can enforce a combination of pickup time window with a source location and delivery time window with a destination location. This option selects objects with a source location matching the geographical source selection attribute and the pickup time window falling into the pickup horizon and objects with a destination location contained in the geographical destination selection attributes and the delivery time window in the delivery horizon. This allows scenarios considering all inbound and outbound objects in the same horizon at a certain location. The freight document stop contains an index time, which represents the earliest time of the other times on stop level, and enables faster selection that can be activated by the parameter USE INDEX TIME FOR SELECTION.

Geographical selection attributes

Transportations within different countries or regions can often be planned independently of one another, enabling a geographical decomposition of the global transportation planning scenario into multiple local and independent planning parts. Geographical selection attributes allow you to select objects based on geographical criteria, which you can maintain in the SOURCE LOCATIONS, SOURCE ZONES, DESTINATION LOCATIONS, and DESTINATION ZONES tabs, as shown in Figure 6.13.

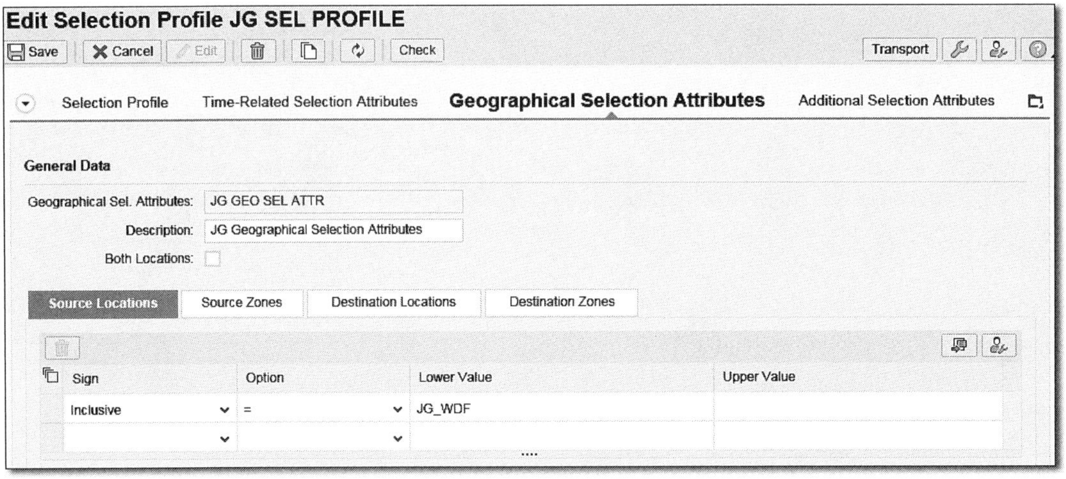

Figure 6.13 Selection Geographical Attributes

You can enter several criteria for each tab. All source locations and locations in source zones marked as INCLUSIVE are relevant, but the corresponding locations marked as EXCLUSIVE are not. The destination locations are determined the same way. The BOTH LOCATIONS checkbox defines whether the selected objects must meet the criteria for their source and destination or whether it is sufficient if only their source or destination matches its criterion.

Additional selection attributes

You can use the additional selection attributes to define more selection criteria explicitly or via condition-based filtering. The selection criteria are defined the same way as the location selection in the geographical selection attributes, but here you can refer to any other fields available in the queries for freight units, freight orders, freight bookings, transportation units, forwarding orders, DTRs, and OTRs. Multiple criteria for the same object and field are combined with logical `OR`, and the criteria for different fields are combined with logical `AND`. Additionally, you can specify whether planned objects will be included or excluded, and whether objects blocked for planning will be included or excluded.

Performance of Condition-Based Filtering

Condition-based filtering is run after all other selection criteria. All other criteria are evaluated first, resulting in an intermediate selection. Then the hits in the intermediate selection are processed according to the specified condition, which may be much more time-consuming than explicit selection criteria that operate directly on efficient database queries.

Therefore, use as many explicit criteria as possible and avoid condition-based filtering. Only use condition-based filtering after intensive performance testing and if the performance is satisfactory. See also SAP Note 1765952 for hints on improving performance for planning and, in particular, for the use of condition-based filtering.

Planning Profiles and Settings

Planning profile

The planning profile defines numerous parameters that control how planning is to be performed. It consists of six building blocks: capacity selection settings, optimizer settings, planning cost settings, load planning settings, incompatibility settings, and carrier selection settings. Similar to the selection profile's building blocks, all these settings can be maintained independently of each other and of the planning profile, and can therefore be reused in different planning profiles. You can create a planning profile and additional settings by following the menu paths Application Administration • Planning • Planning Profiles • Create Planning Profile and Application Administration • Planning • Planning Profile Settings.

Planning horizon

The planning profile itself contains many parameters grouped into different sections, as shown in Figure 6.14. The PLANNING HORIZON area is identical to the relative horizon definitions already explained in the context of time-related selection attributes. However, the planning horizon defines the horizon in which new freight documents can be created by planning.

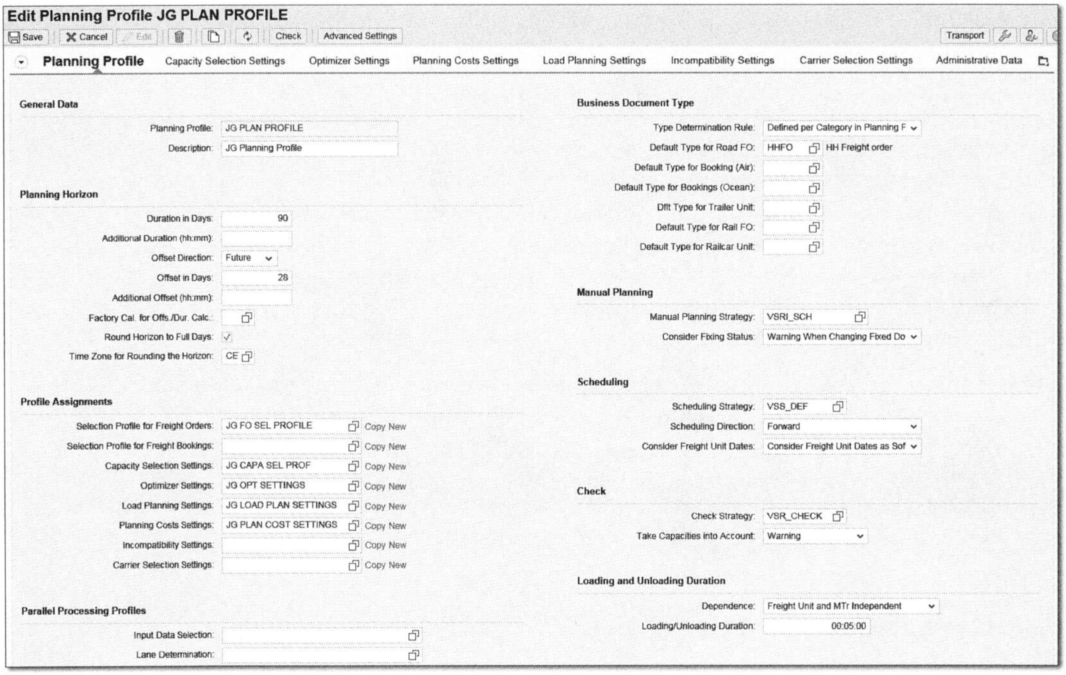

Figure 6.14 Planning Profile

Business document types

The BUSINESS DOCUMENT TYPE area defines the types of the objects that are created by planning. You can either use the default types per category, which all are marked as default in the corresponding type Customizing, use a condition for document type determination, or explicitly maintain the document types per category. You can maintain the document types for road freight orders, rail freight orders, ocean freight bookings, air freight bookings, trailer units, and railcar units. If no explicit document type is specified, the corresponding default type is chosen. If the condition does not specify the document type, then the explicit type definition is used as the fallback.

In the PROFILE ASSIGNMENTS area you can reference these six planning-relevant settings and include selection profiles to select freight orders and freight bookings, which is useful for scenarios in which you want to explicitly include certain freight documents in planning. With the COPY and NEW links, you can copy the assigned profiles and settings or create new ones, respectively.

Profile assignments

The MANUAL PLANNING section defines the manual planning strategy, which you can choose from the available strategies. For example, you can choose `VSRI_SCH` and `VSRI_1STEP` to trigger scheduling and carrier selection, respectively, after each manual planning step, or `VSRI_DEF` for the default manual planning operation without any subsequent step. Additionally, you can define the reaction of the system to your attempt to change a fixed document in the cockpit. Either the system issues a warning (in which case you can continue planning and save the document) or it reports an error (in which case the planning operation cannot be executed).

Manual planning

The SCHEDULING section defines how the automatic scheduling is performed. You can enter the scheduling strategy, for which the default is `VSS_DEF`. You can also specify the scheduling direction—either forward or backward—and define how the time windows of freight units will be considered by scheduling; either both hard and soft time windows are considered, only soft time windows are considered, or time windows are ignored. Refer to Section 6.3.7 for more details on scheduling.

Scheduling

The CHECK section defines the check strategy, for which `VSRI_CHECK` is the default, and the handling of capacity violations. The system issues either a warning (in which case you can continue planning and save the affected document) or an error (in which case the planning operation cannot be executed), or it simply ignores the capacities.

Checks

The LOADING AND UNLOADING section defines how the duration for loading and unloading a freight unit is determined. You can choose from the following options:

Loading and unloading duration

- Independent from freight unit and means of transport
- Dependent on freight unit and independent from means of transport

- Independent from freight unit and dependent on means of transport
- Dependent on freight unit and means of transport

In the first case, you can explicitly maintain the desired duration. For the other three cases, you have to maintain a condition for determining loading and unloading durations. These cases are useful if the loading and unloading depends on the volume, weight, or number of pallets in the freight unit, and it is also possible to model that some means of transport can load and unload the same freight unit quicker than others (e.g., because special loading equipment is on board).

Parallel processing

The PARALLEL PROCESSING PROFILE section allows you to define profiles to control parallel processing for the optimizer input data selection, the transportation lane determination, and the distance and duration determination. Via the menu path APPLICATION ADMINISTRATION • GENERAL SETTINGS • DEFINE PARALLEL PROCESSING PROFILE, you can specify the server group used for parallel processing, the maximum number of parallel processes, the package size that determines the number of relevant objects to be grouped together in a package for parallel processing, and the queue time defining how long the system has to wait for resources to become available for further processing. Parallelization can reduce the runtime for big optimization scenarios.

Capacity selection settings

The capacity selection settings shown in Figure 6.15 select the resources and schedules for the planning scenario at hand. You can define multiple criteria based on attributes of the vehicle resource. The mode of transport can be maintained explicitly, allowing selection of different road and rail resources within one selection. You can use the attribute means of transport to select active or passive vehicle resources. The definition of criteria for schedules is analogous to vehicle resources.

Optimizer settings, planning cost settings, incompatibility settings, load planning settings, carrier selection settings

For the remaining settings, we refer to the sections that describe them in detail. The optimizer settings are described in Section 6.4.2, except for the transportation proposal parameters, which are explained in Section 6.3.6. The planning cost settings cover costs and constraints, which are discussed in Section 6.4.3 and Section 6.4.4, respectively. The incompatibility settings are covered by Section 6.4.4, too. The load planning settings are described in Section 6.4.7; Chapter 7, Section 7.1.4 explains the carrier selection settings.

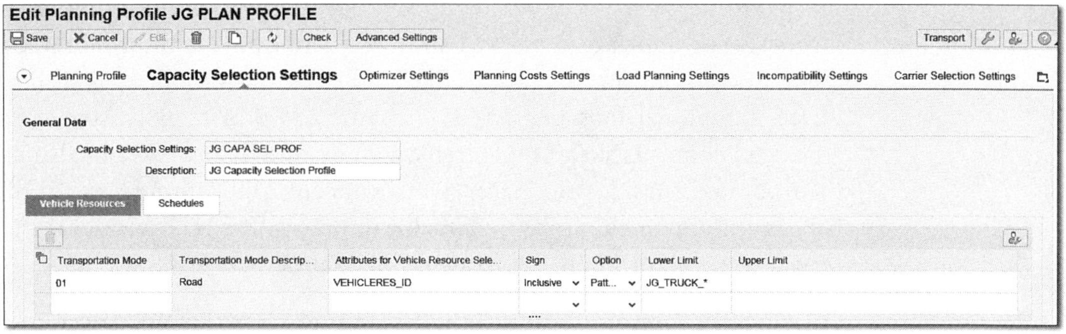

Figure 6.15 Capacity Selection Settings

POWL query for profiles, settings

The menu path PLANNING • WORKLIST • OVERVIEW SETTINGS takes you to a POWL containing queries for all profiles and settings. This provides a good overview and allows you to navigate to the individual profiles and settings and create new ones.

Transporting profiles and settings

Transporting Profiles and Settings

Click on the TRANSPORT button for planning profiles and selection profiles to trigger a (software) transport for them from one system into another. This avoids having to maintain the same profiles in development, quality, and production SAP TM systems, and is less error-prone than maintaining the same profiles multiple times—or once per change in each system.

6.3.3 Configuring the Transportation Cockpit

One configurable user interface for all planning scenarios

The ultimate goal of planning is to define a transportation plan that matches transportation demands with transportation capacities. The transportation cockpit is the central user interface for performing any planning operation. However, due to the vast structural variety of planning scenarios in practice, it is impossible to statically define one user interface that perfectly fits all scenarios of all transportation businesses in the world. Therefore, it is essential to configure the transportation cockpit, adapting it to the planner's needs so that all relevant information is visible and all irrelevant information is hidden.

Configuration by page layouts

The configuration of the transportation cockpit is done by page layouts that allow you to adapt the appearance of the transportation cockpit to the planning scenario at hand. A page layout defines the number of screen areas, their positions and sizes, the tabs contained in the screen areas, and all buttons that are visible in the cockpit's toolbar, as well as in the individual tabs per screen area. In addition to flat lists, more advanced concepts such as hierarchical views and dual views can also be incorporated. Besides the page layout concept, the column sets for all lists and hierarchies can be personalized; that is, the relevant fields can be selected and ordered according to the planner's needs.

Diversity of scenarios

To illustrate the variety of different planning scenarios, we describe a few examples, each accompanied by a dedicated page layout.

Local distribution with own trucks

The planner of a brewery with its own truck fleet that serves only local customers such as restaurants and pubs needs information about freight units, vehicle resources, and road freight orders, maybe accompanied by a map to visualize the current transportation plan. If the planner gets new freight units after he has already created a plan, he may use the full-screen map to check which of the already planned freight orders fits the new freight unit from a geographical perspective. Figure 6.16 shows an example with freight unit stages in the top screen area, road freight orders in the middle screen area, vehicle resources in the bottom-left screen area, and freight order details in the bottom-right screen area, which includes information about the stages of the chosen freight order, the utilizations along the stages, the map, and other information.

Distribution of pallets, loading plan per truck

The planner of a manufacturing company shipping pallets may perform the same tasks as the brewery's planner, but in addition she may need a separate view (including the list of road freight orders with a detailed screen for the load plan per road freight order) to check whether the pallets can be placed according to their and the truck's length, width, and height, and respective axle weight limits. Figure 6.17 shows an example of the load planning with road freight orders in the top screen area and the load planning details in the bottom screen area. The details are displayed in two parts: the load plan list on the left and its three-dimensional visualization on the right. Thus, the planner may use two different setups of the transportation cockpit.

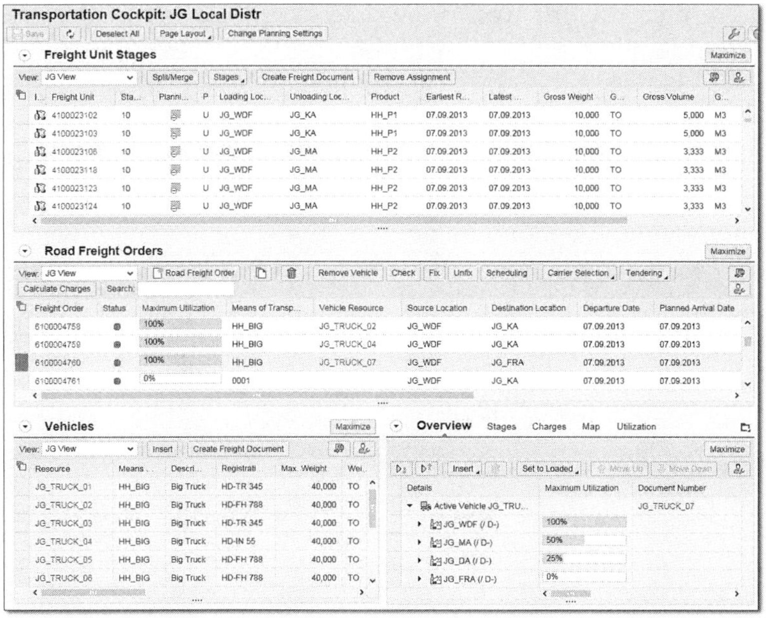

Figure 6.16 Cockpit Local Distribution Layout

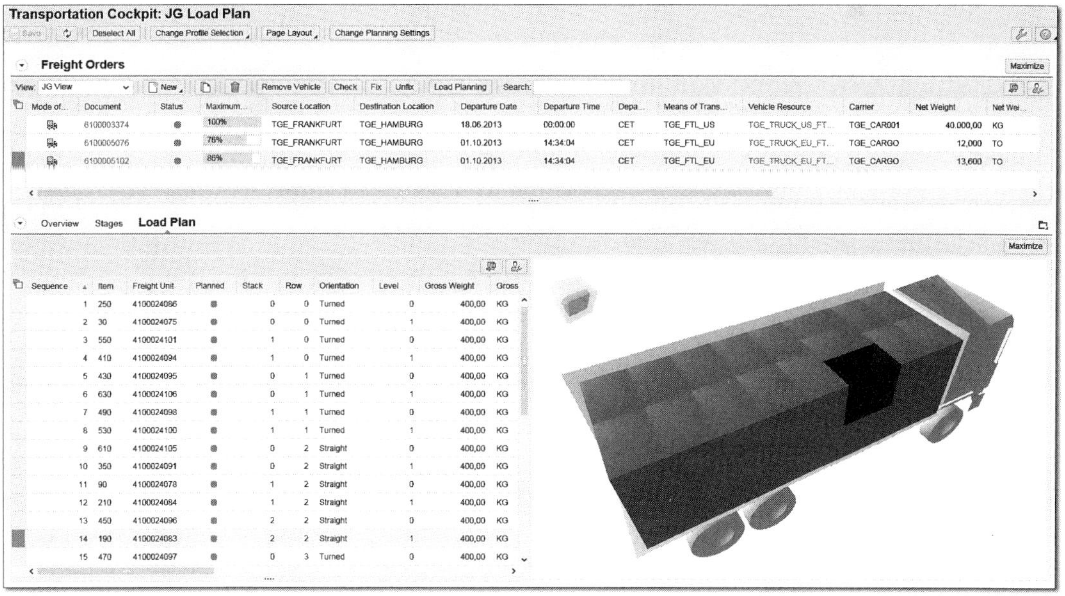

Figure 6.17 Cockpit Load Plan Layout

Road carrier with trucks, trailers

In a road carrier company with its own fleet of trucks and trailers, the planner needs trailer units and trailer resources. If the freight units were already assigned to trailer units a priori by an appropriate FUBR, the planner would not need freight units at all. Figure 6.18 shows an example that includes freight units. Here, the planner sees the trailer units in the top-left corner, the road freight orders in the top-right corner, the freight unit stages in the bottom-left corner, and the available vehicles and trailers in the bottom-right corner.

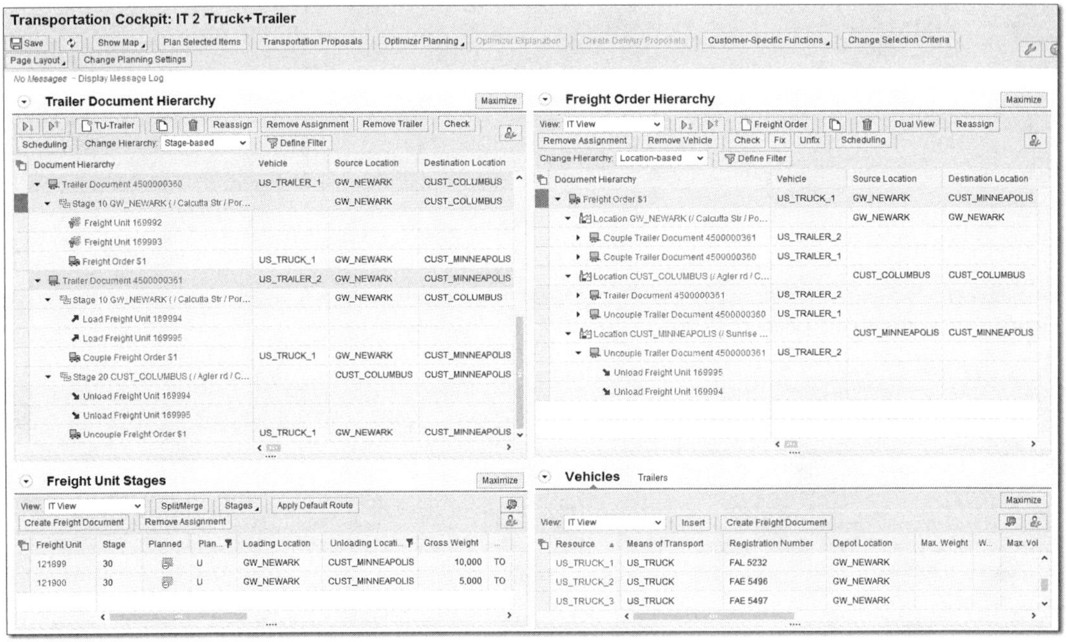

Figure 6.18 Cockpit Truck and Trailer Layout

Global ocean/air freight-forwarding

In a global ocean forwarding company, the ocean freight planner needs freight units, ocean freight bookings, and schedules. Similarly, the air freight planner needs freight units, air freight bookings, and maybe schedules. The planners responsible for the pre-legs and subsequent legs require freight units, road freight orders, and schedules or vehicle resources. Within the same company, different planners need very different setups.

Figure 6.19 presents a layout for the ocean freight planner, in which the top-left screen area contains freight unit stages, the top-right screen area contains the ocean booking hierarchy, the screen area in the bottom left contains schedules, and the bottom-right screen area contains the freight document details, spread over multiple tabs such as MAP, EXECUTION INFORMATION, CARGO MANAGEMENT, and so on. The freight document details are shown for the selected freight document—in this case, the ocean freight booking in the top-right screen area.

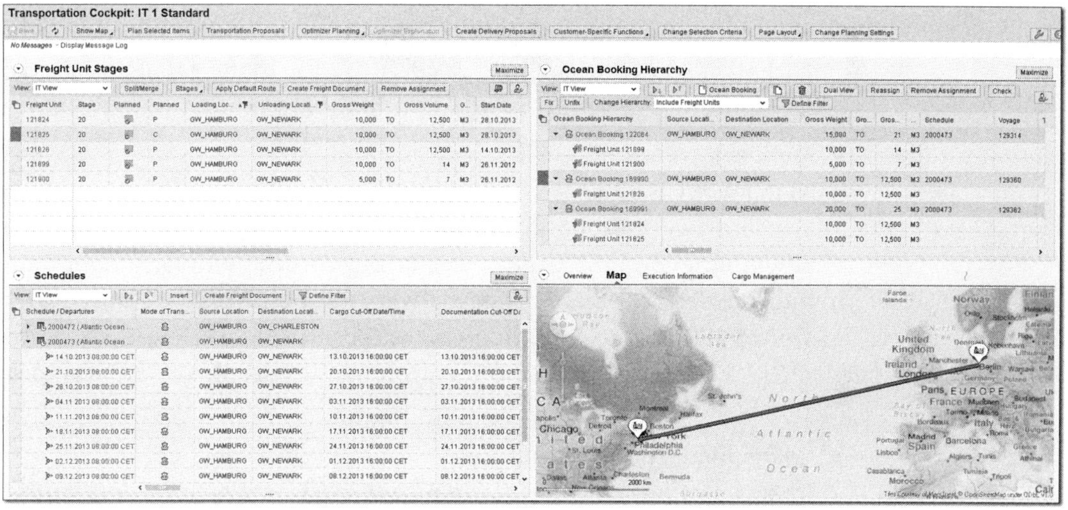

Figure 6.19 Cockpit International Layout

Replanning

All of these planners may want to change their plan by reassigning freight units between freight documents or transportation units. For this, they may need to get a different view of their plan by focusing on the relevant freight documents or transportation units and their substructures. Such requirements are covered by hierarchical views and dual views, which are explained in Section 6.3.4.

Defining a page layout

To define a page layout, follow the menu path APPLICATION ADMINISTRATION • PLANNING • GENERAL SETTINGS • PAGE LAYOUTS and choose one of the following contexts for which you want to define a layout:

- **Transportation cockpit**
  This is the central planning user interface, as just mentioned. Since it offers the biggest range of configuration possibilities, the rest of this section focuses on the transportation cockpit. Configuration for the next three contexts is much more straightforward and is therefore only mentioned briefly, without discussion of details.
- **Carrier selection**
  This layout focuses on freight documents and their details, which is sufficient for carrier selection purposes. From this perspective, these page layouts are special cases of page layouts for the transportation cockpit. Refer to Chapter 7, Section 7.1 for more details on carrier selection.
- **Delivery creation**
  This layout is required for SAP ERP order integration scenarios in which delivery proposals are created in SAP TM for the freight units resembling SAP ERP orders. The layout consists only of freight units and delivery proposals. The delivery proposals can be propagated from SAP TM into SAP ERP, where SAP ERP deliveries are then created. See Chapter 4, Section 4.1.5 for more details on delivery proposals.
- **Transportation proposal**
  This is a semi-automatic planning approach in which layouts consist of transportation proposals and may also include screen areas for preferences and a map. Transportation proposals are described in Section 6.3.6.

Next you define the general values in the LAYOUT area, as shown in Figure 6.20, such as the name of the layout and its validity, which can be valid for all users, a role, or a specific user. You include a transportation proposal page layout, which will be used if you determine transportation proposals out of the transportation cockpit, as described in Section 6.3.6, and you can activate command line planning, which is explained in Section 6.3.5. The VISIBILITY PUSHBUTTON area allows you to define which functionalities are offered as buttons in the toolbar of the transportation cockpit. Section 6.3.5 describes the available functionalities.

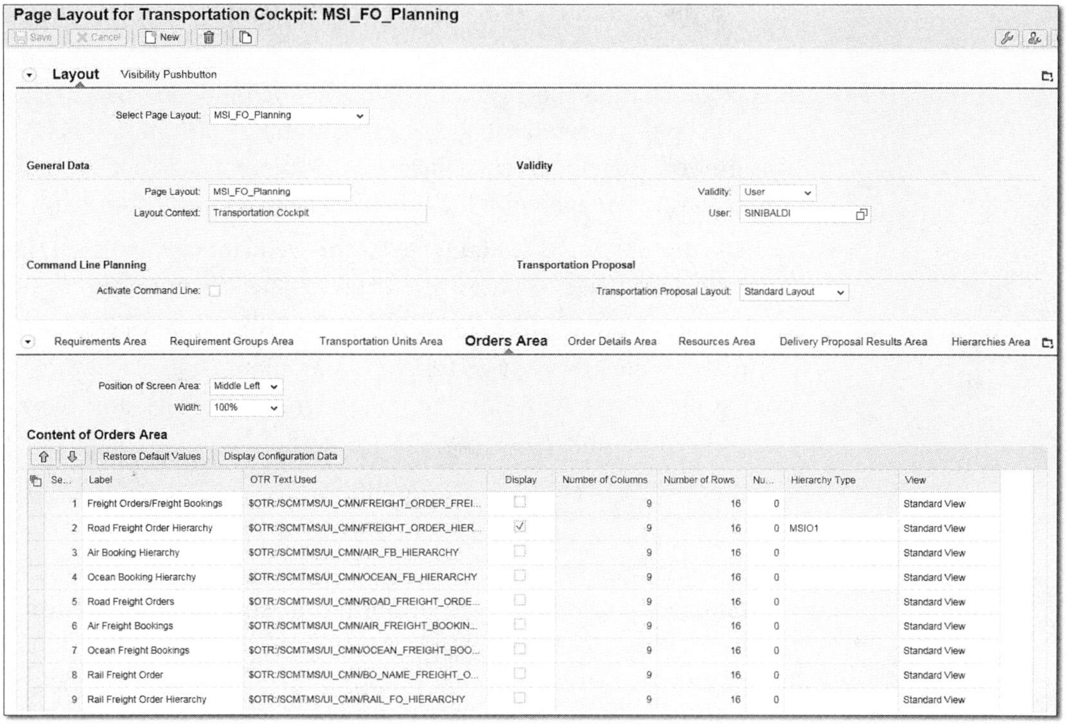

Figure 6.20 Page Layout

Available screen areas, tabs

The page layout offers various screen areas and a set of tabs per screen area, from which you can choose your screen areas and the tabs in them. The following screen areas and tabs are offered:

- The REQUIREMENTS AREA offers tabs for the freight unit stages list and the freight unit hierarchy.
- The REQUIREMENTS GROUP AREA provides the freight unit stage group tab.
- The TRANSPORTATION UNITS AREA offers the trailer unit hierarchy and the railcar unit hierarchy tabs.
- The ORDERS AREA contains tabs for the road freight order list, rail freight order list, ocean freight booking list, air freight booking list, and a combined list for all freight orders and freight bookings. In addition, there are tabs for the road freight order hierarchy, the rail

freight order hierarchy, the ocean freight booking hierarchy, and the air freight booking hierarchy.

- The ORDER DETAILS AREA provides tabs for the following freight document details: overview, stages, carrier ranking, allocation, charges, map, predecessor documents, successor documents, execution information, cargo management, load plan, utilization, and equipment.
- The RESOURCES AREA contains tabs for vehicle resources, trailer resources, schedules, locomotive resources, and railcar resources.
- The HIERARCHIES AREA offers tabs for the freight unit hierarchy, road freight order hierarchy, rail freight order hierarchy, ocean freight booking hierarchy, air freight booking hierarchy, trailer unit hierarchy, and railcar unit hierarchy. All these hierarchies are offered by other screen areas, too. The only purpose of the HIERARCHIES AREA is to enable layouts containing the same hierarchy twice.

Screen area position/width

The screen areas can be placed on a grid with three rows and two columns, and you can select from the options TOP LEFT, TOP RIGHT, MIDDLE LEFT, MIDDLE RIGHT, BOTTOM LEFT, BOTTOM RIGHT, and NOT VISIBLE. The width of one screen area can cover one column (50% width) or two columns (100% width). The system ensures consistent entries; for example, two screen areas can only be placed in one row if they have 50% width each. For each visible screen area, you can choose among the available tabs.

Defining tabs per screen area

You choose the tabs in a screen area by selecting the DISPLAY checkbox and change their ordering with the MOVE UP and MOVE DOWN buttons. It is possible to change the label of the tab and to define the hierarchy type for a hierarchical view. The number of rows used by the MAXIMIZE button can be specified; a value of 0 means that the maximum size of the window is used. If you have created a personalized view for the tab, which defines the number of rows, column set, ordering, and so on, you can set it as the initial view for the tab. Otherwise, the standard view is used. By clicking on the DISPLAY CONFIGURATION DATA button, you can show the association used and change the user interface configuration in order to use your own configuration.

Defining buttons per tab

Clicking on a tab displays its details; you can activate the buttons for different functionalities in the transportation cockpit, as shown in Figure

6.21 for the road freight order list. For tabs that offer the dual view, activating the DUAL VIEW button yields more details to specify the dual view itself, which is described in Section 6.3.4.

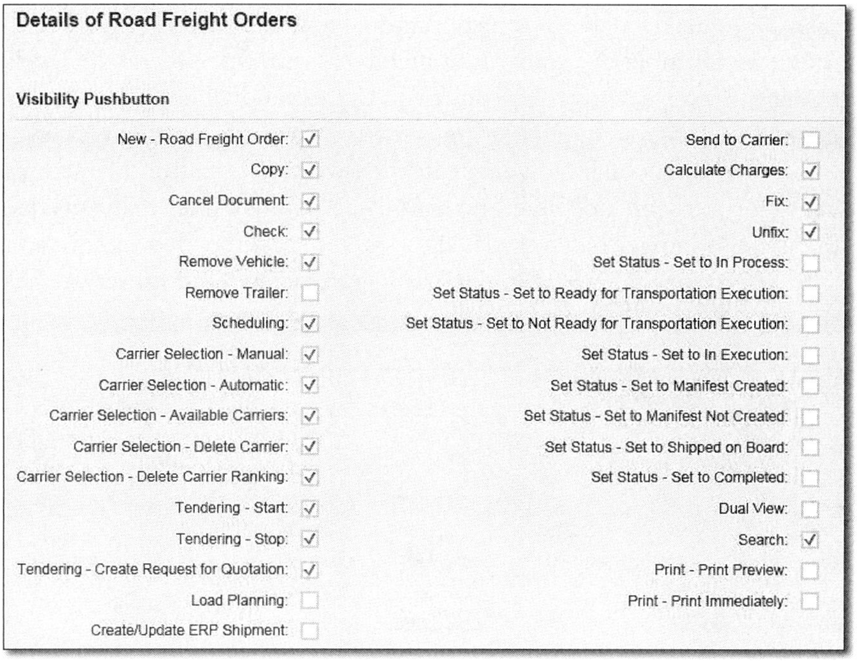

Figure 6.21 Road Freight Order Buttons

6.3.4 Hierarchical View and Dual View

Motivation

Whereas in many planning scenarios the information shown in flat lists for freight units and freight documents is sufficient, several planning scenarios require visibility of the substructures of the business objects in order to make the right planning decisions. For example, if you want to assign an unplanned freight unit to one of the existing freight orders, you need to know the stop sequences of the freight orders to choose the best fit. Another example is to improve the ratio of volume and weight for your air freight bookings, for which it is necessary to see the reassignment potentials of the assigned freight units. Here, you need to know which freight units are assigned to the relevant air freight bookings.

Hierarchical views/ hierarchy types Hierarchical views can be used for freight units, transportation units, and freight documents. The substructure of a hierarchical view is defined by a hierarchy type, which can be configured a priori. Hierarchical views allow planners to dynamically drill down into the substructures of the relevant objects in their transportation cockpit sessions. Figure 6.22 shows an example of a trailer unit hierarchy with three hierarchy levels: trailer unit, stage, and freight unit. You can expand all nodes or selected nodes in the hierarchy and collapse them afterward. Using the CHANGE HIERARCHY functionality, you switch between alternative hierarchical views for the same document type. You can define a filter that considers the nodes on lower hierarchy levels, too. Note that the filtering functionality determines the nodes in the first hierarchy level and preserves their substructures.

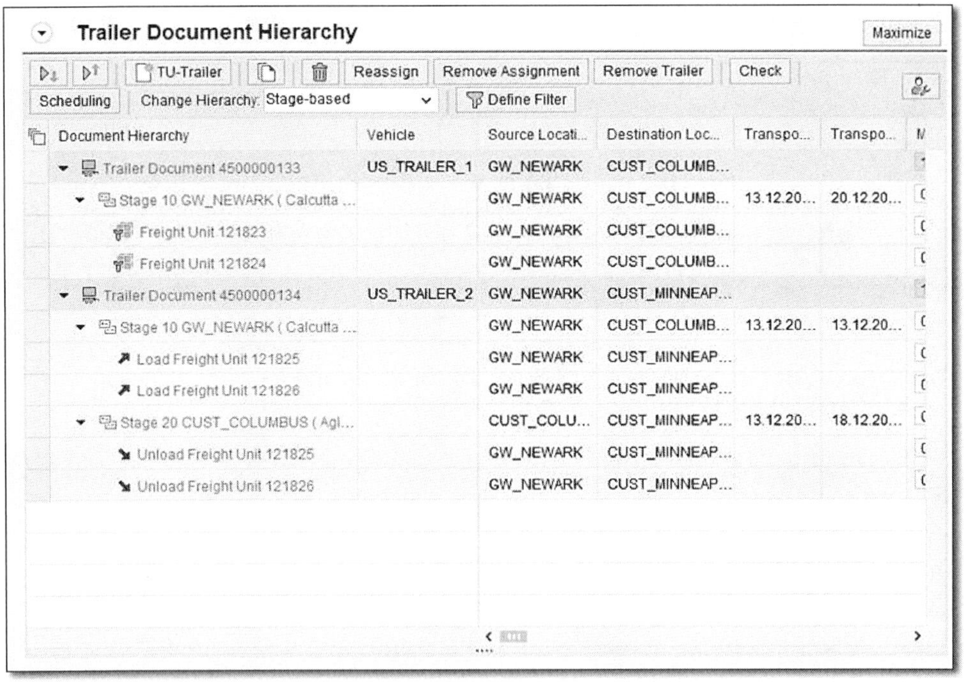

Figure 6.22 Trailer Unit Hierarchical View

Defining hierarchical views To define a hierarchical view, select the IMG menu path SAP TRANSPORTATION MANAGEMENT • TRANSPORTATION MANAGEMENT • PLANNING •

General Settings • Define Hierarchical Views for Freight Documents. You can edit an existing hierarchy type or create a new one. First, you have to choose the hierarchy usage from the following alternatives (see Figure 6.23):

- Transportation Cockpit: Freight Units
- Transportation Cockpit: Transportation Units
- Transportation Cockpit: Freight Orders and Freight Bookings
- Freight Document: Cargo
- Freight Document: Equipment
- House Bill of Lading

The hierarchy usage ensures that only reasonable hierarchy levels can be defined. Then you choose the mode of transport, which also restricts the possible entries for the hierarchy levels. In addition, you can specify whether only items added or removed at the current location are displayed. If a freight order delivers 50 freight units from one source to 10 different locations, you may want to see for each stop only those freight units that are loaded or unloaded at that stop. Alternatively, you can configure the hierarchy view so that all freight units still on the truck are displayed for each stop. Thus, the freight units going to the last destination would be shown at all previous stops.

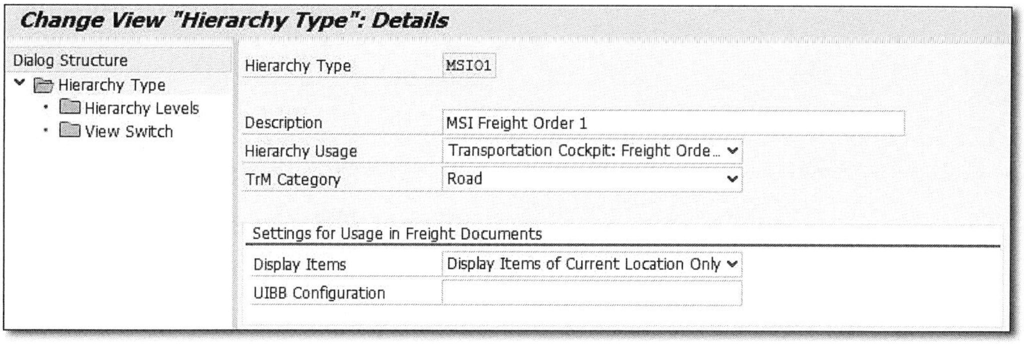

Figure 6.23 Hierarchy Type Header

Hierarchy levels, filtering, grouping

The next step is to define the hierarchy levels, as shown in Figure 6.24. For each hierarchy level you choose the object type and define whether the hierarchy is initially expanded. It is also possible to define the text and icon displayed in the hierarchical view. By activating the filtering option, you can suppress displaying locations at which no freight unit is loaded or unloaded and no trailer unit is coupled or uncoupled. With the grouping functionality, you can aggregate multiple objects into their own hierarchy level based on a grouping criterion that is either explicitly defined by attributes (as shown in the example for the grouping attribute CONSIGNEE_ID) or implicitly defined by a class, which you can introduce to define a more sophisticated grouping logic.

Note that it is even possible to define multiple grouping levels, which is useful if you want to structure your freight units according to destination location and product. If you have hundreds of freight units to be planned manually, this kind of grouping structures the freight units according to your needs, enabling you to quickly get an overview of the different freight unit groups by collapsing all hierarchy levels. You can introduce grouping levels in lower hierarchy levels also (e.g., to structure the freight units assigned to a stop of a freight order by product). Note that you can drag and drop freight unit groups to a freight order, which allows many similar freight units to be handled by one manual planning operation, without the need to care about the individual freight units in the group.

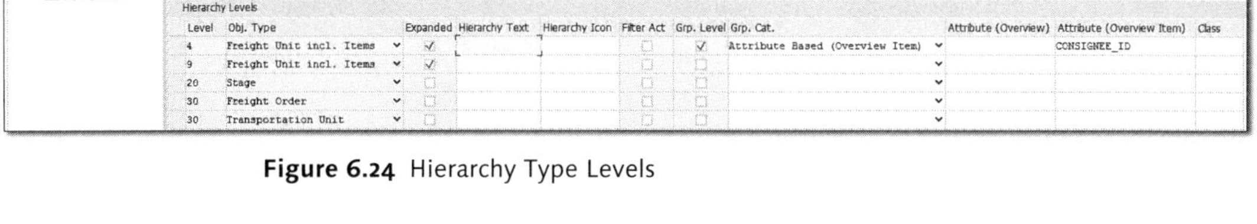

Figure 6.24 Hierarchy Type Levels

View switch

If you want to dynamically switch between alternative hierarchical views, you can define multiple hierarchy types and list the alternatives in the VIEW SWITCH list of one hierarchy type. In the example shown in Figure 6.25, you can switch from the hierarchy type MSIF1 to the alternative hierarchy types FUSTG and MSIF2. In some scenarios, it is useful to see the substructure from a stage-based view or from a location-based view. Here, you could define two hierarchy types and link them by the view switch functionality, making it possible to choose them dynamically from a stage- or location-based view.

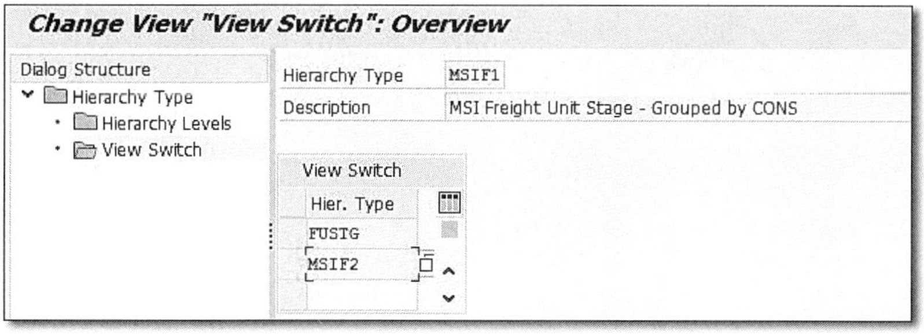

Figure 6.25 Hierarchy Type View Switch

Performance

Performance of Hierarchical Views

A hierarchical view contains much more information than a flat list and therefore may be slower than a flat list. Consider a road freight order hierarchy with locations and freight units on the second and third hierarchy level, respectively. Imagine a scenario with 100 road freight orders, each having on average 5 stops and each stop contains 5 freight units. The fully expanded hierarchical view then contains 2,500 rows, instead of 100 rows in the road freight order list.

The greater level of detail does not come for free. Therefore, if you are defining a hierarchical view, avoid defining too many hierarchy levels and include only hierarchy levels that are very important for your business. We recommend that you do rough estimations of the average number of nodes per hierarchy level first to get an estimate of the expected number of rows for your hierarchical view if all nodes are expanded. If this number is much too big—for example, it contains 10,000 rows—check whether you can cut your scenario into pieces by appropriate selection profiles or whether you can omit hierarchy levels in your definition.

We recommend displaying only items of the current location because displaying items of all locations would yield even more rows in the hierarchical view.

The grouping functionality helps aggregate the objects, but it also creates more rows, so keep this in mind when introducing one or multiple grouping levels.

> Although it is nice to see the full substructure of an object, usually not all columns in the hierarchical view can be meaningfully filled for all rows. While the source and destination location as well as certain times can be meaningfully defined for the freight order level and stage level, the utilization or vehicle resource is determined on the freight order level, and the freight unit's transportation group cannot be easily aggregated into the freight order level. Therefore, we recommend checking whether all intended columns are needed and focusing on the most important pieces of information.
>
> In a nutshell, be aware of the number of columns and rows created by your hierarchy type and chosen columns.

Dual view If you have already created a plan, either manually or automatically, you may want to check it in detail for local improvements, such as reassigning a freight unit from one freight booking to another or moving a whole freight order stop with all its freight units to another freight order. Hierarchical views allow you to browse through the freight documents and their substructures. However, when you have found one freight order that you want to optimize manually by adding or removing freight units, you need to search for another freight order to perform the re-assignment. Usually, you want to keep the first freight order and its substructure visible while identifying the second freight order. Since both freight orders are in the same hierarchical view, scrolling within one hierarchy may hide your first freight order. The dual view overcomes this issue by offering two hierarchies at the same time, which allows scrolling and searching within one hierarchical view while keeping the other hierarchical view constantly visible. The two views show the same information, but from a different angle, so if you change something in one hierarchy, you immediately see the effect in both hierarchies.

Vertical/horizontal dual view Depending on your scenario, you may want to see as many rows as possible or as many columns as possible. The dual view allows you to switch between the vertical view and the horizontal view, as shown in Figure 6.26 and Figure 6.27.

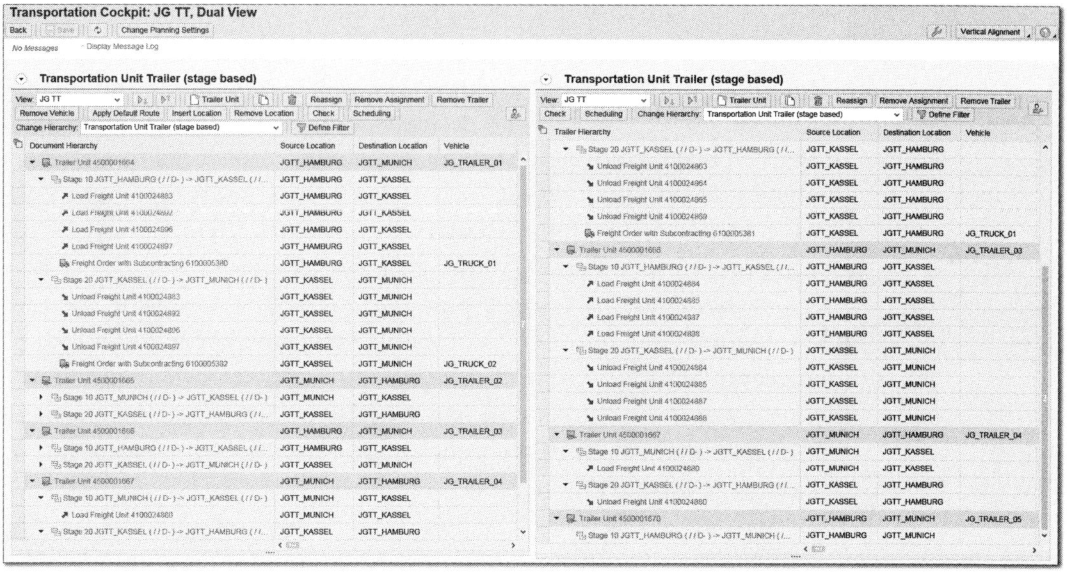

Figure 6.26 Vertical Dual View

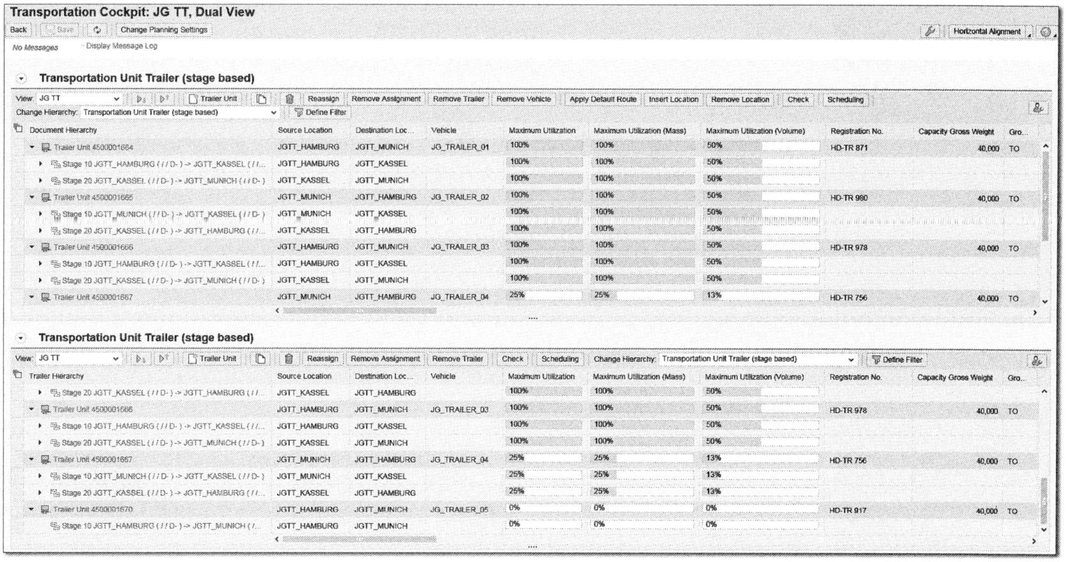

Figure 6.27 Horizontal Dual View

Defining a dual view

Within the transportation cockpit, the dual view is triggered by the DUAL VIEW button, which can be activated in the page layout for the hierarchical views showing transportation units or freight documents. This button has to be activated in the page layout definition for the hierarchical view at hand, as shown in Figure 6.28 for the road freight order hierarchy. If the DUAL VIEW checkbox is selected, a separate DUAL VIEW area is shown below the VISIBILITY PUSHBUTTON area. Here, you can specify the initial alignment of the dual view (either vertical or horizontal) and the hierarchy types for the two hierarchical views in the dual view. If you define only one hierarchical view, it will be used for the second hierarchical view, too. If your layout is user-specific, you can also define the view for the column sets. It is also possible to define the number of rows and columns; the initial value 0 uses the full space available for the page layout.

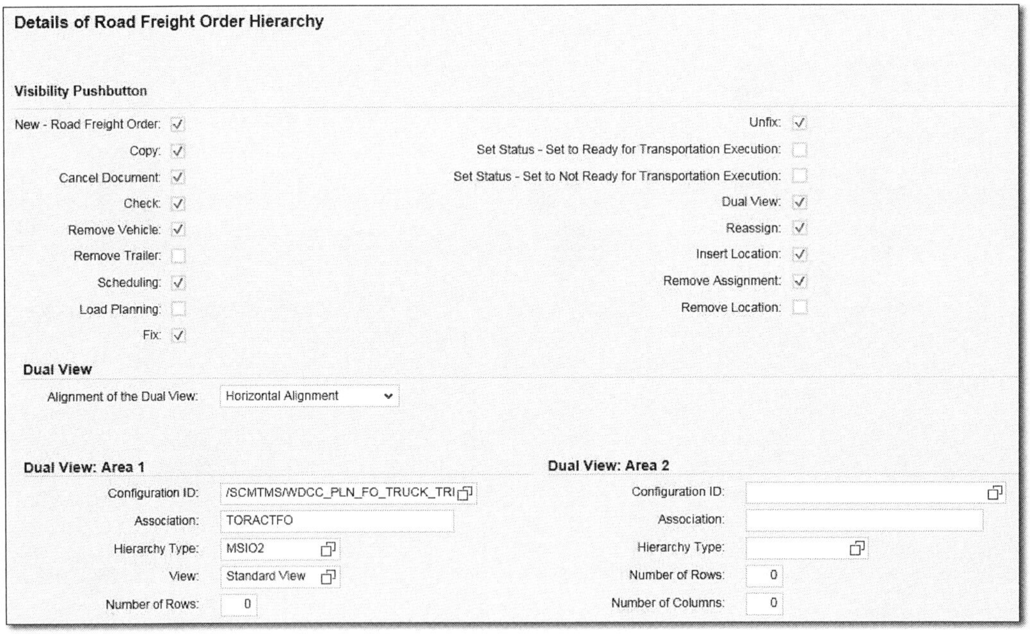

Figure 6.28 Layout Freight Order Details

Two identical hierarchical views and other screen areas

The dual view shows only two hierarchical views and cannot be combined on one screen with other screen areas such as resources or freight units. However, you can define a layout with two hierarchical views and other screen areas by choosing one hierarchical view for the corresponding screen area, such as the road freight order hierarchy, and adding the road freight order hierarchy view again in the HIERARCHY screen area. If you want to enable an easy switch between the horizontal and vertical views, as is built into the dual view, you have to define two layouts: one resembling the horizontal variant and one for the vertical variant.

6.3.5 Using the Transportation Cockpit

The previous sections mentioned the need for configuring the cockpit's appearance and introduced page layouts, hierarchical views, and dual views. Let's turn our attention to how you can use the transportation cockpit by exploring the ways to start the transportation cockpit and the functionalities it offers for navigating through the objects, changing the appearance on the fly, doing manual planning, triggering automatic planning, and performing subcontracting and execution-related tasks.

Entry into the Transportation Cockpit

Entry into transportation cockpit

The menu path PLANNING • PLANNING • TRANSPORTATION COCKPIT takes you to the entry screen for the transportation cockpit. You can choose from two entry possibilities in the VIEW choice: one based on selection criteria and one based on profile and layout sets.

Selection criteria

The entry by selection criteria allows you to define ad hoc selection criteria for the various objects or to choose from previously defined selection criteria (saved searches), as shown in Figure 6.29. In the FREIGHT UNIT STAGES, TRANSPORTATION UNITS, FREIGHT ORDERS, FREIGHT BOOKINGS, RESOURCES, and SCHEDULES sections, you can define multiple criteria for the desired object types, each criterion referring to one attribute and one matching criterion. For example, the first criterion for freight unit stages refers to the source location attribute and the matching criterion IS. You can add or remove criteria by clicking on the plus and minus buttons. The planning profile and the transportation cockpit layout can be selected in the SETTINGS section. Once you have made your choice, click on CONTINUE to start the cockpit.

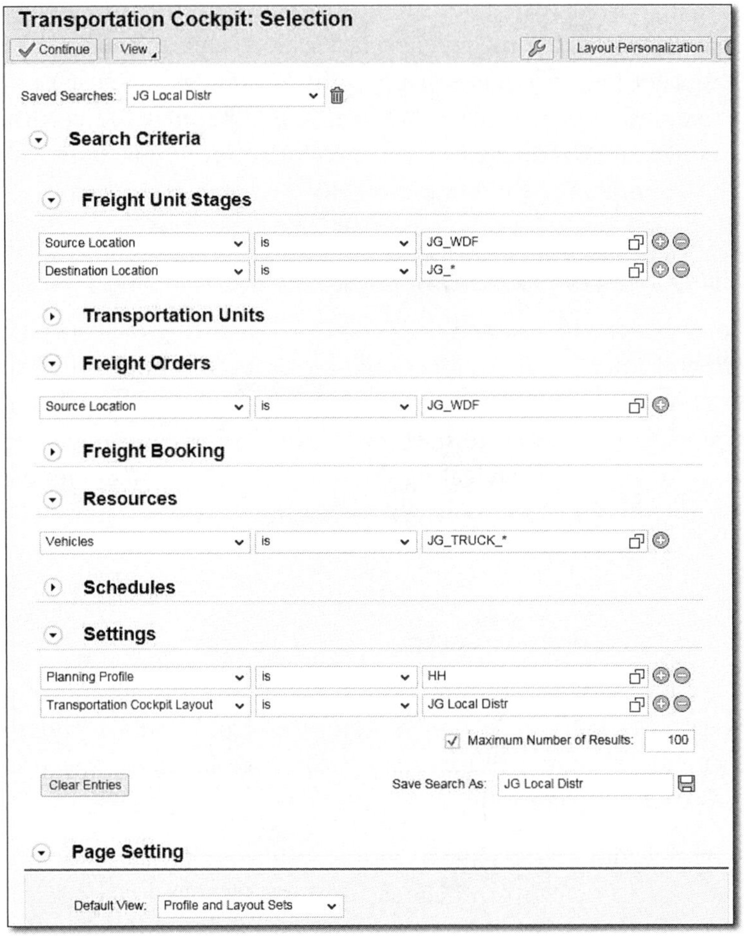

Figure 6.29 Selection Criteria

Profile and layout sets

Figure 6.30 shows the entry by profile and layout sets method. A profile and layout set is defined by a description, a selection profile for freight units, a selection profile for trailer units, a selection profile for freight orders, a selection profile for freight bookings, capacity selection settings, incompatibility settings, a page layout for the transportation cockpit, and a layout for the planning result screen. You can define your own sets, which is helpful if you are using the transportation cockpit for different scenarios. Then you can select one set and start the transportation cockpit as defined by the set. If you have defined multiple sets but you usually use only one set, you can define it as the default set, which frees you from having to select it every time you start the cockpit. For each profile and layout set, the system displays the last time it was used and how many times it was used. Additionally, the system indicates the number of unplanned freight unit stages, which is helpful to compare the workload for a planner using several profile and layout sets per day.

Selection criteria versus profile and layout sets

Comparing these two entry options, the selection criteria are simpler to maintain and very intuitive, allowing a quick start of the transportation cockpit without having to define several selection profiles. The entry with profile and layout sets is more powerful because the selection profiles provide more sophisticated selection and the individual selection profiles for the different objects can be combined and reused by multiple users. Especially for big companies with many users, the profile and layout approach enables structuring users' scenarios and roles on a fine-grained level.

Defining the default entry

Using the Default View choice, you can define whether selection criteria or profile and layout sets will be used as the default. If you always start the cockpit with the same profile and layout set, you can set it as the default profile and layout set. Then choose the Hide Screen option, which suppresses the whole entry screen when you start the transportation cockpit. If you want to see the entry screen again, just start the transportation cockpit and click on the Change Profile Selection button. If your layout does not permit this, change the layout to activate this button.

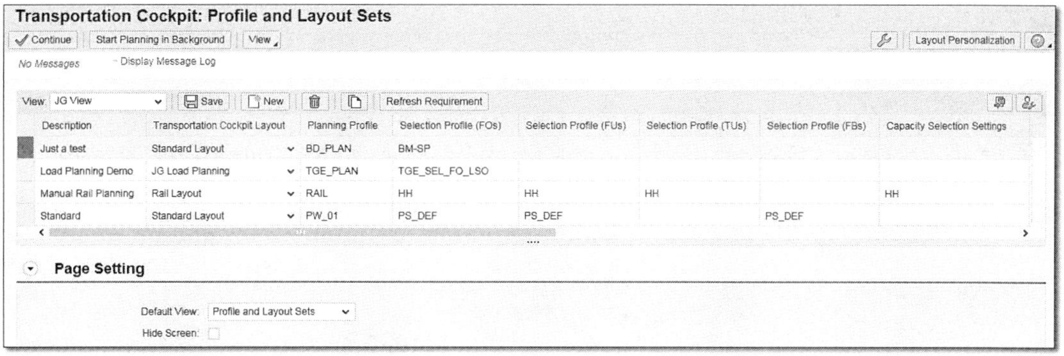

Figure 6.30 Profile and Layout Sets

Entry from POWL queries/business objects

In addition to these entry options, the transportation cockpit can also be triggered from various POWL queries, such as OTRs, DTRs, road freight orders, and rail freight orders. It is also possible to trigger the transportation cockpit directly out of a business object such as the forwarding order.

Transportation Cockpit Functionalities

The transportation cockpit offers a rich set of functionalities that can be triggered by selecting objects in lists and hierarchies, buttons, and drag and drop.

Save and refresh

You can save the results of planning by clicking on the SAVE button, and you can refresh the documents in the transportation cockpit session with the REFRESH button. This may be useful if, for example, execution information from freight documents has been updated outside the transportation cockpit session.

Manual planning operations

The following manual planning operations are offered:

- **Assigning objects to another object**
  You can assign any set of objects to a target object by selecting the objects and clicking on the PLAN SELECTED ITEMS or REASSIGN button or, alternatively, using drag-and-drop. By clicking on the UNASSIGN button, you can revert the assignments. You can also use command line planning, which allows you to enter an assignment command in a text field. For example, the command `5 6 7 - 2` assigns the freight unit stages with indexes 5, 6, and 7 to the vehicle resource with index 2, creating a new road freight order for the vehicle that contains the three freight unit stages.
- **Create freight documents**
  This can be triggered from selected freight unit stages, selected resources, selected schedules, and selected schedule departures by assigning freight unit stages to resources, schedules, or schedule departures, either from scratch or by copying an existing freight document.
- **Create transportation units**
  You can create trailer units and railcar units either based on selected trailer resources and railcar resources, respectively, by assigning

freight unit stages to resources, from scratch or by copying an existing transportation unit.

- **Split and merge freight units**
  You can split and merge by stages and quantities as mentioned in Section 6.1.
- **Apply default route**
  For freight unit stages, trailer units, and railcar units, the system determines a matching default route and creates stages accordingly. See Chapter 3, Section 3.2.6 for more details on default routes.
- **Insert and remove locations**
  You can insert new stops and remove stops from a transportation unit and a freight order.
- **Fix and unfix**
  Freight documents and transportation units can be fixed—that is, they cannot be changed by automatic and manual planning unless they are explicitly unfixed.
- **Cancel**
  You can cancel transportation units and freight documents, which results in unassigning the assigned freight units and transportation units.
- **Remove resource**
  You can remove the resource assignments from transportation units and freight orders.

Manual planning on the map

With the SHOW MAP button, you can visualize all objects or the selected objects or the selected objects with all resources on a full-screen geographical map, which is called the visual transportation cockpit. It is shown in Figure 6.31 for the selected freight orders in a local distribution scenario with a plant in Walldorf, Germany. If you leave the full-screen map by clicking on the BACK button, all results of map-based planning are visible in the original page layout. On the map, the following functionalities are offered:

- **Display profile**
  This allows you to select the planning object types be shown on the map. For example, you may want to focus on freight orders only or freight unit stages.

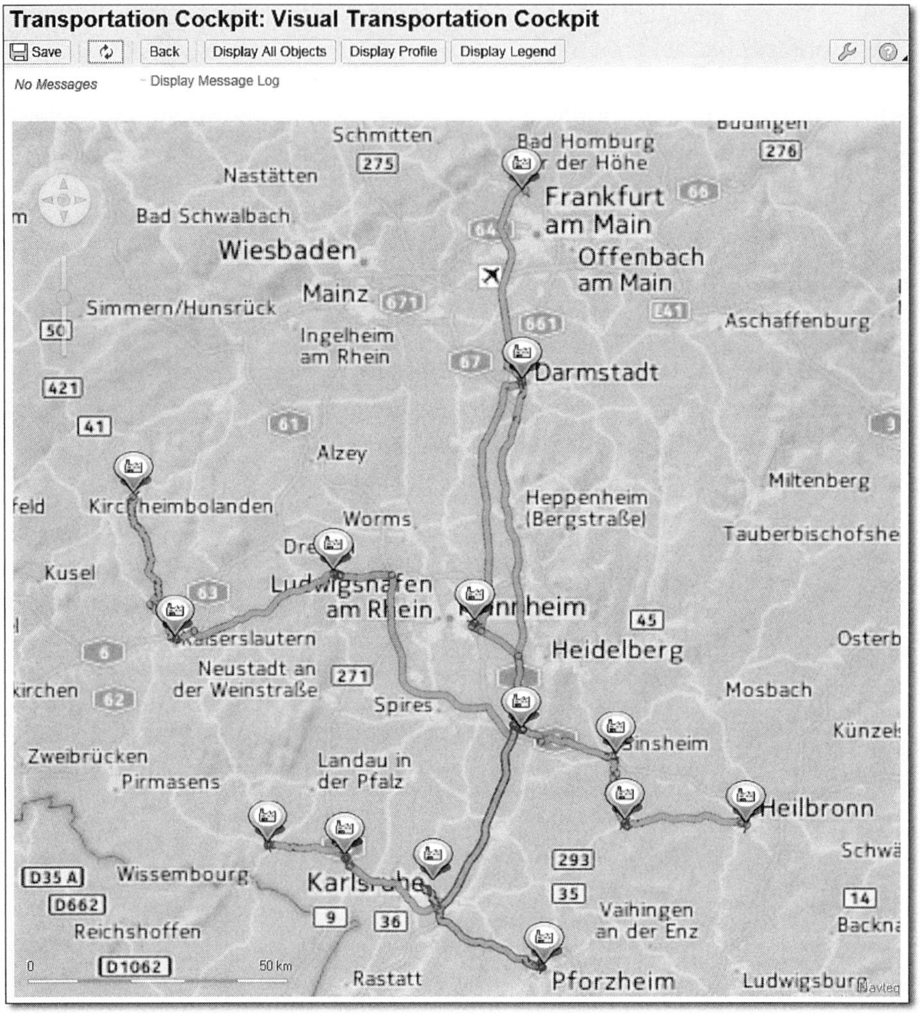

Figure 6.31 Full-Screen Map

- **Assigning objects to one object**
  You can select unplanned freight unit stages and start planning via the context menu. The system then offers the possible assignments of the freight unit stages to the resources, from which you can choose. You can also drag and drop unplanned freight unit stages to vehicle resources or freight orders, and vice versa. It is also possible to assign vehicle resources to freight orders, and vice versa.

- **Freight unit stage split**
  You can split a freight unit stage into two new freight unit stages by dragging and dropping the freight unit stage onto a location, which then serves as a new intermediate transshipment location for the original freight unit stage.
- **Insert new stop into freight order**
  Similar to the previous function, you can also add new stops to freight orders.
- **Multi-object planning**
  If multiple objects correspond to the same arc shown on the map, you can either perform the manual planning operation on one of those objects or on all objects associated with the arc. The latter is a very powerful functionality, because it makes it possible to plan a bundle of freight unit stages with one step on the map.
- **Show transshipment locations**
  In the context menu, you can trigger the display of the relevant transshipment locations for your scenario.
- **Detailed routes**
  Using the context menu, you can switch between displaying the objects on the map as straight lines or detailed (street-level) routes.
- **Address search**
  You can enter an address and show its geographical position on the map. Search results can be cleared afterward.

Automatic planning

You can also trigger automatic and semi-automatic planning from the transportation cockpit:

- **Optimization**
  You can run the vehicle scheduling and routing (VSR) optimizer on all the data in the cockpit, only on the selected objects, or on the selected objects together with all resources in the session. See Section 6.4 for more details on the VSR optimizer.
- **Transportation proposal**
  The system can determine a set of transportation proposals for the selected freight unit stage, from which you can choose one proposal. See Section 6.3.6 for more details.

- **Scheduling**
  The start and end times for all activities represented by selected freight documents and transportation units are scheduled automatically. As defined in the planning profile, the scheduling direction can be forward or backward. Refer to Section 6.3.7 for more details on scheduling.
- **Load planning**
  This triggers the load optimizer for the selected road freight orders. The results are shown in the LOAD PLANNING tab of the freight document details screen area. In the three-dimensional visualization of the load plan, which is based on SAP Visual Business (refer to Chapter 2, Section 2.3.6), you can rotate the truck, zoom in and out, and hide objects in a row, stack, or level to analyze the interior parts of the load that are not visible if you see the full load. See Figure 6.32 for an initial load plan (left), in which first one stack (middle) and then one level (right) is hidden. Refer to Section 6.4.7 for more details on the load optimizer.
- **Optimizer explanation tool**
  This tool helps you understand the input and output data for automatic planning—that is, for VSR optimization, transportation proposals, and the load optimizer. See Section 6.4.6 for more details on the explanation tool.

Figure 6.32 Load Plan Hiding Stack and Level

Adjusting the user interface on the fly

You can adapt the user interface in the same cockpit session as follows:

- **Maximize**
  Using the MAXIMIZE button, you can focus on one screen area that is shown in full window size and then go back to the original layout.

- **Page layout**
  You can switch between alternative page layouts.
- **Dual view**
  You can trigger the dual view, switch between its horizontal and vertical version, and then go back to the original page layout. If you have selected objects in the tab from which you trigger the dual view, only these objects are shown in the dual view.
- **Change hierarchy**
  You can switch between alternative hierarchical views.
- **Change (column) view**
  You can define the set of columns per list and hierarchy and the columns' widths and order and fix the first columns so that they are protected against horizontal scrolling. All this is captured in a personalized view, and you can switch between alternative views that you have defined. The lists also provide a grouping and aggregation functionality, which can be personalized, group objects by predefined criteria, and aggregate certain values by determining the maximum, minimum, average, or sum value for all entities in a group.

See Figure 6.33 for an example of freight unit stages shown in maximized view and grouped by unloading location and product, and aggregation of weight and volume by sum. The bottom line provides the aggregation of all objects in the list.

Navigating through lists, hierarchies

Navigation through the objects in the various lists and hierarchies is possible as follows:

- **Sorting**
  You can sort according to a sequence of columns, and for lists you can store this information in the personalized view.
- **Filtering**
  You can define filters for multiple attributes, and for lists you can store this in the personalized view. For hierarchies, filtering determines the matching objects on the first hierarchy level, considers the content of lower level nodes, too, and preserves the substructure below the filtered first-level nodes.

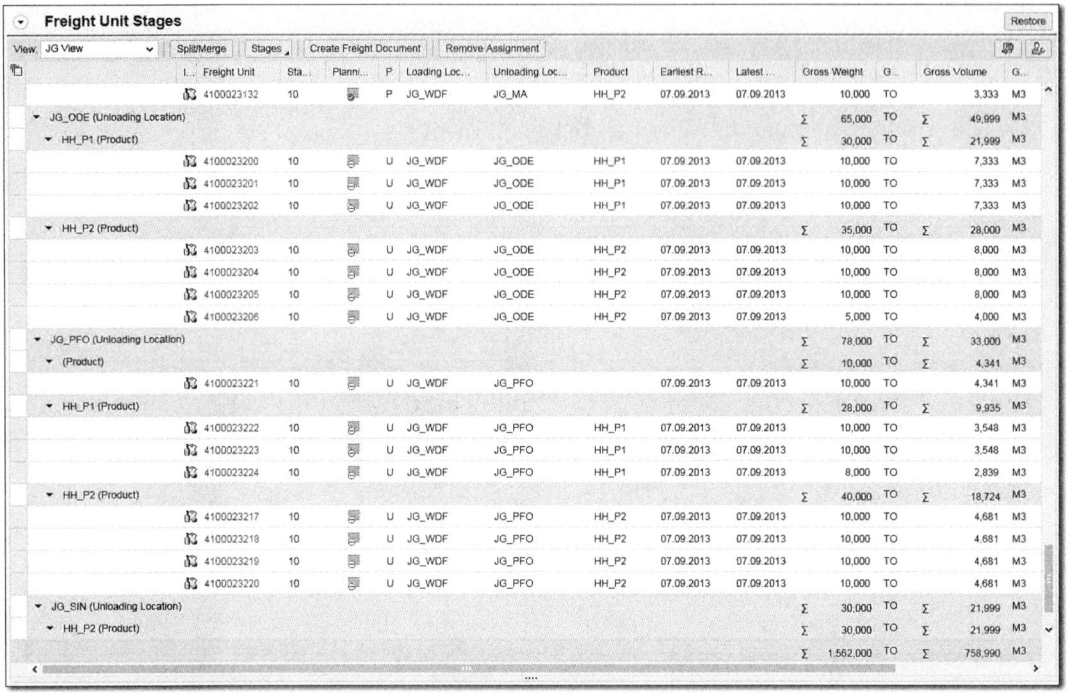

I...	Freight Unit	Sta...	Planni...	P	Loading Loc...	Unloading Loc...	Product	Earliest R...	Latest ...	Gross Weight	G...	Gross Volume	G...
	4100023132	10		P	JG_WDF	JG_MA	HH_P2	07.09.2013	07.09.2013	10,000	TO	3,333	M3
JG_ODE (Unloading Location)										Σ 65,000	TO	Σ 49,999	M3
HH_P1 (Product)										Σ 30,000	TO	Σ 21,999	M3
	4100023200	10		U	JG_WDF	JG_ODE	HH_P1	07.09.2013	07.09.2013	10,000	TO	7,333	M3
	4100023201	10		U	JG_WDF	JG_ODE	HH_P1	07.09.2013	07.09.2013	10,000	TO	7,333	M3
	4100023202	10		U	JG_WDF	JG_ODE	HH_P1	07.09.2013	07.09.2013	10,000	TO	7,333	M3
HH_P2 (Product)										Σ 35,000	TO	Σ 28,000	M3
	4100023203	10		U	JG_WDF	JG_ODE	HH_P2	07.09.2013	07.09.2013	10,000	TO	8,000	M3
	4100023204	10		U	JG_WDF	JG_ODE	HH_P2	07.09.2013	07.09.2013	10,000	TO	8,000	M3
	4100023205	10		U	JG_WDF	JG_ODE	HH_P2	07.09.2013	07.09.2013	10,000	TO	8,000	M3
	4100023206	10		U	JG_WDF	JG_ODE	HH_P2	07.09.2013	07.09.2013	5,000	TO	4,000	M3
JG_PFO (Unloading Location)										Σ 78,000	TO	Σ 33,000	M3
(Product)										Σ 10,000	TO	Σ 4,341	M3
	4100023221	10		U	JG_WDF	JG_PFO		07.09.2013	07.09.2013	10,000	TO	4,341	M3
HH_P1 (Product)										Σ 28,000	TO	Σ 9,935	M3
	4100023222	10		U	JG_WDF	JG_PFO	HH_P1	07.09.2013	07.09.2013	10,000	TO	3,548	M3
	4100023223	10		U	JG_WDF	JG_PFO	HH_P1	07.09.2013	07.09.2013	10,000	TO	3,548	M3
	4100023224	10		U	JG_WDF	JG_PFO	HH_P1	07.09.2013	07.09.2013	8,000	TO	2,839	M3
HH_P2 (Product)										Σ 40,000	TO	Σ 18,724	M3
	4100023217	10		U	JG_WDF	JG_PFO	HH_P2	07.09.2013	07.09.2013	10,000	TO	4,681	M3
	4100023218	10		U	JG_WDF	JG_PFO	HH_P2	07.09.2013	07.09.2013	10,000	TO	4,681	M3
	4100023219	10		U	JG_WDF	JG_PFO	HH_P2	07.09.2013	07.09.2013	10,000	TO	4,681	M3
	4100023220	10		U	JG_WDF	JG_PFO	HH_P2	07.09.2013	07.09.2013	10,000	TO	4,681	M3
JG_SIN (Unloading Location)										Σ 30,000	TO	Σ 21,999	M3
HH_P2 (Product)										Σ 30,000	TO	Σ 21,999	M3
										Σ 1,562,000	TO	Σ 758,990	M3

Figure 6.33 Grouped Freight Unit Stage List

- **Searching**
  In the lists, you can search for freight documents covering a specific location, containing a certain freight unit, or being assigned to a specific vehicle or trailer resource. The matching freight documents are selected in the list.
- **Expanding and collapsing hierarchies**
  You can expand or collapse all nodes or all selected nodes.

Planning session environment

It is possible to adjust the environment for your planning session as follows:

- **Change planning settings**
  This allows adjusting several parameters in the planning session. The initial values stem from the planning profile.

- **Change profile selection**
  This allows going back to the selection screen to adjust the selection and restart the cockpit.
- **Insert**
  You can insert new resources and freight units into the cockpit session.

Subcontracting

In addition to all its planning capabilities, the cockpit can also be used as a work center for subcontracting and execution tasks. You can execute the following subcontracting activities from the cockpit:

- **Carrier selection**
  You can start manual or automatic carrier selection, determine the available carriers and store them in the carrier ranking list, remove the carrier assignment, and clear the carrier ranking list.
- **Tendering**
  You can start and stop the tendering process and create requests for quotation.
- **Subcontracting**
  You can send a freight document to the carrier.

Execution

It is also possible to trigger execution activities for the freight documents at hand:

- **Calculate charges**
  You can trigger charge calculation for the selected freight documents.
- **Printing**
  You can print freight orders or review the print preview first.
- **Set status**
  For freight orders, you can update their status (e.g., *In Process*, *In Execution*, *Completed*) and set the organization interaction status for freight units, as described in Section 6.1.2.
- **Create and update the SAP ERP shipment**
  You can send the freight order information to the SAP ERP shipment. Refer to Chapter 7, Section 7.2.2 for more details on this process.

The system provides the following additional features:

- **Check**
  The system performs a check of the selected transportation units and freight documents.
- **Export to spreadsheet**
  You can export the content of a list into a spreadsheet file.

Flexible sequence of planning steps

There is no predefined sequence in which the various planning activities have to be performed. Therefore, you have full flexibility to use the most efficient and effective sequence of planning steps for your scenario.

For example, you can use the VSR optimizer to create freight documents and then manually adjust the plan. Alternatively, you can create freight documents before the cockpit session via a nightly VSR optimizer background run that assigns freight units to the freight documents or by systematically creating schedule-based freight documents for a certain time period, as described in Chapter 5, Section 5.1.2, and then manually assigning freight units to them. You can also create new schedule-based freight documents in the cockpit by choosing an appropriate schedule and departure.

Resources can be chosen by the optimizer or manually. You can first create road freight orders and then assign freight units and select a vehicle resource. Alternatively, you can assign freight units directly to a resource, thereby creating a freight document.

In truck and trailer scenarios, you may first determine the stages of the trailer units and then assign appropriate vehicle resources, but you could also first define the stages of the vehicle resources and then assign matching trailer units, or let the VSR optimizer determine the whole plan.

6.3.6 Transportation Proposal

Transportation proposal = semi-automatic planning

The transportation proposal engine automatically determines a set of alternative proposals for a given freight unit stage, thereby considering the complete transportation network definition and all the other constraints for automatic planning, as described in Section 6.4. The alternatives are presented to the user for making his choice, as shown in Figure 6.34. Alternatively, the user can refine his preferences or change some parameters and let the system again determine new transportation proposals according to his preferences and parameters. Since the transportation proposal combines automatic planning—for determining the alternatives—and manual interaction by the user choosing from those alternatives, this planning process can be called semi-automatic.

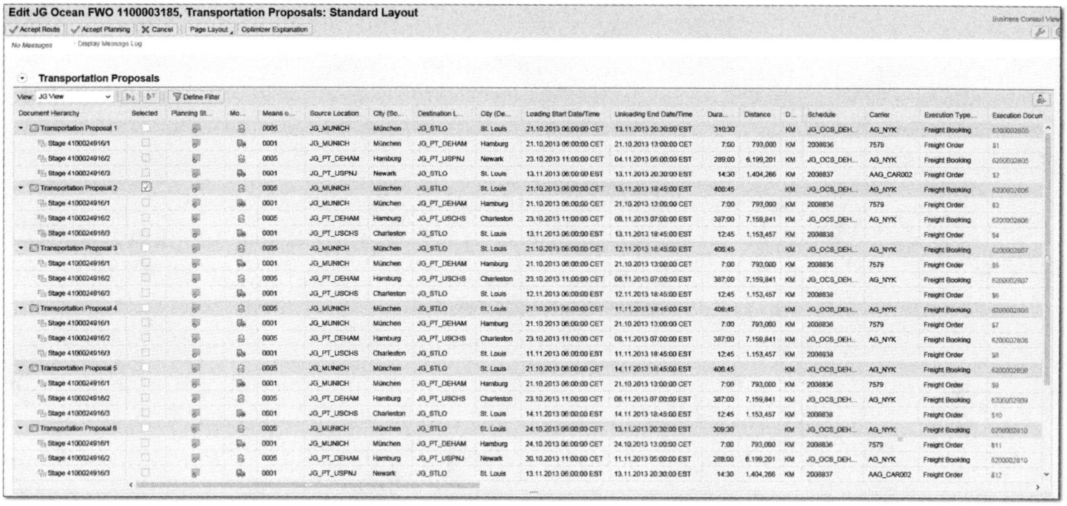

Figure 6.34 Proposal Layout Only Results

Entry from forwarding order

Similar to the transportation cockpit, there are multiple entry options for the transportation proposal user interface. You can trigger it directly from a forwarding order by defining the actual route, as described in Chapter 4, Section 4.2.1, or from the forwarding order POWL query by defining the route for a selected forwarding order. The forwarding order type Customizing allows you to declare the planning profile and the page layout for the transportation proposal. There, you can also define how the selected proposal is copied into the forwarding order, considering either only its route or its whole plan, including start and end times.

Entry from transportation cockpit

In the transportation cockpit, you can select a freight unit stage and click on the corresponding button to trigger the transportation proposal user interface. The manually chosen result is directly applied to the selected freight unit stage in the cockpit.

Another use case: ERP sales order scheduling

The transportation proposal is also used for SAP ERP sales order scheduling, which is described in Chapter 4, Section 4.1.6 and allows the determination of the total transportation duration during sales order entry in SAP ERP—that is, before creating a persistent OTR in SAP TM. In this scenario, the best transportation proposal is automatically chosen, and its dates and times are used to determine the dates and times for the SAP ERP sales order.

Page layout for transportation proposal

The transportation proposal user interface employs the page layout concept, as already mentioned in Section 6.3.3. The page layout for transportation proposals offers the following screen areas:

- The TRANSPORTATION PROPOSAL RESULT AREA contains a hierarchical display of the transportation proposals, structured with proposals on the first level, freight units on the second level (omitted if you call the proposal for one freight unit only), and stages on the third level. This screen area allows you to choose from the proposals, and hence it should be activated in the page layout. You can sort and filter the transportation proposals via the available columns. Collapsing and expanding the hierarchy allows you to compare the proposals on an aggregated level and analyze the stages of the proposals in greater detail, respectively. Personalization of the hierarchy's columns is supported, too, as in the transportation cockpit.
- The MAP AREA shows a geographical map, enabling comparison of the transportation proposals' routes on the map.
- The TRANSPORTATION PROPOSAL PREFERENCES AREA allows you to define preferences that are considered by the automatic transportation proposal determination.

Like the page layout for the transportation cockpit, one to three of these alternative screen areas can be placed on a grid with three rows and two columns, with each screen area covering either two columns or only one column. See Figure 6.35 for an example that shows the results area on top and the map on the bottom.

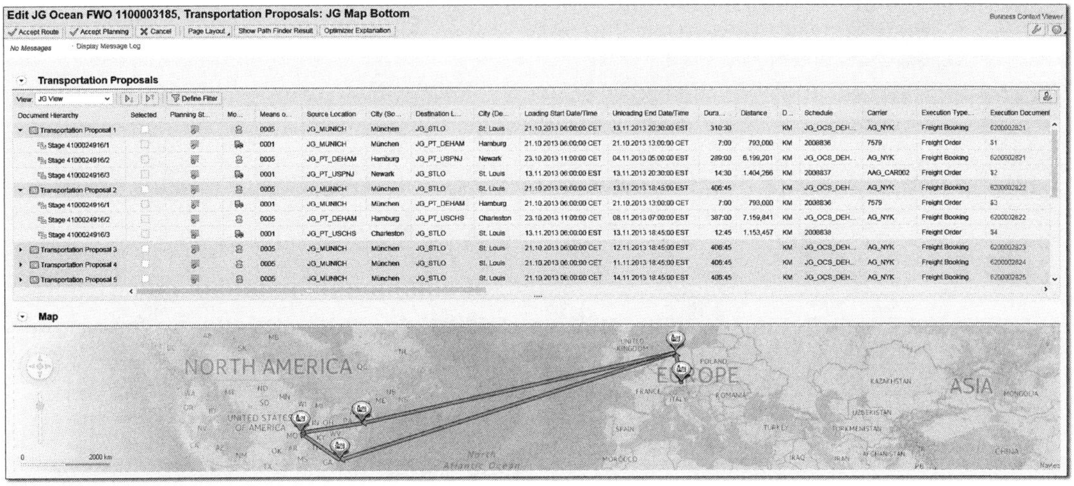

Figure 6.35 Proposal Layout Results and Map

Preferences for transportation proposal determination

The preferences can be entered before triggering a new transportation proposal determination. (See Figure 6.36 for a layout that has preferences on the top and results and map on the bottom.) Suppose that you transport goods from Germany to the United States, and you want them to go via ocean and the port in Hamburg. You can use the STAGES area in the PREFERENCES screen area to define that the transport must contain a stage from the port in Hamburg to the destination location and that ocean will be used as the mode of transport for this stage. Note that the system would still be allowed to split this stage into one main stage and a subsequent stage, which would then result in an ocean main stage and a subsequent stage that could be handled by road or rail, but not the air mode of transport. You can also predefine a sequence of transshipment locations (e.g., to force the system to determine only transportation proposals going via the ports in Hamburg and Newark).

Besides the preferences for transshipment locations and mode of transport per stage, you can also predefine a carrier and dates and times for loading at the source and unloading at the destination. The system determines proposals that have at least one stage covered by the predefined carrier, and loading and unloading does not take place earlier or later than the corresponding preferences.

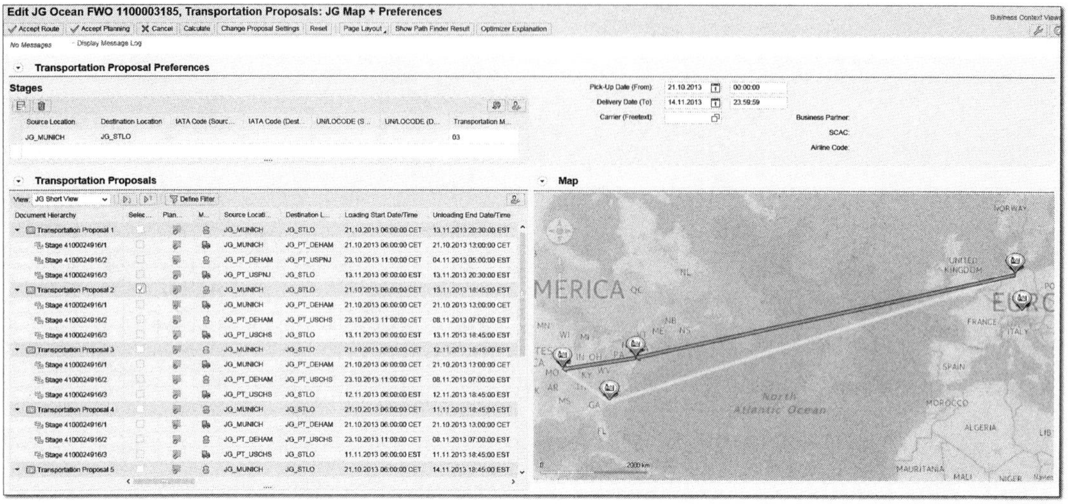

Figure 6.36 Proposal Layout Preferences Results Map

Transportation proposal UI

The VISIBILITY PUSHBUTTON section in the page layout for transportation proposals offers the following buttons:

- ACCEPT ROUTE
  This allows acceptance of only the route of the result; that is, only the stages for the forwarding order or freight unit at hand are stored according to the proposal.
- ACCEPT PLANNING
  This option allows storage of the complete plan of the proposal, including start and end times.
- CALCULATE
  This triggers a new transportation proposal determination.
- CHANGE PROPOSAL SETTINGS
  You get a pop-up with parameters, which you can change for this transportation proposal session. Initially, these parameters are defined as in the planning profile. Overruling these may be useful, perhaps if you are used to considering air freight bookings' capacities, but no satisfactory proposal is found. Then you can relax the booking capacity check to find better proposals, knowing that you can increase the corresponding capacity as necessary later.

- RESET
  This resets the preferences and allows re-starting the transportation proposal determination, which may be required if you tried several preferences but did not find a satisfactory transportation proposal.
- PAGE LAYOUT
  This allows you to switch among alternative page layouts.
- SHOW PATH FINDER RESULT
  This allows you to search for connections and display them on the map, which is helpful for analyzing your network definition. This functionality is also available in the transportation network cockpit and is described in greater detail in Chapter 3, Section 3.2.8.
- OPTIMIZER EXPLANATION
  This tool helps in understanding the input and output data of the transportation proposal determination. It is described in more detail in Section 6.4.6.

Transportation proposal parameters

The optimizer settings contain two groups of parameters dedicated to the transportation proposal. The TRANSPORTATION PROPOSAL SETTINGS section offers the following parameters:

- ACCEPT TRANSPORTATION PROPOSAL
  You can predefine whether only the route of the selected proposal result is stored or the corresponding freight documents are considered, too. Alternatively, you can keep this parameter undefined, letting the user choose between these two options in the transportation proposal user interface.
- PLANNING STRATEGY FOR TRANSPORTATION PROPOSAL
  This allows you to define whether only the transportation proposals are determined (strategy VSR_DEF) or carrier selection is performed for each transportation proposal (VSR_1STEP), as already mentioned in Section 6.3.1.
- MAXIMUM NUMBER OF TRANSPORTATION PROPOSALS
  This parameter defines how many alternative proposals are shown in the user interface. Note that automatic planning may yield fewer proposals, (e.g., if the network or capacity situation does not allow offering the defined number of proposals).

- Do Not Create Transportation Proposals Immediately
  You can control whether the proposals are already determined before you enter the transportation proposal user interface. If you want to maintain transportation preferences first, you can activate this parameter so that proposals are only calculated after you click on the Calculate button.

The Transportation Proposal Preferences section contains parameters to control the diversity of the determined transportation proposals. For each diversity criterion, you can define the relevance to be high, medium, or low, set it as not relevant, or rely on the default behavior in the transportation proposal algorithm. The following parameters are offered:

- Route Variation
  This determines the relevance of how the freight unit is routed through the network. If this aspect is relevant, the system searches for transportation proposals differing in their used transshipment locations, modes of transport, and means of transport.
- Carrier Variation
  This parameter controls whether the system searches for transportation proposals differing in their carriers. For example, if this parameter is relevant but the previous parameter is not, the system would determine the best routing through the network and search for alternative proposals along this routing that use different carriers.
- Departure Date Variation
  Defining this parameter as relevant, the system searches for proposals with alternative departure and arrival dates and times. For example, if the previous two parameters are irrelevant, the system would determine the best routing and carrier and search for alternative departure and arrival dates and times.
- Time Relevance
  In the trade-off of timely delivery versus transportation costs, this parameter defines the relevance of timely delivery. If time relevance is high but cost relevance is not relevant, the system focuses on transportation proposals that meet the freight unit's time windows as closely as possible; if no time windows have been specified, it delivers the goods as early as possible.

- Cost Relevance
  In the trade-off of timely delivery versus transportation costs, this parameter defines the relevance of transportation costs. If cost relevance is high but time relevance is not relevant, the system searches for minimum cost transportation proposals, and compromises on timely delivery are allowed (within the hard time windows of the freight units).

Transportation proposal engine

Note that the transportation proposal determination is performed by a special operating mode of the VSR optimizer, for which technical details are described in Section 6.4.2.

6.3.7 Scheduling

Scheduling determines the start and end times for a set of activities, considering multiple constraints such as a predefined relative ordering among the activities and time windows for the activities. Scheduling can be triggered for one or more selected freight documents.

Freight documents represent activities

A freight order represents a sequence of activities. For example, a freight order moving from location A to B to C may contain two freight units: one from A to B and one from A to C. This freight order represents the following activity sequence:

1. Load first freight unit at location A
2. Load second freight unit at location A
3. Transportation from location A to B
4. Unload first freight unit at location B
5. Transportation from location B to C
6. Unload second freight unit at location C

Note that the freight order could also be represented by an alternative activity sequence obtained by swapping the first two activities. However, the rest of the sequence is predefined according to the freight order's location sequence. It is important to note that scheduling does not change the location sequence of a freight order.

Truck and trailer scenarios For truck and trailer scenarios, coupling and uncoupling activities are considered, too. The coupling activity takes place just before the first joint transportation activity of the coupled truck and trailer, and the corresponding uncoupling activity is scheduled just after the last joint transportation activity. Coupling, decoupling, and transportation activities are relevant for both the truck and the trailer. Freight units loaded into the trailer are represented by the loading and unloading activities of the trailer unit.

Scheduling algorithm The scheduling algorithm can be run forward or backward, which can be set in the planning profile described in Section 6.3.2. Forward scheduling uses the first activity to be scheduled as the anchor, defines its start and end times, and then iterates through the whole activity sequence until the last activity is reached, assigning start and end times for each activity. Backward scheduling uses the last activity to be scheduled as the anchor, defines its start and end times, and then iterates through the activity sequence until the first activity is reached, assigning start and end times for each activity. For both directions, the scheduling algorithm tries to schedule the activities as compactly as possible to avoid idle times between the activities. However, time windows and other constraints may make it impossible to avoid idle times entirely.

Forward scheduling is useful if you want to push the goods out of a depot to minimize your inventory. Backward scheduling aims at meeting the delivery time window as closely as possible, and your transportation happens "just in time" before the delivery.

Scheduling constraints Scheduling considers the following constraints:

- Time windows for loading and unloading activities, as specified for the corresponding freight units and described in Section 6.1.2
- Loading and unloading durations, as specified in the planning profile described in Section 6.3.2
- Coupling and uncoupling durations, as mentioned in Section 6.4.4
- Calendars of the involved vehicle resources for loading, unloading, transportation, coupling, and uncoupling activities
- Calendar resources' and handling resources' calendars and capacities for loading and unloading activities, as described in Chapter 3, Section 3.2.1

- Minimum and maximum goods wait times defined for transshipment locations, as described in Chapter 3, Section 3.2.1
- Scheduling constraints to reflect driving time regulations, as described in Section 6.4.4

Freight unit time windows

The planning profile defines how scheduling will consider the time windows of freight units: either both hard and soft time windows are considered, only soft time windows are considered, or time windows are ignored. Considering all time windows is the most restrictive version and may cause the scheduler to fail if your time windows make it too hard to find a feasible scheduling for the activities at hand. However, not considering the time windows at all may lead to extreme solutions that fully exploit the boundaries of the planning horizon. Therefore, you should carefully specify these parameters and consider the interplay with the planning horizon to ensure that scheduling can produce the desired results. Please also refer to SAP Note 1908165 for handling time windows in the scheduling algorithm and the VSR optimizer.

Scheduling = operating mode of VSR optimizer

Note that scheduling is performed by a special operating mode of the VSR optimizer, for which Section 6.4.2 provides technical details.

6.4 Automated Planning

Zero-click process

In contrast to the previous section, our focus now is on planning processes that do *not* require user interaction. One way of limiting user interaction is to omit it completely, like in the zero-click process described in Section 6.1.3. Freight orders are created automatically instead of freight units upon the creation of their predecessor business documents, and subsequent planning steps such as carrier selection and tendering are triggered by background jobs. Thus, planning in the background is the first topic here.

Optimization

In the zero-click process freight orders are actually rule-based (created using FUBRs). This way, the planning process of creating freight orders (i.e., choosing the right vehicle resource, choosing the shortest path for the vehicle resource, and choosing the best possible time schedule for loading and unloading) is omitted. These tasks can be done manually or interactively, as described in the previous section, by using the VSR optimizer (as standard functionality) or by defining custom planning strategies. How the VSR optimizer creates transportation plans in a structured way is the main topic of this section.

Finally, automated planning is also supported in other transactions outside the planning domain. The select schedule functionality to create stages based on default routes directly in the forwarding order (as described in Chapter 3, Section 3.2) or the transportation proposal functionality (as described in Section 6.3.6) are examples of automated planning that requires some user interaction. The direct shipment option (see Section 6.1) is a means of using "real" costs in the automated planning step.

When we elaborate on planning in the remainder of this chapter, only the first planning step that creates the freight order is addressed. Subsequent planning steps such as carrier selection and tendering are discussed in detail in Chapter 7.

6.4.1 Background Planning and Vehicle Scheduling and Routing

Background planning

Background planning can be initiated by starting Transaction /SCMTMS/BACKGRD_PLAN from the menu path Application Administration • Background Reports, scheduling the report /SCMTMS/PLN_OPT, or clicking on the Start Planning in Background button in the transportation cockpit. Figure 6.37 shows the configuration options of this report. The Selection Profile (FUs) field determines the *scope*—that is, which freight units (stages) are selected for planning in the background, whereas the Planning Profile field determines the *method*—that is, the planning strategy (e.g., whether carrier selection should be included with planning strategy VSR_1STEP), available capacities, planning costs, and constraints.

Background planning is used when no or limited user expertise is required or even desired for creating a reasonable transportation plan. Thus, it can be used in simple planning scenarios (e.g., to group freight units by temperature requirements and assign them to a suitable resource). Both grouping of freight units with similar temperature requirements and their resource assignment can be expressed by the definition of incompatibilities, which then implies that freight units are sorted and assigned properly. In this type of scenario, automated planning takes the burden of manual assignments from the transportation planner, and a high volume of freight units and freight orders can be processed in a short amount of time.

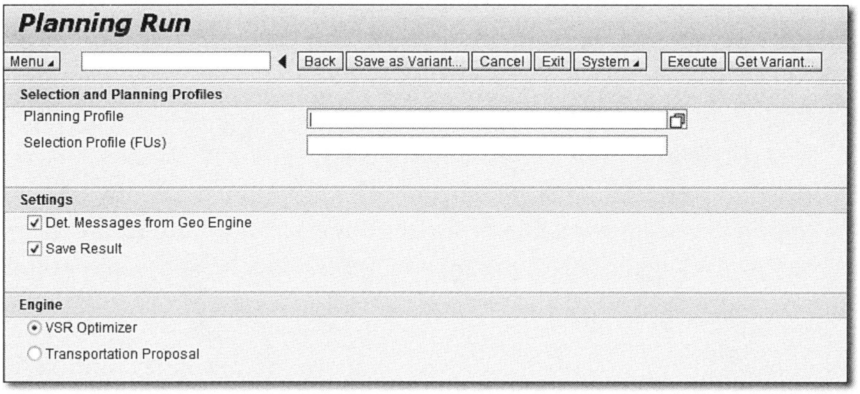

Figure 6.37 Background Planning

However, background planning can also be used in rather complex scenarios (e.g., if the consolidation of freight units into freight orders involves complex routing and scheduling decisions) because many freight units with different delivery windows based on fixed appointments have to be delivered to many destination locations with a limited number of vehicle resources. In this case, the decision situation is rather complex because of the combinatorial nature of the planning problem; the VSR optimizer can explore many more alternatives to find the best possible solution than a human planner can in the same amount of time. For this kind of scenario, background planning can be scheduled during the night, and the transportation planner can check and adapt the results in the morning upon arriving to work.

Vehicle scheduling and routing

To enable the automatic creation of a systematic transportation plan, a suitable framework needs to be defined that allows the maintenance of the objectives, rules, and constraints to which the planning result should adhere. The VSR optimizer engine provides this framework. For this purpose the VSR optimizer requires all relevant data for this task:

- **Freight units**
  Freight units (precisely, freight unit stages, because the different stages of a freight unit can be planned independently of each other in a transportation chain) are selected directly via the selection profile.
- **Transportation capacities**
  Transportation capacities are selected via the planning profile and can be vehicle resources or schedules (selected via capacity selection settings), existing freight orders (selected via the selection profile for freight orders), or freight bookings (selected via the selection profile for freight bookings).
- **Master data**
  Finally, the relevant master data (i.e., the transportation network consisting of information about locations, transportation zones, transportation lanes, and transshipment hierarchies) is retrieved based on the selected freight units and transportation capacities.

During automated planning, the system has to consider dependent objects to be able to keep all objects consistent and at the same time consider the effects of parallel processing (e.g., if other users or reports try to change the same objects in parallel).

Context determination

The selection of dependent objects is called *context determination*. It is relevant, for example, in a scenario in which new freight units are assigned to existing freight orders. In this case, the existing freight orders (created either manually or automatically) already have freight units assigned. These freight units are also relevant, because their constraints (pickup and delivery windows) or properties (required temperature conditions in transportation) need to be respected if the freight order is adapted. The same is valid in the opposite case. If a freight unit needs to be transported via several stages, some of these stages may already have been planned, so existing freight orders for some stages can apply relevant context information for the current planning scope. Finally, the context information for vehicle resources is determined to limit the timely and geographical availability of a fleet based on already scheduled freight orders.

Objects that have not been explicitly selected but have been retrieved via context determination are only for information purposes; that is, they will not be changed by the VSR optimizer, but only impose additional (side) constraints.

Parallel planning, locking

To keep planning data consistent while several users work in parallel, a locking concept is imposed to allow parallel planning. If different users or processes of the same user try to access the same object (e.g., a vehicle resource) in manual planning or in VSR optimization, a message that the object is locked is issued. Only the user or process that locked the object is allowed to change it. Subsequent processes cannot change the locked object until the lock is released. Freight bookings are exempt from this locking concept. The VSR optimizer or a transportation proposal can both access the freight booking in parallel, with both processes being able to use the remaining capacity of the freight booking. Because saving is done in an asynchronous mode, this can result in overbooking the freight booking.

Locking of Multi-Resources

Recall from Chapter 3, Section 3.2 that multi-resources are often used to represent external vehicle resources in subcontracting scenarios. If these are only available in a limited number for certain means of transport, you can specify in Customizing (Transportation Management • Master Data • Resources • Define Means of Transport) whether these multi-resources will be locked or not. However, the lock applies to all copies of the multi-resources at once.

6.4.2 Configuring Optimizer Settings

Optimizer Settings

The configuration of the VSR optimizer is primarily done in the Optimizer Settings view of the planning profile (see Figure 6.38). Defining a planning strategy is mandatory because the system needs to know whether only the VSR optimizer should be used (VSR_DEF) or a subsequent carrier selection will also be triggered (VSR_1STEP). The freight order building rule decides how freight orders are structured. If the same vehicle resource is scheduled to pick up freight units from several locations and deliver to several locations, different freight orders can be created either based on the load of the vehicle (New Freight Orders when Resource is Empty) or whenever it returns to its depot location (New FO when Resource is Empty and Depot Location Reached).

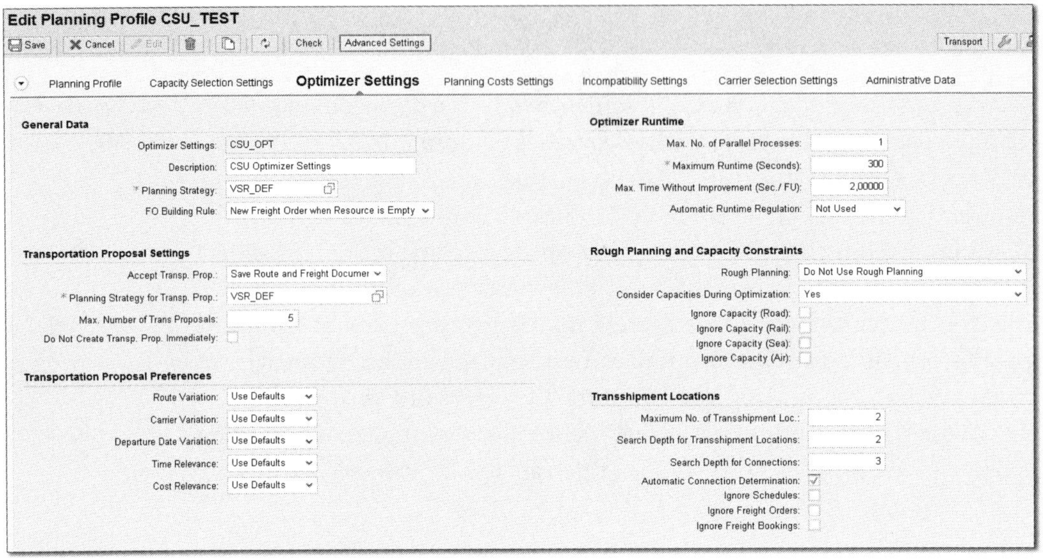

Figure 6.38 Optimizer Settings

Optimizer runtime

You have to define a maximum runtime to specify the amount of time the algorithm will use to calculate the best possible result. The required runtime depends on many factors (e.g., number of freight units to be planned, number of available vehicle resources, and complexity of the transportation network) and has to be determined during testing. It is possible to define a second termination criteria for the VSR optimizer: MAX. TIME WITHOUT IMPROVEMENT (SEC./FU). If the VSR optimizer does not improve the best solution found for the defined amount of time per freight unit, then it is automatically terminated prior to the defined maximum runtime.

Modern hardware can parallelize processes. The MAX. NO. OF PARALLEL PROCESSES field allows you to define how many parallel processes the VSR optimizer is allowed to start. Each process requires one CPU core, so this setting needs to take into account the available hardware as well as the number of parallel users that can run the VSR optimizer at the same time.

The transportation proposal settings and transportation proposal preferences refer to the use of the VSR optimizer for creating transportation proposals, as explained in Section 6.3.6.

Rough planning

Rough planning and capacity constraints deal with exact or rough duration determination in planning (rough planning) and whether vehicle capacities should impose a constraint for planning with the VSR optimizer. For many business processes, it is not important to plan complete, end-to-end transportation in detail. For example, it may be important to find the right flights, but it is known that the airport can be reached within a predefined time interval (e.g., eight hours), and planning for this stage of the journey won't require the same degree of precision. The assignment of a transshipment location to a transportation zone (that is, an entry or exit point into a transportation network) can be used to allow planners to specify a duration that can be used for *rough planning* instead of the exact distance and duration determined based on transportation lanes. This means that a detailed transportation network (transportation lanes) is not required for the pre/on-carriage in this kind of scenario.

Transshipment locations

The parameters for transshipment locations influence the complexity of the transportation network and therefore have a significant impact on the amount of time required to calculate a reasonable planning result. The Maximum No. of Transshipment Loc. field defines the number of transshipment locations any freight unit is allowed to be routed through between its source and destination. It should be as small as possible to limit the number of paths in the transportation network that the VSR optimizer has to consider as possible alternatives.

To ease maintenance of the transportation network, you don't need to define transshipment locations that are part of a schedule explicitly as transshipment locations, but automatic connection determination can be activated in the optimizer settings. This implies that all possible connection points (stops of flight or sailing schedules) are implicitly considered transshipment locations (see Chapter 3, Section 3.2).

Technical configuration

Technical Configuration of the VSR Optimizer

The VSR optimizer is a separate piece of software that has been developed for performance reasons in C++ and not in ABAP. Technically, the optimizer is called via the *remote control and communication framework* (RCCF) using an RFC. You find relevant transactions for the technical configuration at Application Administration • General Settings • Remote Control and Communication Framework.

Edit destinations

In the Edit Destinations transaction (see Figure 6.39) you can define which hardware you have available for the VSR optimizer and the other external engines and how these can be reached. The following applications are other external engines that use the same framework:

- TSPS – Carrier selection
- TVRG – Transportation proposal
- TVSR – VSR optimizer
- TVSS – Manual scheduling
- TVSO – Load optimization

The COMMUNICATION CONNECTION column defines the RFC connection and is defined in Transaction SM59 (TCP/IP CONNECTIONS). The MAX. SLOTS column refers to the number of CPU cores available at the destination server. If these are already in use and an additional optimization run is started, it will be canceled immediately because no hardware is available to process the request. If several destinations (DEST. ID) are defined for the same application (APPL.), the PRIORITY column determines the sequence that the VSR optimizer tries to use for these servers.

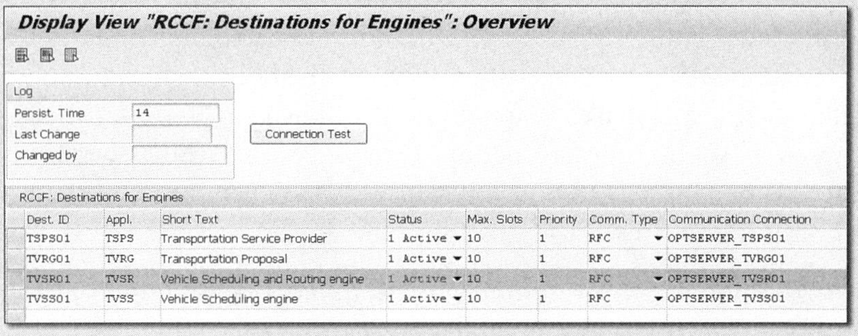

Dest. ID	Appl.	Short Text	Status	Max. Slots	Priority	Comm. Type	Communication Connection
TSPS01	TSPS	Transportation Service Provider	1 Active	10	1	RFC	OPTSERVER_TSPS01
TVRG01	TVRG	Transportation Proposal	1 Active	10	1	RFC	OPTSERVER_TVRG01
TVSR01	TVSR	Vehicle Scheduling and Routing engine	1 Active	10	1	RFC	OPTSERVER_TVSR01
TVSS01	TVSS	Vehicle Scheduling engine	1 Active	10	1	RFC	OPTSERVER_TVSS01

Figure 6.39 Remote Control and Communication Framework: Edit Destinations

Engine Debug configuration

The PERSIST. TIME field specifies how many days the planning logs are kept in the system before they are deleted. The logs can help you analyze the planning result and are explained in Section 6.4.6 in more detail. The logs are also required for SAP Support to reproduce VSR optimizer behavior in case of an error in any of these engines. For this purpose, set TRACE LEVEL to INFO and DUMP LEVEL to 1 in the Engine Debug Configuration transaction (see Figure 6.40).

The correct installation and technical setup of all destinations can be checked in the Version Display transaction. If a version is displayed as VERSION INFORMATION, then everything is fine. Note that the engines are downward-compatible and therefore have to be of the same or a newer version and support pack than the SAP TM application.

The Display Active Sessions transaction provides a monitoring tool for currently running sessions of all engines using the remote control and communication framework. This transaction can also be used to terminate background runs.

Finally, the optimization data can be manipulated prior to the engine run, when the relevant data is sent to the optimizer (e.g., to add additional constraints on the fly), and after the engine run (to adapt the results). This is done in BAdI `/SCMTMS/PLN_PRE_PROC` to preprocess the optimization data and in BAdI `/SCMTMS/PLN_POST_PROC` to postprocess the planning result.

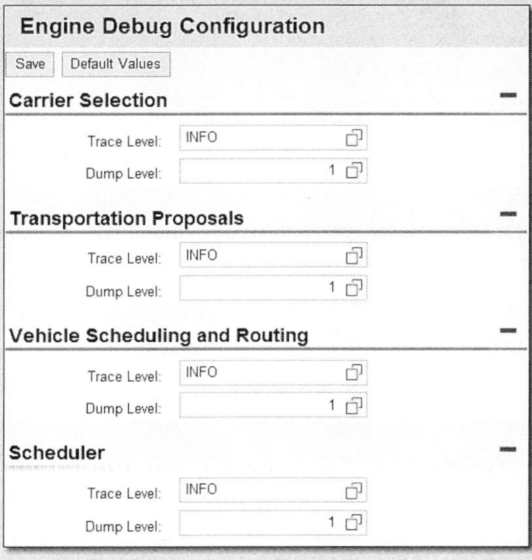

Figure 6.40 Remote Control and Communication Framework: Engine Debug Configuration

6.4.3 Vehicle Scheduling and Routing: Planning Objectives

Planning objective

For the VSR optimizer engine to create a reasonable transportation plan, you need to define an objective. Individual objectives may be different in different implementation projects or even different for different planning situations in the same implementation. This can be illustrated by a simple example.

Assume a simple planning situation with only one source location (plant) and only one delivery location (customer). The customer orders the equivalent of half a truck of goods every day. In this situation, there are two possible solutions from a transportation planning point of view:

1. Every day a truck delivers the goods from the plant to the customer. However, this truck is only half full (or half empty). Though this solution is not very efficient because of the low truck utilization, it provides a high service level because the customer is served exactly to his needs.
2. The customer is served every other day with a full truckload. This solution is very efficient from transportation perspective (100% truck utilization), but the customer service level is poor because the customer is provided 50% of his orders either one day early or one day late based on the delivery pattern.

Which solution is better and will be created from the VSR engine? The answer depends on the individual situation; either one may be preferred. The first solution may be preferred if there are inventory/warehousing constraints at the customer location or if the shelf-life of goods does not allow their storage but forces daily deliveries. The second solution is preferred in competitive situations if cost is the driving decision criteria. The VSR engine can create both solutions, and which one is returned depends on the defined objective. Thus, the relation between different cost elements must reflect the user's business objectives to make it possible to calculate the right solution.

Cost minimization

The VSR optimizer is governed by a cost minimization objective. The cost elements are dependent on the freight units as well as on the transportation capacities. Freight unit-dependent cost elements influence the service level (timely pickup and delivery), whereas transportation capacity-related cost elements influence their efficient usage. In the previous example, the first solution will be returned if the freight unit-dependent costs for earliness and lateness are high compared to the transportation capacity-related cost, because in that case it would be expensive to violate the delivery window for half of the goods. The second solution will be returned if the transportation capacity-related costs are higher.

Let's take a look at which cost elements are considered. Most of them are defined in the PLANNING COSTS SETTINGS screen (see Figure 6.41 and Figure 6.42).

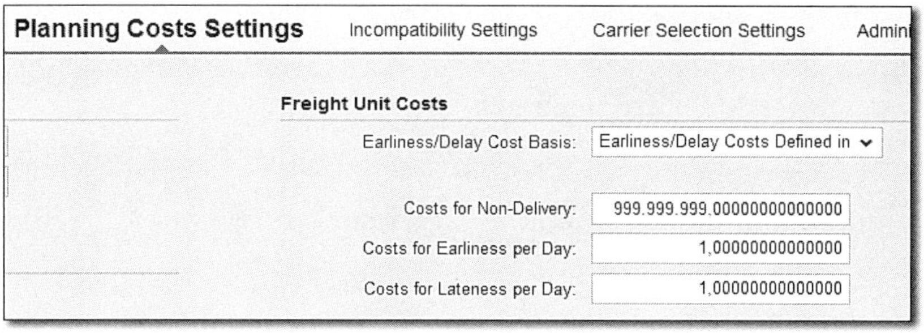

Figure 6.41 Planning Costs: Freight Unit Costs

Freight unit cost

Freight unit costs can be defined either directly in the PLANNING COSTS SETTINGS screen or via a condition (type /SCMTMS/FU_PNLT_COST) that is assigned in that screen. If a condition is used, different costs can be defined based on the characteristics of each freight unit (e.g., different lateness penalties depending on customer priority) or based on whether pickup or delivery windows are violated.

- Costs for non-delivery are incurred in the cost objective if a freight unit is not transported. By default, they are set very high to force the VSR optimizer to find a solution because non-delivery is usually not a valid option.
- Costs for earliness per day apply to both early pickup and early delivery of a freight unit. The duration of the earliness is calculated as the interval between the scheduled time of the pickup/delivery and the requested date to start (see Section 6.1.2).
- Costs for lateness per day also apply to both late pickup and late delivery of a freight unit. The duration of the lateness is calculated as the interval between the scheduled time of the pickup/delivery and the requested date to end (see Section 6.1.2).

Means of transport cost

Means of transport costs define the transportation capacity-related costs and can be maintained directly in the PLANNING COSTS SETTINGS screen (see Figure 6.42) and to some extent also in transportation lane master data (see Chapter 3, Section 3.2).

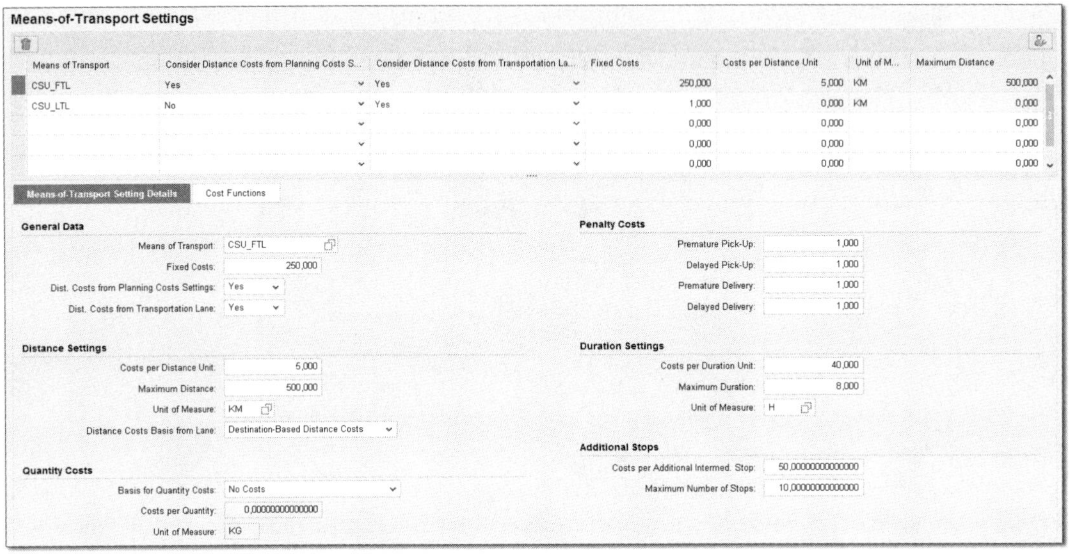

Figure 6.42 Planning Costs: Means of Transport Settings

Means of transport-specific costs can be defined differently for all means of transport in the scope of the planning scenario:

- *Fixed costs* are incurred in the objective function once per freight order. Therefore, fixed costs can be used to minimize the number of vehicle resources used or to select between different available vehicle resources.
- *Penalty costs* are factors that apply to freight unit-dependent earliness and lateness costs. These factors can be used to distinguish earliness and lateness based on the means of transport. For example, violating a pickup window for an ocean transport is much more expensive than for a truck transport.

Distance costs

- *Distance-dependent* costs per means of transport can be maintained in the PLANNING COSTS SETTINGS screen or in the transportation lane (MNS OF TRSP. COSTS field). The actual distance of the freight order is multiplied by the costs per distance unit (see Chapter 3, Section 3.2.3 for the determination of transportation distances). In North America and other parts of the world different concepts for calculating distance-dependent costs exist. In North America, these are known as *destination-based distance costs*, and in Europe, they are *route-based distance costs*.

Figure 6.43 provides an example that illustrates the different calculation concepts, which can even result in different optimal solutions. Destination-based distance costs are calculated by multiplying the actual distance of the complete freight order with the distance cost from the transportation lane between the source and the destination. Route-based distance costs are calculated by multiplying the actual distance of each stage with the cost of the transportation lane for this stage and summing up the costs for the individual stages.

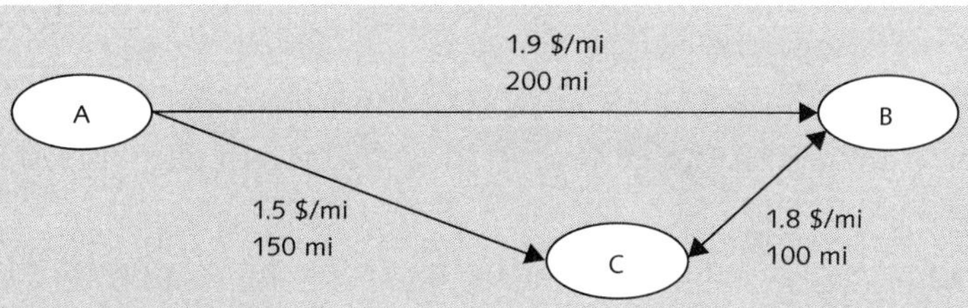

Destination-based cost:
Freight order (A→B→C) = Distance (A→B→C) * Cost (A→C) = 300 * 1.5 = 450 $
Freight order (A→C→B) = Distance (A→C→B) * Cost (A→B) = 250 * 1.9 = 475 $

Route-based cost:
Freight order (A→B→C) = Distance (A→B) * Cost (A→B) + Distance (B→C) * Cost (B→C) = 200 * 1.9 + 100 * 1.8 = 560 $
Freight order (A→C→B) = Distance (A→C) * Cost (A→C) + Distance (C→B) * Cost (C→B) = 150 * 1.5 + 100 * 1.8 = 405 $

Figure 6.43 Destination-Based Cost vs. Route-Based Cost Calculation

Finally, a *minimum cost* can be defined that applies to distance costs calculated for a freight order. A typical example would be a scenario in which freight costs are usually variable based on the distance (e.g., $2/mi), but a minimum of $300 applies because short trips less than 150 miles are not economical with only variable rates. The minimum cost can be defined in the transportation lane master data (Min. MTr Costs field).

Distance costs are usually minimized to lower fuel consumption or reduce operating times of the vehicle resources.

- *Duration costs* refer to the actual use of vehicle resources. They are incurred from its first use (e.g., loading) until its last use (unloading).

The duration is multiplied with the cost per duration unit. Duration costs can be used to minimize operating times, perhaps to create a compact transportation plan with limited idle time.

Quantity costs

- *Quantity costs* based on a unit of measure (e.g., kilogram or cubic meter) can be incurred into the objective function or not (NO COSTS selected in the BASIS FOR QUANTITY COSTS dropdown box). If quantity costs are considered, they can be distance-independent or distance-dependent. For distance-dependent quantity costs, the freight order quantity is multiplied by the actual distance. This calculation is done stage by stage. Additionally, you can define quantity costs universally in the PLANNING COSTS SETTINGS screen or based on the geography in the transportation lane.

  Transported quantities can have an impact on fuel consumption and may therefore be considered distance-dependent cost elements in the objective.

- *Additional stop costs* can be used to minimize the number of stops in a freight order. The costs per additional intermediary stop are incurred for each stop that is not the source or destination of the freight order. For example, in a freight order from A via B to C, the costs per additional intermediary stop would be incurred once for the intermediate stop at B.

Cost functions

- You can define *cost functions* by following the menu path APPLICATION ADMINISTRATION • PLANNING • COST FUNCTION SETTINGS • CREATE COST FUNCTION SETTINGS and assigning them to the means of transport in the PLANNING COSTS SETTINGS screen. A cost function is a stepwise linear function that is referenced to a unit of measure when assigned to a means of transport (see Figure 6.44). The cost function is intended to load vehicles in a more efficient way by associating a cost with the vehicle resource that is dependent on the load of the resource in the referenced unit of measure.

  In the example in Figure 6.44, a load cost of 10,000 is defined if the vehicle is loaded with less than 15,000 kg. From 15,000 kg the load cost is 500 and decreasing linear to 0 at the vehicle capacity of 20,000 kg. With this cost pattern, the VSR optimizer would try to load a vehicle resource at least with 15,000 kg.

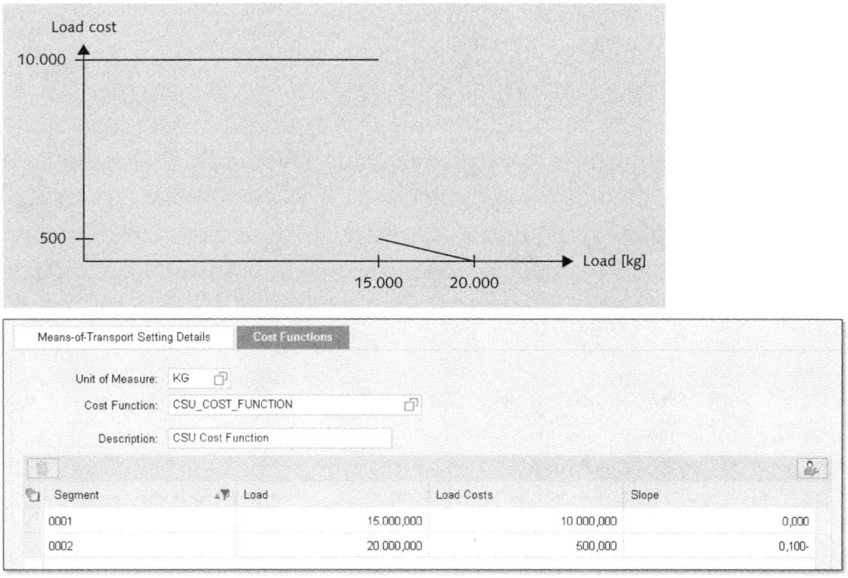

Figure 6.44 Cost Function

6.4.4 Vehicle Scheduling and Routing: Planning Constraints

Planning constraints

The VSR optimizer tries to minimize all different cost elements introduced in Section 6.4.3. The creation of a minimal cost transportation plan also has to adhere to certain constraints. These planning constraints model the physical or execution-related restrictions of the real world in the planning system. For example, because of physical or legal restrictions, trucks can only transport a certain load, which is expressed as a maximum mass, maximum volume, and/or maximum number of pallets that needs to be considered as a capacity constraint in the freight order. Many constraints are available in the VSR optimizer, and we'll go into them. However, it is important to note that good modeling practice is not to constrain a planning problem too much, but to use only those constraints that influence the planning decision.

Capacity constraints

Capacity constraints are important in transportation planning and are considered to be hard constraints; that is, the VSR optimizer will not create any solution that violates capacity constraints. The VSR optimizer would rather decide not to transport a freight unit (and incur a high non-delivery cost) than to violate a capacity constraint. Capacity constraints can be expressed in vehicle resource master data (see Chapter 3, Section 3.3). For each vehicle resource, the VSR optimizer can consider up to eight different capacity dimensions. The prevailing dimensions are mass, volume, and floor space (number of pallets). Only those dimensions are considered in planning that have also been defined as capacity requirements in the freight unit (that is, those that have been defined as planning quantities in the FUBR; see Section 6.1.3). In addition to capacity constraints, the temporal availability defined in the resource master data is considered as a hard constraint by the VSR optimizer.

Capacity constraints are observed on several levels, as illustrated in Figure 6.45. The lowest level is the compartment level. In Figure 6.45, both the vehicle resource and the trailer are physically structured into four compartments (e.g., tanks, C1 to C8). You may need to take this property of the vehicle into consideration, for example, if different liquids that cannot be mixed have to be transported. Each compartment provides a capacity constraint. In addition, each vehicle resource (the vehicle and the trailer) provides a capacity constraint, and finally the vehicle combination can also have a capacity constraint. Capacity constraints on different levels can be defined independently from each other and are considered independently from each other during planning. They are not derived from the other levels.

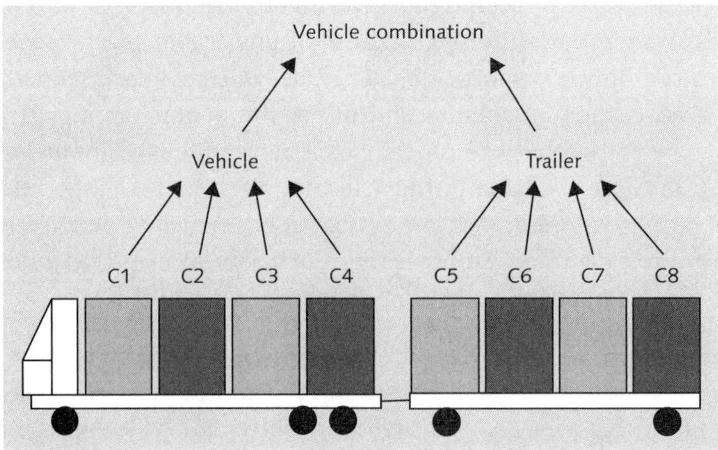

Figure 6.45 Capacity Constraints: Compartment, Vehicle Resource, and Vehicle Combination

Compartments

Compartments are defined in Customizing via the menu path TRANSPORTATION MANAGEMENT • MASTER DATA • RESOURCES • DEFINE COMPARTMENT TYPES in the following way:

1. A compartment type is defined by its capacity (in several dimensions) and attributes. Attributes can be used in conjunction with incompatibilities to control whether only liquids or goods that require special temperature conditions can be loaded into the compartment. In Figure 6.45, different colors indicate that the vehicle consists of four compartments with two different compartment types.
2. In the same Customizing transaction, compartment profiles are defined to combine the compartment types with a vehicle configuration. In Figure 6.45, the compartment profile indicates that the vehicle has two compartments of compartment type 1 and two compartments of compartment type 2.
3. Finally, the compartment profile is assigned to the means of transport (in the same Customizing transaction). In the example in Figure 6.45, the same compartment profile can be assigned to the means of transport representing the vehicle resource and to the means of transport representing the trailer.

Compartments are frequently used in planning scenarios that involve liquids or in retail distribution to meet the temperature conditions of transported goods.

Vehicle combinations

Vehicle combinations are defined in Customizing via the menu path TRANSPORTATION MANAGEMENT • MASTER DATA • RESOURCES • DEFINE MEANS-OF-TRANSPORT COMBINATION in the following way:

1. Means of transport combinations consist of exactly one means of transport that is *not* defined as passive (TRANSPORTATION MANAGEMENT • MASTER DATA • RESOURCES • DEFINE MEANS OF TRANSPORT) and any number of means of transport that have been defined as passive. Any means of transport that is used in a means of transport combination cannot be defined as multi-resource because planning requires tracking individual resources in these scenarios.
2. Similar to compartments, a means of transport combination is assigned capacities (in different dimensions) and attributes.

By planning with the VSR optimizer, each vehicle resource with a means of transport used in a means of transport combination can be used in a vehicle combination and be coupled and uncoupled to other vehicle resources based on any defined means of transport combination. Coupling and uncoupling durations are defined in Customizing via the menu path Transportation Management • Master Data • Resources • Define Coupling/Uncoupling Duration.

In the location master data (Master Data • Transportation Network • Locations • Define Location • TM), you can define how coupling and uncoupling of trailers is to be handled at a location. A few options are available: to allow coupling/uncoupling activities at the location, *not* to allow these activities, to allow them only if freight units are picked up from or delivered to the location at the same time, or whether only trailer swaps are allowed at this location (see Section 6.2.2). The VSR optimizer supports these scenarios and the creation of trailer documents as part of the automated planning process.

Planning with vehicle combinations is used if trailers are frequently exchanged between tractors or if the number of tractors and trailers deviates, because loading and unloading takes a significant amount of time compared to driving, and a tractor can pull other trailers while one is being loaded or unloaded. However, the use of means of transport combinations introduces a lot of complexity into the planning scenario because tractors and trailers have to be planned independently from each other. Therefore, modeling this constraint should be avoided if possible. Note that this feature is limited to truck and trailer scenarios (e.g., one tractor and one or a few passive resources) and is not intended for building trains in a railway scenario.

Capacity constraints are not limited to compartments, vehicle resources, and vehicle combinations. Schedules and freight bookings can also provide capacity in planning.

Decreasing capacities

Decreasing capacities are a capacity constraint that sometimes exists in retail scenarios. If a truck delivers goods to several stores, the goods in the truck are loaded sequentially per store and a separator (thin wall) is used to separate the goods for each store from each other. However, because of the separator, some loading space cannot be used and is lost.

How much space is lost depends on the number of separators—that is, on the number of stores planned to be delivered to with the vehicle. Thus, the available vehicle capacity is not a fixed value, but depends dynamically on the transportation plan. Decreasing capacities are a planning constraint that addresses this situation and defines how much the vehicle capacity is decreased based on the number of stops on the freight order. Decreasing capacities are defined per means of transport via the menu path Application Administration • Planning • General Settings • Decreasing Capacity Settings • Create Decreasing Capacity Settings.

We introduced pickup and delivery time windows in Section 6.1.2. The acceptable start and end dates of both the pickup and delivery time window are hard constraints for the VSR optimizer and are never violated. The requested dates express preferences (soft constraints) and are considered in the objective of the VSR optimizer to incur earliness and lateness costs for freight units (see Section 6.4.3).

Loading and unloading

Pickup and delivery imply that the freight units have to be loaded onto and unloaded from transportation capacities. These activities imply additional constraints for planning. First, loading and unloading take time. Loading and unloading durations can be defined directly in the planning profile if they are independent from the freight unit and means of transport. In this case, the maintained time applies to all loading and unloading activities of all freight units in the scope of planning. However, loading and unloading usually depend either on the freight unit (e.g., freight units can represent a USB stick or a container) or on the means of transport (loading a truck versus loading an ocean vessel) or both. In this case, the loading/unloading duration can be defined using a condition (type /SCMTMS/FU_LOAD_DURA) that is then assigned in the planning profile.

Loading and unloading are activities that require not only the vehicle resource, but may also require availability at the location. Some locations may only be open at certain times (e.g., weekdays from 9 a.m. to 5 p.m.). Other locations may limit the number of parallel loading/unloading activities because of limited ramps or personnel available to load and/or unload vehicles.

To represent time constraints, you can define a calendar resource in master data (Master Data • Resources • Define Resource) and assign it to the location in location master data (Master Data • Transportation Network • Locations • Define Location • Resources). You can assign different calendar resources as operating times for inbound (unloading) and outbound (loading) activities. In addition, you can assign operating times that are specific to a means of transport. If the number of parallel loading and unloading activities is restricted, you define handling resources in master data (Master Data • Resources • Define Resource) and assign it to the location in location master data (Master Data • Transportation Network • Locations • Define Location • Resources). Similar to opening times, you can distinguish between inbound and outbound and also define these constraints as being means of transport specific (see Chapter 3, Section 3.2.1 and Section 3.3 for the definition of master data).

Transshipment locations Section 6.4.2 introduced the maximum number of transshipment locations in a transportation chain as a constraint that drives complexity. Transshipment locations are defined in the transshipment hierarchy (Master Data • Transportation Network • Locations • Assign Transshipment Location; see also Chapter 3, Section 3.2.5). This constraint helps the VSR optimizer limit the number of possible alternatives to transport a freight unit from its source to its destination. However, if the constraint maintained is too low, no feasible alternative may be found. Using transshipment locations (e.g., ports or airports) often implies a second constraint that is the cutoff time at the transshipment location. If a flight takes off at 2:15 p.m., it is usually too late to deliver freight to the airport at 2:10 p.m. Thus, a minimum cutoff time should be respected. On the other hand, delivering freight to the airport one week in advance is also not an option because this may incur additional costs or not be accepted at all. Therefore, minimum and maximum cutoff times can be defined either in the location master data (Master Data • Transportation Network • Locations • Define Location • TM) or in schedules (Master Data • Transportation Network • Schedule • Create Schedule).

Depot location If a fleet is in the scope of transportation planning, then depot or home locations of vehicle resources may be in the scope. If trucks are parked in depot D overnight and need to transport a freight unit from A to B the next day, then the freight order has to include the empty runs from D to A and from B back to D because these stages also require time and incur cost.

Depot locations are maintained in the resource master data (MASTER DATA • RESOURCES • DEFINE RESOURCE) for each vehicle resource. They are only considered as a constraint for means of transport that are flagged as YOUR OWN MTR and are not flagged as MULTIRESOURCE in Customizing (TRANSPORTATION MANAGEMENT • MASTER DATA • RESOURCE • DEFINE MEANS OF TRANSPORT).

Distance, duration, stop limits

If planning local delivery tours from a depot is in the scope of the planning scenario, there may be an additional constraint in place that each tour should last at most eight hours, reflecting the working times of the drivers. This type of constraint can also be considered by the VSR optimizer. Essentially, the optimizer is allowed to impose limits on one or all of the following:

- Total duration of a freight order
- Total distance of a freight order
- Number of stops of a freight order

These limits are separate constraints and can be used independently from defining depot locations. For each means of transport, these limits can be maintained in the means of transport settings of the planning costs settings (see Figure 6.42).

Scheduling constraints

Chapter 3, Section 3.2 explained how transportation duration is calculated based on the transportation distance. However, this is not a very precise calculation and cannot be used universally, which is obvious from even a simple example.

To calculate the transport duration for a container ship from Rotterdam to Boston, you can take the distance (3,200 nautical miles) and divide it by the speed (20 knots/hour) to calculate the expected duration to be 160 hours (approximately one week). The same calculation logic will fail for a truck going from Boston to Los Angeles. The distance (3,000 miles) divided by the truck speed (50 miles/hour) would indicate that the truck would arrive after 60 hours (2.5 days) at its destination.

However, in contrast to the ship, which sails day and night, the truck (driver) has to take some breaks that prolong the transportation duration. If the driver were allowed to drive 10 hours per day, the trip from Boston to Los Angeles would take 6 days instead of 2.5 days. Scheduling constraints can be defined to take those breaks into account. As shown in Figure 6.46, scheduling constraints are defined in Customizing (TRANSPORTATION MANAGEMENT • PLANNING • GENERAL SETTINGS • DEFINE SCHEDULING CONSTRAINTS) in the following way:

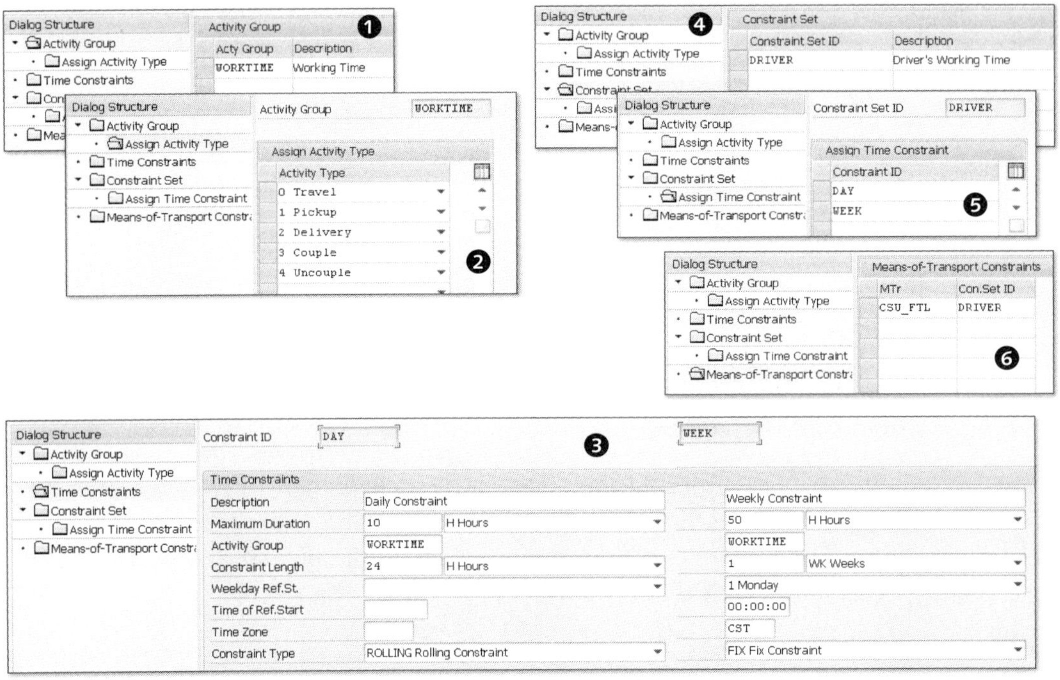

Figure 6.46 Scheduling Constraints

❶ Define an activity group.

❷ Assign activity types to the activity group. The available activity types include travel, pickup (loading), delivery (unloading), couple (vehicle combination), and uncouple (vehicle combination). Depending on the origin or purpose of this planning constraint in a specific scenario, either all activity types or just a subset may be relevant for the scheduling constraint. This step maintains the relevant activity types.

❸ Several time constraints can be maintained in the third step. In the example shown in Figure 6.46, two time constraints are defined. Constraint ID DAY represents a constraint in which the maximum duration of activities of activity group WORKTIME, defined in step ❷, is 10 hours in any 24-hour time interval (constraint length). The constraint type is defined as a rolling constraint; that is, it does not apply per workday, but applies to any 24-hour period. If it were defined as a fixed constraint, it would be possible to schedule activities during the last 10 hours of day 1 and during the first 10 hours of day 2, which would allow 20 hours of uninterrupted driving. The second constraint ID, WEEK, allows 50 hours of activities within one calendar week.

❹ Because both constraint IDs should be considered together, a constraint set is defined in this step.

❺ Both constraints are assigned to the constraint set in this step.

❻ Finally, the constraint set is assigned to a means of transport.

Incompatibilities

We've already mentioned incompatibilities as constraints because they are used in several areas of the application (see Section 6.1.3; Chapter 4, Section 4.1; and Chapter 7):

- Freight unit building
- VSR planning (manual, transportation proposal, optimizer)
- Carrier selection
- Delivery proposal

Incompatibilities define which two objects are not compatible with each other. In VSR planning the following incompatibility types exist:

- **Freight unit—freight unit (vehicle level)**
  Two freight units are not allowed on the same vehicle (e.g., because they may be explosive if mixed in an accident).
- **Freight unit—freight unit (compartment level)**
  Two freight units are not allowed in the same compartment (e.g., chemicals should not be mixed with milk in a tank).
- **Freight unit—freight unit (means of transport combination)**
  Two freight units are not allowed on the same vehicle combination (e.g., because they may be explosive if mixed in an accident).

- **Freight unit—vehicle resource**
  A freight unit is not allowed on a vehicle (e.g., chemicals are not allowed in a truck intended for food transport).
- **Freight unit—transshipment location**
  A freight unit is not allowed to be routed via a transshipment location (e.g., a heavy turbine cannot be routed through ports that do not have cranes to lift it).
- **Freight unit—vehicle compartment**
  A freight unit is not allowed on a vehicle compartment (e.g., chemicals are not allowed in a milk tank).
- **Vehicle resource—vehicle resource**
  Two vehicle resources cannot be coupled, although in general it would be allowed by the means of transport combination definition.
- **Vehicle resource—location (stay level)**
  A long vehicle resource cannot visit certain locations (e.g., because of maneuvering limitations) on its route.
- **Vehicle resource—location (loading/unloading level)**
  Locations that do not have loading and unloading equipment must not be visited by vehicles that do not carry their unloading equipment (e.g., crane, forklift) themselves.
- **Vehicle MTR combination—location**
  A long vehicle combination must not visit certain locations (e.g., because of maneuvering limitations) on its route.
- **Freight unit—booking**
  A freight unit containing hazardous goods cannot be booked on some freight bookings.
- **Freight unit—schedule**
  A freight unit containing hazardous goods cannot be booked on some schedules.

Incompatibilities are defined in incompatibility definitions (Application Administration • Planning • General Settings • Incompatibility Definitions; see Figure 6.47). The incompatibility definition determines the incompatibility area and the incompatibility type. It also defines whether violations of the incompatibility are allowed, will result in a warning, or are forbidden. This decision can be made separately for manual planning and automated planned—perhaps to forbid violations in automated planning but to allow a manual override of the decision, resulting in a warning only. The comparison of the evaluation result of two conditions that are evaluated for the two objects defined in the incompatibility type determines whether these two objects are compatible with each other.

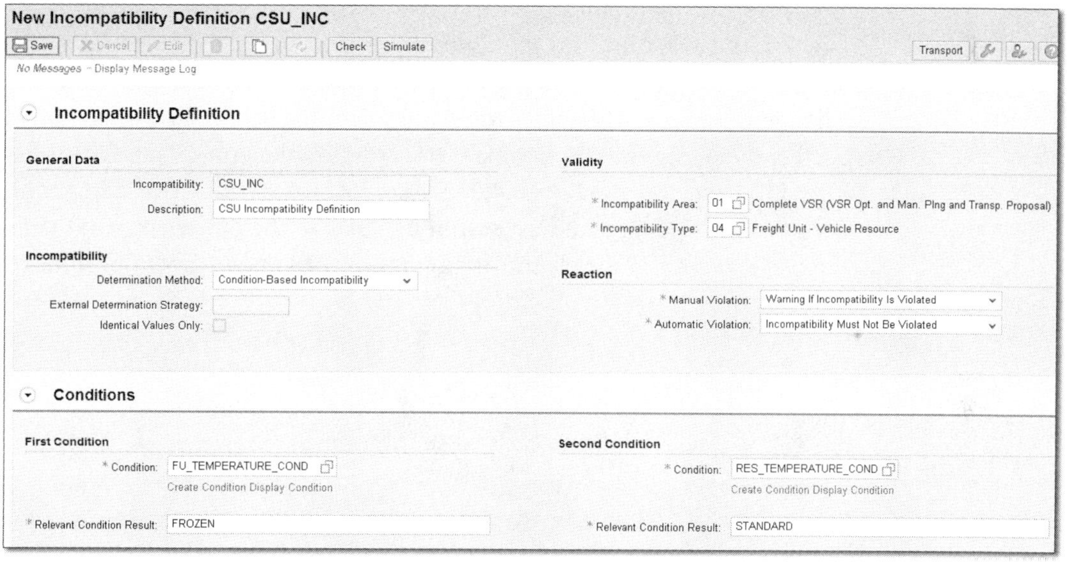

Figure 6.47 Incompatibility Definition

Incompatibility settings (APPLICATION ADMINISTRATION • PLANNING • GENERAL SETTINGS • INCOMPATIBILITY SETTINGS) are a list of incompatibility definitions that are grouped together because all of them should be considered in a specific planning scenario. The incompatibility settings are assigned to the planning profile.

An overview of all defined incompatibility definitions and incompatibility settings is available in a worklist using one of the queries for general settings (APPLICATION ADMINISTRATION • PLANNING • OVERVIEW SETTINGS).

6.4.5 Vehicle Scheduling and Routing: Optimization Algorithm

VSR optimizer algorithm

The VSR optimizer combines the ideas of several metaheuristics in a population-based optimization algorithm that tries to determine reasonable transportation plans within an acceptable runtime. The algorithm is based on the basic principle of evolutionary local search, so a population of candidate solutions is subject to an evolutionary search process by iterative selection and variation. The initial population is created by several insertion heuristics that iteratively assign all freight units to transportation capacities. In the improvement phase, different variation operators either reassign freight units to other transportation capacities, change the routes of freight orders, or adapt the scheduling of activities. In this process, the VSR optimizer tries to minimize total costs defined as the objective (see Section 6.4.3) while respecting all active constraints (see Section 6.4.4). The best solution found within the maximum runtime (see Section 6.4.2) is returned. Figure 6.48 shows how the solution evolves over time.

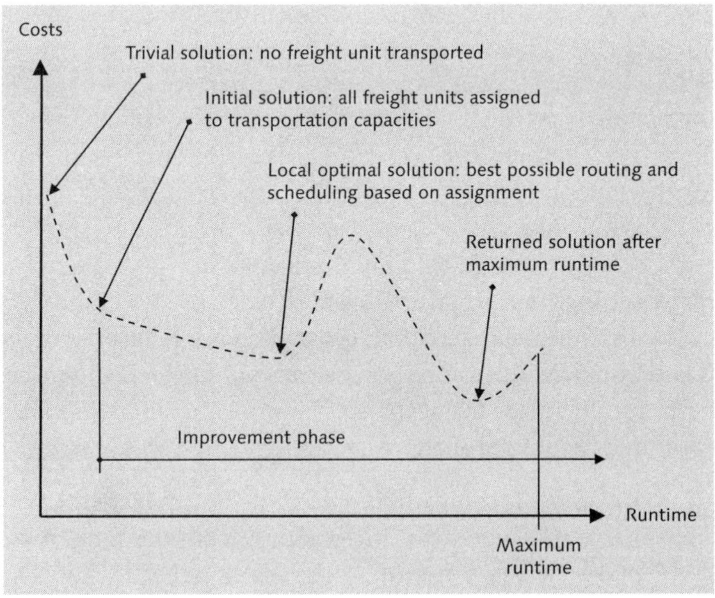

Figure 6.48 VSR Optimizer Runtime Behavior

Complexity of Vehicle Scheduling and Routing Problems

The VSR optimizer is based on an optimization algorithm for the *Vehicle Scheduling and Routing Problem* (VSRP). The VSRP is a combinatorial optimization problem that is NP-complete—meaning that, based on current research, no polynomial time algorithm is known that solves the problem exactly.

In practical terms, NP-completeness means that an exact procedure for large problem instances would require computing time that would be too high to determine the global optimum. That is why the VSR optimizer relies on approximation procedures that achieve an acceptable solution quality in an acceptable computing time.

Algorithmic decisions

In the process of generating solutions, the algorithm has to make the following decisions:

- For each freight unit, it decides whether it is transported.
- If the decision is made to transport the freight unit, the path (direct or via which transshipment locations) in the transportation network needs to be chosen.
- Each stage of the freight unit has to be assigned to a transportation capacity (e.g., vehicle resource, schedule, freight booking).
- For each vehicle assignment, the compartment assignment (if defined) needs to be made.
- For each transportation capacity, the sequence of activities (e.g., loading, travel, unloading, coupling, uncoupling) needs to be determined.
- A date and time need to be determined for each activity according to their sequence.

6.4.6 Explanation Tool

Optimizer explanation tool

Based on the previous section, you can see that the VSR optimizer is based on a very powerful algorithm, but to some extent it is not transparent to the user how the VSR optimizer created a specific result and why. So an explanation tool is provided to help the user understand and analyze the optimization results. To activate logging of the optimizer data for optimization runs, a prerequisite is that the user parameter

/SCMTMS/EXP is set to X for the specific user. You can access the explanation tool interactively by clicking on the OPTIMIZER EXPLANATION button either while working in the transportation cockpit or directly after any interactive optimizer run prior to accepting or canceling the result. Both actions take you to the explanation of your last optimizer run.

Log display

Another way of accessing the explanation tool that is also applicable for background runs is the Log Display transaction (APPLICATION ADMINISTRATION • GENERAL SETTINGS • REMOTE CONTROL AND COMMUNICATION FRAMEWORK • LOG • LOG DISPLAY). This transaction lists and identifies the logs of all engine calls (optimizer, transportation proposal, scheduling; see also Section 6.4.2) via the user that triggered the engine run, the planning profile used, and the date and time of this activity. The explanation tool can be accessed from here for each engine call by clicking on the EXPLANATION TOOL button for the specific run (row of the table).

Explanation Tool: Cleanup

Note that the persistence time for engine logs mentioned in Section 6.4.2 does not relate to the explanation tool. For performance reasons, old explanation data should be deleted regularly with report /SCMTMS/PLN_EXP_DELETE.

Input data

The explanation tool displays all data that is sent to or retrieved from the engine. Let's walk through Figure 6.49, which shows the input data for the optimizer. On the left side of the screen, it is possible to browse in a folder structure through all the tables, which are organized based on the data origin (e.g., freight unit data, transportation network data, etc.). The right side of the screen displays the content of the tables selected on the left. Often not all tables contain entries because not all planning constraints mentioned in Section 6.4.4 are present in each optimization problem. You can click on the SHOW/HIDE EMPTY TABLES button to toggle between the display of all tables or only tables that have at least one entry. Figure 6.49 shows how the scheduling constraints created in Figure 6.46 are transferred to the VSR optimizer. Note that for entries in dimension time (columns MAX. DURATION and CONSTRAINT LENGTH in the BUCKET CONSTRAINTS table), no unit of measure is transferred, but the unit of measure seconds is always used (10 hours corresponds to 36,000 seconds).

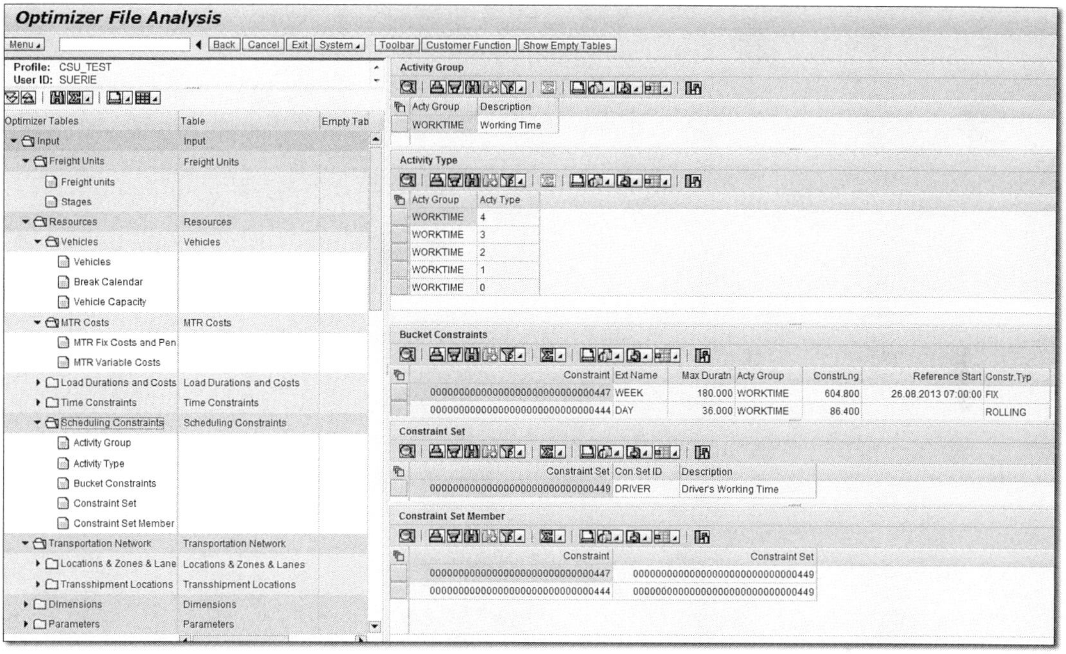

Figure 6.49 Optimizer File Analysis: Input Data

Result data

Figure 6.50 shows how the planning result is displayed in the explanation tool. Again, the available tables are displayed in a folder structure on the left, with the table contents displayed on the right. If freight units could not be planned as shown in this example (freight unit 168386, PLANNING STATUS NOT PLANNED), an explanation would be displayed as FREIGHT UNIT RELATED MESSAGES. In this case, the explanation NO VALID TRANSPORTATION LANE TO DESTINATION LOCATION for location CSU_CUST01 indicates that transportation lane information was missing for the optimizer. This is only a hint to the user because the system cannot know the real root cause. Other possible root causes for this scenario include the following:

- The resources that have been selected for this planning run are defined for a means of transport that is not assigned to the (existing) transportation lane.
- A freight booking or schedule should have been used but was not selected, so the missing transportation lane was not relevant.

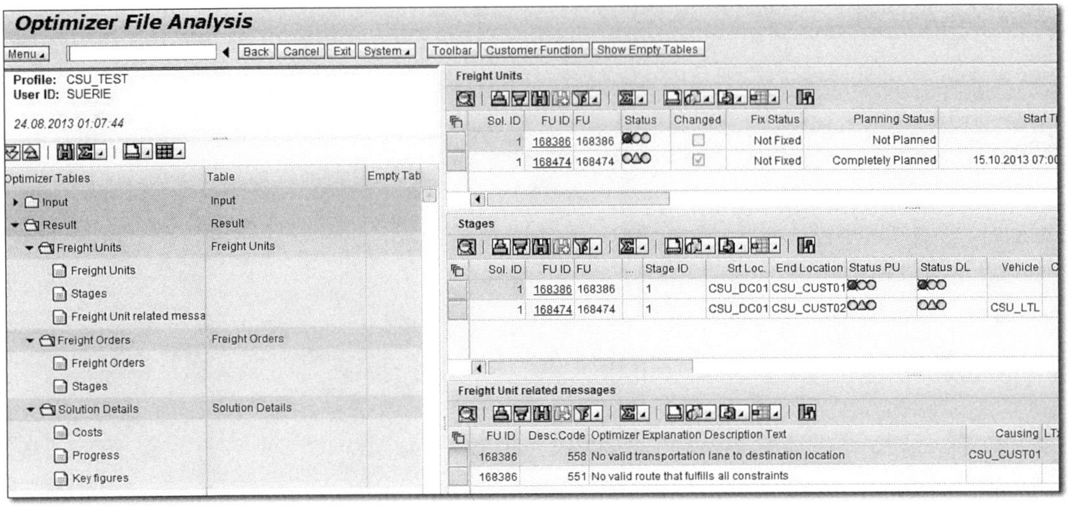

Figure 6.50 Optimizer File Analysis: Result Data

6.4.7 Load Optimization

Load planning Load planning deals with the creation of a load plan, primarily for loading trucks with pallets. For that reason, master data for vehicle resources and dimensions of freight units needs to be available in more detail (see Section 6.1.3 and Figure 6.51). Load planning can be done manually (see Section 6.3.5) or using an optimization engine that is triggered from the transportation cockpit. The optimization engine is fed with freight orders, pallet information from freight units, and truck profiles (axles, size, etc.). Its result is the position of each pallet on a truck.

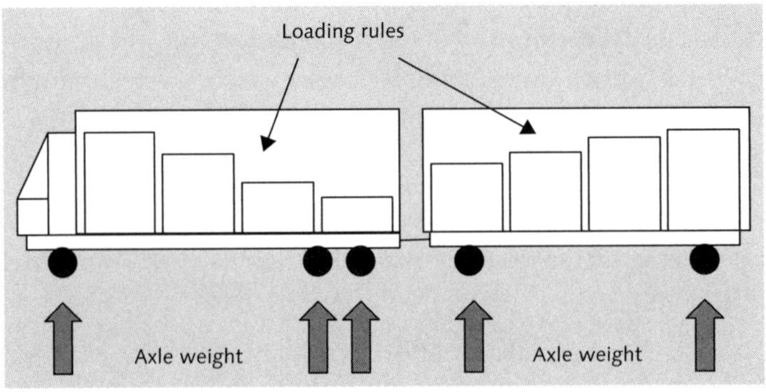

Figure 6.51 Load Planning Rules and Axle Weight

Depending on the goods to be transported, load planning can become important because legal restrictions and safety considerations may require that the load be distributed inside the vehicle resource based on certain rules. For example, in the United States, the bridge formula establishes the maximum weight any set of axles on a vehicle resource may carry on the interstate highway system. These rules have to be observed when loading a vehicle.

Load optimization

The load optimizer is based on an objective function that is to be minimized and some hard constraints that need to be met. Hard constraints relate to the physical attributes of the vehicle resource (maximum gross weight, maximum weight on a split deck, maximum weight on an axle group, etc.) and the physical attributes of the load (pallet dimensions, pallet gross weight, stackability, etc.). The objective is to find a suitable assignment of pallets to the positions in the truck that meets all hard constraints and minimizes the other objective criteria:

- All pallets should be loaded onto the truck. Not loading a pallet results in a penalty cost, which is considered very high because this situation needs to be avoided first.

Load planning rules

- All load planning rules must be met according to their priority. The load planning settings (APPLICATION ADMINISTRATION • PLANNING • PLANNING PROFILE SETTINGS) control the behavior of the load optimizer and allow you to specify load planning rules and to assign priorities to them.

  More than ten load planning rules are available; these include ascending stack height in driving direction and descending stack height in driving direction (see Figure 6.51).

The optimization engine is based on metaheuristics. It runs for a maximum runtime defined in the load planning settings and returns the best found solution within this runtime.

This chapter explored the depths of planning in SAP TM. Starting from the definition and creation of freight units, it explained different planning methods (manual and automatic, interactive, and background) that produce freight orders, freight bookings, or trailer documents. The next chapter focuses on freight orders and how these are subcontracted, tendered, or both.

Transportation execution can be outsourced easily to specialized companies such as LSPs or carriers. Therefore, subcontracting is an important component of a transportation management solution because it leaves the supervision of the transportation process in the control of the company while outsourcing its physical execution.

7 Carrier Selection and Subcontracting

In the previous chapter, we discussed transportation planning as a planning task that deals with the assignment of freight units to available transportation capacities. However, we haven't yet discussed the topic that will provide these transportation capacities: whether they represent a fleet of vehicle resources or whether the objective is to subcontract the execution of freight orders.

So the focus of this chapter is twofold. One focus is freight order management, and the other focus is selecting the best possible carrier through carrier selection and/or tendering processes.

Carrier selection, tendering

The objective of carrier selection is to provide a list of ranked carriers that are available to execute a planned freight order. A broad list of options and constraints is available to streamline the selection of the best possible carrier. These options include expected freight order costs and priorities, as well as incompatibilities, transportation allocations, and business shares representing various contractual obligations.

You can perform carrier selection manually by using an optimization algorithm or by using an auctioning mechanism as part of the tendering process. The tendering process can itself be used as part of the carrier selection process or as a separate process. In the tendering process, individual freight orders can be tendered to one carrier (peer-to-peer tendering) or to several carriers (broadcast tendering) in parallel. The tendering process involves communication with the carrier, in which the carrier can quote prices for the tendered freight order (broadcast tendering) or inform about acceptance or rejection of the freight order (peer-to-peer tendering).

The chapter is structured in the following way: Section 7.1 explains the carrier selection process and its objectives and constraints. It describes the carrier selection settings, which are the available configuration options. Section 7.2 deals with freight order management, freight order configuration, freight order types, and use, mainly in land transportation. This section also talks about special processes using freight orders, such as pickup and delivery freight orders in sea and air transportation, customer self-delivery and pickup, and service orders. Finally, Section 7.3 focuses on freight tendering and tendering process configuration options.

7.1 Carrier Selection

The primary objective of carrier selection is to assign a reliable and cost-efficient carrier to a freight order. This can be done in the background, interactively using manual steps, or by using an automated optimization procedure. The various options are described in the next section.

7.1.1 Process

Carrier selection process

The carrier selection process can be initiated using various methods interactively and in the background. Figure 7.1 shows the carrier selection process when initiated interactively. Carrier selection can be started directly from the freight order user interface or from any worklist that displays freight orders. It can also be started for freight orders inside the transportation cockpit (PLANNING • PLANNING • TRANSPORTATION COCKPIT) or using a transaction dedicated to carrier selection (PLANNING • PLANNING • CARRIER SELECTION). Finally, carrier selection can be executed as part of a planning strategy (e.g., by including method VSR_TSPS like in planning strategy VST_1STEP, which is delivered as a standard planning strategy that combines VSR optimization and carrier selection; see Chapter 6, Section 6.3).

The input data for carrier selection is one or more freight orders (selected interactively in a worklist) or a selection profile by which the freight orders that need to have a carrier assigned are determined. In addition, carrier selection settings are required to specify exactly how carrier selection should be carried out. The carrier selection settings (see Section 7.1.4) are either specified explicitly as part of the definition of a background job (see Figure 7.4) or retrieved indirectly, perhaps because they have been assigned to the planning profile that was used to enter the transportation cockpit.

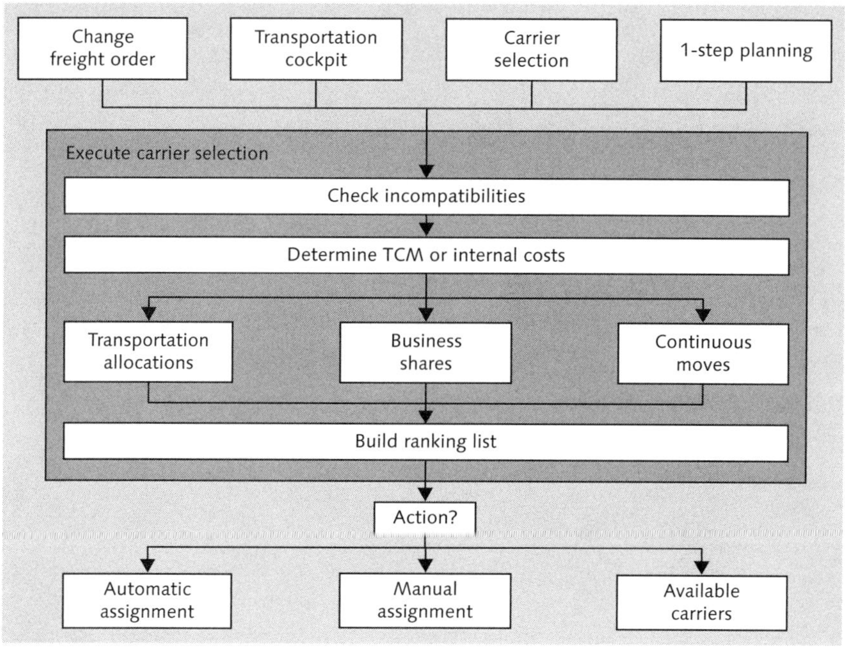

Figure 7.1 Carrier Selection Process

Available carriers

Once carrier selection has been initiated, the system determines the available carriers for each selected freight order. A carrier is considered available if it has been defined as a valid carrier for the means of transport that is used in the freight order for a transportation lane from the source location to the destination location of the freight order. The list of available carriers is reduced by those carriers that are incompatible with the freight order. Incompatibilities arise in a number of ways, such as when a customer does not want to be served by a specific carrier or because the freight units assigned to the freight order contain hazardous goods and the carrier is not allowed to handle these hazardous goods (e.g., because his drivers lack experience, have not been trained for it, or do not possess a legally required permit).

Available Carriers

To determine which carriers are available, the system proceeds as follows:

1. For each stage of the freight order, the carriers defined for the means of transport used in the freight order in the most specific transportation lane that represents each stage are determined. See the top half of Figure 7.2.
2. The carriers defined for the means of transport used in the freight order in the most specific transportation lanes from the source to the destination of the freight order are determined. See the bottom half of Figure 7.2.
3. Depending on whether the OVERALL CARRIER AVAILABILITY flag has been set or not, either those carriers are considered available that have been identified for each stage in step 1 and in step 2 (flag is set) or only those identified in step 2 (flag is not set).

Figure 7.2 illustrates this procedure. If the flag for OVERALL CARRIER AVAILABILITY is set, only carriers 1 and 2 will be considered available, whereas carriers 1, 2 and 5 will be considered as available if the flag has not been checked.

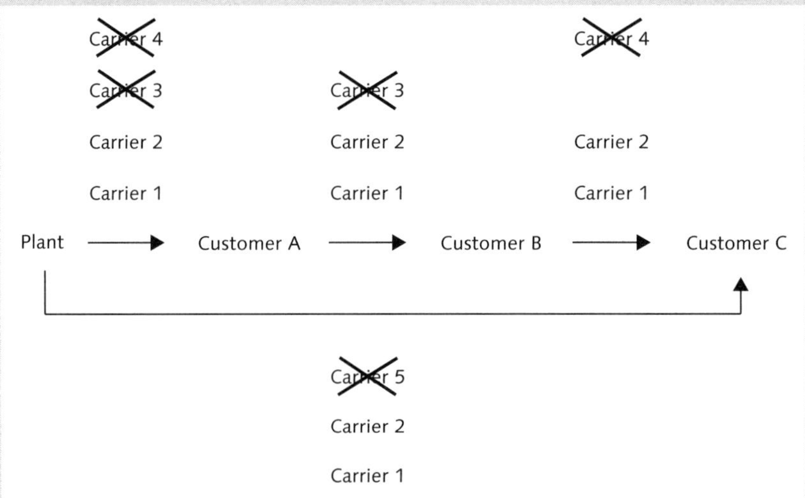

Figure 7.2 Available Carriers

Carrier selection optimization

After eliminating incompatible carriers, it is time to decide how to choose one from among the remaining carriers. Decision criteria can include priorities, internal costs, or transportation charge management costs based on the defined strategy.

Ranking list

The next step is an optimization procedure that takes into account the decision criteria calculated in the previous step as well as all constraints that have been defined. (We outline objectives and constraints in Section 7.1.2 and Section 7.1.3.) The result of this process is a ranking list of all available carriers from the reduced lists based on the decision criteria. Figure 7.3 gives an example of a ranking list. It lists the relevant information—such as the carrier name, SCAC code, expected transportation cost, and priority—that a user will require to make a reasonable assignment. Note that the first-ranked carrier is not necessarily the cheapest because his ranking may have been forced by constraints such as transportation allocations or business shares. All carriers beyond the first position are sorted in descending order of the decision criteria (e.g., cost).

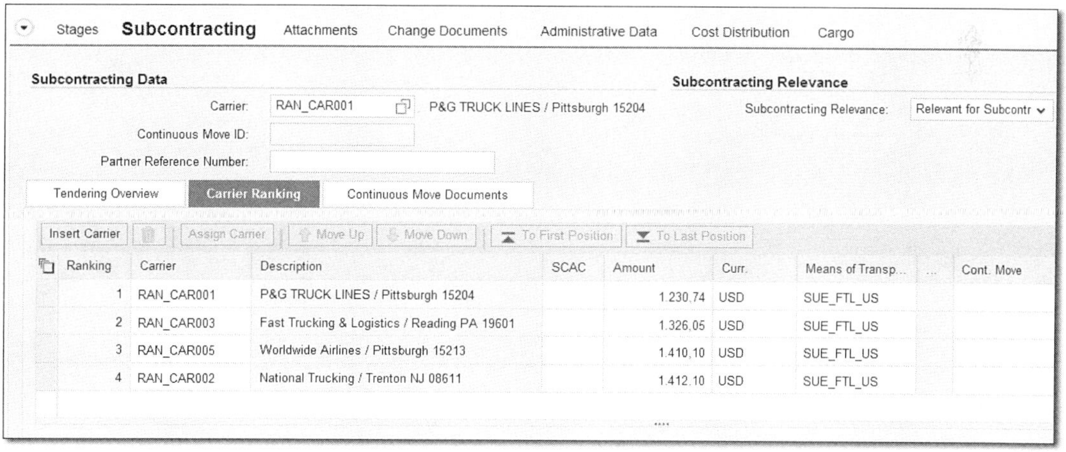

Figure 7.3 Ranking List

Carrier assignment

Finally, the carrier that is ranked first can be assigned to the freight order as part of the automatic carrier selection process, or the assignment can be delegated to a user in a manual process. If carrier selection has not been executed to *determine* a carrier, but only to identify the *available* carriers, the ranking list can be used in a subsequent tendering process (see Section 7.3). If carrier selection is done to initiate a broadcast tendering process, the optimization step can be skipped because the decision criteria (cost) will be determined only then.

In addition to starting manual and automatic carrier selection and determining a list of available carriers, the transportation cockpit also offers interactively the functionality to delete existing carrier assignments as well as to delete existing carrier rankings for a freight order.

Background processing

In the background, the carrier selection process can be initiated in the following ways:

- As part of the transportation planning process with background report /SCMTMS/PLN_OPT and using a planning strategy that includes carrier selection (e.g., VSR_1STEP)
- By scheduling report /SCMTMS/TSPS_OPT_BGD (see Figure 7.4)

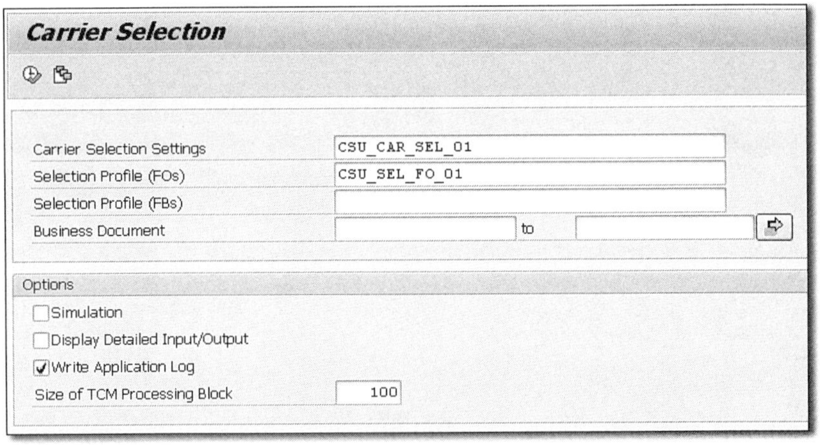

Figure 7.4 Background Report for Carrier Selection

7.1.2 Objective of the Carrier Selection Optimizer

The objective of carrier selection is to assign the most suitable carrier to a freight order, which usually means assigning the cheapest one. However, there may be criteria other than cost involved, such as whether another carrier needs to be assigned to a defined minimum transportation quantity based on contractual obligations (minimum transportation allocation), which can force an assignment even if that carrier is more expensive for a specific freight order.

Carrier selection optimizer

Carrier Selection Optimizer

The carrier selection optimizer converts the assignment problem into a *mixed-integer linear problem* (MILP). Based on the number of freight orders, the complexity of the constraints, and if the runtime specified in the carrier selection settings is carefully chosen, the carrier selection optimizer usually returns the optimal solution. If the runtime is not sufficient to determine the optimal solution, the best solution found within the specified runtime is returned.

Optimizer costs

Automatic carrier selection is based on an optimization algorithm that minimizes costs. The algorithm considers the following cost components:

- **Transportation charges for the freight orders of the assigned carrier**
  The transportation charges of the assigned carrier are often the only relevant cost component. For each freight order, the transportation charges are evaluated for all available carriers. The transportation charges can have two different origins. Either the transportation charges are calculated via transportation charge management (which we cover in Chapter 10) and thus represent the expected real freight cost for a freight order, or via internal costs. Internal costs can be defined in the transportation lane master data, or, if they do not depend on the geography, they can be defined in the carrier profile. You can also allowed use a combination of both, if required.
- **Nondelivery charges for freight orders to which no carrier could be assigned**
  While the objective of carrier selection is to find an assignment of a carrier to each freight order, it's not always possible, and the result is nondelivery. For example, finding a carrier may not be feasible because it would violate the maximum transportation allocation of all possible carriers. Because this situation should be avoided, the system defines a penalty cost for not assigning any carrier to a freight order. This penalty cannot be influenced by the user and is calculated to be prohibitively high in order to outweigh the other cost components and avoid nondelivery charges.

Transportation allocation

- **Penalty charges for violating minimum transportation allocations**
  If minimum transportation allocations refer to a monthly quantity, these may often not be fulfilled in the beginning of the month. Thus, minimum transportation allocations cannot be considered a hard constraint because if there are not enough freight orders waiting for assignment, it is impossible to fulfill all minimum transportation allocations. Therefore, this constraint is considered a soft constraint, and violation of minimum transportation allocations is penalized by costs. Similar to nondelivery charges, this penalty cost is calculated automatically by the system, and it is set higher than the other cost components to avoid violation if possible.

Business shares

- **Penalty charges for noncompliance to business shares (outside negative and positive tolerance)**
  Business shares can be defined to yield a predefined distribution of freight orders among several carriers. This constraint can be used because of contractual obligations or simply to avoid assigning all freight orders to one carrier and therefore becoming too reliant on this carrier. To allow the assignment of carriers that have higher transportation charges, noncompliance with defined business shares is penalized, as outlined in Figure 7.5.

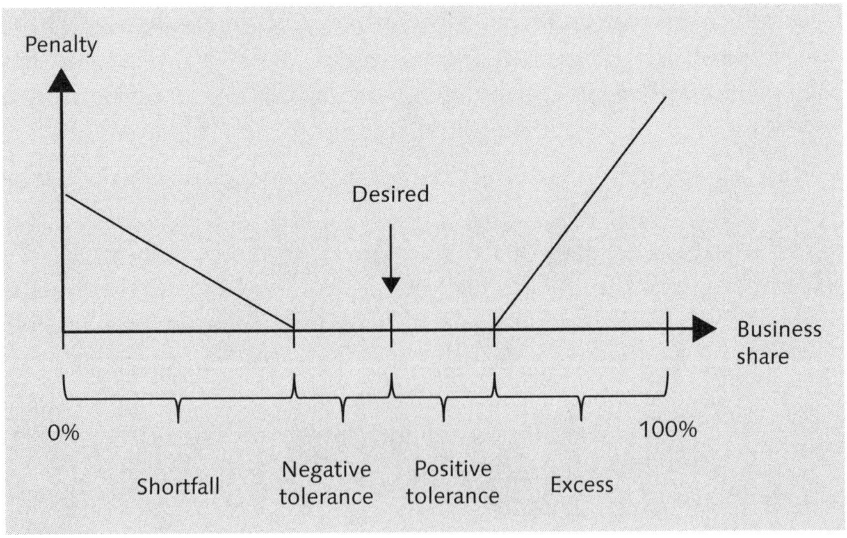

Figure 7.5 Business Shares

Based on a desired business share for a carrier, a negative and positive tolerance can be defined; actual business shares within these tolerances are cost-free. A shortfall of the desired business share minus the negative tolerance or an excess assignment beyond the desired business share plus the positive tolerance is penalized with (different) penalty costs. By allowing the user to specify these penalty costs, the user can influence to what extent the consideration of business shares overrules the assignment of carriers based on transportation charges.

- **Discounts granted by the creation of continuous moves**
  In certain situations, carriers grant discounts to transportation charges. These discounts are also taken into account by the carrier selection optimizer.

Continuous moves

Continuous Moves

Sometimes carriers grant discounts if they are assigned several freight orders for the same vehicle resource because the incentive reduces effort on their end to look for additional freight after having completed a single freight order. This situation is addressed in carrier selection by defining *continuous move* options. When carrier selection allows it, continuous moves are a way to save money on transportation costs.

Continuous moves may be offered for transports between regions that have economic disparities, like mainland Europe and the UK. Because many more goods are transported from mainland Europe to the UK, many trucks have to return empty from the UK. A carrier may therefore be inclined to offer a discount on the transportation charges if he is also offered to transport freight on his return trip.

Figure 7.6 shows two continuous move types.

- In a *simple continuous move*, there is no relation between the destination of the second freight order and the source of the first freight order. Only the destination of the first freight order and the source of the second freight order have to be close.
- In a *round trip*, the destination of the second freight order and the source of the first freight order have to match. In addition, the destination of the first freight order and the source of the second freight order have to be close. A round trip is limited to two freight orders, whereas an unlimited number of freight orders can be combined by simple continuous moves.

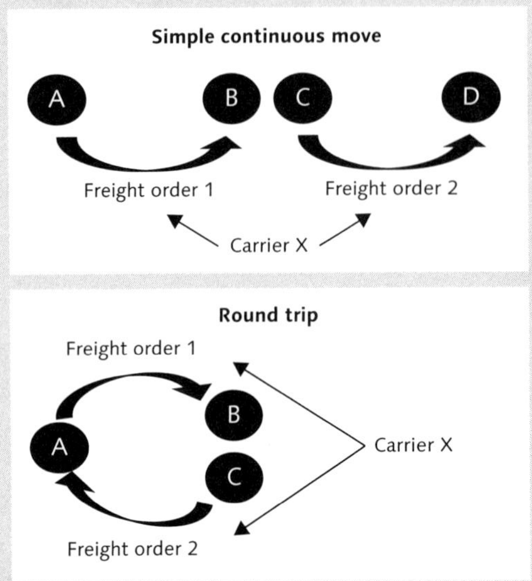

Figure 7.6 Continuous Move Types

Closeness, or proximity, has two dimensions and is defined in the transportation lane.

- One dimension is distance. A continuous move only provides commercial benefit to the carrier if the source of the second freight order is not too far away from the destination of the first freight order. A maximum distance can be defined in the transportation lane master data.
- The second dimension is time. The departure time of the second freight order has to be reasonably close to the arrival time of the first freight order. It does not help the carrier if his vehicle has to wait for a week. Therefore, time constraints for these windows are also defined in transportation lane master data.

Which continuous move types are allowed for a carrier is also defined in the transportation lane and therefore can deviate by geography. In addition, the carrier needs to be marked as eligible for offering discounts on continuous moves in the carrier profile (CONTINUOUS MOVE flag).

If certain freight orders should not be combined in a continuous move—even if these definitions would allow it—this can be specified via incompatibility type 82.

So before the carrier selection optimizer is started, it is evaluated whether two or more freight orders and carrier assignments qualify for continuous moves. These continuous move opportunities and their associated discounts are then provided as additional information to the optimizer, which takes these discounts together with the other cost components (freight charges, penalties) into account when calculating the optimal (cost minimal) solution.

If two or more freight orders build a continuous move, this is identified by a continuous move ID that is visible in the freight order (see the SUBCONTRACTING DATA section of Figure 7.3).

Priorities

While cost is the most obvious objective, carrier selection can also be based on priorities or a combination of priorities and costs as an alternative. Using priorities may be an option if the focus of the implementation project is planning and it is deemed to be too much effort to actually set up transportation charge management or internal transportation costs. If the contractual setup with carriers allows you to express preferences based on geographical information (transportation lane master data), priorities can be used as a selection criteria for carrier selection.

One way of combining priorities and costs is to use the multiplication of both numbers as a decision criteria (*priorities × costs*). Priorities are considered a key performance indicator for carrier reliability. A value smaller than 1 indicates a bonus in the selection because of good customer service, while a value larger than 1 corresponds to a disadvantage, which increases the perceived costs of this carrier. Similarly, you can argue for using the sum of both decision criteria (*priorities + costs*).

7.1.3 Constraints

We recommend considering the following constraints during carrier selection:

- Transportation allocations
- Business shares
- Incompatibilities

In the carrier selection settings, each constraint can be switched on or off individually by geography (transportation lane level) or globally for one carrier selection optimizer run (see Section 7.1.4). Transportation zone hierarchies and means of transport hierarchies are considered with these constraints.

Transportation allocations

Chapter 5 introduced transportation allocations in detail. A *transportation allocation* allows you to define minimum and maximum allocations (capacities) that *need to be* assigned to a carrier (minimum) or that *can be* assigned to a carrier (maximum).

The transportation allocation type customization defines whether a transportation allocation is valid for carrier selection (SAP TRANSPORTATION MANAGEMENT • TRANSPORTATION MANAGEMENT • PLANNING • GENERAL SETTINGS • DEFINE TRANSPORTATION ALLOCATION TYPES). A transportation allocation can be created by following the menu path PLANNING • TRANSPORTATION ALLOCATION • CREATE TRANSPORTATION ALLOCATION and is defined by the following characteristics, which are shown in Figure 7.7:

- A trade lane
  - Its source location or transportation zone
  - Its destination location or transportation zone
  - An orientation (see Chapter 3, Section 3.2.7 for details)
  - A means of transport
- A carrier
- A validity period
- A planning period
- An allocation type

Trade lanes and orientation define the geographical validity of the transportation allocation. Means of transport and carrier identify exactly for whom and for what the transportation allocation is defined. The validity period defines the temporal validity of the transportation allocation. This time period is divided into buckets according to the setting of the planning period (e.g., daily, weekly, monthly, quarterly, yearly, or per scheduled departure).

Unit of measure

Customizing the transportation allocation type also specifies the unit of measure for which the transportation allocation is defined. In the example used in Figure 7.7, gross weight has been defined as the relevant dimension. Therefore, in the transportation allocation you can maintain a minimum and maximum weight for each bucket, as well as display what has already been allocated to the carrier for this transportation allocation (right column of Figure 7.7).

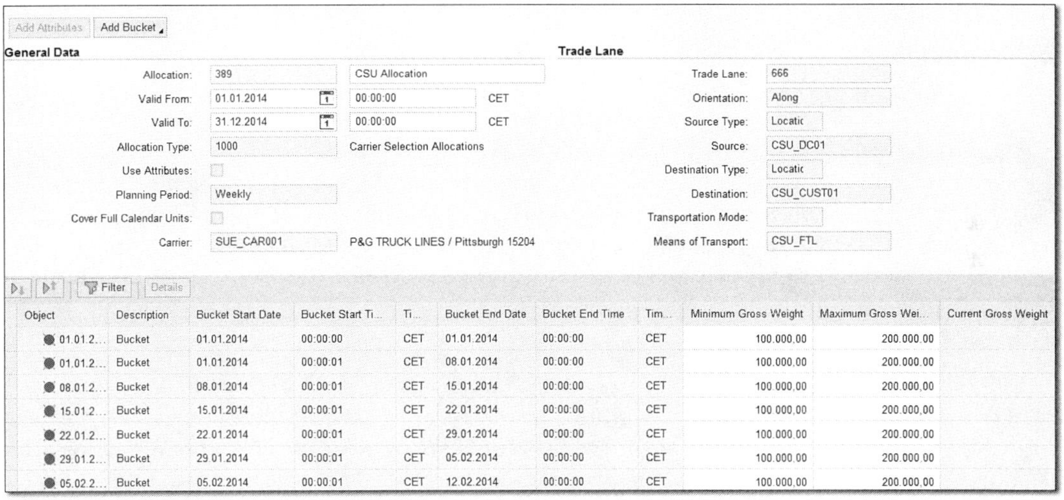

Figure 7.7 Creation of Transportation Allocations for Carrier Selection

Furthermore, in the transportation allocation type you have specified how to account for transportation allocations. The BUCKET OVERLAPPING flag specifies whether only the first date (the requested start date) or the complete period of the freight order allocates capacity. If a freight order covers a period of three days (e.g., a truck driving from Spain to northern Germany) and the planning period for the transportation allocation is daily, this flag determines whether the transportation allocation takes this freight order into account on all three days or only once on the first day.

A freight order is considered relevant for a transportation allocation if all of the following characteristics are met: Usage

- The means of transport in the allocation is the same or superior to the means of transport used in the freight order.
- The requested start date of the freight order falls into the validity period of the transportation allocation.

- The geographical criteria (trade lane and orientation) are met for the source and destination location of the freight order, considering also the transportation zone hierarchy.
- The check for transportation allocations has been set in the transportation lane corresponding to the freight order.

One Freight Order can Consume Several Transportation Allocations

Note that one freight order may consume several transportation allocations because means of transport hierarchies, transportation zones, and transportation zone hierarchies are taken into account when determining the relevant transportation allocations for a freight order. It's not possible to automatically assign a carrier if even one transportation allocation is violated, but you can force the assignment manually.

For example, let's say that a carrier leaving plant P is allowed to take a maximum of ten freight orders per week. A maximum of three freight orders per week is allowed for the same carrier from plant P to warehouse W. If three freight orders from P to W have been assigned to this carrier in a particular week, no additional freight order from P to W can be assigned to this carrier, even though there is an open allocation of seven freight orders outbound from plant P. This assignment would violate the second transportation allocation (P to W).

If a freight order is created or changed, this will update all relevant transportation allocations, meaning that potential violations will be checked. However, an update of the transportation allocation does not have an impact on existing carrier *assignments* to freight orders.

Business shares

Business shares have a lot of similarities to transportation allocations. They can be created by selecting PLANNING • TRANSPORTATION ALLOCATION • CREATE BUSINESS SHARES and are defined by the following characteristics shown in Figure 7.8:

- Trade lanes
  - Source location or transportation zone
  - Destination location or transportation zone
  - Orientation
  - Means of transport
- Validity period

- Business share period
- Positive and negative tolerances
- Penalty costs
- Target business shares per carrier

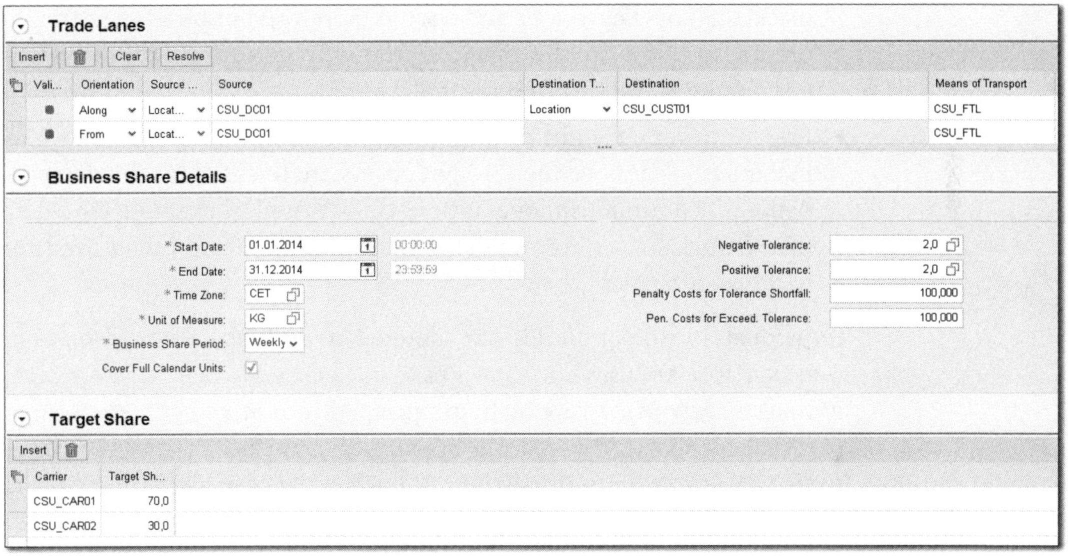

Figure 7.8 Business Shares

Similar to transportation allocations, defining and using business shares takes into account transport hierarchies, transportation zones, and transportation zone hierarchies. When you are calculating the actual business share, you consider not only the orders that are part of the selection for carrier selection, but also those that already have a carrier assigned and that fall into the business share period (historical business share).

The major difference between transportation allocations and business shares is that business shares express preferences of carrier assignments in relative numbers (percentages), while transportation allocations represent capacity restrictions and are defined in absolute numbers. In addition, business shares are represented as soft constraints, and their consideration is based on penalty costs, whereas transportation allocations are hard constraints.

Incompatibilities

For carrier selection, two incompatibility types are relevant:

- **Freight order—carrier**
  Incompatibilities between freight orders and carriers are used if a carrier will not be assigned to a freight order for any reason. Possible reasons are typically driven by the business, such as if a carrier is blacklisted by a customer and should therefore not be used in freight orders for this customer or if a carrier will not process certain goods because he is not certified for it.
- **Freight order—freight order**
  Incompatibilities between freight orders are relevant for the determination of continuous move options only. A typical example would be that a truck's cleaning requirement prevents it from being used for backhaul activities.

Individual incompatibilities are defined as *incompatibility definitions* (Application Administration • Planning • General Settings • Incompatibility Definitions), which are grouped into *incompatibility settings* (Application Administration • Planning • General Settings • Incompatibility Settings) to finally be assigned in the *carrier selection settings* (Application Administration • Planning • Planning Profile Settings • Carrier Selection Settings). This has been described in detail for planning-related incompatibilities in Chapter 6, Section 6.4.4.

Example

Figure 7.9 illustrates in an example of which information is considered for automatic carrier selection.

Although the example does not consider alternative carriers or several means of transport, it shows that a lot of information needs to be processed to adhere to all constraints relevant for carrier selection. In real-world situations, several available carriers or means of transport organized in a hierarchy with transportation allocations or business shares defined on several levels may add complexity to the decision.

The example consists of four circle locations (A, B, C, and D), which are assigned to boxed transportation zones Z1 (A and B), Z2 (B and C), and Z3 (A, B, and C via Z1).

Freight orders are for the following:

- Freight order 1 (FO1) from source **A** to destination **B**
- Freight order 2 (FO2) from source **A** to destination **C**
- Freight order 3 (FO3) from source **D** to destination **C**
- Freight order 4 (FO4) from source **D** to destination **A**

Three transportation allocations are in the scope of this scenario: one from **A** to **Z1** (orientation: along), one from **D** to **Z2** (orientation: along), and one for **D** (orientation: from). In addition, business shares have been defined from **A** to **Z1** (orientation: along) and from **D** to **Z3** (orientation: along).

In this example, the two freight orders originating from **D** (FO3 and FO4) have to respect two transportation allocations and one business share, whereas one transportation allocation and one business share are relevant for the other two freight orders (FO1 and FO2) that start in **A**.

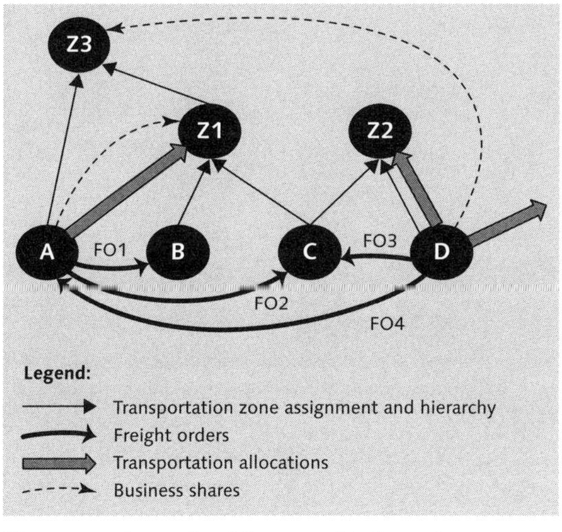

Figure 7.9 Example of Processed Information

7.1.4 Configuration

Required information

The configuration of carrier selection with all its constraints is controlled by many objects. These objects have been explained in earlier chapters:

- Carrier profiles (see Chapter 3)
- Transportation lanes (see Chapter 3, Section 3.2)
- Transportation zones and transportation zone hierarchies (see Chapter 3, Section 3.2)
- Transportation allocations (see Chapter 5 and Section 7.1.3)
- Business shares (see Section 7.1.3)
- Means of transport hierarchies (see Chapter 3, Section 3.2)

Carrier selection settings

One central configuration step is the carrier selection settings (APPLICATION ADMINISTRATION • PLANNING • PLANNING PROFILE SETTINGS • CARRIER SELECTION SETTINGS), which are shown in Figure 7.10; these settings control the carrier selection process and determine which constraints will be used and how. Therefore, setting carrier selection settings for automatic carrier selection is mandatory.

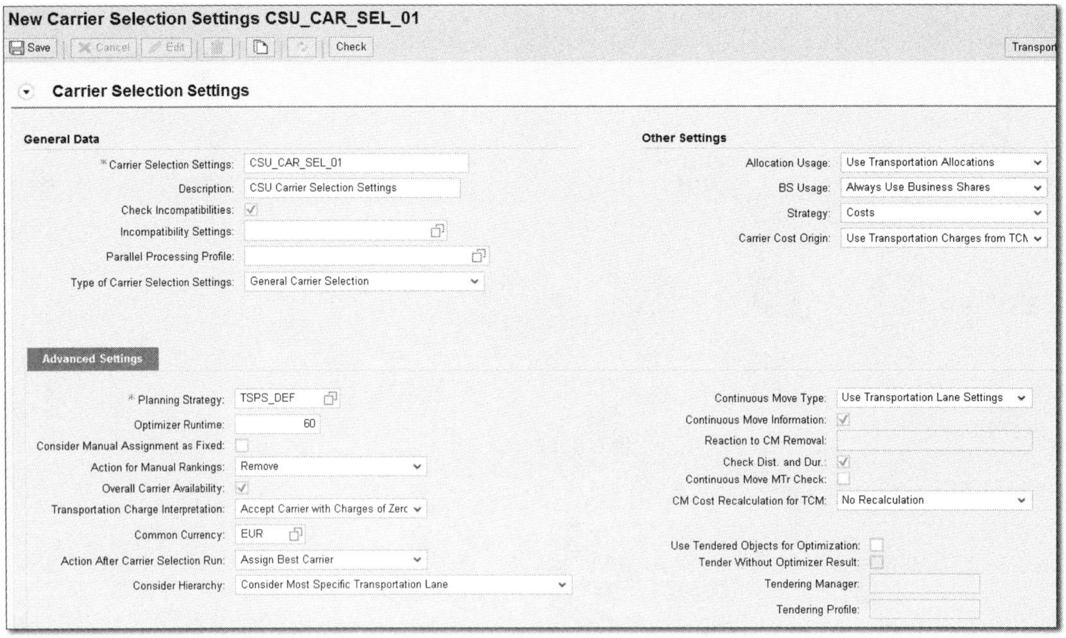

Figure 7.10 Carrier Selection Settings

The following fields are the most important to influence the carrier selection process:

- Type of Carrier Selection Settings
  This parameter defines the purpose of carrier selection. Available options are General Carrier Selection, Carrier Selection for Tendering, and Carrier Selection for Direct Shipment.

- Strategy
  The strategy determines how the objective of carrier selection is calculated in optimization. The objective can be based on cost, priority, cost plus priority, or cost times priority. In addition, the Use Transportation Lane Settings option can delegate this decision by geography to the individual transportation lanes, such that in different geographical areas, different objectives can be pursued (e.g., priority in one transportation lane and costs in another transportation lane).

Objective

- Carrier Cost Origin
  Carrier cost origin defines how costs are calculated. Transportation charges from Transportation Charge Management or internal costs can be used (see Section 7.1.2). The No Cost Determination strategy may be used if carrier selection for tendering searches for the available carriers in a broadcast tendering process (see Section 7.3). Similar to the settings for strategy with the Use Transportation Lane Settings, you can delegate this decision by geography to the individual transportation lanes.

- Transportation Charge Interpretation
  If, for any reason, the transportation charges for a carrier for a freight order are evaluated as zero, this parameter determines how to deal with it. Available options are to either ignore the carrier availability for this freight order or to accept the carrier for this freight order either as the cheapest available carrier or as the most expensive one.

- Check Incompatibilities
  Incompatibilities (see Section 7.1.3) are only checked if this checkbox has been selected.

Constraints

- Incompatibility Settings
  If the Check Incompatibilities checkbox is selected, incompatibility settings define which incompatibility definitions need to be adhered to.

- ALLOCATION USAGE
  This parameter determines the consideration of transportation allocations. Available options are to use transportation allocations, not to use them, or to decide on the transportation lane level.
- BS USAGE (business share usage)
  This parameter determines the consideration of business shares. Available options are to use business shares, not to use them, or to decide on the transportation lane level.
- CONTINUOUS MOVE TYPE
  With this parameter you decide whether only simple continuous moves are allowed, only round trips are allowed, continuous moves are not considered at all, or this decision is made on the transportation lane level (see Section 7.1.2).

Planning strategy

- PLANNING STRATEGY
  The left column of Figure 7.11 shows the default planning strategy for carrier selection TSPS_DEF. The assigned methods illustrate the carrier selection process with all relevant steps from gathering relevant data (carriers, step 20; incompatibilities, step 30; continuous move opportunities, step 40; other constraints, steps 50 to 70) to optimization (step 80).

Method assignment to Strategy

Strategy	Strategy Description	Method	Method Description	Sequence
TSPS_DEF	Default TSPS Strategy	TSPS_PRE	CS: Preprocessing	10
TSPS_DEF	Default TSPS Strategy	TSPS_CARR	CS: Get carriers	20
TSPS_DEF	Default TSPS Strategy	TSPS_INC	CS: Check incompatibilities	30
TSPS_DEF	Default TSPS Strategy	TSPS_CM	CS: Build continuous moves	40
TSPS_DEF	Default TSPS Strategy	TSPS_FILT1	CS: Optimizer restrictions	50
TSPS_DEF	Default TSPS Strategy	TSPS_TAL	CS: Load relevant TAL/BS	60
TSPS_DEF	Default TSPS Strategy	TSPS_FILT2	CS: Process TAL Bucket Restrictions	70
TSPS_DEF	Default TSPS Strategy	TSPS_EXEC	CS: Execute Process	80
TSPS_DEF	Default TSPS Strategy	TSPS_POST	CS: Postprocessing	90

Figure 7.11 Planning Strategy TSPS_DEF

- OPTIMIZER RUNTIME
  This parameter specifies the maximum runtime for the optimizer (in seconds). The optimizer returns the optimal solution prior to this runtime or the best solution found at this runtime.

- CONSIDER MANUAL ASSIGNMENTS AS FIXED
  Manual assignments
  How to deal with manual assignments in automatic carrier selection is an important topic because there is very likely a reason for manual assignments, and they should not simply be overridden. Therefore, manual assignments can be considered as fixed when this checkbox is selected.
- ACTION FOR MANUAL RANKINGS
  A similar question is how to deal with manual rankings. Available options for manual rankings are to keep them, remove them, or keep the carrier only when he is considered available (see Section 7.1.1).
- ACTION AFTER CARRIER SELECTION RUN
  Available actions after the carrier selection run are the automatic assignment of the highest ranking carrier to the freight order or doing nothing (i.e., leaving this decision to a manual process/user, based on the created carrier ranking). In addition, automatic or manual tendering based on the parameters for tendering (tendering profile, tendering manager) can be initiated. The tendering process is explained in detail in Section 7.3.
- CONSIDER HIERARCHY
  This parameter chooses the available carriers. Options include considering only those carriers defined on the most specific transportation lane or considering all transportation lanes to retrieve available carriers. In which sequence hierarchies (based on source location, destination location, and means of transport) are evaluated to determine the most specific transportation lane defined in Customizing activity SAP TRANSPORTATION MANAGEMENT • TRANSPORTATION MANAGEMENT • MASTER DATA • TRANSPORTATION NETWORK • GENERAL SETTINGS FOR TRANSPORTATION NETWORK DETERMINATION.

7.2 Freight Order Management

Freight order

For the most part, freight orders are created as the result of planning, especially in land transportation, whereas freight bookings are used in air or sea transportation. In a subcontracting scenario, the freight order is the document that is sent to the carrier once he is assigned via either carrier selection or tendering (see Section 7.3). In this section we focus on the role of the freight order and its automatic or manual creation from a planning perspective; the next chapter covers the view from the execution perspective, including the printing of freight documents.

7.2.1 Freight Order: Configuration and User Interface

Freight order creation

You can create a freight order either manually or automatically. Choose the manual process if you already know what to order because you regularly create similar freight orders for a carrier. In this case, you would manually enter the relevant information (logistical data such as source and destination locations as well as dates/times and items to be transported) in the freight order user interface, or you would copy an existing freight order and change the required fields in your version.

However, there are different options for automatic creation of freight orders:

- **Result of planning**
  Freight orders can be created as a result of two kinds of planning: manual planning in the transportation cockpit and automatic planning using either the VSR optimizer or transportation proposal functionality.
- **Direct creation based on a transportation requirement**
  Freight orders can be created directly from the FUBR (shortcut planning process, see Chapter 6, Section 6.1.3) or as a result of a direct shipment option (see Chapter 6, Section 6.1.2).
- **Based on SAP ERP shipment integration**
  Freight orders can be created without a predecessor document in SAP TM based on integration of a shipment document from SAP ERP.

Freight order type

The most important settings for the freight order are defined in the Customizing activity of the freight order type, which you can access by following the menu path SAP Transportation Management • Transportation Management • Freight Order Management • Freight Order • Define Freight Order Types. Figure 7.12 shows the freight order type Customizing with all its options.

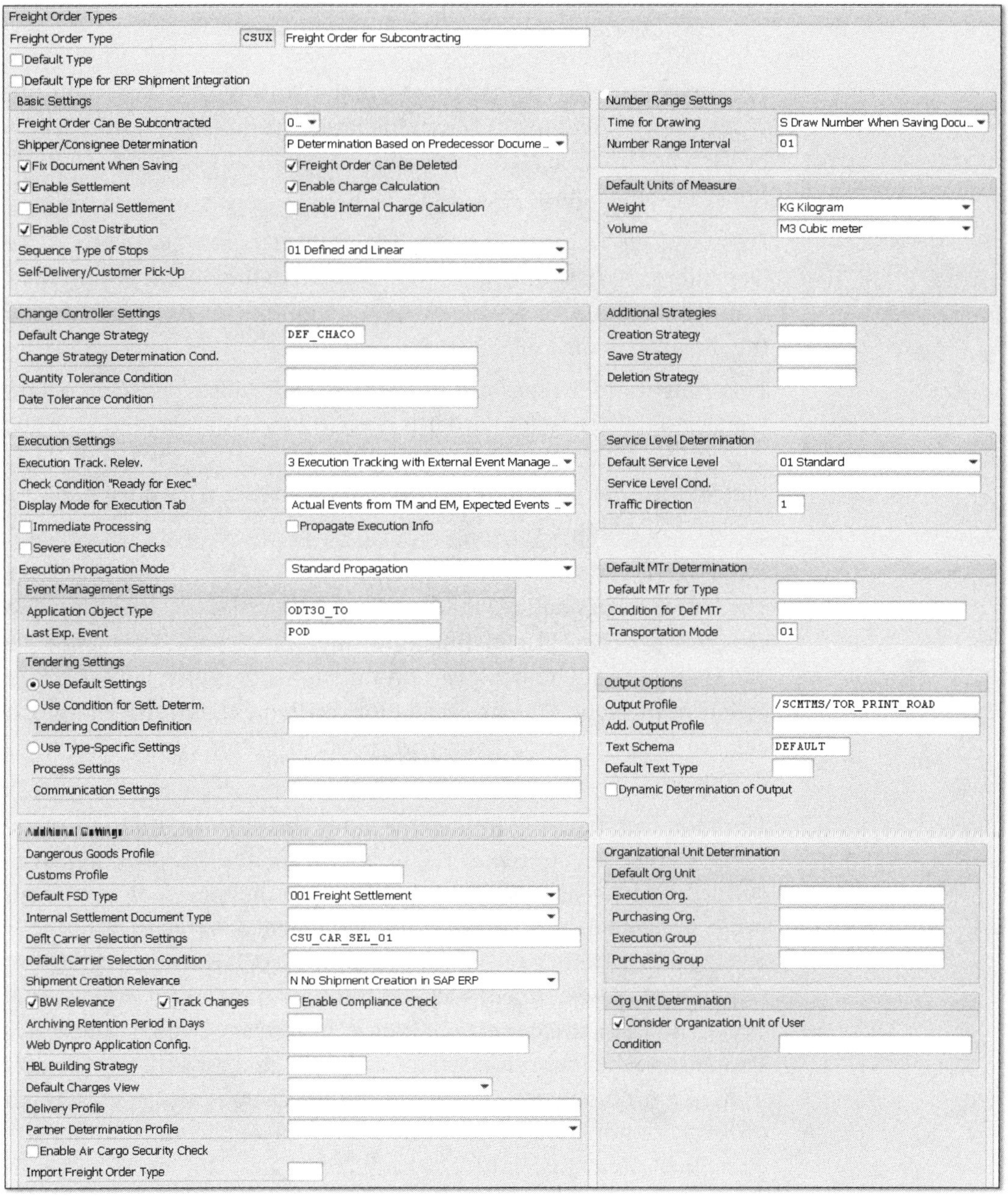

Figure 7.12 Freight Order Type Customizing

Let's walk through the most important settings:

- Freight Order Type
  The freight order type has to be unique with respect to freight unit types, freight booking types, transportation unit types, and service order types because all of these objects technically originate from the same business object: /SCMTMS/TOR.

Default type

- Default Type/Default Type for ERP Shipment Integration
  Depending on how a freight order is created, the freight order type is manually entered, determined by a condition, or determined from the planning profile.

  Two freight order types can be marked as default types. One default type is used if the freight order is created based on a predecessor document in SAP TM and no other freight order type is being determined; the other default type is used if the predecessor document was an SAP ERP shipment.

Sequence type of stops

- Basic Settings
  The basic settings define whether the freight order is relevant for subcontracting. They also define how the shipper and consignee are determined and whether a certain functionality is enabled for freight orders of this type. Charge calculation, settlement, internal charge calculation, internal settlement, and cost distribution can be enabled or disabled.

  The sequence type of stops determines the structural design of the stages of the freight order. Freight orders created via planning have a defined and linear stop sequence because the stages of the freight order represent the route that the assigned vehicle resource is expected to drive (e.g., from A to B, from B to C). However, other stop sequences can be defined, such as nonlinear star-shaped (from A to B, from A to C) or disconnected (from A to B, from C to D). These stop sequences, which physically cannot be executed as such, may be used in freight orders that are only relevant for charge calculation. Freight orders with nonlinear stages can only be created using customer-specific functions or from SAP ERP shipments. No planning activities are allowed for these freight orders.

- Number Range Settings
  The number range settings specify a number range interval and whether a number is drawn immediately or only when the business document is saved.
- Default Units of Measure
  You can specify a default unit of measure for weight and volume.
- Change Controller Settings
  The concept of the change controller was explained in detail in Chapter 2, Section 2.3. The change controller settings specify which change controller strategy is used for freight orders of this type or which conditions are used to determine the change controller strategy.

Strategies

- Additional Strategies
  Similar to the change controller strategy executed for document changes, additional strategies can be defined to be executed at document creation, save, and deletion.
- Execution Settings
  The execution settings deal with the execution tracking relevance of the freight order type and define the application object type that will be created in SAP Event Management if execution tracking is activated. We describe the integration with SAP Event Management in Chapter 8.
- Service Level Determination
  The service level for the freight order is determined based on the default setting or is retrieved via a condition.
- Default MTr Determination
  The means of transport of a freight order is usually determined from the vehicle resource assigned to the freight order. However, if the freight order is not created from planning, the means of transport can be determined from the default value defined here or using the specified condition.
- Tendering Settings
  Settings related to the tendering process are defined here. These settings are explained alongside the freight tendering process in Section 7.3.

- ADDITIONAL SETTINGS
  The additional settings cover attributes that do not fit in any of the other categories. Among others things, you define whether changes are tracked for freight orders of this type, whether these freight orders are relevant for BW integration, and whether and how freight orders of this type relate to SAP ERP shipments.
- OUTPUT OPTIONS
  The output options are explained in detail in Chapter 8.
- ORGANIZATIONAL UNIT DETERMINATION
  For a freight order, the purchasing organization is relevant in subcontracting scenarios because in this case, transportation services are purchased from an external vendor or carrier. Alternatively, the execution organization may be relevant if the freight order is not subcontracted. The relevant organizational unit for a freight order is either determined from the default values, the condition maintained here, or based on the assignment of the user that creates the freight order to an organizational unit.

Freight order user interface

Freight orders can be accessed for editing or displaying either via worklists (FREIGHT ORDER MANAGEMENT • LAND • OVERVIEW FREIGHT ORDERS or PLANNING • WORKLIST • OVERVIEW PLANNING), from the document flow of predecessor or successor documents, from the transportation cockpit (PLANNING • PLANNING • TRANSPORTATION COCKPIT), or via the menu path FREIGHT ORDER MANAGEMENT • LAND • EDIT FREIGHT ORDER. Figure 7.13 shows the user interface for the freight order.

The information stored in the freight order is organized into the following tabs:

- GENERAL DATA
  The content of the general data tab is depicted in Figure 7.13. It gives an overview of the freight order, including logistical information (source and destination location, dates and times, cumulative quantities for the cargo assigned to the freight order), assigned capacities (vehicle resources), and carrier as well as organizational assignments (responsible purchasing organization or planning and execution organization).

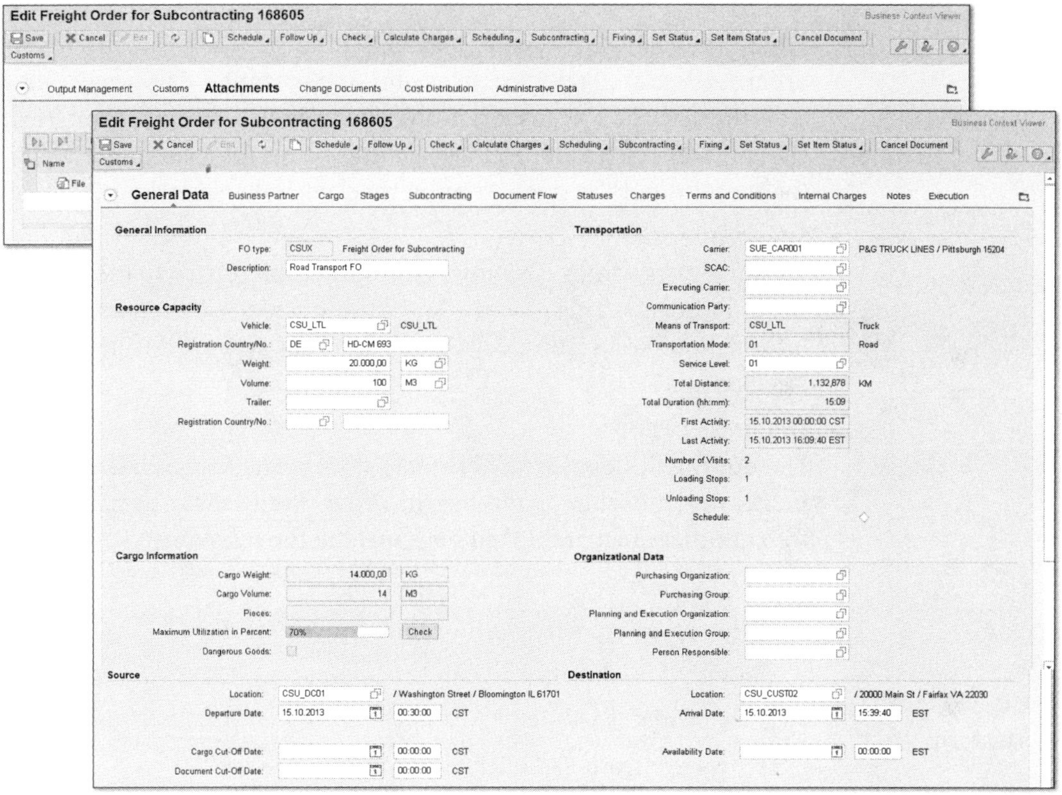

Figure 7.13 Freight Order User Interface

- Business Partner
  The business partners for the relevant business partner roles are displayed here. The most important business partner roles for the freight order are carrier, shipper, consignee, executing carrier, and communication party.
- Cargo
  Logistical data
  All items assigned to any stage of the freight order are displayed on the Cargo tab. All information related to cargo is consolidated here: product details, quantities, dangerous goods information, content identification information, customs information, and notes. From this tab, discrepancies can be reported, documented via attachments, and resolved.

- STAGES
  The STAGES tab displays the logistical information about the freight order. For each stage, distance, duration, planned dates and times, and the assigned transportation requirements (freight units) are shown. Stage-dependent block statuses—one for planning and one for execution of the stage—can be checked and changed on this tab.
- SUBCONTRACTING
  The SUBCONTRACTING tab shows all information related to the subcontracting process. This includes the carrier ranking, continuous move information (see Section 7.1), and tendering process documentation (see Section 7.3).
- DOCUMENT FLOW
  The document flow lists all business documents that have a direct or indirect relationship to the freight order. Predecessor documents of freight orders that are linked here include the following:
  - Freight units
  - Order-based transportation requirements (OTRs)
  - Delivery-based transportation requirements (DTRs)
  - Forwarding orders
  - Deliveries (from SAP ERP)
  - Sales orders, purchase orders, or stock transfer orders (from SAP ERP)

  A successor document would be the freight settlement document. Both predecessor and successor documents are displayed in a hierarchical structure.

Statuses

- STATUSES
  The freight order can be monitored using various statuses:
  - The lifecycle status shows whether the business document is new, in process, completed, or canceled.
  - The fixing status prevents changing existing planning results (e.g., you are not allowed to add additional freight units to a fixed freight order). The reason can be that the freight order has been subcontracted and the carrier has confirmed the requested quantities. Adding freight units at this point could lead to situations in

which the carrier cannot transport the additional (unconfirmed) freight units. For manual planning activities, you can specify to ignore the fixing status in the planning profile.

- The archiving status shows whether the business document has been archived.
- The subcontracting status deals with the current status of the subcontracting process (e.g., whether a carrier has already been assigned), and the corresponding confirmation status shows whether the carrier has confirmed this assignment. Additionally, the invoicing status shows whether an invoice has been received for the freight order.
- The execution status is adapted based on the execution information.
- The manifest status shows whether documents have been created for the freight order.
- The customs status is set based on information from SAP Global Trade Services (see Chapter 9, Section 9.1).
- The transmission-to-ERP status records the transmission of data to SAP ERP as well as transmission dates (first and last). The last confirmation received from SAP ERP is also recorded here.
- Consistency check statuses (for the document and cross-document) report the integrity of the business document data.
- Finally, a freight order can be blocked for planning, execution, and invoicing. This is controlled with three separate blocking statuses available on the STATUSES tab.

Financial data

- CHARGES
  This tab shows the calculation result and all individual charges for external (carrier) settlement. How the charges are determined is the focus of Chapter 10.
- TERMS AND CONDITIONS
  Terms and conditions include information about the incoterms for the freight order and about the value of goods (for customs and insurance).

- INTERNAL CHARGES
  The INTERNAL CHARGES tab shows the calculation result and all individual charges for internal settlement. Internal settlement is covered in Chapter 11, Section 11.2.
- NOTES
  This tab allows the display of freight order texts in different languages.

Execution data

- EXECUTION
  On the EXECUTION tab you can compare expected and actual dates and times of the events for this freight order as well as report events. SAP Event Management integration is explained in Chapter 8, Section 8.2.
- OUTPUT MANAGEMENT
  Print documents or electronic messages can be the result of the output process and are listed here based on their action status (unprocessed actions, successfully processed actions, and errors). Output actions can be generated and triggered from here. More details about this topic are provided in Chapter 8.
- CUSTOMS
  The CUSTOMS tab displays customs-relevant information (see Chapter 9, Section 9.1 for details).
- ATTACHMENTS
  This tab allows you to link files and URLs to the freight order.
- CHANGE DOCUMENTS
  If change tracking has been activated in the corresponding freight order type, the change documents allow you to keep track of who changed what in the freight order.

Cost distribution

- COST DISTRIBUTION
  The COST DISTRIBUTION tab shows how the charges (from the CHARGES tab) are distributed among the cargo transported by the freight order. If goods for two customers (and therefore originating from two sales orders) are transported together, the transportation cost has to be borne by these two accounting objects. How the costs are split is shown here (see also Chapter 11, Section 11.2).
- ADMINISTRATIVE DATA
  The administrative data shows information about the creation and latest change date/time and user.

Freight Orders for Rail

Compared to freight orders for road, which has been the focus of this section, freight orders for rail have some specifics. They can represent a train with one or more locomotives and one or more railcars as items. Railcar items can come from railcar units that are assigned to the rail freight orders (similar to trailer units assigned to road freight orders).

A specific user interface is available for rail freight orders (FREIGHT ORDER MANAGEMENT • RAIL) that offers additional tabs (equipment) and rail-specific fields. For some further details on rail freight orders specific to planning, see Chapter 6, Section 6.2.

Freight order actions

Many actions can be triggered from the freight order user interface. The user can change statuses on the document header level, but also on the item level for the cargo included in the freight order. In addition, planning activities (scheduling the freight order as well as starting the transportation cockpit) and actions related to subcontracting (e.g., carrier selection, tendering, and document transfer to the carrier) can be initiated here. Charge calculation and the creation of a freight settlement document, creation and printing of freight documents, and actions related to the logistical integration with SAP ERP (sending delivery proposals to SAP ERP and creating or updating an SAP ERP shipment document) can be triggered via buttons on the freight order user interface.

Hint

The planning profile includes some important information for the freight order, such as whether capacity violations result in an error or warning and which incompatibility settings need to be watched. Therefore, a planning profile needs to be associated with the freight order even when the freight order needs to be changed in transactions that are not planning-related, such as the freight order user interface.

For that purpose, a planning profile is stored in the PLAN_PROF_KEY field of the root node of the business object /SCMTMS/TOR. This field is populated with the planning profile that has been used to create the freight order (e.g., in the transportation cockpit or via a background planning run) or via evaluation of a condition with condition type /SCMTMS/TOR_PLN_PROF. If no planning profile can be determined, then the defaults are applied (such as no incompatibility check).

7.2.2 Special Processes with Freight Orders

Freight order for pickup/delivery

Freight orders for pickup and freight orders for delivery can be used in an ocean freight process or an air freight process. In an ocean freight process, they are created out of the freight booking for the transfer of containers from the export gateway to the port of loading (freight order for pickup) and for the transfer from the port of destination to the import gateway (freight order for delivery). This process is shown in Figure 7.14. In an air freight process, these orders are created similarly between the gateways and the airline's delivery/pickup address to transport either uniform load devices (ULDs) or loose cargo.

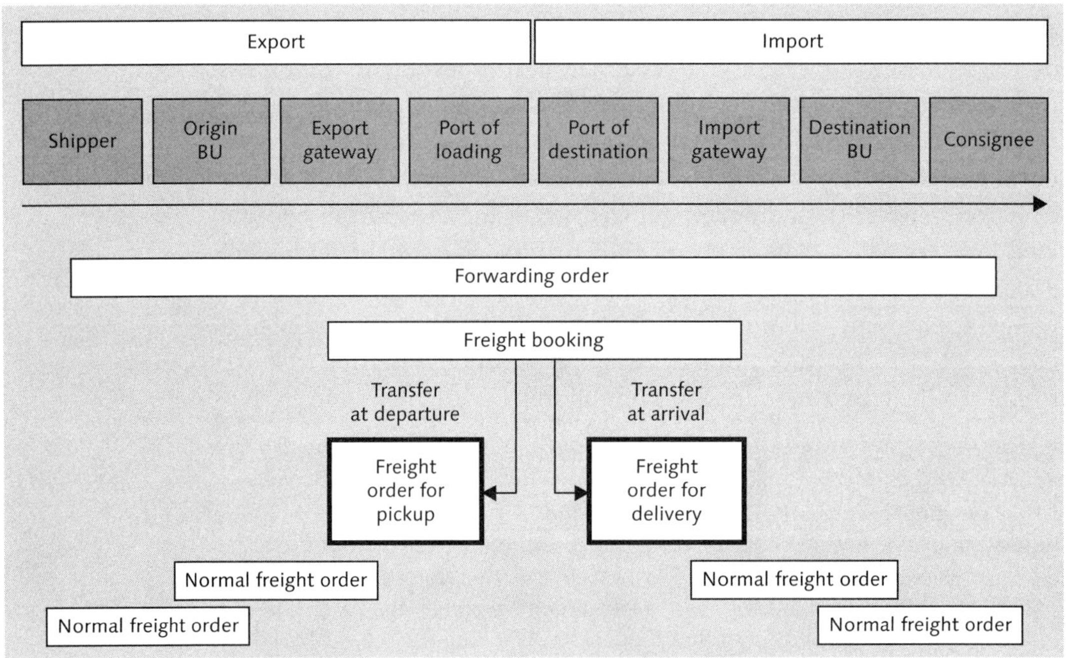

Figure 7.14 Freight Order for Pickup and Freight Order for Delivery

A prerequisite for this process is that the freight booking (ocean or air) has been created from the export gateway to the import gateway. In this situation, the freight orders for pickup and delivery can be created from the freight booking user interface (CARGO tab) or from a freight booking worklist (FREIGHT ORDER MANAGEMENT • AIR • OVERVIEW AIR FREIGHT BOOKINGS or FREIGHT ORDER MANAGEMENT • OCEAN • OVERVIEW OCEAN

Freight Bookings). The freight order type that is used for the creation of these freight orders is defined in the Customizing of the freight booking type (SAP Transportation Management • Transportation Management • Freight Order Management • Freight Booking • Define Freight Booking Types). You can create one or several freight orders for pickup and delivery from each freight booking, but you can also consolidate freight from several freight bookings into one freight order for pickup and delivery.

The advantage of tight integration between the freight booking and its freight orders for pickup and delivery is that changes in one document automatically update the other document. The following information is copied from the freight booking to the freight orders for pickup and delivery:

- Seal information and changes to seals
- Item changes
- Location changes
- Date/time changes
- Adjustment of handling execution status

The following information is copied from the freight orders for pickup and delivery to the corresponding freight booking:

- Change of cargo receipt status
- Reporting/resolution of discrepancies

Customer self-delivery, customer pickup

Freight orders for customer self-delivery and pickup can be created if the LSP's customer organizes either the delivery of freight to the origin business unit of the LSP or the pickup of freight from the LSP's destination business unit (see Figure 7.15). In this case, this stage is not relevant for planning, and a freight order with limited functionality is sufficient for this process. This is indicated in the Customizing of the freight order type (SAP Transportation Management • Transportation Management • Freight Order Management • Freight Order • Define Freight Order Types), which is marked as freight order for customer self-delivery or pickup (see Figure 7.12), and the use of a "light version" of the freight order user interface by defining Web Dynpro application configuration /SCMTMS/FRE_ORDER_SDCP in the freight order type

Customizing. Subcontracting and charge calculation cannot be done with these freight orders.

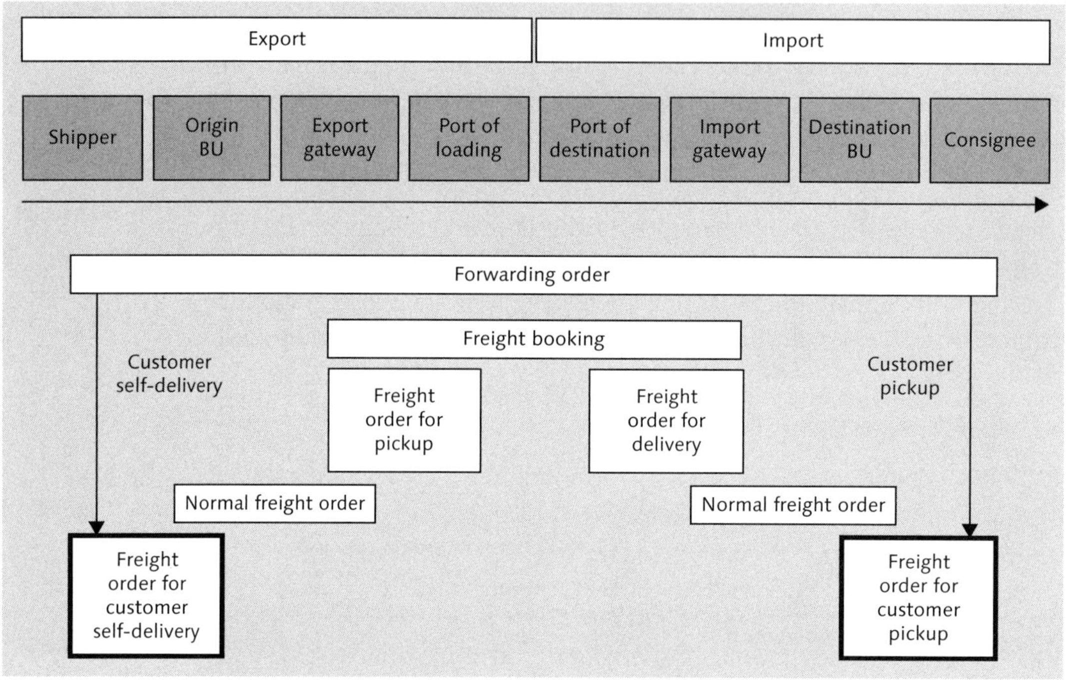

Figure 7.15 Freight Orders for Customer Self-Delivery and Pickup

Freight orders for customer self-delivery and pickup can be created directly from the forwarding order user interface (the STAGES tab) or from a forwarding order worklist (FORWARDING ORDER MANAGEMENT • WORKLIST • OVERVIEW FORWARDING ORDERS) in an air or ocean process. If these freight orders are created based on the forwarding order, it is a prerequisite that the corresponding stages are defined based on a stage type with either category CUSTOMER SELF-DELIVERY or CUSTOMER PICK-UP (in Customizing: SAP TRANSPORTATION MANAGEMENT • TRANSPORTATION MANAGEMENT • FORWARDING ORDER MANAGEMENT • DEFINE STAGE TYPES). In addition, these stages need to be set to NEVER RELEVANT FOR PLANNING in the stage type sequence of the movement type used in the forwarding order (in Customizing: SAP TRANSPORTATION MANAGEMENT • TRANSPORTATION MANAGEMENT • FORWARDING ORDER MANAGEMENT • DEFINE STAGE TYPE SEQUENCE FOR MOVEMENT TYPES).

ERP shipment integration

Freight orders can also be created based on SAP ERP shipments (see Section 7.2.1). Shippers that already have implemented their processes in SAP ERP successfully and want to use functionality in SAP TM that is not available in SAP ERP can take advantage of this option. Use cases include processes involving carrier selection, tendering, or transportation charge management and settlement. A prerequisite for this process is that the technical SAP ERP shipment integration has been configured on the SAP ERP side. When a freight order is being created based on an SAP ERP shipment, the system first determines the freight order type. You can define the freight order type in three ways:

- Customizing using a mapping table between SAP ERP shipment types and freight order types (SAP TRANSPORTATION MANAGEMENT • TRANSPORTATION MANAGEMENT • INTEGRATION • ERP LOGISTICS INTEGRATION • SHIPMENT INTEGRATION • ASSIGN FREIGHT ORDER TYPES TO ERP SHIPMENT TYPES)
- Using a condition with condition type /SCMTMS/FRO_TYPE_SHP
- Using the default type defined in freight order type Customizing (SAP TRANSPORTATION MANAGEMENT • TRANSPORTATION MANAGEMENT • FREIGHT ORDER MANAGEMENT • FREIGHT ORDER • DEFINE FREIGHT ORDER TYPES)

Note that SAP TM cannot create freight orders based on any type of SAP ERP shipment because only certain stop sequence types (see Section 7.2.1) are allowed in SAP TM. Thus, any SAP ERP shipment that creates a freight order needs to respect this limitation. Once created, the freight order can be used for subcontracting purposes (carrier selection, tendering) and updates the SAP ERP shipment with the result of this process.

Creation of ERP Shipments Based on Freight Orders

You can also create ERP shipments based on freight orders, such as when you want to carry out transportation execution in SAP ERP or use SAP EWM for warehouse planning and execution. For ERP shipment creation, several restrictions apply to freight orders:

- Freight orders can only have freight units assigned that originate from DTRs.
- All freight units originating from the same DTR have to be assigned to the same freight order.

- All freight units assigned to a freight order need to have either only inbound deliveries or only outbound deliveries as the source of the DTR.

You can initiate ERP shipment creation from the freight order user interface via FOLLOW-UP • CREATE/UPDATE ERP SHIPMENT, from a freight order worklist, from the transportation cockpit, or via the background report /SCMTMS/TOR_FO_PROC_BATCH.

7.2.3 Service Orders

Service orders

Cleaning of containers, fumigating, and performing security services or documentation are typical examples of services that can occur for items of a freight booking. The *service order* is used to account for and track services, calculate charges, and enable settlement of the charges for services that have been provided for individual items in a freight booking or for the entire freight booking.

Service orders are created from the freight booking user interface (return to Chapter 5, Section 5.4). The SERVICE ORDER tab in the freight booking user interface displays an overview of the service orders created for the freight booking, while similar information for freight booking items is provided on the CARGO tab of the freight booking user interface. The service order user interface (see Figure 7.16) provides the following information:

Service order type

- GENERAL DATA
  The GENERAL DATA tab shows the involved parties (purchasing organization and service provider) as well as the service order type and status information. The service order type is defined in Customizing (SAP TRANSPORTATION MANAGEMENT • TRANSPORTATION MANAGEMENT • FREIGHT ORDER MANAGEMENT • SERVICE ORDER • DEFINE SERVICE ORDER TYPES). In the service order type Customizing, you define options similar to those in the freight order type Customizing. That is, you can enable charge calculation and settlement, define number range settings, define output options, define how organizational units are determined for service orders of this type, and define change controller and execution tracking settings.

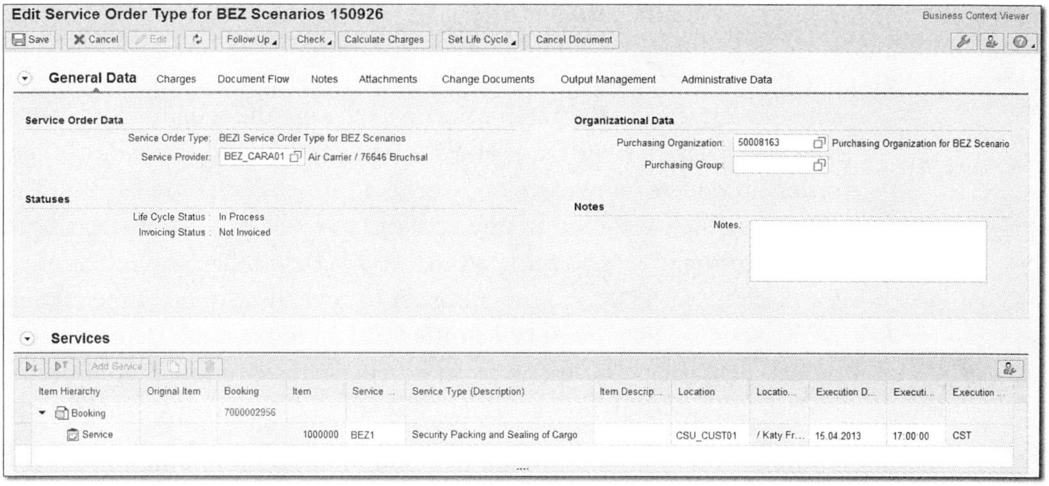

Figure 7.16 Service Order User Interface

- SERVICES

  Service types

  Under SERVICES, the link to the freight booking or freight booking items is displayed as well as the service type(s) that have been or will be provided for the freight booking or freight booking items, their location, and the execution time per service type. Service types are defined in Customizing (SAP TRANSPORTATION MANAGEMENT • TRANSPORTATION MANAGEMENT • BASIC FUNCTIONS • GENERAL SETTINGS • DEFINE SERVICE TYPES). To be able to select a service type for a freight booking or freight booking item, you need to assign the service type to the item category in Customizing (SAP TRANSPORTATION MANAGEMENT • TRANSPORTATION MANAGEMENT • BASIC FUNCTIONS • GENERAL SETTINGS • ASSIGN SERVICE TYPES TO ITEM CATEGORIES).

The CHARGES, DOCUMENT FLOW, NOTES, ATTACHMENTS, CHANGE DOCUMENTS, OUTPUT MANAGEMENT, and ADMINISTRATIVE DATA tabs of the service order user interface provide similar information as the corresponding tabs on the freight order user interface covered in Section 7.2.1.

The service order can be edited or displayed by following the menu path FREIGHT ORDER MANAGEMENT • SERVICE ORDER • EDIT SERVICE ORDER, by selecting the service order from a worklist (FREIGHT ORDER MANAGEMENT • SERVICE ORDER • OVERVIEW SERVICE ORDERS), or via the document flow of a related document (e.g., freight booking).

7.3 Freight Tendering

Tendering process

After you've created a freight order and have potentially identified a suitable carrier, the final step prior to executing the freight order can be tendering. The freight tendering process can be used to tender freight orders to one or more carriers. Numerous process variants exist for this process, which is shown in one variant in Figure 7.17. In this example, the tendering process is initiated in SAP TM by sending a freight request for quotation (RFQ) for a freight order to several carriers. These carriers receive and review the RFQ via different communication channels (web portal, EDI, etc.). The carriers can accept or reject the tendered freight orders and quote prices in a freight quotation. SAP TM evaluates the carriers' responses and awards the freight order to one carrier (carrier 1, in the example). This carrier receives the freight order and acknowledges receipt of the freight order with a confirmation message.

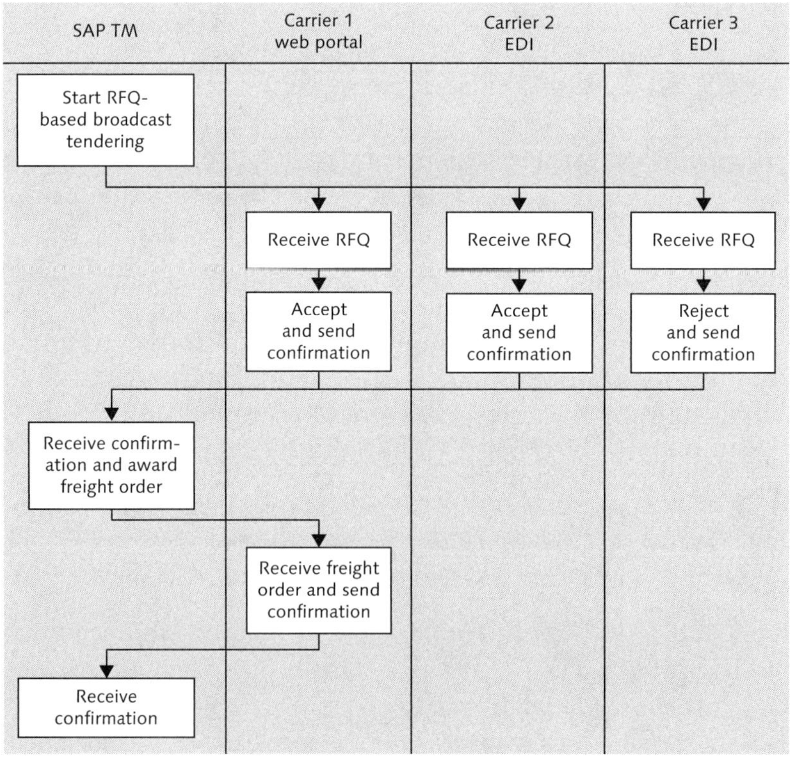

Figure 7.17 Tendering Process Example

The tendering process can be configured in various ways using a tendering profile (APPLICATION ADMINISTRATION • TENDERING • TENDERING PROFILES) as displayed in Figure 7.18.

Tendering profile

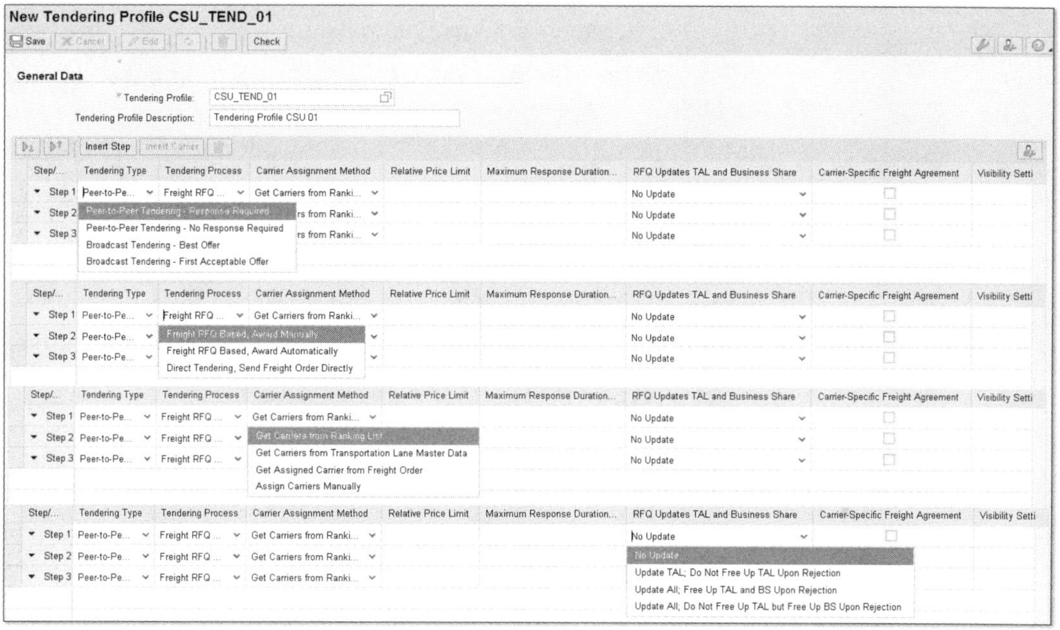

Figure 7.18 Tendering Profile

In the tendering profile you can define the individual steps of a tendering process. For each step you define the following criteria:

Tendering type

- TENDERING TYPE
  The tendering type specifies the auction mechanism that is used in the tendering process. Four available tendering types are offered.
  - PEER-TO-PEER TENDERING—RESPONSE REQUIRED initiates the tendering process with one or more carriers sequentially. This means that the system will wait for a (negative) response from the first carrier before contacting the second (or third and so on) carrier. Response required indicates that the lack of a response is considered a rejection.
  - PEER-TO-PEER TENDERING—NO RESPONSE REQUIRED also initiates the tendering process with one or more carriers sequentially, but

in this case, the system treats no response within the specified duration as acceptance of the tendered freight order. This tendering type can be used if carrier acceptance is the usual and expected behavior and thus reduces the number of messages exchanged between the involved parties.

- Broadcast Tendering—Best Offer is an auctioning mechanism in which several carriers are contacted simultaneously. The evaluation of the different offers is based on the quoted price of the carrier.
- Broadcast Tendering—First Acceptable Offer is also an auctioning mechanism in which several carriers are contacted simultaneously, but here the focus is on a fast response rather than the best price. With this tendering type, the first carrier to quote a price below the specified price limit is chosen.

- Tendering Process
The tendering process can be specified as freight RFQ-based or direct tendering. The difference is that in the first method, a freight request for quotation is sent to the carrier, whereas in the second method the freight order is sent directly. The freight RFQ process always requires an award step (manual or automatic) after the freight quotation has been received from the carrier. The award step determines which carrier will receive the freight order, and the freight order itself is only sent at this stage of the process as a separate message. In the direct tendering process the freight order is automatically awarded to the carrier if he has not rejected it explicitly within the maximum duration.

Carrier assignment method

- Carrier Assignment Method
An important decision in each step of the tendering process is the selection of the carrier(s) that will be included in each step. Four different options are available:
  - The carriers can be taken from the carrier ranking list, which has been created by the carrier selection process (see Section 7.1) or selected manually.
  - Carriers can be retrieved from the transportation lane.

  - The carrier that is currently assigned to the freight order can be used.
  - Carriers are manually assigned to a process step.
- RELATIVE PRICE LIMIT
  The relative price limit is important for the automatic award mechanism of freight orders to a carrier. It is used as a threshold to avoid awarding a freight order based on too high a price quotation. The relative price limit is therefore relevant for all four tendering types.
- MAXIMUM RESPONSE DURATION
  The maximum response duration specifies the time limit within which SAP TM waits for carriers' responses. With tendering type BROADCAST TENDERING—BEST OFFER, it specifies the time after which the best available offer is selected.
- RFQ UPDATES TAL AND BUSINESS SHARE

  Transportation allocation, business shares

  Section 7.1 introduced transportation allocations and business shares as a means of influencing the carrier selection decision based, for example, on contractual obligations. How are transportation allocations and business shares considered in a tendering process? If a minimum allocation of three freight orders per week has been agreed upon with a carrier, how is a rejection of a freight RFQ accounted for? Does this freight RFQ count against the minimum allocation or not? The answers to these questions can be influenced by this setting. The following options are available:
  - No update
  - Update of transportation allocations, but no update (free up) upon rejection
  - Update of transportation allocations and business shares, and update (free up) of both upon rejection
  - Update of transportation allocations and business shares, and update (free up) of transportation allocation, but no update (free up) of business share upon rejection
- CARRIER-SPECIFIC FREIGHT AGREEMENT
  If this checkbox is selected, the price limit is calculated per carrier with his specific freight agreement.

Visibility settings

- Visibility settings

  The visibility settings are defined in Customizing (SAP Transportation Management • Transportation Management • Freight Order Management • Tendering • Define General Settings for Tendering) and determine the following:

  - Whether the price limit is disclosed to the carrier (read-only) or not (hidden)
  - Whether the submitted price can be edited, is read-only, or is hidden
  - Whether the stop dates can be edited, are read-only, or are hidden

  This allows you to specify whether the carrier is allowed to change stop dates of stages of the freight order to accommodate the freight order better into his vehicles' schedules. Any changes made by the carrier need to be reviewed to determine whether they fit into your plans before the freight order is awarded.

In the tendering profile, you can combine different steps. The first step can be a peer-to-peer tendering process with the preferred carrier selected based on carrier selection (see Section 7.1). A broadcast tendering to all available carriers defined in the transportation lane is a second step to quickly make an assignment if the preferred option (step one) failed. Which tendering profile is chosen can be decided manually or is specified in the tendering settings of the freight order type (see Figure 7.19). The process settings and communication settings can be defined per freight order type, based on a condition (with condition type /SCMTMS/TEND), or per using default settings. The tendering profile is part of the process settings defined in Customizing (SAP Transportation Management • Transportation Management • Freight Order Management • Tendering • Define General Settings for Tendering).

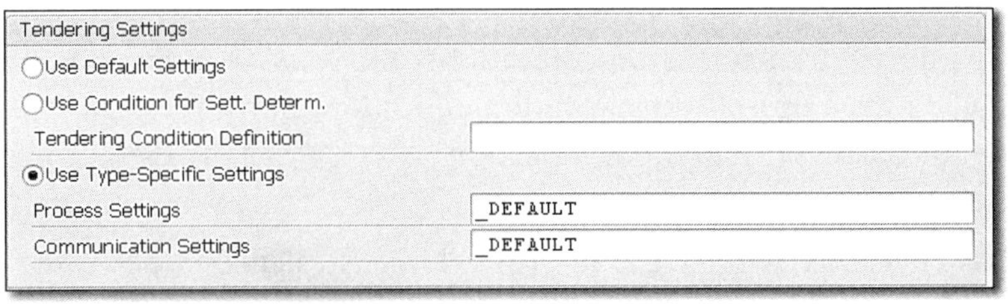

Figure 7.19 Tendering Settings

Tendering Customizing

In addition to the visibility settings, process settings, and communication settings, you can define the following in the tendering Customizing:

- Rejection reason codes
- Email and SMS content, if these communication methods will be used
- Carrier-specific communication settings
- Default settings for the tendering process

The tendering process can be initiated interactively from the freight order user interface or from the transportation cockpit. Additionally, the tendering process can be initiated in the background via report /SCMTMS/TENDBATCH (see Figure 7.20). To be run automatically in the background, report /SCMTMS/TEND_CONT_PROCESS must be regularly scheduled. This report evaluates the responses received from the carriers.

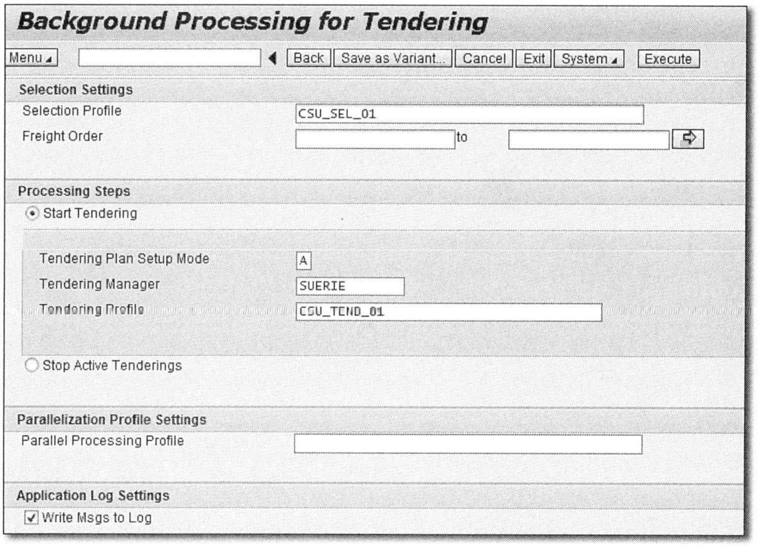

Figure 7.20 Tendering: Background Report

Monitoring

Freight RFQs and freight quotations can be monitored by the responsible user via worklists (FREIGHT ORDER MANAGEMENT • LAND • OVERVIEW TENDERING). Here, alerts are visible in separate queries (FREIGHT QUOTATIONS TO BE REVIEWED, UNSUCCESSFUL TENDERINGS, and TENDERINGS STOPPED DUE TO FO CHANGES).

You can monitor the tendering process for a freight order with its individual steps in the SUBCONTRACTING tab of the freight order user interface. Figure 7.21 shows an example. In the first step, the carrier SH_CARR_US accepted RFQ 510000551, but his submitted price exceeded the price limit. Therefore, he was not awarded the freight order, but a second tendering step has been started with an increased price limit. In the second step, carrier SH_CARR_DE first rejected the offer (RFQ 510000552) because of lack of capacity but later submitted a successful acceptance. If a carrier submits more than one freight quotation, the system will consider only the last freight quotation of each carrier that is submitted within the maximum duration. Carrier SH_CARR_OC, who was also invited in the second step with RFQ 510000553, did not respond.

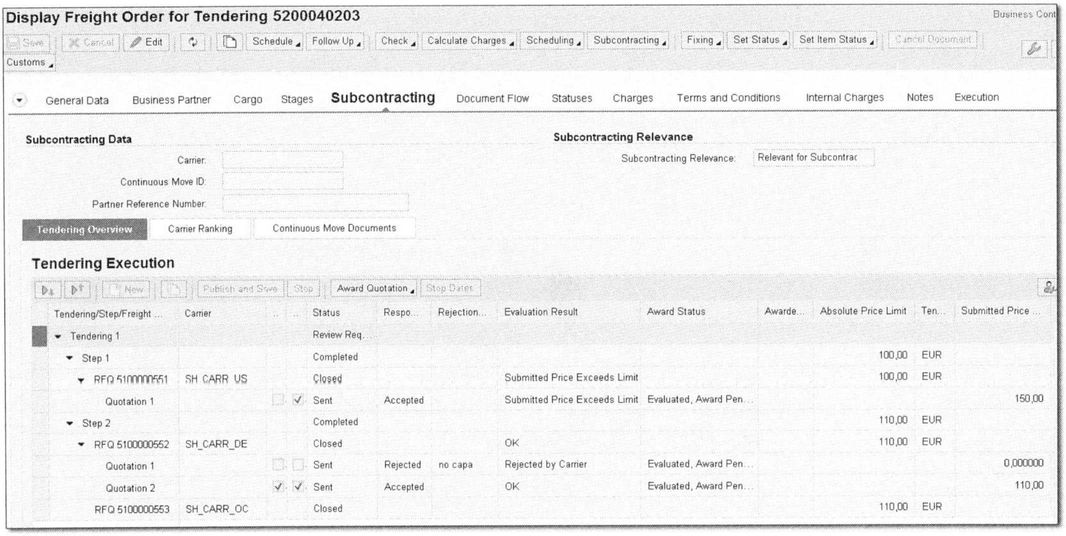

Figure 7.21 Tendering Overview

Connecting to Marketplaces

The freight tendering process expects carriers to communicate directly with the ordering party (the owner of SAP TM). Frequently, marketplaces are used to bring together companies that want to ship goods with carriers that offer these services. With small enhancements, the standard freight tendering process can be utilized to take advantage of freight marketplaces. In this process, the marketplace is specified as the carrier to which the freight RFQ is sent. The marketplace assigns the real carrier, and its response includes the real carrier and price information. Thus, based on the marketplace's response, only the real carrier and its price needs to be assigned to the freight order.

This chapter has dealt with freight orders that are primarily created as the result of planning, especially in land transportation. These can get subcontracted with a carrier selected either via the carrier selection process, the tendering process, or a combination of both. Now that the planning process is complete, the next chapter covers the view on freight orders from the execution perspective, including the printing of freight documents.

SAP Transportation Management and SAP Event Management provide documentation and status management and track and trace capabilities for end-to-end cargo movements.

8 Transportation Execution and Monitoring

Transportation execution and monitoring includes a set of process steps and corresponding functionalities in SAP TM and SAP Event Management that allow you to organize and document activities that involve the physical movement and handling of cargo along the supply chain. The SAP system supports employees working in documentation and freight-handling departments of logistics service providers (LSPs) and shipper's logistics departments to fulfill their responsibilities in a compliant manner and create an audit trail for the actual cargo movement.

Execution and monitoring components

Execution and monitoring of freight movement processes is done using two components of the SAP Business Suite:

- **SAP TM Freight Order Management**
  Before and during execution, the Freight Order Management component allows you to create documents related to the cargo movement and its compliance to document actual quantities and details of the cargo. You can record discrepancies that occur between order capture and planning on the one hand, and physical checking and handling on the other. Moving cargo also results in updates of order and cargo status, such as documenting a loading status either manually out of the SAP TM system or automatically in conjunction with SAP Event Management. The organizational aspect of the cargo movement is supported by export/import handling, which provides you with the handover capability between multiple organizational units along the path of a cargo move.

- **SAP Event Management**
  SAP Event Management is an extremely flexible, efficient, and generic software tool for management and visualization of track and trace processes, status management processes, and key performance indicators based on these process types. You can use SAP Event Management in logistics and beyond to automatically record activity information related to cargo movements, order processes, resource lifecycles, and financial management aspects related to shipments (e.g., cargo prepayment).

In this chapter, we explain the features and background of SAP TM transportation execution and SAP Event Management.

8.1 Transportation Execution

Transportation execution comprises all activities involved with handling and documenting shipments in transit. It is more than just tracking a vehicle on the road, and includes recording any changes to the planned transport and handovers to other business partners or entities.

In the previous chapters we mainly talked about how to use SAP TM to record transportation requirements, how to plan them, and how to subcontract them. Actual pallets, containers, and wagons had not yet been loaded, let alone put into motion.

The Freight Order Management component not only supports transportation management during the planning phase, but is also used after the vehicle has left the loading location.

Before the vehicle actually leaves for the transit, the loading process gathers new information that might be relevant for the shipper or carrier. So far we've been dealing with planned quantities, but now the carrier has to deal with actual quantities, which could differ from the freight quantities ordered. These discrepancies between ordered and actual quantities can be recorded and, depending on the agreement between the ordering party and the carrier, taken into consideration for charge calculation.

In international, multi-modal transports, more than one LSP unit is often involved in the overall transportation. Sometimes importing in the destination country can be very specialized, and a local LSP unit or organization can support the import leg much better than a foreign LSP unit. Therefore, we can observe a handover between the exporting LSP unit (who is also the single point of contact for the customer) and the local importing LSP unit.

In addition, transportation is not only about moving goods from point A to point B. Sometimes it seems like it's more about moving paper or information from point A to point B. Waybills, custom declarations, bills of lading, and other documents need to be generated, printed, and transported with the goods or transmitted in advance. Often legal requirements also need to be respected.

Transportation execution is more than moving goods from one place to another; it involves significant administrative effort, legal requirements, transparency, and organizational interactions. The Freight Order Management component supports all these areas, so we now delve deeper into each of them to show you how you can leverage these requirements with the SAP TM application.

8.1.1 Document Creation

Regardless of which transportation mode is involved, creating, printing, and carrying documents is very important in the transportation process.

Depending on the perspective from which we look at the transportation process, several documents are involved. In this section we concentrate on the creation of bills of lading and waybills.

Bills of lading and waybills

The difference between these two documents is their legal and practical purposes. The *bill of lading* (BL) serves as proof that a contract or order has been issued between a shipper and a carrier stating that certain goods need to be transported. A *waybill* is the more logistical document, listing the goods that need to be transported.

However, in a process involving a freight-forwarder as an agent between an actual carrier and a shipper, the bill of lading and waybill can mean the same thing. The LSP now issues its own bill of lading, called the *house bill of lading*, which is at the same time also called a house waybill. Because waybills are usually seen in the context of the transportation mode, the terminology for waybills is usually used together with the mode of transport (e.g., sea waybill for ocean transports or air waybill for air transports).

If we look at it from a shipper's perspective, for some modes of transport (e.g., ocean) we are obliged to hand over a bill of lading to the consignee that lists all goods to be transported. The bill of lading is a legal and negotiable document enabling the receiver—who is usually the consignee—to claim the goods at port of discharge. On the other hand, the consignee may also sell the goods during ocean transit and hand over the bill of lading to the buyer.

The LSP or carrier who manages the order for the shipper issues the bill of lading. In case of consolidated transportation for different shippers, the LSP will issue himself or receive from the carrier a bill of lading for the consolidation and issue multiple house bills of lading which he provides to the involved shippers.

Making It Simple

Recall that the terminology of waybills is often used in combination with its mode of transport. To make this chapter easier, we will use air transportation as an ongoing example from now on; therefore, we refer to *house air waybills* (HAWBs) and *master air waybills* (MAWBs)

Process controller strategies

House bills of lading can be created from the forwarding order as well as from the freight units, freight orders, or freight bookings. By customizing the forwarding order, you can define how the number of the house bill of lading or HAWB should be put together. We talked about this already in Chapter 4, Section 4.2.1. Nevertheless, you can also define in Customizing how the house bill of lading should be composed by defining a process controller strategy that takes over the job of creating house bills of lading.

Several possibilities are offered with the standard strategies. In general, the house bill of lading is built out of the *items* of a document, not the header data:

- **By shipper and consignee**
  All items containing the same shipper and consignee combination are consolidated into one house bill of lading.
- **By shipper, consignee, and container**
  This method of building a house bill of lading is very similar to the previous possibility. However, this process controller strategy also considers the assignment of the goods into their respective containers. Each container will get its own house bill of lading.
- **By forwarding order**
  If the house bill of lading or HAWB is created by an LSP, all items belonging to the same forwarding order can be put together in one document.
- **By freight documents**
  Again, this scenario is built for LSPs. All items planned on the same document on the main carriage will be consolidated into one house bill of lading.

Which strategy you use and from which document you would like to create the house bill of lading depends on your business case and industry.

Waybill creation

The waybill, on the other hand, is the logistical document passed between two parties cooperating in a transportation business. The waybill document composes the information about the cargo and the transportation route and terms.

The consumer of a transportation service and the provider of that transportation service share a common number range from which the consumer can draw a number himself and give this number to the provider. This number is then a unique referral for both parties of the transportation business.

In SAP TM, the number ranges are stored in *waybill stock IDs* that define agreed upon number ranges.

Waybills in SAP TM can be separated into house waybills and master waybills. The house waybills represent the transportation documents between the sales side of the company using SAP TM and an ordering party. The master waybills represent the purchasing document between the company using SAP TM and a carrier. You can compare this differentiation with freight settlement documents and forwarding settlement documents—they are similar-looking documents, but are built for different parts of the transportation process.

Waybills Are Not a Separate Entity

Other than with freight settlement documents and forwarding settlement documents, waybills are not represented with their own separate entities in the SAP TM system. Waybills only serve as print documents on the forwarding order and freight orders, freight bookings, and freight units. These documents provide the functionality of drawing the right number. However, there is no separate business object designed for the waybills.

Defining number stock types

To create waybill stock IDs, you need to carry out some customizing activities. Remember that the number ranges in SAP TM are stored as waybill stock IDs. To create these waybill stocks, we need to define *waybill stock types*, which we can do in Customizing via the IMG menu path SAP Transportation Management • Transportation Management • Master Data • Waybill Stock • Define Waybill Number Stock Types.

As shown in Figure 8.1, you can customize several details about the waybill number stock before you define the waybill number stock itself. Waybill number stocks are always transportation mode-dependent, so the assignment of a transportation mode is mandatory in this Customizing activity. With the stock category, you specify whether the number stock is used as a waybill number or as a tracking number.

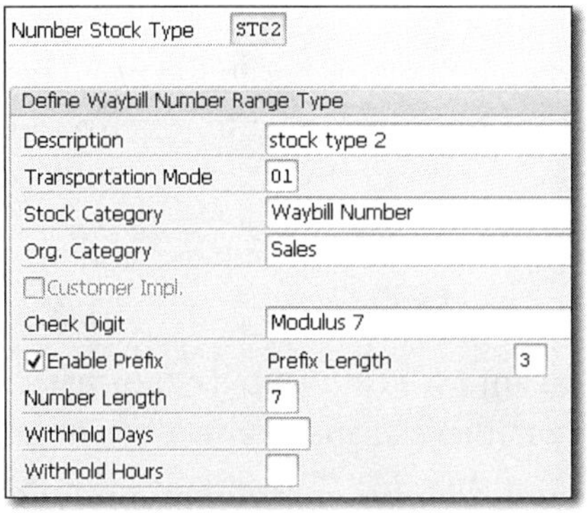

Figure 8.1 Waybill Number Stock Customizing

Waybill Numbers and Tracking Numbers

Technically, there is no difference between a waybill number and a *tracking number*. Both numbers are used to uniquely identify a transport in communication between the transportation service consumer and the provider.

However, the different terms are used in different transportation scenarios. Air, sea, road, and rail transportation use the waybill number; we refer to tracking numbers in parcel scenarios.

Later in this chapter you will see that tracking numbers are maintained the same way waybill numbers are.

Assigning the waybill stock type

The organization category defines whether the waybill number stock we are creating is supposed to be used for master waybills or house waybills. Recall that house waybills are used between the sales side and an ordering party, while master waybills are used between the purchasing side and a carrier. Therefore, the house waybill usually represents the entire transportation ordered, and the master waybill represents only a certain transportation leg.

The lower part of the Customizing activity illustrated in Figure 8.1 addresses how the number should be composed. First, you can define that a carrier-specific prefix should be added by selecting the ENABLE PREFIX checkbox.

Carrier Prefix

The prefix for the carriers is also defined in Customizing. In the Customizing activity found via the IMG menu path SAP TRANSPORTATION MANAGEMENT • TRANSPORTATION MANAGEMENT • MASTER DATA • BUSINESS PARTNER • DEFINE IATA AIRLINE CODES, you can assign a waybill prefix to the carriers.

Airline-specific prefixes

In addition, waybill numbers may have a check digit. SAP TM offers two possibilities for automatically adding a check digit to the waybill number. The waybill number is either calculated with Modulus 7 or Modulus 10. If none of the provided calculation rules fit the specific waybill number stock requirements, you can implement custom implementation using a BAdI.

Calculation with Modulus

Calculation with Modulus is often used in IT. When you use the calculation operation `MOD`, the base number is first divided by the divisor. But it's the remainder after the division that's important, not the result of the division.

For example: 11 `MOD` 3 = 2.

As you would expect, the waybill number stock type also defines the general length of the waybill number. The waybill number is always extended with leading zeros to match the waybill number length defined in the waybill stock type.

In certain processes, master waybill numbers can be returned, such the cancellation of a freight booking for which a waybill number was already assigned. Though the returned number cannot directly be reused by other documents, it needs to be withheld for a certain time because the cancellation might be replicated into other systems as well. Therefore, a withholding time in days and hours can be maintained in the waybill stock type.

Actual waybill number ranges

After defining a waybill stock type, we can create the actual waybill number ranges. In SAP NetWeaver Business Client (NWBC) you can find the corresponding menu entry with the menu path Master Data • General • Overview Waybill Stock. Here you can see a POWL that differentiates between house waybills, master waybills, and tracking numbers.

When creating a new waybill stock with the New button on the top of the POWL list, you will see that you need to select a waybill stock type first. After you've chosen a waybill stock type, the definitions from Customizing are automatically propagated into the waybill stock.

A Short Note on Terminology

In this chapter we use the terms *waybill stock, waybill stock IDs,* and *waybill number ranges.* All these terms describe the same entity in SAP TM: the waybill stock.

Defining waybill stocks

With the waybill stock, you now define the actual number range for the waybill document in the FROM NUMBER and TO NUMBER fields. For air waybills, the waybill stock needs to be assigned a specific airline to represent the airline's prefix, as you can see in the top-right corner of Figure 8.2. However, as you can see in the inset, the waybill stock can be defined among several organizational units (sales organizations for a house waybill and purchasing organizations for a master waybill) and several external parties (ordering parties in house waybills and carriers in master waybills).

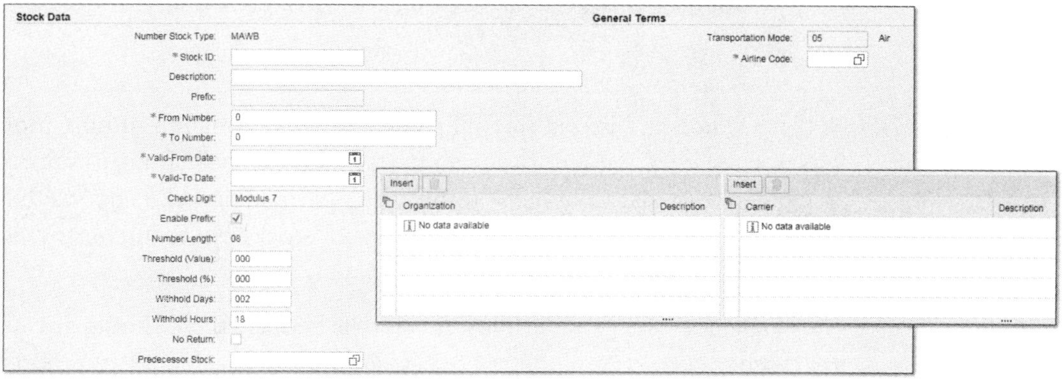

Figure 8.2 Definition of Waybill Stock

Automatically drawing a waybill number

Waybill numbers can automatically be drawn on all SAP TM documents that are related to waybills. In freight orders and freight bookings you can do this with a follow-up activity called DRAW MAWB NUMBER; in forwarding orders you have a separate button in the global toolbar called HBL, where you find the action DRAW HBL NUMBER.

Numbers are drawn for waybills based on the waybill stocks that we defined earlier. For house waybills, the system looks for waybill stocks that the sales organization and ordering party of the corresponding forwarding order assigned. If several waybill stocks are found, a pop-up displays where the user can choose between the different relevant waybill stocks. During searches for waybills stocks, the validity dates of the waybill stock are also considered; only waybills stocks that are currently valid are taken into consideration.

Predecessor Stocks

If several valid waybill stocks represent the required organizational unit/ ordering party combination, you need to manually choose the waybill stock from a pop-up.

However, you can use the predecessor stock functionality to define a priority among waybill stocks that are valid at the same time. You can see in Figure 8.2 that you can insert an ID of another waybill stock as the predecessor stock. If a predecessor stock is maintained, numbers from this predecessor stock are drawn first.

Only when the predecessor stock is exhausted will the next waybill stock be considered.

If there is no valid house waybill stock that represents the combination of sales organization and ordering party, then the system next looks at waybill stocks that only have an ordering party assigned. If no waybill stocks are found in this case either, waybill stocks are then considered that only have sales organizations assigned.

This logic does not exist for master waybills because these should always represent a certain combination of carriers and purchasing organizations.

If the waybill stock is running out of numbers, you can define a threshold value (either an absolute value or a relative value) for your waybill stock. When the use of the waybill stock exceeds the threshold value, the user sees a warning message when drawing another number from the almost-exhausted waybill stock.

Delegating waybill numbers

When a waybill stock that represents a number range agreed upon between an organizational unit and an external party is eventually exhausted, the organizational unit has to approach the external party to agree on a new number stock. However, if the organizational unit foresees that only one more number is needed, this number can be drawn from another waybill stock that was not designed for the relationship between this organizational unit and the external party. This process is called *delegation* of a waybill number. When you access your waybill stock in SAP NWBC, as shown in Figure 8.2 in the lower part of the screen, you can see which waybill numbers have already been used and on which documents they were used.

In the example in Figure 8.3, one waybill number has already been drawn and used from the waybill stock. You could access the freight document directly from here. If a number was returned but the withholding time has not yet expired, the number is still displayed together with the freight document it was previously used on, but the status of the number is *Returned*.

If you want to delegate a number to another organizational unit as just explained, you can also do this in the waybill stock. In Figure 8.3 you can see that there is a DELEGATE button above the overview of waybill numbers. If you click on this button, the next available number from the waybill stock is drawn and added to the list of numbers in the DETAILS area of the waybill stock. The status is set to DELEGATED, and you can now enter the organizational unit to which you would like to delegate the number.

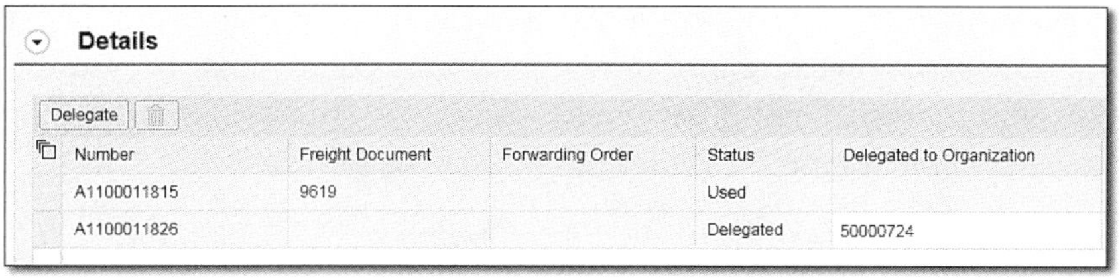

Figure 8.3 Used and Delegated Waybill Numbers

Restrictions on Delegation

Please note that delegating waybill numbers is only possible for master waybills, not for house waybills. Furthermore, the number will only be delegated to another purchasing organization; the carrier has to remain the same.

Once the delegation is entered in the waybill stock, the next automatic drawing of a waybill number takes the delegated number into consideration.

Organizational structures

As you know from Chapter 3, organizational units can be created hierarchically, representing the responsibilities of some organizational units to other units. The organizational hierarchy is considered in the number drawing of waybill numbers as well. In waybill stocks, you can enter not only purchasing organizations as organizational units, but also other functional roles such as forwarding houses and companies.

If different waybill stocks are defined for purchasing organizations and forwarding houses, the automatic number drawing only considers the more specific organizational unit; in this example, this is the purchasing organization. Only if the waybill stock of the most specific organizational unit is exhausted, is the next higher level considered as a fallback solution. With this functionality you can make sure you have some fallback numbers maintained if a waybill stock unexpectedly runs out of numbers.

Printing waybills

Now that we've discussed the process of drawing waybill numbers, let's take a look at the waybill itself. In Customizing of the freight order or freight booking type, you define which documents can be printed out of the corresponding document. Via the IMG menu path SAP TRANSPORTATION MANAGEMENT • TRANSPORTATION MANAGEMENT • FREIGHT ORDER MANAGEMENT • FREIGHT BOOKING • DEFINE FREIGHT BOOKING TYPES (or the definition of freight order types), you can assign two output profiles to the document type, as shown in Figure 8.4.

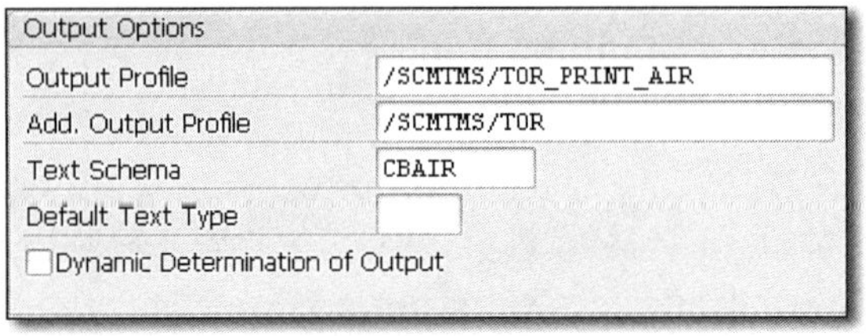

Figure 8.4 Assignment of Output Profiles to Document Type

The output profile defines which documents may be printed for the document. Therefore, you can find output profiles for each supported transportation mode in Customizing. If you want to define your own output profiles, you can do this in the Post-Processing Framework that was discussed in Chapter 2, Section 2.3.3.

Figure 8.5 shows an MAWB in the standard format that SAP TM provides. You can change the layout of this document in Transaction SFP with Adobe Document Server, as we described in Chapter 2, Section 2.3.4.

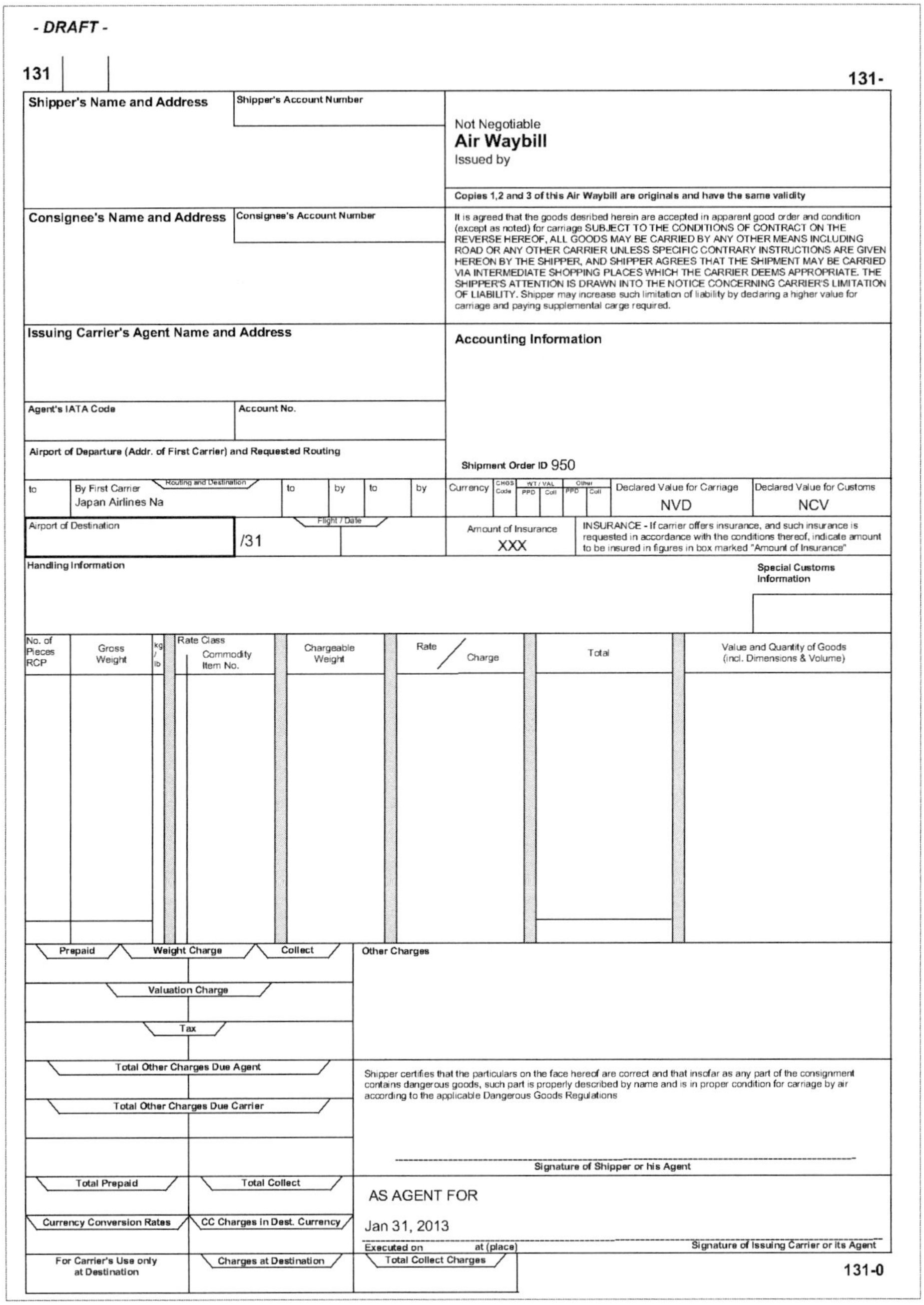

- DRAFT -

131 | | 131-

Shipper's Name and Address | Shipper's Account Number

Not Negotiable
Air Waybill
Issued by

Copies 1,2 and 3 of this Air Waybill are originals and have the same validity

Consignee's Name and Address | Consignee's Account Number

It is agreed that the goods desribed herein are accepted in apparent good order and condition (except as noted) for carriage SUBJECT TO THE CONDITIONS OF CONTRACT ON THE REVERSE HEREOF, ALL GOODS MAY BE CARRIED BY ANY OTHER MEANS INCLUDING ROAD OR ANY OTHER CARRIER UNLESS SPECIFIC CONTRARY INSTRUCTIONS ARE GIVEN HEREON BY THE SHIPPER, AND SHIPPER AGREES THAT THE SHIPMENT MAY BE CARRIED VIA INTERMEDIATE SHOPPING PLACES WHICH THE CARRIER DEEMS APPROPRIATE. THE SHIPPER'S ATTENTION IS DRAWN INTO THE NOTICE CONCERNING CARRIER'S LIMITATION OF LIABILITY. Shipper may increase such limitation of liability by declaring a higher value for carriage and paying supplemental carge required.

Issuing Carrier's Agent Name and Address

Accounting Information

Agent's IATA Code | Account No.

Airport of Departure (Addr. of First Carrier) and Requested Routing

Shipment Order ID 950

to	By First Carrier (Routing and Destination)	to	by	to	by	Currency	CHGS Code	WT / VAL PPD / Coll	Other PPD / Coll	Declared Value for Carriage	Declared Value for Customs
	Japan Airlines Na									NVD	NCV

Airport of Destination | Flight / Date: /31

Amount of Insurance: XXX

INSURANCE - If carrier offers insurance, and such insurance is requested in accordance with the conditions thereof, indicate amount to be insured in figures in box marked "Amount of Insurance"

Handling Information

Special Customs Information

No. of Pieces RCP	Gross Weight	kg / lb	Rate Class	Commodity Item No.	Chargeable Weight	Rate / Charge	Total	Value and Quantity of Goods (incl. Dimensions & Volume)

Prepaid	Weight Charge	Collect
	Valuation Charge	
	Tax	
	Total Other Charges Due Agent	
	Total Other Charges Due Carrier	

Other Charges

Shipper certifies that the particulars on the face hereof are correct and that insofar as any part of the consignment contains dangerous goods, such part is properly described by name and is in proper condition for carriage by air according to the applicable Dangerous Goods Regulations

Signature of Shipper or his Agent

Total Prepaid | Total Collect

Currency Conversion Rates | CC Charges in Dest. Currency

AS AGENT FOR

Jan 31, 2013

Executed on | at (place) | Signature of Issuing Carrier or its Agent

For Carrier's Use only at Destination | Charges at Destination | Total Collect Charges

131-0

Figure 8.5 Example Master Air Waybill

When you want to print a waybill from a freight document such as a freight order or freight booking, click on the PRINT button in the global toolbar of either document. After you've decided whether to display a print preview in PDF format or trigger the actual printout, you choose the document you want to create from a pop-up. The choice of documents depends on the output profiles that are assigned to the document type in Customizing.

House waybills on forwarding orders can be printed and viewed on the OUTPUT MANAGEMENT tab. If the output profile was assigned to the forwarding order type, the document automatically appears on this tab.

8.1.2 Discrepancies

So far we have only discussed transportation process steps that deal with requested and planned quantities. However, in some cases the actual quantity cannot be estimated precisely beforehand. When the transportation execution starts, the actual quantity needs to be recorded as well. This process in SAP TM is called *discrepancy handling*.

Discrepancies are a hassle for transportation execution because they might affect the choice of vehicle resource being used for the transportation or lead to a different charge calculation. Therefore, discrepancies need to be discussed with the shipper before transportation can continue.

The transportation process starts as usual, requested quantities are entered into a forwarding order, and freight units are created out of this document. Once the freight units are planned, the execution of the transportation may begin. The carrier now receives the cargo physically and checks the actual quantities against the requested quantities.

This checking can be done on the CARGO tab of either the freight order or the freight booking, depending on the required document flow. On this tab, which is shown in Figure 8.6, you can see that each item of the freight units or forwarding order is displayed separately. In this example, we have created a one-click scenario, creating a freight order directly from a forwarding order. You can see that the requested quantities are also provided, but the actual quantities are still empty.

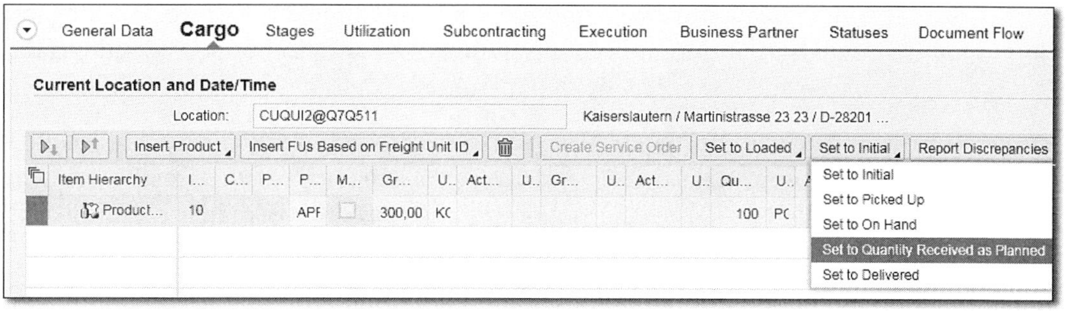

Figure 8.6 Cargo Tab on Freight Order

If the carrier receives the cargo and realizes that the actual quantity matches the requested quantity, you can just fill the actual quantities with the requested quantities by selecting SET TO LOADED • SET TO QUANTITY RECEIVED AS PLANNED from the toolbar above the list of items.

Quantity discrepancies, other discrepancies

However, if the carrier does not receive the cargo as planned, you need to report the discrepancy. In general, we can differentiate between two types of discrepancies:

- **Quantity discrepancies**
  The actual quantities are different from the requested quantities because of a change in the quantity, gross weight, or volume.
- **Other discrepancies**
  Discrepancies that are not caused by a change of quantities are called *other discrepancies*. If the cargo is damaged or documents are missing for the cargo, these events can be recorded as other discrepancies.

When you discover a quantity discrepancy, you can enter the actual quantities in the corresponding fields of the CARGO tab. The system automatically checks the actual quantities against the requested quantities. If a discrepancy exists, it is automatically reported.

Discrepancy types

In Customizing you can define different types of discrepancies. You can find this Customizing activity by following the IMG menu path SAP TRANSPORTATION MANAGEMENT • TRANSPORTATION MANAGEMENT • FREIGHT ORDER MANAGEMENT • DISCREPANCY TYPES.

For the discrepancy type, you can define a tolerance range, meaning that if the actual quantities are within the tolerance range defined, no discrepancy is recorded. The tolerance range is defined as a percentage.

Once a discrepancy is recorded, the actual quantities of the subsequent transportation stages and the actual quantity of the freight units are updated. The carrier now has to discuss the discrepancy with his customer before the execution of the transport may continue. Therefore, once a discrepancy is reported, all transportation stages carrying a freight unit with unresolved discrepancies get a planning and execution block. In discrepancy type Customizing, however, you can specify that the reported discrepancy of a special discrepancy type does not lead to a planning and execution block. If you remove the freight unit with unresolved discrepancies from the freight document, the planning and execution block is also removed.

Resolving discrepancies

In the freight order on the Cargo tab, you can select the item for which you have entered discrepant quantities; in the Details area below the table of items, the reporting of discrepancies was protocolled and a corresponding event (assigned to the discrepancy type in Customizing) was triggered. In this area, you can add notes and attachments to this discrepancy, such as a weighing protocol or photographs of damaged goods.

If the carrier discusses the discrepancy directly with the shipper, you can report the resolution of the discrepancy directly on the Cargo tab. From the toolbar above the list of items, select Report Discrepancies • Set Discrepancies to Resolved, as shown in Figure 8.7.

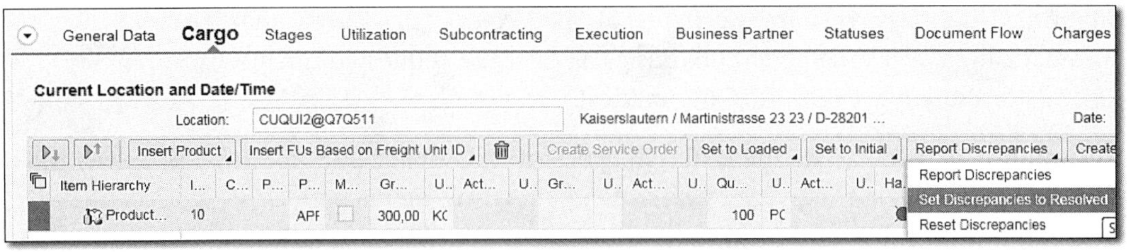

Figure 8.7 Resolving Discrepancies in the Freight Order

Once the discrepancy is set to Resolved, the planning and execution block is removed.

In other cases, the carrier notifies only the freight-forwarder of the discrepancy, and that freight-forwarder has to discuss the discrepancy with the shipper. Therefore, the discrepancy is also propagated to the forwarding order. If you select the item of the forwarding order, you'll see the Discrepancies tab in the Details area of the forwarding order items.

As you can see in Figure 8.8, the freight-forwarder can now also set the discrepancy to *Resolved* by selecting the Discrepancy Resolved checkbox, which leads to the removal of the planning and execution block of the stages in the assigned freight documents.

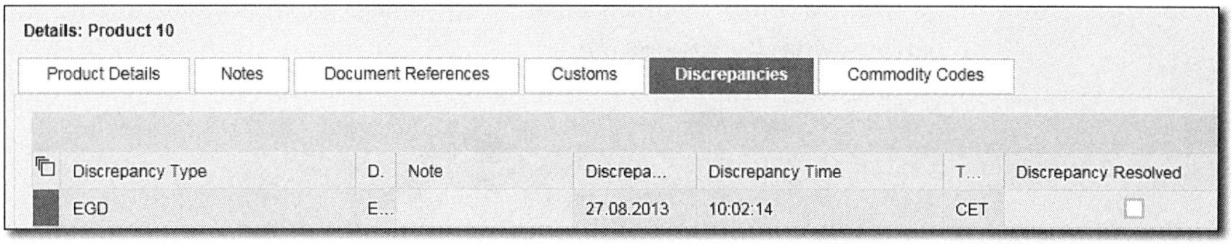

Figure 8.8 Resolving Discrepancies in the Forwarding Order

If the cargo was already damaged when handed over to the carrier, the carrier can report discrepancies as well. As you can see in Figure 8.7, the Report Discrepancies button also offers the possibility to report discrepancies without entering actual quantities in the cargo information of the freight order or freight booking. The processing of other discrepancies is exactly the same as the processing of quantity discrepancies.

If you have reported a wrong discrepancy or the discrepancy was resolved by changing the quantity, you can click on the Reset Discrepancies button to reset the discrepancies so that the actual quantities no longer differ from the requested quantities.

8.1.3 Export/Import Processing

In international, multi-modal transports, several organizational units or even business partners are often involved in the planning and execution of the transport.

Interaction between export/import organizations

It's not uncommon for an export organization to organize the pre-carriage and main leg of the transportation while an import organization deals with the on-carriage of the same transport. Due to customs regulations and special circumstances in the importing country, this makes sense because organizations with local knowledge can participate in the transportation planning, making the transport more efficient and, in most cases, cheaper.

Let's take a deeper look at how the interaction between export and import organizations is established in SAP TM. As an example of the process, we concentrate on an ocean transport ordered at the export organization. The export organization will deal with the pre-carriage and the main carriage, while the import organization will be responsible for the organization and execution of the import transportation leg. Figure 8.9 illustrates this division of labor.

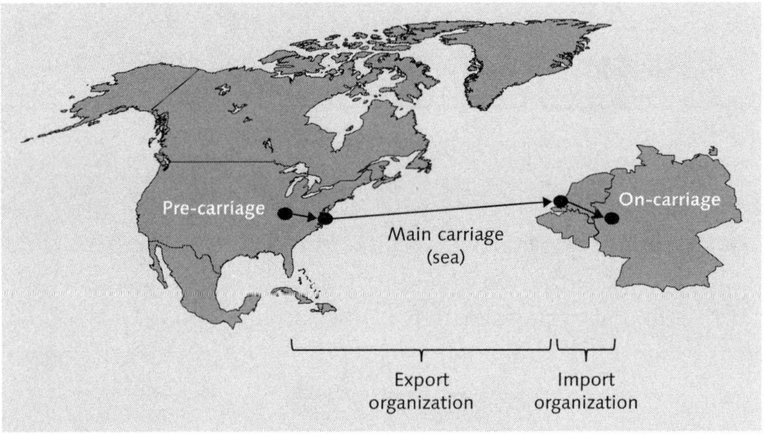

Figure 8.9 Responsibilities in an Export/Import Process

Export and import scenarios

In general, we can differentiate between different scenarios in the export/import process:

- **Full container load (FCL)**
  Since a full container is shipped completely from the shipper to the carrier, there is exactly one transportation request on the export side and one transportation request on the import side. The number of freight documents are also exactly one on each side.

- **Less than container load (LCL)**
  The number of transportation requests on either side still matches. However, since several transportation requests are now consolidated into one container, more than one transportation request is created on either side, but there is still only one freight document on each side.
- **Buyer's consolidation**
  In this scenario, the export side needs to deal with several pickup transports from many shippers and consolidate them at the port. The import side, however, only receives one container that can be delivered as a whole to the consignee. Therefore, the export side creates several transportation requests and one freight document (for the main leg), but for the import side it is sufficient to create only one transportation request.
- **Shipper's consolidation**
  Just like in the buyer's consolidation scenario, consolidation effects should be exploited. In this scenario, the shipper creates only one transportation request (because the pre-carriage and on-carriage are transported in one container) and one freight document, but the import side now needs to create one transportation request and one freight document per consignee (also because one waybill needs to be created per consignee).

As you've seen while browsing through the different scenarios, two SAP TM documents are crucial to using the export and import process: the transportation request (in our case, a forwarding order) and the freight document (in our case, freight booking).

Import/export organizations on forwarding order

If you know the forwarding order quite well, you will be wondering how the information about the export and import organizations is stored on the forwarding order. In the forwarding order, you can define whether the forwarding order should be for export or for import purposes with the Traffic Direction field. If the forwarding order is an export forwarding order, the sales organization of this transportation request is the export organization of the entire transport. If you are viewing an import forwarding order, the sales organization of this forwarding order acts as the import organization for the transport.

On an export forwarding order, the import organization is assigned on the BUSINESS PARTNER tab shown in Figure 8.10. With the party role IMPORT ORGANIZATION, you can assign the business partner created for the organizational unit to the export forwarding order as the import organization. This also applies to import forwarding orders, where you can assign an export organization in the same way.

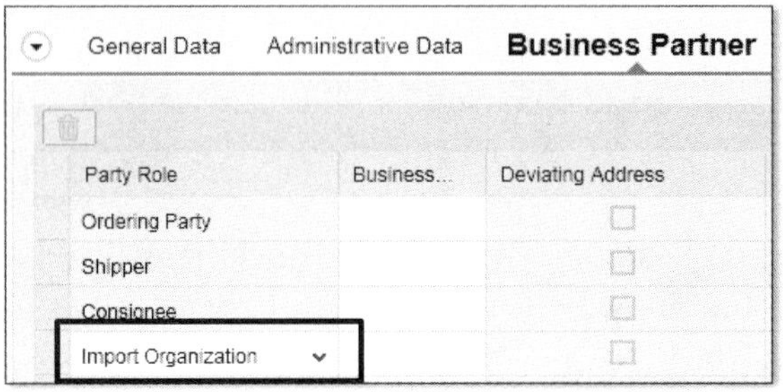

Figure 8.10 Import Organization on an Export Forwarding Order

Business Partners for Organizational Units

So the import and export organizational units have to be created as business partners as well. They also need to be assigned to the organizational units.

There are several ways to create organizational units and business partners and to link these two entities. The easiest way is to ensure that a business partner is automatically created upon creation of the organizational unit. Maintain the following entry in the database table T77S0:

GRPID = HRALX
SEMID = HRAC
GSVAL = X

If you have already created organizational units and business partners separately, you can also assign the business partner to the organizational unit in Transaction PPOME. Manually assign the business partner in the ORG. DATA tab.

The report /SCMB/ORG_CREATE_BP_ASSIGNMENT automatically links organizational units to business partners if both entities have been created already. To establish the link between organizational unit and business partner, the business partner's search term 1 needs to be the name of the organizational unit, and search term 2 needs to be the description of the organizational unit.

Internal and external communication

Let's take a look at how the export and import organizations interact and which documents need to be created in SAP TM. In general, we can differentiate between an internal communication of two organizational units and an external communication. In the internal communication, both organizational units use an SAP TM system and even access the same client; this means that no web service-based information flow needs to be established. In the external communication, the two parties use different transportation management systems (TMSs), but at least one of them uses SAP TM (otherwise, we wouldn't describe this case in this book!).

Internal communication process

Let's first take a look at the ideal case of both organizational units working in the same SAP TM system.

The general transportation process starts as usual with the creation of a forwarding order. In our case, we create an export forwarding order. The traffic direction (which defines whether we are dealing with an export or an import forwarding order) can be entered manually in the document or can be predefined in the forwarding order document type via the IMG menu path SAP TRANSPORTATION MANAGEMENT • TRANSPORTATION MANAGEMENT • FORWARDING ORDER MANAGEMENT • FORWARDING ORDER • DEFINE FORWARDING ORDER TYPE in Customizing. It makes sense to define one dedicated forwarding order type for export forwarding orders and another for import forwarding orders.

As we have already discussed, the export organization acts as the forwarding order's sales organization, while the import organization is an additional party role on the BUSINESS PARTNER tab. For the shipper's consolidation scenario, you might not want to define one import organization for the entire forwarding order, but instead define several import organizations for each item of the forwarding order in the DETAILS area of the forwarding order items, as discussed in Chapter 4, Section 4.2.1.

Once the forwarding order is defined, the export organization can start planning the pre-carriage and the main carriage. The method of planning the two stages is completely up to the planner. We have taken a look at the different ways of planning in Chapter 6.

Because the export and import processes usually take place in ocean or air scenarios, SAP TM supports these processes only when using freight bookings, which can be either an ocean booking or an air booking. This means that to be able to realize the interaction between the export and import organizations, the main leg has to be planned on a freight booking.

The traffic direction of the freight booking must now also be EXPORT, since it was created by the export organization. In addition, we again enter the import organization on the BUSINESS PARTNER tab of the freight booking. The export organization now has to deal with all the necessary process steps regarding the customs declaration for exporting the goods. The freight booking in this case is therefore used as the supporting document for the export declarations but also in general for the capacity reservation at the carrier.

Once the freight booking is set to SHIPPED ON BOARD (in air freight booking, the status is *Uplift Confirmed in Air*), as shown on Figure 8.11, the SAP TM system automatically starts generating the import documents. From a process perspective, this means that the import organization is only notified by the upcoming transportation request after the cargo is loaded onto the vessel or the aircraft.

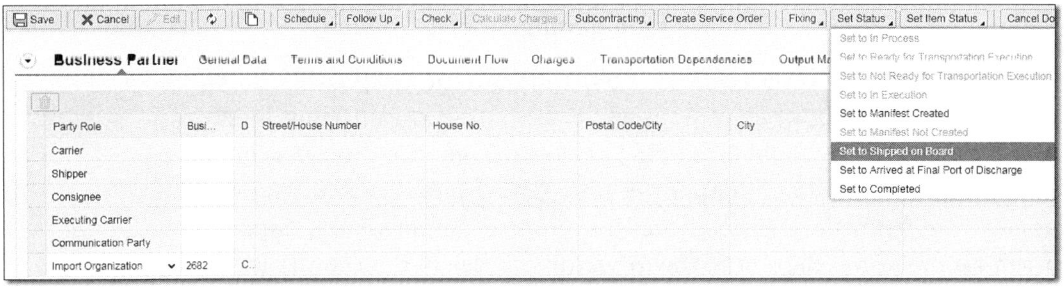

Figure 8.11 Setting the Ocean Freight Booking to Shipped on Board

Automatic generation of import documents

Once the execution status of the freight booking is set to *Shipped on Board*, the automatic generation of import documents is triggered by the Post-Processing Framework (PPF).

To enable the internal communication—that is, create the import documents within SAP TM—the corresponding PPF action has to be assigned and activated for the business partner assigned to the import organization. In the master data of the business partner, follow the SAP NWBC menu path Master Data • General • Define Business Partner and select the Output Management tab. On this tab (shown in Figure 8.12), you need to check whether the corresponding PPF actions are activated for the business partner. For internal communication, the PPF action /SCMTMS/COPY_EXP_BOK_IMP_BOK needs to be activated and assigned; the action /SCMTMS/TOR_BKWBIL_NTF_EXT needs to be deactivated.

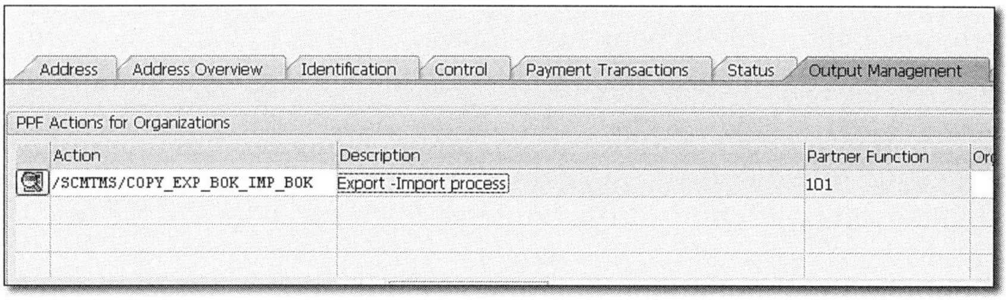

Figure 8.12 Assignment of PPF Action to Business Partner

The import documents are created by a PPF action. For this to occur, import document types for the import freight booking, import forwarding order, and import freight units need to be defined up front. The determination of the correct document types during the automatic creation of import documents can be done in two ways:

Determining import document types

- **Assignment of import document type on export document type**
  In Customizing of the forwarding order type as well as in Customizing of the freight booking type, you can directly assign a corresponding document type for the import document. During the creation of the import document from the export document, the corresponding document type is taken into consideration.
- **Condition-based document type determination**
  SAP TM offers two condition types for the condition-based determination of the import forwarding order type and the import freight booking type:
  - The condition type /SCMTMS/FWO_TYPE can be used to determine the forwarding order type for the import document.

- The condition type /SCMTMS/FRB_TYPE_IMP can be used to determine the import freight booking type.

Both condition types are singleton conditions, meaning they cannot be assigned anywhere in Customizing, and you need to define one global condition for all use cases.

The SAP TM application first creates an import freight booking out of the export freight booking. The import organization can use this import document to carry out the import declarations for customs, and it is also the basis on which to plan the on-carriage.

After the import freight booking is created, the import forwarding order and freight units are also created from the booking. The freight units are now created by the freight unit building rule (FUBR) assigned to the import forwarding order type, so the automatic import document creation actually has nothing to do with the creation of import freight units. This is done independently by the forwarding order functionality.

Relation of Forwarding Orders and Freight Bookings

Remember the beginning of this chapter when we discussed buyer's and shipper's consolidation? For a shipper's consolidation, one import booking is created, but because the items will eventually be delivered to different consignees, several import forwarding orders are created—one for each consignee.

With the import documents created, the import organization can now start with the import declarations and with planning the on-carriage based on the freight units created out of the import forwarding order.

On the import freight booking, the purchasing organization is now automatically the import organization, and the export organization is entered as the additional party role in the Business Partner tab. The same applies to the import forwarding order that was automatically created. The import organization now acts as the sales organization of the forwarding order, and the export organization is shown on the Business Partner tab in the corresponding party role.

Import documents in draft status

The import forwarding order and the import freight booking were created in the draft status. In a document with this status, data cannot be changed, except for the purchasing organization in the freight booking or the sales organization in the forwarding order. The import organization now has to check the forwarding order and the freight booking for completeness; then it can set the status manually to *In Process* when starting the on-carriage planning.

Import Documents and Service Items

In SAP TM, the standard process is to copy only cargo items from the export documents to the import document. Service items are not transferred. However, you can influence the system's copy logic by implementing a BAdI via the IMG menu path SAP Transportation Management • Transportation Management • Business Add-Ins (BAdIs) for Transportation Management • Basic Functions • Export/Import Processing • BAdI: Service Item Processing for Import Forwarding Orders. With this BAdI, you can make changes to the copy logic so that service items are copied to the import document.

External communication scenarios

The external communication scenario is not especially different from the internal communication scenario because the physical and legal process does not differ from the internal communication. Only the use of TMSs is different here.

With the external communication, we need to differentiate between two cases:

- **The export organization uses SAP TM**
  In this case, we can start the process just like in the internal communication scenario because we are going to create an export forwarding order and export freight booking. After that, the information concerning the import is sent out to the import organization's TMS.
- **The import organization uses SAP TM**
  If the import organization uses SAP TM, the transportation request communicated by the customer is recorded in an external TMS. Only after the cargo is loaded into the vehicle executing the main transportation leg is a message sent to SAP TM to create import documents.

We now take a look at both cases, starting with the export organization using SAP TM.

As already mentioned, the export organization—just like in the internal communication scenario—starts creating an export forwarding order and plans the freight units created out of this forwarding order on an export freight booking. Now when execution status is set to SHIPPED ON BOARD, as shown in Figure 8.11, the system needs to react differently.

To enable the different behavior, the PPF actions that we saw earlier in this chapter need to be assigned differently to the business partner of the import organization. Now the PPF action /SCMTMS/COPY_EXP_BOK_IMP_BOK needs to be disabled, and the PPF action /SCMTMS/TOR_BKWBIL_NTF_EXT has to be enabled on the OUTPUT MANAGEMENT tab of the business partner, as shown in Figure 8.13.

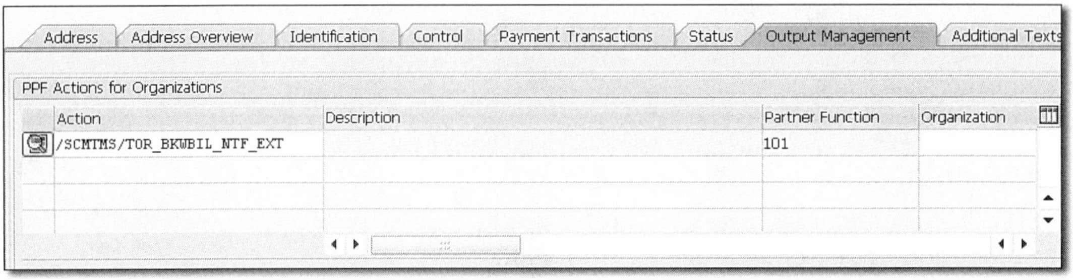

Figure 8.13 Enabling External Communication for Import Organization

Sending export freight bookings

Now a web service is called that sends out all the necessary information to the import organization's connected external TMS. The service interface `TransportationOrderBookingWaybillNotification_out` is sent to SAP NetWeaver PI, where the routing of the message is processed.

System Landscape Setup

To use external communication between the export and import organizations, the external TMS needs to be connected to the system landscape on which the SAP TM system is located. We recommend that you connect the external TMS with the SAP TM system using SAP NetWeaver PI or any other middleware.

Receiving import information

When the import organization uses the SAP TM system, another service interface can be used. The service interface `TransportationOrderBookingWaybillNotification_In` integrates the information from an external TMS into SAP TM and triggers the creation of an import freight booking as well as the creation of an import forwarding order based on the information provided by the service interface.

Let's compare internal and external communication. First, SAP TM supports both scenarios. However, internal communication offers some advantages because export document updates are received seamlessly on the importing side. In addition, internal communication provides more transparency because the import organization, using the same system, can be notified by upcoming imports earlier. They can proactively look for export documents with shipments to the region the import organization is responsible for.

Manually creating import freight bookings

In rare cases, manual creation of freight bookings is necessary. Import freight booking can therefore be created manually in the system. As always with the manual creation of freight bookings, you need to ensure that the necessary information from the export freight booking is correctly copied into the import freight booking. On manually created import freight bookings, you cannot use all the functionalities that you usually use on an export freight booking, such as the following:

- **Subcontracting the freight booking**
  Because planning and execution is done by the export organization, the import organization does not have to do anything about subcontracting the freight booking.
- **Assignment of schedules**
  Like with subcontracting of the freight booking, planning (and therefore also schedule assignment) is done by the export organization.
- **Automatic drawing of master waybill number**
  The waybill number has been negotiated by the export organization and the carrier. Therefore, the import organization is not aware of the number ranges available for the freight booking. However, if the waybill number is already known from the export document, the user can manually enter the waybill number in the import freight booking.

After the import freight booking is manually created, the user can also manually create the import forwarding order as a follow-up action, as shown in Figure 8.14.

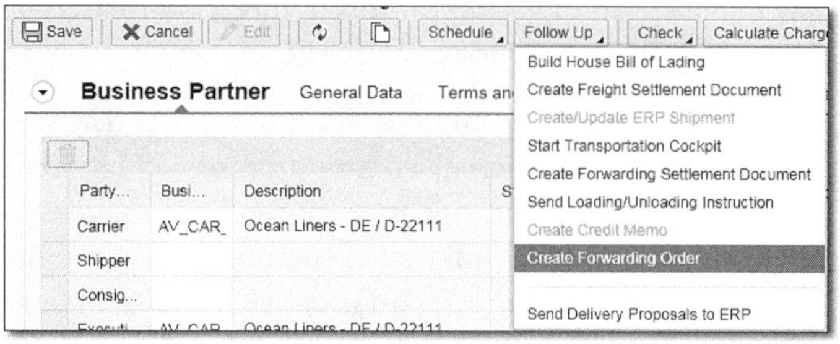

Figure 8.14 Creating an Import Forwarding Order from the Import Freight Booking

8.1.4 Statuses

If you look at the STATUSES tab in the freight order user interface, you will see a huge number of statuses that indicate the progress of the various processes a freight order can handle. As you can see in Figure 8.15, these show not only statuses of the execution of the freight order, but also of process steps that are handled before the actual execution. They include SUBCONTRACTING STATUS as well as statuses that are of a more technical nature, such as FIXING STATUS and ARCHIVING STATUS. However, this section is only concerned with statuses of the execution of the transport.

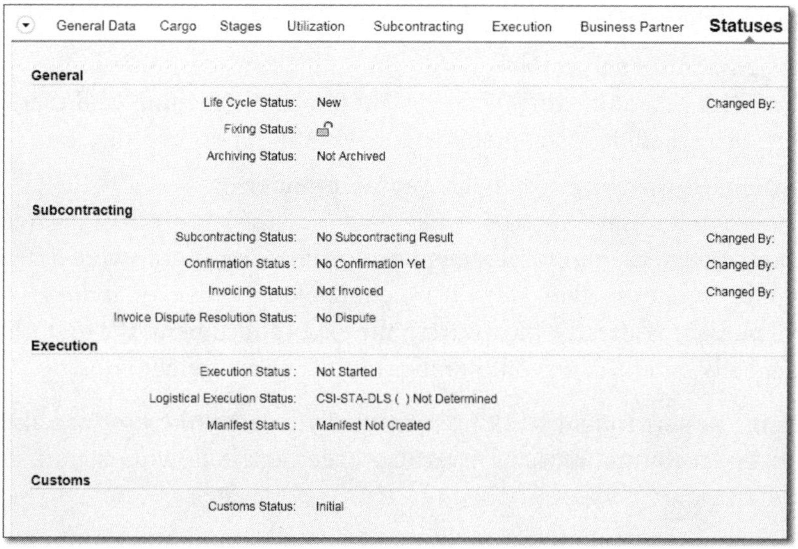

Figure 8.15 Statuses of a Freight Order

Handling and cargo execution status

Among the execution statuses, we can differentiate between two types: the *handling execution status* and the *cargo execution status*.

To understand the difference between these two statuses, let's look at an example. We have a freight booking that is supposed to consolidate different freight units into one container to transport these freight units together on the main leg. The cargo execution status tracks the progress of loading these freight units (meaning package items) into the container. The handling execution status, on the other hand, tracks the progress of the container being loaded onto the vessel.

Since freight bookings can have multiple stops (e.g., in the case of connecting schedules), the handling execution status needs to be tracked on every stop and for every item of the freight booking. If the handling execution status is updated for an item, the same status will be propagated to the subordinate items. The cargo execution status, on the other hand, is only defined on the container item level because it is assumed that the container will only be loaded and unloaded once during the transportation part covered by the freight booking.

Tracking/reporting execution status updates

Both the handling execution status and the cargo execution status can be seen on the Cargo Management tab in the freight booking or freight order. (In air freight bookings, this tab is called Operations.) You might remember this tab from Section 8.1.2.

As you can see in Figure 8.16, the handling execution status is tracked on each item of the freight booking. In an item hierarchy like the one shown in this example, the handling execution status of the container item is then propagated to the package items.

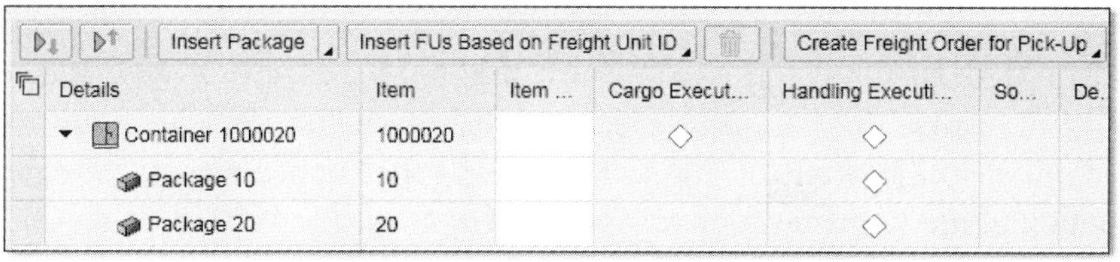

Figure 8.16 Cargo Execution Status and Handling Execution Status in Freight Booking

Keep It Simple

In this chapter we only refer to a very simple example to show the course of the execution statuses. In addition, we refer to the freight document as a freight booking, meaning an ocean freight booking.

Bear in mind that the execution statuses for the freight documents also apply to more complex scenarios and are also used (and work in the same way) on freight orders or freight bookings for other modes of transport.

Handling execution status

For the handling execution status on the item level, SAP TM provides the following statuses:

- *Not Determined*
  This is the initial status after the freight document is created. Once the freight document is ready for execution (we cover this later in this chapter), the initial status changes to the first status in the process
- *Not Loaded*
  Once the freight document is ready for execution, the handling execution status for the items changes to this status because the system now awaits the loading of the container onto the vehicle or vessel
- *Loaded*
  Packed container is confirmed to be loaded to the transporting vehicle
- *Not Unloaded*
  At its destination, the container is still sitting on the transport vehicle waiting for unloading
- *Unloaded*
  The container was unloaded from the vehicle

Since the last three statuses are probably self-explanatory, we need to add here that the handling execution status on the item level always adapts to the current location. This means that in a multi-stop freight booking, the item status changes from *Not Loaded* to *Loaded*, and at the destination location of the first transportation stage it changes to *Not Unloaded* and *Unloaded*. However, if the first transportation stage is finished, the status of the item changes to *Not Loaded* again.

The handling execution status on the stop level represents the statuses on the item level and offers the additional events DEPARTED and ARRIVED. Usually, these two events are reported by SAP Event Management, as we explain later in this chapter. In addition, the handling execution status on the stop level also represents all cargo execution statuses. If the current stop is the first stop of the freight booking, meaning this is where the cargo is loaded into a container, then the handling execution status on the stop level will also show the current progress of the loading of the cargo into the container. If the current stop is an intermediate stop, these statuses will not be shown.

Cargo execution status

Now let's take a look at what cargo execution statuses the system offers. As already mentioned, the cargo execution status is only *defined* on the item level, but is also *shown* on the stop level using the handling execution status if the current stop is a stop where cargo is loaded or unloaded.

SAP TM offers the following cargo execution statues:

- *Not Determined*
  Just like with the handling execution status, the cargo execution status is first set to this initial status before the execution process is started.
- *Cargo Ready for Loading*
  Before the cargo can actually be loaded, it needs to arrive at the loading location. In an ocean scenario, we can imagine that the cargo is brought to the port with trucks, and the container waits for the cargo at the port. Therefore, the container and the subordinate items are only ready for loading when the cargo items have arrived, meaning that the prerequisite freight order has arrived at its destination location.
- *Cargo Not Loaded*
- *Cargo Partially Loaded*
  In an example like Figure 8.16, each subordinate item needs to be loaded into the container separately. If some package items have already been loaded into the container and others have not yet, the container's cargo execution status would be cargo partially loaded.

- *Cargo Loaded*
- *Cargo Ready for Unloading*
  As already mentioned, when the freight booking has arrived at its final destination, the cargo needs to be unloaded from the container again. Please note that this only applies to the last location of the container item; the cargo execution status is not changed on intermediate stops. All statuses regarding the unloading process correspond to the loading statuses, so we will only list the unloading statuses without going into details.
- *Cargo Not Unloaded*
- *Cargo Partially Unloaded*
- *Cargo Unloaded*

Setting the execution statuses manually

The CARGO MANAGEMENT tab on the freight booking is used not only to monitor the current statuses of the cargo items, but also to manually set these statuses. Recall that the handling execution status and the cargo execution status can be linked to SAP Event Management events. Since we cover this in Section 8.2, for now we just focus on setting these statuses manually. This is applicable in many use cases because the loading and cargo loading is often done not by the shipper or LSPs (which created the freight booking in SAP TM), but by the carrier, who does not have access to the SAP TM system and therefore cannot set the execution status. Because of this, the carrier calls the LSP to report the current status of the cargo and container.

If the freight booking is ready for execution, you can start setting the execution statuses in the freight booking.

Readiness for Execution

A freight booking's readiness for execution depends on various factors and information in the freight booking. For example, if a freight booking type is defined in Customizing as relevant for subcontracting, then the freight booking needs to have a carrier assigned to it before the freight booking is ready for execution.

You can check the readiness for execution by selecting CHECK • READY FOR TRANSPORTATION EXECUTION in the global toolbar of the freight booking.

If you want to set the cargo execution status or handling execution status manually in the freight booking, click on the SET TO LOADING button above the item hierarchy displayed in the CARGO MANAGEMENT tab.

As you can see in Figure 8.17, the choice of statuses combines the different statuses of the cargo execution status and the handling execution status. Note also that not all of the statuses are selectable. Which statuses are selectable depends on the current status of the items. For example, if the freight booking has already left the port of loading, you can no longer select any statuses that are concerned with loading cargo into the container.

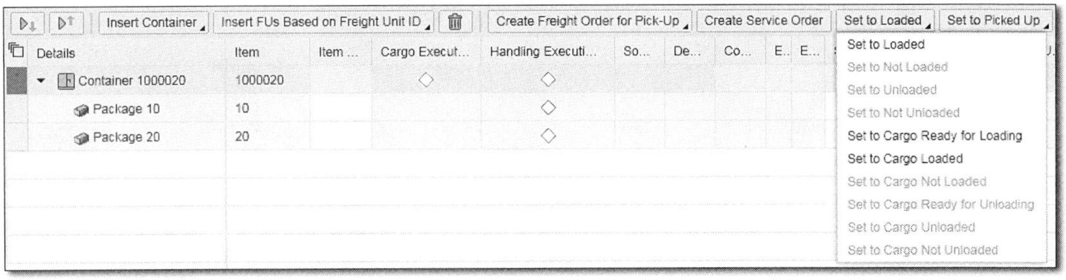

Figure 8.17 Setting the Execution Status Manually

In addition, you can also set the handling and cargo execution status by clicking on the SET ITEM STATUS button on the global toolbar of the freight booking without having to go to the CARGO MANAGEMENT tab. When using this button, you will find the same statuses that you can set as described on Figure 8.17.

The Statuses Need to Be in Order

The statuses have a defined sequential order in which they can appear. Therefore, the cargo must be loaded into the container before the container can be loaded onto the vessel. The same applies at the port of discharge; the container needs to be unloaded from the vessel before the cargo can be unloaded from the container.

This reflects the common use case at ports and airports, where the containers are usually unloaded outside the vessel or aircraft.

Progress of execution statuses

You can see in Figure 8.18 that the execution statuses of the items change depending on the overall execution progress of the freight booking. The vessel leaves the port after the cargo is loaded into the container and the container is loaded onto the vessel. Once the vessel has left the port, the cargo and handling execution status is reset to a status awaiting the next execution at the next port.

Even if the vessel has arrived in the port and the container has been unloaded from the vessel, the handling status is not set to *Unloaded* if the current port is the port where the cargo is unloaded from the container. The cargo needs to be unloaded from the container before the handling execution status turns green, indicating a status of *Unloaded*.

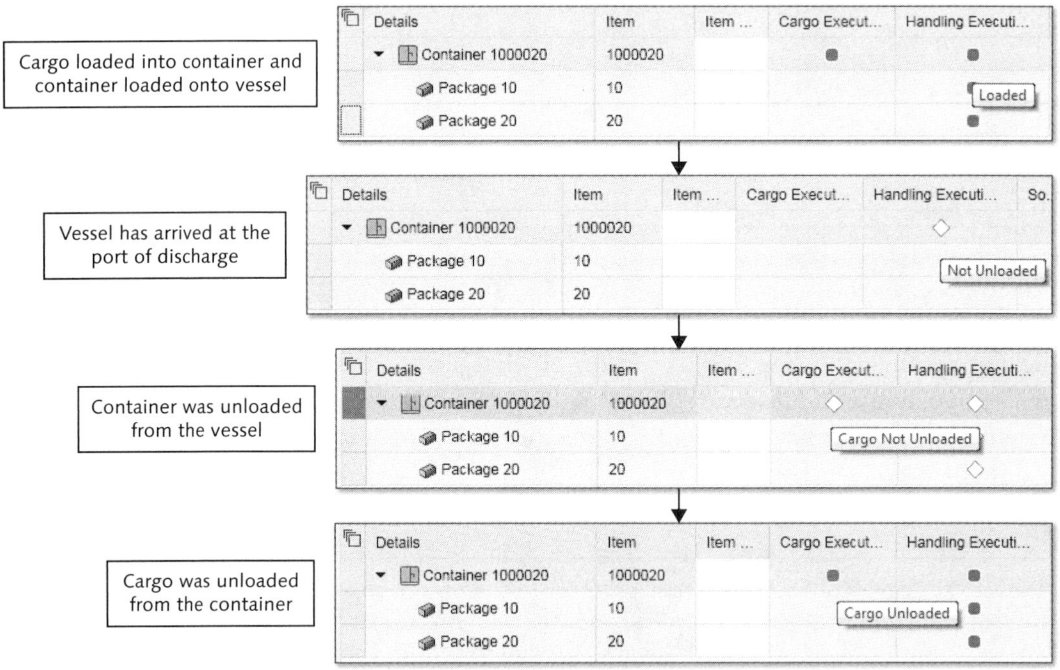

Figure 8.18 Progress of Execution Statuses

Status interdependence

Recall from Figure 8.15 the overall status called execution status, which we haven't yet discussed. The following are some of the most important execution statuses:

- *Not Started*
  This is the initial status of a newly created freight booking.
- *Ready for Transportation Execution*
  All preparations for the shipment have been successfully finalized and the shipment is ready to go.
- *In Execution*
  The resource used in the freight booking has left the source location but has not yet reached the destination where the container and cargo can be unloaded. Think of this as *en route*.
- *Executed*
  The resource has reached its final destination, and the cargo has been unloaded from the container. In Customizing of the freight order type, as described in Chapter 7, Section 7.2, you can specify which event from SAP Event Management is supposed to set the overall execution status to this status.
- *Not Relevant*
  If you have defined in the Customizing of the freight booking type that the freight booking is not relevant for execution tracking, the execution status will always be not relevant.

The overall execution status is influenced by the handling execution level on the stop level. If the handling execution status on the first stop changes to *Loaded* or *Partially Loaded*, the overall execution status will change to *Loading in Process*.

The same applies to the actual transportation of the goods. If the handling execution status has been set to *Departed* (which, as we said, is usually done by an event in SAP Event Management), the overall execution status will change to *In Execution*.

8.2 SAP Event Management

SAP Event Management is a versatile and adaptable tool that manages processes for object and status tracking and tracing, performing collection, and analysis of key performance indicator data. SAP Event Management can be integrated into an SAP and legacy system landscape that communicates with partner systems in a worldwide network.

Core functions

AMR Research breaks SAP Event Management's functions into five core areas:

- **Monitoring**
  Monitoring of processes and objects is based on their status and events that are expected to happen within the process or with the object. Usually, monitoring has certain real-time requirements; there sometimes need to be immediate reactions to occurring or missing events. An example of monitoring is the tracking and tracing of a shipment.
- **Notification**
  Decision makers need to be notified if a process deviates from a planned progression. First, the deviation needs to be detected (which is an outcome of the monitoring function). Then a notification via an appropriate channel has to be raised, giving information about and access to the critical situation (e.g., sending an email that alerts the recipient of a delay in the delivery of a shipment).
- **Simulation**
  In case of process deviations or delays, it can be sensible to simulate different options for recovery or alternative progression. Simulation is a tool for decision making that allows evaluation of the impact of particular actions in terms of complying with customer or internal service level definitions.
- **Control**
  Any situation within a process that is monitored through events or status values can lead to reactions that allow you to control the process itself or dependent activities within a business system. The decision on the type of control required in a particular situation is based on a rule set. An example of control is posting a goods receipt in a distribution system if a customer reports the complete arrival of the goods at his premises.
- **Measure and analyze**
  The planned and actual process data, status, and event information can be used in a variety of ways to identify weak points in processes or to determine key performance indicators of the capabilities of an

organization. This data can be collected by SAP Event Management and evaluated in a business warehouse. An example is the average delay time of deliveries made by a certain carrier.

Configurability of SAP Event Management

Real-world business processes usually entail a variety of process requirements that need to be reflected in an implementation of an event management process. Because SAP Event Management is not bound to any predefined designated business objects (e.g., a sales order), you have flexibility in deciding which object type, process steps, and reactions you want to model and implement in a tailor-made event management process. Yet, because SAP Event Management is the standard tracking and tracing system within SAP logistics applications, many of the processes it covers are connected to a corresponding preconfigured scenario (in SAP TM, enabling plug-and-play integration to process visibility).

Event Management Processes

An event management process is an implementation of a real-world process in an event management system (EMS), where major milestones and characteristics of the process are reflected in the system. The event management process can have different flavors depending on the main emphasis:

- Control processes keep control over a business process.
- Track and trace processes show the current status of a process.
- Visibility processes provide an end to end overview of a process.

Process viewpoints

Even if there are many cases when an event management process can be a straightforward implementation, you can find examples where different objects interact and where different phases or viewpoints have to be taken into account. SAP Event Management can cover this kind of process and can be configured accordingly.

Let's consider two examples of such processes: one that involves ordering and producing cars and another for tracking shipments by shipper and carrier.

In the automotive industry, a car can be tracked from two different viewpoints:

- As a to-be-produced material object that will be sold in the future
- As an order from a customer who wants a made-to-order car

In such a process, either form may occur first (i.e., an order for a specific car that is not produced yet or the production of a specific car that has not yet been ordered by a customer). Because the sequence is not known up front, all of the related objects (customer order or car production order) need to be modeled in an EMS and be able to be instantiated independently of each other.

Still, it needs to be possible at any time to relate the visibility process for one object to the visibility process for the other. It may happen that someone just created a production order of a specific car, and in parallel someone else created a customer order for exactly the same car configuration. In this case, the new production order can be directly assigned to the new customer order, and the processes will be logically merged.

Let's consider the second example. When tracking shipments, we also face situations where we are looking at the same objects from the perspective of a consignee or a carrier. The consignee is focused on his shipment, which may be only one of many on a carrier's truck. He is not entitled to view the truck or any other shipments on the truck addressed to other consignees. The carrier, however, can look at his overall tour with many shipments, one of them being our consignee's shipment. Similar to the car example, the objects (shipments and tour) are closely related to each other but can be independently managed and have different constraints in terms of access rights and visibility.

8.2.1 Event Handlers and Event Messages

Event handler examples

The key business object of SAP Event Management is the event handler. All event management processes are based on at least one event handler, which allows you to define main characteristics, statuses, and steps that need to be tracked and controlled. Event handlers can represent a material object, a process, or a virtual operation. The following are examples of material objects that need to be tracked:

- A pallet that is used as the package for a shipment that needs to be tracked. (You are interested in the shipment, but the pallet carrying it is the object that is identifiable from outside.)
- A container is an asset that needs to be tracked during its complete ownership lifecycle.
- A production device, whose correct operation needs to be monitored and logged.
- A shipment such as an express parcel that needs to be tracked from pickup to delivery.

Examples of more process-specific visibility scenarios include the following:

- A customer order handled in various order processing statuses.
- A payment process that should result in the balance of an invoice.
- A purchase order that needs to be tracked from ordering time to delivery and quality inspection of goods.

Event Handlers for the Car and the Shipment Example

Returning to our two examples, there are usually two event handlers for each of the examples to control the corresponding process:

- For the car order, one event handler would represent the order; a second event handler would represent the production order. Both can be created independently from the other and linked as the process requires it.
- Similarly, for the shipment example, there would be one event handler for the individual shipment and a second one for the truck tour. Again, both event handlers could be linked when applicable.

Event handler lifecycle

Each event handler has a lifecycle that corresponds to the lifecycle of the object or process it represents. An event handler is instantiated by an incident in a business process, which could be related to a certain status (e.g., order accepted) or to the creation of a business object (master data object for a container created). During its lifetime, the event handler processes a variety of events and reacts to them according to a defined rule set. It can be put to sleep and woken up again before being deactivated and finally archived. Table 8.1 lists some examples of typical event handler lifecycles and event counts.

	Event Handler Type (Business Usage)		
Characteristics	Tendering Process	Shipment Tracking	Container Resource Tracking
Lifecycle	2 hours	4 weeks	5 years
Number of processed events	3–5	approx. 20	>10,000

Table 8.1 Examples of Event Handler Lifecycles

Examples of event management scenarios

To underline the flexibility and comprehensive applicability of SAP Event Management, let's consider a few examples of how it is used throughout various SAP industry segments:

- Order management including production monitoring, delivery, and invoice settlement (mill industry)
- Tendering and visibility for logistics execution (high tech industry)
- Distribution processes in a complex environment (industrial machines and components)
- International ocean freight, including customs management (retail industry)
- Returns management (automotive industry)
- Tracking of handling units in logistics outbound processes (LSPs)
- Spare parts and equipment management (aerospace and defense industry)
- Tracking of parcels including hierarchical loading (postal services)
- Rail car management (chemical and mill industry)
- Integration with vehicle management systems (automotive industry)
- Integration with traders and schedulers workbench (oil and gas industry)

High-Performance Tool

SAP Event Management is designed to process scenarios with very large amounts of data. Many large postal companies use SAP Event Management for parcel tracking, where several billion events have to be processed every year. Big data is not new to SAP Event Management.

Starting with SAP Event Management 9.1, it runs on SAP HANA database, which enables in-memory use of event handler and event message data.

Creation of event messages

Event messages are announcements related to real-world processes or objects represented in an event management context. These messages are communicated in a standardized form to SAP Event Management; they carry information to identify the related process, the incident, the time and location of the incident, and further contextual details. Event messages can be created and communicated in various ways:

- Interactive creation by humans (e.g., with a mobile device, scanner, Internet application, or business system)
- Automatic creation by machines (e.g., a technical system, production system, or RFID scanner)
- Forwarded by business systems (EDI or XML messages with business content)

Characteristics of events

We can define events by some essential characteristics:

- **Event type**
  The event type is a definition of the incident that should be reported by the event message (e.g., acceptance of an order, departure of a shipment at a location, proof of delivery of a shipment).
- **Repeatability**
  The repeatability defines whether the event type occurs only once in the context of the current process or whether the same event type can reoccur at the same or another location (e.g., an *arrival event* may happen multiple times during a truck tour as several customers are visited).
- **Expected event date/time**
  Expected date/times define a point in time or a timeframe when an event should happen; an earliest and latest point in time may be assigned to the expected event.
- **Expected message date/time**
  Even if an event is expected to happen within a certain time frame, it may be reported via an event message at a different time. The

expected message date/time or time frame is a characteristic that can be defined as a benchmark for reporting compliance.

- **Actual event date/time**
  Once the event is reported, the actual date and time of event occurrence are defined. At this point they can be measured against the expected date and time.

Event categories

Besides the fact that an event management process usually has initialization and termination events, the various events happening in the course of the process can be assigned to four event categories. Event categories are determined as part of the monitoring function of SAP Event Management and lead to different behavior in terms of notification, simulation, control, and analysis. Figure 8.19 shows an overview of the event categories in the context of an event handler lifecycle, which moves from left to right.

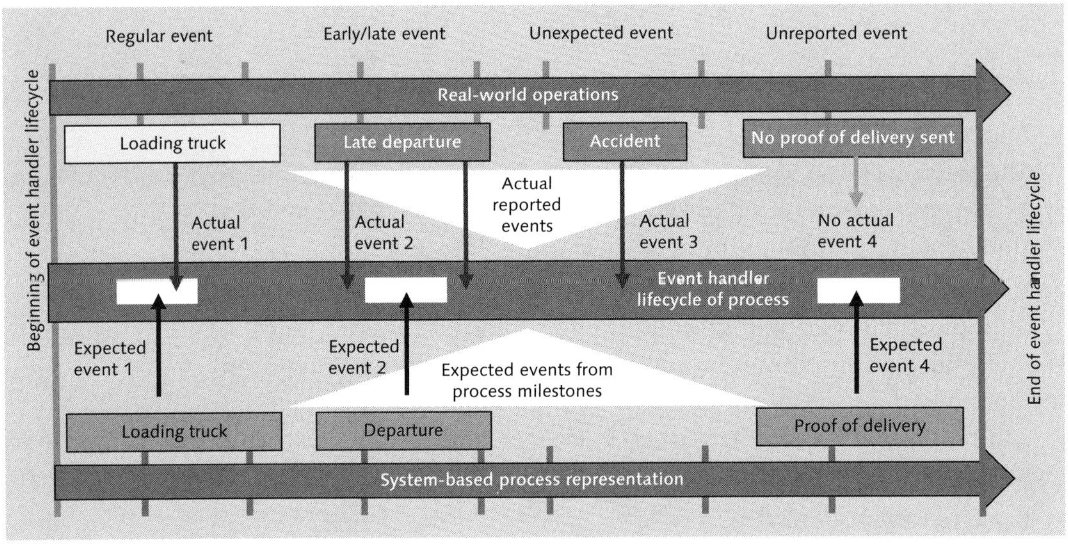

Figure 8.19 Event Types and Event Handler Lifecycle

We can divide the example events in Figure 8.19 into four event categories:

- **Regular events**
  A *regular event* is defined by a milestone in a business process that is reflected in the definition of the expected event and its expected time

frame. The actual event occurs within the expected time frame and is reported within the expected message time frame.

Say, for example, that a container should arrive at a terminal between 9:00 a.m. and 10:00 a.m. The confirmation is expected until 1:00 p.m. The container arrives at 9:43 a.m., and the confirmation (event message) is sent at 10:35 a.m.

In some cases, a regular event can occur without a defined expected time frame (e.g., it can happen at any time but it needs to happen at least once).

- **Early or late events**
  Like a regular event, the *early or late event* is a milestone that is expected to happen within a defined time frame. The actual event occurs either earlier or later than the expected time frame, or it is reported earlier or later than the expected messaging time frame.

  Say, for example, that a container should arrive at a terminal between 9:00 a.m. and 10:00 a.m.; the confirmation is expected until 1:00 p.m. Instead, the container arrives at 10:33 a.m., and the confirmation (event message) is sent at 3:27 p.m.

  A specific real-time reaction to the early or late event is not planned. Instead, the fact that the process is not executed according to the expected milestones is registered and used for analytical and process improvement steps (perhaps to evaluate the quality of service of a business partner).

- **Unexpected events**
  Unexpected events do not have a corresponding milestone in the business process, and there is no expectation that the event is reported. We need to differentiate between two situations:

  - The event simply happens and is registered (e.g., location is reported by a GPS device in a truck).
  - The event indicates a serious problem, and some corrective measures have to be taken (e.g., accident report of a delivery truck).

  In some cases, the situation needs to be evaluated based on further event characteristics (e.g., if a railcar is reported to be in a switchyard in Chicago when it should be en route from Denver to Los Angeles).

- **Unreported events**
  Unreported events are also based on an expected and time-defined milestone, but the actual event does either not happen or an event message is not received within the required time frame. The nonoccurrence of the event is rated as noncompliance and immediately leads to exception handling. This allows at least the raising of a notification to a person who can assess the situation and take further actions. For a well-defined escalation process, an automatic control process can be initiated as the reaction.

8.2.2 Event Management Processes

Initiation of event management processes

An event management process is usually triggered by one of the following sources:

- A transactional object in a business system is created or set to a particular status (e.g., an order that reaches the status accepted).
- A master data object is created in a business system (e.g., truck resource).
- A process in a business system reaches a particular state (e.g., a delivery process reaches the state of goods issued).
- A message from external sources indicates that an event management-relevant process has been kicked off and there are event messages to be expected in the future.
- A process implemented in SAP Event Management starts with manual creation of an event handler for pure event management-based handling. This kind of process is special because SAP Event Management is creatively used to run a business process on the implemented event handlers without a backend business system (e.g., returns management process on SAP Event Management).

Business Objects and Application Objects

In a business system, manifold business objects represent substantial entities of the business processes (e.g., sales order, shipment, invoice, and so on). In the context of event management, *application objects* define a more granular and semantic classification of objects that depends on the individual characteristics of a business object. This is necessary because an event management process may differ considerably depending on what a business object represents. In the context of event management, the business systems are generally also called *application systems*.

As an example, a shipment business object can represent a less than truckload (LTL) truck tour or an ocean full container load (FCL) container shipment. The nature of the business object can only be determined by looking up characteristics such as mode of transport, type of cargo, or an indicator such as shipment type.

Depending on these characteristics, SAP Event Management needs to initiate a different event management process and use an individual type of event handler. Therefore, based on the business object characteristics, the *application object type* is determined to control the event management process.

Event management process

Figure 8.20 illustrates the elements of SAP Event Management that are involved in an event management process. Usually, the process starts in a business system such as SAP TM or SAP ERP, where the mother object or process is created. In the course of running the event management process, the following stages or steps are executed:

❶ A business process is started in the backend business system. At a certain step, the process reaches a status where an event management process needs to be triggered. When the business object is saved, the business object data is handed over to a BAdI layer (SAP ERP) or to the PPF for post-save action handling. In both cases, the business object data is handed over to the SAP Event Management application interface, which is a configurable integration component that manages the communication to SAP Event Management.

❷ The application interface determines the tracking relevance of the business object by checking the relevance of an application object type and analyzing the business object status. If an application object type is relevant for tracking and the object status requires communication to SAP Event Management, the relevant context data is extracted from the business object, and several data packages containing object data in a standardized way are created (e.g., expected events, tracking IDs, and general parameters). In addition, the EMS to be used is determined (it is possible to use multiple EMSs for different purposes). Finally, the data packages are sent asynchronously to the SAP Event Management instance. The same procedure applies in the case of updates or deletions of business objects, where change or delete requests are sent to the EMS.

❸ Legacy systems do not have an application interface (unless they are built on an SAP NetWeaver ABAP stack). Therefore, a legacy system needs to determine the tracking relevance on its own, build the application data packages, and send them over to SAP Event Management using a BAPI or web service call. This integration is not uncommon; one of the largest customer installations of SAP Event Management uses this integration technique.

❹ The control channel in SAP Event Management receives the request from the application interface of the business systems or from the legacy systems. The data received is forwarded to the event controller, which determines if an event handler exists or, if not, which type of event handler needs to be created.

❺ If an event handler is created, SAP Event Management checks to see if messages have already been received. If so, the buffered messages are processed in sequential order. If an event handler already exists, it is first changed according to the new data (e.g., it may contain changed or new expected events). Then the already received messages can be reprocessed to check whether the altered event handler is still compliant with the previously received event messages.

SAP Event Management finally sends back a status protocol of applied change steps to the application systems, where this information is logged in the application log (Transaction SLG1, object SAPTRX, subobject APPSYS).

❻ Event messages can be received from various sources as single messages or as a batch of messages. The messages are sequenced and related to event handlers before being forwarded to event processing by an event controller and rule processor.

❼ The event controller retrieves the event handlers for the event messages to be processed and hands over the individual batches of messages to each event handler.

❽ Each event handler processes its batch of event messages sequentially by use of the rule processor. The rule processor analyzes the received event message and applies the rules of the rule set to the event and event handler context. Based on current and previous data of the event handler, a decision on reactions to reported events can be made.

9 Extraction of event management data for analytics is one sort of reaction. The extracted data on process performance is sent to SAP NetWeaver BW or SAP HANA and stored in InfoCubes or data stores.

10 The web communication layer allows data to be retrieved from one or multiple event handlers and presented in a role-based web interface. Due to its configurability, the web communication layer can be used to present visibility data to customers and provide access for partners or internal employees.

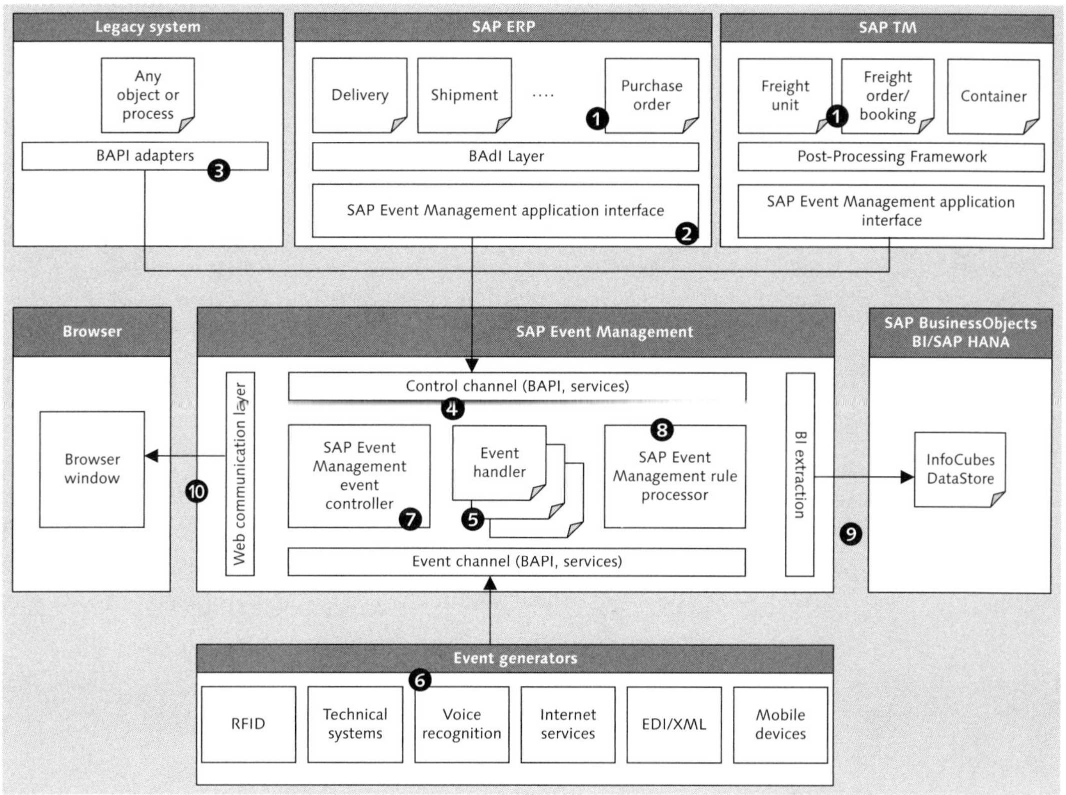

Figure 8.20 Elements of SAP Event Management

8.2.3 SAP Event Management Configuration

SAP Event Management configuration requires changing settings on the SAP application side (e.g., in SAP ERP or SAP TM) as well as the SAP Event Management side, which need to match in some places. In this section we describe how to set up SAP Event Management data extraction and communication in the application interface and how the directly related Customizing has to be done in SAP Event Management. Visibility process configuration components are also explained.

Configuration in SAP Application Systems

The SAP Event Management-related configuration capabilities of SAP application systems are provided by the application interface and are part of the application basis software layer of SAP NetWeaver. Therefore, you can use it in all systems of the SAP Business Suite. The technical integration between the business processes and the application interface is delivered out of the box for various business objects. Table 8.2 provides an overview of the logistics-related SAP Event Management integration objects.

Object	Component	Event Management Usage
Freight unit	SAP TM	Cargo item/container tracking
Freight order/ freight booking	SAP TM	Shipment tracking, master bill tracking
Resource	SAP TM	Equipment tracking
SOP instruction	SAP TM	Tracking of standard operating procedures
Forwarding order	SAP TM	Tracking of order status
Purchase order	SAP ERP MM	Purchase order tracking
Delivery	SAP ERP Logistics Execution (LE)	Delivery tracking
Sales order	SAP ERP SD	Order fulfillment
Shipment	SAP ERP LE	Shipment tracking

Table 8.2 Logistics-Related SAP Business Suite Objects with SAP Event Management Integration

The application interface provides a standardized way of configuring the integration between business processes and the corresponding event handling. Depending on the semantic context of a business object, configuration can control which process in SAP Event Management is fed particular data and expected milestones. The configuration is mainly controlled by definition of business object type and application object type.

Business process types Customizing

The business process type is directly related to the business objects as they are defined in the business object repository of the application system (e.g., a transportation order object). It relates to a technical integration of the business object with the application interface and a list of data tables with object content that can be used for extracting data and events to be sent to SAP Event Management.

You can customize the business process type by following the IMG menu path INTEGRATION WITH OTHER SAP COMPONENTS • EVENT MANAGEMENT INTERFACE • DEFINE APPLICATION INTERFACE • DEFINE BUSINESS PROCESS TYPES. For each business process type, settings for technical data posting (dialog task, V1 update task) and queuing are defined. In addition, a list of data structures is provided that allows you to characterize the data dictionary structure used, and an indication how to evaluate business object changes. The indicator allows you to define, for example, which value in a structure field indicates a newly created object that needs to be communicated to SAP Event Management. Figure 8.21 shows the business process types of an SAP TM 9.1 system and some details of the application table definition of the business process type of the transportation order object.

Application object type Customizing

You can define the settings for application object types via the IMG menu path INTEGRATION WITH OTHER SAP COMPONENTS • EVENT MANAGEMENT INTERFACE • DEFINE APPLICATION INTERFACE • DEFINE USED BUS. PROC. TYPES, APPL. OBJ. TYPES, AND EVT TYPES. Each application object type is directly related to a business process type.

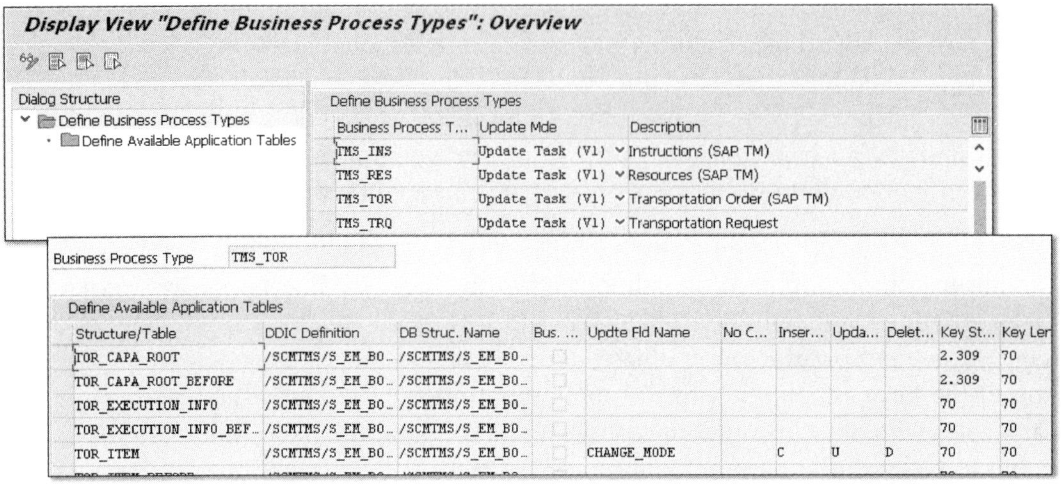

Figure 8.21 Business Process Types of SAP TM and Application Tables of the TOR Object

> **Using Multiple Event Management Systems for Your Visibility Processes**
>
> If you integrate your application systems, you have the option to connect multiple EMSs for different tracking purposes. In application object configuration, you can choose which EMS should handle the processes for a particular application object type (e.g., shipment tracking on EMS A and resource tracking on EMS B).

In the General Data settings of the application object type, you have to define the EMS where the visibility process is started. In addition, you can set the behavior of the application object creation (e.g., if the application object is relevant to trigger an EM communication; you can use this to deactivate an application object type).

The Control Table settings define which of the business object data tables represent the main object. An application object may, for example, be created for a shipment (header level) or for each shipment item (item level). In the first case, the main table would be the objects header table; in the second case, the main table would be the item table, and the header table has to be assigned as a master table. For some objects, deleted records are kept in separate tables. In this case, the table for deleted objects could deviate from the main object table.

On the OBJECT IDENTIFICATION tab, you can configure how the application object ID is compiled. You can extract it from one or two fields of the business object data tables or use a function module for extraction.

The EVENT MANAGEMENT RELEVANCE settings determine when an application object is communicated to the EMS. The determination can be done by either a Boolean function or an ABAP function module. Alternatively, the application object can be set to be always relevant, which triggers a communication to SAP Event Management as soon as a business object is created.

Using ABAP Function Modules in SAP Event Management

In application system and SAP Event Management configuration, there are many places where ABAP function modules are used to either extract data or do some kind of processing (e.g., rules engine).

Many function modules are delivered with the standard software. You also have the option to create your own function modules from templates that are available or as copies of existing modules. If you assign such a function module in Customizing, you first have to create an entry for it in Customizing because the function module name is not directly entered, but a logical name is assigned to it. You can find the assignment via the IMG menu path INTEGRATION WITH OTHER SAP COMPONENTS • EVENT MANAGEMENT INTERFACE • DEFINE APPLICATION INTERFACE • DEFINE SAP EM EXTRACTION FUNCTIONS.

Data extraction from the application object

In the PARAMETER SETUP tab, you can define which data of the business object is handed over to event management to create or update an event handler. There are multiple categories of data to be extracted. Figure 8.22 shows the Customizing screen.

- Tracking identifications (IDs) and code sets are used to identify the event handler when event messages are received. Tracking IDs are usually numbers such as shipment numbers, bill of lading numbers, pallet numbers, or order numbers. The code set associated with the number helps you find the correct event handler (there might be a shipment with number 12345 identified by SHP 12345, and an order with number ORD 12345, where SHP and ORD represent the code set and 12345 represents the tracking ID). Tracking IDs can be extracted via a table field reference or an ABAP function module.

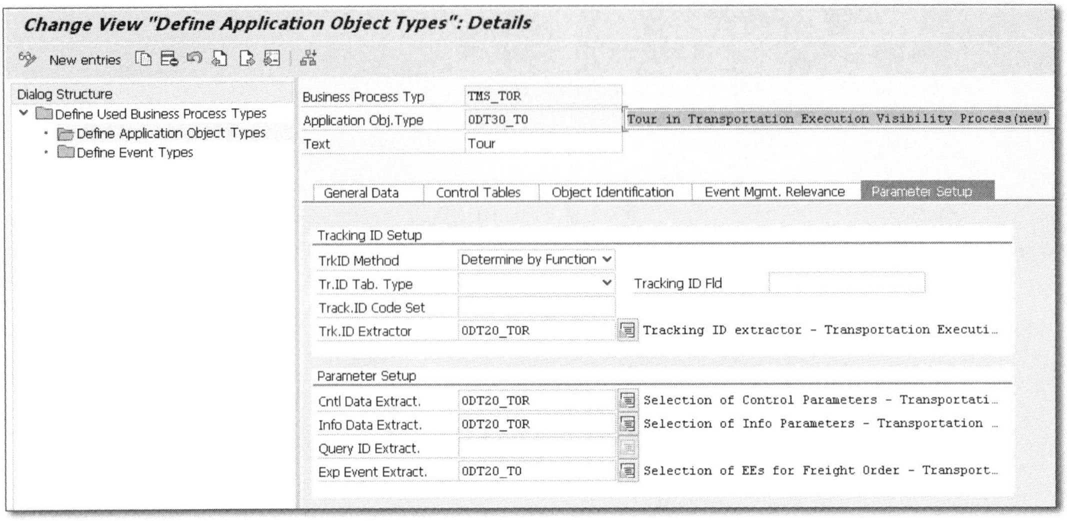

Figure 8.22 Parameter Setup of an Application Object Type

- Control and information data extraction provides containers to hand over any kind of data in the form-name-index value to SAP Event Management. This data is related to the object or process and gives additional information. The control data container holds information that can directly influence the SAP Event Management logic (e.g., an indicator that a shipment contains dangerous goods). The information data container holds additional object characteristics (the name of the truck driver). The index is used to relate several entries belonging to the same group (PRODUCT[1] = "Television", QUANTITY[1] = "200", TYPE[1] = "Yamamoto DXTV-230").
- Query IDs provide the option to assign additional code set/alpha string pairs to an event handler that can be used for data retrieval from SAP Event Management but not for message processing.
- Expected event extraction allows the retrieval of milestones from the application object that are later used to create the expected events in the event handler. Because the extraction is done in an ABAP function module, you have the option to enhance the expected event list by calculating or enriching the milestones given in the business object context. If, for example, a shipment object contains only a departure

date, you can additionally create a loading end date 30 minutes earlier, which allows you to track your internal load dock schedule compliance.

Customizing of event types

For each business object type, you can configure one or multiple event types. Event types allow you to set up event messages to be sent from the application system to a connected EMS in the context of the backend process. An example is the confirmation of goods issue in an SAP ERP delivery, which needs to be registered in the shipping process tracking in SAP Event Management, or setting a load status on a freight order in SAP TM.

Data extraction from the application is done in an assigned ABAP function module that allows you to build the event message context in a very flexible way. Figure 8.23 shows the event types defined for the SAP TM transportation order objects (freight order and freight booking).

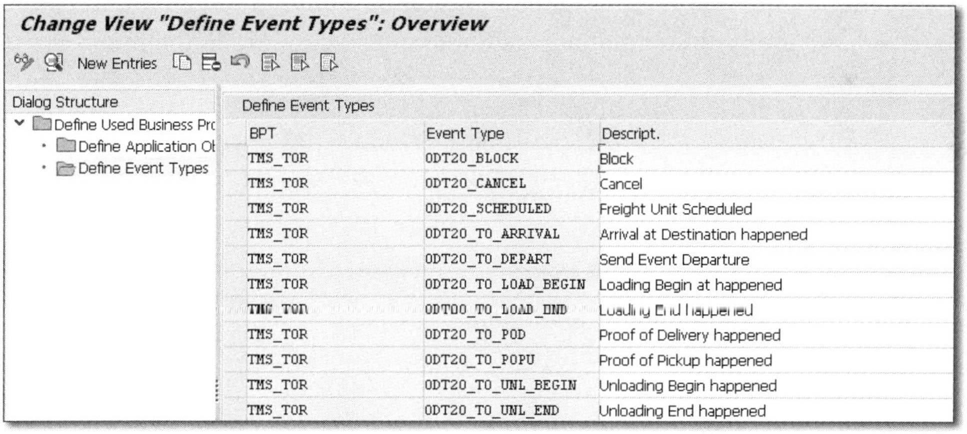

BPT	Event Type	Descript.
TMS_TOR	ODT20_BLOCK	Block
TMS_TOR	ODT20_CANCEL	Cancel
TMS_TOR	ODT20_SCHEDULED	Freight Unit Scheduled
TMS_TOR	ODT20_TO_ARRIVAL	Arrival at Destination happened
TMS_TOR	ODT20_TO_DEPART	Send Event Departure
TMS_TOR	ODT20_TO_LOAD_BEGIN	Loading Begin at happened
TMS_TOR	ODT20_TO_LOAD_END	Loading End happened
TMS_TOR	ODT20_TO_POD	Proof of Delivery happened
TMS_TOR	ODT20_TO_POPU	Proof of Pickup happened
TMS_TOR	ODT20_TO_UNL_BEGIN	Unloading Begin happened
TMS_TOR	ODT20_TO_UNL_END	Unloading End happened

Figure 8.23 Definition of Event Types for the Transportation Order Object of SAP TM

Application log

The standard application log, which you can start via Transaction SLG1, provides detailed information on the success or failure reasons for the activities happening within the application interface. You can access details about the event management relevance of application objects, data extraction and application data, or event transmission to the EMS. The response of SAP Event Management to the application system requests is transmitted back to the application system and stored in the log (e.g., no suitable event handler type could be determined in SAP Event Management). You can see the complete process by accessing the log for object SAPTRX and subobject APPSYS. Figure 8.24 shows an example of an application log.

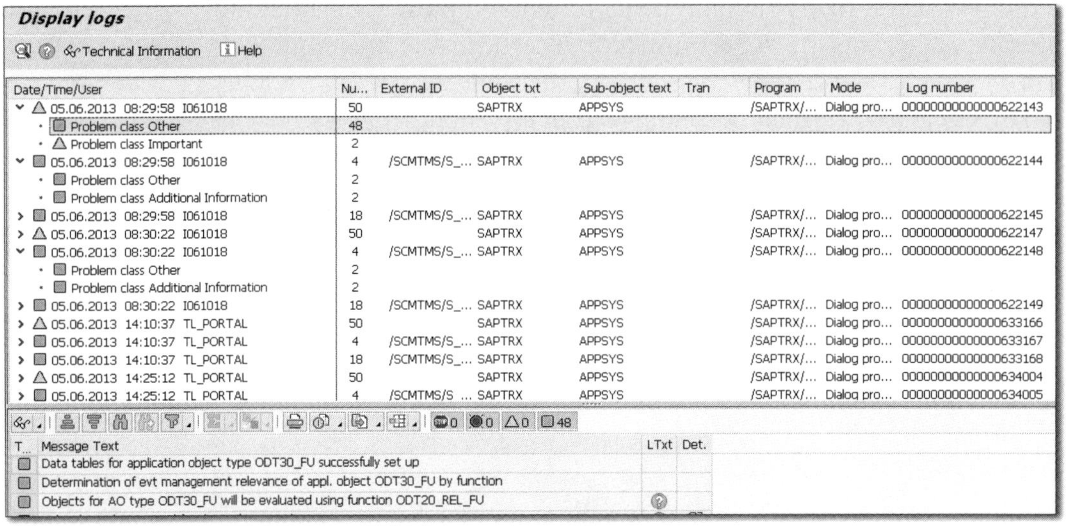

Figure 8.24 Application Log for Application Object Processing

Configuration of SAP Event Management

The configuration of SAP Event Management allows you to define the behavior and characteristics of the visibility processes. The main activities in SAP Event Management configuration are to define the following:

- Event handler types, expected event messages, and status settings
- Event handler type determination and data mapping from the application system data
- Event messages and event codes to be processed
- Rules about how to react to received or missing event messages
- Setup of personalized web transactions for accessing event management data

To allow process synchronization between application systems and SAP Event Management, a few essential settings have to be made to connect application system processes with SAP Event Management configuration. You can find these settings in the IMG by selecting EVENT MANAGEMENT • GENERAL SETTINGS IN SAP EVENT MANAGEMENT.

- Define RFC connections to enable technical communication.
- Define logical systems to identify application systems and EMSs.
- Define application systems to give a name to the systems for which you set up processes in SAP Event Management.
- Define business process types because you need them to synchronize the application system extraction process to the event handler creation process.

Configuration of event handlers

Definition of different event handler types allows you to control the creation, composition, and behavior of individual visibility processes on the event management side. Event handler types are directly related to a business process type, which you need to define in the event handler type settings. Figure 8.25 shows the setup of an event handler type that you can find in SAP Event Management Customizing by selecting EVENT MANAGEMENT • EVENT HANDLERS AND EVENT HANDLER DATA • EVENT HANDLERS • DEFINE EVENT HANDLER TYPES.

The main fields of the event handler type are as follows:

- BUS. PROC. TYPE
  You have to assign the event handler type to a *business process type.*
- PRIORITY and CONDITION
  You can set a *priority* to allow a ranking within event handler type determination. You define a *condition* that specifies the applicability of the event handler type (e.g., only to be used for a specific application object type).
- RULE SET
  The *rule set* defines how the created event handler reacts to incoming or overdue event messages.

- STAT.ATTR.PROF
  Using the *status attribute profile*, you can detail the creation of status fields and set an initial value to it during event handler creation.
- EE PROFILE
  The *expected event profile* summarizes the expected events that are created with the event handler based on application system milestones or other expected events.
- AUTH. PROF. ID
  Using the *authorization profile*, you can define who has access to which part of the event handler data.
- EH UPD.ACTY 1 and EH UPD.ACTY 2
  The event handler *update activities* allow you to specify ABAP function modules to update data of the application system before creating the event handler or update the event handler after running through the standard event handler creation process. They can be used like traditional user exits.
- EXT. TABLE ID
  The *extension table IDs* for event handlers and event messages allow you to add fields on the header level of event handlers and messages.
- BW PROFILE
  The *BW profile* defines which data is extracted from the event handler to be sent to a connected data warehouse system (SAP NetWeaver BW or SAP HANA).
- CHANGE DOC. ACTIVATION
  Changing document activation and logs allows you to capture additional data for auditing purposes.

Event Handler Extension Tables and Use of Logs

You can use an extension table to extend the event handler header for various purposes. Most important is the ability to create fields that are part of a database index to allow a fast search (e.g., a field for location of last sighting). Each event handler type can have its own extension table.

Logs—especially in the verbose mode—should be used very carefully or mainly for testing when running SAP Event Management in high-performance scenarios or with high data volumes because they may create multiples of the data load of the pure tracking process.

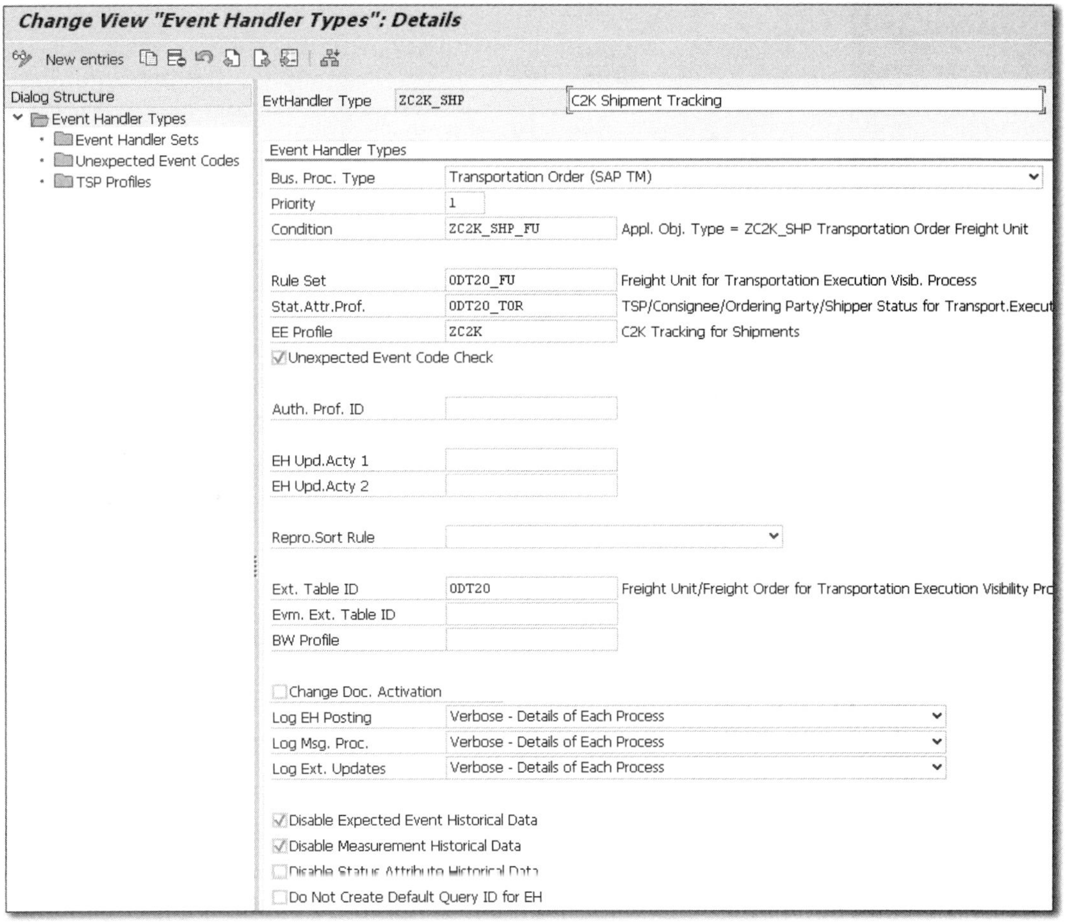

Figure 8.25 Configuration of an Event Handler Type

Example event handler data

Figure 8.26 gives example data of an event handler. The event handler header keeps the references between the application system and SAP Event Management and provides identification and control characteristics for the process. Milestones, tracking IDs, and other attributes are stored in related tables. The header extension table contains important context data for the process. Control and information parameters store more detailed information that can be presented upon request.

Event handler header				
GUID	***Appl. object type***	***Appl object ID***	***Business process type***	***Event handler type***
34eG62fs6DF	SHIPMENT	9926007	TMS_SHPMT	TMS_CONT

Tracking IDs	
Codeset	***ID***
CONTNR	EASU7777779
SHPMNR	9926007
HAWBNR	EASC -2777098-2
MAWBNR	EASC -8755545

Query IDs	
Codeset	***ID***
ORD_NR	35500726
PO_NR	978925007-08

Extension table (System parameters) Z_CONT_EH_TAB				
Origin	***Destination***	***Container type***	***Last sighting***	***Container line***
HAMBURG	SINGAPORE	22GP	ROTTERDAM	EASC

Control parameter		
Name	***Idx***	***Value***
COMMODITY	1	Toys
COMMODITY	2	Furniture
HTS	1	34526700
HTS	2	34584100
WEIGHT	1	3360 KG
WEIGHT	2	7840 KG

Info parameter		
Name	***Idx***	***Value***
VESSEL	1	Northstar
SHIPPING LINE	1	EASC
CUSTOMS AGENT	1	MÜLLER&CO
CONSIGNEE	1	Jun Chen Ltd., Singapore, Main St.1

Expected events				
Event	***Actual time***	***Location***	***Partner***	***Reporting time***
LOADING	03.02.13, 14h00	HAMBURG	HHLA	04.02.13, 12h00
DEPARTURE	03.02.13, 22h00	HAMBURG	EASC	05.02.13, 12h00
SIGHTING	05.02.13, 04h00	ROTTERDAM	EASC	06.02.13, 0h00
ARRIVAL	23.02.13, 17h00	SINGAPUR	EASC	25.02.13, 12h00
UNLOADING	23.02.13, 23h00	SINGAPUR	SGP_PORT	25.02.13, 12h00

Measurements			
Measurement	***Expected***	***Actual***	***Status***
OVERALL WEIGHT	11200 KG	11350 KG	OK

Figure 8.26 Example Data of an Event Handler (Incomplete)

Expected event profiles

An event handler that needs to work with milestones must be assigned an expected event profile. You can define expected event profiles in the IMG by selecting EVENT MANAGEMENT • EVENT HANDLERS AND EVENT HANDLER DATA • EXPECTED EVENTS • DEFINE PROFILES FOR EXPECTED EVENTS.

When an event handler is created, SAP Event Management uses the information in the expected events profile to generate a list of expected events to serve as milestones for processing actual received event messages and as a basis for detecting overdue events. The right side of Figure 8.27 shows an example of an expected event list for an air freight tracking scenario that has been set up to support Cargo 2000-like event processing.

Event groups and dates

The expected events of a profile can be bundled into groups to allow alternative processing of events. One example is a group of events where either an approval or a rejection event is expected. Either of these two events fulfills the requirement to receive an answer for a request and therefore satisfies the necessity to receive an answer within a defined time frame.

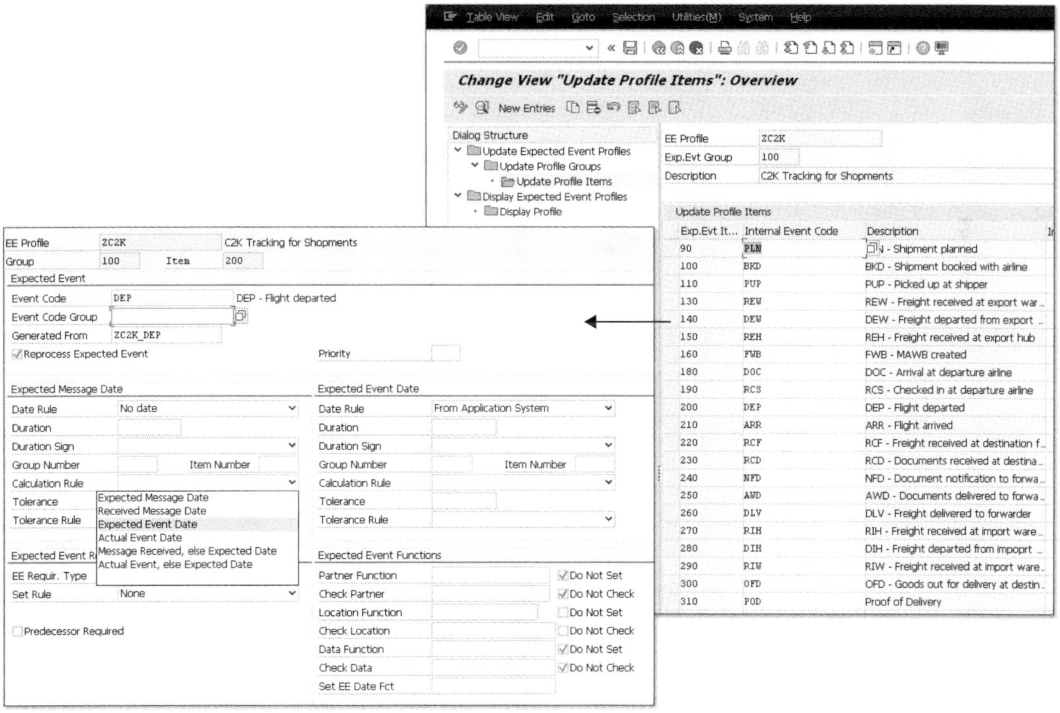

Figure 8.27 Event List in an Expected Event Profile and Detail Settings for Expected Event Generation

The detailed setting for expected event generation (see the left side of Figure 8.27) allows you to define how an expected event is created. You can relate the event to an expected event communicated from the application system. Dates and times can be directly moved or manipulated as required. You also can create an expected event by referencing a previously generated expected event (e.g., loading end is always 30 minutes after loading start). If the event scheduling follows a more complex rule or needs to reference other data, you can use an ABAP function module to determine the correct date and time.

Each expected event can have event and message dates with a defined earliest/latest time frame. If an event code occurs multiple times in a process, the individual event instances can be enriched by event locations or sending partners (e.g., if a departure event occurs multiple times on a distribution tour).

Keeping Track of Original Plans

Within the expected event structure, an event handler keeps track of the expected and the actual event date and time. In addition, an original expected event date and time are kept. Even if a plan changes multiple times, the original plan is retained in the expected event structure.

Status attribute profiles

Because business processes often require quite specific status settings, SAP Event Management provides a tool to define the status of each event handler type individually. You can configure the Customizing for status attribute profiles by selecting Event Management • Event Handlers and Event Handler Data • Statuses • Define Status Attribute Profiles. The profiles, which are assigned to one or multiple event handler types, consolidate a list of status attributes, each offering a status definition, possible status values, and an initial value that is set when the event handler is instantiated. Transition of the status values from one setting to another is accomplished by rules processing with specific status modification activities. Figure 8.28 shows an example of a status attribute profile (SAP TM transportation order event handler) with status attributes for block, delivery, and transportation status.

Event handler type determination

When an event handler is created by a request from an application system, the data extracted and transmitted is used to determine the appropriate type of event handler. The first step is to compare the business process type assigned to the application object against the business process type assigned to event handler types. The two should match. Optionally, the alternative business process type assigned to the application object is compared. Event handler types matching the business process type are now ranked by their priority. Subsequently, the conditions of the event handler types are checked until the first applicable one is found. This event handler type is then used to instantiate the event handler.

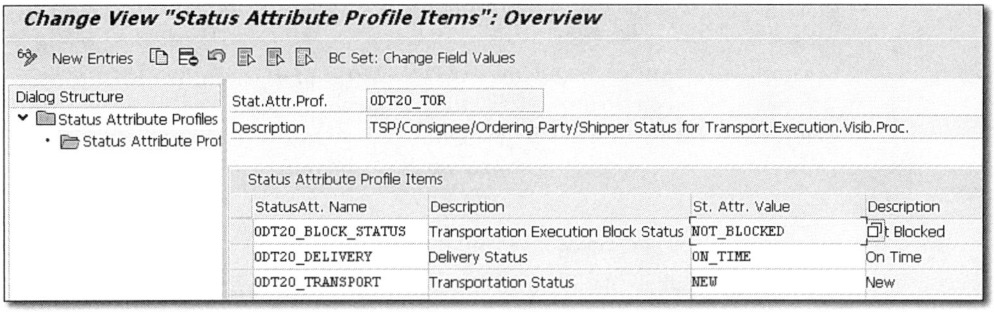

Figure 8.28 Status Attributes of an Event Handler for Transport Order Tracking

Application system parameter mapping

Because not all application systems are structured the same way, you can harmonize individually created parameters in a common process using the parameter mapping functionality. Even in the SAP world, orders of different kinds exist (CRM orders, SD orders, and so on). To avoid a cross-system harmonization of transmitted parameters, SAP Event Management offers a parameter mapping tool to assign parameter entities to a joint EMS-specific naming definition. For example, an ORD_NUM parameter from SAP CRM and an ORDERNUM from SAP ERP can be mapped to an ORDERNUMBER parameter in the event handler. The setup of a corresponding mapping profile in the IMG under Event Management • Event Handlers and Event Handler Data • Parameters • Define Parameter Mapping is mandatory; otherwise, the event handler cannot be created, and you will find a mapping error in the application log. In a simple case, the mapping profile just defines that all parameters are routed through the way they are received from the application system.

8.2.4 Event Messages and Event Processing

Event messages

An event message is a structured set of data that conveys information about the what, when, where, who, and why of a real-world incident to SAP Event Management. SAP Event Management uses this information to identify potential event handlers as receivers by comparing the tracking ID of the message with the tracking IDs of the event handlers. All active event handlers with a matching tracking ID and code set get a feed of the event message.

Technical Processing of Event Messages

Event message processing is mass-enabled to support high volumes of event messages to be passed to SAP Event Management with a single transmission. All received raw event messages are first saved in the database before being forwarded to message processing, which can be done either synchronously or asynchronously.

Message processing then packages all received messages by tracking ID, sorts the packages by actual event time stamp, and pushes each package into the event handler update process. In the end, the processing status and logs are saved with the messages. The event handler update process assigns the internal event code, checks the feasibility of message processing, and finally executes the rule processing for each message.

Event codes

Events codes characterize the purpose of an event message. They are divided into an external and internal view. Because there are many standards that define how to report a specific incident (e.g., EDIFACT, ANSI X.12, Rosetta Net), the external view needs to be flexible. Therefore, you can define external event codes and a mapping rule to harmonize and transform them into an internal view defined by internal event codes.

You configure the event codes in Customizing under Event Management • Event Handlers and Event Handler Data • Codes. There are multiple settings to define external and internal views and the possibility to group them. Figure 8.29 shows the definition and grouping of internal event codes.

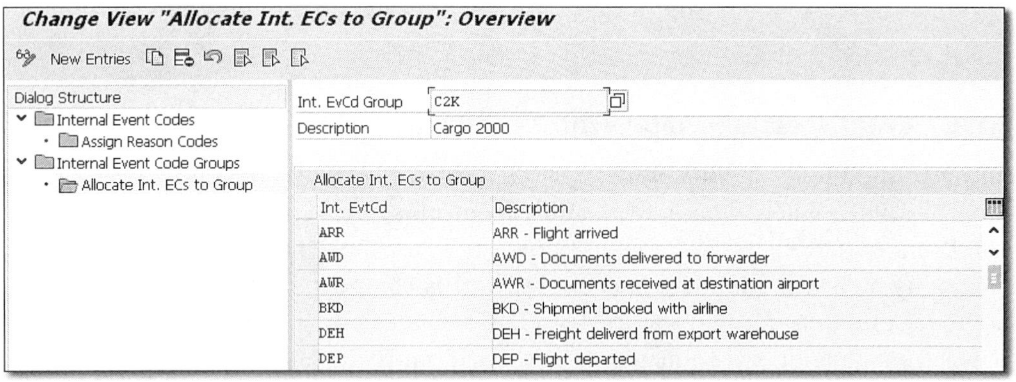

Figure 8.29 Internal Event Code Definition and Grouping

Rule sets

Event message processing is organized through the definition and use of rule sets. A rule set is a sequence of activities that are conditionally applied upon receipt of an event message or if an expected event is overdue with respect to either an expected event or message data.

A rule set is directly associated with an event handler. It can be a comprehensive list of rules, where each rule contains a condition under which it is executed and a definition of which activity should happen if the condition is true or false. The activity may be an ABAP function module, a method, or a procedure containing multiple other activities (see Figure 8.30). You also can define a next rule that should be executed based on the result of a called activity, which could be either true, false, or an error.

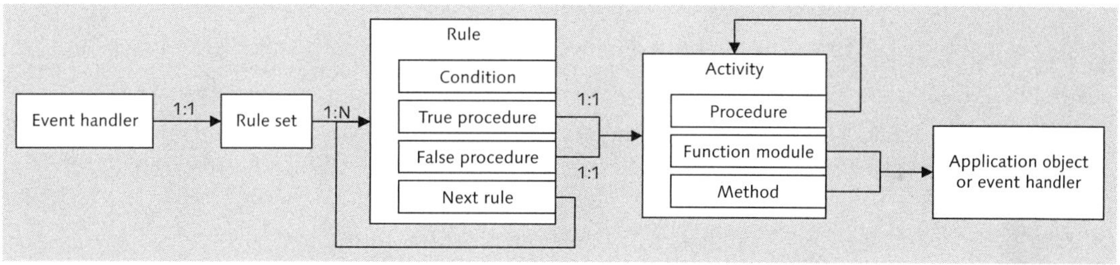

Figure 8.30 Association between Event Handler, Rule Sets, Rules, and Activities

Using Rules to Build an Event Handler Hierarchy

If you need hierarchical relations of event handlers (e.g., a tracked box in a tracked container), inheritance of tracking IDs allows you to easily indicate this reference. As soon as the box is packed into the container, the container event handler passes its tracking ID to the box event handler so that each message for the container event handler is now also processed by the box event handler (same tracking ID). This can be easily managed by rules and activities in the SAP Event Management activity repository (see the IMG under EVENT MANAGEMENT • GENERAL SETTINGS IN SAP EVENT MANAGEMENT • FUNCTIONS, CONDITIONS, AND ACTIVITIES IN SAP EVENT MANAGEMENT). Once the box is decontainerized, the box event handler deactivates its relation to the container event handler and can then be tracked on its own.

You define rule sets in Customizing under EVENT MANAGEMENT • REACTION TO EVENT MESSAGES • DEFINE RULE SETS. Here you can manage rule sets and define the single activities and logic in it. You also have access to activity definitions, multitasking activities (procedures), and rule conditions.

To provide a better overview of complex rule sets, you can also display rule set details, which takes you to a screen where the complete rule set with all its rules, conditions, and activities is displayed.

Figure 8.31 shows the rule set maintenance and the rule set details display. In the details display you can identify the procedural structure of the rule set, for example, in the LOADING_BEGIN section, where the following steps are done:

1. Check if the event is LOADING BEGIN.
2. Check if the event handler was created from an SAP TM freight unit.
3. Check that the event was not sent multiple times (i.e., that it had not been reported before).
4. Send the event time stamp to SAP TM for a freight unit update.

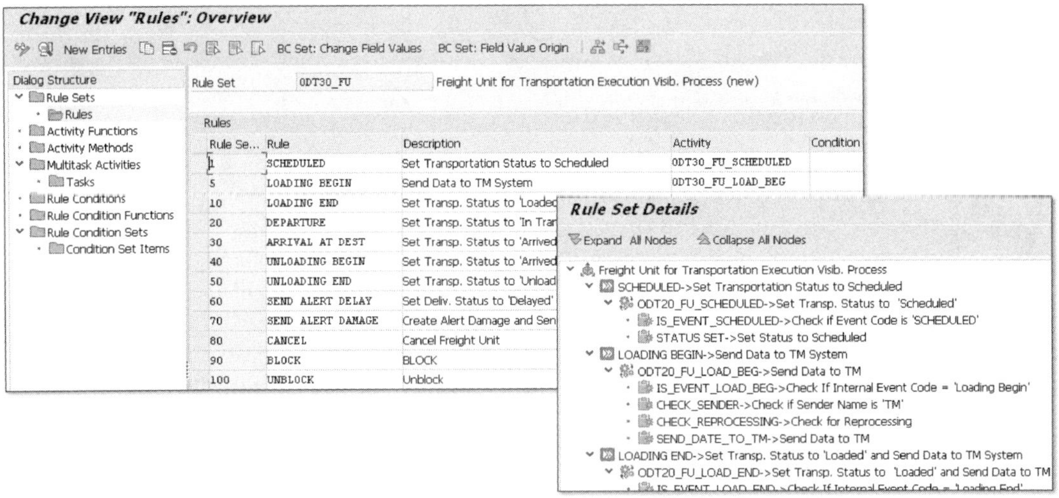

Figure 8.31 Rule Set Maintenance and Details Display

Definition of solutions, scenarios

To ease and support the development and distribution of SAP Event Management settings, you can use solutions and scenario definitions to package all settings done for a particular scenario implementation on the

SAP Event Management or application system side. In the IMG, the setup can be found under EVENT MANAGEMENT • SOLUTIONS AND SCENARIOS. Here you can assign event handler types, rule sets, extension tables, parameters and conditions, functions, activities, web interface transactions, users, and document flow to a scenario. Each scenario can then be exported from SAP Event Management as a business configuration (BC) set, which allows you to import it into another EMS.

8.2.5 End User Interaction, Lists, and Background Processing

SAP Event Management offers different methods of user interaction:

- Role-based web end user interaction to retrieve data from SAP Event Management or interact based on event messages
- List processing for power users to execute maintenance, retrieval, and inspection tasks
- Transactions for simple event message input

Role-based web interface

The role-based web interface allows you to define web transactions for event handler data retrieval and posting of event messages. It can be used to provide tracking information to end users and customers or partners and offers them the possibility to take part in the event reporting. An example is a transaction for parcel tracking, where a customer can follow the status of his parcel and the receiver can send a proof of delivery event if he receives the package in the correct condition.

Because the web transactions are role-based, you can assign different authorizations to each role and define the kind of event handlers and event messages and also the details a user is allowed to see. In addition, the event reporting feature can be authorized or restricted. You can find the setup of the web interface in the IMG under EVENT MANAGEMENT • EVENT MESSAGES, STATUS QUERIES, AND WEB INTERFACE • WEB INTERFACE, where you can define the web interface transactions, configure the visibility of data and authorization, and assign users or roles to the web interface transactions. Figure 8.32 shows an example of a web interface transaction where drilldown capabilities and a map display have been configured to do freight unit tracking.

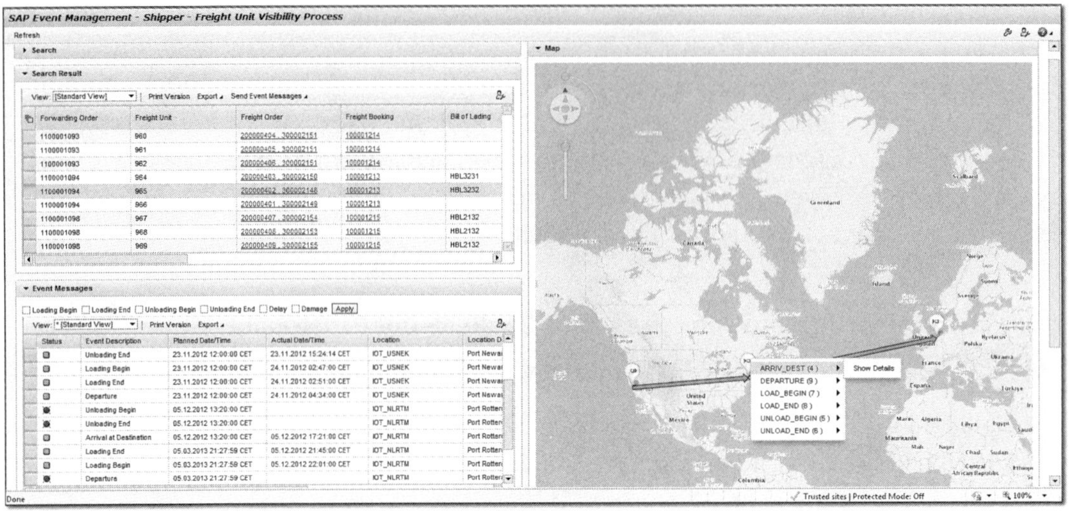

Figure 8.32 Web Interface of SAP Event Management with SAP Visual Business Map Display

Monitoring and event handler lists

With the lists in SAP Event Management, you can either retrieve information as a professional user or control the processing of data. Control processing can either be done interactively, when a user starts the corresponding transaction, or as a background process started by a batch job to regularly do data processing. The following list processes are provided in SAP Event Management:

- **Event handler list (interactive)**
  Using the event handler list, you can find and retrieve event handlers from SAP Event Management. You can drill down into the event handler overview and the detail display, which offers an in-depth overview of all event handler data (see Figure 8.33).

  From the event handler details, you can also update event handler data, which is a functionality that should be used only for maintenance purposes because it may corrupt data consistency.

- **Unprocessed message list (background or interactive)**
  The unprocessed message list allows you to process received messages that could not be processed (e.g., due to a locking situation). If an event handler receives two independent messages within a millisecond time frame, the second message may find the event handler

locked while processing the first message. In this case, the unprocessed message processor can be scheduled and run regularly to resolve this situation.

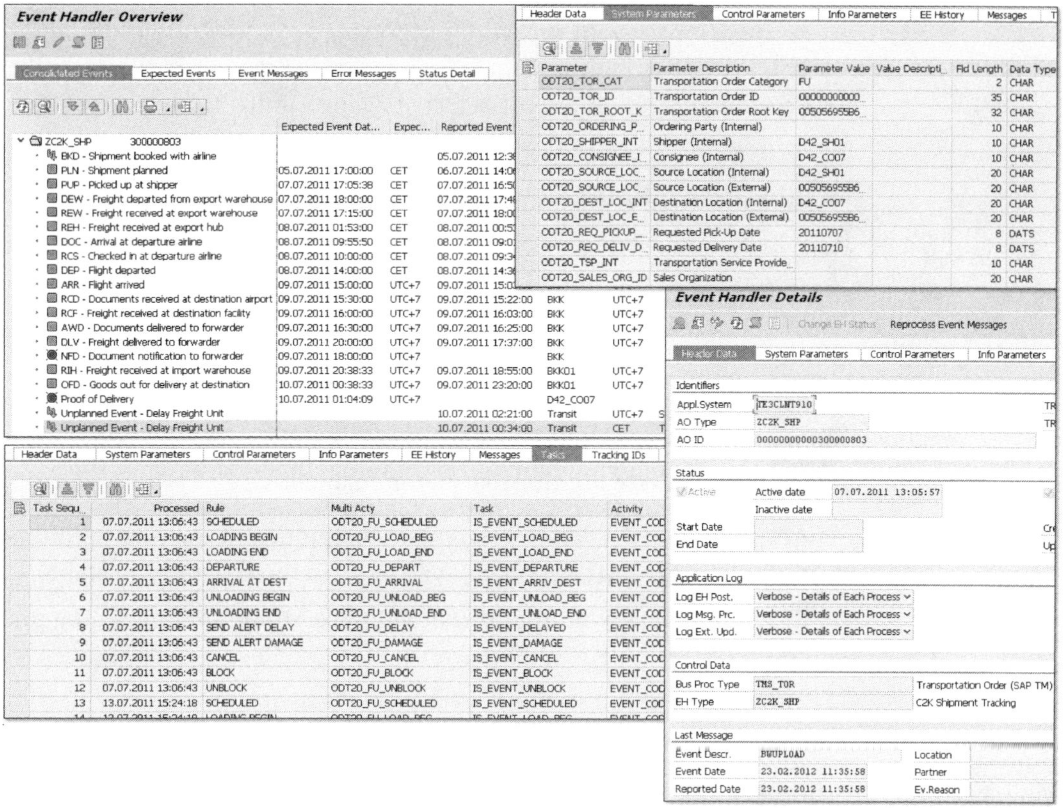

Figure 8.33 Event Handler Overview for a Cargo 2000-like Air Freight Tracking with Details of the Event Handler

- **Expected event overdue processing (preferably background)**
  The expected event overdue processing should be done regularly, best triggered by a batch job in less than an hour time frame (usually 5–10 minutes, depending on the on-time criticality). The processing checks if any expected event was overdue and raises an exception processing of the corresponding event handlers that can be handled in the rule set.

- **Event message reprocessing (interactive)**
  In situations where event handler rules need to be tested, reprocessing can support the process so that not every test requires new event messages.
- **Status list for event message processing**
  This status list shows the processing status of the event messages that have been received for the selected event handlers.

8.2.6 Preconfigured Event Management Content

SAP Event Management comes with preconfigured visibility processes. As described in Section 8.2.3, the SAP TM application objects forwarding order, freight unit, freight order, resource, and SOP instruction are connected to the application interface.

For most of these application objects, there is a corresponding preconfigured process, which you can use out of the box just by enabling it in Customizing. There are three steps for enablement:

1. Activate the SAP Event Management integration in the corresponding settings of the SAP TM object type (e.g., in the freight order type).
2. Enable and set up the application interface for the corresponding process (see Section 8.2.3).
3. Enable and set up the SAP Event Management process (see Section 8.2.3).

Visibility processes for SAP TM

The following visibility processes are ready to use, as shown in Table 8.3.

Visibility Process	Application Object	Event Handler Type
SOP instruction tracking	ODT30_INS	ODT30_INS
Freight unit tracking	ODT30_FU	ODT30_FU
Freight order tracking	ODT30_TO	ODT30_TO
Resource tracking	RES30_RESOURCE	RES30_RESOURCE

Table 8.3 Visibility Processes and Their Implementations

SOP instruction tracking is automated for the management of standard operating procedures. It supports the alerting and overdue management for the instructions created in SAP TM. If a forwarding order or freight unit has instructions with deadlines that need to be tracked, SAP Event Management will take over the correct setting of the overdue status of the corresponding instruction.

Freight unit tracking allows you to manage the scheduling, loading, unloading, departure, and arrival dates of a freight unit. Statuses such as blocking, delay, and damage can also be handled. Freight unit tracking is very useful for doing end-to-end tracking of a freight unit such as a package, pallet, or container. Actual events reported to SAP Event Management are fed back into the actual dates and times of SAP TM.

Freight order tracking supports the visibility on a consolidation level because it can be used, for example, to track trucks, ocean master bills of lading, and MAWBs, which are represented by the freight order or freight booking. If you have multiple consolidated moves in a transportation chain, the freight order event handler usually covers only one of the legs, whereas the freight unit is end-to-end. Events tracked for freight orders are the same as for freight units. In addition, events for proof of pickup, proof of delivery, and customs clearance are handled.

Resource tracking deals with visibility on the resource level; that is, the system tracks the asset represented by the resource from the beginning of its lifecycle until its end. Events handled are sighting, arrival, departure, and damages.

Cargo 2000-like scenarios

Based on the standard freight unit tracking scenario, it is possible, for example, to configure a tracking process that supports events processing in a Cargo 2000-like scenario that is used in air freight tracking. This can be achieved by renaming the events. In addition, an expected event extractor has to be created on the SAP TM side (ABAP function module). This extractor names the created expected events according to Cargo 2000 naming and derives the corresponding dates correctly from the freight unit dates and times. The event handler overview in Figure 8.33 shows the events that have been created from the extractor. This scenario is not part of the standard SAP TM or SAP Event Management delivery.

8.2.7 Integrating SAP Event Management with Other Components

In Section 8.2.2, we described the integration of application systems (which includes SAP TM) to SAP Event Management for creating event handlers or posting event messages. On top of this very fundamental integration, SAP Event Management offers a variety of additional integration points with SAP components and legacy systems.

Integration between SAP TM and SAP Event Management

There are two additional integration points between SAP TM and SAP Event Management:

SAP Event Management and SAP TM integration

1. The first is posting and update of SAP TM-relevant data from SAP Event Management to SAP TM as a result of rule processing. In many scenarios, receipt of an event needs to update corresponding data in SAP TM. If, for example, an event handler has an expected event for arrival at a destination that is derived from an SAP TM freight unit arrival date, then the rule processing can call an activity, which updates the SAP TM freight unit actual date upon receipt of the arrival event message (see Figure 8.34).

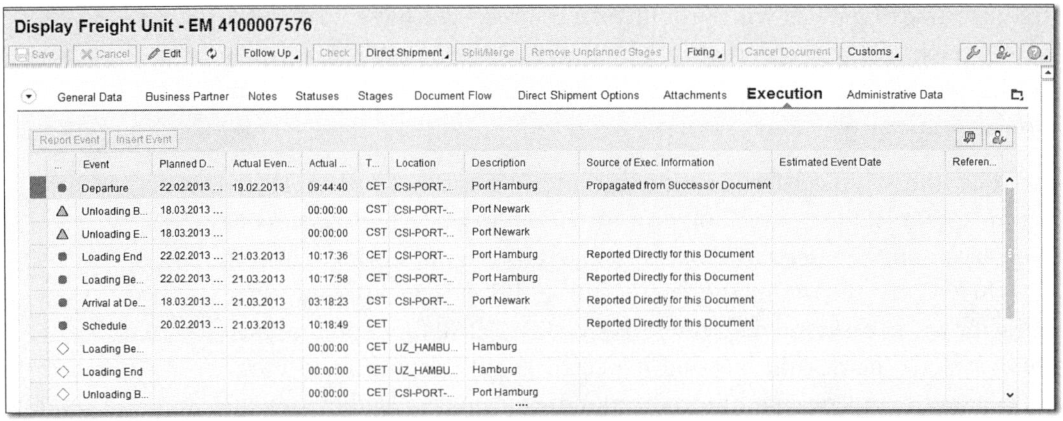

Figure 8.34 Event Message Data Displayed in SAP TM Freight Unit Context

2. SAP TM (and also SAP ERP) can retrieve event handler data for displaying inside the application context. With this integration, a user

can see the event handler status without calling a web transaction. The data is displayed as part of the current transaction (e.g., in the EXECUTION tabs). Additionally, SAP TM also offers the sending of event messages to SAP Event Management upon manual setting of the actual dates, for example, in the freight unit maintenance. Figure 8.34 shows an example of displayed event message data in an SAP TM freight unit context.

Integrating SAP Event Management with Other Systems

BAPIs, IDocs, and services

SAP Event Management offers three important interfaces for communication, which can be used as business application interfaces (BAPIs), enterprise web services like SAP NetWeaver PI, or intermediate documents (IDocs). The interfaces can also be used to integrate with non-SAP systems such as legacy systems running in a customer's landscape.

Table 8.4 shows the interfaces that exist in SAP Event Management.

Interface Function	Type	Implementation
Create, update, and deactivate event handlers	BAPI	/SAPTRX/BAPI_EH_POST
	IDoc	EHPOST01
	Enterprise service	TrackedProcessRequest_In
Receive event messages	BAPI	/SAPTRX/BAPI_ADDEVENTMSG_02
	IDoc	EVMSTA02
	Enterprise service	TrackedProcessEvent-NotificationCreateRequest_In
Retrieve event handler and event message data	BAPI	/SAPTRX/BAPI_EH_GET_DATA
	Enterprise service	TrackedProcessByIDQuery-Response_In

Table 8.4 Interfaces of Event Management

Integration with Other SAP Systems

Due to its flexibility and universality, SAP Event Management can also be connected and integrated with many other SAP components. The following integration cases have been done within SAP environments either as prototypes, as part of standard product, or as custom development projects:

- Integration of SAP Event Management with SAP Global Trade Services (GTS) for tracking customs approval status.
- Integration of SAP Event Management with SAP Extended Warehouse Management (EWM) to track the detailed movements of items in a warehouse or yard.
- Integration of SAP Event Management with SAP CRM Incident Management to create incidents to be handled by customer service in the event of critical situations in the supply chain that effect service-level agreements for handling of customer cargo. This integration is done as a custom development add-on.

This chapter gave an overview on how to execute and monitor transportation processes. In the next chapter, we offer insight into transportation compliance handling in terms of legal and trade compliance and dangerous goods regulations.

Functional capabilities in the SAP portfolio can support your daily processes that deal with transportation-related legal, environmental, and trade compliance. In this chapter, we focus on two SAP applications: SAP Global Trade Services and SAP Environment, Health, and Safety Management.

9 Transportation Compliance

Highly globalized markets and integrated supply chains offer many opportunities to companies acting in international economies. But since every coin has two sides, international business can involve risks for supply chains and the environment, and numerous underlying national and international regulations. Dealing with such hurdles in an effective manner without neglecting a company's core business is a challenge for many businesses. SAP applications for transportation compliance support you in overcoming these obstacles.

Administrative Efforts of Handling Shipments in Global Markets

An international shipment can easily entail the creation of up to 35 documents and interaction with 25 business partners, including authorities, logistics service providers (LSPs), banks, and security authorities. More than 600 local laws have to be followed, depending on the countries of origin and destination, transit countries, and kind of goods.

Areas of transportation compliance

Transportation compliance describes the expectation of abiding by local and international laws and regulations in transportation and logistics. The following list categorizes the different areas of transportation compliance, in line with SAP software applications:

- **Import and export management**
  Every country checks incoming and outgoing cargo regarding customs obligations and ensures the conformance with national and international trade laws. This requires the creation of export/import

declarations to authorities, embargo checks, and sanctioned party screenings for each foreign shipment. In addition, trade finance services can fall under this category to ensure payment of goods via a letter of credit. Countries and economic zones such as the North Atlantic Free Trade Agreement (NAFTA), European Free Trade Association (EFTA), and the Association of Southeast Asian Nations (ASEAN) have individual regulations and require individual interfaces for data exchange to conform to import/export compliance. The United States, for example, requires a full list of all incoming goods on each container ship and aircraft for Import Security Filing (ISF) via the Automated Manifest System (AMS). Other import/export systems that require a data exchange between companies and customs authorities include the following:

  - U.S. Automated Export System and Automated Broker Interface
  - Brazil Nota Fiscal Electronica
  - European New Computerized Transit System (NCTS)
  - German ATLAS System

- **Special customs procedures**
  The usage, categorization, and further processing of goods can entail various exceptions to regular import/export customs requirements, which are called special customs procedures. Goods that remain in-bond, for example—that is, remain in a warehouse before being exported again—are not subject to customs duty. Inward/outward processing prevents custom obligations if goods are being exported, processed abroad, and reimported. Only the value-added is subject to customs duty. The same applies for raw materials that are imported and exported again after processing or assembly.

- **Trade preference management**
  Countries have preferences about importing goods from certain countries and prohibitions against others. Trade preference management is the process of managing the eligibility of products for reduced import duty rates. This includes both the processing of long-term vendor declarations that empower a company to claim duty reductions and the issuance of declarations for the company's customers.

- **Dangerous and hazardous goods**
  Dealing with dangerous and hazardous goods exposes companies to additional risks and therefore responsibilities. Multiple legal regulations have to be followed depending on a categorization of products according to both national and international law. This includes specific requirements for documentation and operational limitations such as mixed loading prohibitions and averting the use of wrong container types, trucks, or even aircrafts with certain goods. Entire import bans or quantity restrictions can be applied to certain dangerous and hazardous goods classes.

Figure 9.1 shows an example process flow for a shipment with a few compliance regulations that need to be followed.

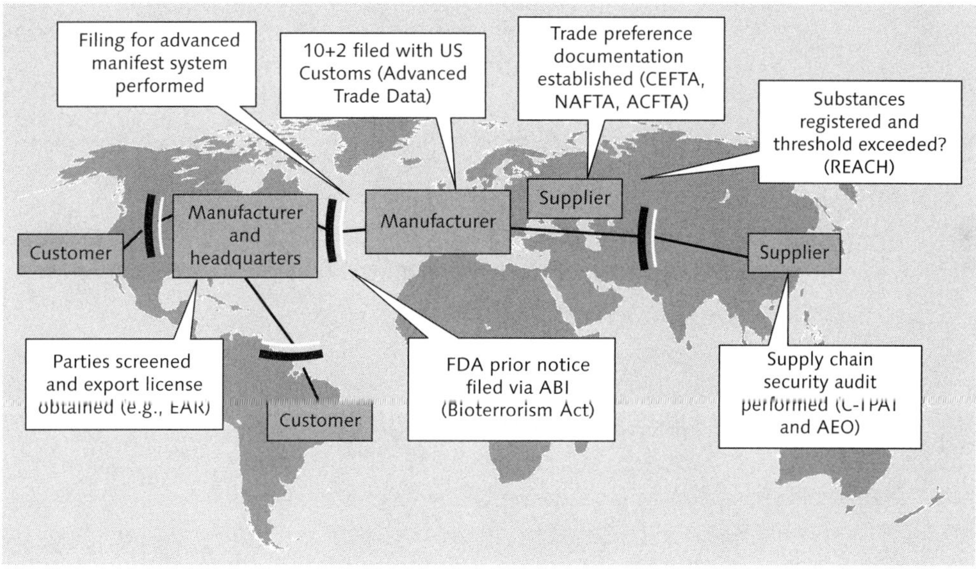

Figure 9.1 Example Process for Global Trade and Transportation Compliance

SAP GTS, SAP EHS Management

The two core applications in the SAP portfolio that support your business in dealing with challenges regarding transportation compliance are SAP Global Trade Services (GTS) and SAP Environment, Health, and Safety (EHS) Management. Both applications are integrated into SAP TM and contain individual functional components:

SAP GTS is part of SAP's Global Risk and Compliance (GRC) portfolio. It is integrated into SAP ERP, SAP Customer Relationship Management (CRM), SAP Extended Warehouse Management (EWM), SAP Event Management, and SAP TM.

SAP EHS Management is part of the SAP Business Suite and is a component of SAP ERP. It offers extensive functionality, such as employee health and safety management, product safety and stewardship, and environmental and product compliance, as well as registration, evaluation, and authorization of chemicals (REACH) and dangerous and hazardous goods management.

Note

Both applications contain additional functionality that exceeds the transportation and logistics focus of this book and is therefore not covered extensively. We focus on the integration of SAP TM with SAP GTS and SAP EHS Management.

It is important to mention that SAP TM has its own functional components to support trade compliance for specific scenarios. For example, strict legal compliance regulations apply to air freight shipments, such as known shipper checks, cargo screening, and prohibition checks between dangerous goods and aircraft types. These functional components are discussed in Chapter 15 in an air freight scenario overview.

The upcoming sections describe the capabilities and use of SAP GTS (Section 9.1) and SAP EHS Management (Section 9.2).

9.1 SAP Global Trade Services

Flexible system architecture

SAP GTS is the most extensive application in the SAP portfolio for trade compliance management. It can be used very flexibly in your system architecture; you can implement SAP GTS on top of an existing SAP ERP system, either as its own technical client or within an existing client of an implemented system. Alternatively, you can install it on any other existing SAP application that runs on the SAP NetWeaver stack or even install it standalone to support your legacy processes.

Business functionality

SAP GTS is highly integrated into the SAP landscape, especially into SAP ERP and SAP TM. In this area SAP GTS has five functional components to support your business:

- **Export management**
  SAP GTS helps you to streamline complex export processes to ensure faster delivery to your customers. SAP GTS automates your interactions with authorities due to its numerous interfaces with customs applications. This supports minimizing delays at national borders while ensuring compliance with relevant regulations and mitigating financial risks.
- **Import management**
  Similar to the export process, you can expedite customs clearance for import shipments and reduce costly buffer stocks, aiming for just-in-time inventory management. SAP GTS allows you to easily classify products, calculate duties, streamline electronic communication with customs authorities, ensure import compliance, and manage letters of credit.
- **Trade preference management**
  You can achieve savings by making the most of international trade agreements with your vendors, determine the eligibility of your products for preferential customs treatment, and issue certificates of origin to your customers.
- **Special customs procedures**
  This functional component supports you in managing your in-bond customs warehouses. You can drive process efficiency for duty reliefs regarding inward and outward processing and decrease costs by referring duty obligations.
- **Special regional procedures**
  Businesses operating in the European Union (EU) can use SAP GTS to manage and calculate the restitution for the export of common agricultural products (CAP) out of the EU with capabilities to assign securities, manage export licenses, maintain recipes, and calculate and apply for refunds.

The following section details the general capabilities of SAP GTS and how it supports your business for trade compliance. In Section 9.1.2 we show whether functionality interacts with processes in SAP TM or is used only in SAP ERP scenarios. Last, we dive deep into the functional scenarios available in the integration of SAP TM and SAP GTS.

9.1.1 Functionality Overview

Let's take a closer look at specific SAP GTS functionality to construct a clear picture of how you can use it in a standard implementation.

Import and Export Management

Import and export classification

The first key functionality in Import and Export Management is the import/export classification. Companies are required to ensure that they have licenses for certain goods and declare their license codes when filing customs declarations with authorities. SAP GTS allows you to assign import/export control classification number codes to your existing materials from your SAP ERP system (products in SAP GTS). This functionality is performed in SAP GTS and is mainly used for shippers because most LSPs do not maintain product masters for goods they ship for their customers.

Import and export compliance

The second core functional component, import and export compliance, is supported by three capabilities: sanctioned party list screenings, embargo checks, and license management. You can use complex checks to screen your business partners (vendors, customers, etc.) for denied parties. This functionality is highly integrated into transactions in SAP ERP and SAP TM: SAP ERP sales and purchase orders, SAP ERP Financial Accounting (to prohibit financial interactions), SAP TM Order Management and Transport Execution, SAP ERP Human Capital Management (to check potential new employees), and so on.

Embargo checks follow the same concept. Transactional documents in SAP ERP (sales orders, purchase orders, and payment transactions) or in SAP TM (forwarding orders, freight orders, and freight bookings) are screened and blocked if they match an embargoed country. For both embargo checks and sanctioned party lists, you can use SAP GTS to manage all blocked documents centrally. Last, you can maintain and use license catalogs to store legal licenses for transporting specific goods and integrate the licenses into transactional documents. This capability is only used in the integration between SAP ERP and SAP GTS.

Customs services management

Third, SAP GTS has a strong functional component that supports you in actually generating and processing customs declarations as well as interacting and communicating with customs authorities. If you work as a shipper, you will benefit from the goods classification of your materials and product master data. If you work as a freight-forwarder, you will collect the majority of customs-relevant information directly from the shipper/consignee. Again, SAP GTS is integrated into both SAP ERP and SAP TM. For orders in SAP ERP (e.g., sales orders, purchase orders) and inbound/outbound shipments, SAP GTS checks the customs relevance based on the product classification of the goods that you plan to ship. With SAP TM, the customs check happens in SAP GTS based on freight orders/freight bookings or forwarding orders. The corresponding document is blocked for further processing. You can navigate to SAP GTS to create the actual customs declarations and required documentation. SAP GTS offers numerous interfaces to local authorities, with standardized e-filing formats.

Letter of credit

The last process categorized under import/export management is the handling of a letter of credit. When you import goods, you can generate or print a letter of credit directly from the SAP GTS system. You can assign your transactional documents, such as purchase orders in SAP ERP, to the letter of credit. In SAP GTS you will find a central cockpit to monitor all transactional documents affected by a letter of credit that is in process. The letter of credit functionality checks SAP ERP documents and is used a lot by shippers.

Trade Preference Management

Long-term vendor declaration

SAP GTS Trade Preference Management is a functionality that supports you in identifying any products that you procure regarding preference handling by customs authorities. Especially within NAFTA and the EU, companies can be granted preference handling and lower customs duties for specific goods procured from developing countries. This process can take on a certain complexity, especially if you procure multiple components or semi-finished goods that you assemble and use for production. In this case, each part needs to be traced back to its country of origin and feature a *long-term vendor declaration* (LVD). The goal of SAP GTS trade preference management is to maintain your LVDs to automatically mark all your sales and purchase orders and shipments in SAP ERP that are applicable for automatic preference handling. In addition, SAP GTS allows you to generate and issue certificates of origin for your customers. SAP GTS Trade Preference Management is used only by shippers and does not include any interaction with SAP TM. It mostly exchanges information between SAP GTS and your SAP ERP system.

Special Customs Procedures

Besides numerous obligations and regulations in international trade, authorities still support your international business with certain customs exceptions. SAP GTS supports three key processes to benefit from such duty exemptions.

Customs warehouse

First, SAP GTS supports you in your customs warehouse activities to store duty-unpaid goods. This is permitted in two situations. If you keep cargo in-transit before it is shipped to another (possibly yet unknown) country, no duties apply. Alternatively, you can keep a stock of duty-unpaid goods if they are being processed for free circulation with a prescribed end use for industrial assembly. For both cases, SAP GTS is tightly integrated into SAP ERP and SAP EWM for warehouse processes. When receiving inbound deliveries based on SAP ERP purchase orders, SAP GTS can automatically detect whether they are to be treated as in-bond goods. SAP GTS helps you in its integration with SAP ERP and SAP EWM to manage stocks of in-transit and duty-unpaid products.

Free circulation management

Second, you can manage your processes of free circulation with prescribed end use and processing under customs control. The goods you have imported have a higher duty obligation then a finished product they might be used for in an assembly or industrial process. Free circulation under prescribed end use supports the process of letting a product be manufactured or assembled by a third party after you import it. It is crucial for you as the importer to always maintain the link between the imported goods and the finished product that grants the lower customs duty. In this process, SAP GTS supports identification of goods that qualify for such treatment, monitoring and management of stocks, and warehouse movements and interactions with authorities to calculate duty deductions. It is integrated with both SAP ERP and SAP EWM.

Inward and outward processing

Third, you can declare goods for outward and inward processing. This process aims at you as a manufacturer, exporting products into another country for processing and reimporting them for assembly/disposition (outward processing). You only have to pay customs duties for the newly added product components/value. The import process works the other way around. You have to declare products for inbound processing if you intend to re-export them after processing. Again, SAP GTS is the core application that supports this process, integrated into billing documents and orders in SAP ERP. Figure 9.2 summarizes the process of inward and outward processing.

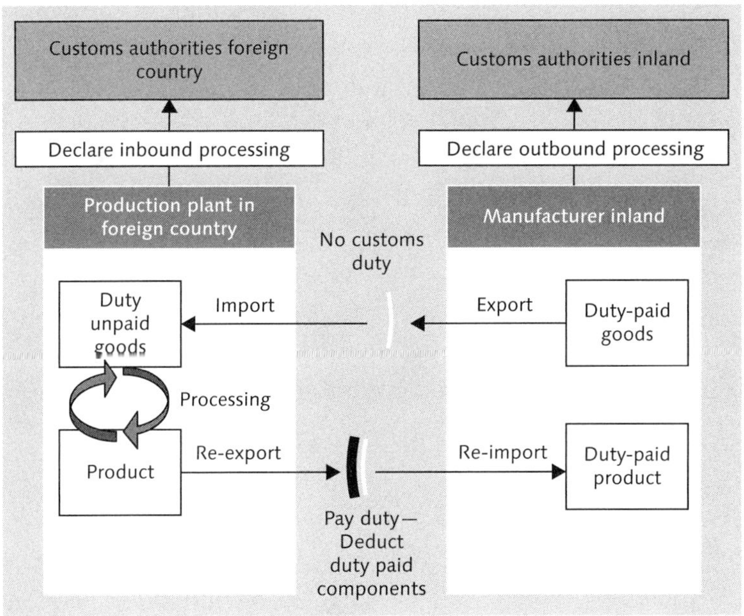

Figure 9.2 Inward and Outward Processing

Special Regional Procedures

Restitution management

SAP GTS also supports regional customs procedures. So far this capability focuses on requirements from the EU, so it handles identification and processing of restitution for certain farm and food products exported from the EU. The EU grants specific subsidies to companies exporting agricultural products to compensate for differences in EU prices and the world price. The legal basis is given in the Common Agriculture Policy (CAP).

SAP GTS supports you in a clear process. The products that comprise each order, shipment, or billing document created in your SAP ERP system are compared against your restitution master in SAP GTS. Once a product has been identified as eligible for subsidies, SAP GTS calculates the restitution amount and generates the appropriate customs declarations. Again, this is a functionality for shippers without any interaction with an SAP TM system.

Excise duties

We want to highlight two specific use cases where SAP GTS supports your business dealings with regional customs procedures. Both apply according to European law. First, all goods subject to specific excise duties, such as alcohol or tobacco, require specific customs handling. SAP GTS supports you in transporting these products in duty suspension through electronic communications with the customs authorities.

Electronic compliance reporting

Second, it is an EU requirement to declare all goods movements when trading between European countries. Based on the measured flows of commodities, governments and the EU can collect valuable statistical information. SAP GTS supports you in classifying affected shipments and declaring goods movements to Intrastat or other channels for statistical declarations. The generation of customs declarations or payment of customs duty is consequently not required.

9.1.2 Integration of SAP GTS and SAP TM

Integration to SAP ERP vs. SAP TM

The previous section explained how each functional component of SAP GTS is integrated into feeder systems, such as SAP ERP, SAP TM, SAP EWM, and SAP Event Management. Table 9.1 summarizes the SAP applications that are integrated to the different functional components of SAP GTS. In bold you can see the main functional components GTS supports, such as Import and Export Management and Trade Preference Management. We have listed the integration points of the more granular functionalities as part of import/export management as well.

Functional Components	SAP ERP	SAP EWM	SAP TM	SAP EM
Import and Export Management				
▸ Import/export classification	X			
▸ Customs services	X		X	X
▸ Letter of credit	X			
▸ Embargo checks	X		X	
▸ Sanctioned party list screening	X		X	
Trade Preference Management	X			
Special Customs Procedures	X	X		
Special Regional Procedures	X	X		

Table 9.1 Functional Components Using SAP GTS

In the following sections we explain the three functional integrations between SAP GTS and SAP TM as introduced in Table 9.1. In Section 9.1.3 we dive deep into embargo checks and sanctioned party screenings. Section 9.1.4 and Section 9.1.5 offer details about export and import customs services.

Integration of master data, transactional data

Let's shift our focus to the functionality that is used in the integration between SAP TM and SAP GTS only. From a transportation perspective, we consider SAP ERP as a pure back-end system that stores a materials master and generates orders (from the shipper's perspective) and a financials and billing solution. In this section we differentiate between the integration of master data and transactional data. Master data is replicated from SAP ERP. The transactional integration is enabled for two major technical objects in SAP TM: the transportation request (TRQ) and the transportation order (TOR).

Integration of Master Data

As a first step in the integration between SAP GTS and SAP TM, you need to synchronize the master data from your back-end SAP ERP system to SAP GTS. SAP provides an asynchronous master data interface based on the SAP Application Link Enabling (ALE) technology. Both the vendor and customer master from SAP ERP are replicated to the business partner in SAP GTS. Materials master data from an SAP ERP system creates the product master data in SAP GTS. Last, you can integrate the organizational data to SAP GTS. After a successful integration of master data, you need to enrich records with additional SAP GTS-specific information such as product classifications (to identify goods based on customs authority definitions), product groupings (which allow logical grouping of goods with equal customs requirements), and so on. Additional master data needs to be maintained directly in SAP GTS because it is used solely in the SAP GTS system and not replicated from SAP ERP; this includes country groupings (assignment of country of departure and country of destination to country groups), customs list (import/export classification numbers), licenses for import/export, and sanctioned party lists.

Integration of Transactional Data

User interface

Recall from Chapter 2, Section 2.2 that SAP TM runs entirely on SAP NetWeaver Business Client (NWBC) technology. Figure 9.3 shows the standard user interface of SAP GTS in the SAP NWBC. Based on the capability to search any transaction, even Customizing and other back-end transactions can be executed from SAP NWBC.

Let's focus now on the transactional integration between SAP GTS and SAP TM. All transactional data between SAP TM and SAP GTS is exchanged via SAP's service-oriented architecture (SOA) using SAP NetWeaver Process Interface (PI). Alternatively, it is possible to define point-to-point (P2P) connections without going via SAP NetWeaver PI but still facilitating the PI Enterprise Service Repository. To set up the integration between SAP TM and SAP GTS, you have to configure the communication basics first. This procedure technically links your SAP TM feeder system to SAP GTS. You need to define the logical system, set up RFC destinations, and define the distribution model for the ALE.

Standard SOA Services for SAP TM-SAP GTS Integration

SAP delivers six standard SOA services for transactional integration:

- `ExportDeclarationSUITERequest_Out` (and `_In`)
  Enables SAP TM to create customs declarations in SAP GTS.

- `ExportDeclarationSUITECancellationRequest_Out` (and `_In`)
  Cancels a customs declaration from SAP TM.
- `ExportDeclarationSUITEConfirmation_Out` (and `_In`)
  Transmission of status information from SAP GTS to SAP TM.

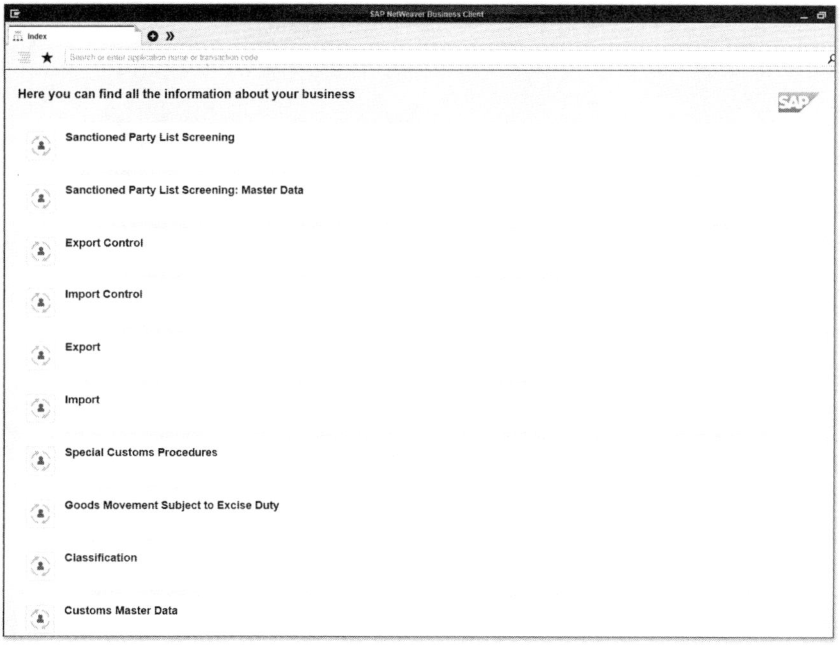

Figure 9.3 User Interface of GTS in SAP NetWeaver Business Client

Because you have already been introduced to the different objects in SAP TM in previous chapters, such as freight orders, forwarding orders, and so on, you can see that these major transactional documents are interfaced with SAP GTS for specific functionality. Figure 9.4 shows which objects can be used for which functionalities in the integration to SAP GTS.

The forwarding order, for example, can be validated against a sanctioned party list because this is where you would maintain your shipper, consignee, and so on. In the upcoming sections we introduce the details of the functionalities in the transactional context. It is important to highlight that the integration between SAP TM and SAP GTS for customs services is currently only available in the standard for export customs services and export compliance. The functionality for import customs declarations and transit procedures is currently not integrated into SAP GTS but can be enabled via a Business Add-In (BAdI).

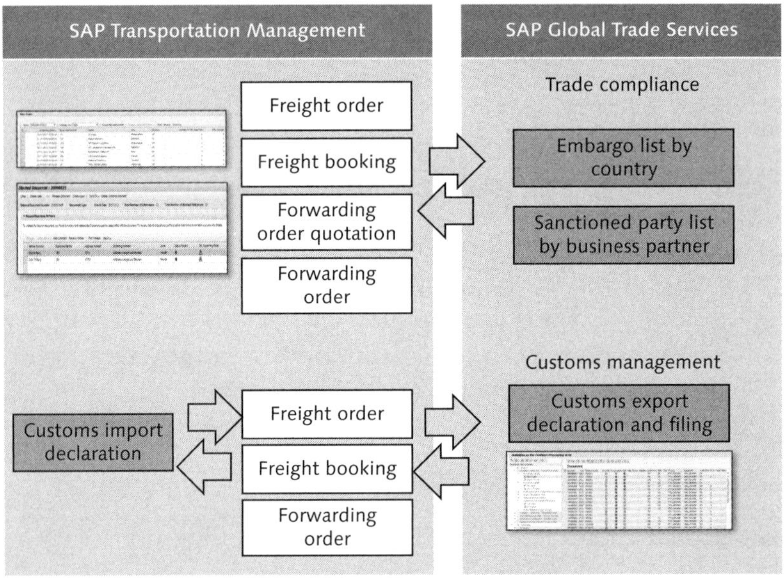

Figure 9.4 Transactional Interface between SAP TM and SAP GTS

9.1.3 Export Compliance for Sanctioned Party Screening and Embargo Checks

Business background

Section 9.1.1 introduced the general concept and functionality of sanctioned party screenings and embargo checks. Let's now dive deeper into the details of the integration between SAP GTS and SAP TM. Various historical events, especially the terrorist attacks of September 11, 2001, led to stricter legal regulations to monitor and blacklist companies if they violate local law. Different countries and legal authorities (e.g., the Office of Foreign Asset Control in the United States) have issued blacklists that name all businesses and individuals with whom it is prohibited to have a business relationship, including importing and exporting goods. Companies are now legally obliged to check their business partners against government blacklists of sanctioned parties. The challenges in this process are that companies can have thousands or even millions of records of business partners with various spellings and addresses. Blacklists frequently change and require updates, which makes manual maintenance very difficult, if not unfeasible.

Manufacturer Blacklisted until 2018 for Intending to Violate U.S. Trade Regulations

In 2008, the manufacturer Winter Aircraft Products SA of Madrid was blacklisted for a period of 10 years by the U.S. Department of Commerce, Bureau of Industry and Security. The company had tried to transship aircraft parts subject to U.S. regulations to Iran without the correct license or informing the suppliers about the final destination. This serves as an example of both possible sanctions caused by negligent trade compliance and how companies might end up being blacklisted in a sanctioned party list.

Functional process

It is similarly relevant for both shippers and freight-forwarders to validate business partners and check destination or transit countries against sanctioned parties and embargo lists that appear in your orders and freight documents. Figure 9.5 gives an overview of the entire process for both a shipper and an LSP.

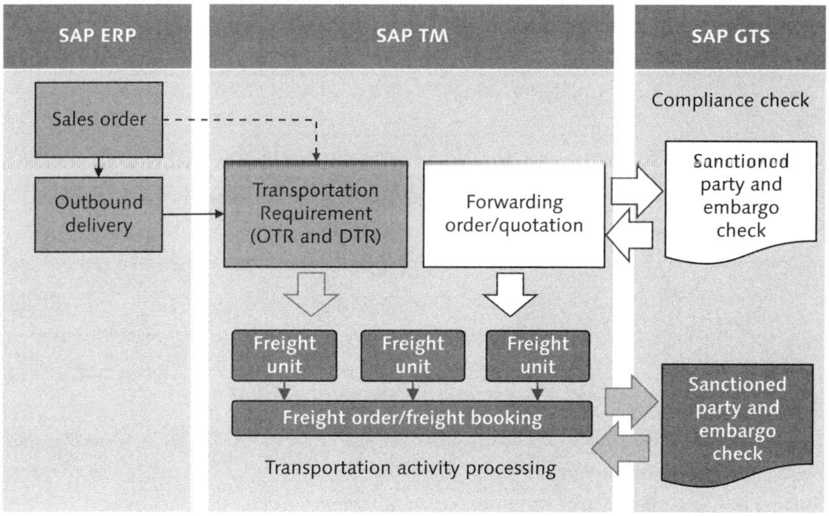

Figure 9.5 Embargo and Sanctioned Party Checks with SAP TM and SAP GTS

In the white boxes you can see the LSP-specific objects, such as the forwarding order/quotation. The shaded boxes illustrate the shipper scenario. In the dark boxes you can see the common objects that are relevant for both an LSP and a shipper, such as the freight units.

As a shipper, you can start the process for compliance checks by creating regular orders and deliveries in SAP ERP that generate a transportation requirement (OTR or DTR) in SAP TM, as described in Chapter 4, Section 4.1. Alternatively, as a freight-forwarder or carrier, you would start the process by creating a forwarding order directly in SAP TM.

Screening of forwarding order

Figure 9.6 shows that an LSP performs the compliance checks directly when capturing a forwarding order or when providing a quotation to a customer. This allows you to prevent the creation and execution of an order very early in the process. After a forwarding order is created, it is automatically blocked for execution with the reason of Compliance Check Required.

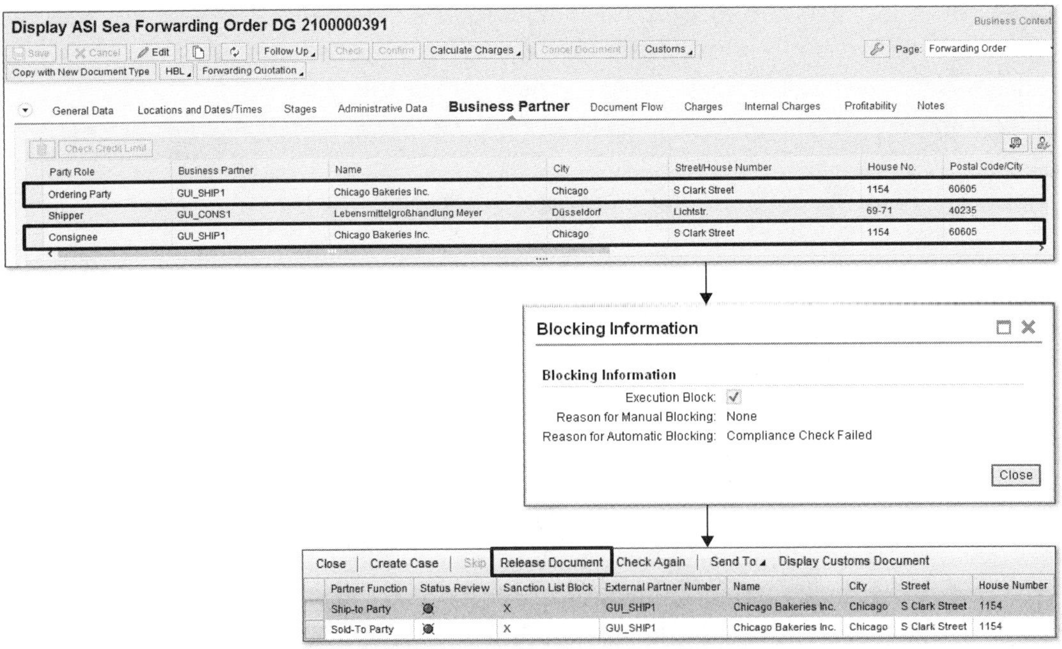

Figure 9.6 Sanctioned Party Screening in Forwarding Order

After the document is saved, the transactional interface with SAP GTS transfers the business partners and country information for the validation. SAP GTS returns the compliance status as either not compliant or compliant. If a business partner or country is detected as blacklisted or under an embargo, you can cancel the forwarding order or manually release the business partner or document as compliant in SAP GTS to release the execution block in the forwarding order. In Figure 9.6 you can see the compliance management between a forwarding order in SAP TM and SAP GTS.

Freight order or freight booking screening

If you work as a shipper, you do not use the forwarding order. Instead, you can perform the compliance checks after planning your transportation from either the freight order or the freight booking. As an LSP, you can use the same functionality to validate freight orders and bookings. After you have successfully created a freight order or booking, the compliance status will be set to NOT CHECKED, and the execution block will be activated. Only after you save the document is the interface to SAP GTS triggered. SAP GTS again validates the business partners and countries against blacklists and embargos. The document is released or blocked, depending on the screening result. You can now handle the exceptions of blocked freight orders or bookings in SAP GTS. You can monitor all blocked documents and business partners and manually release them with the appropriate reason code (e.g., a mistaken identity) by following the menu path SAP NWBC • SANCTIONED PARTY LIST SCREENING • BLOCKED BUSINESS PARTNERS • RELEASE SELECTED ENTRIES.

Customizing in SAP TM

In SAP TM very few customizing settings are required to enable embargo checks and sanctioned party screenings. You need to configure the technical integration with your SAP GTS system, as described in Section 9.1.2. In addition, you must select the ENABLE COMPLIANCE CHECK checkbox for the following document types: forwarding order type, forwarding quotation type, freight order type, and freight booking type.

Customizing in SAP GTS

To enable the sanctioned party screening and embargo checks in SAP GTS, you have to undertake various Customizing activities. All the configurations you need to make in SAP GTS are made in the following Customizing menu paths: IMG • GLOBAL TRADE SERVICES • COMPLIANCE MANAGEMENT • SANCTIONED PARTY LIST SCREENING and IMG • GLOBAL TRADE SERVICES • COMPLIANCE MANAGEMENT • EMBARGO CHECK. Let's highlight a few key configurations:

- Maintain the Customizing table in Activate Business Partner at Business Partner Function Level. This setting enables the sanctioned party checks for different business partner functions.
- To define the check rules for the sanctioned party screening (e.g., using address information and keywords), maintain the settings in Define Control Procedure for Address Comparison.
- You have to assign the defined check rules to a legal regulation, such as the German Foreign Trade Regulations. You can specify which types of sanctioned party lists should be considered, as well as audit trails and notification workflows. To do this, specify the settings in Control Settings for Sanctioned Party List Screenings.
- To enable embargo checks, implement the Customizing table Define Types of Agreements and Assign Determination Procedure for Active Legal Regulations. You also need to maintain the table Activate Legal Regulations, and enter, for example, the United Nations Embargo Regulation. For each legal regulation, you can select Control Settings for Embargo Checks to assign business partner groups that are relevant for checks.

SAP GTS Configuration Guide for Compliance Management

You can find a detailed description of all GTS Customizing activities in the *GTS Configuration Guide for Compliance Management* at *service.sap.com/swdc* by selecting Release & Upgrade Info • Installation & Upgrade Guides • Analytics • Governance, Risk and Compliance • Global Trade Services • Release 10.1 • GTS10.1 Configuration Guide for Compliance Management.

9.1.4 Export Customs Services

To successfully process your international shipments, as both a freight-forwarder and a shipper, SAP TM offers integration with SAP GTS for export customs services. You can check whether your outgoing shipments are customs-relevant, depending on the origin and destination country of the shipment. You cannot process a shipment without appropriate legal documentation, so the application blocks orders and shipments until appropriate documentation has been issued. SAP GTS supports the generation of the customs declaration and the electronic communication with authorities via EDI. SAP GTS supports the creation and printing of export documentation, based on customs (e.g., legally required documentation), compliance (legally required certificates), and customer-specific requirements (individual documents).

Figure 9.7 shows the process overview for both a freight-forwarder (top row) and a shipper (bottom row). As the central application, SAP GTS is used the same way for shippers and freight-forwarders. The major difference is that a freight-forwarder triggers the generation of a customs declaration directly from the order, whereas a shipper starts the generation of customs declaration from a freight order or booking.

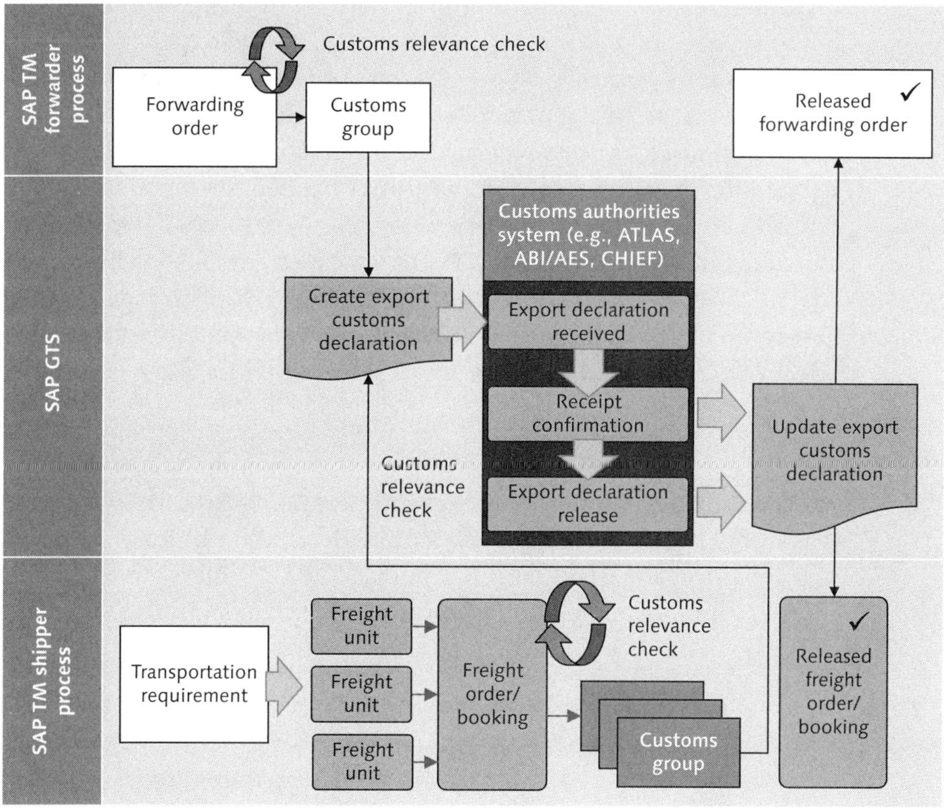

Figure 9.7 Process Overview for Export Customs Services

The first activity in export customs processing is to check whether the forwarding order or freight order/booking requires customs documentation

Customs relevance check

at all. For shipments within the European Union, for example, no export customs declaration except Intrastat reporting (standard with SAP GTS) is required. When you save your document, the customs relevance check is automatically performed directly in SAP TM. If the result is customs-relevant, you see that the order/booking is automatically blocked from execution for the reason CUSTOMS DECLARATION CHECK REQUIRED.

Export Customs Declaration Management

Customs groups

Once a freight order/freight booking or a forwarding order has been identified as customs-relevant, you need *customs groups*. The reason is that each freight order/booking or forwarding order can contain multiple consignees or goods sellers and you might need to generate a separate customs declaration for each one. The freight unit items are grouped together in a customs group. Each customs group created in SAP TM generates one export customs declaration in SAP GTS. You will see that the generation of customs groups happens automatically and is flexibly configurable. At least one customs group is required to generate an export declaration in SAP GTS. Each object—the forwarding order, freight order, and booking—has a CUSTOMS tab to manage customs groups and show the latest statuses. To generate the relevant groups, select CUSTOMS ACTIVITY • BUILD CUSTOMS GROUP.

Generation of export customs declaration

After the automated or manual generation of a customs group(s) in SAP TM, you can create the export customs declaration by selecting one or multiple customs groups and choosing CREATE EXPORT DECLARATION. It is possible to automatically generate the declaration in SAP GTS based upon transmission of the customs groups. SAP GTS offers you the ability to monitor all incoming and generated declarations in the SAP NWBC operative cockpit by selecting EXPORT • OPERATIVE COCKPIT • DISPLAY OUTBOUND ACTIVITIES.

Operative cockpit

The comprehensive operative cockpit allows you to closely monitor all inbound and outbound messages with the customs authorities, including the management of errors or interruptions in the process. It is used for the communication on both the import and export side. In our example, we use the operative cockpit for the monitoring of the export customs declaration displayed in Figure 9.8. Documents listed under status

Open have not been transmitted, whereas status *In Process* represents that a message has been sent but no answer has been received yet. Figure 9.8 shows also the process of a blocked forwarding order or freight order/booking and the generation of customs groups, which eventually triggers a customs declaration document in the SAP GTS export customs operative cockpit. SAP GTS offers several standard web services for the integration with legal authorities. In the COMMUNICATION tab of the details of each document a message log is visible. Once the communication has been sent, and a successful answer is received from the authorities, then a reference number (e.g., the movement reference number, or MRN) is returned to both the export declaration in SAP GTS and to the customs groups in the forwarding order or freight order/booking. The customs status in SAP TM changes to *Approved*. Only now is the SAP TM document released from the execution block. For a rejection or hold, the corresponding documents in SAP TM stay blocked with an appropriate status.

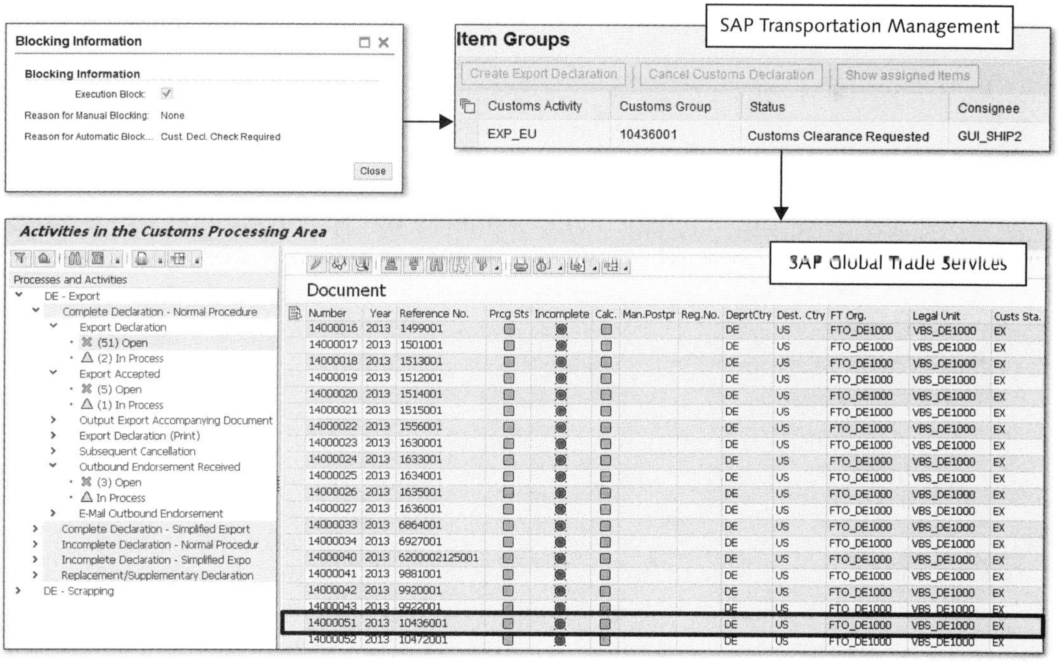

Figure 9.8 Generation of Export Customs Declaration

Configuration for Export Customs Management and Services

Customizing in SAP TM

To enable the process of export customs integration between SAP TM and SAP GTS, you have to perform settings on both the SAP TM and SAP GTS side.

Process controller settings

SAP delivers a standard service (TM_GT_GRP) that splits your forwarding orders, freight orders, and freight bookings in SAP TM into customs groups. This is required in case you have to generate more than one declaration per document (e.g., in cases where you ship to two consignees). You need to assign this service to a strategy in Customizing in order to use it via SAP Transportation Management • SCM Basis • Process Controller • Define Strategy.

You can define multiple ways to split the items of a freight order or freight booking. There are two standard methods provided to group the items: per consignee and per LSP. This is defined in a method that you can access by following the menu path SAP Transportation Management • SCM Basis • Process Controller • Define Method.

The last step is to assign the method to a strategy. As a result, the strategy combines the service and method for splitting and grouping items. We use the strategy later and assign it to a customs profile.

Customs relevance check and statuses

The customs relevance check defines whether an order or booking in SAP TM requires customs clearance. This differentiates all national shipments from international transports. The customs relevance check can be defined via the menu path SAP Transportation Management • Transportation Management • Basic Functions • Global Trade • Define Customs Relevance Checks. You must maintain the status values that you want to be displayed depending on the processing status of a customs procedure (e.g., *Requested*, *Approved*, and so on). Follow the menu path Transportation Management • Basic Functions • Global Trade • Define Cumulation of Customs Statuses.

Customs activity and profile

A customs activity combines some of the settings we described, such as the customs relevance check and the process controller strategy. These settings determine what status an object (e.g., freight order) needs to have in order for TM to trigger the check of customs requirements. Last, the customs activity contains the list of all status values related to the customs process that will be displayed in your documents. Figure 9.9 shows the configuration in SAP TM. The Customizing table can be found via the menu path SAP Transportation Management • Transportation Management • Basic Functions • Global Trade • Define Customs activities and profiles. As just described, the last step is to assign a customs profile to forwarding order, freight order, and booking types.

Customizing in GTS

On the SAP GTS side, various settings are required to enable the export customs functionality, but only few are required to set up the integration with SAP TM. Besides the technical integration described in Section 9.1.2, you first need to maintain a mapping between the organization's SAP TM partner number to the SAP GTS foreign trade organization. Second, you have to map your SAP GTS business partner functions to SAP TM partner functions via the menu path Global Trade Service • General Settings • Partner Structure • Define Partner Functions. Last, you need to map the SAP TM packaging material types to SAP GTS package types.

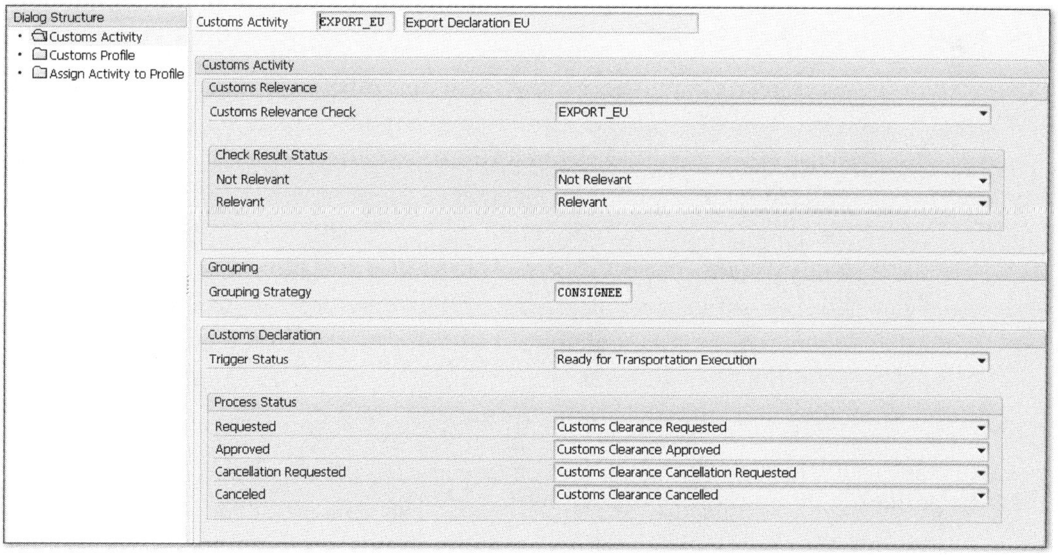

Figure 9.9 Customs Activity

> **GTS Configuration Guide for Customs Management**
>
> You can find a detailed description of all SAP GTS Customizing activities for export declaration and integration with SAP TM in the *GTS Configuration Guide for Customs Management*. The configuration guide can be downloaded from *service.sap.com/swdc* by selecting RELEASE & UPGRADE INFO • INSTALLATION & UPGRADE GUIDES • ANALYTICS • GOVERNANCE, RISK AND COMPLIANCE • GLOBAL TRADE SERVICES • RELEASE 10.1 • GTS10.1 CONFIGURATION GUIDE FOR CUSTOMS MANAGEMENT.

9.1.5 Import Customs Services

To ensure end-to-end transportation across countries, more than an export declaration is required from a customs service perspective. You also need an appropriate import declaration to declare goods and determine customs duty. Alternatively, you can issue a transit procedure if goods are not declared at the border and will be transported duty-unpaid or under bond. The importing country can have very specific requirements for the content and procedure for import declarations and transit procedures.

It is very important to emphasize that this functionality in the standard SAP TM 9.1 release does not have any integration to SAP GTS or other customs management applications. In other words, it is executed completely standalone and manually in SAP TM. A BAdI is provided to implement a custom integration, which we cover shortly. The functionality of the import customs declaration is aimed solely at LSPs and freight-forwarders acting as the customs broker for a shipper. Consequently, the functionality is only enabled for the freight unit object, in combination with the import forwarding order. The transit procedure also involves your freight documents.

Import customs declaration

When generating an import forwarding order in SAP TM for a customer, you first need to select the IMPORT DECLARATION BY LSP checkbox in the GENERAL DATA tab. If a customer always requires import customs clearance, you can maintain this checkbox on the business partner as well, which defaults it to the forwarding order. Per forwarding order item (e.g., products and containers), you can find the CUSTOMS tab. In this tab, it is necessary to specify what customs activity needs to be performed for the forwarding order item. You can choose between a transit procedure or, in our case, the import declaration. You also need to specify in which location the cargo is imported to a country. Select CUSTOMS • CREATE IMPORT DECLARATION in the top panel of the forwarding order to generate the customs groups, for example, based on the customs location. As soon as the customs groups are generated, the forwarding order is blocked for execution because a processing and approval of the import declaration is required. The appropriate customs groups are visible in the freight units. For each customs group in the freight unit, you need to set the status to *Customs Declaration Approved*, as you receive the actual approval from the customs authorities. As mentioned earlier, the entire generation of the customs declaration—as well as its submission to and handling by customs authorities—is currently not supported. Once the status for all freight unit custom groups is set to *Customs Declaration Approved*, the forwarding order is released from the execution block.

Transit procedure

As an alternative to the import customs declaration, you can generate a transit procedure. Navigate to the CUSTOMS tab in the forwarding order items and set the inbound customs activity as TRANSIT PROCEDURE. You can enter relevant data, such as commodity codes. In the STAGES tab, you can mark each transportation leg (each stage) that requires a transit procedure for each freight unit in the forwarding order. After you have planned your forwarding order and assigned the freight units to freight orders and bookings (refer to Chapter 6), SAP TM automatically marks the same stages in the freight documents as TRANSIT PROCEDURE RELEVANT. The freight documents that are affected can be blocked for execution. You can open and close transit procedures to indicate whether the goods are currently in transit (e.g., between the border to a trade union and the transit end location). To open a transit procedure, navigate to the freight document and select CUSTOMS • OPEN TRANSIT. Customs groups are generated based on the settings in Customizing. You can manually interact with the authorities per customs group and as soon as the transit is approved, change the status to *Open Transit Approved*. You can also store the customs document number and movement reference number manually; these are provided by the authorities. The freight document is released from an execution block.

A similar process can be followed to close the transit procedure: Choose CUSTOMS • REQUEST UNLOAD. SAP TM generates a new customs activity and customs group, which can be processed manually to reflect the offline interaction with authorities. Each customs group's status can be changed to *Transit Unload Approved*. In addition to the capability of opening a transit procedure from a freight document, you can also trigger it directly from a freight unit. The closure still happens from the freight document.

Customizing To enable import customs management in SAP TM, you have to perform configuration in Customizing. First, you need a forwarding order type with a traffic direction of import. Set CUSTOMS HANDLING to AUTOMATIC. Only forwarding orders that have been set to automatic customs handling are considered for import customs. All other required settings are very similar to the export declaration from Section 9.1.4.

Another requirement is to set up the process controller by using an import declaration strategy (`IMP_FU` or `IMP_LSP`) and a method (`GT_B_IM_FU`). The transit procedure must be set up the same way (e.g., strategy `TRA_OPEN`, `TRA_CLOSE`, and `TRA_OPENFU`). To do this, follow the menu path IMG • SAP TRANSPORTATION MANAGEMENT • SCM BASIS • PROCESS CONTROLLER. You must also implement the customs relevance check, a customs activity, and a profile, similar to the export process (refer to Section 9.1.4 for details). A key difference from the export process is that you need to assign the customs profile to a freight unit type. To enable both the transit procedure and import customs declaration, you need to generate two separate custom activities and assign them to a customs profile.

Extensibility To enable automated communication with customs authorities via a customs management application such as SAP GTS, you can implement a BAdI provided in standard SAP TM. You can find the BAdI in SAP TM Customizing via the menu path TRANSPORTATION MANAGEMENT • BUSINESS ADD-INS (BADIS) FOR TRANSPORTATION MANAGEMENT • GLOBAL TRADE • DECLARATIONS.

9.2 SAP Environment, Health, and Safety Management

Cargo moved in shipments has to comply with dangerous goods regulations. Dangerous goods—also known as hazardous materials, hazmat, or DnH (dangerous and hazardous)—include solids, liquids, and gases that can harm people, other organisms, or the environment. Shipments moving these dangerous goods can therefore have a direct influence on the environment, the health of people that are in contact with them, and the safety of all surrounding material objects. Recall the 2013 crash of 70 tank railcars filled with crude oil in Quebec and the associated death toll as a reminder of the seriousness of this subject.

Dangerous goods risk factors

The risk of moving dangerous goods is based on multiple factors and characteristics:

- The characteristics of the material itself and its impact on the environment; these include whether it is flammable, explosive, corrosive, radioactive, oxidizing, toxic or asphyxiating, pathogenic or allergenic, or biohazardous in nature
- The risk of inappropriate transportation or handling, which includes improper or damaged means of transport, unsuitable transportation routes, and unqualified, unreliable, or over-worked personnel
- Risk factors due to external influence such as terrorism

In SAP TM and SAP ERP you can perform checks to ensure safe and compliant transportation of dangerous goods. These checks reflect the international and national regulations regarding dangerous goods transport and depend on transportation mode, transit countries, and so on. SAP Environment, Health, and Safety (EHS) Management provides a framework to deliver certain checks and is used in combination with SAP TM. SAP EHS Management services are part of the SAP SCM Basis layer and can be used from any application in SAP SCM. SAP EHS Management consists of several components:

- Basic data and tools
- Product safety
- Occupational health

- Industrial hygiene and safety
- Waste management
- Dangerous goods

From this set of components, basic data, tools, and dangerous goods are integrated into SAP TM.

9.2.1 Dangerous Goods Regulations

Dangerous goods regulations differ by mode of transport, issuing country or region, and activity category or material status. For example, warehouse handling rules for flammable materials in the United States may differ from regulations that determine flammable materials transportation rules in Swiss tunnels.

Global dangerous goods regulations

Several globally valid dangerous goods agreements and regulations are relevant for transportation. The following are the most important ones:

- **International Maritime Dangerous Goods (IMDG) Code**
  The IMDG has been defined by the *International Maritime Organization* (IMO) to standardize terminology, packaging, labeling, and markings of dangerous goods and advice on stowage, segregation, handling, and emergency reactions for dangerous goods transport on vessels.
- **IATA Dangerous Goods Regulations (DGR)**
  The *International Air Transport Association* (IATA) is the governing body of airlines and published the fifty-fourth edition of the DGR in 2013. The DGR is a guide to safely shipping cargo or passenger luggage by air.
- **ICAO Safe Transport of Dangerous Goods by Air**
  The *International Civil Aviation Organization* (ICAO), which is an association of countries, has defined joint rules for safe air cargo transportation in the annex 18 of their international standards and recommended practice (SARP).

U.S. and European dangerous goods regulations

- **Hazardous Materials Regulations (HMR) of the Code of Federal Regulations (CFR)**
  The HMR is regulated by the U.S. Department of Transportation

(DOT) in Title 49 CFR Parts 171-180. The HMR is applicable to transportation of hazardous materials in interstate, intrastate, and foreign commerce by aircraft, railcar, vessel, and motor vehicle.

Some European regulations are also commonly used:

- **Regulations concerning the International Carriage of Dangerous Goods by Rail (RID)**
  The RID, issued by the *Intergovernmental Organization for International Carriage by Rail* (OTIF), regulates dangerous goods processes on rail cargo for European, Middle Eastern, and some North African countries.
- **European Agreement concerning the International Carriage of Dangerous Goods by Road (ADR)**
  ADR regulations are targeted at transnational transport of dangerous goods in Europe and were launched by the *United Nations Economic Commission for Europe* (UNECE).
- **European Agreement concerning the International Carriage of Dangerous Goods by Inland Waterways (ADN)**
  Similar to ADR, ADN regulates the transport of dangerous goods on inland waterways in Europe.

9.2.2 Dangerous Goods Classification

Dangerous goods classification is to some extent harmonized through the UN Recommendations on the Transport of Dangerous Goods with the goal of making it easy to understand what kind of cargo is transported and what hazards are entailed without having in-depth knowledge of chemistry or physics. Dangerous goods are divided into the following classes:

UN dangerous goods classes

- Class 1: Explosive substances and articles
  - 1.1: Substances and articles that have a mass explosion hazard
  - 1.2: Substances and articles that have a projection hazard but not a mass explosion hazard
  - 1.3: Substances and articles that have a fire hazard and either a minor blast hazard or a minor projection hazard or both, but not a mass explosion hazard

  - 1.4: Substances and articles that present no significant hazard
  - 1.5: Very sensitive substances that have a mass explosion hazard
  - 1.6: Extremely insensitive articles that do not have a mass explosion hazard
- Class 2: Gases, including compressed, liquefied, and dissolved under pressure gases and vapors
  - 2.1: Flammable gases (e.g., butane, propane acetylene)
  - 2.2: Nonflammable and nontoxic, likely to cause asphyxiation (e.g., nitrogen, CO2) or oxidizers (oxygen)
  - 2.3: Toxic (e.g., chlorine, phosgene)
- Class 3: Flammable liquids
- Class 4.1: Flammable solids, self-reactive substances, and solid desensitized explosives
- Class 4.2: Substances liable to spontaneous combustion
- Class 4.3: Substances that, in contact with water, emit flammable gases
- Class 5.1: Oxidizing substances
- Class 5.2: Organic peroxides
- Class 6.1: Toxic substances, poison
- Class 6.2: Infectious substances, biohazardous materials
- Class 7: Radioactive material (e.g., uranium, plutonium)
- Class 8: Corrosive substances (e.g., acids and alkalis)
- Class 9: Miscellaneous dangerous substances and articles (e.g., asbestos, airbag inflators, self-inflating life rafts, dry ice)

Detailed material classification is given by the four-digit UN number (currently, the number range is from 0001 to 3600). We recommend referencing UN resources for a detailed list of UN numbers.

9.2.3 Requirements for Dangerous Goods Checks in Transportation

Dangerous goods regulations cover various aspects of dangerous goods handling, movements, accident avoidance, and legal documentation. In transportation, a subset of these requirements is applicable, and they differ slightly from a shipper's and an LSP's perspective.

Shipper's requirements for transportation

Shippers involved in transportation usually have very detailed master data on the products and materials they manufacture or trade. The corresponding master data records are home to the related dangerous goods characteristic that controls the processing and documentation behavior in transportation. For shippers, the dangerous goods master data is maintained in SAP ERP in the material master definition. The applicability of a dangerous goods check for a shipper depends on whether the shipper is executing and planning transportation himself or subcontracting it to an LSP.

In the first case, the shipper is obligated to thoroughly execute all checks required and provide his personnel with appropriate training and transportation documentation, such as a shipper's declaration for dangerous goods or dangerous goods sheets. If the shipper subcontracts complete dangerous goods transportation, he simply needs to provide accurate dangerous goods data for his forwarding orders to the LSP.

LSP dangerous goods requirements

An LSP who actively handles dangerous goods transportation usually does not have detailed master data records for the materials shipped (unless it is a contract logistic business, where the LSP manages the supply chain for the shipper). In this case, the forwarding order needs to bring the correct and detailed classification for the forwarding order items that allow running the transportation of dangerous goods check process.

Checking for dangerous goods in transportation encompasses the following auditing steps (see also Figure 9.10):

Dangerous goods procedures

- Check the forwarding order to determine whether transportation of the ordered items can be done at all. It may not be feasible for the following reasons:
  - There are no appropriate vehicles or trained personal.
  - Certain materials (e.g., ammunition) are not accepted at all.
  - Origin, destination, or transit countries do not allow the materials to be shipped.

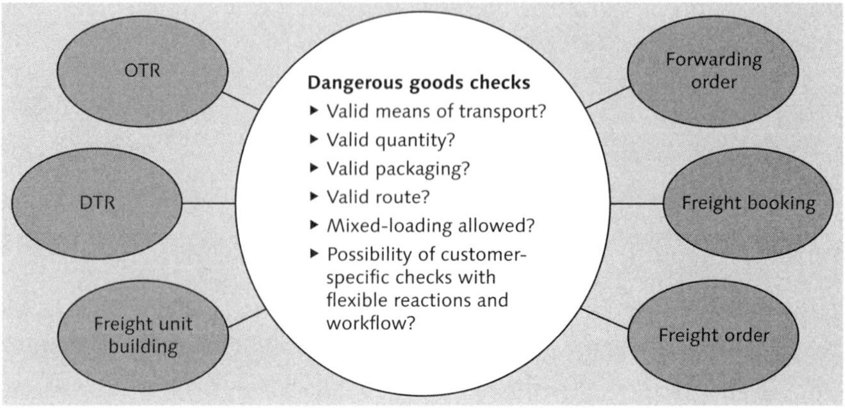

Figure 9.10 Dangerous Goods Checks in SAP TM

- Check a single forwarding order to determine whether quantities, packaging, and shipment properties comply with the regulations:
  - Certain quantities cannot be exceeded.
  - Certain material combinations cannot be shipped together at all or together in one transportation or packaging unit (on a pallet, in less than *x* yard distance, in the same container, on the same truck, etc.).
  - The evaluation of all cargo items must not exceed a certain limit in an evaluation scale (e.g., 1,000 points rule).
  - Certain materials are not allowed to be shipped on certain routes (e.g., flammable liquids in Swiss road tunnels).
- Check a consolidation of cargo or transportation plan details during or after transportation planning. All rules for the single forwarding order are applicable here, too:
  - Respect quantity limits, prohibited combinations, overall evaluation of hazard potential, allowed vehicles, and allowed shipping routes for the consolidated cargo.
  - In multi-stop shipments, loading limits for each stop need to be checked individually. For example, for a container line business, there are certain limits on the number of containers that can be loaded or unloaded in a port.

SAP TM can perform many of these checks fairly well, but for some (such as the load restrictions per port) the use of custom extensions is required.

9.2.4 SAP ERP Master Data for Dangerous Goods and Shipper Handling

Manufacturers and shippers need to maintain their product master data for raw, semi-finished, and finished goods to be procured, used in production, or sold in SAP ERP Materials Management or Sales and Distribution. For each maintained product, you can define additional dangerous goods data as described in Chapter 3, Section 3.1. Figure 9.11 shows an example of a dangerous goods master record (classification data) for a specified regulation (ADR) and a dropdown list of the various tabs offering access to further relevant areas.

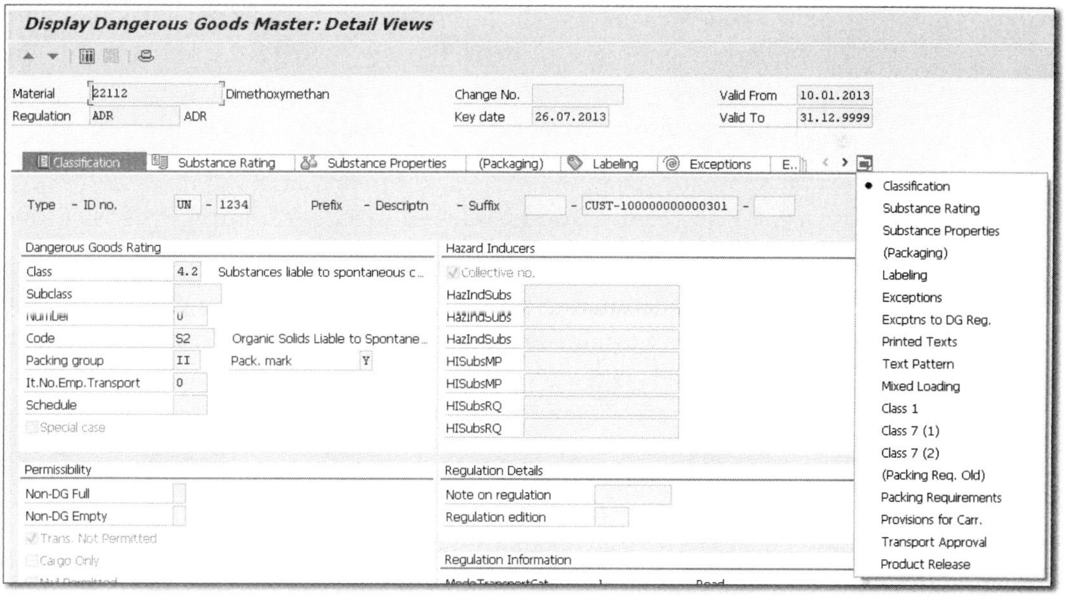

Figure 9.11 Dangerous Goods Master Record in SAP ERP

Dangerous goods checks in SAP ERP Logistics

SAP ERP offers a variety of dangerous goods compliance support for shippers and manufacturers in terms of handling their distribution requirements.

- **Packaging proposals in deliveries**
  In the handling unit functions of delivery, the system can propose allowed dangerous goods packaging codes based on packaging data in the dangerous goods master. In addition, proposing allowed packaging materials is done based on packaging codes and packaging approval specification.
- **Packaging and mixed packing checks within a delivery**
  If incompatible or noncompliant dangerous goods are packed together but have UN numbers that are not allowed to be packed together, the system sends an error message and blocks further processing.

Dangerous goods checks in SAP ERP

- **Mixed loading checks and transit country checks in SAP ERP shipments**
  Shippers using SAP ERP Logistics Execution (LE) transportation functionality may use the dangerous goods checks embedded in the SAP ERP shipment. The handling unit-based mixed packing checks work the same way here as in the delivery. In addition, the checks are implemented across all deliveries assigned to the shipment. If the route and stages of the shipment have been defined, dangerous goods checks can also be done for the transit countries passed en route.

9.2.5 Dangerous Goods-related Integration and Process Flow between SAP ERP and SAP TM

In shipper scenarios and systems, where SAP ERP is the main operational system for production and delivery processes, data integration for dangerous goods is required if you want to use SAP TM for transportation management. There is master data and transactional data integration. The following data transmission procedures are used to synchronize SAP ERP with SAP TM:

Data transmission

- Transfer of dangerous goods master data from SAP ERP to SAP TM.
  - Dangerous goods master data and text phrases are distributed via ALE integration.
  - Material master of SAP ERP is transferred via the Core Interface Framework (CIF).

- Transactional data transmission is embedded in the integration processes between SD sales orders and OTRs and between LE deliveries and DTRs. The required dangerous goods data and master data records are provided by SAP ERP via web services to SAP TM business objects, where they are available as references to the SAP SCM dangerous goods master.

Figure 9.12 shows an overview of the integration principle and the internal processing and integration of SAP TM and SAP EHS Management as part of the SAP SCM Basis layer.

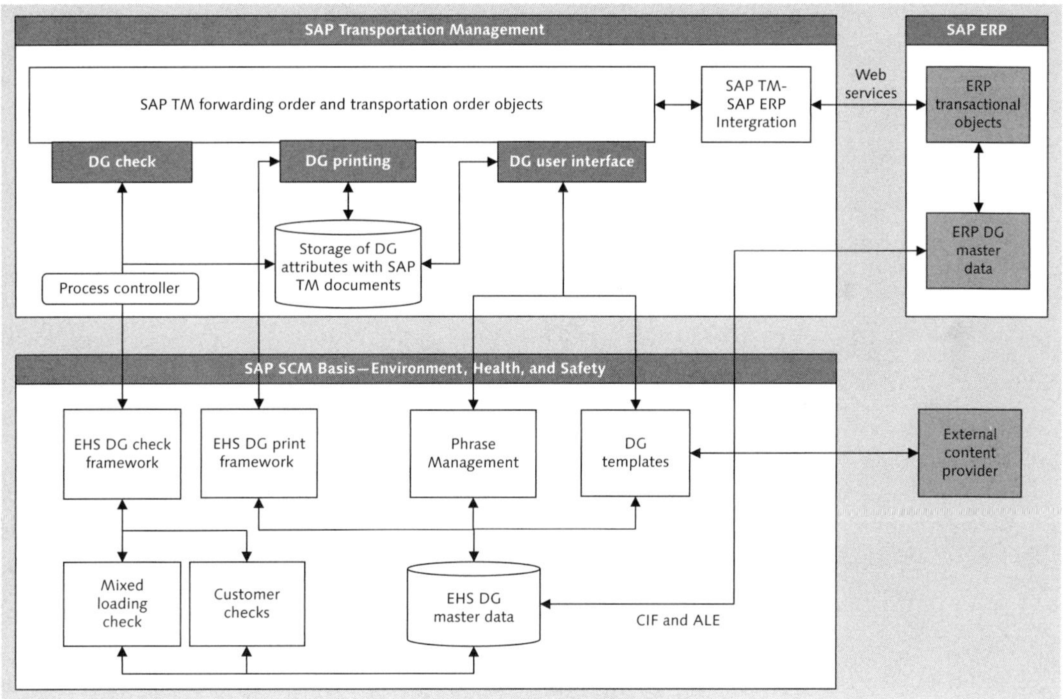

Figure 9.12 Integration between SAP TM and SAP EHS Management

Check process flow

Once material and dangerous goods master data has been transmitted from SAP ERP to SAP TM, transactional processing of forwarding orders and transportation orders (freight orders and freight bookings) can start.

The integration of the checks is technically done in the same way, independent of the ordering or subcontracting side.

The dangerous goods check triggers the SAP EHS Management check framework to call appropriate configured dangerous goods checks, such as mixed loading checks or customer-specific checks. All check modules can relate back to the dangerous goods master data created in SAP EHS Management or transferred from SAP ERP. Dangerous goods paper printing is accomplished via a dangerous goods print framework. The dangerous goods user interface allows you to flexibly configure the tabs for dangerous goods maintenance depending on the applied regulations, use dangerous goods templates for screen layout, and use phrase management for automatic text determination.

Master data integration setup

To set up master data transmission from SAP ERP to SAP TM, follow these steps:

1. Create tRFC ports in SAP ERP and SAP TM using Transaction SM59.
2. Create partner profiles in the SAP ERP system using Transaction WE20. In the partner profiles you have to set up outbound options for the dangerous goods master data IDocs DANGEROUSGOOD04, HAZARDOUSSUBSTANCE01, PHRMAS01, and SYNCHRON.
3. Create corresponding partner profiles in SAP TM for inbound parameters DANGEROUSGOODS, PHRMAS, and SYNCH.
4. Create an ALE distribution model in SAP ERP using Transaction BD64. Maintain a model view for SAP EHS Management with BAPI methods for the following business objects: BUS1078 with method `SAVEREPLICAMULTIPLE`, BUS1091 with method `SAVREPMUL`, and BUS1119 with method `SAVEREPLICATE`.
5. Activate change pointers for the utilized message types in SAP ERP in Customizing via Application Server • IDoc Interface/Application Link Enabling (ALE) • Modeling and Implementing Business Processes • Master Data Distribution • Replication of Modified Data • Activate Change Pointers for Message Types.
6. Transfer dangerous goods master data to SAP TM using Transaction DGP5 in SAP ERP, where you trigger a selective distribution of master data via the Send Dangerous Goods Master Data screen.

9.2.6 Configuration of Dangerous Goods Checks for SAP TM

Customizing of SAP EHS Management

Before you can use the dangerous goods checks of SAP EHS Management, you have to enable the checks in Customizing. If you follow the IMG menu path SAP TRANSPORTATION MANAGEMENT • SCM BASIS • EH&S SERVICES • BASIC SERVICES • SPECIFY ENVIRONMENT, you have to set the environment parameter `DG_SERVICES_ACTIVE` to value X. Figure 9.13 shows the corresponding setting.

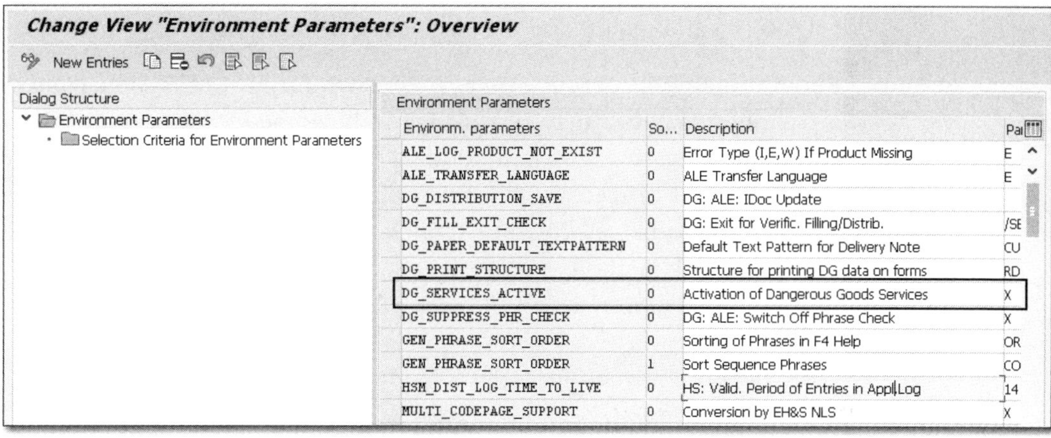

Figure 9.13 Enabling Dangerous Goods Checks for SAP TM

Like in SAP ERP, the SAP EHS Management component in SAP SCM Basis has a variety of settings with which you can configure the behavior of dangerous goods management and phrase management:

- Phrase management settings allow you to define phrase libraries and language definitions for the phrases.
- Dangerous goods management provides setup for all of the different codes, categories, regulations, and classes that build the legal framework.
- Dangerous goods checks provide the configuration for the check rules and the reaction of the system on success or failure of a check.
- Dangerous goods documents allow you to set up the output conditions, formatting rules, and languages for dangerous goods paperwork.

You can find these Customizing settings in SAP TM via the IMG menu path SAP Transportation Management • SCM Basis • EH&S Services. Some additional configuration settings have to be made via SAP Transportation Management • Transportation Management • Basic Functions • Dangerous Goods. Here you can define, for example, the error behavior of the SAP TM-related dangerous goods checks or quantity definition for the 1,000 points rule according to ADR 1.1.3.6.

Let's take a look at two configuration examples: dangerous goods check rules setup and maintenance of segregation rules.

SAP EHS Management dangerous goods rules

Because many dangerous goods checks within SAP EHS Management are based on check rules, which have to be embedded in a rules framework, you need to maintain individual rules. Alternatively, you have the option to load dangerous goods content via a loader so that not everything needs to be set up from scratch. Figure 9.14 shows the maintenance of check methods and the level of granularity for operations (e.g., header level and item level).

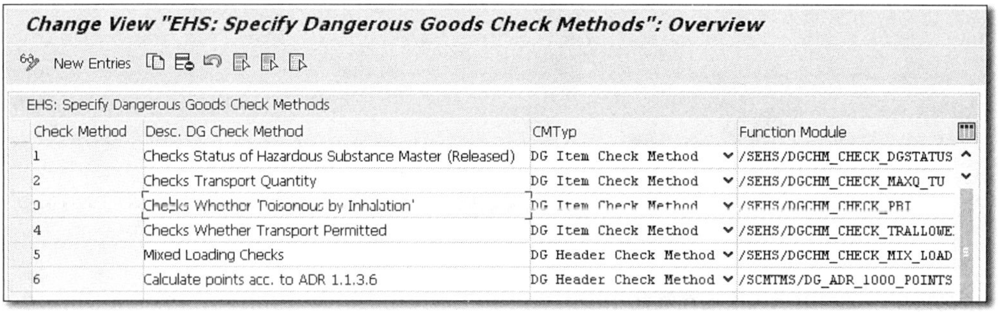

Check Method	Desc. DG Check Method	CMTyp	Function Module
1	Checks Status of Hazardous Substance Master (Released)	DG Item Check Method	/SEHS/DGCHM_CHECK_DGSTATUS
2	Checks Transport Quantity	DG Item Check Method	/SEHS/DGCHM_CHECK_MAXQ_TU
3	Checks Whether 'Poisonous by Inhalation'	DG Item Check Method	/SEHS/DGCHM_CHECK_PBI
4	Checks Whether Transport Permitted	DG Item Check Method	/SEHS/DGCHM_CHECK_TRALLOWE
5	Mixed Loading Checks	DG Header Check Method	/SEHS/DGCHM_CHECK_MIX_LOAD
6	Calculate points acc. to ADR 1.1.3.6	DG Header Check Method	/SCMTMS/DG_ADR_1000_POINTS

Figure 9.14 Dangerous Goods Check Methods Maintenance in SAP TM

You can maintain the check methods in Customizing via the menu path SAP Transportation Management • Transportation Management • EH&S Services • Dangerous Goods Management • Dangerous Goods Checks and Dangerous Goods Documents • Dangerous Goods Checks • Specify Dangerous Goods Check Methods.

Maintenance of segregation rules

Segregation keys are used to define criteria to keep materials apart from each other in stowage, loading, and consolidation. Segregation key definition is done as part of dangerous goods master data maintenance (see Figure 9.11). Segregation keys are maintained on the MIXED LOADING tab. In SAP EHS Management Customizing, you can set up the segregation rules for mixed loading and define the system response if you plan to load two cargo items with particular segregation keys together. The result of rules also depends on the individual dangerous goods regulation and can either be a warning or an error. You can maintain the segregation rules via the IMG menu path SAP TRANSPORTATION MANAGEMENT • TRANSPORTATION MANAGEMENT • EH&S SERVICES • DANGEROUS GOODS MANAGEMENT • DANGEROUS GOODS CHECKS AND DANGEROUS GOODS DOCUMENTS • SPECIFY SETTINGS FOR MIXED LOADING CHECKS • SPECIFY SEGREGATION RULES FOR SEGREGATION KEYS. Figure 9.15 shows the segregation rule maintenance.

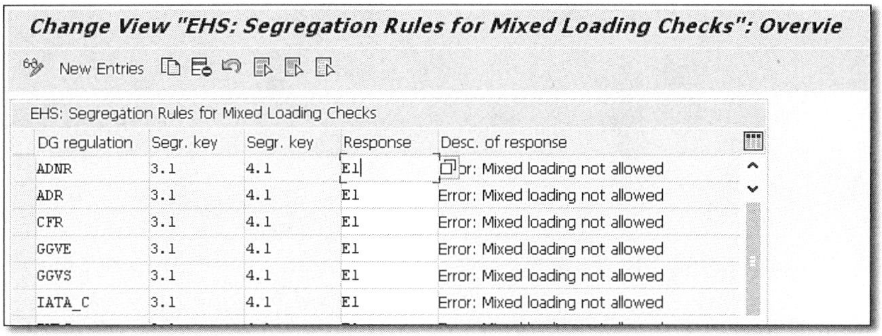

DG regulation	Segr. key	Segr. key	Response	Desc. of response
ADNR	3.1	4.1	E1	...or: Mixed loading not allowed
ADR	3.1	4.1	E1	Error: Mixed loading not allowed
CFR	3.1	4.1	E1	Error: Mixed loading not allowed
GGVE	3.1	4.1	E1	Error: Mixed loading not allowed
GGVS	3.1	4.1	E1	Error: Mixed loading not allowed
IATA_C	3.1	4.1	E1	Error: Mixed loading not allowed

Figure 9.15 Segregation Rule Maintenance in SAP TM

9.2.7 Dangerous Goods Content Loader

Dangerous goods content

The dangerous goods content loader allows you to import SAP EHS Management regulatory content into the dangerous goods master data tables in SAP TM. The content can be found in the software download center of the SAP Service Marketplace (*service.sap.com/swdc*) via the menu path INSTALLATIONS AND UPGRADES • A–Z INDEX • E • EHS REGCONTDANGER GOODS • EHS REGCONTDANGER GOODS 2.0. The content files provided contain dangerous goods regulations, regulation texts such as substance names and phrases, and Customizing data to implement the regulations.

The content loader can be very useful for LSPs to get the system quickly set up with the data and regulations that concern the shippers. The imported content can be used to perform consistency checks on the forwarding order data provided by shippers. In addition, it provides template data for document-based dangerous goods records. Figure 9.16 shows the content load UI, which you can call in SAP TM application under Master Data • Dangerous Goods Management • Dangerous Goods Content Loader.

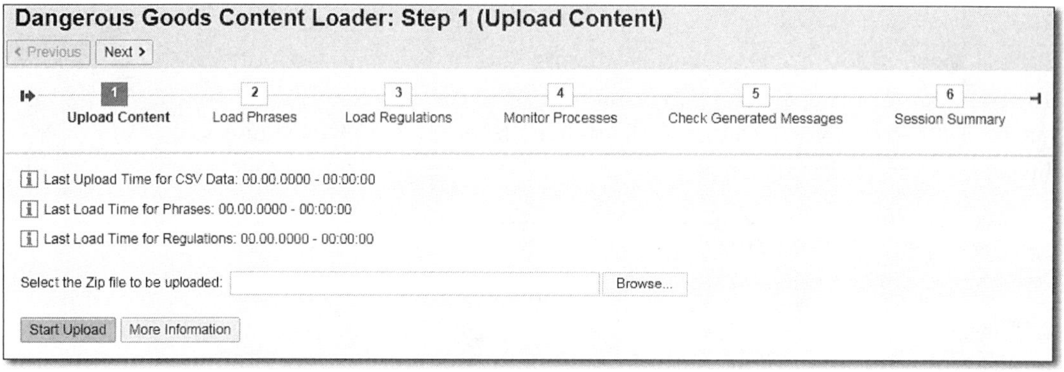

Figure 9.16 Dangerous Goods Content Loader

Content loading procedure

The loading procedure is as follows. After uploading the content zip file, you first load phrase data because this is reused in later stages. Loading progress and success can be always viewed using the monitoring step 4. In step 3, the system displays a list of regulations that are specified in the CSV content files. You can select the regulations to be loaded. Please make sure that you have already created the corresponding regulations to be uploaded in the SAP EHS Management Customizing settings; otherwise, the system prevents you from loading. After loading, you can review the message log and a summary.

9.2.8 SAP TM Application Level Setup for Dangerous Goods Support

As of SAP TM 9.0, you can create individual user interfaces for document-based dangerous goods data that reflect the specific data and phrase requirements of the dangerous goods regulations.

Regulation-specific user interfaces

The user interfaces specified in a UI profile can hold tabular and field data with individual field labels (see Figure 9.17). Fields can be assigned to UI groups. You can define check functions for a UI profile and use them by clicking on the CHECK RECORD button on the forwarding order's DANGEROUS GOODS tab (see also Figure 9.19).

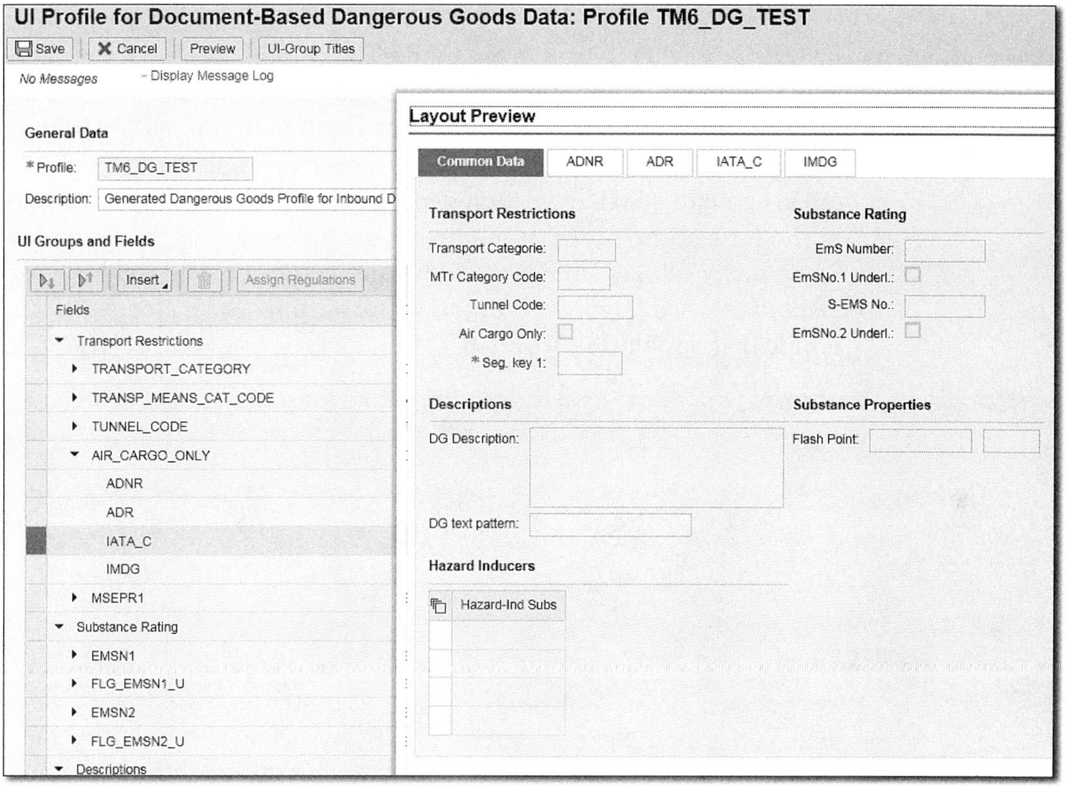

Figure 9.17 Configuration for Dangerous Goods UI Profile

You can find the tool for defining dangerous goods UI profiles in the SAP TM application menu via APPLICATION ADMINISTRATION • GENERAL SETTINGS • DEFINE UI PROFILE. The assignment of the UI profile to the forwarding order happens on the forwarding order item level. You can set the dangerous goods UI profile in the Customizing of the item types for forwarding order management.

9.2.9 Dangerous Goods Data and Checks in SAP TM

As described in Section 9.2.3, dangerous goods data definition and dangerous goods checks happen mainly on the forwarding order and transportation order level (freight order and freight booking). Additionally, dangerous goods are considered in freight unit building and transportation planning.

Dangerous goods data in forwarding orders

In the forwarding order, most of the dangerous goods data is made available on the item level. The header contains only status and summary information such as the dangerous goods indicator, ADR points, or an ADR exemption indicator. Recall from the previous section that you can assign individual, regulation-dependent UI profiles to the item types of the forwarding order. If multiple regulations are assigned, you will see as many sub-tabs under the forwarding order item as you have regulations defined. Figure 9.18 shows a forwarding order item with a UI profile that has only an ADR tab.

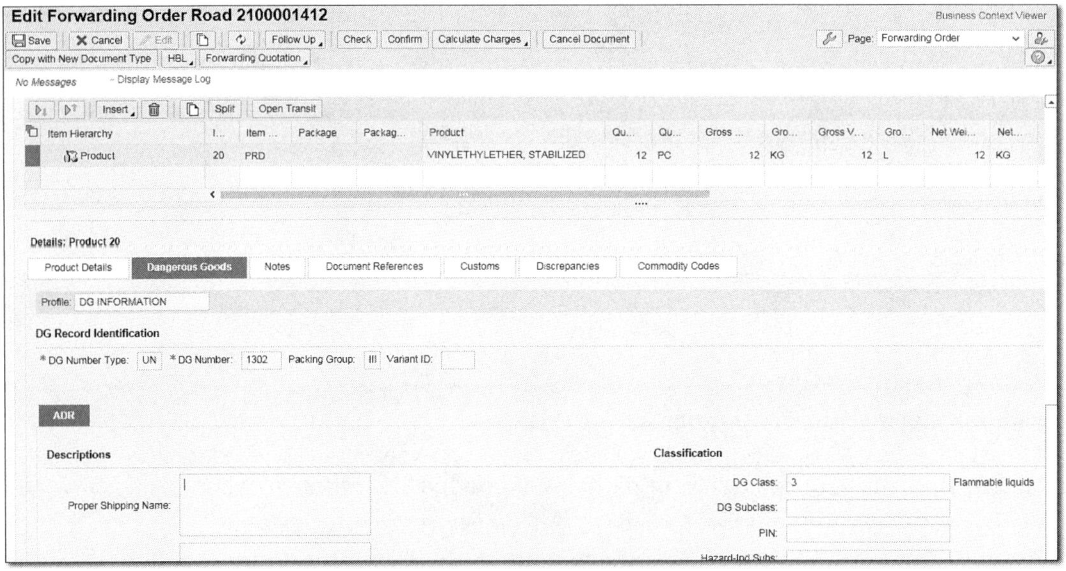

Figure 9.18 Dangerous Goods Item Details Screen in a Forwarding Order

If you invoke the check functionality in the forwarding order, the configured dangerous goods checks are also executed, and the result of the checks is displayed in the message log of the forwarding order. Figure 9.19 shows a check result where a mixed loading check found incompatibilities concerning the UN class assignment of the two forwarding order items and a prohibited combination of segregation keys.

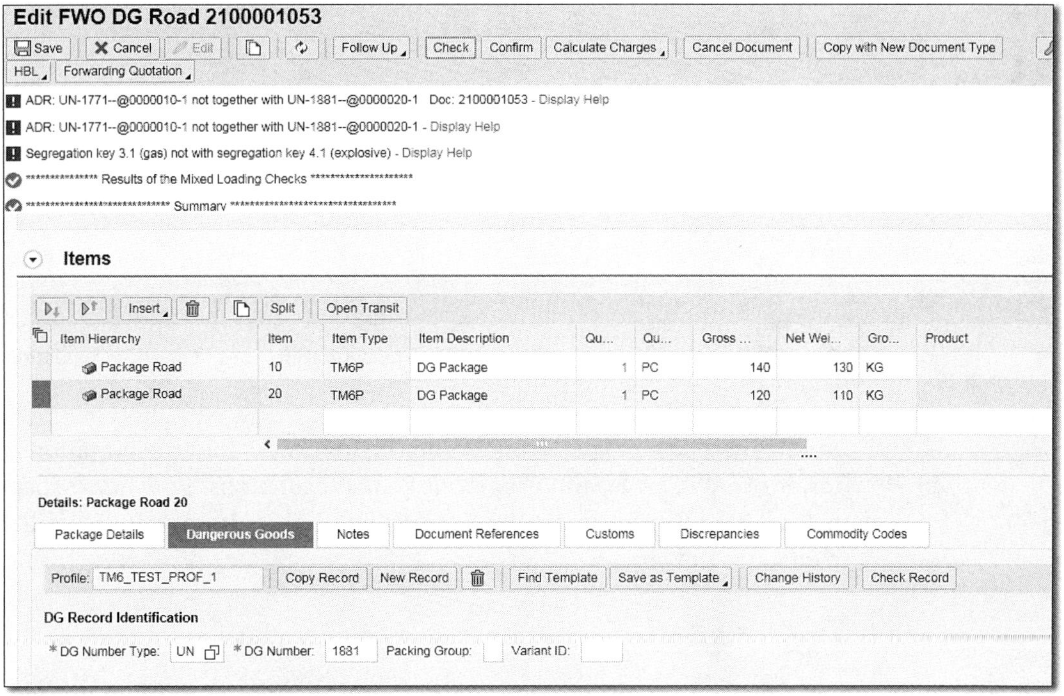

Figure 9.19 Mixed Loading Check Result in a Forwarding Order

When you start planning in the transportation cockpit (see Chapter 6), the mixed loading checks are applied to the consolidated items of the freight units that you plan to be moved together in one vehicle. Material compatibilities are inspected in any mode of transport. Additionally, the system summarizes the ADR points of all cargo items that should go on a truck in road transportation. Figure 9.20 shows the result of the application of the ADR 1,000 points rule to an LTL freight order with the warning that the limit is exceeded.

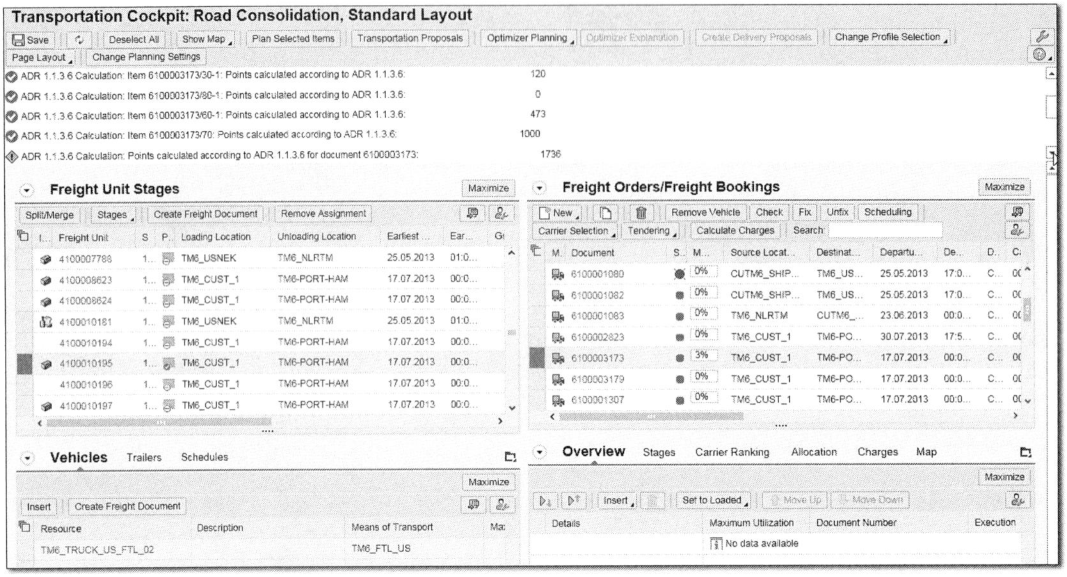

Figure 9.20 1,000 Points Rule Result in the Transportation Cockpit after Consolidation Planning

In this chapter we explained the compliance checks for transportation in terms of legal, trade, and dangerous goods regulations. In the next chapter, we will introduce service products, tariffs, customer and supplier contracts, and charge calculation.

Understanding the flexible contract management and charge calculation capabilities in SAP TM will benefit your supply chain management. Learn how to maintain vendor and customer contracts, store complex rate agreements, and calculate charges in your operational documents.

10 Transportation Charge Management

Managing transportation charges is a core process requirement for shippers and logistics service providers (LSPs). From a shipper's perspective, it is necessary to maintain contracts with suppliers and to ensure full visibility of transportation costs. As a freight-forwarder and carrier, you equally need transparency of your supplier contracts and costs. In addition, it is a key requirement for your commercial processes to maintain and manage customer contracts and have visibility of your profitability.

The Transportation Charge Management component of SAP Transportation Management (TM) is a very powerful and flexible engine; it allows you to maintain transportation contracts and fully integrate them into your operational processes. This core benefit of SAP TM means that contracts are centrally maintained and will be cardinally used for the automated charge determination in your freight documents: forwarding orders and quotations, freight orders, freight bookings, and service orders. Consequently, SAP TM Transportation Charge Management supports your charge calculation processes across all modes of transport for shippers, freight-forwarders, carriers, and railways. SAP TM itself also allows all service providers to work with service products and standard operating procedures.

In this chapter we introduce to you the complexities of SAP TM Transportation Charge Management in four different sections. As with your user experience, with the setup of a scenario using SAP TM Transportation Charge Management, we will start by setting up the master data objects in Section 10.1. We will explain each master data element and their core Customizing settings. With this foundation, we dive into the second section and apply the defined master data to orders and bookings.

When determining rates in transactions, the logical step is to execute contract determination to find the correct master data such as agreements that were previously set up (Section 10.2). Following the chronological order, it subsequently calculates the charges per the agreement. We use Section 10.3 to dive deep into the actual charge calculation logic from agreements, offering various scenarios and possible outcomes of a successful charge calculation as illustration.

In Section 10.4, we branch away from using contracts operationally, and introduce the capabilities of SAP TM Transportation Charge Management to support more strategic customer and vendor management. This chapter explains the latest functionalities of quotation and procurement management, as well as how this enables you to negotiate and set up customer and vendor agreements.

Architecture

Figure 10.1 shows the big picture of the architecture in SAP TM Transportation Charge Management. In the center of the picture, you can see the master data repository of Transportation Contract Management. It allows you to store customer contracts the SAP TM forwarding agreements and vendor contracts—the SAP TM freight agreements. Furthermore, it allows you to maintain SAP TM internal agreements used for internal settlement processes. Finally, the SAP TM Transportation Charge Management master data contains the setup of a product master, the service product catalog.

The forwarding agreement is fully integrated with forwarding orders for the calculation of customer charges. A forwarding agreement contains not only the contracted rates, but also payment terms, service products, and standard operating procedures. Besides the forwarding order, you can calculate the charges for forwarding quotations. Your freight agreements are stored in a similar structure as forwarding agreements that contains complex rate tables, service products, and payment terms. The freight agreement is integrated into freight orders, freight bookings, and service orders to calculate your transportation costs.

The charges which are calculated in the SAP TM Forwarding Order and SAP TM Freight Management components are settled with the SAP TM Transportation Charge Settlement component. We will describe the details of the settlement process in Chapter 11. In Figure 10.1, you can see the charge settlement in the bottom layer and include the settlement for internal charges.

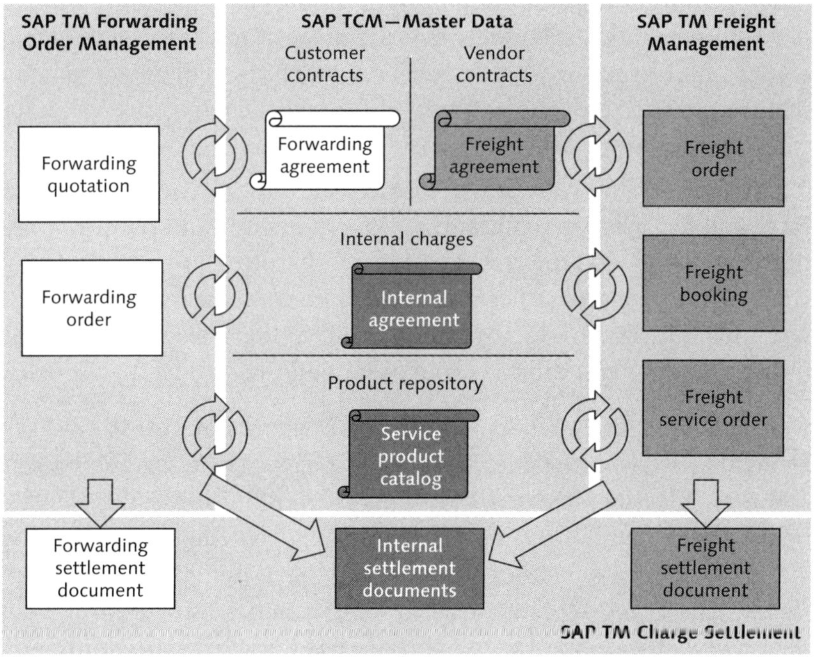

Figure 10.1 Overview of SAP TM Transportation Charge Management

We now dive into each master data element of SAP TM Transportation Charge Management. In Section 10.2, we delve deeper into the contract determination logic and integration to order documents. In Section 10.3, the focus is on the actual calculation logic SAP TM offers for calculating charges.

10.1 Charge Calculation Master Data

Let's walk through the SAP TM master data objects maintained in a central repository, the Transportation Contract Management component:

1. Forwarding and freight agreements serve as customer and vendor contracts, and contain charges and rates, optionally based on service products.
2. Internal agreements represent rate agreements between your own internal organizations for the internal settlements process. You can maintain standard costs for intercompany and intracompany charges.
3. *Transportation charge calculation sheets* (TCCSs) have a direct cardinality to agreements. Each agreement has at least one TCCS that lists all charge types (e.g., surcharges, basic freight charges) of an agreement. It also strongly steers the behavior of the actual charge calculation for each charge type in order documents.
4. Rate tables are the central place where all the actual values of your rates and charges are maintained. We can again find cardinality between the TCCS and the rate tables. Each charge calculation sheet can contain multiple rate tables. Each rate table contains several scales. If you imagine a rate table as a spreadsheet, the scale would be the headlines of the columns (i.e., the weight of cargo, the distance, etc.).

Transportation Contract Management

The left side of Figure 10.2 depicts the hierarchical relationship between agreements, TCCSs, rate tables, and scales. On the right, we apply these concepts to a business example to show how you can use this flexible structure in ocean freight. Here, the forwarding agreement is a customer contract and can have, among other things, the scope of multiple trade lanes as items of the agreement. Each trade lane has its own TCCS with different global and country-specific charges (e.g., basic ocean freight, bunker fee, Suez Canal fee, etc.). For each charge, an applicability rule or condition can be maintained. In addition, each charge is assigned a rate table that contains all the actual rates. A rate table is constructed based on multiple scales, such as origin, destination, equipment, and commodity.

Last, SAP TM Transportation Charge Management allows you to maintain a configurable service product catalog. You can set up your own service products based on individual service items, which can represent value-added services or customer-specific services. Using service products when setting up agreements and order documents is entirely optional. You can assign standard operating procedures to the service products to operationalize them.

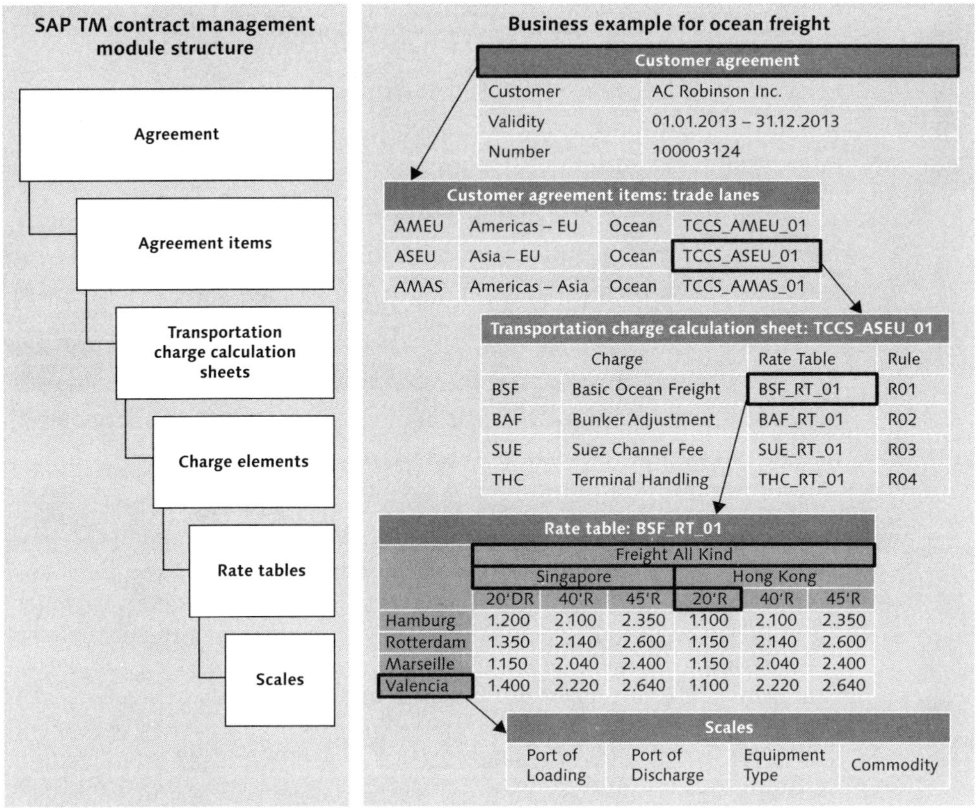

Figure 10.2 Transportation Contract Management Component with Business Example

Now let's look more closely at each individual master data object.

10.1.1 Contract Management with Agreements

With SAP TM, you can overview and query all charge-relevant master data objects and search through multiple different queries, including rate tables, TCCSs, service product catalogs, and much more. Navigate in SAP NWBC to Master Data • Charge Management and Service Product Catalogs • Overview Charge Management and Service Product Catalogs. In the so-called master data cockpit, you can query all your agreements.

- The *forwarding agreement* represents a customer contract. This is of great relevance to all LSPs.

- The *freight agreement* represents a carrier contract with the rates of your vendors, which is relevant for both shippers and LSPs.
- The *internal agreement* is used for internal settlements. Agreements can represent both short- and long-term relationships between your company and business partners.

The great benefit of this functionality is that you can search for agreements based on the most granular components, such as a single rate table entry or source destination, and SAP TM highlights all agreements that contain this specific rate table line.

As you will see in this chapter, the contractual rates are not the only part of an agreement. All agreements can be used for the agreement determination and charge calculation in SAP TM.

External agreements

Note

In addition to maintaining agreements yourself, with SAP TM you can interface with external agreements in third-party systems via a web service. Rates that are stored in contracts residing in an external system can automatically be pulled into SAP TM and used for rating and settlement. If you want to work with external agreements, you still need to maintain an equivalent in SAP TM that contains no data, but the reference ID of the external agreement and a *calculation method type* of external system that triggers the interface during charge calculation.

For more details on calculation methods, refer to Section 10.3.1.

Forwarding Agreement

Creation, general header data

You can create a forwarding agreement by navigating to SAP NWBC • Forwarding Agreement Management • Forwarding Agreements • Create Forwarding Agreement. You need to specify a forwarding agreement type, which is to be defined in Customizing, and then assign the agreement to one or multiple sales organizations.

In Chapter 3, you learned the essentials about the organizational structure. The sales organization represents a customer-facing organization that issues forwarding orders. If you work as a freight-forwarder, your individual freight stations might have autonomous agreements with your customers. Alternatively, it is possible to maintain multiple sales organizations for an agreement. Besides the sales organization, you need to store the business partners who are permitted to use the contract. You have two options:

1. Define exactly one business partner (your contracting party).
2. Maintain a list of business partners in a table format. All parties are permitted to book forwarding orders under this contract.

For example, the second option is used to reflect a shipper association or the affiliates of a shipper. In addition, you can choose to maintain a *business partner hierarchy* (BPH). Each member of the BPH can use the rates of the contract. This is important if you maintain an affiliate or partner structure of your customers in the *business partner hierarchy master*. Last, it is mandatory to maintain the validity dates of a contract.

Versioning

Another attribute on the header of a forwarding agreement is the version number. This new feature of TM 9.1 supports with both regulatory filings and keeps track of changes in agreements. You can generate new versions of an agreement by clicking on the GENERATE NEW VERSION button. A deep copy of the entire contract with all rates is generated. A history of the different versions is also available.

Figure 10.3 shows an example of the general header information of a forwarding agreement. This contract is maintained for one business partner only, but is valid for three sales organizations. To prevent use while agreements are in maintenance, you have to activate each agreement. As a result you see IN PROCESS in the AGREEMENT STATUS field.

Forwarding agreement items

Items in the agreement line differentiate the scope inside an agreement. Figure 10.4 shows an example of agreement line items. There are a few important requirements to take into consideration here. You must assign one TCCS to each agreement item, and each forwarding agreement has to have at least one line item. When executing the charge calculation for forwarding orders, the system picks up the rates from a line item in the determined agreement only if the defined scope between your forwarding order and the agreement line item match. We call this scope for an agreement item a *precondition*.

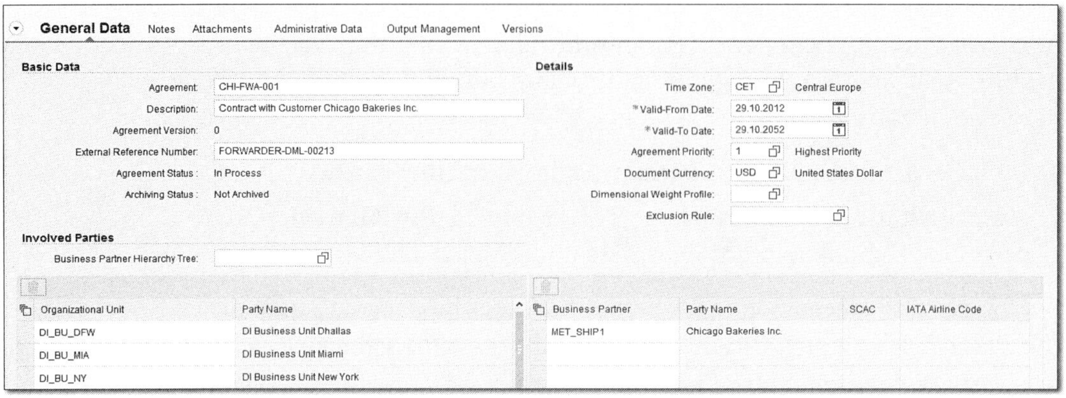

Figure 10.3 General Data of Forwarding Agreement

Consider these examples of agreement line items and preconditions:

1. Maintain only one line item per transportation stage category (e.g., pre-, main, and on-carriage, as shown in Figure 10.4). As a result, an item would only be used to determine trucking charges for an individual stage (pre-carriage) in the forwarding order.
2. Agreement items can also represent service products; we discuss this further in Section 10.1.4.
3. Mode of transport, movement type, and service level are examples of preconditions. If you do not want to maintain services as proper master data, note that the service level is a simplified concept of a service product catalog.

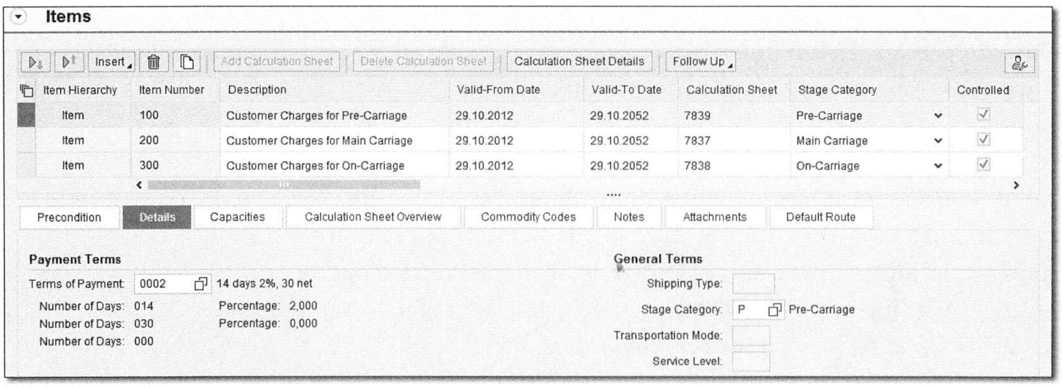

Figure 10.4 Forwarding Agreement Item

You can enhance agreement item preconditions through Customizing. On the forwarding agreement line items you can also maintain the *settlement basis*. This setting determines how your settlement documents are generated (e.g., if you want to generate an invoice based on all goods loaded on a resource). The standard setting is per forwarding order. You can find details on the settlement process in Chapter 11.

The tabs in the forwarding agreement item provide more information, such as the DETAILS and CALCULATION SHEET OVERVIEW tabs. In the CALCULATION SHEET OVERVIEW tab, you can see the information from the TCCS once it has been assigned to the item. The PRECONDITION and COMMODITY CODES tabs allow you to store more conditions for the TCCS determination, which we detail in Chapter 10, Section 10.2. Last, you can maintain capacities that you have agreed upon with a customer. The freight agreement has stronger capabilities to manage your agreed vendor capacities. Consequently, we provide you with more details on capacities on the freight agreement, especially in Chapter 5, Section 5.2. SAP TM cannot yet maintain customer space or capacity allocations. The capacities in the forwarding agreement are majorly used for reference purposes and in an RFQ process, as described in Section 10.4.1, but are not yet operationally used.

Notes and attachments

Both the agreement header and items have other useful tabs, such as the NOTES section, where you can maintain free text clauses and other text elements. You can also upload any document to the agreement, such as a signatory page or a legal filing confirmation.

Output management

For proper contract management, you need to be able to print agreements and define other output modes, such as email or fax. For this, define the settings for output management as part of the Post-Processing Framework (PPF) in Customizing via the menu path CROSS APPLICATION COMPONENTS • PROCESSES AND TOOLS FOR ENTERPRISE APPLICATIONS • REUSABLE OBJECTS AND FUNCTIONS FOR BOPF ENVIRONMENT • PPF ADAPTER FOR OUTPUT MANAGEMENT • MAINTAIN OUTPUT MANAGEMENT ADAPTER SETTINGS. We covered this in Chapter 2, Section 2.3.3.

After you have configured output management for agreements, you can generate actions in the OUTPUT MANAGEMENT tab by selecting GENERATE •

Actions Including Condition Checks. Figure 10.5 shows some examples of configured actions for agreements. A core difference from the order documents (e.g., forwarding order, freight booking) is that you do not maintain the output profile on the document type (agreement type).

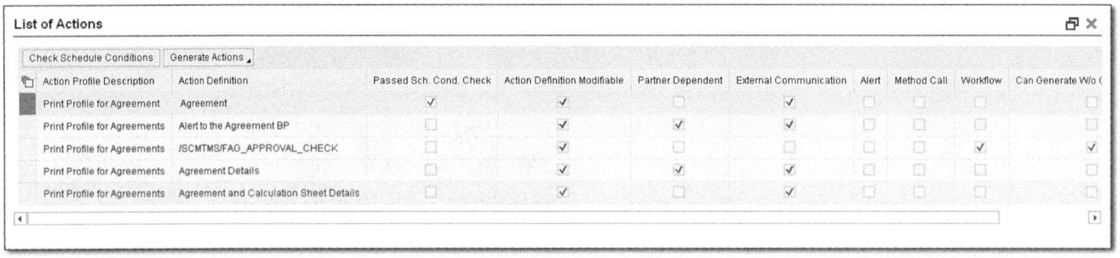

Figure 10.5 Output Management

Forwarding agreement type

Let's highlight the key Customizing settings required for using a forwarding agreement. The forwarding agreement has its own type, which you need to configure. You can maintain the type in SAP TM Customizing via the menu path Transportation Management • Master Data • Agreements and Service Products • Define FWA and Service Product Catalogue Types.

The agreement type is a key setting; for example, it controls the user interface layout. You can select the Multiple Parties checkbox to maintain more than one business partner and organizational unit, as shown in Figure 10.3. In the same Customizing section, you can specify preconditions for the agreement items such as the stage category, movement type, and so on, as just described.

Item type

The forwarding agreement items have their own type; you can maintain them by following the menu path Transportation Management • Master Data • Agreements and Service Products • Define FWA and Service Product Item Types. The forwarding agreement item types need to be assigned to the forwarding agreement type.

Let's turn our attention to another agreement.

Freight Agreement

Creation and general data

You can create a freight agreement by navigating to SAP NWBC • Freight Agreement Management • Freight Agreements • Create Freight Agreement. You need to specify a freight agreement type, which affects, for example, the user interface and the number range that is assigned. You define the freight agreement type in Customizing. The freight agreement is of great relevance for both shippers and LSPs because they use it to store vendor contracts with trucking companies, airlines, ocean lines, railways, and freight-forwarders.

Let's take a closer look. Note that the object structures of the freight and forwarding agreements are very similar.

Figure 10.6 shows the header section of a freight agreement. The appropriate organizational unit here is the purchasing organization, which actually procures the services from a vendor. Your contracting party is the carrier itself. As with the forwarding agreement, you can maintain multiple parties—both carriers and purchasing organizations.

The Details section, too, is very similar and contains the validity dates of the vendor contract, the document currency, and an optional Exclusion Rule field in case you want to exclude an agreement for specific charge calculations (see Section 10.2.2 for more details on rules and conditions). Like forwarding agreements, freight agreements support versioning.

Other tabs

All other tabs in the freight agreement have similar functionalities as the forwarding agreement. You can add notes and store attachments, and you can access the entire capabilities of the output management to print a document or trigger a workflow, email, or alert.

Freight agreement items

Similar to the forwarding agreement, you need to create freight agreement items. Each item is assigned one TCCS (or a determination rule) and has its own validity. Use preconditions in order to define which item will be used in a charge calculation for freight orders and freight bookings.

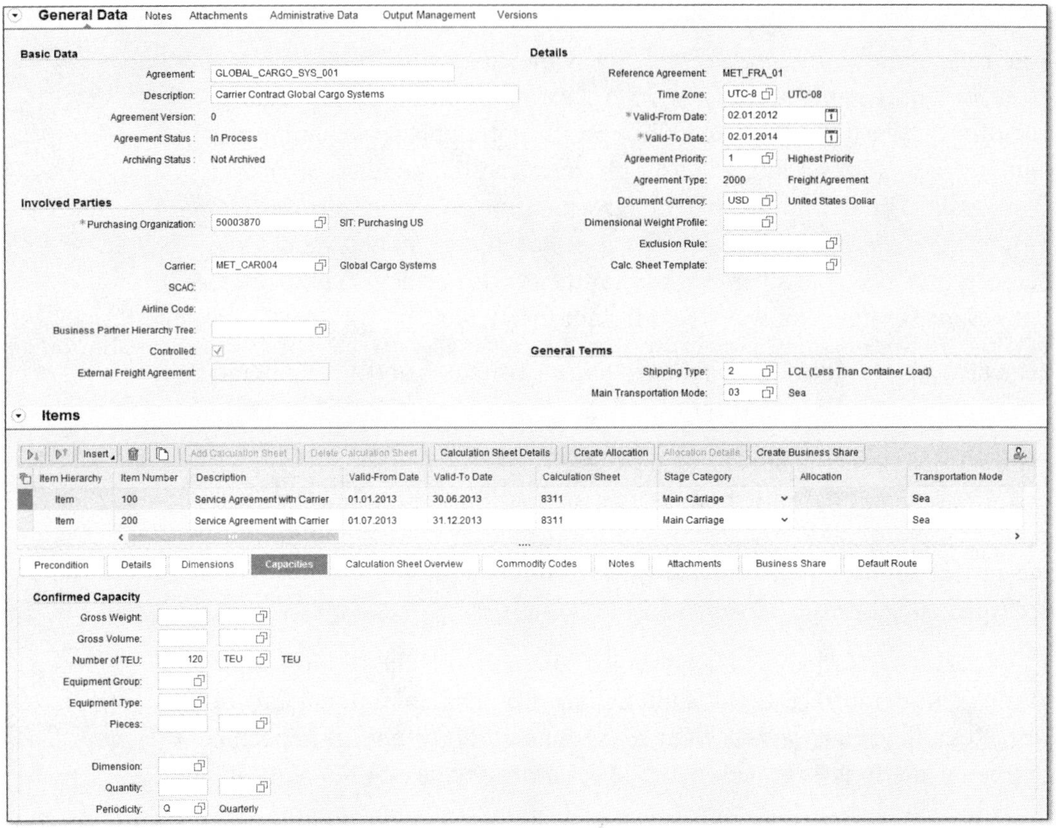

Figure 10.6 Freight Agreement

Similarly to the forwarding agreement, you can use standard freight agreement item preconditions. In addition, you can use the freight agreement items to maintain different service products you have agreed to procure from your vendor. This requires setting up service products, which we cover in Section 10.1.4.

Capacities in agreements

Another functionality of the freight agreement items is to maintain *capacities*. When negotiating a service contract with vendors, especially freight-forwarders and carriers, it is a common business practice to agree not only on rates, but also on capacities to be shipped under a contract. This allows carriers better forecasting and managing capacities. Higher capacities can be associated with discounts on the rates for the shipper or LSP.

In Figure 10.6, you can see the DETAILS tab for capacities. By clicking on the CREATE ALLOCATIONS button in the AGREEMENTS LINE ITEMS table, you can generate an agreement allocation out of such capacities. In order to enable this capability, you need to assign an agreement allocation type to your freight agreement types. Besides generating an SAP TM agreement allocation from a freight agreement, it is also possible to generate a business share from a freight agreement in SAP TM 9.1. This capability allows you to distribute your total demand in capacity between your suppliers. This capacity share is used in carrier selection. Refer to Chapter 5 for more details on allocations and business shares and to Chapter 7 for carrier selection.

Freight agreement type

The freight agreement type is the core required Customizing setting. You can find it in SAP TM Customizing by following the menu path TRANSPORTATION MANAGEMENT • MASTER DATA • AGREEMENTS AND SERVICE PRODUCTS • DEFINE FREIGHT AGREEMENT TYPES. In addition to the basic settings on number ranges and the UI layout, you can specify an allocation type for the capacities functionality and other settings concerning workflow and timing rules.

Freight agreement item types

Next to the freight agreement type, you need to maintain item types in Customizing (TRANSPORTATION MANAGEMENT • MASTER DATA • AGREEMENTS AND SERVICE PRODUCTS • DEFINE FREIGHT AGREEMENT ITEM TYPES). Multiple preconditions can be assigned to the item types. The item types are assigned to agreement types.

Internal Agreements

The last agreement we examine is the internal agreement. This object is important to LSP and trucking companies' management of internal settlements. For example, a freight-forwarder can use internal agreements to maintain the internal rate agreements between freight stations and purchasing organizations. Each purchasing organization could have its independent standard rates, which would be charged internally to the sales organizations generating the revenue. We go into further detail about business scenarios of the internal settlement process in Chapter 11, Section 11.2.

Alternative Use of Internal Agreements

In implementation projects, the internal agreement has been modeled differently from the internal charge settlement process. Customers decided to maintain internal agreements to store their internal standard cost of shipments, which contains many more cost components than just the purchasing cost for the transportation service. Such additional cost components can contain the cost of operations, container evacuation, or overhead cost. Since this data flows into the profitability analysis of the forwarding order, it gives an even more precise insight into the margin.

General data

The structure of the internal agreement is almost the same as the forwarding and freight agreements. In the header section you maintain the purchasing organization that is offering internal services to other organizations.

The unique aspect of the internal agreement is that you also need to maintain the agreement partner. For example, a purchasing organization grants the same internal rates to all sales organizations in the same country but different rates to stations in other countries. As a result, a table is provided to maintain the involved parties for each internal agreement.

Agreement items

The internal agreement items are very similar to forwarding agreement/freight agreement items. Functionalities such as preconditions, commodity codes, notes, and attachments are supported. Capacities cannot be stored on the items, assuming that internal organizations would not negotiate on capacities that need to be consumed by internal sales organizations. You need to assign a TCCS or a determination rule to each agreement item.

Customizing

You need to define an internal agreement type before you can use this functionality. Maintain internal agreement types by following the menu path Transportation Management • Master Data • Agreements and Service Products • Define Internal Agreement Types. You also need to configure the internal agreement item types and assign them to internal agreements: Transportation Management • Master Data • Agreements and Service Products • Define Internal Agreement Item Types.

10.1.2 Transportation Charge Calculation Sheet

As discussed in the previous section, each agreement in TM needs to contain a TCCS that stores the charge types and assigned rate tables. This setup is mandatory for using the charge calculation for orders and bookings.

The TCCS contains all charges you want to bill to your customers or invoice with your vendors as part of an agreement. Each agreement needs to have at least one TCCS assigned to it; agreements have a heavy hand in steering the behavior of the charge calculation logic, which we describe in Section 10.3. It is tremendously important because in the TCCS you map rate tables to each charge, and it contains the logic for how to apply charges and the sequence in which they need to be calculated.

TCCS Structure

Creation of TCCS

To generate a TCCS, you generally have two alternatives:

1. **Embedded TCCS**
   You can create the TCCS directly from an agreement. Notice the ADD CALCULATION SHEET button in both Figure 10.4 and Figure 10.6. This action automatically generates a TCCS in the background and assigns it to the agreement item. By clicking on the CALCULATION SHEET DETAILS button, you jump directly into the TCCS for maintenance.
2. **Standalone TCCS**
   You can create a TCCS separately and assign it to agreement items later. Navigate to SAP NWBC • MASTER DATA • CHARGE MANAGEMENT AND SERVICE PRODUCT CATALOGS • CALCULATION SHEETS • CREATE CALCULATION SHEET.

The TCCS contains very little header information, as you can see in Figure 10.7. Most important here is the CHARGE USAGE field. Unlike an agreement, the TCCS can be generically used for customer, vendor, or internal agreements.

TCCS line items

The main focus of the TCCS is the item table. Each item represents a charge that the system is going to evaluate and use in the charge calculation for orders and bookings. In the example in Figure 10.7, we have maintained a business scenario for ocean freight charges with a basic sea freight, currency adjustment factor, bunker charges, terminal charges, and two additional service charges, such as below deck stowage. We call all these charges generically *charge types* in SAP TM, and they are maintained in Customizing.

This flexible structure can be used for any other mode of transport or business model because you can define your own charge types. Each charge type needs to be assigned either a fixed rate amount or, most commonly, a rate table like in our example. You can also maintain charge types as percentage values in rate tables or directly in the TCCS and reference them in another charge type. Each line item in the TCCS has a details section with multiple tabs, such as RATE, BASIC DATA, and CLASSIFICATION, where you can maintain important settings for the charge calculation logic and define how the charge types are used and calculated. We spend more time on the setup and further use of charge types and item details in the next section of this chapter.

Figure 10.7 Transportation Charge Calculation Sheet

Templates

Now let's introduce the *transportation charge calculation sheet template*, which you can create and maintain by following the menu path SAP NWBC • MASTER DATA • CHARGE MANAGEMENT AND SERVICE PRODUCT CATALOGS • CALCULATION SHEET TEMPLATES. The TCCS template contains the most important elements of a regular TCCS. When generating a new TCCS, you are always asked to optionally specify a template that will automatically populate all specified charge types and other settings into the new TCCS. You can also assign a TCCS template to an agreement type in Customizing. As a result, that TCCS template is used when you generate a TCCS from an agreement.

TCCS Setup and Customizing

Charge types

The most important Customizing activity in the context of a TCCS is to maintain charge types (IMG • TRANSPORTATION MANAGEMENT • BASIC FUNCTIONS • CHARGE CALCULATION • BASIC SETTINGS • DEFINE CHARGE TYPES). Figure 10.8 shows a list of charge types on the left, and gives the details of one selected charge type on the right.

Figure 10.8 Charge Type Customizing

You can define whether a charge type should be used as a positive or negative value (e.g., for discounts) and whether it is an absolute or percentage value. For either setting, you can specify that both are valid and enter the details in the TCCS line item details. The rounding profile is optional and can be directly maintained in the TCCS line item.

Item details for charge calculation

The details of each TCCS line item can contain many settings based on Customizing entries evaluated by the system during charge calculation. When we refer to the TCCS line item details, we refer to selecting one specific charge type. We dive deep into the impact of such settings on the charge calculation logic itself in Section 10.3, but we give you a brief introduction here. Keep in mind that most of these settings are optional and are only required if you want to model specific use cases.

Let's start by describing the BASIC DATA tab, which is where you specify the charge type of each item. You can mark a charge type as the leading charge type. If the system identifies more than one charge calculation sheet as valid during charge calculation, it checks the leading charge of each TCCS to determine which one to use. You have the possibility to disable this determination feature in the SAP TM charge profile and simply randomly use one valid agreement (see Section 10.2.2, where we introduce the charge profile). You can assign a rounding profile—which you have to configure earlier in Customizing—to round calculated charges up or down. A commonly used feature is the manual charge calculation where a user can specify a rate amount in the transactional document. By enabling the MANUAL CHARGE ITEM checkbox in the TCCS, the charge type will automatically be determined in the charge calculation but the rate value is entered by the user directly in the forwarding order, freight order, booking, or service order.

Dimensional weight profile

As a shipper or freight-forwarder, you'll find that the chargeable weight is very important in your charge calculation logic. In SAP TM you can specify your own dimensional weight profiles in Customizing to calculate the correct chargeable amounts (e.g., 1 kg:6000 cm^3 for the metric system or 1 lb:166 in^3 in the imperial system). The customized profile can be assigned in the TCCS item details. To specify a profile in Customizing, navigate to IMG • TRANSPORTATION MANAGEMENT • BASIC FUNCTIONS • CHARGE CALCULATION • DATA SOURCE BINDING FOR CHARGE

CALCULATION • DEFINE DIMENSIONAL WEIGHT PROFILE. Besides maintaining the dimensional weight profile per charge type in the TCCS line item details, you can also maintain it on the agreement line item itself. Last, you can maintain it on the agreement header in the GENERA DATA tab.

Calculation method

An optional but very powerful setting is the calculation method. You can specify logic for how a charge should be calculated. Whenever you expect the system to simply read a rate from a rate table, no specific calculation method is required. But if a computation *is* needed, such as for break weights in air freight, SAP provides a few standard calculation methods. You can flexibly develop your own calculation methods as an enhancement, maintain them in Customizing, and plug them into the TCCS by following the menu path IMG • TRANSPORTATION MANAGEMENT • BASIC FUNCTIONS • CHARGE CALCULATION • ENHANCEMENTS OF CHARGE CALCULATION ENGINE • DEFINE CALCULATION METHOD.

Resolution base and grouping rule

The last setting we want to highlight is the resolution base and grouping rule. Both are of great importance in the charge calculation. The resolution base defines on which level you apply a charge in your order document (e.g., per container, per package, or once per document [header]). The grouping rule allows the system to calculate one charge for multiple items. You could specify the grouping rule as destination. The system groups all packages or containers in a freight order and takes the total weight in order to calculate the charges. This can lead to a different calculation result.

The next section we highlight is the RATE area. The associated rate table information of a charge type is stored in this tab.

Rate tables and amounts

You always have the option to maintain either a fixed amount with a currency or a rate table. The most common scenario is to store a rate table because it offers much more flexibility. You can either associate exactly one rate table to the charge type or implement a *rate table determination rule* (e.g., per origin country). As the name indicates, this capability allows you to define a Business Rules Framework plus (BRF+) rule and maintain it in this tab. During charge calculation, the system executes the condition and finds the correct rate table. More details on the implementation of charge-related conditions can be found in

Section 10.2.3. Each of your country organizations might be in charge of maintaining its own country-specific surcharges and hence have its own rate tables for the same charge type (a terminal handling charge).

Calculation rule, calculation base

The calculation rule is another important setting on the RATES tab. A calculation rule determines how to calculate a charge based on a fixed rate of the charge. You can maintain calculation rules in a similar manner when using rate tables. For example, you might maintain a charge per distance (kilometers), in which case, you have to multiply this rate by the actual distance a vehicle has traveled, which is stored in your order document.

The actual rule needs to be specified by a *calculation base*, which specifies the data access to the order document (in our example, this is ACTUAL_DIST for the actual distance). If you maintain a rate table for the line item, the calculation rule needs to be defined in the scales of a rate table itself and becomes irrelevant in the TCCS. Refer to Section 10.1.3 for more details on scales and the calculation base.

The next tab in the TCCS item details is the CLASSIFICATION tab. Here you indicate, for example, whether a charge is a statistical charge. Statistical charges are used in the charge calculation like any other charge, but they are not settled in the invoicing process. In the same tab, you can also find the tax indicator. In this tab you can specify a charge type as mandatory by activating the appropriate checkbox. In the event that a charge is mandatory, but it is not possible to calculate a rate in an order or booking, then an error message is displayed to the user. Without the checkbox, only a warning message is shown.

Charge preconditions

The PRECONDITIONS tab provides an important functionality for the charge calculation logic. This setting allows you to maintain flexible conditions when a charge should be used. You might have charges for which you only apply for a certain geographical scope or for certain customers. You can maintain a precondition for a surcharge in three ways:

1. Use the standard TRADE LANES section to maintain geographical conditions, such as FROM, TO, or WITHIN.
2. Specify that a charge is only applicable to specific partners. The standard business partner master data is used for this functionality.

3. Assign the precondition rule for the calculation sheet item, which is the most flexible way to implement your own BRF+ rule.

Business Examples for Conditional Charges

In the air, and especially in ocean freight business, you can find numerous rules and conditions for various charges. We want to highlight a few examples:

- All ocean freight transports with the destination Spain have to pay a mandatory banking charge that is a percentage value of the amount to be paid in Spain.
- In India, a mandatory service and education tax has to be paid for every inbound and outbound shipment.
- For all shipments going to the United States, a mandatory surcharge has to be paid for legal filings. Other surcharges apply if a ship crosses the Panama or Suez Canal.

Similar examples in air freight apply, such as specific terminal handling charges, screening charges to x-ray cargo, and other security surcharges that are based on origins and destinations.

Let's turn our attention to the functionality of the rate tables, where the actual values of your charges are stored.

10.1.3 Rate Tables

Think of rate tables in SAP TM as the heart of the Transportation Charge Management component.

Rate tables contain all rates that are required for the charge calculation in a flexible format. A rate table can never be used standalone. It must be assigned to a TCCS, which itself is assigned to an agreement. You can maintain rate tables both manually and by using upload and download capabilities, as well as mass update functions. A rate table always needs to be designed before you can define its dimensions, called scales.

Creation and Structure

You have two options for creating a rate table.

Creation of rate tables

First, you can generate it from a TCCS. You mark one line item in the TCCS item table and choose ADD RATE TABLE to generate a new rate table in the background. As we described earlier in the chapter, a rate table is always assigned to the TCCS item that contains a charge type. Optionally, you can maintain a rate table template in the TCCS item details that can pre-populate the new rate table.

Second, you can generate and maintain a standalone rate table by navigating to the MASTER DATA • CHARGE MANAGEMENT AND SERVICE PRODUCT CATALOGUE • RATE TABLES. You are asked to optionally specify a rate table type.

General data The rate table has a GENERAL DATA section, as you can see in Figure 10.9. A rate table can be used to store selling, buying, or internal rates. As a result, you need to maintain CHARGE USAGE as a mandatory field. It is possible to set the usage for all three purposes.

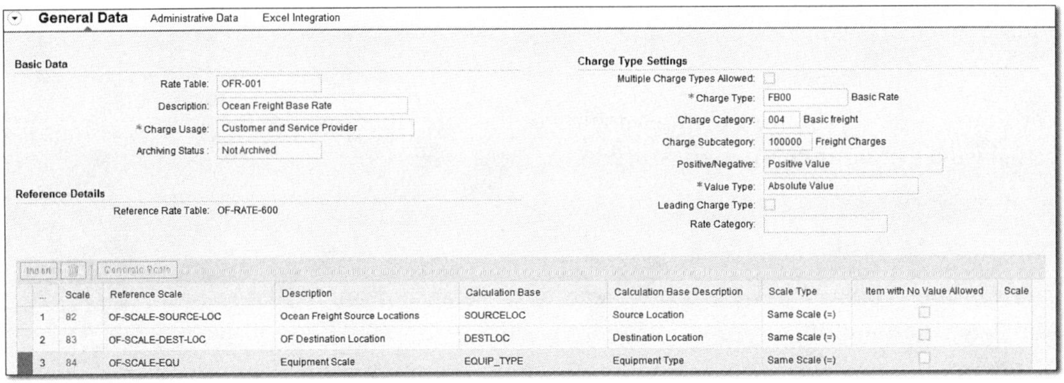

Figure 10.9 Rate Table General Data

Another important section is the CHARGE TYPE SETTINGS area. The dimensions for storing the rate values can be very different for each charge type. As a result, you maintain a rate table for each charge in most scenarios. Consequently, you need to maintain the charge type in the rate table. If you generated the table from the TCCS, the charge type is automatically pulled from the TCCS item. The MULTIPLE CHARGE TYPES ALLOWED checkbox enables the use of one rate table for multiple charge types. In this case, you could maintain the charge type as its own dimension (scale).

Scales in rate tables

The key section in the General Data tab is the Scales area. This is the most important step to actually design your rate table and specify the rows and columns it should contain. In the given example, you can see a table at the bottom of the screen where three scales are maintained. You can write a description for each scale, but the Calculation Base Description column is the best way to identify a dimension; in our example, this is source location (origin), destination location (destination), and equipment type. In the upcoming subsection on scales, we provide more details.

Dates and Values

Let's zoom in to the Dates and Values tab shown in Figure 10.10. Notice that the actual rate table is empty. You need to generate one entry in the header item table with a valid-from date, a valid-to date, and a currency. Then the specified dimensions in the form of the rate table scales are visible as columns.

Figure 10.10 shows the rate table with the same three scales as defined in the General Data tab. Here you can maintain your rates, which are used in the charge calculation. You have the option to toggle between different layouts of the rate table to display or hide the descriptions and notes. You can easily imagine how rate tables can grow enormous in terms of row size, so use the filter functionality to display only a specific set of rows.

Before you can use a rate table for charge calculation, you have to release each header item for use.

Scale items, scale types

After you have successfully generated the rate table and the corresponding scales, you need to maintain the scale items before you can start entering rates. Scale items are actual values maintained against each scale. You can enter scale items directly row by row in the Rates tab by clicking on the Insert button, or you can maintain the entire list of scale items in the Scale Items tab. In our example, we add different load and discharge ports, as well as equipment types and scale items. Based on the *scale type*, you have the option to maintain tiered scale items (e.g., distances, weight, or volume). The *same scale* (=) is commonly used for nonnumeric values or if an exact match is expected. The *to scale* (<=) and *base scale* (>=) are used for tiers.

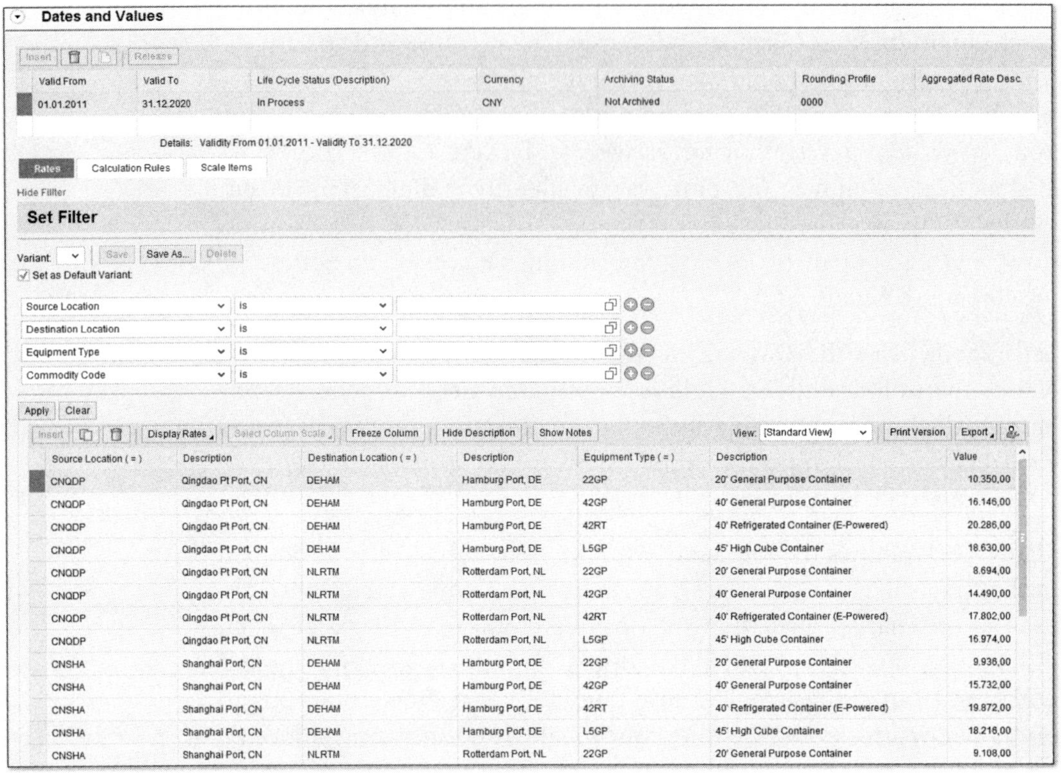

Figure 10.10 Rate Table Dates and Values

Calculation base for calculation rules

In Section 10.1.2 we mentioned that you can maintain a calculation rule in the TCCS or directly in the rate table. Recall that a calculation rule is used in the charge calculation logic to multiply a retrieved value from the rate table with a certain data field. You could specify, for example, that the chargeable weight of a cargo item needs to be multiplied by the rate in the rate table by changing the settings in the CALCULATION RULES tab.

Scales

Calculation base for scales

Scales are mandatory, despite being in the lowest layer in the transportation charge calculation master data in the actual dimensions of a rate table.

Think of a scale as a shell that needs to be designed for a specific purpose. For example, you wouldn't find a scale for equipment type, source location, or chargeable weight. A scale is always generic, but you define the SAP TM data element that it is associated with it by assigning the *calculation base*.

When we speak of a scale, we're mostly referring to the calculation base that actually defines the dimension, such as EQUIP_TYPE for the equipment type, SOURCELOC for a source location or CHRG_WEIGHT for the chargeable weight. The calculation base has two use cases: to generate a scale and to maintain a calculation rule.

Construction of scales

You can generate scales directly from a rate table by clicking on the GENERATE SCALE button, but first you need to lay the groundwork with a few mandatory Customizing settings. Figure 10.11 shows the step-by-step scale generation process.

Let's walk through each of these steps.

❶ **Define the scale base in Customizing.**
The initial Customizing setting is to define the scale base, which is the scale's technical foundation. In the scale base you can define the technical field a scale value is stored in, whether a scale is a numerical value, whether a currency is required, and which dimension can be maintained for the unit of measure (mass, volume, density, etc.). You can customize the scale base via the IMG menu path TRANSPORTATION MANAGEMENT • BASIC FUNCTIONS • CHARGE CALCULATION • DATA SOURCE BINDING • DEFINE SCALE BASE.

❷ **Specify calculation base in Customizing.**
You know that the calculation base is used both to maintain calculation rules and to specify the actual scale. You can maintain a calculation base in SAP TM Customizing via the menu path IMG • TRANSPORTATION MANAGEMENT • BASIC FUNCTIONS • CHARGE CALCULATION • DATA SOURCE BINDING • DEFINE CALCULATION BASES. Note that you need to assign a scale base to the calculation base. You can specify whether a calculation base is an absolute or a relative value. In addition, you can implement new calculation bases with BAdIs and conditions. Other attributes of the calculation base are technical settings, such as those that specify database fields.

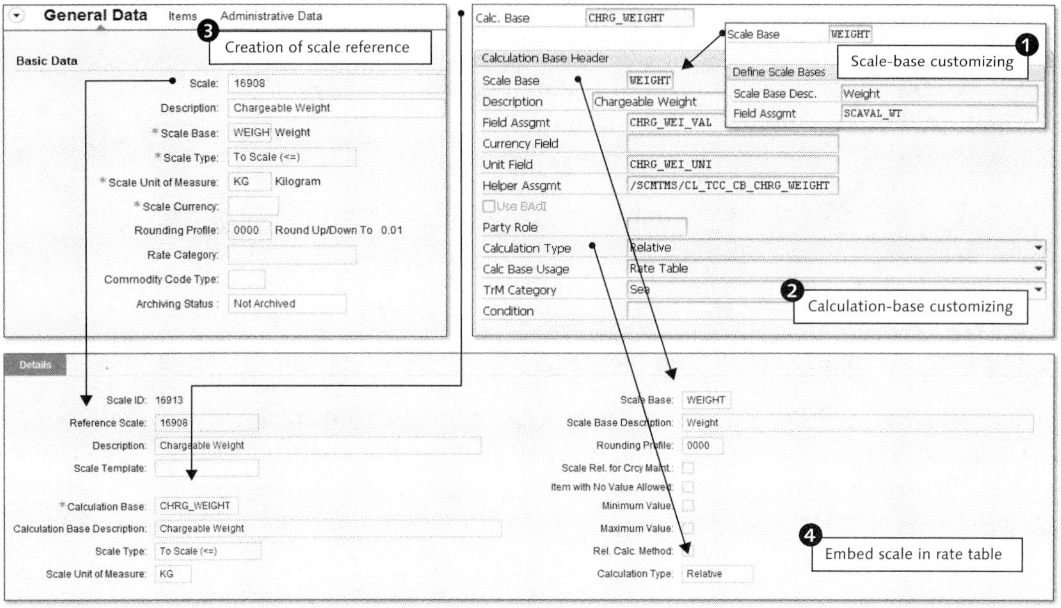

Figure 10.11 Structure to Generate a Scale

❸ **Define reference scale (optional).**
You can create a standalone scale in SAP NWBC and assign it later as a reference scale in a rate table. Navigate to SAP NBWC • Master Data • Charge Management and Service Product Catalogs • Scales. You can also maintain scale templates. Most important when generating a standalone scale is that you have to specify the scale base, but the calculation base is always assigned in the context of the rate table.

❹ **Embed the scale in the rate table.**
The last step returns to where we started to generate the scale in the rate table master data. As shown in Figure 10.11, you have to choose a calculation base as a mandatory setting after you have clicked on the Generate Scale button. All settings specified in the calculation base are populated in the scale, especially the noneditable Scale Base and Calculation Type fields. Optionally, you can assign the reference scale in order to populate the specified fields from the reference, such as a scale unit of measure, scale currency, and the scale type. As you might have noticed, both the calculation base in the

new scale and the reference scale contain a scale base. If the two scale bases do not match, an error will be displayed.

Note that you only have to perform the first three steps once as part of the initial implementation system setup, which means that the process of creating a rate table with scales as master data in a common business environment is less complicated. You can even skip this last step if you work with rate table templates, which we recommend.

Maintenance Tools

The manual creation and maintenance of large rate tables can be a very tedious process. Especially in the environment of large LSP or carrier contracts, a rate table can contain tens of thousands of rows. SAP TM offers two mass update tools to ease this maintenance process, but a certain level of manual effort will still be required.

Rate table upload/download

First, SAP TM offers integration with Microsoft Excel to upload or download entire rate tables. You have the option to define a reference rate table with all required scales in SAP TM and download it to Excel. Then you can maintain the rate values and even change the scale items in the spreadsheet. When uploading the file back to SAP TM, the system validates the newly added or changed scale items against the master data. Alternatively, you can start directly in Excel to create a fresh rate table. You can define a name for the rate table, a rate table template that should be used when uploading it to SAP TM, and the number of rate tables to be generated. You can use the following report to mass download rate table templates for maintenance in Excel: /SCMTMS/TCC_RATE_MASS_CREATE.

Mass update function

In the ocean freight industry, it is a common business practice to regularly (e.g., quarterly) adjust basic ocean freight charges with general rate adjustments and surcharges with batch updates. Such mass updates of rates usually impact multiple rate tables across numerous agreements. Other industries have similar requirements, such as the air freight industry, trucking, and railway businesses. Shippers and LSPs are equally affected on the buying side because they have to update increased/decreased rates in their vendor contracts. SAP TM offers a mass update report /SCMTMS/RATE_MASS_UPDATE for such use cases, as shown in the top left of Figure 10.12.

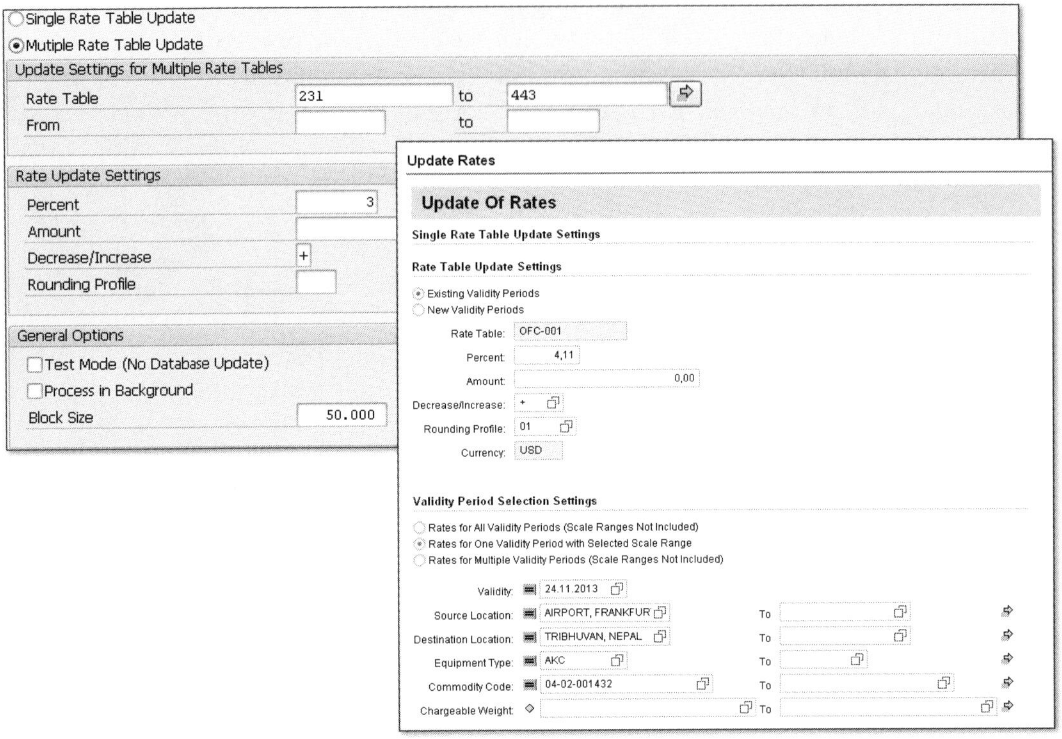

Figure 10.12 Rate Table Mass Update Function

In the mass update report, you can select numerous rate tables and apply a markup or deduction or an absolute value. Unfortunately, it is still not possible to specify the scope for a mass update. It is a common industry practice to update only certain entries in a rate table across multiple agreements. The current functionality always updates *all* entries in a rate table and disallows specifying certain rows based on origin, destination, commodity, equipment types, and so on. In order to cater to the specific requirements needed for the exact scope in terms of rate table line items, a second mass update functionality is supported. You can launch this second mass update directly from a rate table in the master data by selecting UPDATE RATES in the top panel. The updates are limited to this rate table, which is a clear disadvantage. Refer to the bottom-right corner of Figure 10.12, where you can find an example of the rate table update functionality.

The advantage is that you can specify the scope for the rate table line item updates based on the existing scales in a rate table. The scope parameters in the function are dynamically shown based on the maintained scales in the rate table. In our example, you can find five different scope parameters (e.g., source location and destination location) based on the scales maintained in the rate table.

10.1.4 Service Products and Standard Operating Procedures

The professional services industry, which includes marketing, consulting, and the business-to-consumer service industry, has been kicking around the concept of packaged service products for a long time. For LSPs, especially carriers and freight-forwarders, the concept of modular and packaged service products is increasingly prominent. Offering non-freight-related, complimentary services can provide multiple benefits for an LSP, such increasing revenue potentials, achieving cost savings due to standardization, and establishing a unique selling proposition.

With SAP TM, you can maintain service products and service items as master data. A *service item* is a granular component and can be combined with a basic freight transportation service such as container fumigation or reduced cutoff. A *service product* is a bundle of multiple service items (e.g., expedited air freight or cold-chain services). You can define a service product catalog and use it when setting up agreements and generating orders. SAP TM can also operationalize the production of services with standard operating procedures.

Let's look into these options now.

Services in Agreements and Orders

Using service product catalogs

You can maintain service product catalogs to create service products as bundles of multiple service items. You can use these service products as the basis to generate customer contracts representing service agreements. Remember that a service agreement is nothing more than a forwarding agreement that contains service products.

You can work with service products and service items in your transactions in two ways:

1. Generate a forwarding order directly from a service agreement. The new forwarding order automatically inherits all service items, which are bundled under the service product in the service agreement.
2. Generate forwarding orders and insert service items. It is not mandatory to maintain a service product catalog. A limitation of the forwarding order is that you cannot insert an entire service product but only service items. Defaulting an entire bundle of services is only possible when generating the forwarding order directly from the agreement, as recently described.

Creating service product catalogs

Now let's introduce the creation of a service product catalog. You can generate a service product catalog in SAP NWBC by following the menu path MASTER DATA • CHARGE MANAGEMENT AND SERVICE PRODUCTS • SERVICE PRODUCT CATALOGS. In the structure, the catalog is similar to a forwarding agreement, but you do not maintain any agreement parties, validities, or capacities, as you can see in Figure 10.13. Most important, you can create service products in the item table.

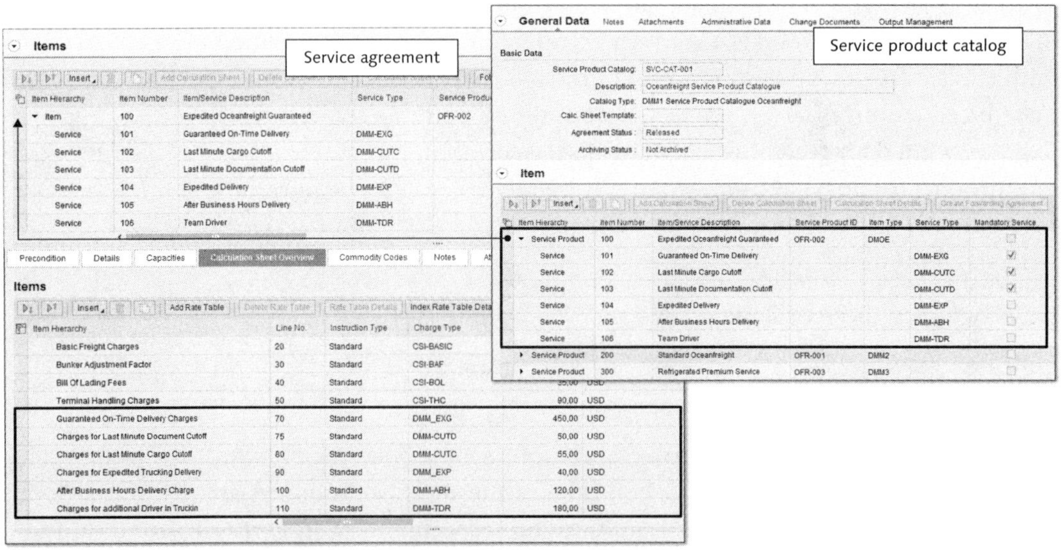

Figure 10.13 Service Product Catalog and Service Agreement

Select INSERT • SERVICE PRODUCT and assign a service product item type as specified in Customizing.

You can now give the product its own ID and bundle service items under the product by selecting INSERT • SERVICE. It is possible to directly assign a TCCS to each service product. In the TCCS, you can include all the charge types that are relevant to calculate the charges for the individual service items. It is not possible to assign a charge type directly to a service product (e.g., to offer a price for an entire bundle).

You can maintain more than one service product catalog in order to distinguish the scope of service products with preconditions in Customizing. Figure 10.13 depicts the creation of a service agreement using a service product from a product catalog.

Creation of service agreement

In a service agreement you can insert a service product in the item table. You can search through your repository of service products for each service catalog and select a suitable product. All service items and charge types are pulled into the agreement. Choose FOLLOW UP • CREATE FORWARDING ORDER to generate an order directly from the agreement to default a service product into the order.

Service products in a forwarding order

When generating a forwarding order, you can enter service items in the item table. You can assign services either to single containers only or to all containers as a header service. It is possible to browse for a service product from a service agreement or catalog, but you cannot default all associated service items to the order. You will always have to choose each service item. Once you execute the charge calculation, the rates for service items are calculated based on the maintained amounts in the service agreement.

Settings for Services in SAP TM

Customizing for service products

To generate a service product catalog and to use services in orders, you need to define a repository of service types in Customizing first, together with various other settings. Service types are the most granular items (e.g., value-added services that can be bundled under a service product or that are particularly used in orders).

1. **Service types**
   You need to set up your individual repository of service types, such as fumigation, providing generator sets, customs brokerage, container cleaning, expedited delivery, and GPS cargo tracking. To do this, navigate to the Customizing menu path IMG • TRANSPORTATION MANAGEMENT • BASIC FUNCTIONS • GENERAL SETTINGS • DEFINE SERVICE TYPES.
2. **Forwarding order item types**
   Recall from Chapter 4, Section 4.2 that in the forwarding order you can use item types mainly to enter cargo items, containers, railcars, and trailers. To also be able to enter service items in the forwarding order item table, you need to maintain a new item type in Customizing by following the menu path IMG • TRANSPORTATION MANAGEMENT • FORWARDING ORDER MANAGEMENT • DEFINE ITEM TYPES FOR FORWARDING ORDER MANAGEMENT. Maintain at least one entry with the item category SERVICE, and assign this item type to your forwarding order type in Customizing by following the menu path IMG • TRANSPORTATION MANAGEMENT • FORWARDING ORDER MANAGEMENT • FORWARDING ORDER • ASSIGN ITEM TYPES TO FORWARDING ORDER TYPES.
3. **Service product item types**
   Besides the granular definition of your individual service types, you need to establish a repository of your service products to generate the list of all the service products you want to offer. You can establish the repository by following the menu path IMG • TRANSPORTATION MANAGEMENT • MASTER DATA • AGREEMENTS AND SERVICE PRODUCTS • DEFINE FWA AND SERVICE PRODUCT ITEM TYPES. In this setting, you can define the description for your service product, such as guaranteed on-time delivery, expedited services, or temperature control deluxe.

   It is crucial now to assign the list of granular service types to the service product item type, which could be bundled under it. It is important to understand that this is not the actual bundling of services, which happens in the master data.

   Optionally, you can specify preconditions that influence whether a service product can be used with the scope of an order or agreement.
4. **Service catalog type**
   The service catalog is technically a forwarding agreement object, so you need to generate a service catalog type that influences the number

ranges and user interface design and contains templates and defaults. You can create a service catalog type in Customizing via the menu path IMG • Transportation Management • Master Data • Agreements and Service Products • Define FWA and Service Product Catalogue Types. To enable the use of service products in the service catalog, you need to assign the possible service product types to the catalog types in the same Customizing path.

Enable charge calculation for services

You have now made the basic Customizing settings to enable service products in SAP TM. To allow a charge calculation for services in forwarding orders, you need to map your service types to charge types. You must maintain at least one similar charge type per service type, but, alternatively, you can map a service type to multiple charge types. Navigate to Customizing in SAP TM to maintain the mapping (IMG • Transportation Management • Basic Functions • Charge Calculation • Basic Settings • Assign Charge Types to Service Types).

Standard Operating Procedures

Definition

Imagine an end-to-end transportation process of an LSP, freight-forwarder, or carrier. You will find that numerous tasks and actions need to be performed to execute a single shipment. Such tasks can include anything from taking and validating an order, planning and executing the transportation, issuing transportation documentation, performing customs clearance, and then finally invoicing a customer. In addition, specific agreements in customer contracts might outweigh the standard handling of a shipment and require additional services.

Standard operating procedures support consistent and structured handling of all such tasks by generating instructions to be executed from user-specific worklists, which ensures process standardization, compliance, and clear guidance for staff. Figure 10.14 gives an example of a worklist with various instructions for one forwarding order.

Instructions and instruction sets

In SAP TM, you can define instructions to represent executable tasks for a user, such as arrange fumigation, generate customer invoice, verify shipping instructions, perform cargo screening, and so on. The example in Figure 10.14 includes checking for green light relevance and completing shipment data.

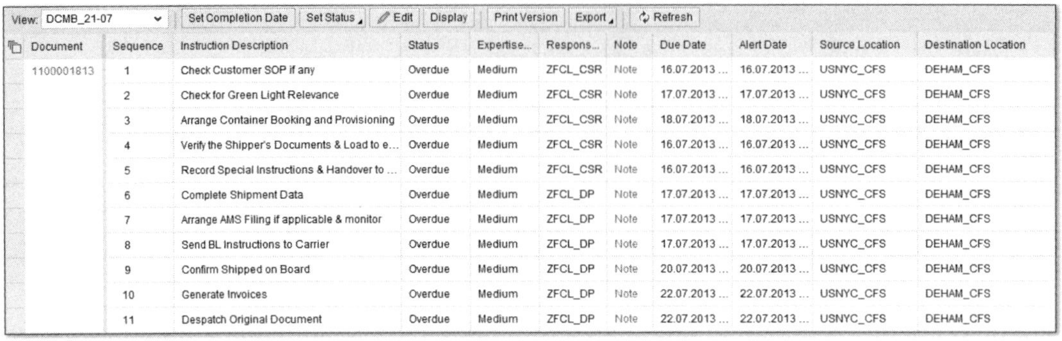

Document	Sequence	Instruction Description	Status	Expertise...	Respons...	Note	Due Date	Alert Date	Source Location	Destination Location
1100001813	1	Check Customer SOP if any	Overdue	Medium	ZFCL_CSR	Note	16.07.2013 ...	16.07.2013 ...	USNYC_CFS	DEHAM_CFS
	2	Check for Green Light Relevance	Overdue	Medium	ZFCL_CSR	Note	17.07.2013 ...	17.07.2013 ...	USNYC_CFS	DEHAM_CFS
	3	Arrange Container Booking and Provisioning	Overdue	Medium	ZFCL_CSR	Note	18.07.2013 ...	18.07.2013 ...	USNYC_CFS	DEHAM_CFS
	4	Verify the Shipper's Documents & Load to e...	Overdue	Medium	ZFCL_CSR	Note	16.07.2013 ...	16.07.2013 ...	USNYC_CFS	DEHAM_CFS
	5	Record Special Instructions & Handover to ...	Overdue	Medium	ZFCL_CSR	Note	16.07.2013 ...	16.07.2013 ...	USNYC_CFS	DEHAM_CFS
	6	Complete Shipment Data	Overdue	Medium	ZFCL_DP	Note	17.07.2013 ...	17.07.2013 ...	USNYC_CFS	DEHAM_CFS
	7	Arrange AMS Filing if applicable & monitor	Overdue	Medium	ZFCL_DP	Note	17.07.2013 ...	17.07.2013 ...	USNYC_CFS	DEHAM_CFS
	8	Send BL Instructions to Carrier	Overdue	Medium	ZFCL_DP	Note	17.07.2013 ...	17.07.2013 ...	USNYC_CFS	DEHAM_CFS
	9	Confirm Shipped on Board	Overdue	Medium	ZFCL_DP	Note	20.07.2013 ...	20.07.2013 ...	USNYC_CFS	DEHAM_CFS
	10	Generate Invoices	Overdue	Medium	ZFCL_DP	Note	22.07.2013 ...	22.07.2013 ...	USNYC_CFS	DEHAM_CFS
	11	Despatch Original Document	Overdue	Medium	ZFCL_DP	Note	22.07.2013 ...	22.07.2013 ...	USNYC_CFS	DEHAM_CFS

Figure 10.14 Worklist for Standard Operating Procedures

For each instruction, you can also define which user role needs to execute this task and provide notes and descriptions. To bring these instructions in the proper sequence and context, you need to group them in instruction sets. Instruction sets serve as a framework to define due dates for tasks to appear before or after a specific event. You can define, for example, that a task needs to appear in a user's worklist 12 hours before cargo cutoff. You can also define time-dependent alerts for a user in case a task has not been executed in time. After an initial setup, SAP TM automatically generates instructions in a user's worklist after a forwarding order has been generated, dependent on the scheduled time. You can generate instructions and instruction sets in the SAP TM Customizing via the menu path IMG • Transportation Management • Basic Functions • Instructions • Define Instructions and Instruction Sets.

Assignment of instruction sets

Instructions are always stored on the forwarding order, where you can find the entire list of relevant instructions in the Instructions tab. You have four options for how to assign an instruction set to ensure it is appropriate for the forwarding order:

1. **Forwarding order type**
   The most generic assignment allows you to assign instruction sets to forwarding order types. When using the specific forwarding order type, the instructions are pulled into the forwarding order.
2. **Item type per forwarding order**
   The instruction set is based on the item type in the forwarding order.

3. **Stage type**
   An instruction set is pulled into a forwarding order if a specific stage type is used.
4. **Service type**
   When you work with the service product catalog, you can assign an instruction set to each service type. This ensures the compliant execution of services you sold to a customer.

The assignment of instruction sets is performed in SAP TM Customizing via the menu path IMG • TRANSPORTATION MANAGEMENT • BASIC FUNCTIONS • INSTRUCTIONS • ASSIGN INSTRUCTION SETS.

10.2 Contract Determination

The creation of master data in SAP TM Transportation Charge Management allows you to store and manage your customer and vendor contracts. The real added value in SAP TM comes into play when you use the integration of agreements into your transactional documents, such as orders and freight documents. It is possible to automatically calculate the charges of forwarding orders, quotations, freight orders, freight bookings, and service orders in SAP TM based in the contract determination logic. This functionality is the key to automatically finding and determining an appropriate contract with your business partners.

There are numerous ways to set up this determination logic because it is tremendously flexible and so can serve all industries and modes of transport. In this section we introduce the basics of defining the *contract determination*. In Section 10.3, we cover charge calculation logic, setup, and preconditions.

Note

It is important to highlight the key difference between the contract determination in this section and the charge calculation logic in Section 10.3: contract determination finds a correct agreement, agreement item, TCCS, and charge type. As soon as this level is reached, the charge calculation logic determines the actual value from a rate table of a charge type and how to use this value in charge calculation.

10.2.1 Contract Determination Logic

SAP TM contract determination follows a specific logic to find the correct agreement and charges. It is triggered with the charge calculation by clicking on the CALCULATE CHARGES button in a forwarding order, quotation, or freight document. The following logic is executed:

1. Determine an appropriate forwarding, freight, or internal agreement.
2. Determine each item of the appropriate agreement(s).
3. Check all qualifying TCCSs.

Forwarding/freight agreement determination

Let's break this down further. SAP TM first searches for a forwarding agreement based on the business partner and sales organization. It uses the purchasing organization for a freight agreement determination. The business partner in the transactional document (e.g., the role of the ordering, prepaid, or collect party or the carrier in a freight document) has to be present as a business partner or in the business partner hierarchy assigned to the agreement. In addition, the calculation date in the order or freight document has to be within the validity of the agreement. Based on Customizing settings, you can further influence the contract determination. If multiple agreements qualify based on a similarity in the described attributes, SAP TM has three ways to narrow down to one valid agreement:

- Select randomly one agreement. This behavior can be activated in the SAP TM charges profile (refer to Section 10.2.2 for more details).
- Check whether priorities are maintained on the header of the identified agreements in tab GENERAL DATA. Use the agreement with the highest priority.
- Validate whether a leading charge type exists in the agreements. Use the leading charge type to check for available rates (e.g., in an assigned rate table) in each qualifying agreement. You can de-activate this behavior in the SAP TM charges profile (refer to Section 10.2.2 for more details).

Internal agreement determination

A slightly different approach is used for the determination of an internal agreement. The determination logic uses the purchasing organization from a freight order or booking to search for an internal agreement with a matching organization. Only when it is modeled as a freight-forwarding house does SAP TM recognize a matching organization.

The next step of the determination is to validate the business partner in the internal agreement against the business partner of the sales organization in each forwarding order loaded to the freight order or booking. The rest of the determination follows the same logic as for the forwarding agreement and freight agreement.

Agreement item determination

The determined agreement can contain multiple agreement items. Before checking the assigned TCCS, SAP TM automatically checks the attributes in the items, such as the following:

- Shipping type (e.g., less than container load, full truck load, ULD)
- Transportation mode (e.g., sea, air, rail, truck)
- Stage category (e.g., pre-, main, or on-carriage)
- Movement type (e.g., CFS to CFS, gateway to gateway, door to CFS)
- Means of transport (e.g., standard truck, cold chain truck)
- Service level, such as those of a carrier (e.g., express, standard, cold chain, or fumigation)

TCCS determination

After one or multiple agreement items are determined, an appropriate TCCS needs to be identified and validated. This can be influenced by the charge types contained in the TCCS and preconditions. The attributes of an order or freight document need to fulfill all preconditions maintained in the TCCS. You can use the standard preconditions that are provided, such as trade lanes and business partners with partner roles. You can also store a list of commodities and use them as a pre-condition for a TCCS. Consequently, SAP TM can validate a specific business partner role from the order or freight document against the TCCS precondition and check that the geographical scope matches or the commodity being transported exists in the contract.

In addition to these three standard preconditions, it is possible to refer to the business rules framework (BRF+ in Chapter 2, Section 2.3) for more complex condition expressions. In Section 10.2.3 we elaborate more regarding the use of preconditions for contract determination.

Leading charge type

Charges can be calculated upon successful determination of one TCCS. It is still likely and possible that multiple calculation sheets qualify, containing many charge types. For each contract determination run, SAP TM always selects one suitable TCCS, so it's not possible to use charges from various matching calculation sheets. As a result, SAP TM evaluates only the leading charge type of each TCCS to determine whether it will be used.

You have the option to de-activate the determination based on leading charge types, as introduced earlier. This will result in a random selection in case multiple TCCSs are determined. When you model your TCCS structure, try to always use just one leading charge type per TCCS; this could be the basic ocean freight charge, the main air freight charge, or the basic freight charge, depending on your mode of transport.

For each TCCS and leading charge type, SAP TM now checks whether a rate is available in the rate table. If yes, the TCCS is used. If multiple TCCSs apply, one of them is picked randomly. We cover the actual rating and charge calculation per individual charge type of the determined TCCS in the next chapter.

Calculation level

There are three ways to determine how often the contract determination logic runs for each forwarding order/quotation or freight order/booking. This behavior is steered by the *calculation level* in the SAP TM Customizing:

1. **Calculation at the header level**
   The contract determination logic runs only one time on the header of the transactional document and identifies exactly one agreement with one matching TCCS.
2. **Calculation at the item level**
   SAP TM runs the logic one time per main item (e.g., a container) in the order/freight document. SAP TM generates charge items in the CHARGES tab of the order or freight document for each item.
3. **Calculation at the stage level**
   Determine an agreement for each stage in a forwarding order, freight order, or freight booking. This is specifically required in all business scenarios with incoterms or wherever different legs require using

charges from different agreements. This calculation level is very useful because it can coexist with the header level calculation. SAP TM always tries to determine one agreement for the header charges and another agreement for each stage.

Forwarding agreement determination with incoterms

It is possible to combine a stage level and a header level calculation. To do this, choose the stage level calculation base but maintain a separate agreement line item; you need to have selected the HEADER-LEVEL CHARGE checkbox.

If you work as a freight-forwarder or LSP and your customer orders are subject to incoterms, then you need to choose the stage calculation level in the calculation profile. Enter the incoterm and prepaid and collect agreement party in your forwarding order. SAP TM populates the correct prepaid or collect business partner on each stage of the transportation, based on your incoterm and stage settings. Consequently, SAP TM uses the business partner stored on each stage to determine the agreement and agreement item for each stage. As a result, it is possible to determine one forwarding agreement for the prepaid party and another forwarding agreement for the collect party.

Release Your Agreements and Rate Tables

It is easy to forget this small step when setting up SAP TM Transportation Charge Management (which leads to there being no results in the rating), but remember that both agreements and rate tables have to be released before they can be used. This contingency prevents you from using rates while they are still in negotiation or setup. So remember to always change the lifecycle status of the rate table header items and the agreement header to *Released* once the maintenance is completed.

Agreement determination rule

The calculation level needs to be defined in the SAP TM Customizing as part of the calculation profile that is an important setting for the contract determination. As an alternative to the SAP TM standard contract determination logic, you can design your own agreement logic and implement it as part of an agreement determination logic in the calculation profile.

10.2.2 Setup of Contract Determination

Before you use the automated SAP TM contract determination logic, you must make a few customizable settings. First, customize the required preconditions for both the agreement and agreement item determination in SAP TM Customizing. For agreement preconditions, navigate to the agreement type configuration (see Section 10.1.1) and select SPECIFY PRECONDITIONS. Similarly, for agreement item preconditions, click on the SPECIFY PRECONDITIONS button in the freight agreement item type or forwarding and service product catalog item type. Only the preconditions defined here are available in the application.

Calculation profile

A key setting for the contract determination is the *calculation profile*, which contains the *calculation level*. This steers the behavior of both contract determination and charge calculation. You can find the profile in SAP TM Customizing via the menu path IMG • TRANSPORTATION MANAGEMENT • BASIC FUNCTIONS • CHARGE CALCULATION • DEFINE CALCULATION PROFILE.

The calculation profile also contains the calculation date type. The calculation date type specifies which date to use to check the validity of an agreement, calculate the charges, and determine the exchange rate date. Numerous other settings can be made in the calculation profile, which we discuss more in Section 10.3. Figure 10.15 shows the settings in the calculation profile.

Charges profile

It is possible to maintain multiple calculation profiles. To enable SAP TM to pick the correct calculation profile, you need to assign it to a *charges profile*. The charges profile is a very important setting for the agreement determination and is depicted in Figure 10.15. It is possible to maintain two calculation profiles in the same charges profile. This is only useful for LSPs working with forwarding houses. In the charges profile, you can also de-select the EVAL. ALL AGREEMENTS checkbox, as displayed in the figure. SAP TM would randomly determine one valid agreement with an appropriate business partner and organization, as we introduced earlier in Section 10.2.1. If the checkbox is enabled, all agreements which have an appropriate business partner and assigned organization are being checked for valid rates based on the leading charge types and priorities.

Besides the calculation profile the charges profile contains the *settlement profile* and the *cost distribution profiles*, which we introduce in Chapter 11. You need to assign the charges profile to an organizational unit in the SAP TM master data. This gives you flexibility in determining agreements and settling your charges.

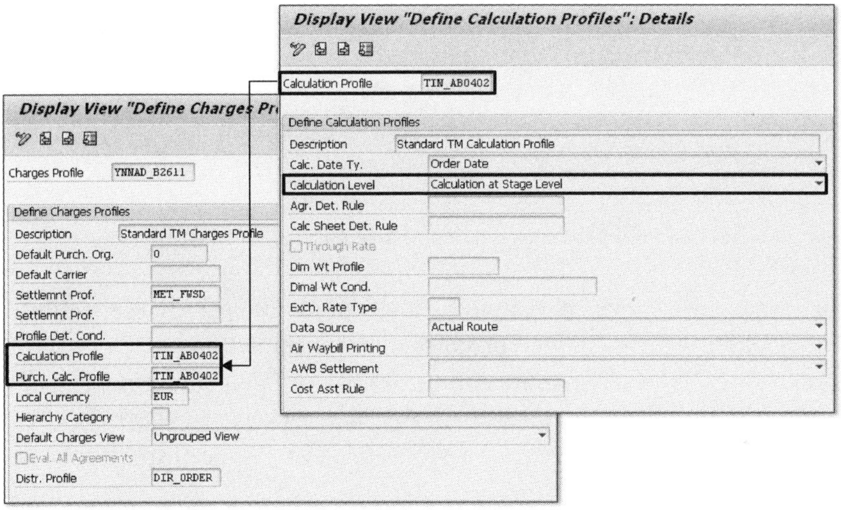

Figure 10.15 Calculation and Charges Profile

10.2.3 Preconditions for Contract Determination

To enhance the flexibility of both the contract determination and charge calculation logic in SAP TM, you can specify condition expressions using BRF+ (refer to Chapter 2, Section 2.3). SAP TM has several areas where you can enable the use of such a condition.

Data access definition

The basic principle is identical for all BRF+ rules: All attributes in a transactional document—such as the forwarding order, freight order, or booking—can be used to design a rule and for validation against a BRF+ decision table. In the event that an attribute is stored on a line item level (e.g., a rate table line item), it has to be available as a calculation base. Attributes that are used as elements in a rule table and for the validation between transactional data and the table structure are called *data access definitions*. Figure 10.16 shows an example of a BRF+ condition table for a TCCS determination rule.

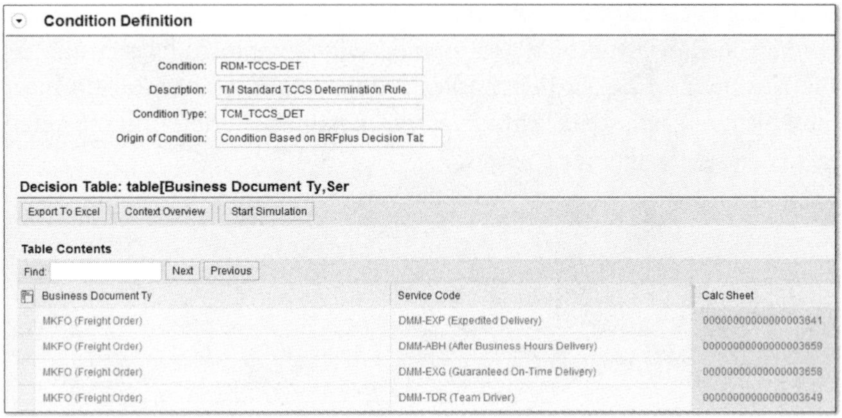

Figure 10.16 Condition for Transportation Charge Calculation Sheet Determination

BRF+ for contract determination

In this example, we maintained a determination rule that specifies a TCCS based on the service product of a carrier with a specific freight order type. You can generate conditions for charge management in SAP NWBC by following the menu path Application Administration • General Settings • Conditions • Create Condition. You can specify the following custom conditions:

1. **Agreement determination rule**
   The entire logic of how to perform the contract determination logic can be bypassed using a BRF+ condition. You can implement your own custom logic, and the system will ignore the standard. Alternatively, you can still use the standard contract determination logic but apply an additional filter. The system uses an exclusion rule to narrow down the remaining agreements to one result.
2. **Rate table determination rule**
   In standard SAP TM, you can assign a fixed amount to either a charge type in the TCCS or a rate table. Alternatively, you can implement a rate table determination rule, perhaps to differentiate multiple tables by country of origin or destination. In a scenario where your charges are maintained on a regional or country level, you can assign each country its own rate table for each charge type. In the TCCS only one charge type is required. This charge type points to multiple rate tables via a BRF+ rule.

3. **TCCS determination rule**
   As recently described, you can specify a specific rule to determine a TCCS, perhaps one based on a specific business partner acting as an affiliate or named account.

BAdIs versus BRF+ Rules

In the standard SAP TM Transportation Charge Management component, there are various enhancement spots where a BAdI or a BRF+ rule can be implemented. You can find the available BAdIs in SAP TM Customizing via the menu path IMG • TRANSPORTATION MANAGEMENT • BUSINESS ADD-INS (BADIS) FOR TRANSPORTATION MANAGEMENT • MASTER DATA • AGREEMENTS/RATE TABLE/ CALCULATION SHEETS.

The actual logic developed in BAdIs has a slight performance advantage compared to BRF+. A BRF+ rule, in essence, is very easy to implement because it is configured purely in decision tables and can be flexibly maintained. Whether to enhance the SAP TM Transportation Charge Management via a BAdI or BRF+ rule is an individual decision based on the complexity and required flexibility of the enhancement.

10.3 Charge Calculation Logic in Selling, Buying, and Internal Invoicing Processes

In the previous section, you learned how SAP TM automatically determines one or multiple agreements for the charge calculation. The agreement determination logic identifies a matching agreement, agreement item, and TCCS.

In this section we focus on the pure logic of how charges are calculated per charge type in a TCCS. The charge calculation process can vary wildly in your industry, the type of charge, and the mode of transport. In ocean freight, for example, there can be hundreds of surcharges that are all calculated differently, such a bunker adjustment factor formula or a charge for out-of-gauge cargo. Similarly, in air freight, there are complex requirements for charge calculation, such as a charge calculation for unit load devices (ULD rating) with pivot rates. LSPs calculate charges based on chargeable weight. In trucking there can be controversial requirements for charges, such as fuel adjustment factors.

Consequently, SAP TM offers a large variety of functional components to support multiple business scenarios and industries. We introduce the general principle of how charges are calculated in SAP TM in Section 10.3.1. In order to introduce SAP TM's strong capabilities to manage air freight charges, we have dedicated Section 10.3.2 to the rating for air freight. Last, in Section 10.3.3 we dive into several business scenarios in which SAP TM has specialized capabilities for particular industries.

10.3.1 Charge Calculation Logic for Selling and Buying Processes

Trigger for charge calculation

Charge calculation in SAP is triggered by the contract determination. You can calculate charges from any freight and service order, freight booking, forwarding order, and quotation. You need to ensure that the charge calculation is enabled in the relevant transactional document type in Customizing. The charges can be calculated manually by selecting CHARGES • CALCULATE charges in the top panel of any transactional document or from a POWL. In addition, you can enable an automated charge calculation in the document type Customizing of orders, quotations, and freight documents. In addition, SAP TM offers two reports that allow you to mass calculate charges. You can find the reports by navigating to SAP NWBC • APPLICATION ADMINISTRATION • BACKGROUND PROCESSING • CALCULATE CHARGES FOR FREIGHT ORDERS and CALCULATE CHARGES FOR FORWARDING ORDERS. Last, the SAP TM charges are automatically calculated when you generate a settlement document (depending on the strategy for settlement documents in Customizing; refer to Chapter 11) or recalculated in credit memos.

General Processing Logic

Rate determination

Let's rewind for a moment. The final result of the contract determination is one charge item per calculation level (e.g., per stage or item) and a corresponding TCCS. Each determined TCCS has a leading charge type with an applicable rate.

The SAP TM charge calculation now evaluates each charge type in each TCCS in a sequential order by generating a calculation table in the background:

1. Each charge type in a TCCS needs to be qualified to determine whether it is required for the calculation. You can maintain preconditions for each charge in the TCCS by using SAP BRF+, as described under the contract determination. The order or freight documents are used only if they meet the conditions of the charge type, such as a specific trade lane/origin-destination pair or business partner. You can also refer to a more complex BRF+ rule for a precondition.
2. Now SAP TM determines the rate amount stored on each charge type in the TCCS. If no rate table exists but there is a single amount, SAP TM uses this one amount and stores it in the calculation table. If a rate table does exist, SAP TM accesses the rate table assigned to the charge and checks for a valid rate table header item. One valid item is used for the charge calculation.
3. It is important to remember that a rate table consists of scales with calculation bases such as source location, destination location, distance, or equipment type. The combination of the scale items in one row of the rate table are validated against the attributes of each charge item element from the transactional document (e.g., for each stage or item). If more than one appropriate table row is found that matches the scale item attributes, only one rate value will be randomly selected by the system. The system now shows the rate amount in the document, influenced by different calculations.

   We should emphasize that some calculations have already been performed at this point. The total rate amounts have simply been retrieved and stored in a background table, invisible to a user. Three major settings that you have the option of maintaining determine the calculation of the final amount: the resolution base, the calculation rule, and the calculation method, as introduced in Section 10.1.2.

Resolution base

The resolution base is a key setting which can optionally be maintained for each charge type in a TCCS. You can find it when marking a charge type in the item details, under BASIC DATA in the RESOLUTION BASE section. The calculation resolution base determines on which level a rate needs to be applied for each charge type. In other words, it specifies the number of resulting rows in the CHARGES tab of an order/quotation or freight document for each charge type. The following are the most important resolution bases in SAP TM:

- **Root**
  This resolution base results in exactly one row in the CHARGES tab. It can be used for charge types that are not calculated on a per container or per item level, as in a documentation fee, bill of lading charges, or airway bill charges for each order or booking. If you leave the resolution base blank, root is the default which will be used.
- **Container (similar to ULD)**
  The container resolution base generates one row in the CHARGES tab for each container in a forwarding order/freight order or booking. It is the most common base for air and ocean freight charges. In air freight, it can also be used for ULDs. Since multiple containers with different attributes—such as weight, volume, commodity, equipment type, and so on—can be contained in an order and a booking, the determined rates can vary per container. Consequently, SAP TM determines a separate rate for each container from a rate table.
- **Package**
  A charge item element is generated and the rate calculated for each package in a transactional document. This resolution base can be used for charge types, such as commodity handling, loading and unloading, and for LTL charges.
- **Product**
  You can generate a charge item element for each product to differentiate product-dependent transportation charges.
- **Service**
  When you work with service items, you can apply the service resolution base for charge types that are mapped against service types. SAP TM generates one charge item element per service. If you assign a service multiple times, each to a different container, SAP TM generates a charge item element per service, per container. Alternatively, it is possible to use the package or container resolution base for a charge type mapped to a service type. As a result, you can enter a service just once as a main item in the forwarding order item table and not for each package/container, but SAP TM still generates one charge item element per service, per package/container.
- **Active and passive resource**
  Especially in trucking and for railways, the calculation based on resources is a common business practice. For example, you can maintain

the passive resource resolution base, and SAP TM generates one charge item element in a forwarding order for each trailer the freight units are assigned to for execution.

When you work with resolution bases, remember that the charges are retrieved from a rate table or charge type amount for each charge line. This is a core difference from the rate determination without resolution bases. The calculation level we introduced in Section 10.2.1 has a strong impact on the resolution base. If a calculation base is chosen as item level but the resolution base of a charge type in a TCCS is selected as root (header), then the change will still be applied for the item (e.g., a container) and not the header (forwarding order). This means that the calculation level overrules the resolution base. Resolution bases cannot be customized, but are provided as part of the standard SAP TM installation.

Resolution base grouping rules

SAP TM generates one charge item and calculates the rates for each resolution base, as described. Grouping rules allow you to cluster different charge item elements with identical charge types and the same resolution base to calculate a consolidated rate. The individually calculated rates per charge item element are summed up to the charge item.

Another grouping functionality is called the *collective rate*. By using grouping rules based on specific calculation bases or conditions it is possible to actually sum up the values of a defined calculation base for all charge item elements in an object. The rate table lookup is based on this summed up value for the calculation base.

Let's look at an example. In a trucking freight order, you deliver goods to various destinations. The resolution base is the package, so you can calculate the charges per package. The calculation base is the gross weight of the packages, but you have agreed with your trucking provider that charges are calculated based on the total weight of all packages per truck load with the same destination. When calculating the charges, SAP TM would first sum up all the weights for the same charge item element based on the grouping rule (here: destination). Second, the lookup from the rate table is executed with the consolidated weight. This might result in a cheaper or more expensive rate compared to an individual rating of each package.

Calculation rule

In the charge calculation logic, the calculation rule greatly influences how a rate is calculated. To construct a calculation rule you need to define three components: the calculation base, the price unit, and the unit of measure.

From Section 10.1.2 you learned that many calculation bases are available, such as source location, equipment type, and gross weight. When constructing a calculation rule, you can only use calculation bases with a numerical value (e.g., gross weight) because SAP TM multiplies or divides this value by the price unit. If you do not use a rate table, the calculation base stored on the charge type is used to multiply or divide a rate amount by a price unit. For example, you can maintain a charge type with the calculation base distance and a rate of 45 Euros per 100 kilometers. In the rate amount you can enter 45 with currency EUR. The price unit has to be 100, and the calculation rule unit of measure is kilometers. As a result, SAP TM uses the actually traveled distance, for example, 2,000 km from a freight order, divides it by 100 and multiplies by 45. The result is a rate of 900 Euros, which is shown in the charge item element in the freight order charges.

Calculation rule in rate table

If you use rate tables, you can maintain multiple calculation rules for more complex business scenarios.

Consider another example. Your ocean carrier has different rates for LCL cargo, depending on origin, destination, and equipment type. The rates are applied per volume of cargo in the container. Consequently, you need to maintain four pieces of information in your rate table with four calculation bases: the source location, destination location, and equipment type are used for scales in the rate table and store a rate per cubic meter. Figure 10.17 shows the setup of a rate table with a calculation rule.

You maintain a calculation rule with the calculation base volume, the price unit 1, and the unit of measure cubic meters (CBM). The level of the calculation rule is maintained as RATE TABLE in the APPL. LEVEL OF CALC. RULE column because you expect to apply the rule on the rate retrieved from the rate table. When issuing a booking, SAP TM finds the correct rate for each CBM in the rate table and multiplies it by the actual volume of the cargo. The resolution base can be maintained as container. Alternatively, you can implement calculation rules that impact single numerical scale items only. This is only possible if your scale item is a relative value and you maintain the corresponding price unit for each scale in the SCALE ITEMS tab.

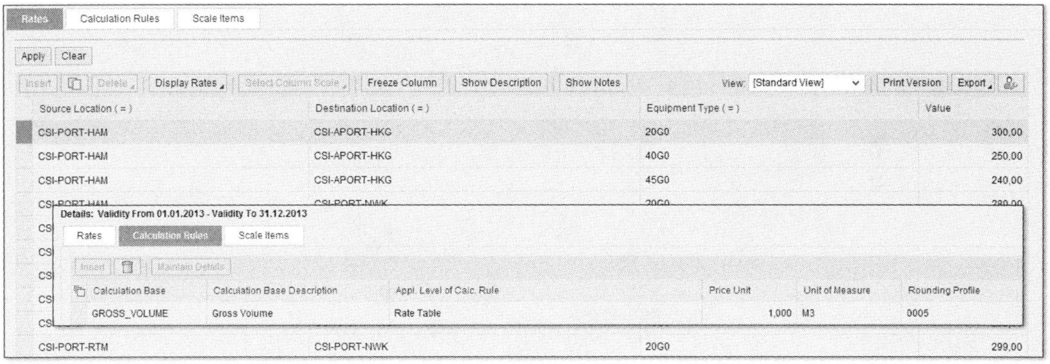

Figure 10.17 Rate Table with Calculation Rule

Calculation method

Calculation methods offer you great flexibility to develop and implement any custom logic that is required for the calculation of your rates (interface SCMTMS/IF_TC_CALC_METHODS). Remember from Section 10.1.2 that you can generate a new calculation method or use the existing SAP TM standard methods. As explained, enabling a calculation method means completely ignoring the standard charge calculation logic. A calculation method is plugged into the charge type in the TCCS.

The following SAP TM standard calculation methods are available:

1. The standard calculation method is used to follow the exact charge calculation logic as described in this chapter. If you do not maintain a calculation method in the charge type, the standard method is used.
2. *Clipping* is a commonly used calculation method in the air freight industry. When you are working with tiered rates in a rate table, the clipping method takes a value that is relevant for the charge calculation (e.g., volume) and applies each tier until the total value is reached or exceeded. At the same time, the clipping method is accumulating the calculated charge.

For example, let's assume you have a charge based on the calculation base gross weight. The cargo you want to transport weighs 11 tons. In a standard calculation method, the calculated rate would result in 11 × 14 = $154 USD based on the rate table in Table 10.1. The result shown in the table is achieved based on clipping.

Tier	Rate	Calculated Amount	Value
Up to 1 ton	$35 USD absolute	$35 USD	1
Up to 5 tons	$20 USD per ton	$80 USD	4
Up to 10 tons	$16 USD per ton	$80 USD	5
Up to 12 tons	$14 USD per ton	$14 USD	1
Sum		$209 USD	11

Table 10.1 Clipping Result

3. Deficit weight rating is a calculation method used especially in U.S. land transportation. If you have a tiered rate table, and the weight of a cargo item is not high enough to reach the tier with cheaper rates, SAP TM sums up the weight of similar cargo from different freight classes and rounds up the weight to reach the next available tier. The charges for all cargo items are calculated based on this cheaper tier. The deficit that was added is rated with the cheapest freight class rate in the appropriate tier and is added to the total rate.

   Table 10.2 and Table 10.3 show an example of the deficit weight rating. The weight of freight class A is 350 kg, and the weight of freight class B is 600 kg. As depicted in the rate tables, the total rate with the standard calculation method is $20,200 (that is, 700 + 13,200) USD. With the deficit weight method the rate is $21,000 (that is, 5,250 + 3,750 + 12,000) USD because the total weight is rounded up by 250 kg to 1,200 kg to achieve the next cheapest tier. Each freight class is rated in its individual tier. In addition, the delta of 250 kg is rated with the cheapest rate in all tiers across the two freight classes.

4. The external agreement calculation method refers to a rate table/agreement maintained outside SAP TM, as introduced in Section 10.1.1. The rates are retrieved from the external agreement without calculating them in SAP TM.

Tier	Rate	Standard Amount	Deficit Weight Amount
Up to 500 kg	$20 USD per kg	$7,000 USD	
Up to 800 kg	$19 USD per kg		
Up to 1,200 kg	$15 USD per kg		$5,250 USD $3,750 USD (15 × 250)

Table 10.2 Rate Table 1 with Freight Class A: 350 kg

Tier	Rate	Standard Amount	Deficit Weight Amount
Up to 500 kg	$25 USD per kg		
Up to 800 kg	$22 USD per kg	$13,200 USD	
Up to 1,200 kg	$20 USD per kg		$12,000 USD

Table 10.3 Rate Table 2 with Freight Class B: 600 kg

There are three more SAP TM standard calculation methods: internal charge calculation, air freight standard, and air freight weight break rating. We introduce these in upcoming sections.

Charge Calculation Results and Charge Estimation

Charges results screen in forwarding order

So far we have put a lot of emphasis on the options for how to configure the contract determination and charge calculation. Now let's highlight what the final results look like, depending on a different setup. After a successful calculation of the charges, you can find the results in the CHARGES tab of your transactional document.

Figure 10.18 shows an example of calculated charges for an FCL ocean freight forwarding order. You can see one charge item because the calculation level in the corresponding calculation profile has been maintained as the header. The various charges are listed below the charge item. Some charges such as the basic freight charge and bunker adjustment factor are calculated per container because the corresponding charge types have the resolution base container. The bill of lading fee and terminal handling charge apply only once on the document because no resolution base is assigned to them and they are consequently calculated on the root. In this example, we did not maintain any calculation rules.

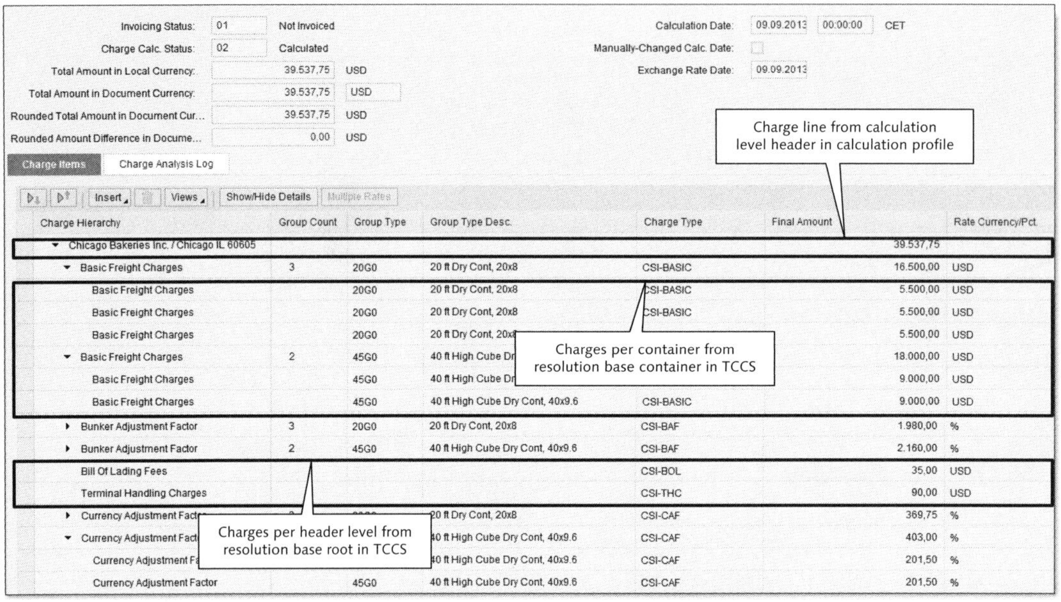

Figure 10.18 Charges Results in Forwarding Order

Manual charges

In the CHARGES RESULTS screen of both forwarding orders and freight orders and bookings, you have the option to change the view. In the example in Figure 10.18 we chose a grouped view, which aggregates all similar charge types for similar resolution bases (in our case, the similar containers). In addition, you have the option to manually add charge items by choosing INSERT • CHARGE LINE. For manually added charges, you need to specify the charge type and the rate amount.

Charge Calculation Details and Log

For clarity, SAP TM displays the results of the contract determination and charge calculation in a detailed section for each charge type and a separate charge calculation log, as shown in Figure 10.19. You can see step-by-step how the charges were determined—that is, using an appropriate agreement: MET_FWA_24 and TCCS 9149. You can see that the rate table DMM_24 has been used to resolve the CSI-basic charge and that some charges applied container as the resolution base.

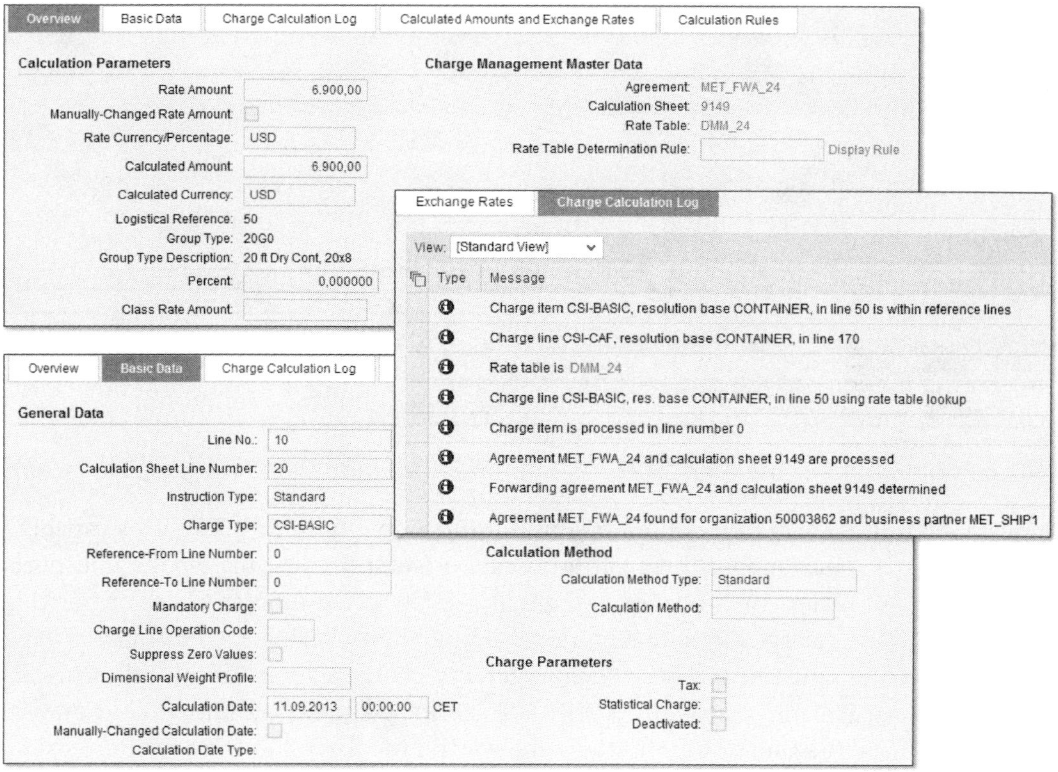

Figure 10.19 Charge Calculation Log

Charges Results screen

Let's introduce another example of the charge calculation that shows the charge determination results for a freight order. In this example (see Figure 10.20), we have maintained an agreement with a trucker. The freight order is transported from the customer warehouse at the origin via two stops in hubs, and on to the final destination. You can see that three charge items are generated based on the calculation level stage in the calculation profile. A calculation rule is implemented to calculate the basic freight charge based on the actual distance traveled between each location per stage. In the Rate Amount column, you can see the per kilometer rate, which is multiplied by the actual distance in the Quantity column and converted to a different currency to arrive at the final amount.

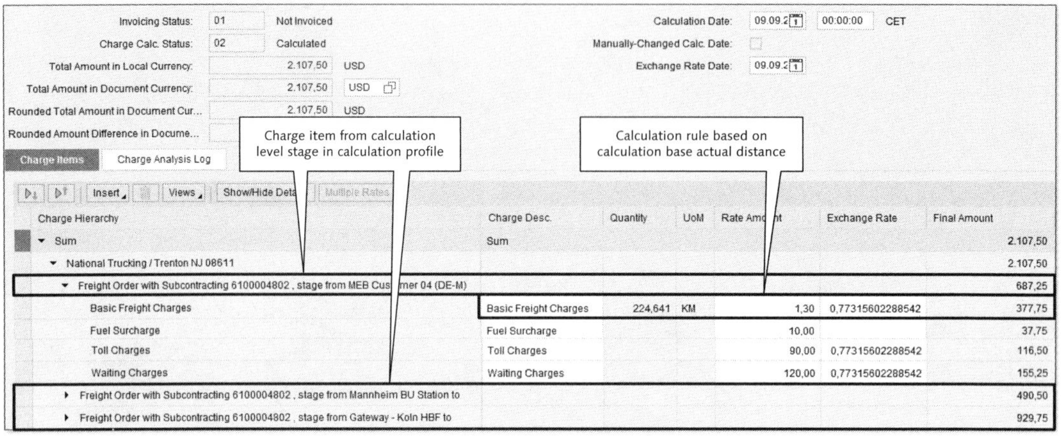

Figure 10.20 Charges Results in Freight Order

Charge estimation

Charge estimation is a useful functionality in SAP TM. This capability can support a freight-forwarder or shipper in estimating the transportation cost or revenue without actually generating an order or booking. Charge estimation exists for both the forwarding agreement and freight agreement sides. You trigger the charge estimation by following the menu path SAP NWBC • FREIGHT ORDER MANAGEMENT • ESTIMATE CHARGES. It works the same way on the forwarding order side.

Technically, the object used is similar to a forwarding or freight order, but the results are not persisted and saved. In addition, it is possible to skip entering a business partner for the charge estimation on the customer (forwarding order) side. Instead, you can maintain a generic agreement without business partners, which is used for the estimation. The user interface is a simplified forwarding order and forwarding order screen. For the charge estimation on the vendor (freight agreement) side, you can use the standard carrier selection (refer to Chapter 7, Section 7.1) if you are not sure which carrier contract to use.

Customizing and conditions

To implement the charge determination logic, most settings have to be made directly in the master data objects, such as agreements, TCCSs, and rate tables. The Customizing activities are minimal.

Let's walk through the Customizing settings in SAP TM. Generate a calculation profile and assign it to a charges profile (refer to Section

10.2.2). The calculation level in the calculation profile is very important for the agreement determination. You can maintain additional settings in the calculation profile that we highlight in Section 10.3.2. In the charges profile, you can maintain a default charges view that is, for example, grouped as shown in Figure 10.20. You can work with rounding profiles to round rate amounts up or down. You can configure a rounding profile in SAP TM Customizing via the menu path IMG • Transportation Management • Basic Functions • Charge Calculation • Rounding. You can assign the rounding profile, perhaps to a rate table item or to a charge type in the TCCS. A very important capability of SAP TM is implementing BRF+ condition rules to specify the charge type applicability. You can maintain flexible rules in BRF+ to manage the complex requirements for surcharges.

10.3.2 Charge Calculation for Air Freight

Complexity

SAP TM Transportation Charge Management offers comprehensive capabilities for the calculation of air freight charges for freight-forwarders. There are multiple methods and variants for calculating charges in air freight. Since this book focuses on the bigger transportation management picture, we do not cover the details of the pure air freight charge calculation or attempt to cover all available functions and setup, but instead aim to convey a general overview of the capabilities in SAP TM. In Chapter 15, you will find more information for an end-to-end air freight process. If you are an SAP customer or partner, you can find a comprehensive implementation guide for air freight with SAP TM on the SAP service market place at *http://help.sap.com/transportationmanagement90*. Navigate to Configuration and Deployment Information • SAP Service Market Place • Scenario Configuration Guides and Scenario Descriptions • LCL Air Freight.

IATA vs. contract rates

SAP TM supports two ways of calculating air freight charges: rating based on the International Air Transport Association (IATA) or rating based on individual contract rates. The IATA globally standardizes processes in air freight and acts as a service organization for the majority of airlines for charge calculation and settlement processes (see Chapter 11, Section 11.1.7). One of the services the IATA offers is to calculate and manage cargo rates in its own database, representing over 100 airlines.

SAP TM is capable of using the same methods and tools that the IATA uses for charge calculation to determine the cost for a freight-forwarder. The core principle is that as a freight-forwarder, when setting up agreements with airlines, you can configure your own freight agreements or base them on the IATA logic. Similarly, when maintaining contracts with your customers, you can charge based on IATA or individual contract terms. For both IATA and contract rates, three rate categories are supported: unit load device (ULD) rates, special commodity rates (SCRs), and general cargo rates (GCRs).

Upload of TACT Rates

SAP TM has the capability to upload the Air Cargo Tariff (TACT) into SAP TM database tables to enable a charge calculation based on IATA. TACT rates represent a generic tariff available to any customer of an airline.

You can upload TACT rates via a report in SAP TM. Navigate to Transaction SE38 in the SAP GUI and run the report /SCMTMS/TACT_RATE_UPLOAD. You can select a file containing the TACT rates. SAP TM generates the appropriate entries in the database with the report /SCMTMS/TACT_RATE_PROCESS.

Unit load device rates

A common way of calculating rates is to have different weight tiers with flat charges per ULD type. This rating logic can be used by both the IATA and if rates are contractually agreed. If the chargeable weight exceeds a weight tier, an over-pivot rate is applied and added. SAP TM calculates the chargeable weight of a ULD and applies the flat rate of the next higher weight tier. This rate is compared against the flat rate of the next lower weight tier, plus the delta in chargeable weight, multiplied by the over-pivot.

Let's illustrate this using the simplified example shown in Table 10.4.

ULD Class	Weight Tier for Chargeable Weight	Rate
04	300	$20,000 USD
04	400	$22,000 USD
04	500	$25,000 USD
	Over-pivot	$40 USD

Table 10.4 ULD Rating Example

Let's say that the ULD in a freight booking has a chargeable weight of 450. The flat rate would be $25,000. The lower weight tier rate is $22,000 plus the over-pivot rate of 50 × 40 = $2,000, which results in $24,000. The lower rate of $24,000 is rated. IATA provides the weight tiers for each ULD class and the flat rates/over-pivot rates for each carrier. The rates are normally dependent on the source and destination IATA cities. This logic is implemented in SAP TM via the air freight weight break calculation method, as introduced in Section 10.3.1.

Special commodity rates

Pure commodity rates are rarely used in air freight. They can be maintained as rates per chargeable weight and are simply multiplied by the chargeable weight of the cargo with a similar commodity class (see example in Table 10.5). It is possible to also apply tiers for the commodity rates, but no over-pivot or comparison logic is applied. A pallet of chocolate on a flight from Frankfurt to Singapore has a chargeable weight of 350. The rate calculated would be 350 × $5 = $1,750.

Origin	Destination	Weight Tier for Chargeable Weight	Commodity Chocolate	Commodity Gold
Frankfurt	Singapore	>100	$6 USD	$30 USD
Frankfurt	Singapore	>300	$5 USD	$25 USD

Table 10.5 Commodity Rating Example

General cargo rates

Finally, the general cargo rates are always used if no specifically discounted ULD or commodity rate is found. Three rate classes are used for the calculation: minimum charge, normal rate, and quantity rate (see example in Table 10.6). The logic is that a normal rate is multiplied by the chargeable weight of the cargo. If it is lower than the minimum, the minimum rate is applied. If the chargeable weight of the cargo is higher than the normal rate weight, the quantity rate of the appropriate tier is used. For example, a piece of cargo on a flight from London to Newark has a chargeable weight of 80. The calculated rate is 80 × $8 = $640 USD.

Origin	Destination	Class	Weight Tier	Rate
London	Newark	Minimum	-	$124 USD
London	Newark	Normal	<50	$9.5 USD
London	Newark	Quantity	>50	$8 USD
London	Newark	Quantity	>100	$7 USD

Table 10.6 General Cargo Rating Example

Logic for charge calculation

When creating an air freight booking, for example, you can specify which rate category to use. SAP TM can also automatically determine the rate category with a specific sequence. Figure 10.21 depicts an automatically rated air freight booking. The charges for airway bill printing and carrier settlement are listed in two separate charge items.

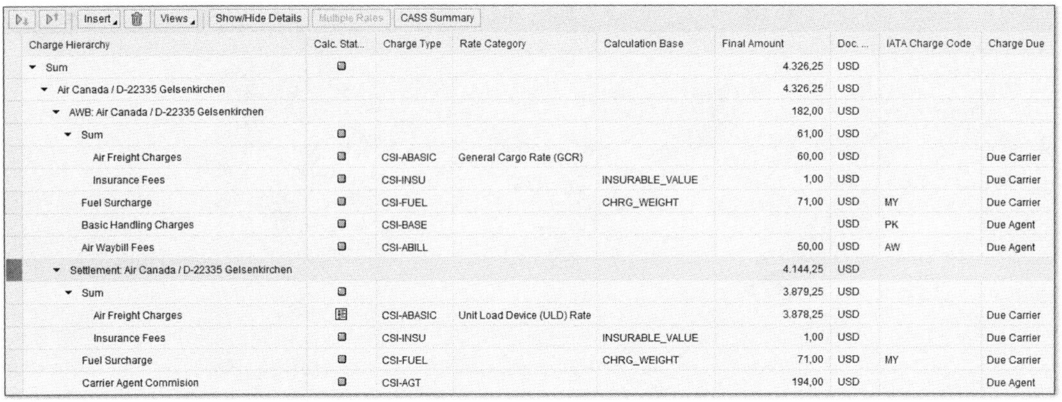

Charge Hierarchy	Calc. Stat...	Charge Type	Rate Category	Calculation Base	Final Amount	Doc. ...	IATA Charge Code	Charge Due
Sum					4.326,25	USD		
Air Canada / D-22335 Gelsenkirchen					4.326,25	USD		
AWB: Air Canada / D-22335 Gelsenkirchen					182,00	USD		
Sum					61,00	USD		
Air Freight Charges		CSI-ABASIC	General Cargo Rate (GCR)		60,00	USD		Due Carrier
Insurance Fees		CSI-INSU		INSURABLE_VALUE	1,00	USD		Due Carrier
Fuel Surcharge		CSI-FUEL		CHRG_WEIGHT	71,00	USD	MY	Due Carrier
Basic Handling Charges		CSI-BASE				USD	PK	Due Agent
Air Waybill Fees		CSI-ABILL			50,00	USD	AW	Due Agent
Settlement: Air Canada / D-22335 Gelsenkirchen					4.144,25	USD		
Sum					3.879,25	USD		
Air Freight Charges		CSI-ABASIC	Unit Load Device (ULD) Rate		3.878,25	USD		Due Carrier
Insurance Fees		CSI-INSU		INSURABLE_VALUE	1,00	USD		Due Carrier
Fuel Surcharge		CSI-FUEL		CHRG_WEIGHT	71,00	USD	MY	Due Carrier
Carrier Agent Commision		CSI-AGT			194,00	USD		Due Agent

Figure 10.21 Charge Calculation in Air Freight Booking

The charge calculation logic is implemented based on two calculation methods: The air freight weight break rating and air freight standard rating can be used for all scenarios. It is also possible to combine the ULD and GCR rate categories, which you might do for mixed loose cargo and ULD cargo scenarios.

1. If you're transporting ULDs, the booking needs to contain a ULD with weight and volume information. SAP TM checks for a ULD rate based on chargeable weight.

2. If no rate is found or the cargo is loose/packages (but not consolidated in an air freight ULD), the SCR is calculated. The commodity code in the booking for the cargo is required.
3. If no specific commodity-based rate is maintained, the generic rate calculation is used as a fallback option.

Handling codes, direct shipments

When you work with contract rates, you can flexibly change this sequence. You can enter specific handling codes in air freight bookings that can indicate, for example, that a cargo screening is required, dangerous goods are being transported, or other services need to be performed. Handling codes can be used to determine additional charge types that need to be included in the calculation. Finally, you have the option to calculate the charges of a direct shipment by pulling the cost calculated in the main air freight booking into a forwarding order (use the calculation method AIR_COST).

To configure SAP TM Transportation Charge Management according to air freight charges, you need to apply a few settings. In Customizing, the following configurations are required. Let's highlight the core differences from a regular charge calculation as described in Section 10.2 and Section 10.3.1.

1. TACT rates
   - ULD rate types are maintained and mapped against ULD types: IMG • Transportation Management • Basic Functions • Charge Calculation • Air Freight Settings • Define ULD Rate Types and Map ULD Types to ULD Rate Types.
   - TACT rates are uploaded.
2. A calculation profile is maintained with the calculation level header and the air waybill printing and air waybill settlement definitions. The calculation profile is assigned to a charges profile of an organization or business partner.
3. A dimensional weight profile is assigned to the calculation profile or the agreement/agreement item.
4. The charge types you use have the transportation mode category air.

Master data settings

Besides pure Customizing, it is essential to set up your agreements, TCCSs, and rate tables accordingly. Consider this summary of the possible settings:

1. The agreement contains a dimensional weight profile. Alternatively it is stored on the calculation profile.
2. The charges in the TCCS are set up for air freight:
   - The rate categories are maintained as either ULD, SCR, or GCR.
   - The rate type is selected as either contract or TACT.
   - Charge type classifications and IATA charge-due definitions are maintained.
3. One of the two air freight calculation methods—air freight weight break and air freight standard—is assigned to the charge types.
4. Rate tables contain the appropriate rate category for a corresponding charge type.

You can also use other functionalities introduced in this chapter, such as a multi-rate hit or minimum/maximum rates, as well as working with preconditions in the air freight charge calculation.

10.3.3 Business Scenarios for Charge Calculation

In Section 10.3.1 we introduced the basic principles of contract determination and charge calculation in SAP TM, but we haven't yet explained the capabilities and numerous functionalities of SAP TM for specific industries and business scenarios. In this section we dive into multiple scenarios to get a deeper understanding of industry-specific capabilities.

Charge Calculation for Freight-Forwarders

Dimensional weight profiles

In the freight-forwarding industry, especially in air freight and LCL cargo, it is a very common business practice to maintain rates not per weight or volume, but instead per the chargeable weight. The chargeable weight is a combination of weight and volume based on a specific factor. For the calculation of rates, it is common to use the higher value out of weight versus chargeable weight.

Let's look at an example. In air freight, the space of an aircraft and the weight that a plane can carry are limited. Consequently, the perfect piece of cargo to meet both the total volume and weight limits of the plane weighs 1 kg per 6,000 cubic centimeters. In example 1 in Table 10.7, the chargeable weight used for rating is 83.33 kg. In example 2, the chargeable weight is 47 kg.

Example	Volume (in cm^3)	Divided by	Dimensional Weight	Actual Weight
1	500,000	6,000	83.33 kg	60 kg
2	240,000	6,000	40 kg	47 kg

Table 10.7 Air Freight Weight Scenarios

You can maintain dimensional weight profiles to store the specific weight:volume ratios in the SAP TM Customizing via the menu path IMG • TRANSPORTATION MANAGEMENT • BASIC FUNCTIONS • CHARGE CALCULATION • DATA SOURCE BINDING • DEFINE DIMENSIONAL WEIGHT PROFILE. You can assign the dimensional weight profile to a calculation profile or on the header/line items of an agreement to use it for the charge calculation and even in the charge details.

Through rate, cross rate

In a regular rating scenario for customers of a freight-forwarder, you can calculate charges per transportation leg. SAP TM checks for a rate for each start location and destination location for each leg when using the resolution base stage.

In the ocean freight industry, however, it is a common business practice of carriers to charge a freight-forwarder an end-to-end rate from origin to final destination. This is called the *through rate*. Alternatively, any other combination of stage-dependent rating is possible based on *cross rates*, meaning that one rate is maintained for the pre-carriage, another rate for the main carriage, and a third rate for the on-carriage.

To pass this concept on to customers of the freight-forwarder, SAP TM supports the through rate/cross rate calculation logic for forwarding orders and quotations. This calculation logic always tries to find the most direct rate with as few individual stages as possible. If no through rate is found, the logic systematically breaks up the end-to-end route into pieces of mixed stage categories (such as pre-carriage and main carriage) to find cross rates including two stage categories. If no cross rate is found, the logic checks for a rate for each stage category. If a rate is not maintained for each stage category, then each start-destination pair is analyzed for a rate. Figure 10.22 shows how SAP TM breaks down the legs of the tour.

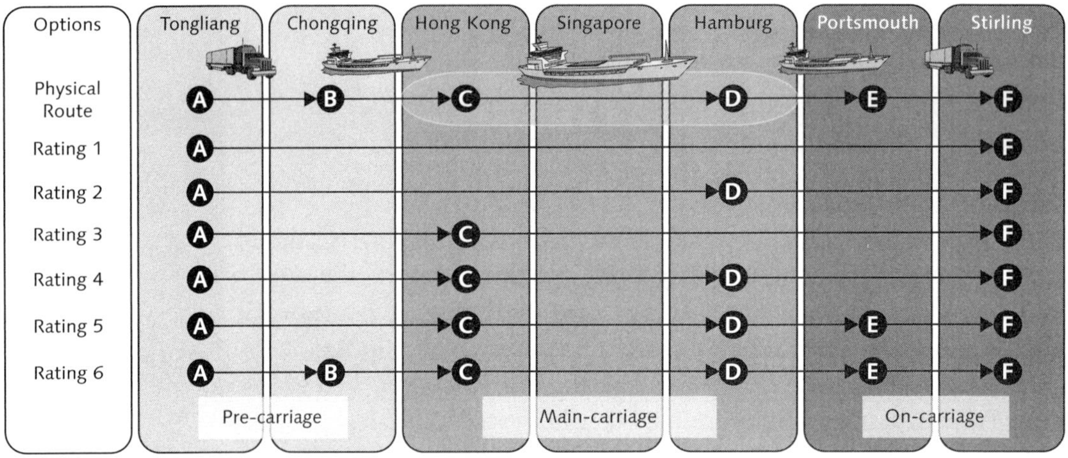

Figure 10.22 Through Rate and Cross Rate Concept

The trip starts with a charge calculation, which checks the rate table for a rate from Ⓐ to Ⓕ. If no rate is found, a second call to the rate table is performed to check for a rate from Ⓐ to Ⓓ and from Ⓓ to Ⓕ, and so on. To enable the through rate/cross rate calculation logic, you need to select the THROUGH RATE checkbox in the calculation profile in Customizing (refer to Section 10.2.2). The calculation level needs to be maintained as the stage level. Note that if a rate is not found for specific legs, SAP TM performs a partial determination where a rate is determined (e.g., only for main and on-carriage).

Uncontrolled transportation

A common business scenario for a freight-forwarder is to handle uncontrolled shipments. A customer uses rates for the main carriage that are directly agreed upon with the carrier. A forwarder can be requested to perform invoice verification for the carrier. Consequently, the charge calculation for both the forwarding order and freight booking are still performed by the forwarder to validate a received invoice. You can maintain a forwarding agreement item, deactivate the CONTROLLED flag, and enter both the carrier and external agreement ID. In the freight agreement with the carrier, you need to mention the uncontrolled agreement party and the external agreement ID.

Additional Capabilities of Charge Calculation

Besides the specific functionality in SAP TM Transportation Charge Management for freight-forwarders and in air freight, SAP TM also supports dedicated rating scenarios in trucking and rail. The following are specific functionalities for trucking: Trucking rating

- Maintain an index rate table for each charge type with a regular rate table in order to calculate fuel surcharges based on indices
- Charge calculation based on active and passive resources via freight orders and trailer units

SAP TM has a sharp focus on railway companies and has been investing in developments to support dedicated functionality for railways. In the area of charge calculation, the following functionalities are available: Railway rating

- **Interline shipments with rule 11**
  Charges can be calculated for each independent stage of the railway move in an interline shipment that considers multiple rail carriers.
- **Railcar charge calculation with day-of-week pricing**
  It is possible to maintain rates in absolute numbers or markups in percentages depending on the delivery date.

Attributes that can be used to calculate rates for resources (e.g., railcars or trailers) will be enhanced with vehicle weights, number of railcars, number of axles, and so on.

Charge Management for Rail Carriers and Parcel Businesses

It is planned to develop additional functionalities in SAP TM to support courier, express, and parcel services and railway businesses. Specific developments are planned in the charge calculation for the parcel and railway industries, but these have not been finalized yet.

A very useful functionality in SAP TM 9.1 is the cost pull capability, through which you can pull the actual transportation cost from a freight Cost pull order or booking into a forwarding order for customer charge calculation. For example, you can maintain markups in percentages that are applied on top of the transportation cost when calculating the customer charges. As you'll see in Chapter 11, Section 11.2, you have to configure the cost distribution before you can use this functionality.

When setting up your TCCS in the forwarding agreement, you need to enter a line item with the new instruction type COST. In the basic data of this line item, assign the calculation method `COST_PULL`, which enables the underlying logic of retrieving cost information from a freight document and using it for charge calculation on the forwarding order side. You can specify whether all charge types or only dedicated charge types from a booking will be retrieved, as well as whether they will be applied as an aggregated lump sum or shown individually as in the booking. To apply a markup, you can enter additional charge types under the cost line item in the TCCS with percentage values.

Multiple rate hit

When calculating charges from a rate table, SAP TM always checks for an entry that exactly matches all attributes of a rate. In a scenario where information in the forwarding order or freight booking is missing, SAP TM can determine a rate with partial information. In other words, if SAP TM does not find one exact match for a rate because a scale value is not provided in a document, it shows all available options as a multiple rate hit so the user can decide. Two scenarios are supported:

1. You can maintain rates with empty scale items in a rate table next to the regularly maintained line items. SAP TM uses these rate lines with blank values as fallback options in case no specific hit can be achieved. The scale you want to leave blank has to be maintained as a wildcard (*). In a freight booking, for example, you might specify a certain port of discharge (POD). If SAP TM does not find a rate for this particular POD, it automatically uses the rate maintained in the rate line with the wildcard for POD. If no POD is given at all, SAP TM shows all options, including the wildcard.
2. If multiple regular rate table line items are defined based on origin, destination, or equipment type, then you can also achieve a multirate hit. For example, the equipment type is not specified in a forwarding order. Consequently, SAP TM does not calculate a charge for each affected charge type but proposes all options for user selection. You

can choose the particular charge type and click on the MULTIPLE RATES button in the CHARGES tab. A pop-up window opens that shows the options from the rate table, as shown in Figure 10.23.

Multiple Rates For CSI-BASIC

Multiple Rate

Amount	Currency	Source Location	Description	Destination Location	Description	Equipment Type	Description
11.200,00	USD	CSI-PORT-HAM	Port Hamburg	CSI-PORT-NWK	Port Newark	45G0	40 ft High Cube Dry Cont, 40x9.6
8.500,00	USD	CSI-PORT-HAM	Port Hamburg	CSI-PORT-NWK	Port Newark	42G0	40 ft Dry Cont, 40x8.6
6.900,00	USD	CSI-PORT-HAM	Port Hamburg	CSI-PORT-NWK	Port Newark	20G0	20 ft Dry Cont, 20x8

OK Cancel

Figure 10.23 Multiple Rate Hit

The multiple rate hit is limited to absolute calculation bases and cannot be used with numerical values. In addition, this functionality cannot be used in all automated charge calculations (e.g., via batch jobs). A selected rate is considered a manual rate and is not recalculated automatically in the event of changes to the transactional document.

Minimum and maximum rates

Another very useful functionality is to force a minimum or maximum rate to be determined during charge calculation. You can maintain a minimum and maximum value on both the charge type and rate table scale levels. If the result during charge calculation is below the minimum rate, SAP TM uses the minimum, and vice versa for the maximum rate.

Event-based charges

In SAP TM 9.1, a new integration with SAP Event Management has been developed for charge calculation. You can now apply charge types dependent on events. This live integration to SAP Event Management calculates event-based charges in the rating logic for forwarding orders and freight orders/bookings. For example, a charge is included in the charge calculation if an expected event has been reported, an unexpected event occurs, or an event becomes overdue. To enable event-based charges, you need to create an event profile and map charge types to events in the SAP TM Customizing via the menu path IMG • BASIC FUNCTIONS • CHARGE CALCULATION • BASIC SETTINGS • DEFINE EVENT PROFILES.

Internal Charge Calculation

Scenarios

Internal agreements in SAP TM represent agreements between organizational units of the same company. It is possible to use internal agreements, TCCSs, and rate tables to calculate internal charges. This is similar to the regular charge calculation logic. Internal charges are used purely for the purpose of internal settlements. We look at the internal settlement process later in the book, but for now let's focus on the calculation of internal charges as part of the SAP TM charge calculation logic.

A scenario in the trucking industry is that a purchasing organization of a freight-forwarder cross-charges the cost of procured transportation services against a forwarding order from a sales organization. The internal charges are visible in the INTERNAL CHARGES tab of the forwarding order. In this scenario, you have two options for how to apply internal charges. First, you can maintain actual internal rates in the internal agreement. When the internal charge calculation is triggered from the forwarding order, rates are determined from the internal agreement. The determination logic is similar to the regular charge calculation logic in Section 10.3.1. Second, you can use the actual cost of transportation based on the freight order or booking to calculate the internal charges. The cost can be distributed to the transported forwarding orders. The logic of cost distribution in SAP TM is described in Chapter 11, Section 11.2. An internal agreement is still required, but you do not need to maintain any rates on the agreement. The rates are retrieved directly from the freight documents (e.g., freight order). You need to use the internal charge calculation method to retrieve the rates.

Here's an example: A purchasing organization of a trucking company uses the resources (e.g., trailers) of another organization with a resource pool. The organization with the trailers cross-charges the cost of using these resources against a freight order from the purchasing organization. The internal charges are visible in the INTERNAL CHARGES tab in the freight order:

- Internal charges based on the forwarding order
- Internal charges for freight orders

Using the first option for this trucking scenario, an internal agreement is required that lists the organization with the resource pool and the purchasing organization as partners. You need to maintain internal rates that represent the standard cost incurred by the resource-owning organization. The resolution base of the internal charges needs to be maintained as a resource. After you have assigned forwarding orders to a freight order that contains a trailer, you can trigger the internal charge calculation. The rates are retrieved from the internal agreement following the same logic as described in Section 10.3.1. If multiple forwarding orders are assigned to different trailers in a freight order, you can use the cost distribution to apportion the internal charges for each trailer to the forwarding orders planned on the freight order.

10.4 Strategic Customer and Vendor Contract Management

SAP TM offers entirely new capabilities for long-term procurement and selling decisions and an integration with SAP Customer Relationship Management (CRM) for contract management. Earlier releases of SAP TM offered limited functionalities for selling and procuring transportation services, mainly to support operational and ad hoc based scenarios, but this has improved. We summarize these new capabilities under the umbrella of strategic customer and vendor management.

Architecture

Figure 10.24 shows the architecture of the strategic customer and vendor contract management. The architecture shows a clear distinction between a strategic customer and vendor management, symbolized by the light section on the left and the white section on the right, separating the figure vertically. Similarly, we first introduce to you the strategic customer management in Section 10.4.1. Illustrated by the gray box in the top left corner of the figure you can see that the opportunity management is now integrated with TM. Furthermore, you can replicate the SAP TM service product master data in SAP CRM and use it in the process of generating opportunities, while having the standard capability of managing prospects and leads.

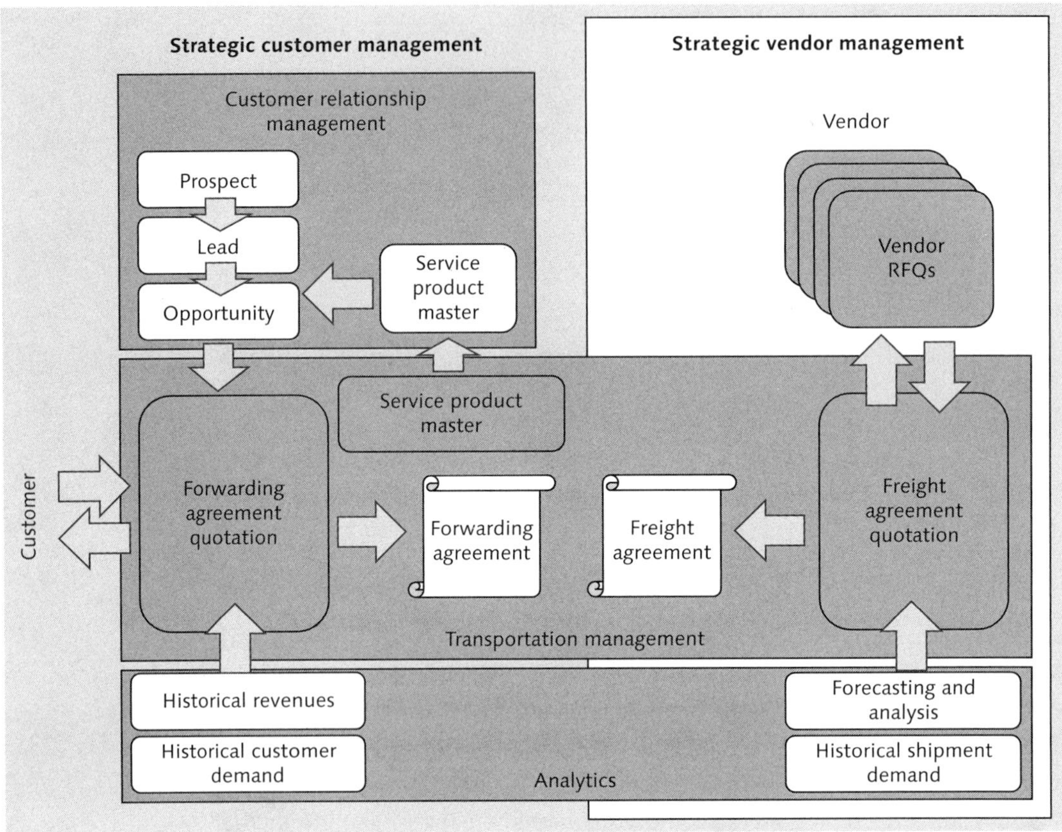

Figure 10.24 Architecture of Strategic Contract Management

It is possible to directly generate the newly introduced SAP TM forwarding agreement quotations (forwarding agreement request for quotation [RFQ]) from SAP CRM. To respond to customer quotations, the forwarding agreement RFQ allows you to analyze historical demand and revenues with customers to support pricing decisions.

This integration to SAP Analytics is displayed in the bottom layer of the figure. As shown by the arrow leading from the forwarding agreement RFQ to the forwarding agreement, you can transform a successfully won RFQ into a forwarding agreement. On the right side of the picture you can see the capabilities for a strategic freight procurement which we focus on in Section 10.4.2. This includes integration with SAP analytics solutions to analyze historical demands and even perform forecasting based on different strategies and what-if scenarios. You can generate a freight agreement quotation, which serves as an RFQ, shown in the center right.

It is possible to perform a carrier ranking and carrier analysis directly from the freight agreement quotation. You can generate multiple vendor RFQs for publishing and collect various responses. In addition, you can compare answers from your vendors, both manually and via an optimizer engine, to award carriers. Finally, you can generate a freight agreement as a new contract from a freight agreement quotation.

10.4.1 Strategic Customer Management

Because SAP TM offers new integration with SAP CRM for both master data and transactions, you can use a harmonized service product catalog across SAP TM and SAP CRM for the generation of opportunities in SAP CRM and a direct integration with forwarding agreement quotations. You also have the capability of executing various SAP TM transactions directly from SAP CRM. The forwarding quotation process allows you to analyze and price RFQs issued from your customers. You can use built-in analytics with the Business Context Viewer to analyze demands and revenues. The process that is supported for customer contract management is still very basic and manual compared to the complex industry requirements. For successful deals, you can directly convert quotations into forwarding agreements.

Integration into SAP CRM

Quotation with service products

It is a very common business practice for freight-forwarders and carriers to have sales employees who work a lot with customers and at the client site. Consequently, they rarely use back-end applications other than an SAP CRM system, which supports the core sales business. With this new integration of SAP TM into SAP CRM, you can fulfill this requirement, and a sales user does not need to navigate to a separate SAP TM system. You can use your standard SAP CRM functionalities, such as campaign management and prospect, lead, and opportunity generation, which are now fully integrated with contract quotations via the forwarding agreement RFQ.

The actual integration starts with an opportunity. In an SAP CRM opportunity, you can choose a sales organization, business partner, and the newly introduced SAP TM service products that will be quoted. You can select a service in the opportunity item table by selecting New • Product ID. The service product ID is equal and replicated from your service product ID in the service product catalog in SAP TM. As a result, your service offering is synchronized across the different systems.

Opportunity integration to SAP TM

Once an opportunity is prepared, you can create a forwarding agreement quotation by selecting Create Follow-Up • TM FWA RFQ. Before a new forwarding agreement RFQ is generated, you can choose between one and multiple service products that will be priced in the agreement quotation. Eventually, a new forwarding agreement RFQ is created, and it contains all relevant fields from the opportunity, such as the business partner, sales organization, and SAP TM service products. Additional attributes from the service product catalog in SAP TM are supplemented in parallel (e.g., all service items that are part of a service product). You can now save the forwarding agreement RFQ, and it is available in SAP TM.

Customizing

To enable forwarding agreement quotations, you need to make a few settings in SAP TM and in SAP CRM.

- First, you need to create a forwarding agreement RFQ type in the SAP TM Customizing by following the menu path IMG • Transportation Management • Master Data • Agreement RFQs • Define Forwarding Agreement RFQ Types.
- Second, it is mandatory to set up the SAP CRM opportunity types and transaction types. Navigate to SPRO • Customer Relationship Management • Transactions • Basic Settings • Define Transaction Types.
- Last, map your forwarding agreement types to the opportunity types that are contained in SAP CRM transaction types: SPRO • Customer Relationship Management • Transactions • Settings for Opportunities • Integration with SAP Transportation Management • Assign Forwarding Agreement Quotation Types to Transaction Types.

Note

To populate the business partner from the SAP CRM opportunity to the SAP TM forwarding agreement quotation, you have to replicate the same business partner data from SAP ERP Central Component (ECC) to both SAP CRM and SAP TM. In addition, you need to use the same organizational structure in both systems to retrieve the sales organization as specified in the opportunity.

Service product master replication

Replicating the service product master data is a very important task in your integration with SAP CRM. The service products are available for the SAP CRM opportunity only after correct setup. It is important to note that the integration is limited to service products and does not include an entire service product catalog, as described in Section 10.1.4. Additional data such as service items are referenced only when you generate the forwarding agreement quotation from SAP CRM.

Now let's dive into the few configuration steps that are required:

1. **Enable SAP TM item types.**
   Activate the ENABLE CRM flag in the SAP TM Customizing and mark SAP CRM as the mapping system by following the menu path IMG • TRANSPORTATION MANAGEMENT • MASTER DATA • AGREEMENTS AND SERVICE PRODUCTS • DEFINE FWA AND SERVICE PRODUCT ITEM TYPES.
2. **Initial and delta replication.**
   Download the initial service product master from SAP TM via Transaction R3AS in SAP CRM and choose the newly introduced load object `TM_SERVICE_PROD`. This new object specifies in which SAP CRM database tables the new information needs to be stored. In addition, you need to choose the source and destination sites that contain SAP TM and SAP CRM as the logical systems for the integration. After you execute the download program, your master data is initially replicated. Whenever you create, change, or delete a service product in SAP TM, this change will automatically be populated to SAP CRM.
3. **Define category and hierarchy.**
   You must assign a category and hierarchy to the SAP TM service product. First, define a category and hierarchy in SAP CRM Customizing, and then assign the hierarchy and category to the product and service object in SAP CRM. You can find this setting in CRM Customizing via

the menu path CROSS-APPLICATION COMPONENTS • SAP PRODUCT • PRODUCT CATEGORY • ASSIGN CATEGORY HIERARCHIES TO APPLICATIONS.

4. **Map hierarchies against service types.**
   The last required Customizing activity is the mapping of the SAP CRM hierarchies to your service items.

Other integration points

In addition to the integration of the SAP CRM opportunity to the SAP TM forwarding agreement RFQ, you can launch other SAP TM functionalities directly from SAP CRM. This is not a real data interface but a remote launch of an SAP TM transaction to avoid switching between systems. Figure 10.25 shows an example of an SAP CRM user interface that is customized for a sales representative and depicts all the transactions that can be accessed directly from SAP CRM.

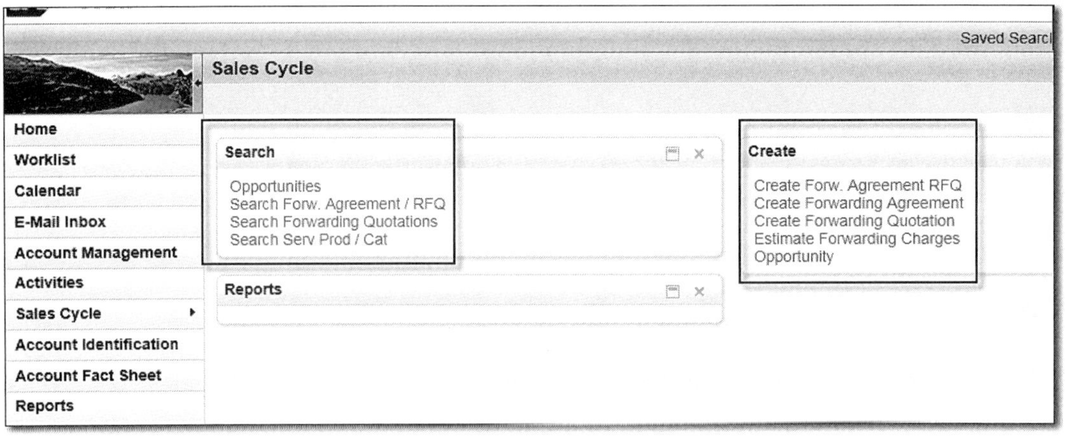

Figure 10.25 SAP CRM User Interface for Strategic Customer Contract Management

Forwarding Agreement Request for Quotation and Analytics

Creation

When a shipper requests a quotation from you as an LSP or carrier, there are two ways you can generate an RFQ. First, you can use the new SAP TM-SAP CRM integration to create the RFQ from an existing opportunity in SAP CRM, as introduced in the previous subsection. Second, you can generate and edit forwarding agreement RFQs from the new section in the forwarding agreement menu tree in SAP NWBC, called FORWARDING AGREEMENT RFQS. The forwarding agreement RFQ header section contains one or multiple sales organizations and customers. This can be used, for example, if a shipper wants to include multiple affiliates or members of a shipper association. You can also maintain various validities and deadlines for the quotation and potential agreement. The rest of the forwarding agreement RFQ follows the same logic as the forwarding agreement: In the item table you can reflect several service products and service items. You can generate the TCCS containing all the different charge types and rate tables.

Supported process flow

The quotation process for customer management can reach a high level of complexity for a carrier or freight-forwarder, especially if a requested scope includes multiple trade lanes and service products and thousands of origin-destination pairs. Submitted RFQs can easily comprise tens of thousands of line items with different equipment types, commodities, and origin-destination pairs. A big challenge is to upload such requests into the SAP TM system in order to allow the generation of a quotation. In SAP TM you are required to first generate a forwarding agreement quotation manually. You can use the standard rate table import capabilities to upload a customer RFQ once you have manually converted it into an SAP TM format.

Pricing and preparation of surcharge structures with the TCCS also have to be done manually. A very useful functionality of SAP TM is the analytics integration via the Business Context Viewer. This supports your pricing decisions with historical revenues and demand breakdowns. Figure 10.26 shows an example of an embedded demand report for four business partners in the RFQ.

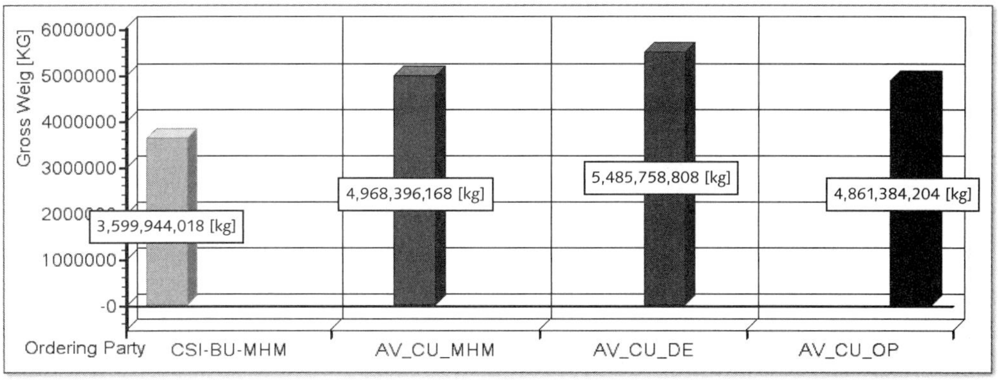

Figure 10.26 Historical Demand for Customer in Forwarding Agreement RFQ

After completing the manual pricing and preparation, you can submit a quotation to your customer, which includes an internal approval workflow. To enable multiple negotiation rounds, you can create different versions of a forwarding agreement RFQ. Finally, you have the option to mark each agreement item as accepted or rejected, depending on the customer response. In a positive outcome you can then generate a forwarding agreement.

Customizing and structure

The forwarding agreement RFQ technically has the forwarding agreement object as the underlying data structure. The structure and user interface are very similar for both business objects, which enables an easy conversion of an agreement RFQ to the actual agreement. The forwarding agreement RFQ has its own document type that needs to be maintained in SAP TM Customizing. Navigate to IMG • Transportation Management • Master Data • Agreement RFQs • Define Forwarding Agreement RFQ Types. You can generate a forwarding agreement quotation and store it for any future RFQs.

10.4.2 Strategic Freight Procurement

Strategic freight procurement is executed from the perspective of a shipper or LSP, requesting capacities and rates from a carrier or freight-forwarder. The SAP TM functionality for strategic freight procurement focuses on mid- and long-term planning and procurement decisions. The intention is to support and enable a quotation and contract management process to establish strategic freight agreements.

The strategic freight procurement process begins by analyzing and planning future demand based on historical data. A *forecast* is the foundation for procurement decisions about freight capacities. Once capacities are forecasted and planned, the actual vendor selection and quotation process can be executed.

Earlier releases of SAP TM lacked the functionality that supports this process. The freight order and freight booking tendering functionality is limited to single shipments only, as discussed in Chapter 7. The innovative aspect of strategic freight procurement is that it supports an entire RFQ process from analyzing historical demands, to generating forecasts, to managing a request for a quotation until the vendor is awarded the contract and a contract is created.

In SAP TM we differentiate between two supported functional processes: the procurement planning and forecasting capabilities based on analytics, and quotation management with SAP TM, as you can see in Figure 10.27.

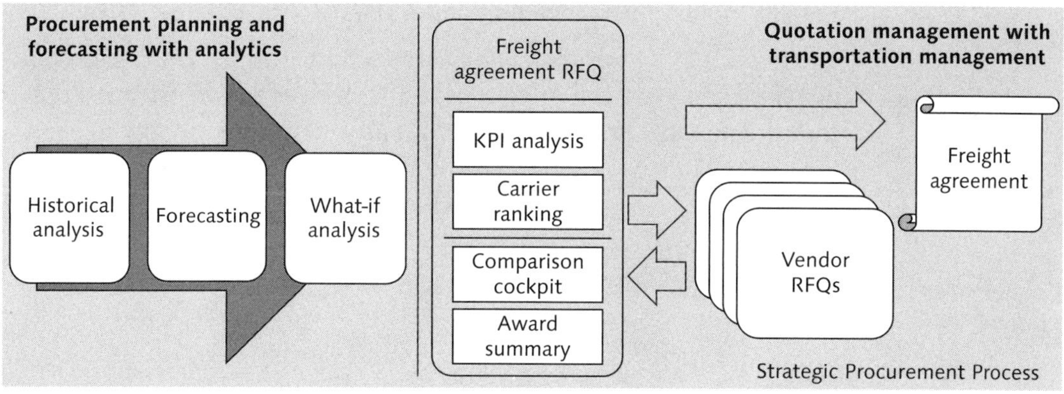

Figure 10.27 Overview of Strategic Freight Procurement

1. **Planning and analytics**
   This process is fully executed outside SAP TM in SAP BusinessObjects Business Intelligence (BI). The bases for procurement planning are historical shipments from SAP TM: freight orders and freight bookings that are stored in SAP NetWeaver BW. These historical demands are the foundation for generating forecasts in SAP BusinessObjects BI, using strategies such as trending and smoothing. Once you have successfully generated a volume forecast, you can use this information for the freight agreement quotation process.

2. **Freight agreement quotation process**
   You can generate freight agreement requests for quotation to subcontract freight capacities based on agreed rates. You can use the forecasted volumes as an input to define capacities you want to request from your vendors. The freight agreement RFQ allows you to define scope, requested capacities, charge structures, and service products that will be contracted. A carrier-ranking tool allows you to short-list potential carriers and forwarders. You can generate vendor

RFQ documents and publish them individually to providers. To award a carrier or service provider, you can use a comparison optimization tool and then generate freight agreements.

Let's take a closer look at each of these components, beginning with freight procurement planning and analytics.

Freight Procurement Planning and Analytics

The procurement planning and analytics functionality has three components that support you in your sourcing decisions: You can analyze historical demands, generate forecasts, and consider alternative scenarios with what-if analyses. All three functionalities are part of the SAP Transportation Procurement Cockpit, which runs completely in SAP BusinessObjects BI. You can access the SAP Transportation Procurement Cockpit from SAP NWBC by navigating to ANALYTICS • STRATEGIC FREIGHT MANAGEMENT.

Historical analysis

Once you have started SAP BusinessObjects BI, you see a table that shows historical shipments and breaks down your shipped containers, weights, volumes, and planned and actual costs. Furthermore, you have a list of filter parameters in the top panel that you can apply to drill deeper into the report. For example, you can specify the scope for the report from a total of containers to a breakdown per carrier and trade lane. In addition, the time horizons can be flexibly adjusted. A clear advantage of the SAP BusinessObjects BI analytics solution is that you can flexibly adjust a report to your needs by setting filters and dragging and dropping additional dimensions in the table. This allows a very intuitive navigation and capabilities for drill-down. Figure 10.28 shows one example of a specified scope for a historical demand report.

Forecasting and version comparison

In the second tab of the Transportation Procurement Cockpit for PLANNING, you can generate capacity forecasts. First define a planning version and describe the scope for the historical data. You can specify the historical basis with start and end dates and whichever trade lanes and origins and destinations you want to include. As a result, you see the historical demand based on your selection. If you select the forecast, you are prompted to specify the forecast period you want to generate.

					Number of Documents	Container Count	Gross Weight in KG	Gross Volume in m3	Planned Costs	Actual Costs
Source Location	Destination Location	Carrier	Equipment Type	Calendar Year		TEU	KG	M3	EUR	EUR
Port Newark	Port of Rotterdam	AV_CAR_002	22G0	2012	23	8.864,000	222.900.000	111.450.000	445.291,00	453.971,00
			22H0	2012	27	9.080,000	228.798.000	114.399.000	541.917,00	553.537,00
			42G0	2012	20	8.048,000	202.394.000	101.197.000	368.780,00	376.155,00
			42G1	2012	17	5.968,000	150.458.000	75.229.000	341.740,00	349.042,00
			Result		87	31.960,000	804.550.000	402.275.000	1.697.728,00	1.732.705,00
		AV_CAR_004	22G0	2012	37	5.632,000	143.196.000	71.598.000	569.592,00	590.470,00
			22H0	2012	37	6.496,000	164.356.000	82.178.000	578.529,00	598.353,00
			42G0	2012	40	6.288,000	159.810.000	79.905.000	643.693,00	666.413,00
			42G1	2012	43	8.160,000	205.808.000	102.904.000	651.620,00	674.104,00
			Result		157	26.576,000	673.170.000	336.585.000	2.443.434,00	2.529.340,00
Overall Result					244	58.536,000	1.477.720.000	738.860.000	4.141.162,00	4.262.045,00

Figure 10.28 Historical Demand per Carrier and Equipment Type

After you confirm the scope, the actual demand projections are computed, including the number of TEUs and the projected cost. You can generate various forecast planning versions using different strategies to calculate the projections. To compare the different forecast planning versions that you have created, you can use the forecast version analysis. Select the versions you want to compare via the filter capabilities, and you will see the overview of projections for each version. It is possible to export generated forecasts into a Microsoft Excel spreadsheet.

Figure 10.29 shows a forecast for a specific origin and destination pair for 2013. Two forecast versions are shown.

Forecast strategies

One essential task is to define the logic of the actual strategy being used to generate the forecast. In SAP BusinessObjects BI you can specify your own planning functions, each with a different forecasting strategy. You can define planning functions in your SAP BusinessObjects BI system using Transaction NRSPLAN. Several standard strategies are available and can be used in a new function, such as simple exponential smoothing, linear regression, seasonal exponential smoothing, and trend-seasonal exponential smoothing. After you've defined and copied a new planning function and assigned a strategy, you need to establish a planning sequence as part of the same transaction. Once a new planning function and sequence are defined, they can be used for forecasting runs.

What-if analysis

After a successful forecast, you can manually alter the projected results. Click on the ENABLE CHANGE button to enter deviating container amounts or weights and volumes. If you select SIMULATE, the system automatically calculates impacts on other parameters, such as the projected cost. You can save the simulated versions.

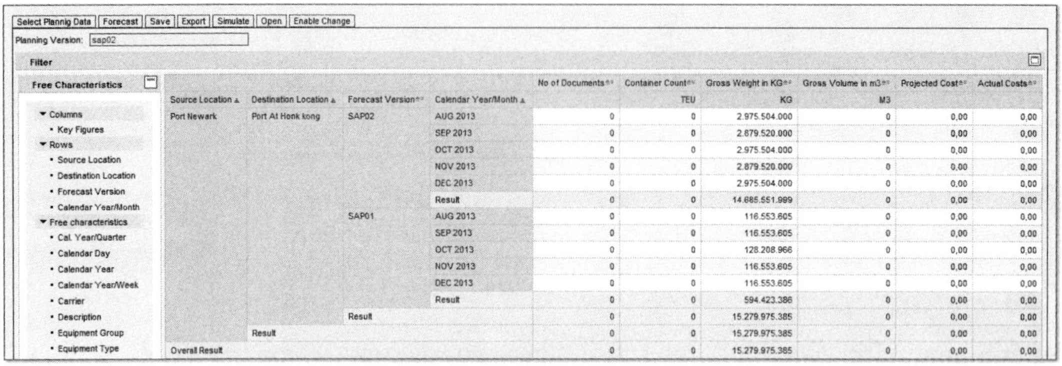

Source Location	Destination Location	Forecast Version	Calendar Year/Month	No of Documents	Container Count	Gross Weight in KG	Gross Volume in m3	Projected Cost	Actual Costs
					TEU	KG	M3		
Port Newark	Port At Honk kong	SAP02	AUG 2013	0	0	2.975.504.000	0	0,00	0,00
			SEP 2013	0	0	2.879.520.000	0	0,00	0,00
			OCT 2013	0	0	2.975.504.000	0	0,00	0,00
			NOV 2013	0	0	2.879.520.000	0	0,00	0,00
			DEC 2013	0	0	2.975.504.000	0	0,00	0,00
			Result	0	0	14.685.551.999	0	0,00	0,00
		SAP01	AUG 2013	0	0	116.553.605	0	0,00	0,00
			SEP 2013	0	0	116.553.605	0	0,00	0,00
			OCT 2013	0	0	128.208.966	0	0,00	0,00
			NOV 2013	0	0	116.553.605	0	0,00	0,00
			DEC 2013	0	0	116.553.605	0	0,00	0,00
			Result	0	0	594.423.386	0	0,00	0,00
		Result		0	0	15.279.975.385	0	0,00	0,00
	Result			0	0	15.279.975.385	0	0,00	0,00
Overall Result				0	0	15.279.975.385	0	0,00	0,00

Figure 10.29 Forecast with Version Comparison

Freight Agreement Quotation Process

Freight agreement quotation creation

In the previous section you analyzed historical demands and generated a capacity forecast. Knowing your future demand is a key input to take strategic procurement decisions. In this section we will use both capacity information and past vendor data to issue an RFQ to shortlisted vendors. The goal will be to sign a long-term freight agreement with the highest ranked and cheapest service provider(s). This process is mainly driven by the new freight agreement quotation, which you need to generate to start a new RFQ.

In SAP NWBC you can find a new section for agreement quotations by following the menu path SAP NWBC • Freight Agreement Management • Freight Agreement RFQs. You are prompted to enter an RFQ type. The RFQ type is a new Customizing setting that you can find in SAP TM Customizing via the menu path Transportation Management • Master Data • Agreement RFQs and Quotations • Define Freight Agreement RFQ Types. The RFQ type specifies number ranges, activates approval workflows, and contains important settings for carrier ranking and the comparison cockpit, as we describe in the following sections. After you have generated the freight agreement RFQ, you can specify basic header information, such as the purchasing organization(s) responsible for the RFQ, validity dates, the deadline for the quotation, and the desired contract duration, as shown in Figure 10.30.

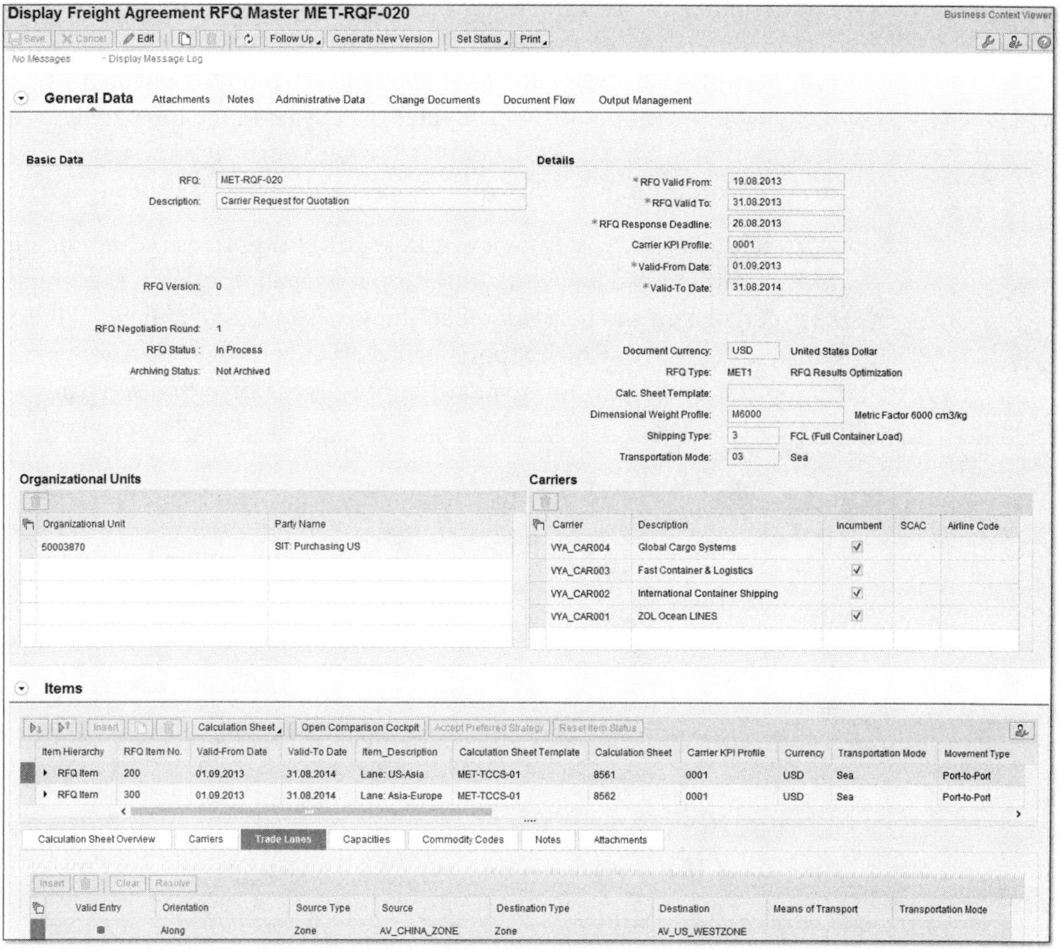

Figure 10.30 Freight Agreement Request for Quotation

The CARRIERS section is used to define the list of vendors you want to consider for the quotation. It does not represent the short list to which you actually are going to *submit* the RFQ, but instead shows the long list of possible vendors. To specify the scope for your RFQ, you need to generate RFQ items. An RFQ item can, for example, resemble different geographical trade lanes, a different set of commodities, or different service products you want to request. You can also maintain a budget against each agreement quotation line to limit your maximum spent.

Vendor KPI analysis

To choose the carriers and freight-forwarders you want to include in the actual submission of the RFQ, you can use the new key performance indicator (KPI) analysis functionality. This capability is enabled by the Business Context Viewer, which you can activate in the top panel on the user interface. The Business Context Viewer is displayed in a side panel to show contact information directly in the user interface of the RFQ.

With KPI analysis, you can view different KPI reports based on the historical shipment data with your suppliers and toggle, for example, between pickup and delivery reliabilities, as shown in Figure 10.31. You can also design your own reports based on your own KPIs. Another KPI supported in the SAP TM standard queries is the invoice discrepancy.

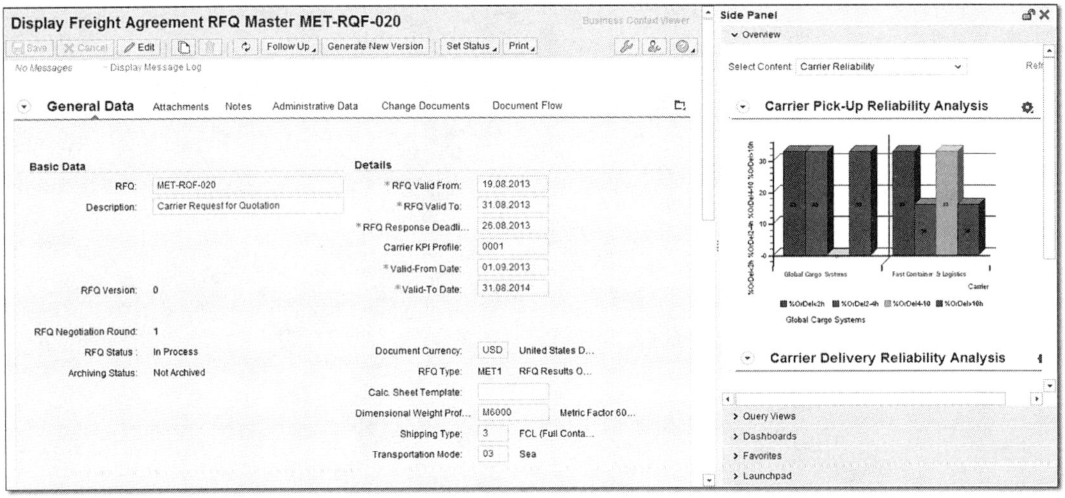

Figure 10.31 Vendor KPI Analysis

Note

For more information concerning the Business Context Viewer, refer to Chapter 12, Section 12.1.

Carrier ranking

In addition to a pure analysis, SAP TM offers a ranking functionality for your vendors based on a configurable carrier KPI profile. You can set up a profile in SAP TM Customizing and assign it as the default profile to the freight agreement RFQ type. Alternatively, you can select it in the RFQ header or line item directly in SAP NWBC.

To set up a profile, navigate to the Customizing in your SAP BusinessObjects BI system by following the menu path SPRO • INTEGRATION WITH OTHER SAP COMPONENTS • TRANSPORTATION MANAGEMENT • DEFINE KEY PERFORMANCE INDICATORS AND PROFILES. In the profile you can specify your desired KPIs, and you can set up weighting factors to multiply your defined KPIs differently and construct a score for each carrier. By comparing the scores with each other, you can select the best-performing vendor. Figure 10.32 shows the results of a carrier ranking with an assigned ranking profile of 0001.

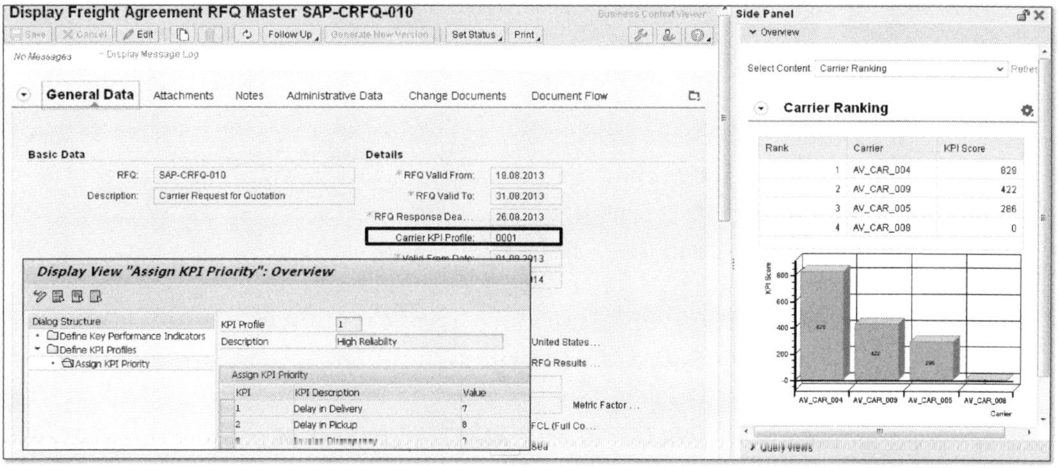

Figure 10.32 Carrier Ranking

Request line items

Once you have successfully short-listed your vendors based on the KPI analysis and carrier ranking, you can proceed to quotation preparation. An important step in the preparation is to define the surcharge structure with which you want your vendors to comply. Like with a regular freight agreement, you can assign a TCCS to each line item and add charge types. The list of charges you select is used for the submission to the vendors. To define the scope of your RFQ with detailed request line items, generate a rate table for each charge type in SAP TM. For charges where you expect a fixed lump sum across all line items, you do not need to assign a rate table (e.g., a currency adjustment factor). Usually, you generate a rate table at least for the basic freight charge and the key surcharges, depending on your mode of transport; for example, enter a rate table for the bunker adjustment factor, security surcharge, and fuel adjustment factor.

You can enter the scope for the RFQ by adding lines in the rate table (e.g., per origin, destination, equipment type, and commodity). The rate table structure is flexible; you can use various scales and calculation bases to model the rate table for the scope of your RFQ. This rate table for the basic freight charge is the central input tool for your vendors to reply with a rate for each request line item. Each rate table you assign to a surcharge generates output that your vendors need to fill in.

Figure 10.33 shows a prepared rate table with a carrier response for a basic freight charge.

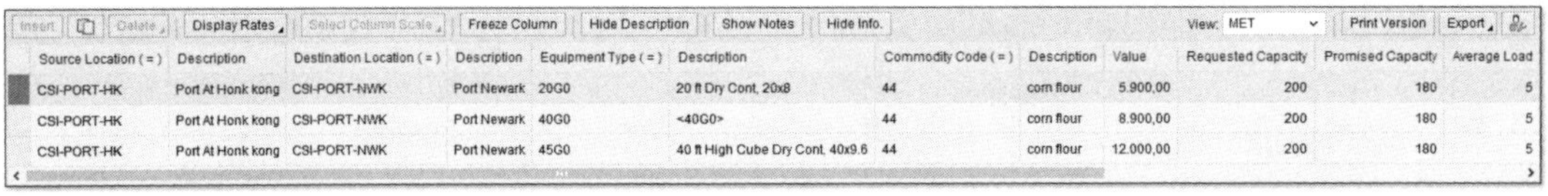

Figure 10.33 RFQ Rate Table for Basic Freight Charge

Capacities

In addition to rates and surcharges, another vital piece of the freight agreement RFQ is the capacities. In the last section we covered how to generate a demand forecast; now we bridge the gap and walk through how to include the capacity information in your RFQ. By selecting BUSINESS CONTEXT VIEWER in the top panel of your RFQ, you can directly access the demand and forecast reports, as shown in Figure 10.34.

You can select the HISTORICAL DEMAND AND FORECAST content in the side panel of the Business Context Viewer. This allows you to view the historical demands you have shipped with your carriers directly in the RFQ. It is possible to personalize the chart and change the parameters, shown in this example by different trade lanes, origin-destination pairs, mode of transport, or time buckets.

You can also view the forecast you generated as part of the procurement planning and forecasting process. You can display your own forecasted capacity demands with the specific carriers in the Business Context Viewer by selecting the content of FORECASTED VERSION ANALYSIS. You can filter the demands (e.g., by the geographical scope). It is even possible to switch between different forecast versions that you created prior to the freight agreement RFQ by entering a different forecast version in the RFQ line item table.

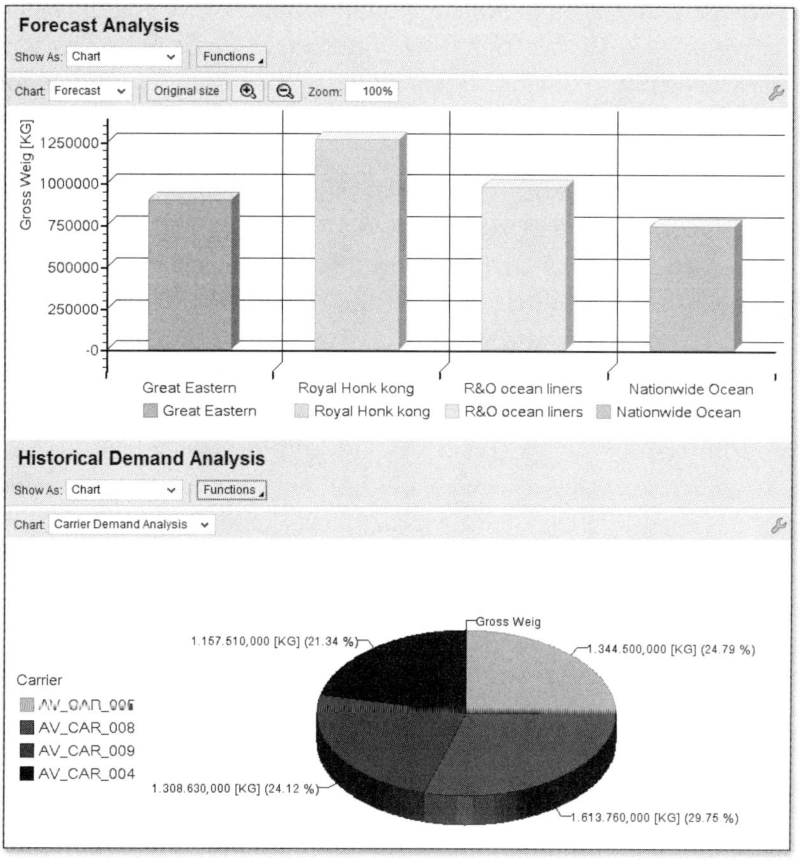

Figure 10.34 Historical Demand and Forecast

Once you've decided what capacities to request, you need to enter your decision in the freight agreement RFQ. You can request capacities both by RFQ line item and in the more granular rate table lines. In the RFQ line items you can specify your requested capacities in various types, perhaps per weight, volume, or TEU per defined time bucket (e.g., per month, per year). A vendor can respond with promised capacities and, after a negotiation, with a confirmed capacity that can be agreed upon.

In addition, you can enter requested capacities for each origin-destination pair in the more detailed rate table items, which is depicted in Figure 10.33. The rate table lines were enhanced in SAP TM 9.1 to store requested, promised, and confirmed capacities.

Publishing carrier RFQs

Once you have successfully prepared the RFQ, but before you publish it to your vendors, SAP TM supports a standard approval workflow that you can enable in the freight agreement RFQ type; it generates a vendor RFQ document for each carrier that you included in your shortlist. Each vendor RFQ has a similar layout and user interface as the master RFQ. It inherits all RFQ line items, the TCCS for each line item, and the rate tables. Only one carrier is stored in the header of the vendor RFQ because at is unique for each vendor. The vendor RFQ document is mainly used to either upload or input the carrier responses. The fields for promised capacities and offered rate values are editable for input only in the vendor RFQ. You can use the standard SAP TM output management functionality that is enabled for RFQs to generate output files. If you want to learn more about SAP TM output management, please refer to Chapter 2, Section 2.3. In addition, you can use the rate table Excel integration to generate RFQ spreadsheets and submit to your vendors.

RFQ response

Once your vendors have replied with rates and capacities, you can use the Excel integration for rate tables to upload all results into your vendor RFQ documents. You need to perform this task once per rate table because there is no mass upload or download capability available in SAP TM. After a successful upload or manual entry, mark each agreement RFQ line item as *Responded*. Clicking on the SUBMIT button in each vendor RFQ makes the rates and capacities available in the initial master RFQ so you can analyze and compare the results. In the master freight agreement RFQ, you can see the status of each vendor RFQ in the document flow, as shown in Figure 10.35.

Comparison cockpit, spend analysis

A key functionality in the RFQ process is the comparison of the results of the vendors to help you decide on the capacities you are going to source for each vendor. In the master RFQ you can select one line item and click on the OPEN COMPARISON COCKPIT button.

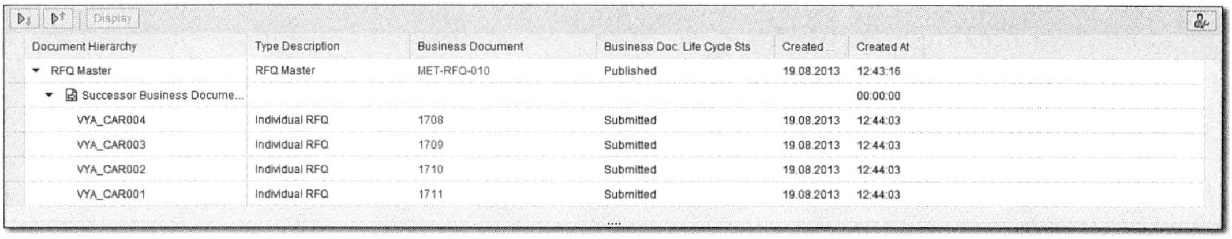

Document Hierarchy	Type Description	Business Document	Business Doc. Life Cycle Sts	Created ...	Created At
▾ RFQ Master	RFQ Master	MET-RFQ-010	Published	19.08.2013	12:43:16
▾ Successor Business Docume...					00:00:00
VYA_CAR004	Individual RFQ	1708	Submitted	19.08.2013	12:44:03
VYA_CAR003	Individual RFQ	1709	Submitted	19.08.2013	12:44:03
VYA_CAR002	Individual RFQ	1710	Submitted	19.08.2013	12:44:03
VYA_CAR001	Individual RFQ	1711	Submitted	19.08.2013	12:44:03

Figure 10.35 Document Flow in Master Freight Agreement RFQ

The comparison cockpit is divided into four sections, which are shown in Figure 10.36.

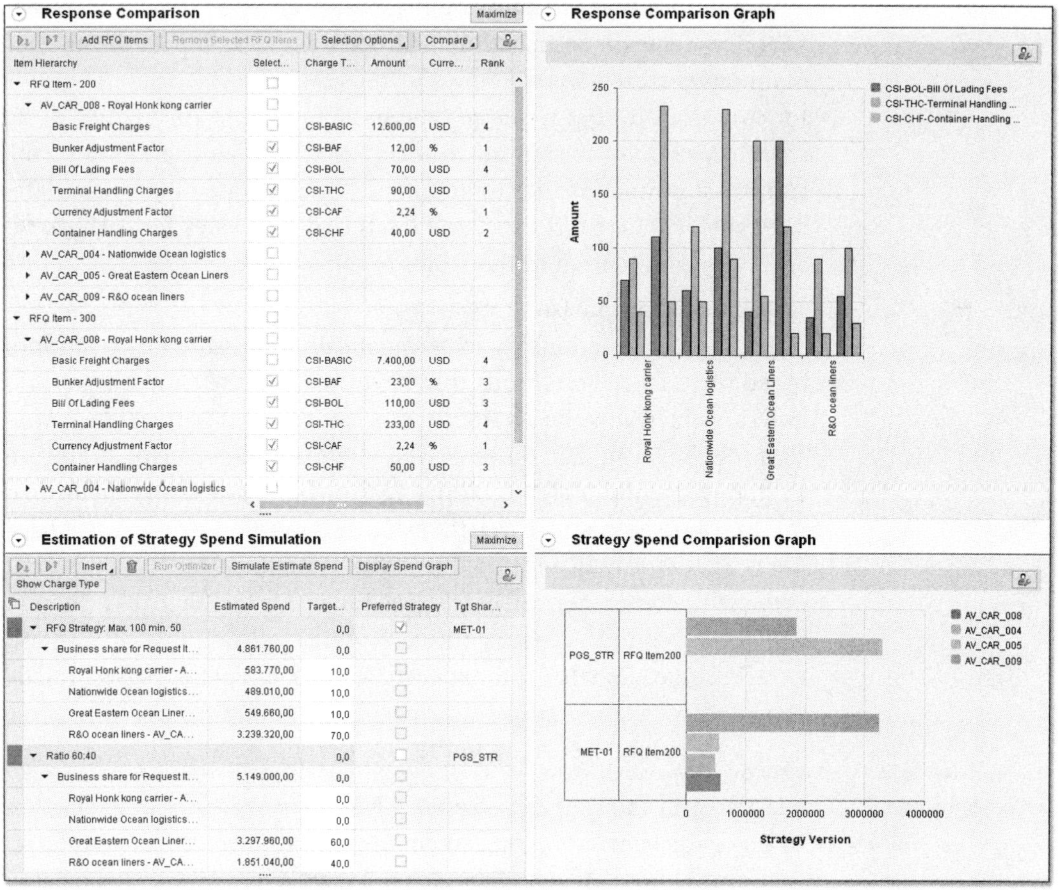

Figure 10.36 Comparison Cockpit

There are two ways to compare the vendor offers:

1. **Manual comparison**
   With manual comparison, you use the RESPONSE COMPARISON section and RESPONSE COMPARISON GRAPH to select the vendors that have replied and submitted their responses. You can see the breakdown of their charges with amounts. Alternatively, you see rate tables assigned to each charge type if those are maintained. You click on the SELECTION OPTIONS button to select multiple charges of the same type for each carrier to compare them. When you select COMPARE • RESPONSES, the graph in the right section is compiled and you can switch between multiple chart types to help you make your decision. You can also add RFQ line items for comparison. SAP TM does not support the analysis of amounts stored in rate tables or the display of them in the graphs for comparison. If any charge type has a rate table assigned to it, the charge type does not appear in the graph. You can use the rate table comparison functionality to automatically generate one consolidated rate table in Microsoft Excel and list the offered rates for each carrier against each line item in the table.
2. **Automated comparison**
   SAP offers the functionality to automatically suggest a cheapest option based on defined conditions that have to be met (e.g., a minimum of two carriers have to be awarded). You can add these conditions before running the optimizer. Alternatively, define a target share strategy in Customizing, and assign it to the agreement type via the menu path IMG • TRANSPORTATION MANAGEMENT • MASTER DATA • AGREEMENT RFQS AND QUOTATIONS • DEFINE TARGET SHARE STRATEGIES. The outcome of the optimizer run is a suggestion of a target share as a percentage of the required capacity for each vendor. When you select SIMULATE ESTIMATE SPEND, the expected cost is calculated based on the provided rates and the given capacities. You can compare multiple strategy versions displayed in the graph in the right section.

Carrier awarding

The last step in the freight agreement RFQ process is the awarding and creation of the contracts. You can choose one strategy as your preferred option and choose ACCEPT PREFERRED STRATEGY in the RFQ line item table. The AWARD SUMMARY tab is available to compare the business share between the vendors and your budget against the actual cost. Last, you can create new agreements for the awarded carriers or amend existing agreements with the new rates and capacities.

With the successful generation of a new freight agreement based on a carrier RFQ, we have now also concluded an entire walkthrough of the SAP TM Transportation Charge Management component. If you recall where we started with Chapter 10, you will see that we have reached the same point again of setting up a forwarding and freight agreement as master data. Let's take this as the chance to summarize and reflect on the big picture of how the various functionalities are interlinked with each other.

We started by setting up the comprehensive master data in SAP TM in Section 10.1.1, which contained forwarding and freight agreements, internal agreements, TCCSs, rate tables and a service product catalog. You have learned about the relationship between agreements, TCCSs, charge types and rate tables as well as the integration with a service product catalog. We pointed out that each forwarding, freight, and internal agreement contains TCCSs, which store all relevant transportation charge types. To store the actual customer and freight rates, you can use flexible rate tables.

We then introduced to you, step by step, to how the defined contract rates are being used in orders and bookings. In Section 10.2, we highlighted the agreement determination logic which is triggered from the forwarding orders or freight orders, freight bookings, and service orders. SAP TM Transportation Charge Management finds the correct agreement based on the contract determination functionality.

In Section 10.3, we dove deep into the actual logic for charge calculation. Corresponding calculation sheets with assigned rate tables are used for the actual calculation logic. We described to you the general approach and very industry-specific charge calculation capabilities and scenarios.

Last, we changed the angle from an operational to a more strategic perspective. We introduced in Section 10.4 the latest capabilities of SAP TM 9.1 surrounding procurement planning and forecasting and sales demand analysis capabilities. We provided a comprehensive overview of the selling and procurement management functionalities for both customers and vendors.

As displayed in our high level picture, once we have done our job from a rating perspective, corresponding settlement documents need to be generated and processed to generate customer bills and perform carrier invoice verifications. This will be the focus of Chapter 11.

This chapter introduces the basics of charge settlement, integration with SAP ERP billing and invoicing, and an alternative integration with SAP Convergent Invoicing.

11 Charge Settlement

Our last chapter introduced to you the capabilities of SAP Transportation Management (TM) charge calculation. This gives you the ideal background to now dive into the process of charge settlement.

Settling charges comes with three different concepts that we want to differentiate in this chapter:

1. **Settlement of supplier freight services**
   Regardless of whether you work as a shipper or freight-forwarder, you will always procure transportation capacities with incurring costs. Both shippers and logistics service providers (LSPs) procure transportation capacities from airlines, ocean liners, railways, and trucking companies. Even for carriers, it is common to procure complementary transportation services (e.g., trucking or railway services). Similarly, costs can occur when you work with agents or alliance partners from a carrier perspective and with freight-forwarders as shippers.
2. **Settlement of customer charges**
   The settlement of customer charges is of significant relevance for LSPs and carriers (such as railways, trucking companies, and ocean liners). Billing customers for the provided transportation services is part of the core business model. This can be a very straightforward process or become increasingly complex. The settlement of charges is very dependent on the incoterms of a shipment. The incoterms define whether freight charges need to be paid by the shipper or the consignee. Consequently, different settlement documents and invoices are sent to the involved shipper(s) to pay for prepaid charges, and the involved consignee(s) to pay the collect charges, according to the

agreed incoterms. Certain customers have preferences about how to split and group invoices, which can become a challenge for logistics companies. The logistics departments of shippers can benefit from the same functionality assuming that they might act as an LSP/carrier to external customers. But as a shipper, your logistics department is probably a cost center and does not bill customers.

3. **Internal settlement**
   A shipper's logistics department might bill other internal departments—let's call them internal customers—for transportation services. For freight-forwarders and trucking companies, this is a very common business practice. LSPs usually organize themselves in such a fashion to increase operational efficiency. They have gateways and hubs which are responsible for the optimization of costs for procurement and generating a profit through consolidation services. In such situations, LSPs tend to operate as internal service providers for their own business units. Incurred cost and profit is commonly settled/shared among the organizations. A selling branch/booking office of an LSP might sell transportation services to a customer, but multiple hubs and gateways are responsible for procuring and providing capacities along the global supply chain.

Integration with SAP ERP SD and MM

Before we describe the different sections, let's clarify the overarching principle and commonality across the three ways of settling charges. If you want to use this capability of SAP TM, you have to use an SAP ERP system as the back-end system, regardless of whether you're working as a shipper with order integration or as an LSP or carrier with an (almost) standalone SAP TM system. The standard SAP TM solution uses the existing capabilities of the SAP ERP Sales and Distribution (SD) billing application and SAP ERP Materials Management (MM) for invoicing. In either case, it is always required to calculate freight charges based on SAP TM's charge calculation functionality before triggering a settlement.

Integration with SAP ERP Convergent Invoicing

We don't want to miss the chance to introduce to you an alternative approach to implementing your charge settlement process for billing. In major customer implementations, especially for LSPs and carriers, a different architecture has been chosen. Instead of using SAP ERP SD for the billing process, SAP Custom Development has developed integration with SAP Convergent Invoicing. This application, which is part of SAP

Billing and Revenue Innovation Management (BRIM), originated from the telecommunications industry. In the telecommunications market, it is common for providers to process enormous amounts of billing information based on very small units, such as minutes of telephone calls, that all have to be invoiced based on different tariffs. This flexibility is enabled by Contract Accounts Receivable and Payable (FI-CA) and consequently can be used in transportation processes to split invoices based on very specific customer requirements.

11.1 Charge Settlement and Billing

Architecture

We have highlighted the resemblance of the settlement process for billing to invoice settlement. For both billing and invoicing, a standard integration with a back-end SAP ERP system is required. You can find more commonalities when you look at the architecture. Figure 11.1 illustrates the standard integration and flow of documents between SAP TM and SAP ERP, with a focus on SD and MM.

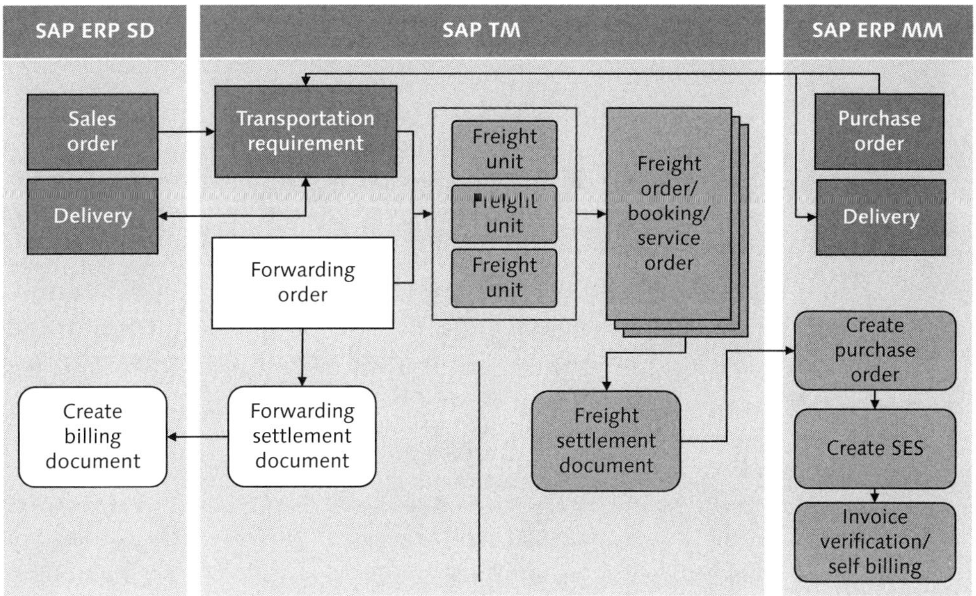

Figure 11.1 Integration of SAP TM to SD and MM for Charge Settlement

The diagram is applicable for both shippers and LSPs or carriers. The light boxes are the documents you use as a shipper: your transportation requirements created based on SAP ERP shipments and orders. The white boxes show the forwarding order document that is relevant for you as a freight-forwarder or carrier. (Recall both options from Chapter 4.) The dark boxes are the common documents that are relevant for both freight-forwarders and shippers and that are required for the settlement process. For both the billing of freight charges to your customers and the settlement of procured services, you generate an SAP TM *settlement document*. The settlement document contains all billing- and invoicing-relevant information, such as the invoicing parties, calculated charges, and currencies. It can be considered as a draft invoice/bill.

Two different business objects differentiate between the customer settlement and the invoicing to service providers: the forwarding settlement document and the freight settlement document. For the billing side, the forwarding settlement document directly generates a billing invoice in SAP ERP SD. The freight settlement document triggers the creation of a purchase order in SAP ERP MM, and a *service entry sheet* (SES) is created. Both documents are generated in the background to eventually enable posting of accruals, invoice verification, or self-billing.

Section 11.1.1 to Section 11.1.3 introduce the settlement of customer charges based on the SAP TM forwarding settlement document and the integration with SAP ERP SD. In Section 11.1.4 you learn about industry-specific scenarios for customer settlement. From Section 11.1.5 to Section 11.1.7 we dive into the opposite side, freight settlement and invoicing via SAP ERP MM. We also highlight industry-specific freight settlement scenarios in Section 11.1.7. Last, in Section 11.1.8, we introduce SAP TM *credit memos* for corrections of settled amounts.

11.1.1 Creating Forwarding Settlement Documents

The forwarding settlement process is highly integrated with and dependent on the way we calculate the charges in the forwarding order. You will quickly see that a multitude of scenarios and requirements have come from customers concerning how their invoices are created and structured. As a result, SAP TM supports a variety of functionality to serve specific scenarios. We introduce to you step by step the different scenarios, including both the business background and the most important Customizing settings.

Incoterm-based Forwarding Settlement Documents

Our process starts with creating the forwarding settlement document, assuming that we have already generated and executed a forwarding order. There are generally three ways to generate a forwarding settlement document:

1. Navigate to a forwarding order and select FOLLOW UP • CREATE FORWARDING SETTLEMENT DOCUMENT from the top navigation panel.
2. Run a batch report, which periodically generates forwarding settlement documents.
3. Collectively or individually generate forwarding settlement documents from a personal object worklist (POWL) by selecting CREATE FORWARDING SETTLEMENT DOCUMENT • INDIVIDUAL or COLLECTIVE.

Mass Creation of Forwarding Settlement Documents

SAP TM offers a standard report that you can run manually or automatically in the background. In SAP NetWeaver Business Client (NWBC), follow the menu path APPLICATION ADMINISTRATION • BACKGROUND REPORTS • CREATE FORWARDING SETTLEMENT DOCUMENTS. Alternatively, navigate to Transaction SE38 in SAP GUI and run report /SCMTMS/CFIR_CREATE_BATCH.

Incoterm-based settlement

The first question to answer is how many forwarding settlement documents are generated when the creation is triggered. For a freight-forwarder, customer invoices are generated based on global incoterms. From Chapter 4 you learned how to maintain and use incoterms in a forwarding order. Then, in Chapter 10 you learned that incoterms can significantly influence the way we calculate charges. Now we show you how incoterms influence the generation of forwarding settlement documents. Figure 11.2 shows an example of a settlement process based on incoterms.

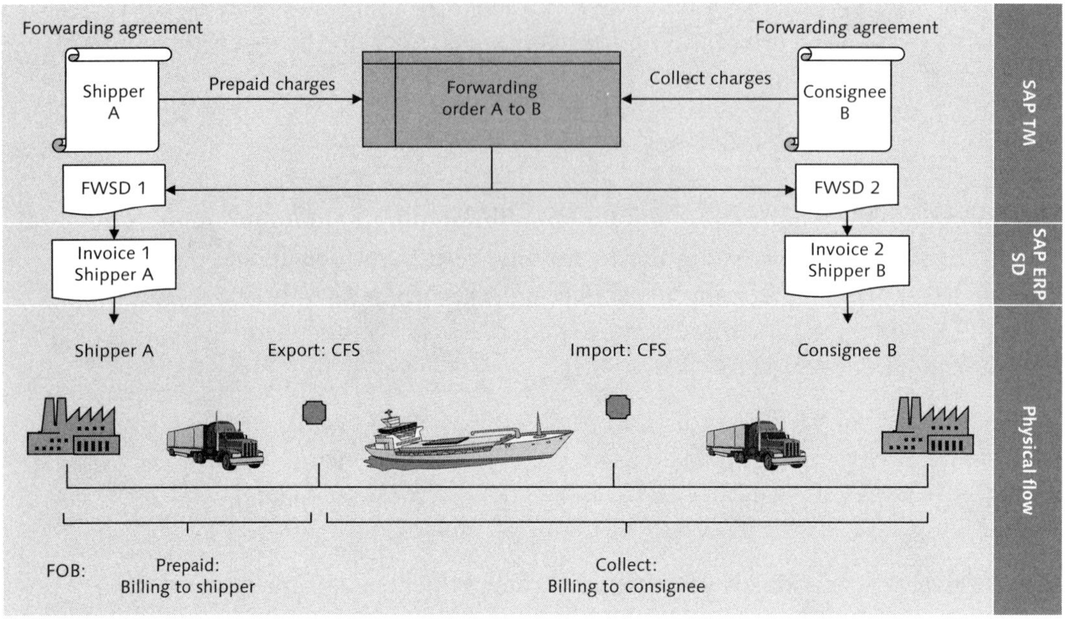

Figure 11.2 Settlement Process Based on Incoterms

Selection of settlement party

Let's assume you have a forwarding order with the incoterm free-on-board. The shipper is responsible for paying all charges until the port of loading, including the export customs declaration. The consignee pays all charges from the port of loading until the final destination, including insurance and import customs. We have two forwarding agreements, one with the shipper and one with the consignee. Assuming we have made all the required settings, you should see the following system behavior after navigating to the forwarding order and selecting FOLLOW UP • CREATE FORWARDING SETTLEMENT DOCUMENT. The system prompts you with a user selection to define which forwarding settlement document to create. This is shown in Figure 11.3.

The system recognizes that two forwarding settlement documents will be created based on our configuration settings. The first business partner represents the prepaid party, as maintained in the BUSINESS PARTNER tab in the forwarding order. The second business partner is the collect party that has to pay the charges from the port of loading. You now choose whether to create just one settlement document for either business partner or both at once. From the forwarding order, a minimum of two settlement documents are generated in total because both the consignee and the shipper pay a proportion of the overall charges. The first forwarding settlement document contains the charges of the shipper who is responsible for paying the prepaid charges from origin to port of loading. The charges are calculated for each transportation stage.

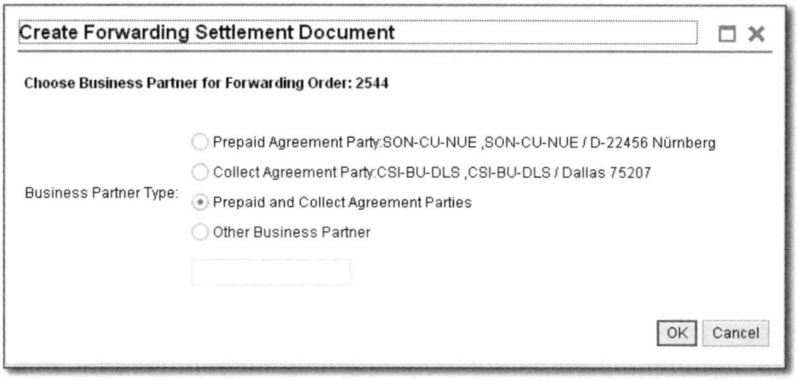

Figure 11.3 User Selection When Creating a Forwarding Settlement Document

The second forwarding settlement document contains the collect charges to be paid by the consignee. The invoicing status in the forwarding order is updated to *Partially Invoiced* if not all settlement documents have been generated. The status changes to *Invoiced* after all forwarding settlement documents have been generated. After the successful creation of multiple forwarding settlement documents based on incoterms, you see an item table of forwarding settlement documents that contains all the settlement documents that were created. You can now select each document to drill down into the details, which we cover in Section 11.1.2.

Note

It is possible to manually create additional forwarding settlement documents by selecting Other Business Partner. This functionality has been developed for exceptional and unplanned cases only, where a user wants to bill a customer manually for charges that weren't retrieved from the customer's forwarding agreement. These additional settlement documents do not impact the invoicing status of a forwarding order.

Import/export-based settlement

It is important to mention that in an export/import scenario, the settlement of customer charges can be based on two forwarding orders for one shipment: the export and import forwarding orders. As you learned in Chapter 4, Section 4.2, SAP TM can generate an export and an import forwarding order for international shipments. As a result, the export organization of a company in the exporting country might only be responsible for settling the prepaid charges to a shipper. The responsible importing organization consequently settles the collect charges with the consignee. Depending on the incoterm, you generate one forwarding settlement document to the shipper from the export forwarding order and a second forwarding settlement document to the consignee from the importing forwarding order. In addition, you can cross-charge the incurred transportation cost between the export and import organizations for an equal split. We detail this scenario of internal settlements in Section 11.2.2.

Settlement documents for canceled orders

If a customer cancels a forwarding order, it is still possible that the costs occurred and consequently the charges need to be applied. You can create a forwarding settlement document for forwarding orders that are in the *Canceled* lifecycle status. You have to create the forwarding settlement document manually. It is not yet possible with the standard SAP TM installation to calculate the charges automatically for canceled orders. As a result you have to manually add them in the settlement document, which is created as an empty shell. Once you have entered the charge types and actual amounts manually, you can follow the standard settlement process flow.

Basic Customizing in SAP TM

Let's take a look at the basic settings that are required to achieve this system behavior. For the creation of forwarding settlement documents, you need to take a few basic Customizing steps in the forwarding order management and charge calculation areas. You need to have defined the incoterms in SAP SCM Basis and maintained the default agreement party roles for stages. The forwarding order you created stores the actual incoterm. In addition, you need to have defined the resolution base as STAGE in your calculation profile (see Chapter 10).

The following Customizing settings are required in charge settlement:

1. **Settlement document type**
   When using SAP TM settlement functionality, you must define at least one forwarding settlement document type in the Customizing and assign it to your forwarding order type by following the menu path SAP TRANSPORTATION MANAGEMENT • TRANSPORTATION MANAGEMENT • SETTLEMENT • FORWARDING SETTLEMENT • DEFINE FORWARDING SETTLEMENT DOCUMENT TYPES. The forwarding settlement document type specifies the number range, output profile, and multiple default settings.
2. **Settlement profile**
   Second, you need to maintain a settlement profile via the IMG menu path SAP TRANSPORTATION MANAGEMENT • TRANSPORTATION MANAGEMENT • SETTLEMENT • DEFINE SETTLEMENT PROFILE. Figure 11.4 shows a standard settlement profile in SAP TM. The settlement profile is the key configuration that influences the behavior of a forwarding settlement document. A settlement profile can be directly assigned to a business partner. Alternatively, it can be assigned to the newly introduced charges profile that is mapped to an organizational unit in SAP TM.

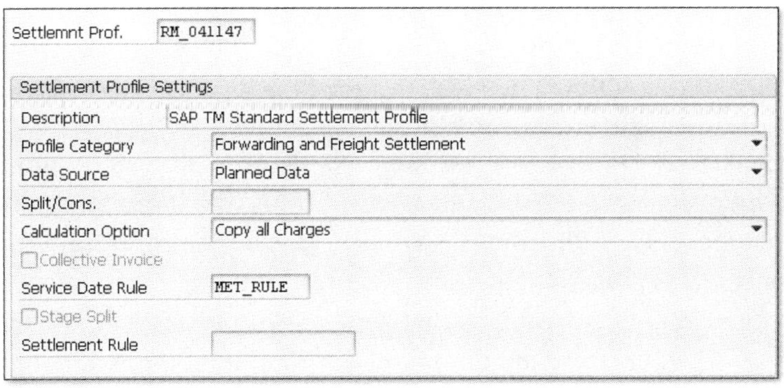

Figure 11.4 Settlement Profile

In the following Customizing path you need to assign your settlement profile to a charges profile: TRANSPORTATION MANAGEMENT • BASIC FUNCTIONS • CHARGE CALCULATION • BASIC SETTINGS FOR CHARGE CALCULATION • DEFINE CHARGES PROFILE. If the settlement profile is assigned to both the business partner and the charges profile, the priority lies with the business partner. The settlement profile is used for both freight and forwarding order settlement documents. An additional capability of the settlement profile is the STAGE SPLIT checkbox. Alternatively to the described incoterm scenarios, selecting this checkbox causes that a forwarding order generates one settlement document per leg for the entire transportation process, independent of the incoterm and involved parties.

3. **Process controller**
   Third, you must set up the process controller for the purpose of creating settlement documents. SAP TM offers two standard methods for the settlement process: (1) Split the forwarding order based on incoterms to generate multiple forwarding settlement documents, and (2) consolidate multiple forwarding orders to generate one settlement document. In addition to the TM standard capabilities, you can develop your own process controller methods and assign them to a strategy that will be used in the charge settlement process and steer the creation process of a forwarding settlement document. Custom logic will not replace the standard methods.

Collective Creation of Forwarding Settlement Documents

In addition to the generation of settlement documents based on incoterms, which allows you to split a forwarding order and generate multiple settlement documents, you can also create one forwarding settlement document across multiple forwarding orders. You can select multiple forwarding orders in a POWL and create a collective forwarding settlement document. One document is created for your selection as long as the basic settlement data is the same. You can configure your own strategies to determine the attributes that have to be similar across forwarding orders to generate a common forwarding settlement document. Examples of such attributes are bill-to party, payer, ordering party, payment term, credit segment, and source or destination location. To activate this grouping logic, you do not need to assign a split/consolidation strategy in the settlement profiles. A default logic is available in standard SAP TM. A customer specific logic can be optionally assigned to the settlement profile which runs in addition to the default in standard TM (see Figure 11.4).

Trailer- and Route-based Settlement

Trailer-based settlement

Billing the customer based on the distance traveled between origin and destination is a common business practice in the trucking industry. Recall from Chapter 5 and Chapter 6 that a trailer can be used to pick up multiple forwarding orders across different origins and drop the cargo off at multiple destinations. In this case, you can invoice one bill to your customer for multiple forwarding orders that are transported on the same trailer. This gives you an alternative way of generating one forwarding settlement document for multiple forwarding orders.

A key aspect of the trailer-based settlement is that the charge calculation logic is different from a regular rating scenario. The charges are calculated for the entire group of forwarding orders in a trailer, and not per individual order. They are treated as one virtual forwarding order document and the consolidated weights, volumes, and so on are used for the charge calculation. (Refer to Chapter 10 for more details on trailer-based rating.) Consequently, you can use a trailer-based settlement process to generate one settlement document per trailer document per customer. For more details of the trailer document, see Chapter 6, Section 6.2.

In cases where the trailer contains forwarding orders from multiple customers, multiple forwarding settlement documents are created. The settlement document has a slightly different layout of the ORDERS section than regular forwarding settlement document, as shown in Section 11.1.2. You can see the trailer document used for the settlement creation as a line item. Select the line item to see the forwarding orders that were loaded on the trailer as the line item details. These scenarios are also referred to as execution-based settlement because the cargo execution information plays a vital role in the settlement process.

Settings

To enable trailer-based charge calculation and creation of settlement documents, one additional setting is required: The trailer-based settlement indicator needs to be selected in the line item details of your forwarding agreement that you want to use to calculate the charges of the corresponding business partner. This will automatically serve as the main item for charge calculation; use of a specific resolution base isn't required.

Note

For a trailer-based settlement process, you can only use the calculation levels header and stage as possible settings for the charge determination. From Chapter 10 you know that the calculation level, which is maintained on the calculation profile, defines how many forwarding agreements are used when resolving the charges. The calculation level item would not make sense in this business context because the system would try to retrieve one forwarding agreement per item (e.g., per package of an order).

In the trailer scenario, the opposite is the case. We want to collectively settle all the packages from one customer that are loaded on the same trailer, based on the distance a trailer has traveled. Consequently, there is no need to use a different forwarding agreement for one package or another because the agreed-upon rate based on the distance would be stored in one forwarding agreement of the business partner only.

Freight order-based settlement

Another functionality of SAP TM is to generate forwarding settlement documents based on the freight order they are loaded to (which is called the route-based settlement). It is similarly targeted at trucking companies. SAP TM allows you to generate a forwarding settlement document for all forwarding orders on the same freight order. The distinctive feature we want to highlight here is that you can use your freight document (and the data contained) as the basis for the settlement to your customers. This bypasses the individual settlement of forwarding orders, similar to a trailer-based settlement. For example, you could combine the route-based settlement with trailer-based settlement in a scenario where a forwarding order is transported on a first stage in a freight order without trailer assignment and for a second leg in a trailer object. Consequently, two forwarding settlement documents are generated per customer—one for the stage of the freight order and one for the stage of the trailer. This functionality can be activated by selecting the route-based settlement basis in the line item details of your forwarding agreement.

11.1.2 Structure of Forwarding Settlement Documents

The forwarding settlement document contains all billing-relevant information inherited from a forwarding order. Think back to Chapter 4, Section 4.2 about forwarding order management and to Chapter 10 about charge management; a lot of the fields and information will already be familiar to you. Therefore, now we want to highlight the most crucial information with respect to the billing process.

General data

The GENERAL DATA tab contains the type of your settlement document, which determines, among other things, the screen design (Web Dynpro application configuration, as described in Chapter 2, Section 2.2). In addition, this tab contains the sales organization, which is determined based on the sales organization from the forwarding order. The incoterms are available based on the forwarding order as well. They can have a major impact on the creation of the settlement document. There is also a section listing the payment terms from the customer's forwarding agreement. Payment terms can be maintained in Customizing of SAP SCM Basis via the menu path MASTER DATA • BUSINESS PARTNER • DEFINE TERMS OF PAYMENT. In addition, there is the invoice date. You can manually change some information in the forwarding settlement document, such as the invoice date and payment terms.

Business partner

To be able to issue a bill to your customers, maintain at least the bill-to party and the payer. Both business partner types are defaulted from the forwarding order. You can use business partner determination rules to automatically determine which parties will be maintained for which party role. For more details on how to maintain and use business partner determination profiles, please refer to Chapter 4, Section 4.2. You can overwrite the address of a bill-to party in the forwarding settlement document if you want to send the invoice to a different address. In this case, the settlement and invoice creation in SAP ERP considers the different address to generate the billing document.

Orders

The forwarding settlement document is created based on actual forwarding orders. As a result, you can always see which forwarding orders were settled in a forwarding settlement document. You might only see one forwarding order assigned to the forwarding settlement document, but remember that we described how to collectively settle multiple forwarding orders in one settlement document. Therefore, the ORDERS tab provides you with the list of all forwarding orders you grouped into a settlement document. For a trailer-based or freight order-based settlement, you will see the freight order(s) or trailer document(s) in the ORDERS tab, containing the forwarding orders as line items. If you resolve the charges based on the calculation level of forwarding order items, you can see the list of forwarding order items in the ORDERS tab.

Service date

In the ORDERS tab you might notice the *service date;* this can be different from order to order. When settling the charges to a customer, you are probably confronted with a legal taxation requirement to only activate your revenues when a service has actually been rendered. SAP TM supports this requirement with the service date. It allows you to set up flexible rules to define at what time you consider the service of a forwarding order fulfilled. The date is transferred with your settlement document to the SAP ERP system for the actual billing (mapped to the *service rendered date* in SD billing). You can define such rules in the SAP TM Customizing by following the menu path SAP TRANSPORTATION MANAGEMENT • TRANSPORTATION MANAGEMENT • SETTLEMENT • DEFINE SERVICE DATE RULES AND RULE PRIORITIZATION.

SAP offers you the choice between multiple transportation-related dates that are stored in your forwarding order and even in the freight order (e.g., expected start date of first stage main carriage or actual end date of loading). You can maintain multiple dates that are considered as the service date and assign priorities between them if a date is not available in a forwarding order. You can assign the service date rule to a settlement profile. It will be converted to a different time zone. For further flexibility, a BAdI is offered to plug in your own service date rule (`/SCMTMS/BADI_FCP_SRV_DATE` of enhancement spot `/SCMTMS/ES_FCP`).

Charges

The CHARGES tab contains the core information on the forwarding settlement document, which gives you the comprehensive overview of all charges that need to be billed to a business partner. All charges are retrieved from your forwarding order. Which charge items actually appear in the CHARGES tab depends on how you have configured the charge calculation and settlement process.

You have the same ability to switch the view of the charges and display them grouped or ungrouped, similar to the charge calculation of the forwarding order (described in Chapter 10). If you use a collective forwarding settlement document to settle multiple forwarding orders, then the charges would appear as a sum for each charge type across all forwarding orders. In a standard settlement scenario based on incoterms, the charges are calculated per stage. As a result, you see a breakdown of the charges per stage in each settlement document (for both the shipper and consignee). You can see two forwarding settlement documents that were created to bill the charges to the prepaid and collect party, based on incoterms in Figure 11.5.

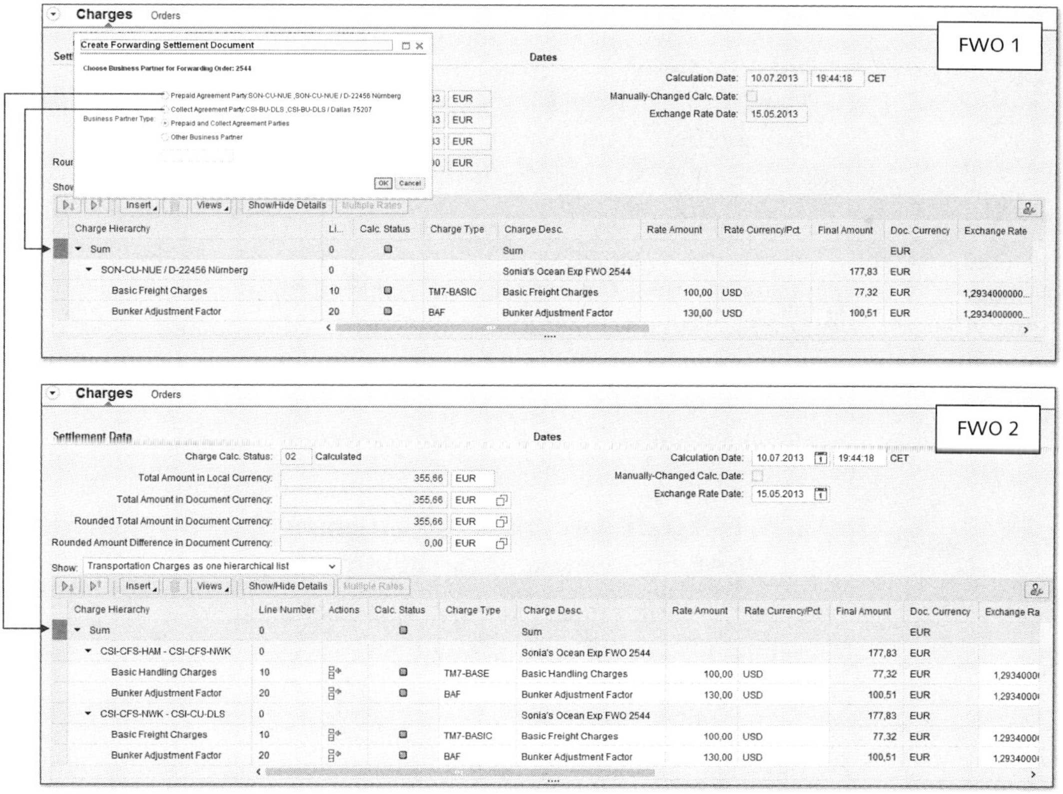

Figure 11.5 Charges Tab for Prepaid and Collect Forwarding Settlement Document

The first settlement document is for the shipper—the prepaid party. In our example, this is a customer in Nürnberg, Germany. The second settlement document (below the first one) is for the consignee (the collect party) who is a customer in Dallas, Texas, the United States.

Statuses

After successful generation of a forwarding settlement document, the actual billing process and the required replication to SAP ERP SD still have to be accomplished. The STATUSES tab gives a comprehensive overview of the various process steps from creation of a forwarding settlement document until the final billing in SAP ERP SD. There are two statuses that ensure visibility of the progress of the entire settlement process. We provide you with example values of both statuses as we describe the integration to SAP ERP:

1. **Lifecycle status**
   This status describes what step was performed last, along the entire billing process. After the initial creation of a new forwarding settlement document, the status is set to *In Process*.
2. **Confirmation status**
   Based on the integration between SAP TM and SAP ERP, multiple activities are performed when transmitting a settlement document to SAP ERP. Hence, a separate status is provided that meticulously tracks the interaction between the two systems. The initial status after the creation of the forwarding settlement document would be *Not Yet Invoiced* in SAP ERP.

11.1.3 Integrating Forwarding Settlement Documents with SAP ERP Sales and Distribution

You have learned throughout this chapter that SAP TM uses the SAP ERP SD billing application to invoice customers. In Section 11.3 we highlight an alternative way of invoicing your customers by integrating SAP TM with SAP Billing and Revenue Innovation Management.

For now, let's dive into the details of the SAP TM-SAP ERP SD integration for billing.

Basic Customizing in SAP ERP

To enable the SAP TM-SAP ERP integration, you have to customize your SD billing system. Moreover, you are required to set up mapping rules between SAP TM and SD billing. Let's start with the essential settings in SAP ERP SD billing:

SD condition types

1. As a first step in SD billing, you need to maintain *condition types*. Condition types are used to steer the logic of how the charges for a sales document are retrieved and calculated. In the integration to SAP TM, you have to define a condition type for each charge type in SAP TM. In SAP ERP Customizing, follow the menu path SALES AND DISTRIBUTION • BASIC FUNCTIONS • PRICING • PRICING CONTROL • DEFINE CONDITION TYPES • MAINTAIN CONDITION TYPES.

Pricing procedures

2. You must create a *pricing procedure*. You are required to assign the condition types to the pricing procedure that you have just defined. The pricing procedure is responsible for defining which condition types are used when creating an invoice. You can maintain hundreds of different condition types, but when generating a billing document only the ones that were assigned to a pricing procedure will be picked up. You can logically group the condition types in a pricing procedure (e.g., by origin port charges, sea freight charges, and destination port charges in an ocean freight scenario). You can maintain multiple pricing procedures because they are assigned for each sales organization or customer. In SAP ERP Customizing, follow the menu path SALES AND DISTRIBUTION • BASIC FUNCTIONS • PRICING • PRICING CONTROL • DEFINE AND ASSIGN PRICING PROCEDURES • MAINTAIN PRICING PROCEDURES.
3. Last, you need to assign your pricing procedure to a sales organization. A sales organization can be assigned multiple pricing sequences. Navigate to SAP ERP Customizing by selecting SALES AND DISTRIBUTION • BASIC FUNCTIONS • PRICING • PRICING CONTROL • DEFINE AND ASSIGN PRICING PROCEDURES • DEFINE PRICING PROCEDURE DETERMINATION.

Now let's focus on the mapping between SAP TM and SAP ERP SD billing:

Mapping

1. It is essential in SAP ERP SD to map the charge types from your SAP TM system to the SAP ERP SD condition types you created earlier.

Before you can do the mapping, you need to create the actual charge types in SAP ERP, similarly to the defined charge types in SAP TM. Hardly any information is stored on the charge types in SAP ERP. Navigate to the following Customizing menu path in SAP ERP to perform the settings: INTEGRATION WITH OTHER SAP COMPONENTS • TRANSPORTATION MANAGEMENT • INVOICE INTEGRATION • BILLING • DEFINITION FOR TRANSPORTATION CHARGE ELEMENTS • DEFINE CHARGE TYPES.

2. Next, navigate to the following Customizing menu path in SAP ERP to perform the actual mapping of the SAP charge types you have now created in SAP EPR and the SAP ERP condition types: INTEGRATION WITH OTHER SAP COMPONENTS • TRANSPORTATION MANAGEMENT • INVOICE INTEGRATION • BILLING • ASSIGNMENT OF TRANSPORTATION CHARGE ELEMENTS • ASSIGN CONDITION TYPES.
3. Optionally, you can set up categories and subcategories, much like the SAP TM Customizing. Categories are mostly used for reporting purposes, and they need to be assigned to charge types. You can find this setting via the menu path INTEGRATION WITH OTHER SAP COMPONENTS • TRANSPORTATION MANAGEMENT • INVOICE INTEGRATION • BILLING • DEFINITION FOR TRANSPORTATION CHARGE ELEMENTS • DEFINE CATEGORY CODES and DEFINE SUBCATEGORY CODES.

Mapping of organizational units

4. In addition to the mapping of charge types, you need to map your sales organizations as defined in SAP TM to sales organizations in SAP ERP. Here you define the following settings for each SAP TM sales organization:
   - Logical SAP TM system
   - Mapped SAP ERP sales organization, division, and distribution channel
   - Billing and sales document types
   - Pricing procedure

Functional Process Flow

Billing simulation

After you have successfully created the forwarding settlement document in your SAP TM system, you can execute a billing simulation in SAP TM before transferring it to SAP ERP SD. The charm of this capability is that you combine the logistical information from SAP TM with the tax data available in SAP ERP. Click on the PREVIEW INVOICE button in the top panel of the forwarding settlement document. SAP TM transfers the relevant logistical information to SAP ERP. SAP ERP SD generates a PDF output file that you can view from SAP TM for each forwarding settlement document. This synchronous processing is not visible to users. When selecting the preview functionality, the actual PDF output file will be visible immediately.

Configuration of billing simulation

To enable this feature, you need to set up the corresponding output types in SAP ERP Customizing via the menu path SALES AND DISTRIBUTION • BASIC FUNCTIONS • OUTPUT CONTROL • OUTPUT DETERMINATION • OUTPUT DETERMINATION USING THE CONDITION TYPE TECHNIQUE • MAINTAIN OUTPUT DETERMINATION FOR BILLING DOCUMENTS • MAINTAIN OUTPUT TYPES.

Because the logic is active and implemented in standard SAP TM, it does not require a particular Customizing. It is possible for you to enhance the logic (e.g., in case you have additional SAP TM fields that need to be sent to ERP). A provided BAdI on both the SAP ERP and SAP TM sides is available for such enhancements. On the SAP ERP side, use the enhancement `TM_BIL_IMPL01` for enhancement spot `BADI_SD_BIL_PRINT01`. In SAP TM, the enhancement `/SCMTMS/BADI_FWSD_PRINT_SIM` needs to be implemented for enhancement spot `/SCMTMS/ES_FWSD`. This is the same enhancement spot that is used for the regular printing of SD invoices.

Once the forwarding settlement document is consistent and ready to be transferred to SAP ERP, you can click on the SAVE AND TRANSFER button in the top panel of the forwarding settlement document. Alternatively, you mark multiple forwarding settlement documents and execute the action from a POWL. You can also use a batch job to transfer the data to SAP ERP.

Mass Transfer of Forwarding Settlement Documents

SAP TM includes a standard report to mass create forwarding settlement documents. You can use the same report to transfer your created forwarding settlement documents to SAP ERP SD billing en masse. In SAP NWBC, follow the menu path APPLICATION ADMINISTRATION • BACKGROUND REPORTS • TRANSFER FORWARDING SETTLEMENT DOCUMENTS. Alternatively, navigate to Transaction SE38 in SAP GUI and run report /SCMTMS/CFIR_CREATE_BATCH.

Settlement with stage calculation

A billing document is created directly in SAP ERP. Depending on your charge calculation level in SAP TM, the corresponding billing items are generated in SAP ERP. If you did a header calculation of charges, the corresponding billing document in SAP ERP contains one billing item. If you calculated the charges per transportation stage, one billing item is created per stage.

Integration process and statuses

In Figure 11.6, you can see the overview of the most important statuses of a forwarding settlement document. After a forwarding settlement document has been created and saved in SAP TM in step ❶, the lifecycle status is set to *In Process*. As we have mentioned, it is possible to change certain information in the forwarding settlement document—such as an alternative address for a business partner—without causing inconsistencies. If you change billing-relevant information in the forwarding order, the two documents will not be consistent anymore. This might happen if, for example, the route of a forwarding order needs to be changed due to a missed cutoff, such as a missed aircraft or vessel departure, and it results in a recalculation of charges.

Only if the document is consistent will the lifecycle status be updated to *Ready for Invoicing in SAP ERP* ❷. Without consistency, you cannot trigger the transfer to SAP ERP, and the forwarding settlement document will be marked with an INCONSISTENT flag in the STATUSES tab. As soon as you trigger the transfer of the forwarding settlement document to SAP ERP in step ❸, the lifecycle status changes. The confirmation status shows *No Confirmation from ERP* until the interface returns a message. In the event of a failed transfer, it depends on the error handling whether a message is returned to SAP TM.

If it is possible to fix a transfer error directly in SAP ERP—say, because of locking issues or a financial period closure—then the transfer can be re-processed in SAP ERP. Until the re-processing, the lifecycle and confirmation status in SAP TM still remain as *Transferred for Invoicing to SAP ERP* and *No Confirmation from ERP* 3A. Only in the event that the error cannot be fixed 3B or if the transfer was successful 3C, both the lifecycle and confirmation status are updated in SAP TM accordingly. Similarly to the inconsistencies that occurred before transferring the forwarding settlement document to SAP ERP, you might still have to change invoicing-relevant data in the forwarding order.

Consequently, you can cancel the forwarding settlement document in both SAP TM and SAP ERP in step 4. The lifecycle and confirmation status are updated accordingly in steps 4A and 4B. It is still possible to cancel the forwarding settlement document by creating a credit memo, which we describe in more detail in Section 11.1.8.

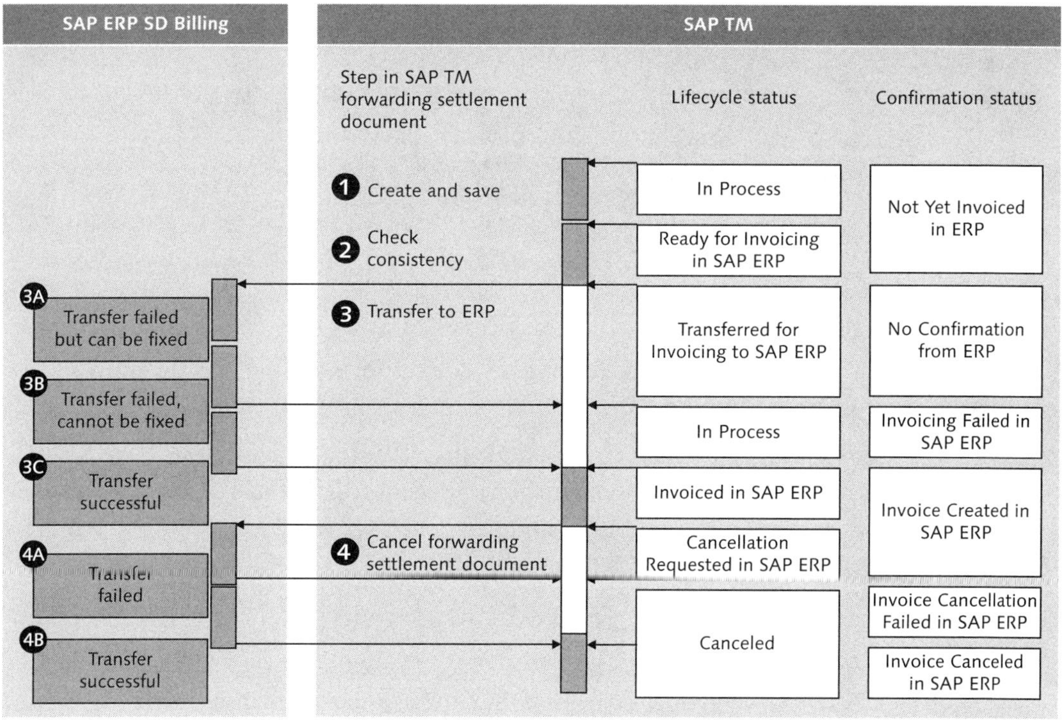

Figure 11.6 Functional Process for SAP TM-SAP ERP SD Billing Integration

Technical Integration

The transfer of a forwarding settlement document from SAP TM to SAP ERP SD is processed via the SAP NetWeaver PI. Alternatively, it is possible to set up the connection via SAP Web Services Reliable Messaging (WSRM).

Standard SOA Services for SAP TM-SAP ERP SD Integration

SAP delivers eight standard SOA services for transactional integration:

- `CustomerFreightInvoiceRequestSUITERequest_Out_V1` (and `_In`)
  Send a request to create or change a forwarding settlement from SAP TM and receive it in SAP ERP.
- `CustomerFreightInvoiceRequestSUITEConfirmation_Out_V1` (and `_In`)
  Send the confirmation and status change for customer invoices from SAP ERP and receive them in SAP TM.
- `CustomerFreightInvoiceRequestSUITECancellationRequest_Out_V1` (and `_In`)
  Send a request to cancel a forwarding settlement from SAP TM and receive it in SAP ERP.
- `CustomerFreightInvoiceRequestSUITESimulate_Out` (and `_In`)
  SAP TM offers to show a preview of an invoice in SAP TM. Send the forwarding settlement from SAP TM to SAP ERP. Receive a PDF of the SAP ERP billing document.

11.1.4 Business Scenarios in the Forwarding Settlement Process

You have learned so far about the standard capabilities of the SAP TM forwarding settlement process. Because different companies can have entirely different billing requirements, we want to highlight the specific scenarios SAP TM supports.

Prepayment Scenario

In the freight-forwarding and ocean liner industries, it is a common business practice to grant customers a certain credit limit. To minimize risks, there are cases where customers are not conceded any credits. A *cash customer* has to pay the invoice for all his prepaid charges of the transportation in advance. In such a scenario, you allow creation of a forwarding order for a cash customer but prevent it from being executed until the payment has been received. As a result, for the system to block the forwarding order, you need to indicate that the specific business partner is a cash customer.

Customizing and master data

When setting up your customer contracts, the forwarding agreements, you can maintain a payment term PRE-PAYMENT. Alternatively you can store this payment term directly in the business partner master data of your customer. In the Customizing for payment terms it is possible to define a block reason (e.g., PAYMENT NOT RECEIVED). Whenever you have a contract with a cash customer, you can assign a payment term containing the block reason PAYMENT NOT RECEIVED to the agreement or the business partner record. After a forwarding order has been created for the cash customer, you can calculate the charges. Based on the charge calculation from the forwarding agreement, the payment term is retrieved into the forwarding order and the document is automatically blocked for execution. The same behavior is invoked if you store the payment term on the business partner and as soon as you assign the business partner to the forwarding order. To still allow the settlement of this order, the creation of a forwarding settlement document is possible despite the block status of the forwarding order. As a result, you can follow the forwarding settlement process as described and create an invoice.

Unfortunately, this feature is not yet fully integrated into the SAP ERP SD billing component. The receipt of the actual payment from the customer in SAP ERP does not automatically release a blocked forwarding order. Instead, a user has to manually navigate to the BUSINESS PARTNER tab in the affected forwarding order and select the checkbox for the corresponding business partner (e.g., select PAYMENT RECEIVED for the shipper). As you learned in Chapter 4, Section 4.2.1, SAP TM is integrated with the credit limit management application. SAP TM automatically deactivates the credit limit check once the payment term is marked as PRE-PAYMENT.

Settlement for Buyer's Consolidation

Business background

A buyer's consolidation is a very common service that freight-forwarders and carriers offer to their customers. A buyer's consolidation is a scenario for an LSP, where the LSP picks up cargo from multiple origins (e.g., manufacturers in different countries). The cargo is transported to a container freight station (CFS) and consolidated into one container. From the CFS onward, the cargo travels in the same container until the final destination. It is relevant for the settlement process that a customer—in this case, a consignee—needs to be invoiced for the consolidated amount and not for each individual forwarding order. As a result, you can generate one forwarding settlement document for all forwarding orders in the same container in a freight booking. Each container in the freight booking (CARGO MANAGEMENT tab) can be marked with the value B in the CONSOLIDATION CONTAINER field. You can create the forwarding settlement document directly from the freight booking or from one of the consolidated forwarding orders. As a result, the combined charges for all forwarding orders that are consolidated in this container are calculated.

Creation of settlement documents

The actual creation of the forwarding settlement documents depends on the incoterms: If the incoterm is maintained as free-on-board, multiple forwarding settlement documents are generated for the pre-carriage legs. One forwarding settlement document per forwarding order is generated because every forwarding order has a different shipper. For both the main carriage and on-carriage, one consolidated forwarding settlement document is generated for the collect agreement party, containing all forwarding orders that were consolidated in the same container in the freight booking. The freight booking is shown as a reference in the ORDERS section. In the example of the incoterm, where the consignee pays all charges, only one forwarding settlement document is generated on the importing side. This document contains all charges for the pre-carriage based on the individual forwarding orders and the charges for main and on-carriage based on the consolidation in the freight booking.

Export versus Import Forwarding Orders

Please keep in mind that since SAP TM 9.0 you always have two forwarding orders in an international shipment: one export forwarding order and one import forwarding order. You generate all settlement documents for the shippers from the exporting side and the settlement document for the consignee from the import forwarding order. Depending on the incoterm, an exporting organization might settle its charges with the importing organization, to be compensated for any occurred cost. This is always required when the shipper does not pay for pre-carriage and main carriage.

We provide a deeper insight into the settlement process between exporting and importing organizations in Section 11.2.2 for internal settlements.

Customizing

The charge calculation follows a specific logic to find the correct forwarding agreement item to retrieve the charges for the forwarding orders:

1. Set the calculation level to STAGES so that the consolidated charges are only calculated for distances the cargo actually traveled together in the container.
2. You can maintain an own forwarding agreement item and select the BUYERS CONSOLIDATION checkbox for it. In the creation of the settlement document, the charges are picked up from this item in the customer's forwarding agreement for all stages that were transported in a consolidated container.
3. If no such item is maintained, SAP TM retrieves the rates from an agreement item with shipping type full container load (FCL) or full truck load (FTL) but never less than truckload (LTL) or less than container load (LCL).

Sales organization for buyer's consolidation

You also need to follow a specific process in the sales organization. Each forwarding order that has been consolidated into the buyer's consolidation might have been captured by a different sales organization. If all sales organizations across the forwarding orders are equal, the buyer's consolidation functionality can be used. If the organizations in the forwarding orders deviate, SAP TM uses the purchasing organization of the consolidated freight booking. It is crucial that this organization is maintained as a forwarding house so it can be used as a sales organization as well.

One forwarding settlement document is generated for all forwarding orders. The ORDERS tab in the forwarding settlement document lists the freight booking(s) where the buyer's consolidation has been marked and that contain the corresponding freight units. In the DETAILS section the container item is displayed together with the forwarding orders.

Settlement for Shipper's Consolidation

Creation of settlement documents

We can apply the concepts introduced in the buyer's consolidation scenario to a shipper's consolidation scenario.

In this scenario, a shipper sends consolidated goods in a full container, for both pre-carriage and main carriage. In the importing country, the container is deconsolidated, and the cargo items are separately transported to multiple consignees. Depending on the incoterms, SAP TM generates one settlement document to group all pre-carriage and main carriage charges in one forwarding settlement document that is charged to the shipper. The charges for the delivery legs are only included in the same forwarding settlement document in the case of an incoterm, where the shipper pays for all charges. Otherwise, separate invoices are generated for each delivery leg to settle with each consignee separately.

Settings

Whether multiple forwarding orders are shipped consolidated as a shipper's consolidation is determined by SAP TM. You have to select the container in the CARGO MANAGEMENT tab of a freight booking to mark the particular container as a shipper's consolidation. This triggers the SC flag to be set in each forwarding order, which can also be done manually. The rest of the settings are similar to the buyer's consolidation. For example, you must mark at least one agreement line item as a shipper's consolidation item.

Flexible Invoicing

We have introduced to you various ways to generate a forwarding settlement document. In the latest release of SAP TM 9.1, you have even more flexibility in how to generate the settlement documents. This capability strongly supports LSPs and, in particular, rail carriers. The flexible invoicing allows you to cluster your relevant charge types into *settlement groups*. An example is that a railway might decide to invoice all charges from the main transportation in one invoice, all accessorial services of an order in a second invoice, and finally, any peripheral charges in a third invoice. Once you create your settlement document, an enhanced pop-up opens. You can choose for which party and settlement group the forwarding settlement document should be created, as shown in Figure 11.7.

You can trigger the generation of the settlement document for either of the two business partners and any group of charges individually. In addition, the batch report to generate forwarding settlement documents has been enhanced to generate forwarding settlement document(s) only for a specific settlement group. You can still use the collective creation of forwarding settlement documents from a POWL. All selected forwarding orders are combined in one forwarding settlement document for the same settlement group. You can create a settlement group in SAP TM Customizing via the menu path Transportation Management • Settlement • Define Settlement Groups and Settlement Rules.

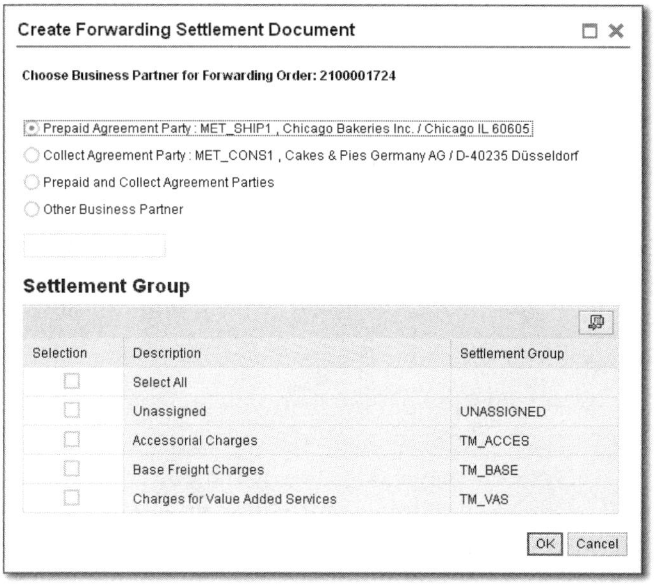

Figure 11.7 Creation of Forwarding Settlement Document with Flexible Invoicing

Multiple settlement groups can be clustered in a *settlement rule*. In a settlement rule you can define more time-relevant parameters for the settlement document generation, such as a billing schedule and the service date rule. To use the settlement rule, which contains the settlement groups, you need to assign it to your settlement profile.

11.1.5 Creating Freight Settlement Documents

We have introduced the settlement of your customer invoices. Now let's elaborate on the opposite: the settlement of your supplier invoices. This functionality is used by any company procuring freight-related services, regardless of whether you work as a shipper, forwarder, or carrier. The invoicing of supplier/service provider bills always originates from your freight documents—namely, a freight order, freight booking, or service order. The freight settlement document can be described as the draft invoice that is generated in SAP TM. It is used to store all invoicing-relevant information from your freight orders, freight bookings, and service orders. The freight settlement document is the document that triggers the interface to SAP ERP for verifying the actual invoice. You will see many similarities in the process of the forwarding settlement document.

Functional Process Flow

Creating the settlement document follows the same process as the forwarding settlement side. First, you can create a settlement document by navigating to your freight documents (e.g., a freight order) and selecting CHARGES/SETTLEMENT • CREATE SETTLEMENT DOCUMENT. Second, you can create the settlement document from a POWL. Third, you can use the standard batch job for mass creation: /SCMTMS/SFIR_CREATE_BATCH.

We start by answering the question of how many settlement documents are created when you trigger the settlement document generation. In a standard scenario, the system creates one freight settlement document for the entire freight order, service order, or freight booking. The reason for this is that each order or booking can have exactly one supplier who performs the services. Hence, it is expected that you receive one invoice per freight order, service order, or booking from this supplier. After the successful generation of a freight settlement document, the invoicing status of the originated freight document is updated.

Additional agreement parties

Alternatively, perhaps you have more than just one service provider as part of a freight order or freight booking. Besides the pure transportation service from a carrier, additional services might be performed by a customs broker or a provider of container cleaning or fumigation. You can capture such additional parties in the BUSINESS PARTNER tab in your freight order or freight booking. When you trigger the settlement document creation, you can select one or multiple parties in a pop-up window. You can create the freight settlement document for the carrier, for one or more additional parties, or for both. The charge calculation (see Chapter 10) determines one freight agreement and one transportation charge calculation sheet (TCCS) per party. If you store the charges for the different services in two separate TCCSs, the system can automatically calculate the charges and uses one TCCS per party. The only restriction applies in case a business partner plays a hybrid role with two freight agreements (one for main carrier rates, one containing rates for specific services). In other words, if a service provider plays a role as the main carrier in a first freight booking but interacts as an additional agreement party in a second freight booking, there is no determination rule available to decide which of the two freight agreements to use.

Collective settlement documents

On the forwarding settlement side, you can collectively create forwarding settlement documents, such as from the POWL. The generation of a freight settlement document works in a similar fashion. SAP TM tries to combine all selected freight orders or freight bookings based on splitting criteria into one or multiple freight settlement documents. The key criteria that needs to be equal among the different freight orders or bookings include the carrier, invoicing party, payee, payment term, and document currency.

Basic Customizing

The essential Customizing in SAP TM will sound familiar to you because, again, it is very similar to the forwarding settlement side. Some additional settings are required on the SAP ERP side, which we describe in Section 11.1.7. Consider these essential Customizing steps to enable the freight settlement process:

1. **Freight settlement document type**
   You can define the freight settlement document type in the SAP TM via the Customizing menu path TRANSPORTATION MANAGEMENT • SETTLEMENT • FREIGHT SETTLEMENT • DEFINE FREIGHT SETTLEMENT DOCUMENT TYPE. The type contains the number range that is being assigned to newly created freight settlement documents. You can also enable/disable the cost distribution function that we introduce in Section 11.2. You have to assign the freight settlement document types to your freight order and freight booking types.
2. **Settlement profile**
   You can use one settlement profile for both the forwarding and freight

settlement side. You need to assign the settlement profile to a business partner or, alternatively, to a charges profile.

3. **Process controller**
   As an optional step, you can implement an alternative way to create and group or split freight settlement documents in the creation process. SAP provides a standard method, but you can also develop your own methods and strategies, which you can assign to your settlement profile.

11.1.6 Structure of Freight Settlement Documents

The freight settlement document has a very similar structure to the forwarding settlement document. It comprises eight tabs that store the invoice-relevant information inherited from the freight order, service order, or freight booking.

General data

The GENERAL DATA tab shows the sum of the invoicing amount, the payment terms, and the organizational data, such as the purchasing organization that was responsible for procuring the services. You can also see the expected invoicing date, which you can change manually. Once the invoice has been received in SAP ERP, and after invoice verification, you see the verified invoice amount in the freight settlement document.

Charges tab

In the CHARGES tab, you see the list of all charge items inherited from the corresponding freight order. If you collectively created a freight settlement document for multiple freight orders or freight bookings, you will see the sum of the charges per freight order. The view depends on your settings in the calculation profile in SAP TM Transportation Charge Management. If you calculated the charges with calculation level stages, then you will see the details of cost per order and cost per stage. You can see an example of the freight settlement document CHARGES tab in Figure 11.8.

Figure 11.8, for example, lists the charges per cargo item from an associated freight order. Figure 11.9 shows a different example. Here, the charges are grouped per transportation leg (stage) of, for example, a freight booking for ocean freight. Two stages are visible and expanded to show the details of the applicable charges and surcharges per leg.

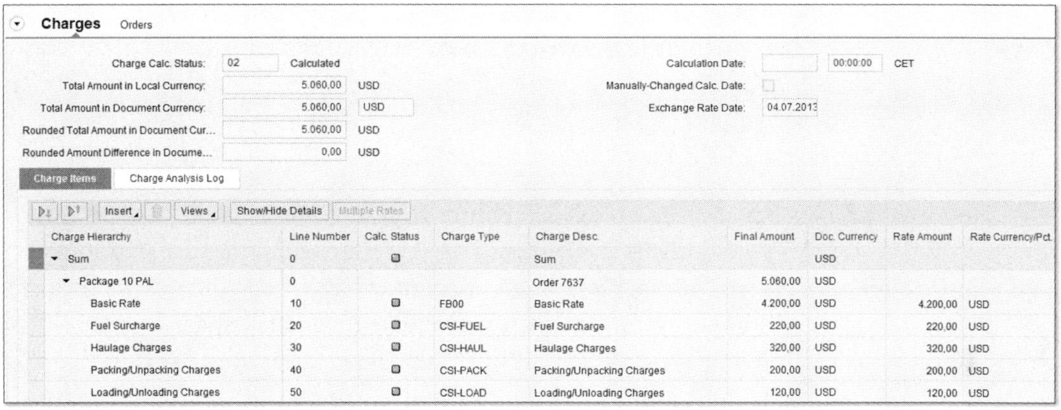

Figure 11.8 Charges Tab in Freight Settlement Document

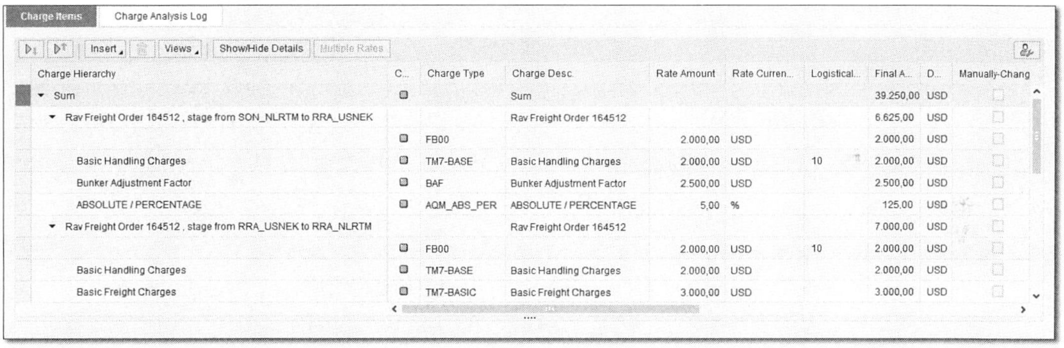

Figure 11.9 Details of Charges per Stage in Freight Settlement Document

Business partner

The BUSINESS PARTNER tab lists all relevant parties for the invoicing. For the SAP ERP settlement process, you require at least the invoicing party and payee. You can include other parties in the settlement document, such as the carrier, which is automatically inherited from the freight order. Similar to the forwarding settlement side, you can assign a business partner determination profile to your freight settlement document type to steer the correct assignment of involved parties to the freight settlement document.

Orders tab

Depending on how the freight settlement document was created, it might contain multiple freight orders or freight bookings that are grouped into one settlement document. All such orders are listed in the

ORDERS tab. If you created a settlement document for each stage based on a freight order with nonlinear stages, and you activate the STAGE SPLIT checkbox in the settlement profile, then you will see the freight order and the stage details of the invoiced stages. This is determined by the calculation level of your calculation profile.

Other freight settlement document tabs

The DOCUMENT FLOW tab in the freight settlement document always gives you an overview of the related predecessor and successor business documents of the settlement document, such as the freight order or freight booking for which it was created. This is especially valuable in the integration with SAP ERP and the generation of the succeeding documents, such as the SES and the purchase order. Other tabs in the freight settlement document are the NOTES tab, where you can enter information as free text, the ADMINISTRATIVE DATA tab, and the CHANGE INFORMATION tab, as well as the ATTACHMENTS and OUTPUT MANAGEMENT TABS. In the STATUSES tab you can see the latest lifecycle and confirmation statuses, which again, supports the integration with SAP ERP. Finally, the COST DISTRIBUTION tab supports a scenario for shippers with an SAP ERP integration for order management. We describe the details of this functionality in Section 11.2.

11.1.7 Integrating Freight Settlement Documents with SAP ERP Materials Management

The freight settlement process is integrated into the SAP ERP Materials Management component. Like in the corresponding integration of SAP TM to SAP ERP SD billing for the forwarding settlement side, SAP decided to integrate to an existing application for invoice verification and the actual posting of accruals and cost in SAP ERP. In this subsection, we highlight the required Customizing settings for the integration, the functional process flow, and some technical basics. It is essential that in SAP ERP we create both an SES and a purchase order.

Basic Customizing in SAP ERP for Service Entry Sheet and Purchase Order Creation

To set up your integration between SAP TM and the ERP MM system, there are a few essential Customizing steps. Let's start with describing the Customizing required to set up the SAP ERP MM and SAP TM integration to generate an SES and a purchase order. A purchase order is created to provide an order reference in your SAP ERP system. The SES comprises all the individual services provided by your service provider. It also contains additional information, such as descriptions of your services. An SES is generated with reference to the purchase order.

We don't go into details on the SAP ERP MM side of the configuration but will instead focus on the essentials to enable the SAP TM integration. First, you have to create a service master for the transportation services. The service master in SAP ERP is a standardized list containing all items that a company might procure in SAP ERP MM. The SAP ERP service master needs to be maintained as master data, and you have to map your SAP TM charge types to the service master in Customizing. The valuation class on the service master can optionally be used in General Ledger (GL) account determination.

Charge types in ERP

You must map your charge types created in SAP TM to SAP ERP, which means that as a prerequisite, you've already defined your charge types in SAP TM, as described in Chapter 10. This Customizing activity can be found in SAP ERP Customizing by selecting INTEGRATION WITH OTHER SAP COMPONENTS • TRANSPORTATION MANAGEMENT • INVOICE INTEGRATION • INVOICING • DEFINITION FOR TRANSPORTATION CHARGE TYPES • DEFINE CHARGE TYPES. You can also define category and subcategory codes and assign them to the mapped charge types. This Customizing step is equal to the forwarding settlement integration, which we described in Section 11.1.3; you only have to maintain the mapping once.

Service master records

Now you need to assign your charge types to service master records and account assignment categories in SAP ERP. To execute this step, you have to ensure that you have maintained entries in your service master record, as described above. Navigate to the Customizing menu path INTEGRATION WITH OTHER SAP COMPONENTS • TRANSPORTATION MANAGEMENT • INVOICE INTEGRATION • INVOICING • ASSIGNMENT OF TRANSPORTATION CHARGE TYPES • ASSIGN SERVICE MASTER RECORD AND ACCOUNT ASSIGNMENT CATEGORY. Based on this setting, SAP ERP can relate your SAP TM-specific charges to actual procurement service master records.

The account assignment category also plays an important role in the determination of accounts, cost objects, and eventually for SAP ERP FI/CO postings.

Mapping of organizational units

Another required setting is the mapping of your SAP TM organizational units to the purchasing organizations in SAP ERP. You can find the settings by following Customizing menu path INTEGRATION WITH OTHER SAP COMPONENTS • TRANSPORTATION MANAGEMENT • INVOICE INTEGRATION • INVOICING • MAPPING OF ORGANIZATION UNITS • ASSIGN ORGANIZATIONAL UNITS FOR PURCHASING. The purchasing organization on the SAP ERP side is just as important as in SAP TM; it impacts account determination and invoice verification in SAP ERP. In addition to the mapping of organizational units, you can also map the SAP TM purchasing organization against SAP ERP internal orders or cost centers. This setting is very important for posting freight costs to the correct internal cost centers. Next to the Customizing setting, you have to activate the `TCM_SE_TORINVPREPREQ_PROC` BAdI. A specific method is provided for the `ACCOUNT_ASSIGNMENT_MODIFY` enhancement. (We dive into the scenario for intra- and intercompany settlements in Section 11.2.)

Basic Customizing in SAP ERP for Invoice Verification and Automated Postings

The previously mentioned settings will allow you to generate the SES and purchasing order in an existing SAP ERP MM system. There is also the capability to allow automated posting of accruals and invoice verification with message integration with the SAP TM freight settlement document. To enable these functional components in the SAP ERP MM and Controlling (CO) system, additional Customizing settings are required. We provide you with a quick synopsis of the key Customizing settings.

Additional Information

For more information on the ERP setup we recommend *Materials Management with SAP ERP: Functionality and Technical Configuration* by Martin Murray (3rd edition, SAP PRESS 2011) or *Integrating Materials Management with Financial Accounting in SAP* by Faisal Mahboob (2nd edition, SAP PRESS 2012).

In order to enable the automated account assignment, you need to first define a mapping for your impacted company code, cost elements, and cost centers. Navigate to ERP Customizing and follow the menu path CONTROLLING • COST CENTER ACCOUNTING • ACTUAL POSTINGS • MANUAL ACTUAL POSTINGS • EDIT AUTOMATIC ACCOUNT ASSIGNMENT. The next important step is to configure the automatic postings so that your SAP ERP system is capable of automatically posting to the correct GL accounts. This has to be maintained per impacted *chart of accounts* in the SAP ERP system of each company code. Navigate to SAP ERP Customizing via the menu path MATERIALS MANAGEMENT • VALUATION AND ACCOUNT ASSIGNMENT • ACCOUNT DETERMINATION • ACCOUNT DETERMINATION WITHOUT WIZARD • CONFIGURE AUTOMATIC POSTINGS to enable inventory postings (Transaction GBB) for your impacted chart of accounts.

Enable automatic account determination

To allow workflow-supported invoice verification, you need to create *event type linkages*. This setting is vital to trigger messages to the SAP TM freight settlement document, enabling the visibility of the actual invoicing status. You have to implement the workflow WS53800006 in the ABAP Workbench. The workflow generates the `InvoiceNotification_Out` message, which shows whether an invoice has been posted successfully or canceled in SAP ERP.

Event type linkages

The SAP ERP and SAP TM systems are technically integrated via SAP NetWeaver PI, as on the forwarding settlement side. SAP provides a standard integration scenario that needs to be enabled in the SAP NetWeaver PI system: `TM_ERPInvoiceIntegration`. Eight messages are supported in a standard SAP ERP system for the integration between SAP TM and SAP ERP:

SAP TM and SAP ERP technical integration

- `TransportationOrderSUITEInvoicingPreparationRequest_Out` (and `_In`)
  Send a request to create or change a freight settlement from SAP TM and receive it in SAP ERP.
- `TransportationOrderSUITEInvoicingPreparationConfirmation_Out` (and `_In`)
  Send the confirmation of received supplier invoices from SAP ERP and receive them in SAP TM.

- `TransportationOrderSUITEInvoicingPreparationCancellation-Request_Out` (and `_In`)
  Send a request to cancel a freight settlement document from SAP TM and receive it in SAP ERP.
- `InvoiceNotification_Out` (and `_In`)
  Send the invoice notification from your SAP ERP system and receive it in SAP TM.

Functional Process Flow

Status management in freight settlement document

After you have configured the integration to your SAP ERP system for the freight settlement process, you are almost ready to transfer a generated document to SAP ERP MM. In Figure 11.10 you can see the different statuses of a freight settlement document. After you have created a settlement document, the lifecycle status is set to IN PROCESS ❶. Before transferring it now to SAP ERP, you can perform an automated consistency validation by clicking on the CHECK button on the top panel in the freight settlement document ❷. If no inconsistencies are detected, the lifecycle status is automatically changed to READY FOR ACCRUALS. It is always required to have charge items in the settlement document; otherwise, you cannot transfer the freight settlement document to SAP ERP.

Creation, transfer in batch mode

If you generated your freight settlement documents collectively with a batch job, the system automatically performs the consistency check. In the batch program, you can also generate and transfer the freight settlement document in one step. In a more manual approach, you can click on the SAVE AND TRANSFER button in the freight settlement document or from a POWL and trigger the interface to SAP ERP MM. The corresponding lifecycle status will be TRANSFERRED FOR ACCRUALS and the confirmation status will be NO CONFIRMATION FROM ERP ❸. If all settings are correct, the system automatically performs three actions in SAP ERP:

1. Generate a purchase order in SAP ERP MM.
2. Create an SES based on the purchase order.
3. Post the accruals based on the purchase order and SES.

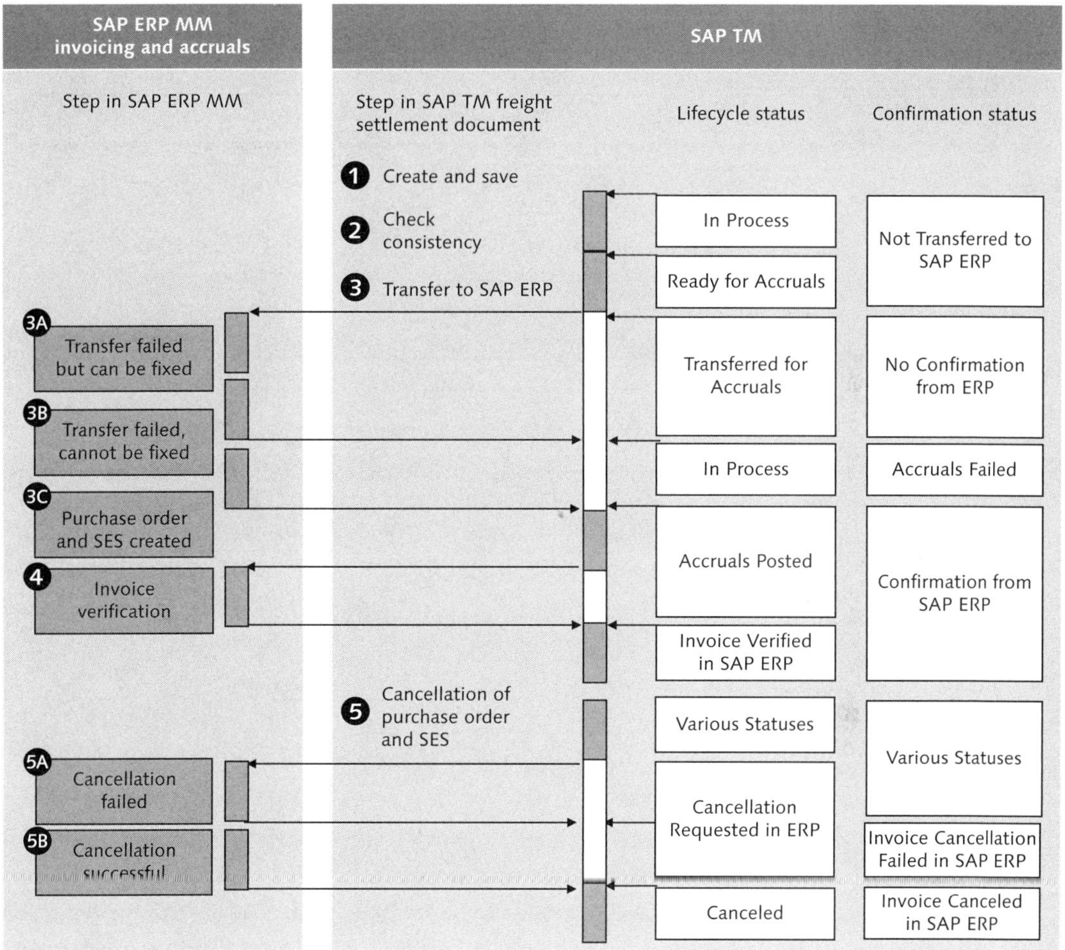

Figure 11.10 Statuses in the Freight Settlement Document

Purchase order and service entry sheet

The purchase order is required for purely technical reasons. The SES enables the posting of accruals and invoice verification in SAP ERP. Figure 11.11 shows a freight settlement document and the related purchase order in SAP ERP. The document flow in the freight settlement document shows the generated purchase order, SES, and the invoice once it has been verified. The purchase order contains all the charge types as service line items, based on the mapping introduced in Customizing: the carrier, statuses, and other references from the freight settlement document.

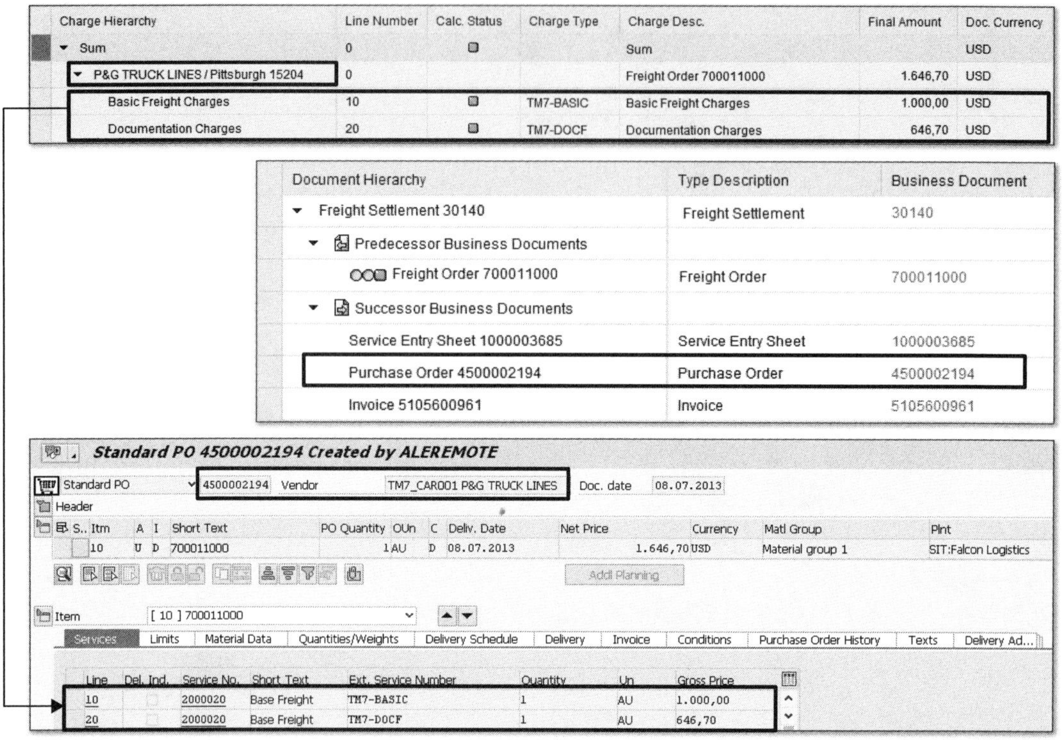

Figure 11.11 Creation of SAP ERP Purchase Order

Transfer and invoice verification

If it has not been possible to post the accruals, the lifecycle and confirmation status in the freight settlement document stay unchanged as long as an error can be corrected in the SAP ERP side. While the lifecycle status is *Transferred for Accruals*, it will no longer be possible to manipulate any data in the freight settlement document in TM **3A**. If the failed accruals posting cannot be fixed, the lifecycle status is automatically updated to IN PROCESS. The confirmation status indicates ACCRUALS FAILED **3B**. If no discrepancies are detected, the lifecycle status is changed to ACCRUALS POSTED, and confirmation status to CONFIRMATION FROM ERP **3C**.

After you have successfully transferred your freight settlement document to SAP ERP MM and the lifecycle status shows ACCRUALS POSTED, it will be possible again to manipulate the data in the freight settlement document in SAP TM. Any changes require a re-transmission of the freight settlement document to SAP ERP, which triggers a few additional steps. First, the SES posted earlier is reversed. Second, items of the SES and purchase order are deleted. Last, the regular process flow is similarly followed as described earlier; a new purchase order and SES are created, and the accruals are posted again. Once the actual invoice from the supplier is received, you can perform the invoice verification in SAP ERP with Transaction MIRO or enable customer self-billing (Transaction MRRL) ❹. Once the invoice verification has been performed, the SAP TM freight settlement document is updated with a corresponding lifecycle status of INVOICE VERIFIED IN SAP ERP. The purchase order and SES in SAP ERP are locked for any manual changes or manipulation. You can issue a credit memo to correct the charges in case of any discrepancies. We elaborate on the credit memo process in Section 11.1.8.

You can only cancel a purchase order and SES from SAP TM by canceling the freight settlement document itself. Keep in mind that even this does not work in all statuses of the freight settlement document: TRANSFERRED FOR ACCRUALS or INVOICE VERIFIED IN SAP ERP. The cancellation action triggers another status change in the SAP TM freight settlement document 5A. If the cancellation request was not successful, the freight settlement document is updated with confirmation status INVOICE CANCELLATION FAILED IN SAP ERP 5B. For a successful cancellation, the status changes to INVOICE CANCELED IN SAP ERP.

Business scenarios for freight settlement process

We have introduced you to the standard capabilities in the freight settlement process. Now we want to highlight some industry- and scenario-specific functionalities.

Settlement for Freight Orders with Nonlinear Stages

In Section 11.1.5 you learned that you can create one settlement document per transportation document or combine multiple freight orders and bookings and generate one freight settlement document.

An alternative way of creating your settlement documents is specific to a scenario for shipping companies with order integration into an SAP ERP system. This scenario supports the creation of one freight order for a star-shaped route with *nonlinear stages*. In other words, the transportation stages are not attached to each other, and a truck returns to an unloading point before picking up more freight from a different location. For example, a truck might pick up goods from a factory A and drop off the cargo at container freight station Z. In the same freight order, you could pick up goods at warehouse B and drop them off, again at container freight station Z, and so on. This scenario can be used when a shipper wants to create a freight order based on ERP shipments to transport goods for a less than truckload consolidation or deconsolidation scenario. As a result, a shipper expects to receive multiple invoices from the supplier for each transportation leg. Consequently, a freight settlement document is required for each transportation leg.

Customizing

To enable this feature, you have to select the STAGE SPLIT checkbox in the corresponding settlement profile. In addition, you have to ensure that the sequence type in the freight order or freight booking type is selected as NON-LINEAR (STAR-SHAPED). Last, you must set the calculation level as STAGES in the corresponding calculation profile for charge management.

Note

This scenario is only available to shippers with order integration with an SAP ERP system. Thus, you would always have an SAP ERP shipment but no forwarding order. Planning and optimization for such freight orders is not supported and has to be done manually, as mentioned in Chapter 6, Section 6.2. Consequently, it is not required or possible to generate freight units for such shipments.

Settlement Scenarios for Air Freight

Cargo Accounts Settlement System

In the air freight business, cargo is often handled through a freight-forwarder or booking agent. To standardize the billing process, most airlines outsource their billing services to the International Air Transport Authority (IATA). The IATA has introduced the Cargo Accounts Settlement System (CASS) as the clearing house for this purpose. Airlines send their air waybills together with other settlement-relevant data to CASS. Based on the air waybill data, such as weights, volumes, charges, and cargo details, CASS generates a monthly or bimonthly billing cycle and lists all charges and credits per air waybill for each freight-forwarder. This billing document is called the cargo sales report (CSR). A freight-forwarder can download the report or introduce an interface via EDI (message CARGO-IMP FCI). The CSR is used by the forwarder for invoice verification, which can be done manually or electronically.

When an error is detected by either side, the freight-forwarder or the airline, and a billed amount is too high or too low, corrections are required. Because the IATA billing cycle happens monthly or bimonthly, corrections can only be made for an already paid invoice in the subsequent billing cycle. In this case, the IATA supports airlines with two documents:

1. The cargo charges correction advice document (CCAD)
2. The debit credit memo (DCM)

Let's consider an example. A freight-forwarder has been invoiced for the weight of his cargo. The airline detected that the actual cargo weight was higher than originally stated. As a result, the airline issues the deviation to the IATA and submits a cargo CCAD. The CASS includes the correction amount as part of the new billing cycle in the cargo sales report to the freight-forwarder. The old amount paid is reversed (similar to a credit memo process), and the corrected amount is added.

The Cargo Accounts Settlement System

Since the IATA introduced CASS in 1999, more than 500 airlines and ground handling companies have joined the CASS billing system. Today the total billing value handled through CASS exceeds $33 billion every year.

Cargo charges correction advice

Now let's introduce what SAP TM offers to support this billing process for freight-forwarders. First, you follow the regular freight settlement process as elaborated in this chapter. A freight settlement document is generated for one or multiple air waybills from the freight booking. You can generate one collective freight settlement document for the same airline for all freight bookings in one month and thereby simulate the cargo sales report in your system.

The integration with your SAP ERP system follows the standard settlement process. A purchase order and an SES are created that contain the appropriate charges. After the receipt of the CSR from IATA, in your SAP ERP MM system you can perform the invoice verification of the accrued charges versus the received invoice amounts and pay the settled amount.

Let's take an example and assume a discrepancy has been detected after the forwarder made the payment. The airline triggers the CCAD to the IATA. As a result, the forwarder receives the corrected settlement amount with the next cargo settlement report. To reflect this discrepancy in your system as the freight-forwarder, follow these steps:

1. Navigate to your originally created and already completed freight booking that generated the trouble-making air waybill.
2. It is possible to correct the freight booking despite its status. You might change the weight of the cargo to the correct weight per your agreement with the airline.
3. Because the freight settlement document has already been generated for the freight booking, you cannot correct or change the existing freight settlement document. You can generate your own cargo correction advice document instead. The CCAD serves as a simulation of the actual CCAD and contains the originally posted amount from your old freight settlement document, the delta amount that needs to be corrected, and the final billing amount.
4. The CCAD needs to be transferred to your SAP ERP system, where a new purchase order is generated. The purchase order stores both the old settlement amount with a reverse indicator and the corrected amounts.
5. Based on the new purchase order, two SESs are generated:
   - First SES: Post the former (wrong) settlement amount based on the reverse indicator.
   - Second SES: Post the accruals for the fully corrected settlement amount.

You are now ready to perform a new invoice verification based on the actual cargo settlement report from the IATA.

Note

The CCAD is a different category of freight settlement document, but technically it is the same object. You cannot create a CCAD if you have not yet generated the freight settlement document or if it has not yet been set as status *Accruals Posted*. The reason is that before the freight settlement document has been generated or the accruals posted, you can still include all corrections in the first settlement document.

11.1.8 Credit Memos

Billing customers and settling supplier invoices may not always be straightforward. Sometimes bills to your customers have errors, and invoices from your suppliers can be too high or too low, perhaps because the cargo measures, weights, or volumes deviated between an order and the actual amounts. In addition, any delays in the transportation chain can cause errors and require you to decrease your bills or receive deductions from your suppliers.

The *credit memo* is a document in SAP TM that you use to post corrections of incorrect settlement amounts. A credit memo can be generated for the forwarding settlement document to give a credit of a wrongly billed amount to a bill-to party. The other way around, you can generate a credit memo for the freight order or freight booking to trigger a deduction of an invoiced amount from your supplier. You have learned in this chapter that settlement documents cannot be changed once they have a certain lifecycle status. Once a billing document is created in SAP ERP, you can still change or cancel the forwarding settlement document, which transfers an updated forwarding settlement document and reverses the old billing document in SAP ERP. Nevertheless, many customers still prefer to use the credit memo for any corrections, especially from a document process flow and audit visibility point of view. The same behavior is relevant for the freight settlement document.

Credit Memos for Forwarding Orders

The credit memo on the forwarding order side is used by freight-forwarders and carriers to correct originally invoiced amounts on a billing document. You can generate a credit memo by navigating to your forwarding settlement document and selecting FOLLOW UP • CREATE CREDIT MEMO button. This is also possible directly from a POWL. Remember that a credit memo can only be created once your forwarding settlement document has been transferred to SAP ERP and a billing document has been generated.

Creation of credit memo

With SAP TM you can generate credit memos for both individual forwarding settlement documents and collective settlement documents that contain multiple forwarding orders. After you have generated a credit memo, you will see that it has the same structure as the forwarding settlement document. The major difference is that the CHARGES tab contains three columns listing the charges. In Figure 11.12 you can see that in addition to the invoiced amount, a credit amount can be entered. In this column, you can enter all deductions from the settlement amount. It is possible to generate multiple credit memos for one settlement document. As a result, the CREDIT REMAINING AMOUNT column shows the remaining billing amount. SAP TM validates and ensures that you cannot give a higher credit than the invoiced amount.

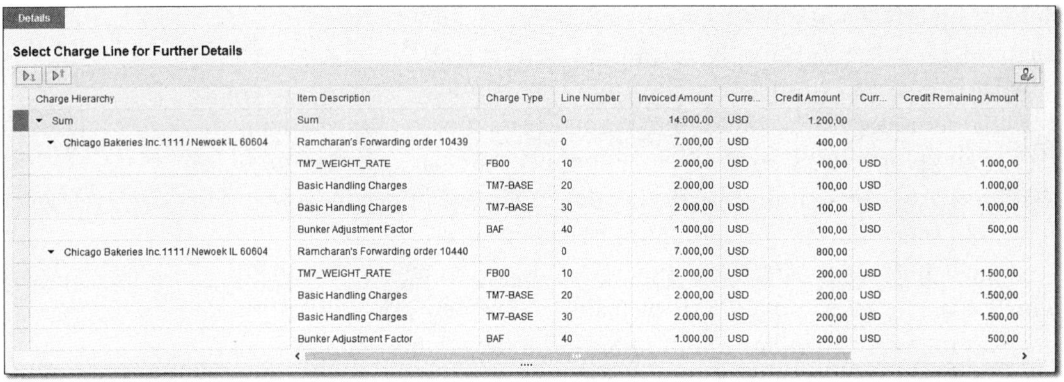
Details

Select Charge Line for Further Details

Charge Hierarchy	Item Description	Charge Type	Line Number	Invoiced Amount	Curre...	Credit Amount	Curr...	Credit Remaining Amount
Sum	Sum		0	14.000,00	USD	1.200,00		
Chicago Bakeries Inc.1111 / Newoek IL 60604	Ramcharan's Forwarding order 10439		0	7.000,00	USD	400,00		
	TM7_WEIGHT_RATE	FB00	10	2.000,00	USD	100,00	USD	1.000,00
	Basic Handling Charges	TM7-BASE	20	2.000,00	USD	100,00	USD	1.000,00
	Basic Handling Charges	TM7-BASE	30	2.000,00	USD	100,00	USD	1.000,00
	Bunker Adjustment Factor	BAF	40	1.000,00	USD	100,00	USD	500,00
Chicago Bakeries Inc.1111 / Newoek IL 60604	Ramcharan's Forwarding order 10440		0	7.000,00	USD	800,00		
	TM7_WEIGHT_RATE	FB00	10	2.000,00	USD	200,00	USD	1.500,00
	Basic Handling Charges	TM7-BASE	20	2.000,00	USD	200,00	USD	1.500,00
	Basic Handling Charges	TM7-BASE	30	2.000,00	USD	200,00	USD	1.500,00
	Bunker Adjustment Factor	BAF	40	1.000,00	USD	200,00	USD	500,00

Figure 11.12 Charge Details of Credit Memo

Statuses of credit memo

Prior to the transfer to SAP ERP, you can use the preview invoice if you want to simulate a preview of the credit memo in SAP ERP together with the tax information. Like the settlement document, you have to ensure consistency in the credit memo before you can transfer it. After you have successfully transferred the credit memo to SAP ERP, and a new document is generated of type credit memo. The lifecycle status of the credit memo is updated after it has been successfully transferred SAP ERP: *Transferred for Credit Memo Creation in SAP ERP*. It is also possible to cancel a credit memo, which changes its status to *Cancellation Requested in ERP* and *Canceled* if successful.

Customizing

To use the credit memo for the forwarding settlement document, you need to perform a few Customizing steps:

1. A credit memo has its own document type that contains the number range and the reason code for a credit memo. You can define a credit memo type in SAP TM Customizing: TRANSPORTATION MANAGEMENT • SETTLEMENT • FORWARDING SETTLEMENT • DEFINE CREDIT MEMO REASON CODES AND TYPES FOR FORWARDING SDS. The credit memo type can be assigned to your freight settlement document type.
2. In the same Customizing path, you can maintain reason codes and assign them to the credit memo. A description of a reason code could, for example, be DELAY IN TRANSPORT or DEVIATING CARGO MEASURES. The reason code generally has two characters with a free text field for a description.
3. All other Customizing settings for the forwarding settlement document on both the SAP TM and SAP ERP sides have to be in place. In addition, you need to map the reason codes from the SAP TM credit memo to SAP ERP order reason codes. Navigate to the SAP ERP Customizing by following the menu path INTEGRATION WITH OTHER SAP COMPONENTS • TRANSPORTATION MANAGEMENT • INVOICE INTEGRATION • BILLING • ASSIGN TM CREDIT MEMO REASON CODE TO ORDER REASON CODE.

Credit Memos for Freight Orders

Now that you are familiar with credit memos on the forwarding settlement side, you will recognize that the process is very similar for the freight settlement documents. You can use a credit memo if your service provider, freight-forwarder, or carrier charged you too much for transportation services.

Creation and transfer to SAP ERP

You can create a credit memo from your original freight order, booking, or service order and include any changes in the document. If, for example, your freight-forwarder charged you based on weight but the cargo items were not as heavy as detailed in the freight booking, you can change the correct weight directly in the freight booking. The credit memo automatically posts the delta amount. To create the credit memo, select FOLLOW UP • CREATE CREDIT MEMO in the top panel of the document. SAP TM generates the credit memo automatically. Similar to the charges correction advice document, every credit memo generates a new purchase order in SAP ERP. In contrast to the CCAD process, the purchase order has only one line as the returns item with the correction amount. Based on the purchase orders, an SES is created that posts the corrected amount.

Customizing

In the Customizing for credit memos on the freight settlement side, you need to execute steps similar to the forwarding settlement side. You need to create a credit memo type and reason codes. The credit memo type for freight settlement can be created via the Customizing menu path TRANSPORTATION MANAGEMENT • SETTLEMENT • FREIGHT SETTLEMENT • DEFINE CREDIT MEMO REASON CODES AND TYPES FOR FREIGHT SDs. In the credit memo type you can define how changes in the invoicing status of the freight order, freight booking, and service order should be handled. The invoicing status of the documents changes automatically to *Completely Invoiced* as soon as you have successfully generated a freight settlement document with the full amount, as calculated for the charges. You can activate the INFLUENCE INVOICING STATUS checkbox in the credit memo type. When generating a new credit memo, the invoicing status of a freight order, booking, or service order is reset to *Not Invoiced*. If you deselect the INFLUENCE INVOICING STATUS checkbox, the creation of a credit memo does not have an impact on the invoicing status. In the same Customizing path, you can maintain reason codes and assign them to the credit memo.

11.2 Cost Distribution and Internal Settlement

Working as a shipper or LSP, you might encounter challenges in apportioning transportation costs to the correct organizational units, accounts, or sales orders. SAP TM offers two capabilities to support automated distribution of transportation charges and the settlement of these between different organizations: cost distribution and internal settlement.

The SAP TM cost distribution functionality supports both shippers and LSPs. It gives you the capability to distribute incurred costs from freight orders and freight bookings in SAP TM to the individual items they contain. For example, you might have a freight order for trucking that contains several pallets as items from multiple forwarding orders or SAP ERP orders or deliveries. The cost distribution ensures an apportionment of the cost for the entire freight order down to each individual cargo item. You can use different apportionment rules (e.g., based on the weight of each cargo item). With this functionality we need to differentiate between shippers and LSPs:

1. **Cost distribution for shippers**
   First, you can manage your transportation costs for inbound deliveries by assigning charges to the materials valuation component in your SAP ERP MM system. This enables calculations with actual costs from your SAP TM freight order or freight booking documents. Second, you can gain better visibility of your sales order profitability by assigning transportation charges for outbound deliveries to the correct orders and company codes. In SAP ERP Controlling and Profitability Analysis (SAP ERP CO-PA) you have the capability to include any transport-related charges in your sales order profitability analysis.
2. **Cost distribution for LSPs**
   You can distribute incurred costs from your procured transportation services and capacities (freight orders or freight bookings) to other organizational units. The distribution can be performed completely in SAP TM, and this enables an internal settlement process.

The internal settlement process is a very common business practice among LSPs and trucking companies. Remember from Chapter 10, Section 10.3.3 that there are two different scenarios for how an internal settlement is supported:

1. The purchasing organization of a freight-forwarder cross-charges the cost of procured transportation services against a forwarding order from a sales organization.
2. A purchasing organization of a trucking company uses resources (e.g., trailers) of another organization with a resource pool. The

organization with the trailers cross-charges the cost of using these resources against a freight order from the purchasing organization.

In Section 11.2.2 we focus on the first scenario and elaborate on the complexity and interdependency in an overall settlement process. The basic concept is that a sales organization (e.g., a booking office or a freight station) capturing orders also invoices the customer and eventually generates the revenue. A planning or purchasing organization (a gateway or container freight station) procures the capacity with a carrier and settles or pays the charges. In an internal settlement process, the purchasing organization (gateway/CFS) settles its charges to the sales organization (booking office/freight station) to be compensated for the internal service of providing capacities and handling a shipment. This enables a profit center structure where each organizational unit can interact autonomously with their own goals and targets. The benefit of such a business model is that a natural competition evolves between different planning and purchasing organizations. This allows differentiation between well-managed organizations and unprofitable units. Especially if you offer end-to-end international transportation services, a shipment can easily involve five or more independent organizational units that settle costs with each other.

11.2.1 Cost Distribution

So cost distribution is for both shippers and LSPs. The concept of apportioning cost is similar in both scenarios, but how we use the cost information is different. In either case, you break down the cost of a freight order or booking to the order document items (e.g., sales order items and forwarding order items). A shipper can use the cost distribution for inbound shipments to perform product costing and pricing. The transportation cost is automatically used in SAP ERP MM materials valuation. For all outbound shipments, the distributed cost can be used for SAP ERP CO-PA. This allows improved visibility regarding actual profitability. For an LSP, the distributed cost can be used as the basis for an internal settlement process. The apportioning of cost is always done per delivery or cargo item.

Functional Process Flow

Now let's detail the functional process flow for the cost distribution. When you distribute cost, the upstream process is similar to a regular handling of an order. Figure 11.13 shows the document flow between SAP TM and SAP ERP.

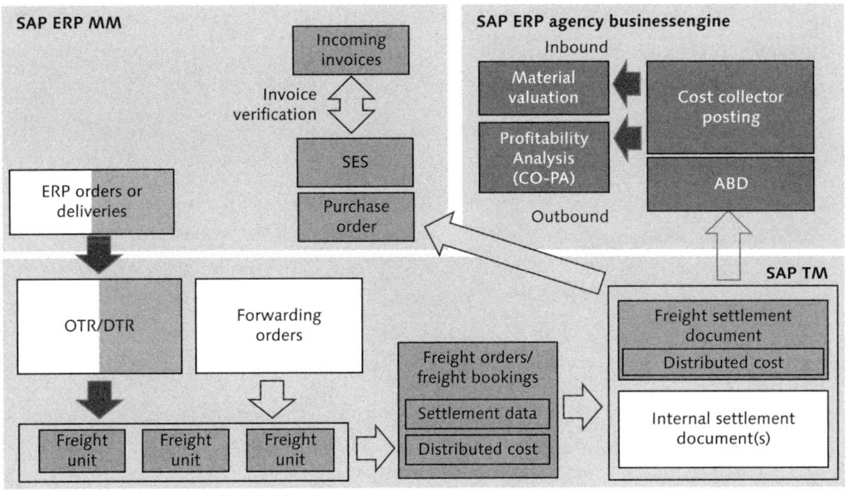

Figure 11.13 Overview of Cost Distribution for Shippers and Freight-Forwarders

To allow a better overview, we have used the same formatting for all documents relevant to a shipper, all documents for a freight-forwarder, and the joint objects. For a shipper, a delivery-based transportation requirement (DTR) or order-based transportation requirement (OTR) is generated based on an SAP ERP document (see Chapter 4 for more details). For a freight-forwarder, you create a forwarding order. Notice that the process does not deviate from any other standard transportation scenario. Freight units can be planned and assigned according to freight orders and freight bookings. The cost distribution takes place when you generate a freight or internal settlement document based on the data available in a freight order or freight booking. The distributed cost is stored and visible in the Cost Distribution tab of the freight orders and freight bookings. After you have calculated the charges and generated the settlement document, the corresponding cost is distributed automatically.

Cost distribution for shippers

For a shipper, it is the freight settlement document that contains the distributed cost and offers visibility in its own tab, as you can see in Figure 11.14. The transportation cost is split and broken down into each individual item of an order. In our example, the bases were two SAP ERP delivery items. You can also see the distribution percentage that was assigned to each item. The cost distribution is executed for every charge type. In this example, the distribution rule is the gross weight of the items. Other rules are available, as we cover in the upcoming subsection on basic Customizing.

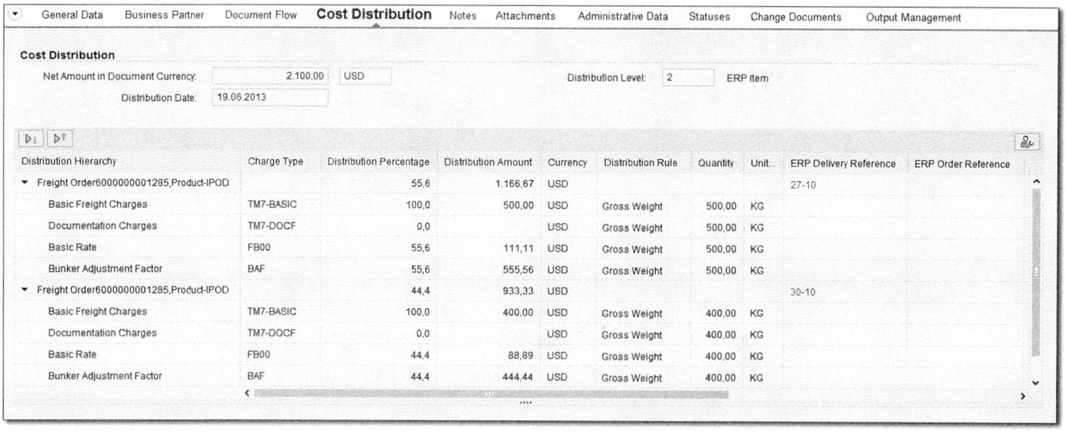

Figure 11.14 Distributed Cost in Freight Settlement Document

Cost distribution for freight-forwarders

For a freight-forwarder, the cost information of a freight settlement document is not relevant. The reason is that a freight-forwarder does not have a use for distributed cost information in SAP ERP. Nor is material valuation required, which targets only shippers, nor a profitability analysis of SAP ERP orders, because a freight-forwarder does not have orders or deliveries in SAP ERP. (We further detail the use of the distributed cost in internal settlement documents for freight-forwarders in Section 11.2.2, but for now we want to dive a little bit deeper into the integration with SAP ERP for shippers.)

SAP ERP integration for shippers

For a shipper, the next step is to trigger the transfer to SAP ERP. Figure 11.14 shows that three documents are generated in SAP ERP, subsequent to triggering the transfer:

1. An SAP ERP purchase order for posting the freight order/booking charges as you are used to from a regular settlement process.
2. The SES (also similar to a regular settlement process) to enable postings of accruals and invoice verification.
3. An agency business document is the central document in the agency business application, which is used as a very flexible posting in SAP ERP.

SAP ERP Agency Business Application

The SAP ERP Agency Business Application is a component of the SAP ERP system that was first introduced in SAP ERP R/3 4.0a. You can find it by navigating in your SAP ERP menu to LOGISTICS • AGENCY BUSINESS. The Agency Business document itself does not interfere with accounts payable or account receivable documents created by your service purchase order or SES. It only posts to expense GL accounts or material accounts.

Agency Business Application

4. The SAP ERP Agency Business Application integrates with various other components in SAP ERP, namely, Material Ledger and Material Valuation in MM, as well as Financial Accounting and CO-PA. You can flexibly configure your use of the created agency business document (ABD) in SAP ERP. The ABD stores the most significant information from the predecessor documents from SAP ERP, such as the sales order, purchase order, and delivery. From SAP TM it inherits the majority of the information from the freight settlement document, including the distributed cost information, the corresponding freight order or freight booking, and even a reference to the newly created purchase order of the transportation services.

Shipper Scenarios for Inbound and Outbound

You know that the agency business document can be flexibly configured. Nevertheless, SAP has foreseen two major use cases for the cost distribution of a shipper.

SAP ERP MM Material Valuation

The first scenario supports you in an inbound process; it starts by generating a purchase order or inbound delivery in SAP ERP. If you follow our standard process flow in SAP TM according to Figure 11.13, you can eventually generate a freight settlement document containing the distributed cost. In SAP ERP, the purchase order for the transportation services, the SES, and an ABD are generated. Figure 11.15 shows the creation of the ABD based on the freight settlement document with type billing document. From the ABD you can now post the financial document to the material account or corresponding price difference account for the materials. The material ledger document is updated based on the freight cost posting.

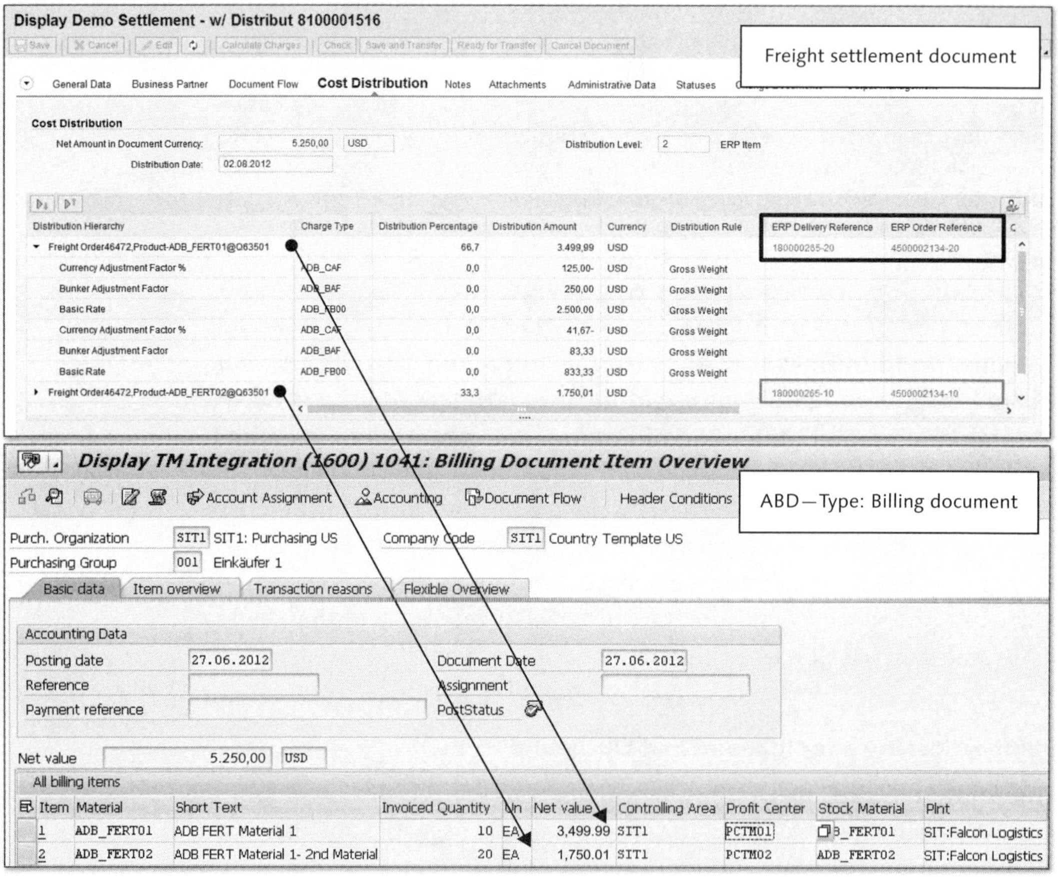

Figure 11.15 Generation of Agency Business Document from Freight Settlement Document

Outbound process with CO-PA

The second use case supports an outbound shipment. Similar to generating a standard scenario, you can generate, for example, a sales order in SAP ERP and eventually a freight settlement document for the freight order in SAP TM (see details in Section 11.1.5). The freight settlement document generates the purchase order and SES again. The ABD is also generated again. This time it is generated with type payment document and is used to post the freight cost of a sales order item to a GL expense account. The details of the ABD include a reference to the original sales order, service purchasing order, freight settlement document, and freight order. The corresponding cost collector in the form of the sales order item is also visible.

You will find that the ABD is generally blocked for any postings, both in Material Valuation and GL Account posting. Transaction WLFLTM (or WLFLTM2) allows you to display all existing ABDs based on various query parameters. To release the document, you have three options:

1. Batch report with Transaction WLFK (Release ABD to Material Valuation).
2. Batch report with Transaction WLFF (Release ABD to GL Account posting).
3. Integration with SAP Event Management. You can use the flexible SAP Event Management framework to define your own rules for releasing ABDs.

Basic Customizing

We need to answer a key question at the beginning of the cost distribution process. How is the cost actually distributed to the order items in the freight order or freight booking? You can define different rules for how to distribute the charges. SAP offers four ways for distribution: gross weight, net weight, gross volume, and distance times weight. You can define your own logic via a BAdI (`/SCMTMS/TCD_DISTRIB_RULE`). This BAdI allows a lot of flexibility, so you can implement your own methods for cost distribution.

Cost distribution profile in SAP TM

The distribution rule needs to be added in a mandatory *distribution profile* in SAP TM. This profile contains both the distribution rule and the level, which has to be the SAP ERP item for shippers or the forwarding order for freight-forwarders. You can configure the profile in SAP TM Customizing via the menu path Transportation Management • Basic Functions • Cost Distribution • Define Cost Distribution Profiles. The cost distribution profile needs to be assigned to a charge profile, just like any other profile relevant to the settlement or charge calculation process in SAP TM Customizing. To do so, follow the menu path Transportation Management • Basic Functions • Charge Calculation • Basis Settings • Define Charges Profile. In addition, you need to enable cost distribution in your corresponding freight settlement document types. Freight-forwarders are required to select the Cost Distribution checkbox in the freight order or freight booking types.

Customizing in SAP ERP

To enable the integration of the cost distribution with SAP ERP, a few Customizing steps are required on the SAP ERP side:

- Map SAP TM charge types to SAP ERP condition types. SAP Agency Business Application can use either the SD or MM pricing procedures and conditions for the mapping. In standard SAP ERP, the MM side is used because these settings are anyways mandatory for the material valuation. As a result, we recommend that you map the SAP ERP MM condition types to SAP TM charge types.
- Define ABD types for both billing documents and payment documents.
- Define your calculation schemas. For example, use standard schema WT0001 for postings in the billing document and WT0002 for the payment document.
- Configure your settings for the automated creation of the ABD. This is a major setting to enable the SAP ERP system to seamlessly generate the ABD: Integration with Other SAP Components • Transportation Management • Invoice Integration • Mapping for Cost Distribution • Posting via Agency Business • Basic Settings for Creation of Agency Documents.

For a detailed description of the required Customizing in SAP ERP, please refer to additional sources, such as the configuration guide: SAP Transportation Management—Basic Settings and Integration for SAP ERP.

11.2.2 Internal Settlement

Let's return to the big picture. Section 11.2 focuses on the internal settlement process for an LSP. In the last subsection, we highlight the principles of internal settlement for a trucking provider.

Business scenarios for internal settlement

A purchasing organization cross-charges the cost for a shipment to a sales organization, which generates the actual revenue with a customer. Let's take a few examples of different business practices:

- In the freight-forwarding industry, it is very common for CFSs/gateways to procure their own capacities with ocean liners or airlines. Capacities are procured and internally provided. A CFS/gateway charges an internal rate to the exporting organization. The incoterms play a major role because an exporting organization might only partially settle the charges with the customer.
- Importing organizations in freight-forwarding are generally in charge of any inland transportation via rail or truck in the importing country. This means that an importing organization can settle the charges of the transportation cost to the exporting organization.
- In the trucking industry, it is a common business practice for a service center to own a pool of resources such as trucks and trailers. The service center provides other internal sales or planning organizations with the required capacities. In this case, the sales or planning organization is charged internally by the service center.
- SAP TM supports the internal settlement process for both intracompany and intercompany charge settlement. Intracompany charge settlement happens between branches or divisions of the same company code; for example, a gateway in Newark, USA, settles the transportation cost to a freight station in New York, USA. An intercompany settlement takes place between legal companies with independent company codes (e.g., if a CFS and export station are in separate countries).

Intracompany versus intercompany settlement

Internal Settlement Process for LSPs

Let's use an example to illustrate an intracompany settlement process with a cross-company shipment execution. Section 11.1.1 covered the settlement process based on incoterms. Now let's complicate this picture to represent an authentic example of the interaction in the settlement process between internal organizational units. We summarize the lessons learned in this chapter and include both a forwarding and freight settlement document creation.

In Figure 11.16, shipper 1 in England transports cargo to consignee 1 in Singapore. Shipper 1 and consignee 1 have arranged the incoterm free-on-board (FOB). As a consequence, shipper 1 is responsible for paying for the transport to the port of loading, and the consignee pays for the rest of the transportation, including customs duty. Any mode of transport and business scenario can be supported, but our example is based on ocean freight (so in parentheses we highlight the equivalent for air freight).

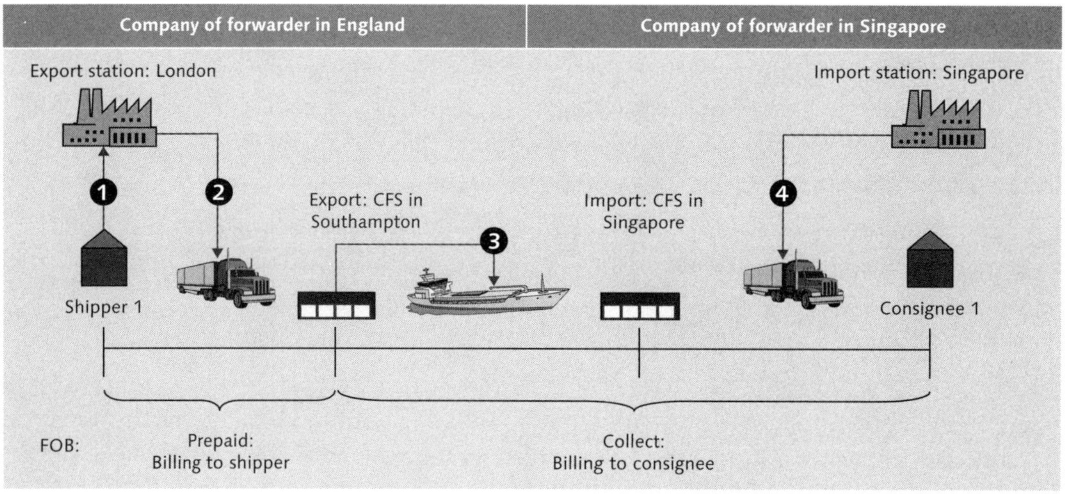

Figure 11.16 Business Scenario for Settlement Process in SAP TM

Example process flow

In step ❶, you receive an export forwarding order in the export station in London. The shipper contacts the local forwarding office to arrange for the transport. Shipper 1 is listed as the prepaid agreement party. Consignee 1 is listed as the consignee, but the importing station in Singapore is marked as the collect agreement party.

In step ❷, the export station in London arranges for a pickup service (a local trucker perhaps) to transport the goods from the customer to the CFS (gateway) in Southampton. A freight order is generated.

In step ❸, the CFS (gateway) in Southampton has already procured capacity on a vessel (on a plane), leaving on a voyage (departure) from Southampton to Singapore with an ocean liner (airline), reflected by a freight booking. As you learned in Chapter 4, Section 4.2, the generation of an export freight booking triggers the generation of an import freight booking and consequently an importing forwarding order.

In step ❹, once the vessel (plane) arrives in Singapore, the importing station is informed and generates another freight order for the pickup of the freight in the CFS (gateway) and delivery to Consignee 1.

In this example, a total of seven settlement documents are generated. Figure 11.17 shows the creation of the settlement documents in SAP TM.

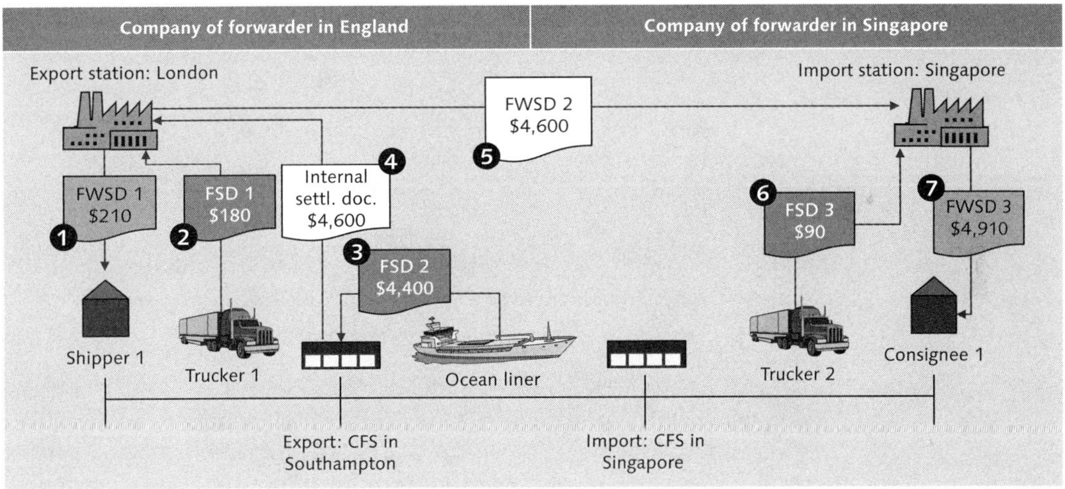

Figure 11.17 Creation of Internal Settlement Documents

Intracompany settlement process

Let's walk through this figure. Based on the free-on-board incoterm, the exporting station in London creates forwarding settlement document 1 for the prepaid party (Shipper 1) from the export forwarding order ❶. These are the only charges the shipper has to cover. The same exporting organization also arranges the pickup and has to settle the charges for the trucker. Freight settlement document 1 is generated for the pre-carriage ❷. The container freight station (gateway) in Southampton generates a freight settlement document to settle the main leg charges with the ocean liner (airline) ❸.

Generation of internal settlement document

An internal settlement document is generated between the CFS (gateway) and the exporting station, again based on the export forwarding order ❹. The intracompany settlement document recovers the cost for the procurement organization, in our example the CFS (gateway) Southampton of the main leg. To calculate the internal charges, in the top panel of the forwarding order, select CALCULATE CHARGES • CALCULATE INTERNAL CHARGES.

The creation of the intracompany settlement document can be executed manually from a POWL that lists all freight bookings/freight orders that require intracompany settlement. To find this POWL, navigate to SAP NWBC and follow the menu path FORWARDING SETTLEMENT • WORKLIST • OVERVIEW FORWARDING SETTLEMENT DOCUMENTS • ALL FORWARDING ORDERS FOR INTERNAL SETTLEMENT. When you select one freight booking or freight order, all the affected forwarding orders are listed. You can click on the CREATE INTERNAL SETTLEMENT button to generate the internal settlement document. Alternatively, you can use a batch job for report /SCMTMS/CFIR_CREATE_BATCH to generate the internal settlement documents.

Note

The POWL only shows relevant freight bookings or freight orders for the generation of an internal settlement document if you have specified a corresponding settlement rule. A settlement rule is defined in Customizing and steers the behavior for generating internal settlement documents based on the execution status of a freight booking or freight order. In our example, we assume the settlement rule is *Executed*. Consequently, the internal settlement document is only ready for generation after the freight booking of the main leg is fully executed.

In our example, the agreed-upon incoterm is free-on-board. Consequently, the exporting organization transfers the cost to the importing organization in Singapore, which bills Consignee 1 as the collect payer for the main leg service ❺. This settlement document FWSD 2 is considered an external settlement document and is generated as a regular forwarding settlement document. In step ❻, the importing organization generates FSD 3 for the delivery charges. Last, in step ❼, FWSD 3 is generated based on the import forwarding order to bill the collect charges to the consignee 1.

Intercompany settlement process

We want to highlight the difference between the intracompany settlement process and the flow for intercompany settlements. Let's assume our CFS (gateway) is located in a different country—say, Scotland, instead of England. In this case, the sales organization in London and the CFS (gateway) in Scotland are assigned to different company codes. The internal settlement between the CFS (gateway) and the station are considered an intercompany settlement. The major change in the process lies in the integration with SAP ERP. You would not see a difference in the actual internal settlement flow in SAP TM.

Table 11.1 summarizes the different cost and revenue components of each organization for our intracompany settlement example.

Export Station London		Export CFS Southampton		Import CFS Singapore		Import Station Singapore	
Cost	Rev.	Cost	Rev.	Cost	Rev.	Cost	Rev.
$180 pickup charges	$210 prepaid charges	$4,400 main leg charges	–	–	–	$90 delivery charges	$4,910 collect charges
$4,600 main leg intrac. settl.	$4,600 main leg charges	–	$4,600 main leg intrac. settl.	–	–	$4,600 Main leg intrac. settl.	–
–	$300	–	$200	–	–	–	$220

Table 11.1 Cost and Revenue per Organizational Unit

Internal cost from agreements

There are generally two ways to calculate the internal rates that are to be settled internally. First, you have the choice to set up internal rates in an agreement between the sales and purchasing organizations. These rates represent the standard cost for a purchasing organization. The standard cost is an average price of procuring transportation services and can even include a profit margin. As a result, it is possible to handle purchasing organizations as profit centers that compete with each other to offer the best rates for capacities in the organization. You can maintain such rates in internal agreements, as described in Chapter 10.

Distributed actual cost

Recall the introduction to cost distribution for LSPs in Section 11.2.1. Based on the cost distribution functionality, you can determine the actual cost of a freight order or freight booking, allocated to the forwarding order items. You have the option to use this actual cost, which needs to be paid to the service provider for an internal settlement between the purchasing organization and the sales organization. The core difference from the standard cost scenario is that instead of an average of the cost, the billing-relevant amounts of the actual service provider charges are used. For consolidated shipments, the cost distribution supports the apportionment of the cost to all affected forwarding orders.

Basic Customizing and Master Data in SAP TM

The functionality of internal settlements in SAP TM is well integrated with and dependent upon other master data and several settings. In this section we highlight the essentials of setting up an internal settlement process.

The basis for every settlement process is the charges and the setup of SAP TM Transportation Charge Management. Please refer to Chapter 10, Section 10.2 to learn what basic settings are required. You can directly assign a calculation profile to a business partner of type sold-to party or to the regular charges profile. If both are maintained, the profile assigned to the business partner takes precedence.

As with the regular settlement process in SAP TM, you need to define a settlement profile. You can use an existing profile from the regular settlement process that is assigned to a charges profile, as you are used to from Section 11.1. Remember that the settlement profile of the purchasing organization will be picked up. Like for the calculation profile, you could assign a settlement profile to a business partner in the type sold-to party, which would take precedence.

You need to define an internal settlement document type in Customizing. Navigate to Transportation Management • Settlement • Forwarding Settlement • Define Forwarding Settlement Type. The forwarding order type needs to be enabled for internal charge calculation and settlement in Customizing.

You can define a specific time you want an internal settlement document to be ready for generation. This is achieved by an internal settlement rule that you can assign to each stage type of a movement type. Navigate to TRANSPORTATION MANAGEMENT • FORWARDING ORDER MANAGEMENT • DEFINE STAGE TYPE SEQUENCE FOR MOVEMENT TYPES.

Most important, you must have maintained an internal agreement between the purchasing organization and the sales organization. As introduced Chapter 10, one internal agreement can be valid between one or many purchasing organizations and one or many sales organizations. If you want to base the internal charges on the standard cost, you have to maintain charge types with amounts or internal rate tables.

> **Note**
>
> Be sure to model your purchasing organizations with the organizational unit function forwarding house. The reason is that a purchasing organization interacts as a sales organization internally when settling internal charges to another organizational unit. In our example, this would be the case for the CFS (gateway) in Southampton.

Cost distribution

To use the actual transportation cost for internal settlements, you require a slightly different setup. The way you generate your internal settlement documents and the integration to SAP ERP are similar. You have performed the setup of the cost distribution for LSPs, as introduced in Section 11.2.1 and all the Customizing steps. The key difference lies in the creation of an internal agreement. You still need to define an internal agreement, but no rates or rate tables. Instead, you set the CHARGE USAGE field of the TCCS as INTERNAL. In addition, you must maintain one charge type with the calculation method 9—INTERNAL CHARGE CALCULATION. The cost is picked up from the corresponding execution documents, freight order, and freight booking.

Integration with SAP ERP

Functional integration

Once you have successfully generated the various settlement documents, they need to be transferred to SAP ERP. For the internal settlement, we need to distinguish between the intra- and intercompany settlement document. The intracompany settlement document does not generate a billing document in SAP ERP SD; a regular forwarding settlement document does. Instead, reposting would take place in SAP ERP. Remember that the cost of procuring a service is always posted to the purchasing organization (in our example, the CFS in Southampton). When you transfer the intracompany settlement document to SAP ERP, the settled cost in the forwarding order from the according freight orders or bookings is reposted from the purchasing organization (the CFS in Southampton) to the sales organization (the exporting station in London). As a result, it is possible to map the primary cost to a cost center or an internal order. We already discussed how the intercompany settlement document has a different sort of integration with SAP ERP. It simply generates a regular billing document in SAP ERP.

Let's take a high-level look at the Customizing steps required to set up SAP ERP. Please refer to the SAP ERP Configuration Guide for more details.

Customizing in SAP ERP

- The standard integration for settlement needs to be configured to enable the intercompany process, as described in Section 11.1.1 and Section 11.2. In particular, a mapping of SAP TM sales organizations to SAP ERP organizations is required.
- You need to map your sales organizations and purchasing organizations to the corresponding cost centers or internal orders in SAP ERP. This is required for intracompany settlements where the charges are being posted from the original purchasing organization of a freight order or booking to the sales organization of the forwarding order.

  Navigate to SAP ERP Customizing by selecting INTEGRATIONS WITH OTHER SAP COMPONENTS • TRANSPORTATION MANAGEMENT • INVOICE INTEGRATION • BILLING (INVOICING) • MAPPING OF ORGANIZATIONAL UNITS • ASSIGN TM SALES (PURCHASING) ORG. UNIT TO INTERNAL ORDER/COST CENTER. Keep in mind that the terms *sales organization* and *purchasing organization* are similar in Customizing compared to the freight booking, freight order, and forwarding order: the purchasing organization is the sender of the intracompany settlement document, and the sales organization is the receiving organization.

- In addition, you are required to assign your charge types to primary cost elements in SAP ERP. Navigate to SAP ERP Customizing by selecting INTEGRATIONS WITH OTHER SAP COMPONENTS • TRANSPORTATION MANAGEMENT • INVOICE INTEGRATION • BILLING • INTRACOMPANY SETTLEMENT • ASSIGN TRANSPORTATION CHARGES TO COST ELEMENTS.

Available BAdIs

- You have the flexibility to enhance the methods for intracompany scenarios based on available BAdIs. This gives you the flexibility to develop a custom logic for determining cost elements, cost centers, and internal orders. The available BAdI is called forwarding settlement document BAdI `TCM_SE_CFIRSUITE_RQ` with the method `INBOUND_PROCESSING` (parameter `CS_CO_DOC`). Similarly for intercompany settlement, you can implement a BAdI to influence the assignment of cost objects (e.g., cost centers): BAdI `TCM_SE_CFIRSUITE_RQ` and method `INBOUND_PROCESSING` (parameter `CT_KOMFKGN`).

Internal Settlement Process for Trucking Businesses

In addition to the process of internal settlements for LSPs based on forwarding orders, as described for the international shipment scenario, you can generate internal settlement documents for resources. This functionality supports trucking companies that own trucks and trailers, for example. A procurement organization arranges a trucking service by using the truck and trailer of another organization that owns the resources. An internal agreement can be maintained between the procurement organization and the organization that owns the resources. These rates are used to generate internal settlement documents between the two organizations. After a successful internal charge calculation, you can generate the internal settlement document directly from the freight order by selecting CHARGES/SETTLEMENT • GENERATE INTERNAL SETTLEMENT DOCUMENT.

The settlement integration with SAP ERP works similarly as described in the last section. The resource-owning unit is the sales organization that sends the internal settlement document. The purchasing organization using the truck and trailer is the receiver of the internal charges. Based on the sales organization, the purchasing organization, the cost centers, or internal orders are derived.

11.2.3 Profitability Analysis

Profitability analysis in SAP TM is used to provide an LSP or carrier with a margin per forwarding order. It is not applicable for shipper solutions. The foundation of the profitability analysis lies in the charge management and settlement, which was introduced earlier in this chapter and in Chapter 10. It accumulates all revenue components in a forwarding order and calculates the delta against all cost components that were incurred as part of the shipment execution in order to calculate the profitability of a shipment.

The SAP TM forwarding order has a PROFITABILITY tab. You can trigger the calculation of the profitability by selecting CALCULATE CHARGES • CALCULATE PROFITABILITY.

Figure 11.18 shows an example of the results in a forwarding order with calculated cost, revenue, and profitability.

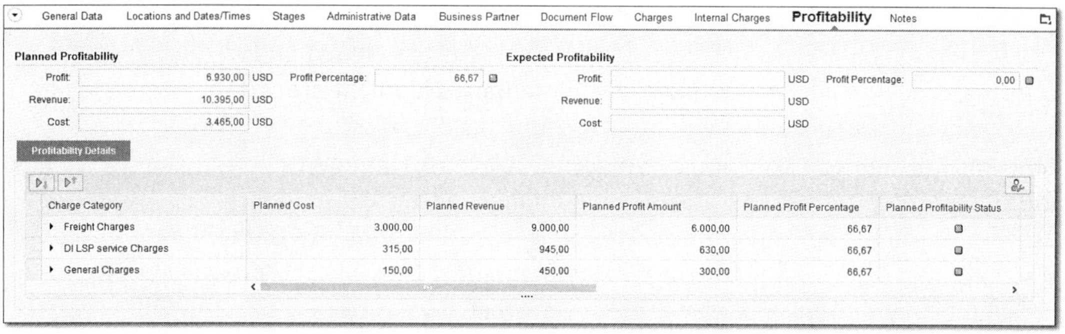

Figure 11.18 Profitability Analysis in Forwarding Order

Calculation logic for profitability

To calculate the profitability, you first need to calculate revenue and cost. The revenue is calculated based purely on the charge calculation for forwarding orders, which are based on forwarding agreements. The cost components are derived from three sources: freight orders, freight bookings, and internal rates. Consequently, you first need to set up the SAP TM Transportation Charge Management component as a prerequisite, as described in Chapter 10. You need to make sure that the charge calculation for forwarding and freight orders and bookings has been executed. Both the cost distribution and internal charge calculation need to be set up and executed as described in Section 11.2.1 and Section 11.2.2 to enable the profitability analysis.

Which cost or revenue components are included in a profitability analysis is not configurable. SAP TM generally uses all charge types of revenue and cost for the comparison. SAP TM calculates both a planned and expected profitability. The major difference between expected and planned profitability is that the latter is based on the charge calculation in forwarding orders, freight orders, and bookings. The expected profitability is calculated based on the settled amounts in the forwarding, internal, and freight settlement documents.

11.3 Settlement with SAP Billing and Revenue Innovation Management

Recall from Section 11.1 that customer billing done with SD billing and FI-CO is quite flexible. It supports full control of integration and also localizations to special billing and taxation standards of a variety of countries worldwide.

Flexibility in billing

However, for companies selling logistics services, the billing flexibility of SAP ERP SD is sometimes not sufficient. Cargo and logistics customers need end-to-end solutions that allow them to efficiently execute invoice services according to service-level agreements (SLAs) defined in customer contracts. Services need to be invoiced the way customers expect or demand in their SLAs. Additionally, an LSP often needs to deal with bad credit history, short and non-payments, credit and collection issues, freight payments to be prepaid in cash before cargo moves, and disputes raised by customers. All these processes need to be handled efficiently with integration into the customer service workplace. Figure 11.19 shows the process steps and the business focus of an end-to-end process concentrating on financial tasks.

Revenue recognition

A second issue for a worldwide LSP is the recognition of revenue in line with legal regulations. If shipments span multiple countries or even continents, or if payments have to be made by multiple parties, you need to properly manage revenue recognition.

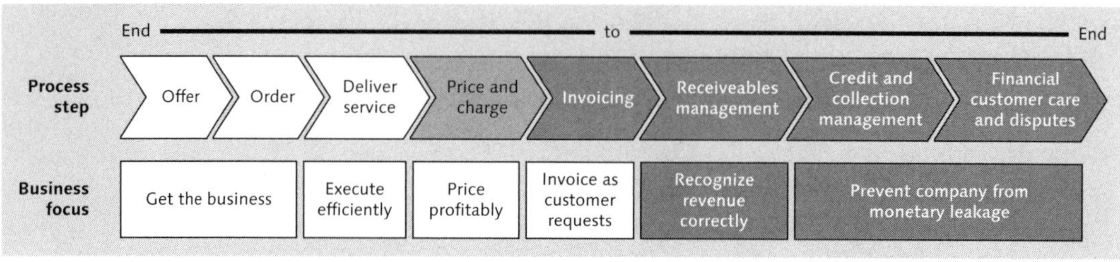

Figure 11.19 End-to-End Process with Billing Focus

Usually, the revenue for a service may only be recognized upon progression of the shipment in the supply chain. Depending on local legal regulations, the following situations may occur:

- A customer orders a rail move across the United States. The payment term is prepaid. The invoice is sent to the customer five days before the pull of the railcars, and the customer pays one day before the pull. The carrier is only allowed to recognize the received revenue upon delivery of the railcar at the destination.
- A customer orders a multi-modal FCL ocean shipment with truck inland haulage in the country of origin, an ocean leg, and inland haulage in the country of destination. Pre-carriage and the ocean move are prepaid by the shipper; the destination haulage is collected from the consignee before delivery of the containers. The carrier can recognize the shipper's revenue in multiple steps. Revenue for origin haulage can be recognized upon delivery of the container at the port of loading. Ocean carriage revenue can be recognized 50% after half of the trip is done; another 50% is recognized once the container arrives in the port of discharge. Consignee revenue is recognized after the container has been delivered in the destination country.

SAP TM integrates with SD, SAP BRIM

The required flexibility in billing and the ability to handle financial customer care and revenue recognition as described above is part of SAP Billing and Revenue Innovation Management (BRIM). SAP TM has the capability to integrate with SAP BRIM via its subcomponent Convergent Invoicing (CI). The technical integration is currently handled by a custom development Repeatable Custom Solution (RCS) and is planned to be available as feature of standard SAP TM in the future. Figure 11.20

shows the integration options. SAP TM creates settlement documents from calculated charges as basis for invoicing. In the PRICE AND CHARGE process step, invoicing and the subsequent financial steps can be either done via SD billing or via SAP BRIM.

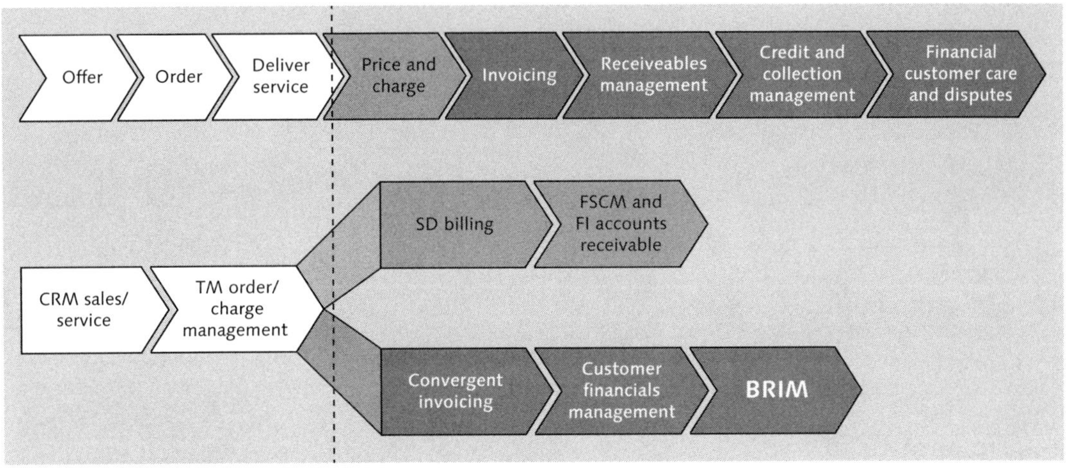

Figure 11.20 SAP TM Integrates with SAP ERP SD and to SAP BRIM for Billing

Looking at the extended billing and collection requirements of LSPs, SAP BRIM is an important building block in their transportation processes. About 90% of LSPs using SAP TM as core solution for their business chose SAP BRIM as their customer billing solution.

SAP BRIM functionality overview

SAP BRIM offers a rich portfolio of services around customer invoicing and receivables management. Figure 11.21 shows an overview of the major abilities of SAP BRIM.

Let's walk through each of these supported functions:

- **Invoice convergence**
  SAP TM creates settlement documents with multiple invoice items. If such a settlement document is sent to SD billing, the whole settlement document is used to create an invoice. With Convergent Invoicing from the SAP BRIM solution, a billable item is created for each charge item of the settlement document. Subsequently, customer profiles (contract account settings) can be used to control a rules engine

that assembles the individual customer invoices from the available charge items. In this case, the content of the invoices is not controlled by agreement and settlement rules in SAP TM, but by contract accounts in SAP BRIM. You can easily use this, for example, to converge fuel surcharges over a longer time while invoicing freight revenue immediately per shipment.

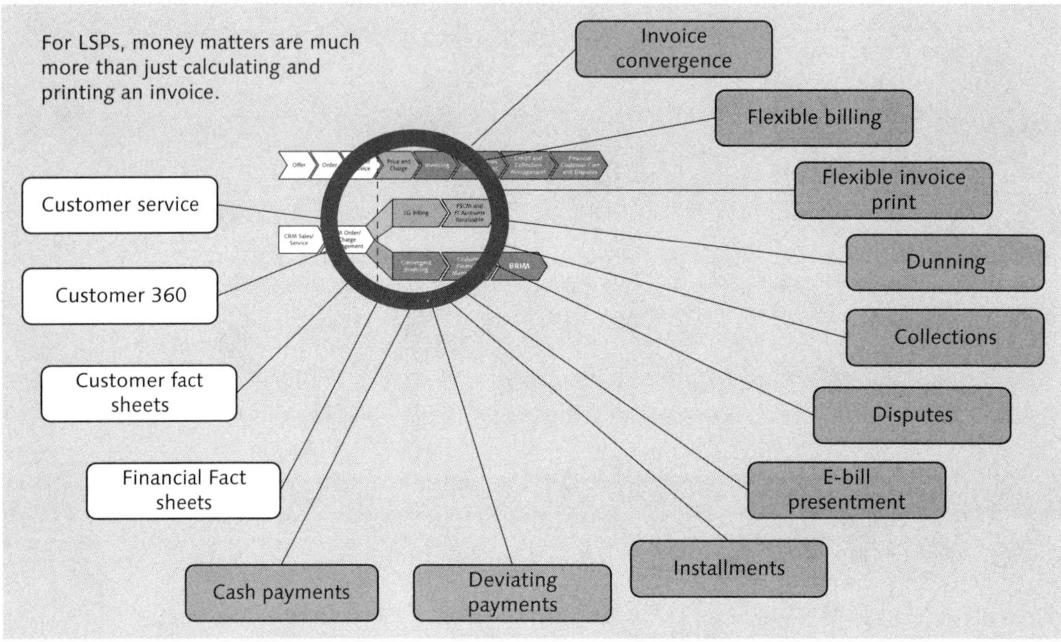

Figure 11.21 Features of SAP BRIM and its Integration into SAP CRM Customer Service

- **Flexible billing and invoice printing**
  Bill creation and communication (technology and paper forms) are also controlled by contract account settings, enabling you to bill according to customers' service requirements.

  The invoice may contain a plethora of logistics and customer context data. In SD billing, this context data is very limited. SAP BRIM, however, offers a powerful concept about how any kind of logistical context data can be stored with each billable item. Therefore, you have all logistics data available to sort, filter, and group items for invoicing.

- **Dunning and collections**
  Dunning and collection mechanisms are very flexible in SAP BRIM. Payment clearing rules and clearing worklists support efficient processing of open receivables items.
- **Invoice and dispute management and customer service**
  Invoice correction capabilities and dispute management are highly flexible in SAP BRIM. Integration into the customer service workplace and the interaction center of SAP CRM is provided. A customer service representative can work directly with invoices in SAP BRIM and manage customer disputes on the charge item level. This has the huge advantage that only the disputed part of an invoice needs to stay open, and the rest can be paid. If, for example, only one item in an invoice with 100 items is disputed, the other 99 items can be collected.
- **E-bill presentment**
  SAP BRIM uses SAP Financial Supply Chain Management functionality for e-bill presentment. A customer can directly access his bills in a portal.
- **Deviating payments and installments**
  If a customer pays an amount that deviates from his invoices or if he pays in installments, these payments can be easily processed against the open receivables.
- **Cash payments**
  In logistics services, especially for ocean cargo or outside of Europe or the United States, cash payments are still a common way of managing receivables. SAP BRIM offers a cashiering functionality for cash desk operation.
- **Customer information**
  To be fully informed about the customer, the customer service workplace offers 360-degree analytics and fact sheet functionality that gives you an overview of customer contracts, orders, payments, disputes, credit history, and other details.

SAP BRIM can do invoicing for SAP TM, but at the same time also process billing data from other system (even legacy systems) and converge multiple receivables streams into joint invoices. You can bring freight invoice data from SAP TM together with warehouse billing data from a legacy warehouse management system and accessorial data delivered from SAP Event Management, which is used to calculate demurrage. You even can use the SAP BRIM component SAP Convergent Charging (SAP CC) to calculate additional prices before going into the invoicing step.

Integrating SAP TM to SAP BRIM

The process integration between SAP TM and SAP BRIM is triggered from the SAP TM side. Figure 11.22 shows how the process is executed. In SAP TM, customer draft invoices (SAP TM forwarding settlement documents) are created based on the charge information in forwarding orders. For a logistical context, the example shows a container shipment from Shanghai, China, to San Pedro, USA. The charge items are sent to SAP BRIM, where billable items (BITs) are created for each charge item. Based on invoicing rules, the available BITs are then converged into three invoices: one for freight charges, one for terminal charges, and a third for additional fees such as a security surcharge.

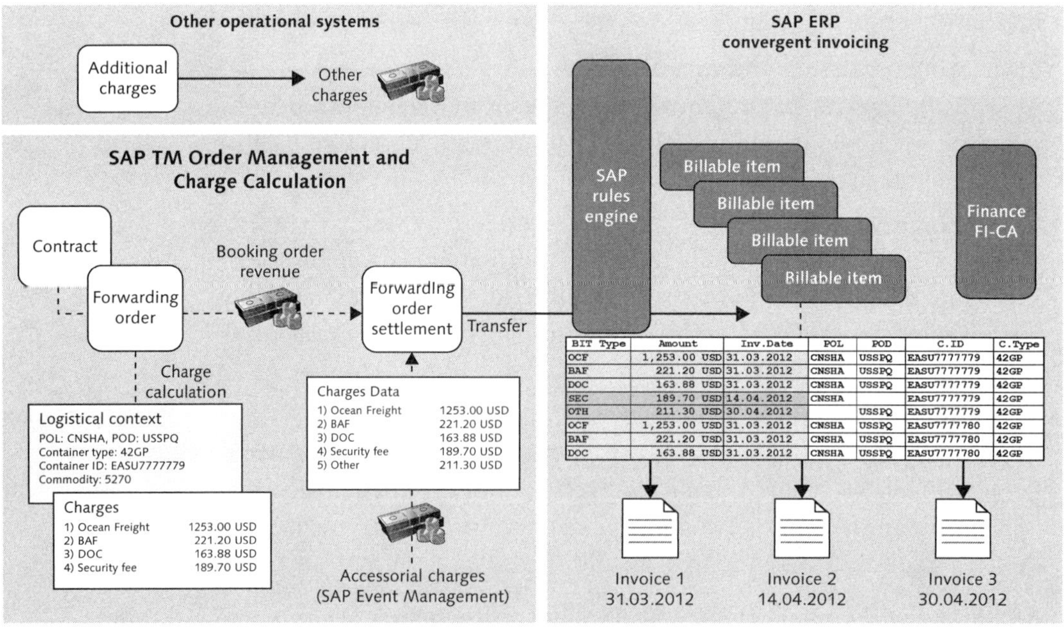

Figure 11.22 Logistics Invoice Creation with SAP BRIM

Accounting is finally done in SAP BRIM Contract Accounting (FI-CA), which is a subledger accounting system in SAP ERP that integrates into the general ledger.

This chapter provided an overview on charge settlement, integration to SAP ERP billing and invoicing, and integration to SAP BRIM as alternative way of customer billing for LSPs. In the next chapter, we offer insight into SAP TM integration to analytics, SAP EWM, and collaborative technologies like portal and mobile access.

If we apply the axiom that a chain is only as strong as its weakest link to business software, we would conclude that a business solution is only as strong as its weakest integration.

12 Integration with Other Components

Though it is usually overlooked and underestimated, integration is one of the most important factors to be considered in a supply chain environment.

You already know that SAP Transportation Management (TM) can map the complex business processes of logistics service providers (LSPs) and shippers. Frequently, such business processes have to be represented across system boundaries. Furthermore, a large volume of data is gathered and added to the business process to satisfy the requirements of carriers, authorities, and customs. Such data is imported electronically and/or automatically ascertained by the system. Nowadays, the volume of data generated during a transportation process is immense, so it is not always easy for an end user or enterprise to draw conclusions from a success or failure at the document or aggregated level.

In this context, SAP TM not only offers strong integration capabilities with analytics solutions from SAP, but also uses the potential and opportunities of mobile devices. We cover these in these in the first part of this chapter.

Integration with other SAP components

When it comes to supply chain operations, SAP TM offers both strong implementation capabilities with third-party solutions and seamless integration with other applications from the SAP Business Suite:

- SAP ERP
- SAP Extended Warehouse Management (EWM)
- SAP Global Trade Services (GTS)
- SAP Event Management

In the previous chapters we have learned that tight integration with other SAP components is mainly used to facilitate transportation planning and execution in SAP TM. The integration of SAP ERP orders and deliveries is represented by transportation requirements in SAP TM (see Chapter 4), while SAP GTS is used to perform compliance checks and customs processing (see Chapter 9). SAP Event Management is used to track and monitor events and statuses for various business objects within the supply chain. The integration with SAP Event Management has been explained in Chapter 8. Integration with analytics is explained in the next section.

Integration with SAP EWM

But from end to end, supply chain management goes far beyond transportation planning and execution. The fulfillment process also uses warehouse operations to source, store, and handle products within the transportation network. The second part of this chapter explains the process integration with SAP EWM.

Collaboration is important and typically uses the latest technology. The third section explains how mobile applications and portal access support the external collaboration in SAP TM.

Analytics and mobility

Available SAP Documentation

The SAP Solution Manager content documentation and installation guides for SAP TM 9.0 offer a comprehensive level of detail and configuration steps to show how the current release of SAP TM can be integrated with other components of the SAP Business Suite.

To learn more about the integration with SAP NetWeaver BW and or SAP EWM via SAP ERP, refer to the installation and upgrade guides at *http://service.sap.com*. Log on with your service user and follow the menu path INSTALLATION & UPGRADE GUIDES • SAP BUSINESS SUITE APPLICATIONS • SAP TM • SOLUTION MANAGER CONTENT DOCUMENTATION.

Two integration guides are relevant for this chapter:

- Integration of SAP NetWeaver BW
- Integration of SAP EWM via SAP ERP

12.1 Analytics

Because they make fact-based decisions, enterprises can benefit greatly from detailed analyses that identify profitable and less profitable transportation and shipping routes. They can use historical data as a basis for learning for the future, structuring their business differently, negotiating with new customers, or plotting shipping routes. A consolidated view of all data from day-to-day activities can give LSPs an important competitive edge. Analyses that connect logistical data with financial data are an important corporate management tool in competition-intensive and low-margin logistics.

In this scenario, analysis-oriented information systems are used to support both planning and strategic processes. They provide enterprises with current and historical data. These systems are frequently based on a data warehouse in which relevant data is collected, formatted, and made available. The core of a data warehouse is a (usually relational) database.

Analytics in SAP TM

In this section we explain how analytics and the integration with SAP TM help decision makers to answer business-related and specific questions.

When it comes to analytics we usually distinguish between the following:

- Role- or user-specific worklists
- Embedded contextual analytics
- Ad hoc reporting and queries
- Graphical dashboards

SAP NetWeaver BW

User-specific worklists have already been explained as part of the user interface technology in SAP NWBC (please refer to Chapter 2, Section 2.2). With SAP NetWeaver BW, SAP provides a solution that includes all of the required components for setting up data warehouse architecture for transportation reporting—embedded, contextual, or using graphical dashboards. In addition to the basic technology for data retention, the system provides all of the essential components for evaluating the transportation-relevant data stored in SAP NetWeaver BW (i.e., reporting tools, data mining methods, and an option for a portal connection). This chapter focuses on analytical reporting, integrating SAP TM with SAP NetWeaver BW, which content is provided, and how the data will be provisioned.

Operational data provisioning

For operational data provisioning, as of SAP TM 9.0, most of the documentation of data sources and BI content has been moved from the BI content documentation to the SAP library for SAP TM. Therefore, in accordance with the SAP Help for SAP TM, we follow the same structure here:

- Business intelligence content (Section 12.1.1)
- Operational reporting and data provisioning (Section 12.1.2)
- Business Context Viewer (Section 12.1.3)

We assume that you are familiar with the main concept of SAP NetWeaver BW and therefore provide some further details about how this data is extracted from the source system, and how—and in which context—the data can be used for operational reporting.

Reporting in general

In the transportation industry, reporting ranges from structural data analysis in day-to-day activities and complex analyses of individual cost components, through to determining key performance indicators (KPIs) for use by corporate management. It demands knowledge and access to data from various operational systems such as transportation, tracking, and financial accounting systems and sometimes even purchasing systems.

In practice, system landscapes are complex, and cross-system enterprise reporting must be possible. Furthermore, both current transaction data and historical data are required to achieve the best possible results. In many enterprises, employees still integrate and prepare data by consolidating Excel tables at the end of the month and generating charts. This is not only a laborious process, but one that is prone to errors. Furthermore, it can be performed by data warehouse systems like SAP NetWeaver BW.

Reporting with SAP NetWeaver BW

A reporting system must permit both incoming and outgoing access to data, irrespective of the data format. SAP TM uses SAP NetWeaver BW to create analyses and reports. SAP NetWeaver BW can be used to process data from SAP systems, legacy systems, web services, and flat files. The data is then displayed either in tabular form or in a meaningful dashboard that consists of traffic lights, meter displays, or charts. SAP NetWeaver BW allows the relevant people within an enterprise to obtain the latest data at the press of a button, thus allowing them to make informed decisions and manage day-to-day activities accordingly.

Business intelligence

The latest technologies in business intelligence use encapsulated enterprise services to extract data from very different systems and display it in an easy-to-configure user interface. Therefore, reporting is no longer just programming; it also involves business process design, which can be performed by almost all employees because drag-and-drop functionality can be used to configure both the data extraction and display levels.

Reporting examples

Let's consider a brief example that shows how reporting based on transactional transportation data can be used to monitor indicators in different enterprise areas. This carbon dioxide example is based on data obtained directly from SAP TM, but it can also be based on or enhanced with data obtained from flat data structures or legacy applications.

Dashboards

Figure 12.1 shows a dashboard that provides the user with a great deal of information. The example concerns carbon dioxide transportation emissions and the analysis by region and transportation mode (land, air, and ocean) on a monthly or quarterly basis. At the press of a button, the user can view the extent to which each zone accounts for total costs, individual and total distance, as well as the contribution to the carbon dioxide footprint. Data can be visualized as a graphic or explored in table format. In addition, the user can identify the total and average carbon dioxide emission by weight and volume.

Integration with SAP NetWeaver BW

Let's apply this SAP TM example to explain the provided BI content of SAP TM and the integration with SAP NetWeaver BW in much greater detail.

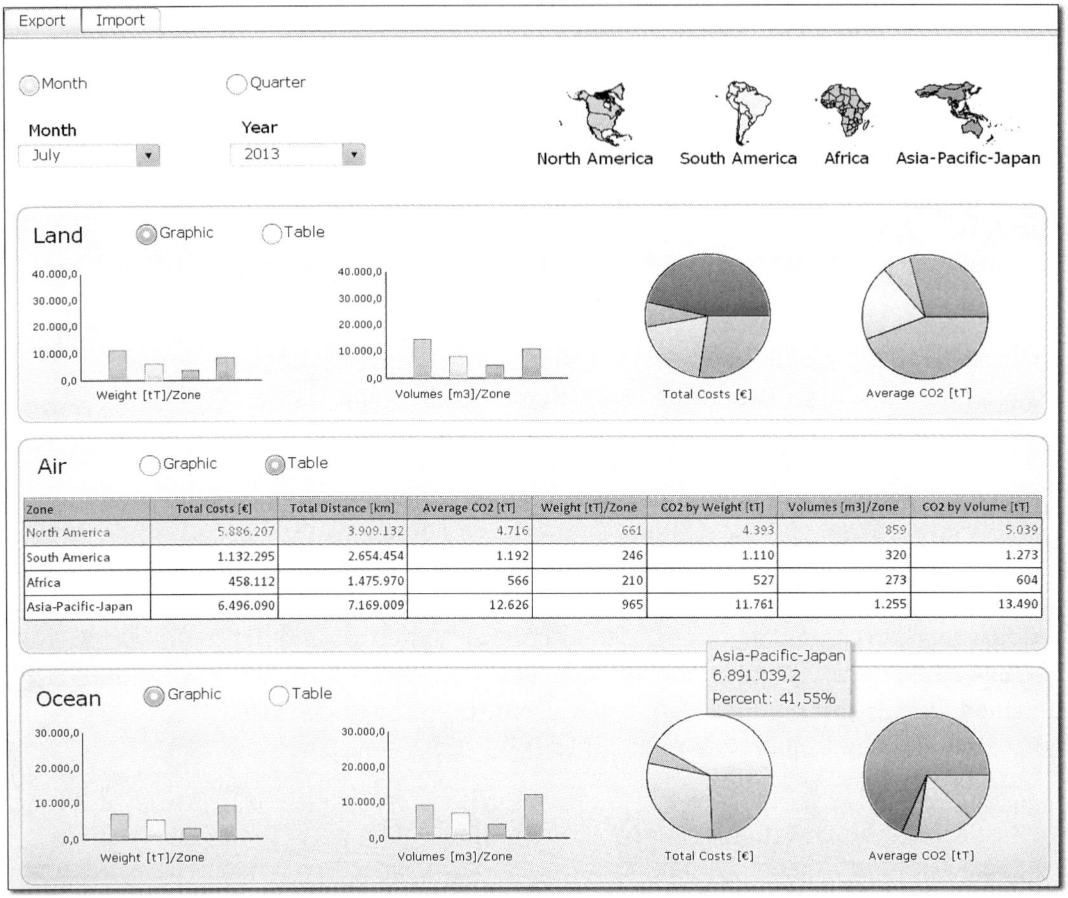

Figure 12.1 Sample Dashboard to Consolidate Business Information

12.1.1 Business Intelligence Content

SAP provides delivered content within SAP NetWeaver BW and delivered extractors for transferring data from SAP TM to SAP NetWeaver BW to further analyze transportation-related data.

Supporting business decisions

SAP TM supports you in managing the physical transportation of goods from one location to another. The BI content of SAP TM empowers decision makers and employees with the queries and analytic tools to evaluate, analyze, and interpret transportation-related business data.

In this section we explain the principles of data extraction from the SAP TM source system and data sources, and the content that provides the data basis and analytical foundation to support and answer daily questions:

- How much weight or volume is transported per trade lane, business share, or transportation allocation—and how long was the distance or duration of individual transports on average?
- How many containers have been shipped from one location to another—and what is the percentage of dangerous goods shipments compared to the overall number of shipments per trade lane?
- What are the average transport costs per trade lane, carrier, shipper, or consignee—and how reliable was the shipper, carrier, or consignee?

Business Intelligence Content

To learn more about the business intelligence content for SAP TM, and for a full list of data sources, visit the SAP Online Help Portal for SAP TM: *http://help.sap.com*. You will find the relevant information at SAP LIBRARY • SAP BUSINESS SUITE • SAP TRANSPORTATION MANAGEMENT UNDER ANALYTICS • BUSINESS INTELLIGENCE CONTENT.

Data Extraction from SAP TM

After examining the application examples and economic background of key performance indicators and reporting, we now continue with a look at the integration of logistics applications with SAP NetWeaver BW.

In general, SAP logistics applications run either via the user interface or through planning or batch reports; a data backup generally takes place before the end of the transaction. Once the backup is completed (i.e., when the transaction is obviously finished and new data has been updated), we need to extract the data relevant to characteristic and key figure determination. Figure 12.2 shows the principle integration process.

Let's walk through it.

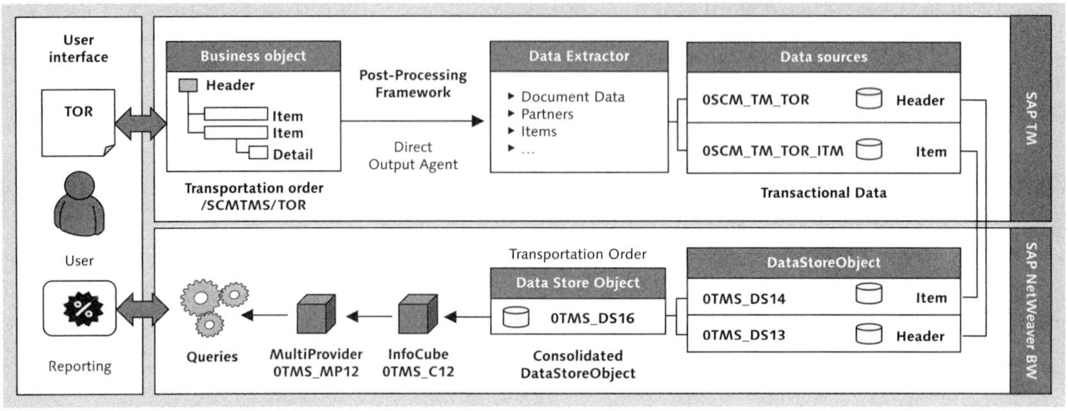

Figure 12.2 Integration with SAP NetWeaver BW

Previous releases of SAP TM

In previous releases of SAP TM, the action definitions and processing types that refer to data extractors are defined in the Post-Processing Framework (PPF) action profile for the shipment request, shipment, and shipment order. Here, a method call is referenced in each case. If you want to enhance or change the standard extraction in any way, you can replace this extractor with your own extractor. The data extractor is only called if the PPF condition for the extraction action is identified as being a true condition. You can define this condition in Customizing for the PPF conditions. The standard condition defined here is such that the business objects obtain the status *Completed*, meaning that no further changes are made. It generally makes the most sense to use the *Completed* status if the data of the processes that are currently running does not have to be included in the SAP NetWeaver BW analyses.

Adapter settings in SAP TM 9.0

In the current release of SAP TM, however, SAP NetWeaver BW integration is based on outputs that do not involve the PPF. Although the configuration is still part of the PPF adapter configuration, output determination and relevance are based on the direct output agent relevance of the business object.

You can maintain this relevance and direct output configuration in the PPF settings for extracting data for SAP NetWeaver BW in the Customizing of SAP TM. In the IMG for SAP TM, select CROSS-APPLICATION COMPONENTS • PROCESSES AND TOOLS FOR ENTERPRISE APPLICATIONS • REUSABLE OBJECTS AND FUNCTIONS FOR BOPF ENVIRONMENT • PPF ADAPTER

FOR OUTPUT MANAGEMENT • MAINTAIN OUTPUT MANAGEMENT ADAPTER SETTINGS. Figure 12.3 shows the Customizing screen for the delta upload configuration of a transportation order.

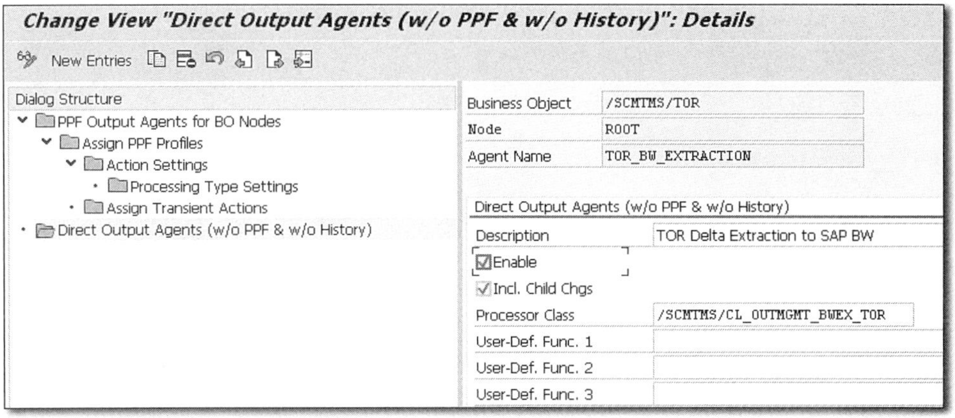

Figure 12.3 Transportation Order: Delta Upload Configuration

General Settings in SAP TM

Before you can use SAP NetWeaver BW for reporting, you have to connect SAP TM with the reporting system. You can find the relevant settings in the IMG for SAP TM by following the menu path INTEGRATION WITH OTHER SAP COMPONENTS • DATA TRANSFER TO BUSINESS WAREHOUSE • GENERAL SETTINGS. Here you enter the necessary parameters to specify the source system and the maximum size and rows of the data packages that will be transferred into SAP NetWeaver BW. These settings are mainly cross-client.

DataSources

Before the data extraction and integration can be configured, use Transaction RSA9 to transfer the application component hierarchy. The necessary DataSources are then delivered as part of the business content of SAP TM and can be activated with Transaction RSA5. These DataSources belong to the source system (here, that is SAP TM), and contain all relevant fields that can be transferred to SAP NetWeaver BW.

Figure 12.4 shows Transaction RSA5 in SAP TM. The relevant DataSources for master data, texts, and transactional data can be found under hierarchy node 0SCM_TM_DATASOURCES. In this context we have to distinguish between an *initial* or *delta upload* from SAP TM to SAP NetWeaver BW.

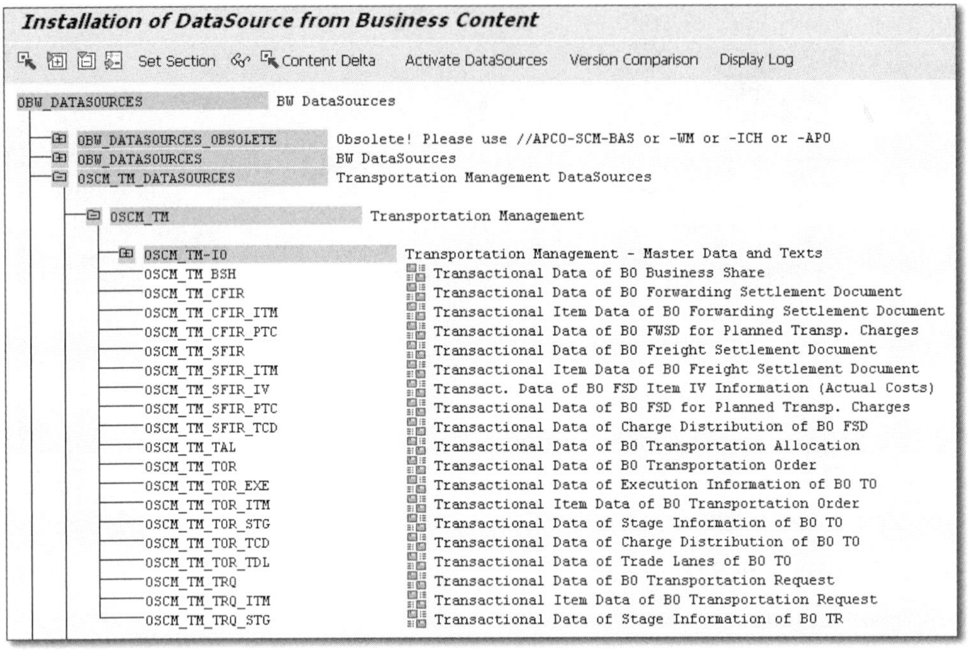

Figure 12.4 Business Content Installation in SAP TM

Initial upload

The initial upload of data to SAP NetWeaver BW can be initialized with or without a physical data transfer. The initial data upload of existing data with data transfer is done using setup tables in SAP TM. Once the data in these setup tables has been deleted, they can be filled with data from the relevant SAP NetWeaver BW extractor. You can find the initialization of these setup tables, the relevant transactions to fill them with existing data, and the execution of the data upload into the SAP NetWeaver BW system in the SAP Easy Access Menu of SAP TM: ADMINISTRATION • INITIAL DATA UPLOAD TO SAP NETWEAVER BW (see Figure 12.5).

Finally, use Transaction RSA6 to perform an extraction test and display the data records of a specific DataSource.

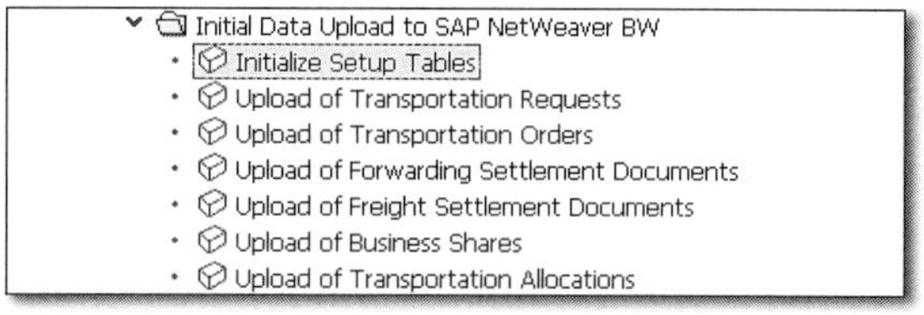

Figure 12.5 Initializing and Filling Setup Tables for Initial Upload

Delta upload

Recall that settings in the PPF are used to control the process of extracting data from SAP TM. For the standard delivery, the following business objects have been integrated into SAP NetWeaver BW:

- Transportation order
- Forwarding settlement document
- Freight settlement document
- Transportation request
- Business share
- Trade lane
- Transportation order execution
- Transportation allocation

As part of normal storage of one of these business objects, processing is triggered in the PPF (see Figure 12.2). In previous releases of SAP TM, once the PPF condition has been successfully checked, the data is extracted using an extraction method specifically built for the business object. This method then makes the data available in a DataSource so that it can be transferred to SAP NetWeaver BW. In the current release of SAP TM, the extraction relevance is based on the settings made for the *direct output agent configuration* (see Figure 12.3).

Analytics is mainly based on the extraction, processing, and visualization of business data. In the next section, we describe both the provisioning and extraction of data from SAP TM and how it is ultimately used for reporting.

12.1.2 Data Provisioning and Operational Reporting

The data is loaded from the SAP TM system to SAP NetWeaver BW into the Persistent Staging Area (PSA). In SAP NetWeaver BW, the Data-Source information is stored in DataStore objects, which then fill Info-Cubes with information that is subsequently used by the MultiProviders from the various sources of data. Once the data has been made available in the MultiProvider, you can use different queries to access the information in SAP NetWeaver BW and then combine this information with other available data, specifically for the purpose of conducting analyses.

Data Provisioning

Before it can be stored in SAP NetWeaver BW, the content is systematically structured on the basis of the technical relevance of the content that is supposed to be mapped. This structure in turn determines and provisions the available options for transportation analytics.

> **Note**
>
> In this book, we can't provide a full introduction to the principles and fundamentals of data provisioning and modeling using SAP NetWeaver BW. However, to familiarize you with the core elements of data retrieval, we provide a brief overview of how SAP TM data is ultimately retrieved and provisioned. For a comprehensive overview, practical examples, and further insight, we recommend *Data Modeling in SAP NetWeaver BW* by Frank K. Wolf and Stefan Yamada (SAP PRESS, 2010).

To become familiar with the core elements of analytics, you need to understand the definition of core objects and structures and how these InfoProviders provide a basis for the provisioned queries in SAP NetWeaver BW.

InfoProvider

An InfoProvider is an SAP BusinessObjects BI object that can be used as a source of data for a query. SAP divides InfoProviders into two categories: those that physically contain data (in a database table) and those that do not. The data objects that contain InfoProviders are InfoCubes, DataStoreObjects, and InfoObjects. Those that do not contain any data themselves refer to other InfoObjects and/or systems for data. Among these are the VirtualProviders, InfoSets, and MultiProviders.

Four of these objects are an important part of the provided standard content and are a basis for the provisioned queries in SAP NetWeaver BW (see Figure 12.6):

- InfoObject
- DataStoreObject
- InfoCubes
- MultiProviders

Let's take a look at these four in more detail. To understand the integration and relationship between these objects, reconsider the example provided in Figure 12.2.

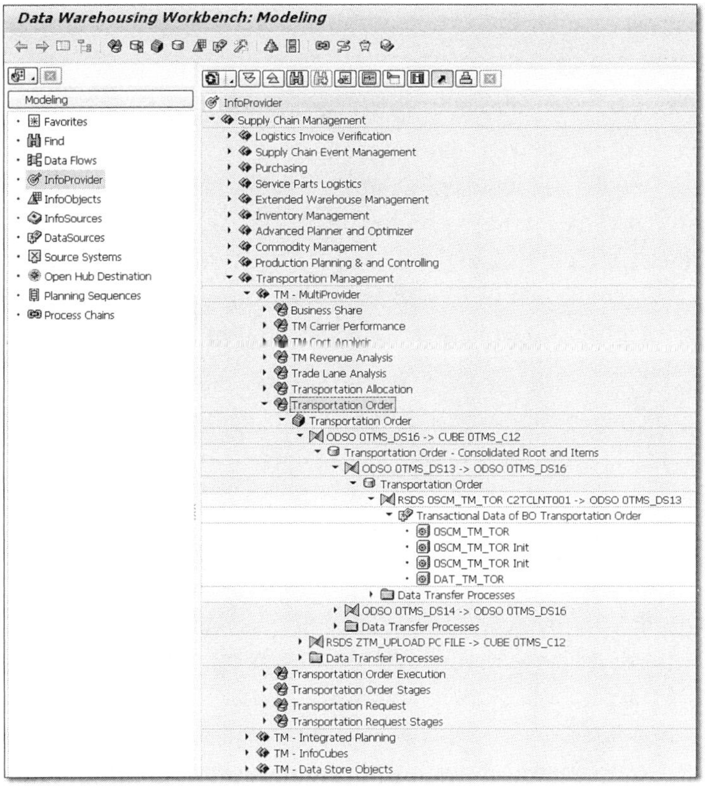

Figure 12.6 Data Warehousing Workbench in SAP NetWeaver BW

In addition, to learn more about the relationship, characteristics, and how the data is ultimately modeled in SAP NetWeaver BW, take a look at the Data Warehousing Workbench in SAP NetWeaver BW via Transaction RSA1 (Figure 12.6) to start the discussion.

InfoObject An InfoObject is a basic building block in SAP BusinessObjects BI data modeling. It acts as a definition for the data (metadata), describing elements such as the type and length of the data, but it can also contain data itself. InfoObjects are divided into KeyFigures (which contain numerical values, amounts, currencies, etc.) and characteristics (texts and IDs of objects). A characteristic, which contains other InfoObjects (either KeyFigures or other characteristics) as attributes, can be used as an InfoProvider.

DataStoreObject A DataStoreObject (DSO) is a store for transactional data that is described by InfoObjects. Some of the InfoObjects are defined as key fields, and the rest contain the "trunk" of the data. Data records with an identical key part can be summed or overwritten.

There are three types of DSOs: standard, write-optimized, and direct-update. In the context of transactional reporting of SAP TM data, the term always refers to the standard DSO. Physically, the standard DSO consists of three database tables: one for new data (has additional, technical key elements), an active data table (only the semantic key), and a change log. DSOs can act as InfoProviders alone but usually are part of an InfoSet or act as an intermediate step in updating the data into an InfoCube (see Figure 12.2).

Figure 12.7 shows the DSO `0TMS_DS14` for the transportation order line items in SAP NetWeaver BW. It contains all data fields from the corresponding data source `0SCM_TM_TOR_ITM` in SAP TM. Together with the header data from the business object `/SCMTMS/TOR` (transportation order), it is consolidated into the DSO `0TMS_DS16` (see Figure 12.2).

InfoCubes From a reporting perspective, an InfoCube is possibly the most commonly used InfoProvider for reporting purposes. It is built from multiple InfoObjects that are grouped together into *dimensions*. The transaction data is stored in a *fact table* to which these dimensions are linked. InfoCubes in SAP NetWeaver BW are either standard or real-time. In this chapter, the term always refers to standard InfoCubes.

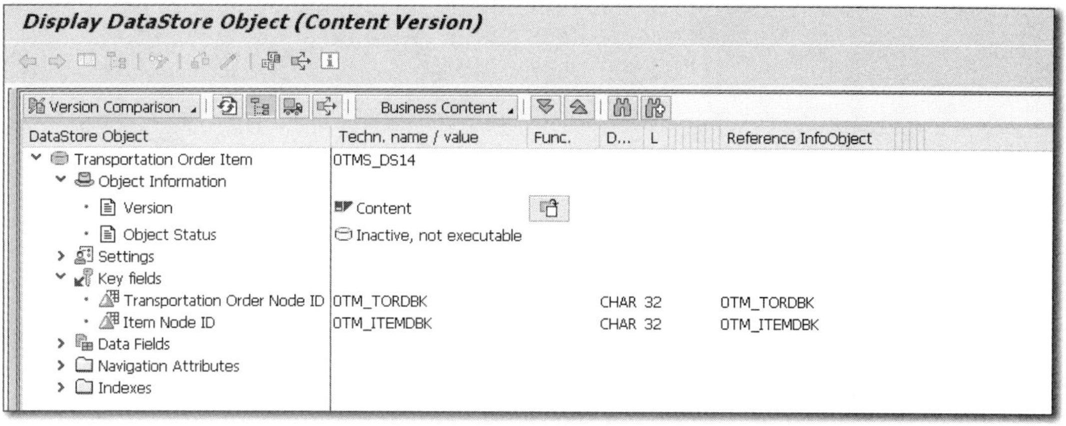

Figure 12.7 DataStore Object: Transportation Order Item

An InfoCube always has three technical dimensions (unit, time, and package) in addition to the user-definable dimensions. The unit dimension contains the units or currencies of the key figures, the time characteristics are included in the time dimension, and the package dimension contains technical characteristics relevant to each data load.

In SAP NetWeaver BW, these InfoCubes exist for virtually all transactional data of SAP TM business objects. Table 12.1 shows which InfoCubes are provided and used for which data.

InfoCube	Description
0TMS_C11	Transportation request
0TMS_C12	Transportation order
0TMS_C14	Planned costs
0TMS_C15	Actual costs
0TMS_C16	Revenue
0TMS_C18	Transportation order execution
0TMS_C19	Business share
0TMS_C20	Transportation allocation

Table 12.1 InfoCubes for SAP TM

InfoCube	Description
0TMS_C21	Trade lanes
0TMS_C22	Transportation request stages
0TMS_C23	Transportation order stages

Table 12.1 InfoCubes for SAP TM (Cont.)

MultiProviders

Other than the basic InfoCubes, several other InfoProviders are available in SAP NetWeaver BW. As an example, MultiProviders are a special type of InfoProvider that can combine data from several InfoProviders. They are then available for reporting. Like InfoSets and VirtualProviders, MultiProviders themselves are simply based on a logical definition and do not contain any data. Their data comes exclusively from the InfoProviders on which they are based. MultiProviders can be based on any combination of the following InfoProviders:

- InfoCubes
- DSOs
- InfoObjects
- InfoSets
- Aggregation levels

Therefore, if a MultiProvider allows you to run reports using several InfoProviders, then it is used for creating reports on more than one InfoProvider at a time. Each of the MultiProviders contains information about the data package, time stamps, units, locations, document numbers, partners, and other dimensions, as well the business object's KPIs. The MultiProviders are defined the same way as the aforementioned InfoCubes.

MultiProviders are often used "on top of" other InfoProviders to simplify the structure for end users and to join two InfoProviders of similar structure together (see Figure 12.2). A classic example is joining two InfoCubes that are identical in design, but one contains actual values and the other contains planned values. The MultiProvider allows easy creation of queries, where these two sets of values can be compared. Table 12.2 lists the MultiProvider definitions for SAP TM.

MultiProvider	Description
0TMS_MP11	Transportation request
0TMS_MP12	Transportation order
0TMS_MP13	SAP TM cost analysis
0TMS_MP16	SAP TM revenue analysis
0TMS_MP18	Transportation order execution
0TMS_MP19	Business share
0TMS_MP20	Transportation allocation
0TMS_MP21	Trade lane analysis
0MTS_MP22	Transportation request stages
0MTS_MP23	Transportation order stages

Table 12.2 MultiProviders for SAP TM

Standard Reporting Content

After the data has been placed in a MultiProvider, you can access information in SAP NetWeaver BW with various *queries* and combine this information with other available data to perform targeted analyses. These analyses can be displayed in spreadsheets, reports, or dashboards.

Queries

You can use queries to analyze transportation processes. The queries delivered with SAP TM are typically based on the information extracted from the transportation request and transportation order business objects and the MultiProviders described in the previous section.

Figure 12.8 describes the transportation-specific queries available in SAP NetWeaver BW. If you want to enhance any of the existing queries, you can use your own queries, which are based on the MultiProviders in SAP TM and other application areas.

SAP standard content

In addition to these InfoProviders and queries, SAP also provides standard reporting content. The COST ANALYSIS FOR TRANSPORTATION MANAGEMENT dashboard example in Figure 12.9 shows of how data can be made available, visualized, and used as a basis for your own reports.

Multi-Provider	Query	
Transportation request	0TMS_MP11_Q0001 0TMS_MP11_Q0002	Customer analysis Party/location analysis
Transportation order	0TMS_MP12_Q0001 0TMS_MP12_Q0002	Carrier analysis Transport analysis
Transportation cost analysis	Planned/actual costs per ...	
	0TMS_MP13_Q0001 0TMS_MP13_Q0001_1_EXT 0TMS_MP13_Q0001_2_EXT 0TMS_MP13_Q0001_3_EXT 0TMS_MP13_Q0001_4_EXT	Transport Carrier Consignee Shipper Transportation mode
	Costs per ...	
	0TMS_MP13_Q0003 0TMS_MP13_Q0004	Number of units shipped per period and mode of transport Carrier
Transportation revenue analysis	0TMS_MP16_Q0001 0TMS_MP16_Q0002	Planned revenue per customer Planned revenue for top 5 customers
Transportation allocation	0SCM_TMS_MP20_Q0001	Carrier by transportation allocations
Business share	0TMS_MP19_Q0001	Business share by carrier
Trade Lane analysis	0TMS_MP21_Q0001 0TMS_MP21_Q0002 0SCM_TMS_MP21_Q0002	Transported weight per trade lane Trade lane analysis per transport Carrier analysis by trade lane
Transportation order execution	0TMS_MP18_Q0001) 0TMS_MP18_Q0002 0TMS_MP18_Q0003 0TMS_MP18_Q0004	Delivery reliability Expected events reliability Unexpected events Duration analysis

Figure 12.8 SAP TM Queries and Data Sources

Cost analysis for SAP TM

You can use this dashboard in SAP TM to do many things:

- Get an overview of costs from the perspective of a carrier, consignee, or shipper
- Display key figures such as actual costs, planned costs, or cost savings

- Access this information for all transportation modes and shipments for any month in the last two years
- Display additional information about the associated source and destination locations

The system extracts the key figures from the SAP NetWeaver BW system when you choose a country in a particular region. The dashboard is intended for managers in the transportation and logistics areas.

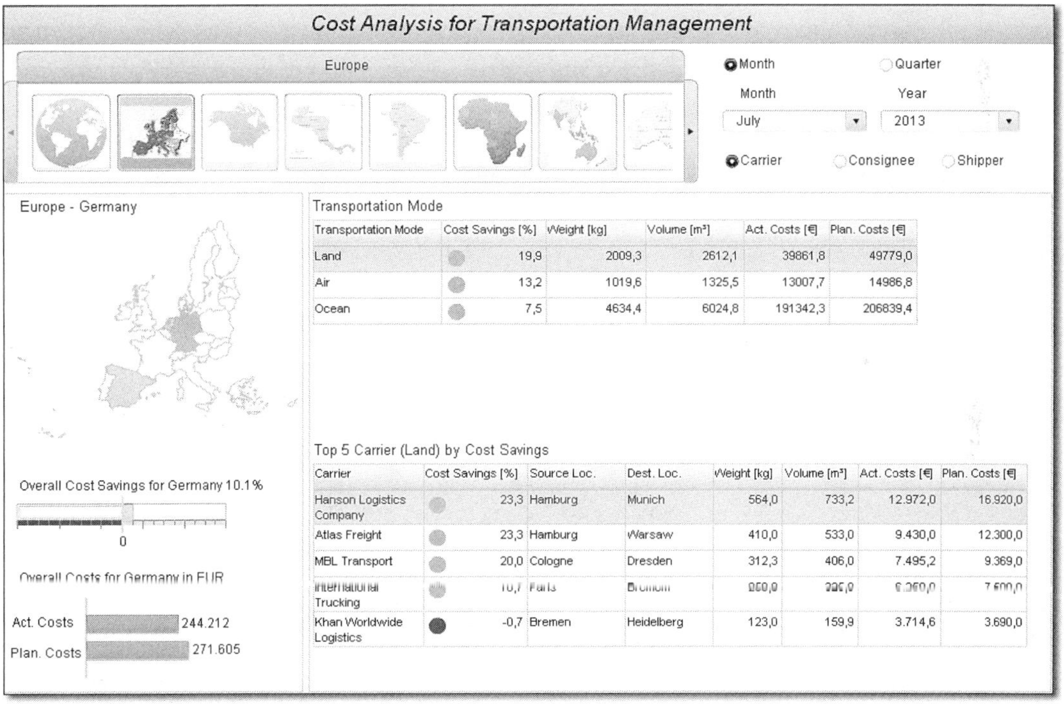

Figure 12.9 Dashboard: Cost Analysis for Transportation Management

12.1.3 Embedded Analytics with Business Context Viewer

Typically, three levels of analytics support decision making in organizations:

- At the corporate level, analytics mainly supports the strategy concerning the direction, composition, and coordination of various business activities within a large and diversified transportation network.

- At the business level, the strategy relates to the creation of competitive advantage.
- At the operational level, analytics usually supports a combination of resources, processes, and competencies to put a strategy into effect or be competitive.

Ad hoc decision support

To support real-time operations and ad hoc decisions, business analytics should offer the crucial ability to integrate different data sources and turn them into valuable insights (see Figure 12.10). The value is mainly based on the ability to collect, analyze, and act on data when the information is needed. To support ad hoc decision making, SAP TM provides standard integration with the Business Context Viewer.

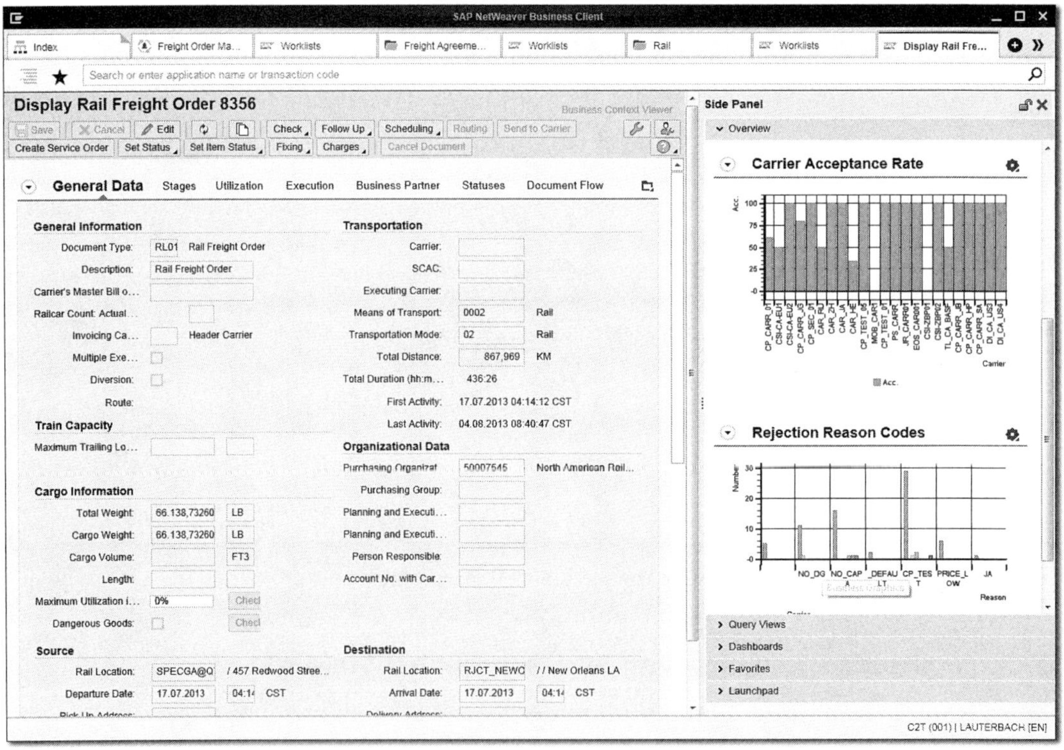

Figure 12.10 Freight Order Detail with Activated BCV Side Panel

Side-bar reporting

Using side-bar reports, full-screen analytical views, and embedded contextual analytics, the Business Context Viewer is a generic framework that allows SAP Business Suite applications to integrate different kinds of additional information into the context of their applications. This information can then be analyzed according to the user's business needs.

Data Provisioning

The Business Context Viewer is not exclusively linked to SAP TM, but it is available to all SAP Business Suite applications. It is mainly a technical framework that uses Web Dynpro ABAP as an underlying UI technology to get and bring data to the UI. This data can either result from individual ABAP coding where relevant information is directly selected from the database or be retrieved from the Business Object Processing Framework (BOPF; refer to Chapter 2 to learn more about the technological foundation of SAP TM). In addition, other data providers can pass information to the user interface, such as InfoSets, web services, workflows, enterprise search, and SAP NetWeaver BW queries, which were mentioned in the previous section (see Figure 12.6). The analyzed data is then shown in a side panel either as a table or Web Dynpro Business Graphics or as an embedded dashboard.

Information about the Business Context Viewer

Although SAP delivers some preconfigured examples that can be a basis for individual reporting, the Business Context Viewer is mainly a technical component that can be used to define and model individual reporting requirements.

Unfortunately, we can't dive deep into the technical details of how the Business Context Viewer is integrated and enhanced or how individual content can be visualized here, so we recommend that you refer to SAP Community Network (SCN) at *http://scn.sap.com*. Use the free text search function and search for "Business Context Viewer". You will find additional background information, cook books, and how-to guides.

You can also consult SAP Help to learn more about the configuration and display of data and how to use search connectors and data providers to retrieve transportation-relevant information. Go to *http://help.sap.com* and follow the menu path SAP Business Suite • Product Lifecycle Management • SAP Portfolio and Project Management • Processes and Tools For Enterprise Applications.

Basic Configuration SAP TM

You know that the Business Context Viewer is a generic framework that can be used by several applications at the same time, so it's necessary to separate its configuration from any other configurations that might exist in the system. SAP TM uses a specific combination of application and business document that identifies both the hosting application and the processed business object.

Prerequisites

To display SAP TM content in the Business Context Viewer, you first have to install and configure the Internet Graphics Service (IGS) for SAP TM. To access transportation-relevant data from the different data provider you then activate the Business Context Viewer Main Application, which is provided via business function `/BCV/MAIN_1` and is part of the SAP Business Suite Foundation 7.03 (SAP_BS_FND 703). In addition, to specify for which business documents and categories Business Context Viewer content should be available, you have to activate the Business Context Viewer side panel link in the main application.

Business Context Viewer Customizing

You can find these settings in the Customizing of SAP TM by following the IMG menu path SAP Transportation Management • Transportation Management • Basic Functions • User Interface • Define Settings for Business Context Viewer (see Figure 12.11).

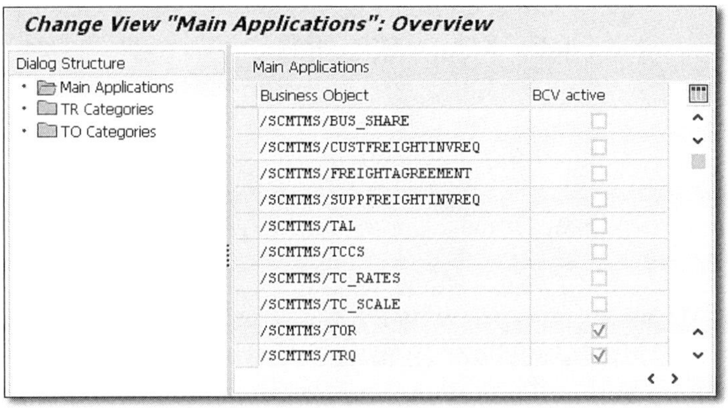

Figure 12.11 BCV Applications

Here you can activate the Business Context Viewer link for the related main applications, and maintain for which business objects the Business

Context Viewer content will be shown. In this example, the Business Context Viewer is active for `/SCMTMS/TOR` and `/SCMTMS/TRQ`. If you would like to activate the Business Context Viewer for transportation requests or transportation orders, you can select the relevant business document category in TR CATEGORIES and TO CATEGORIES, respectively. In this context, you can select the following transportation requirements:

- Order-based transportation requirement (OTR)
- Delivery-based transportation requirement (DTR)
- Forwarding order
- Forwarding quotation

In addition, you can also specify the Business Context Viewer relevance of the individual business document categories such as freight booking, freight units, service orders, freight orders, and freight order copies for tendering.

12.2 SAP EWM Integration

We have seen in Chapter 4 that SAP ERP orders and deliveries are seamlessly integrated into the transportation process and are represented as transportation requirements in SAP TM. They are the basis for transportation planning, where requirements are consolidated into freight orders under consideration of predefined conditions such as volume, desired arrival time, or compatibility of means of transport, and the goods to be transported (see Chapter 6).

SAP ERP transportation requirements

SAP TM facilitates transportation planning and execution, and enables the controller to run a transportation planning based on SAP ERP sales order requirements or outbound deliveries. This tight integration allows SAP TM to continuously react to changes to orders and deliveries that occur in the SAP ERP system and supports the optimization of transportation costs and efforts in a flexible and optimized way. From a cargo handling perspective, the results of transportation planning have a direct impact on warehouse logistics and operations.

Warehouse logistics

A *warehouse* is typically defined as a structural unit with all resources and organizational provisions necessary for the execution of processes connected to inventory and warehouse management, including the organizational units involved with goods receipt and shipping. In a nutshell, warehouse *logistics* involves the storage, maintenance, and handling of goods in warehouses, while warehouse *management* (or warehousing) is in the realm of materials management, a process where goods are temporarily stored or rerouted to a different channel in the network. In this context, we can make a distinction between two basic warehouse types:

- Transshipment warehouses and hubs
- Distribution and supply centers

Transshipment warehouses and hubs

From a transportation logistics perspective, the results of transportation planning influence the warehouse-internal processes such as the retrieval, staging, and provisioning of goods and cargo. In this context, a transshipment warehouse houses goods for short periods between the transloading from one means of transport to another. These hubs are typically operated by LSPs or companies that need to cross-dock and commission prepicked freight to ultimately distribute it within the supply chain network.

Distribution and supply centers

From a distribution perspective, a logistics chain contains several warehouses between a raw materials source and the end customer. Thus, on each individual level of the logistics network, stock is established to either strike a balance between fluctuating requirements and inward goods movement times or to secure material availability due to insecurity about future requirements. With the aid of stock procurement, a disconnection between materials procurement and materials use in production or distribution processes is achieved.

A *supply warehouse* is typically part of the production operation and is used to store raw materials, supplies, and semi-finished products that are needed for either production or distribution to and between other stock-keeping facilities. Stock-keeping fulfills the primary goal of ensuring material availability and represents a reaction to any insecurities in the materials procurement and production planning processes. In this context, distribution logistics primarily concerns sales processes that generally begin with a customer ordering materials and indicating a desired delivery date. A sales order is generated with this information. Depending on the delivery date, shipping activities are initiated such that the materials reach the customer in a timely manner.

The composition of the flow of goods is changed in the *distribution warehouse*; these are usually classified into central, regional, and local distribution warehouses, depending on the area they serve. Central distribution warehouses are generally referred to as central warehouses, while decentralized regional or local distribution warehouses are called *supply* or *delivery warehouses*. In supply warehouses, goods from various suppliers are collected and distributed to one or more production or retail operations. In delivery warehouses, goods from production are stored before being delivered.

Integrating transportation and warehouse logistics

Warehouse logistics takes over the task of commissioning and material provision. As soon as the materials have left the warehouse, a goods issue is booked to update stock and inventory management values. In this context, transportation planning determines when and where to stage what. From an execution perspective, the integration of transportation planning results in warehouse management processes.

In the following section we introduce both warehouse operations based on SAP Extended Warehouse Management in Section 12.2.1, integration with SAP ERP in Section 12.2.2, and standard integration scenarios with SAP TM in Section 12.2.3. The integration of these systems enables smooth outbound processing with an optimized warehouse internal process according to the transportation planning result.

SAP integration guides

Recall that detailed integration guides with step-by-step introductions and process configuration are available. In addition to these guides, we explain the relevant process flow of these integration scenarios for outbound processing. We use screenshots of a sample process to document and explain the most relevant integration points for OTRs and DTRs.

Let's begin by moving our focus from warehouse management in general to SAP EWM in particular. The administration of and transparency regarding existing materials is essential to making precise statements about the availability of a material. Goods movements are usually caused by procurement, distribution, and the associated goods receipts and issues or through stock transfer.

12.2.1 Introduction to SAP EWM

SAP has been providing warehouse management functions since the release of SAP R/3 2.0. Thus, it can look back on more than 16 years of experience in warehouse management and countless successful implementations. Ever since the first SAP R/3-based versions and leading right up to the current SAP SCM-based systems, functionality has been continually expanded and adjusted according to customer demands. In addition to Warehouse Management (WM) as part of SAP ERP, in 2005 SAP introduced the considerably more efficient SAP Extended Warehouse Management (EWM), which is based on SAP Supply Chain Management (SCM).

Originally under the umbrella of SAP Spare Parts Management, today SAP EWM is an independent application that can be used in any warehouse environment and also be integrated with SAP TM. SAP EWM was developed for complex warehouse and distribution centers with several different products and a high document volume, and in contrast to WM, it offers many new and expanded functions and business objects.

To better understand the SAP TM integration scenarios, and how SAP TM documents are mapped and integrated to their corresponding counterparts in SAP EWM, we first briefly explain the most relevant SAP EWM documents and terminology for a standard, outbound scenario. We then use this outbound scenario as the basis for the described integration scenarios.

SAP ERP outbound delivery

From an SAP EWM perspective, the process starts with an SAP ERP outbound delivery: the central document in goods issue. It typically represents a follow-up document to a sales order but can also be created directly, without reference to a preceding document in SAP ERP. The physical shipment, which forms the completion of a goods issue procedure, thus begins with the generation of an outbound delivery document in SAP ERP. Whether it is necessary to distribute this document to a decentralized SAP EWM system depends on the WM warehouse number that is allocated to a delivery item. The system checks to determine whether the warehouse number indicates a decentralized warehouse. If an outbound delivery is relevant to processing in SAP EWM, the document is replicated in the decentralized warehouse system (SAP EWM), where it ultimately generates an outbound delivery request.

In the context of SAP EWM, the outbound delivery records the goods that are delivered to a goods recipient. From an SAP ERP perspective it describes the process of picking goods, reducing the storage quantity, and finally shipping the goods to their final destination (determined by a ship-to party). The outbound delivery process begins with goods picking and ends when the goods are delivered to the recipient. In the SAP system, this operation is represented by the outbound delivery document that is generated during the following activities:

- Goods shipment based on a sales order
- Stock transfer order
- Goods return to the vendor

SAP EWM outbound delivery request

Once the SAP ERP outbound delivery is replicated to the decentralized warehouse system (here, SAP EWM), an outbound delivery request is created. This document basically contains the same information and has the same structure as an outbound delivery in SAP ERP and is activated upon successful replication. The outbound delivery request is an SAP EWM document in the outbound delivery process that contains all relevant logistics data from the initiator of the outbound delivery process (such as the sales order). The outbound delivery request is used to automatically create an outbound delivery order.

SAP EWM outbound delivery order

The outbound delivery order is an SAP EWM document that contains all the relevant data required for triggering and monitoring the complete outbound delivery process. This process starts with the first planning activities for the outbound delivery and extends to the loading and shipping of the goods. The outbound delivery order is either used as a preview for planning upcoming warehouse activities or to create the outbound delivery.

Handling unit

Goods are typically packed and handled on pallets. In this context, the handling unit (HU) represents a physical unit consisting of packaging materials and the products they contain. HUs have a single, scannable identification number.

Packaging material

In the context of SAP EWM, packaging materials are products that are intended to surround or contain products to pack. Products to pack can be packed into or onto the packaging material. The packaging material can be a load carrier. From a warehouse perspective, some examples of the most important packaging materials are crates, boxes, containers, wire baskets, and pallets.

Transportation unit

As soon as the materials have been packed, the resulting packages (which are HUs) can be loaded on transportation units (TUs), and the goods issue is then posted. The system can also automatically determine the staging area and door for goods issue in advance, which is typically based on the route determined when the outbound delivery order was generated. The goods issue can be posted at various times.

Technically, this TU is a special kind of a HU and is used to reflect the truck, trailer, or container that is used for shipping. A TU is assigned to delivery items and therefore contains all relevant information (e.g., carrier, license plate of the truck, and the information regarding which items are supposed to be loaded on the truck). Because loading is an optional step, the goods issue can be posted without previous loading, depending on system settings. Alternatively, posting can be done after loading or, if the Yard Management functionality is used, at the latest after the TU has left the warehouse grounds. When the goods issue is booked, SAP EWM informs Inventory Management in SAP ERP of the change in stock.

12.2.2 Integrating with SAP ERP

Decentralized warehouse solution

SAP EWM is a decentralized warehouse management system. Like SAP TM, which is also decoupled from SAP ERP, and as an autonomous application of SAP SCM, SAP EWM requires integration with an SAP ERP system for master and transactional data. The SAP EWM release in SAP SCM 2007 provided the opportunity to centrally operate SAP EWM from within SAP ECC 6.0 as an add-on and thus use it as an integrated warehouse management system with an SAP ERP system.

Integration with SAP ERP

SAP EWM was developed for complex warehouse and distribution centers with a variety of products and a high document volume. That is why the design of this new, decentralized warehouse management system has special emphasis on the flexible mapping of warehouse-internal processes. The extent of functions has been expanded considerably in comparison to the existing WM and supplemented by a number of warehouse structure elements.

Like for SAP TM, the technical link between SAP EWM and SAP ERP and such functions as the transfer of inbound and outbound deliveries between the systems takes place in real time via defined interfaces. These interfaces enable the seamless integration of both systems by distributing, altering, and returning data relevant to delivery. Inbound and outbound processing is performed asynchronously based on the sequence stored in the inbound and outbound queues. In the event of an error like a missing network connection, this queue saves all transfers and allows processing to continue seamlessly as soon as the error has been located and eliminated. The queue enables the real-time and bidirectional exchange and processing of information.

From a technical standpoint, close integration of SAP EWM and SAP ERP is achieved via interfaces, while process integration is primarily done via organizational data. As in the case of integration of the WM system, in SAP EWM organizational allocation of warehouse numbers is initially achieved by their allocation to specific plant/storage location combinations. Stock management is always executed in SAP ERP, and all quantities of materials stored in an SAP EWM warehouse are allocated on the plant/storage location level. However, the assigned WM warehouse number is identified in the system settings as a decentralized warehouse. It informs the system that warehouse logistics-related processing is to be conducted in a decentralized warehouse management system, and the corresponding document is replicated to the integrated SAP EWM system. This SAP ERP warehouse number is assigned to an SAP EWM warehouse number in the SAP EWM system settings

Stand-alone SAP EWM

From a purely technical standpoint, SAP EWM can also be operated as an autonomous system without direct connection to SAP ERP or a non-SAP system. However, when it comes to integrating SAP TM—depending on the project base of the interfaces to be established and the processes to be implemented—standard integration with an SAP ERP system is preferred. In this section, to illustrate the SAP TM integration and processes, we assume that the SAP EWM system is linked to an SAP ERP system, and that close integration of the existing standard interfaces is established.

Integration with SAP TM

The integration of SAP TM 9.0 with SAP EWM currently runs via SAP ERP using SAP ERP shipments as an intermediate step. As of now, data is not directly sent between SAP TM and SAP EWM. The release of SAP TM 9.1 will support direct integration between SAP TM and SAP EWM without creating shipments in SAP ERP.

12.2.3 Standard Integration Scenarios

To optimize transportation costs and efforts in a flexible and optimized way, SAP TM supports transportation planning either based on sales order requirements or outbound deliveries. As part of transportation planning, the results influence the warehouse-internal processes such as staging. The integration of the SAP TM transportation planning result into SAP EWM warehouse management processes is therefore very beneficial because it enables smooth outbound processing with an optimized warehouse-internal process according to the transportation planning result.

Warehouse operations in SAP EWM

From a logistics perspective, goods issues serve the controlled reduction of stock. Such stock reduction is executed based on sales and distribution processes. So from the viewpoint of distribution logistics, a goods issue represents the completion of a shipping procedure to the customer and serves as an interface between internal and external logistics. Like a goods receipt, the information flow in the goods issue process is primarily based on spatial circumstances, the materials to be picked, and individual process requirements in the warehouse. These requirements can vary from warehouse to warehouse, and may even be dependent on the goods recipient.

For this core process, SAP EWM offers a variety of design options and enables the integration of individual process steps. In the following scenario descriptions, we focus on the main integration points and therefore do not illustrate the basic goods issue process in SAP EWM.

Recall that the integration of SAP TM 9.0 with SAP EWM currently runs via SAP ERP. For the following examples, data is not sent directly between SAP TM and SAP EWM. We now explain this integration and process flow for the following standard integration scenarios:

- Order-based integration
- Delivery-based integration
- Integration with planning in SAP EWM

The business background always concerns sending ordered goods to an external customer, while the transport is executed by an external carrier and the goods are stored in warehouse. In this context, SAP TM is used for transportation planning, while SAP ERP is used for delivery execution and shipment processing. Finally, SAP EWM is used for warehouse planning and execution.

Order-based Integration

The order-based integration scenario is an integrated warehousing and transportation process that sends ordered goods via an external carrier to external customers from a warehouse managed with SAP EWM. Transportation is planned in SAP TM, while warehouse activities are planned and executed in SAP EWM. This process is based on SAP ERP sales orders and outbound deliveries. It involves SAP ERP-SAP TM order integration, delivery integration, and shipment integration, as well as SAP ERP-SAP EWM delivery and shipment integration.

Process overview

You already learned that SAP TM plans transportation on the basis of sales orders created in SAP ERP, and SAP TM then proposes and creates deliveries based on dates and quantities according to transportation planning results, taking into account transportation constraints such as resource availability and transportation durations. The delivery proposals are sent to SAP ERP.

In SAP ERP, this information is used to create deliveries, which are then be forwarded to SAP TM and SAP EWM for further processing. In addition, as a result of the transportation planning in SAP TM, a freight order is created automatically, which then triggers the creation of a shipment in SAP ERP and a TU in SAP EWM. In SAP EWM goods are usually picked by waves directly into shipping HUs (pallets). You then stage the shipping HUs and load them into a truck before posting the goods issue. As a result of the goods issue, the deliveries are updated in SAP ERP for later billing, and the freight order in SAP TM is updated for a later freight cost settlement. Figure 12.12 illustrates the process flow and guides you through the steps of this order-based integration.

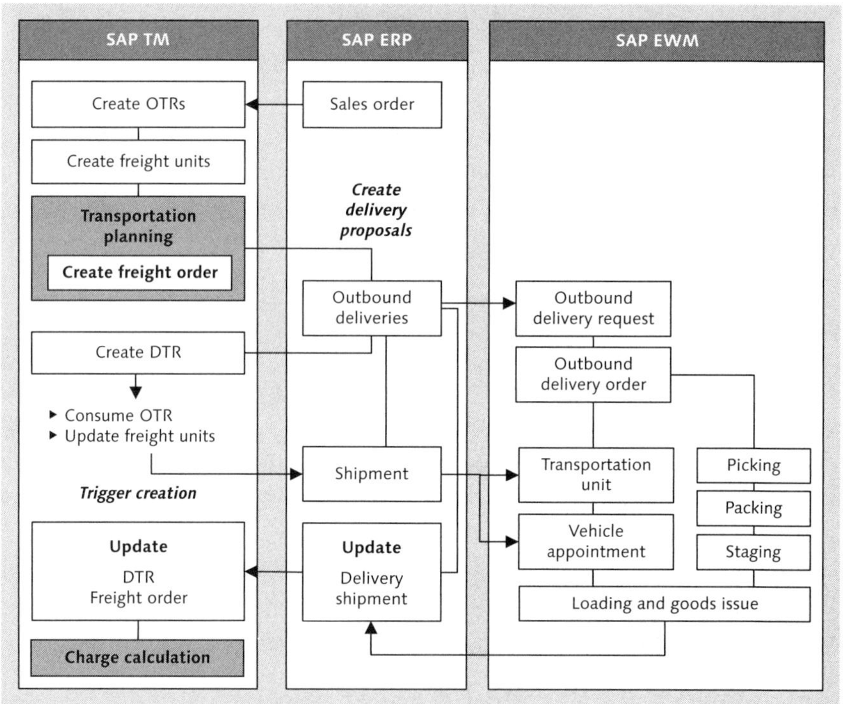

Figure 12.12 Order-based SAP EWM Integration Process Flow

Order and transportation requirements

The process starts with a sales order in SAP ERP and a planned delivery date. In the example shown in Figure 12.13, this is SALES ORDER 1117. With the creation of the sales order in SAP ERP and the integration to

SAP TM, the system automatically creates an OTR and freight units in SAP TM. The OTR contains all relevant information from the SAP ERP order document, such as product information, the requested quantity, and the mapped source and destination locations.

Based on that information, the system automatically creates freight units, which are then the basis for transportation planning and optimization in SAP TM.

Figure 12.13 Order-based Transportation Requirement in SAP TM

Freight units, freight orders

Transportation planning is executed for transportation requirements. For products that are supposed to be transported together across the entire transportation chain, these requirements have been automatically split into freight units. During transportation planning, the freight units are then assigned to freight documents. In our example, OTR 100002159 has triggered the creation of freight unit 6410001010 in Figure 12.14.

Transportation planning results in the creation of freight orders, which are shown in Figure 12.15, and is the basis for shipment and delivery creation in SAP ERP. In SAP TM the freight order can be the basis for tendering and carrier selection (refer to Chapter 7). Once the carrier has been assigned, SAP TM automatically triggers the creation of an outbound delivery in the connected SAP ERP system and communicates the freight order number to the selected carrier.

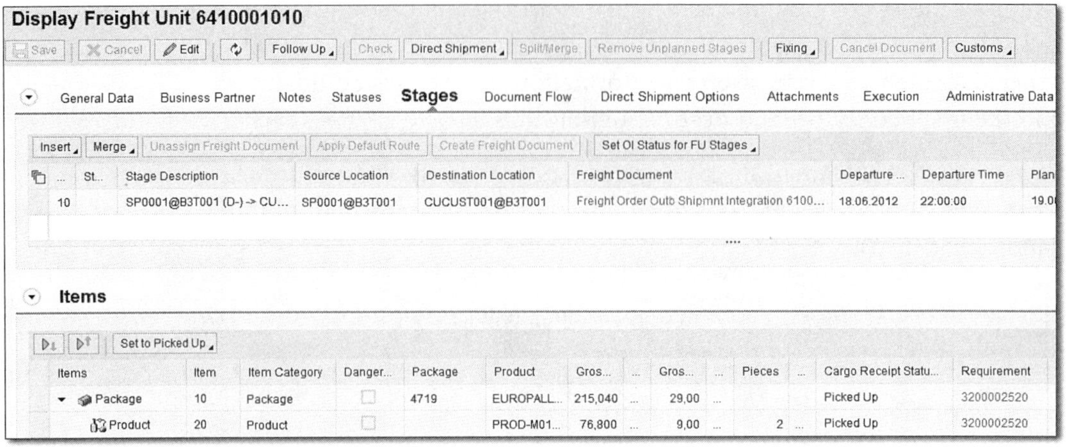

Figure 12.14 Freight Unit in SAP TM

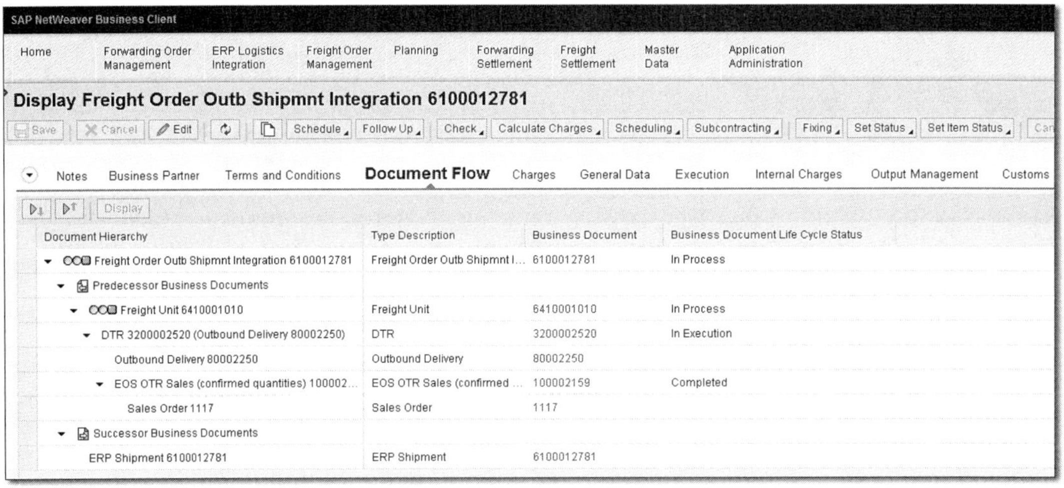

Figure 12.15 SAP TM Freight Order: Document Flow

In our example, the SAP TM freight order that has been created is 6100012781. Figure 12.15 shows the document flow of the freight order so far, as well as all of outbound delivery 80002250, which has been created in SAP ERP in Figure 12.16.

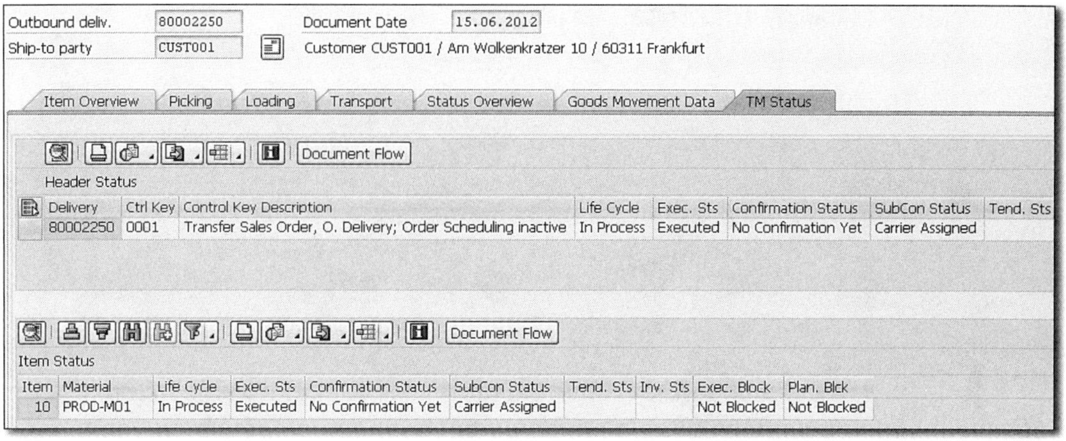

Figure 12.16 Detail of an SAP ERP Delivery with SAP TM Status

Outbound delivery in SAP ERP

Once the transportation requirement has been planned and the freight order has been created, outbound delivery creation in the connected SAP ERP system is triggered. Sending the proposals to SAP ERP ultimately creates SAP ERP outbound deliveries. These deliveries are based on the transportation planning results and therefore take into account planning constraints such as resource availability, distances, durations, and consolidation and dates. This step is executed either automatically as a background job or manually by the planner. The delivery numbers are sent back to SAP TM to update the document flow. This communication is triggered via output processing for the delivery header in SAP ERP.

In the SAP ERP system, the delivery is created with the proposed delivery date from SAP TM. As soon as the delivery has been created, SAP ERP sends the deliveries to SAP TM and SAP EWM for further processing, ultimately creating two things:

- DTRs in SAP TM
- Outbound delivery requests in SAP EWM

Delivery-based transportation requirements

SAP TM automatically creates DTRs based on the transferred deliveries from SAP ERP. Recall that the freight units have been assigned to the OTRs resulting from the SAP ERP sales order, which is now "consumed" by this DTR. Figure 12.17 shows DTR 3200002520, which has been created for SAP ERP delivery 80002250.

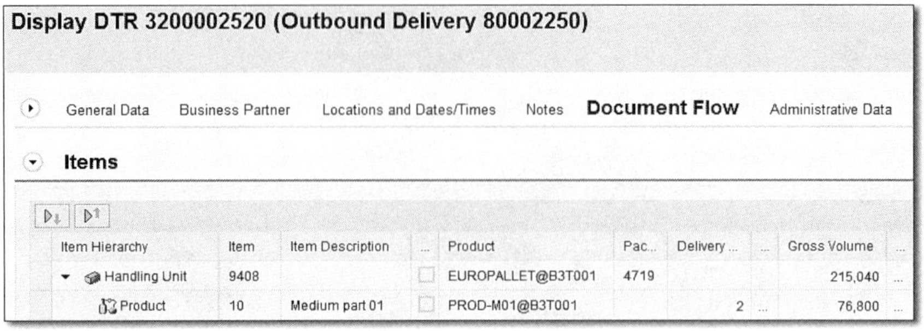

Figure 12.17 Delivery-based Transportation Requirement in SAP TM

Figure 12.18 shows SAP ERP delivery 80002250 with its current document flow. You can monitor all of these steps. From here, you can also navigate to the source systems to access the individual documents.

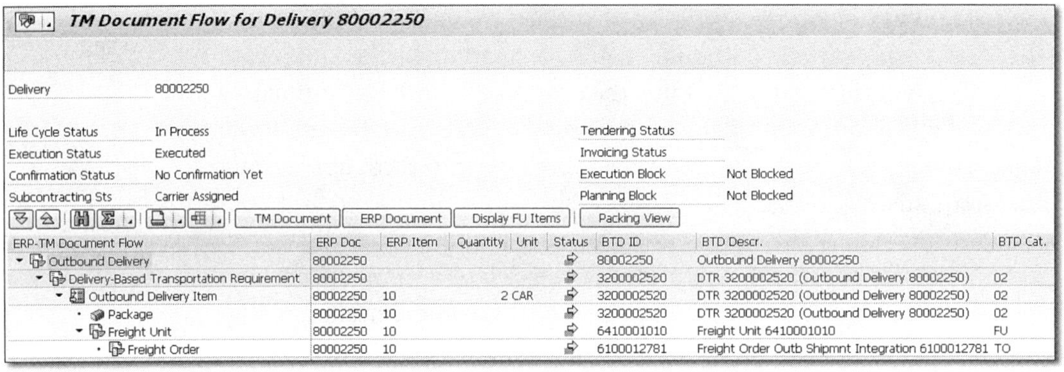

Figure 12.18 ERP Delivery: Document Flow with SAP TM Freight Order

Outbound delivery order in SAP EWM

SAP ERP sent the outbound deliveries to SAP TM to adapt transportation planning and to SAP EWM for warehouse planning and execution. In SAP EWM, the SAP ERP delivery creates an outbound delivery request. The automatic activation of an outbound delivery request immediately creates an outbound delivery order, which acts as the actual warehouse request and initiates the goods issue process in SAP EWM.

The outbound delivery order contains data assumed from the preceding document and all necessary information to trigger the goods issue process and monitor it accordingly (see Figure 12.19). From the perspective of warehouse management, the outbound delivery order represents a worklist that is only completed when the picked materials have been loaded and shipped. When the outbound delivery request is activated and the outbound delivery order is automatically generated, SAP EWM begins all of the procedures required to supply the document with the necessary information and map the process in accordance with the settings in the system based on process- or layout-oriented storage control. We describe the basics of SAP EWM storage control when we explain the warehouse execution.

So far, the outbound delivery order in SAP EWM is still blocked for processing. It will be released as soon as the transportation planning process in SAP TM has been completed and SAP EWM has been updated with the final planning results. Figure 12.19 shows the reference documents of outbound delivery order 310000003422 created for delivery 80002250.

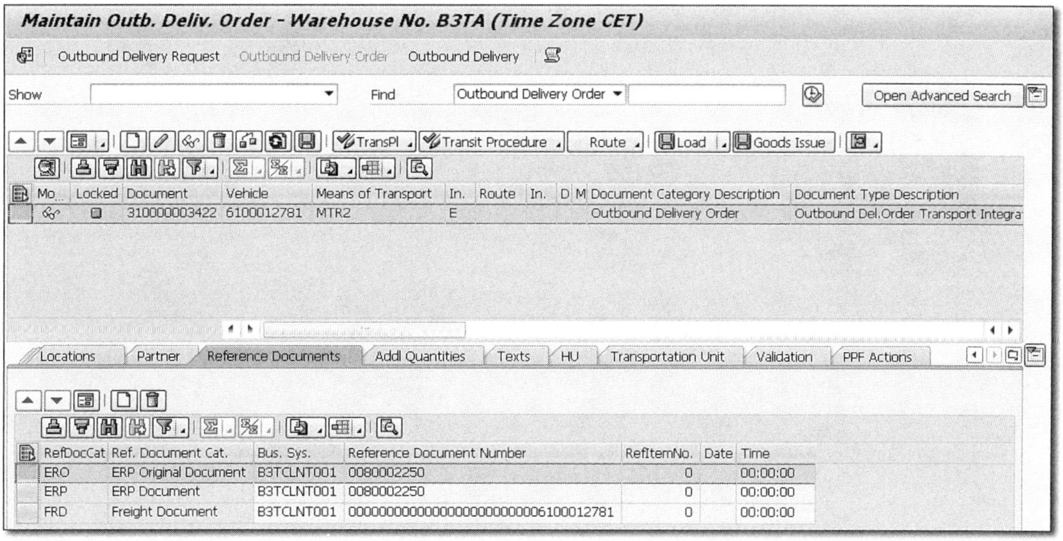

Figure 12.19 Outbound Delivery Order in SAP EWM

Shipments in SAP ERP

The transportation planning in SAP TM is completed once the transportation capacity of the freight orders has been fully planned and all relevant transportation requirements have been assigned. SAP TM adapts transportation planning results and sends the corresponding freight documents to SAP ERP to trigger the creation of shipments (see Figure

12.20). SAP ERP assigns the outbound deliveries to the shipments and transfers the shipments to SAP EWM, where the outbound delivery orders are still blocked for processing.

Transportation units in SAP EWM

Sending the SAP ERP shipments to SAP EWM creates a TU and automatically assigns the outbound delivery orders. With this assignment, the outbound delivery orders are unlocked und updated with the related carrier information from SAP TM. Unlocked outbound delivery orders with TUs are now the basis for warehouse execution in SAP EWM.

The SAP TM freight order now exists as shipments in SAP ERP and as a TU in SAP EWM. To easily identify the same object across all three systems, they all share the same document number. Figure 12.20 shows SAP ERP shipment 6100012781, which has exactly the same number as the SAP TM freight order and the SAP EWM transportation unit that has been referenced with its external key in the SAP EWM outbound delivery. In addition, you can also see the assignment and reference to SAP ERP delivery 80002250 at the stage detail view.

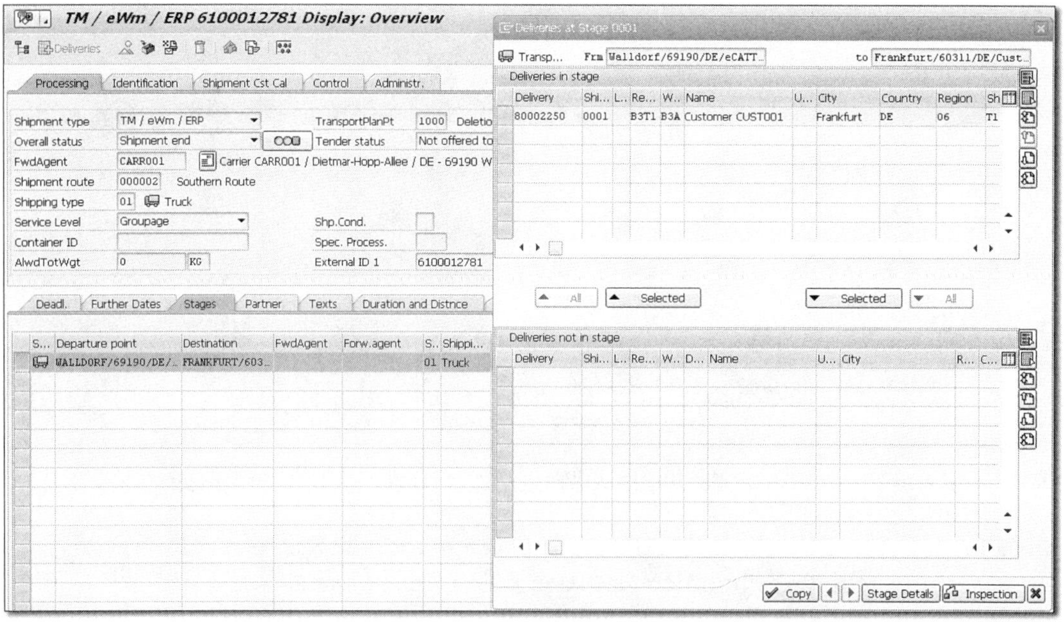

Figure 12.20 SAP ERP Shipment: Stage Detail with Assigned SAP ERP Delivery

Warehouse execution in SAP EWM

The warehouse-internal process starts when the deliveries are unblocked and waves are created and released. When the outbound delivery orders are created, the system can automatically assign them to waves. This assignment to a picking wave is based on the shipping information, and an optional but useful step is to optimize the picking efficiency and to monitor and track the outbound process.

In this context, wave management groups a certain kind of warehouse request into a wave. Regardless of when a wave is to be released (i.e., the point in which warehouse tasks can be processed), wave management generates the necessary work packages in the form of warehouse orders and their corresponding warehouse tasks. The criteria for wave management can be kept in wave templates and can be based on shipping information.

Wave management, task creation

Wave generation and release to create warehouse tasks are typically done automatically. To automatically allocate warehouse tasks to waves, criteria from the delivery can be used. Therefore, waves are created by combining warehouse request items based on common activity areas, routes, or materials in wave picks. The individual criteria and attributes for wave generation can be stored in the system as specifications and serve as the infrastructure for automatic or manual creation of waves.

Therefore, in operational practice, physical picking usually starts with grouping delivery items into waves. In doing so, SAP EWM can even take existing conveyor systems into account. Due to the automated processing, this flexibility enables optimized grouping of the outgoing deliveries and drastically reducing picking times.

Wave management not only generates warehouse tasks, but also groups them into warehouse orders. Warehouse orders are executable work packages for warehouse employees that are generated according to the warehouse order generation rules stored in SAP EWM. The warehouse order contains the warehouse tasks or inventory items to be executed. In our case, these can be the tasks to pick the goods and move them to the staging area.

Warehouse process types

When the outbound delivery order is generated, the system first determines a warehouse process type and route and then conducts a rough estimate of the picking storage bin (source storage bin) based on a picking strategy. At the time, the staging area (destination storage bin) is determined using a so-called warehouse process type. This parameter not only controls whether warehouse tasks should be immediately confirmed and which warehouse door should be determined, but also determines the relevant steps in process-oriented storage control.

Storage control

Individual operational needs and spatial circumstances in the warehouse determine the processes in the warehouse, stock movements, and the process steps to be performed. In real life, it is rare that the material flow is the same for all products and in all areas within a warehouse. Several people and resources are involved in the warehouse processes. Pallets might need to be deconsolidated in the goods receipt area or consolidated in the goods issue area. SAP EWM offers storage control for flexible, tailor-made control of the material flow across several stations, and to enable cross-resource stock movement through various stations.

The purpose of storage control is to map complex, multiple-level stock movements for put-away and removal and for internal warehouse transfer. Storage control is done in relation to the spatial circumstances along the lines of the predominant warehouse processes and the stock to be moved. Using storage control, SAP EWM can specify the put-away or removal route across several stations in a process- or layout-oriented manner. This allows processes such as counting or deconsolidation in the goods receipt area or packing in the goods issue area to be performed in an automated fashion. Storage control can be executed on multiple levels, enabling material flow via several interim storage bins. Stock movements are controlled via storage processes. Each process and all process steps in goods receipt and issue are allocated to a storage process. The possible goods movement types and the direction of movement are assigned to a storage process via a warehouse process type and an activity.

The determination of the storage process depends especially on product and document information. The storage process is automatically selected by the system when the warehouse request is created, based on the document type, the product, and the delivery priority. By using document characteristics and control indicators in the product master, the storage process can be controlled with great flexibility. For simple goods movements, the storage process itself can contain the storage type and bin from or to which a material is to be moved. For complex movements, it can contain the storage process or process-oriented storage control.

Process-oriented storage control

Process-oriented storage control allows you to map complex picking and put-away processes. The individual steps, such as staging or loading, or the execution of supplementary logistics services, can be adjusted as desired and are allocated to a storage process in the system. The determined procedures and their activities are assumed by the HU to be picked or placed. The HU thus possesses the information regarding which process steps are necessary for picking or warehouse-internal stock movements. That is why process-oriented storage control only works with HUs.

Picking

Picking and moving the goods to the staging area is usually carried out using a radio frequency (RF) device. If RF is used, the warehouse worker logs on as a resource and takes an empty pallet or a wire basket as a *pick HU*. Once assigned to an RF queue, he receives the first pick task for execution and is prompted to create the pick HU for the warehouse order he is currently executing. Once the HU is created, the system can automatically print the HU label, which contains both plain text descriptions and the HU number as a barcode.

The worker typically continues to the source bin and picks the requested product and quantity into the pick HU proposed by the system. To avoid mistakes, the source bin, product barcode, and pick HU are scanned before the picked quantity is entered. With the confirmation of the warehouse task, the worker proceeds with the next task. When the warehouse order has been executed and the last task has been confirmed, process-oriented storage control can prompt him to move the pick HU to the staging area. In addition, if this has been foreseen in storage control, there might be additional, intermediate steps, such as shrink wrapping or packaging.

SAP EWM generally allows one- and two-step picking. It also enables you to integrate the packaging process into picking. This procedure, called *pick and pack*, enables direct picking in a shipping unit. An expansion of this function is the *pick, pack, and pass procedure*, a one-step picking procedure with decentralized picking and static staging. Pick, pack, and pass allows the work to be passed from one resource or employee performing the work to another after part of the picking has taken place. The sequence of warehouse tasks for stock removal can be determined according to various criteria, such as the shortest path. SAP EWM and the corresponding SAP ERP system are closely linked for the picking process. This allows picking specifications such as batches or batch characteristics to be taken into account.

Staging

When all warehouse tasks for stock removal are confirmed, the stock is removed from its source storage bin and moved to the destination storage bin. If a difference in quantity is detected in the process because the picked quantity deviates from the quantity to be picked, then another warehouse task can be created or the quantity to be delivered is adjusted and diminished accordingly. When a second warehouse task is generated, stock removal is not complete until the second warehouse task has also been confirmed. The warehouse worker drops the pick HU at the staging area, scans the staging bin barcode, and thus confirms this step in the system. With the last confirmation, SAP EWM prints a shipping HU label to be attached to the pallet.

Loading

As soon as the materials have been picked and staged, the resulting handling units are loaded and the goods issue can be posted. As mentioned above, the system can also automatically determine the staging area and door for goods issue in advance, which is based on the route determined when the outbound delivery order was generated.

From a process perspective, if Yard Management is used, the truck arrives at the checkpoint. It is identified by the external TU number, which also represents the freight order number in SAP TM (see Figure 12.21). The truck can be automatically assigned to a warehouse door. Alternatively, the door assignment can be done manually be the checkpoint clerk, or—in large yards—by the shipping office.

After the truck has arrived at the warehouse and is docked to the door, the loading of the goods can begin. Loading is also supported by RF. Scanning the door starts the loading of the TU. The ship HU is loaded on the truck, its barcode is scanned, and the system automatically creates and confirms a corresponding warehouse task. Loading is confirmed as soon as all ship HUs have been loaded on the TU.

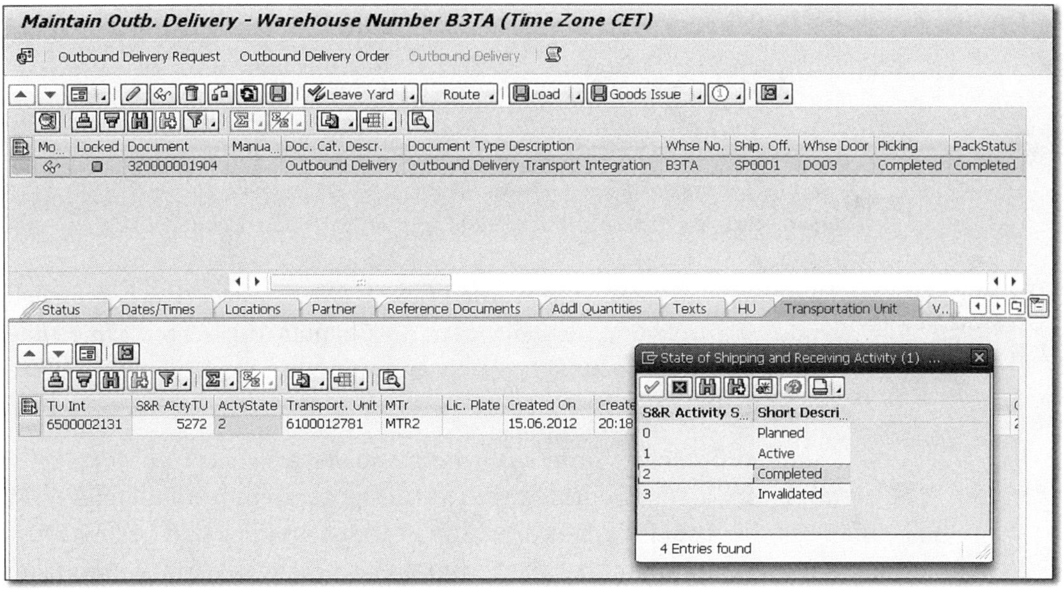

Figure 12.21 Outbound Delivery in SAP EWM

Goods issue in SAP EWM

When the loading is finished, the SAP EWM outbound delivery is created, delivery notes and the waybill can be printed, and the goods issue can be posted in SAP EWM based on the TU. When the goods issue is posted, SAP EWM sends confirmed execution results to SAP ERP.

If a ship HU is missing, the outbound delivery order can be unassigned from the TU before posting goods use. It can be manually assigned to another TU or taken into account during the next transportation planning in SAP TM. Alternatively, the outbound delivery order can be split.

Updating SAP ERP documents

In SAP ERP the corresponding outbound deliveries and shipments are updated with the results of the warehouse execution (e.g., packaging of deliveries, actual quantities and weights, and the goods movement status shown in Figure 12.22). The shipment is also updated with the shipment completion date and precise start date to document the departure of the truck from the warehouse.

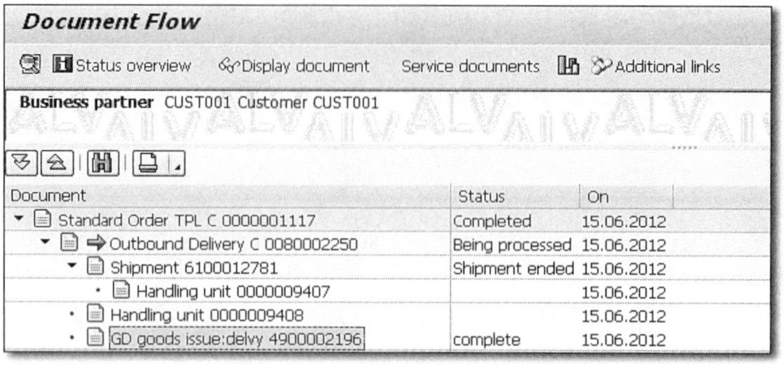

Figure 12.22 ERP Delivery: Document Flow after Warehouse Execution

Updating SAP TM documents

These updates are then forwarded to SAP TM to trigger the charge calculation and freight settlement (refer to Chapter 10). At the same time, SAP TM also completes the DTRs and updates the status of the corresponding freight order.

Outbound delivery orders can be manually assigned to other TUs or taken into account during the next transportation planning in SAP TM. DTRs that could *not* be loaded to the transportation unit in SAP EWM are automatically unassigned from the freight order and can be taken into account for the next transportation planning.

When the carrier confirms the arrival of the truck at the customer site, the freight order is completely executed and confirmed. The freight order confirmation is communicated to SAP ERP and automatically updates the SAP ERP shipment with the end date reflecting the arrival date of the truck.

Cancellation scenario

Once the picking and staging is completed, the shipping HUs are in the staging area and ready for loading. However, there might be business or technical reasons for stopping execution and canceling transportation units and shipments that have already been assigned to outbound delivery orders, perhaps because of major delays, accidents, or the inability to physically load goods to the TU. In most cases, the carrier calls the shipping office to cancel the appointment during or at the end of the picking and staging steps in the warehouse.

Cancellation of TU appointment

Once the cancellation of the arrival of the truck has been communicated to the shipping office, the TU appointment has to be canceled. The cancellation of the order-based scenario can be manually executed in SAP EWM and typically starts with unassigning the outbound delivery orders from the TU appointments. After that, the TU appointments can be deleted.

SAP EWM forwards the TU cancellation to SAP ERP, where the corresponding shipment is automatically deleted and a message is sent to SAP TM to cancel the freight document. After the freight order has also been canceled, the DTRs are reopened and eligible for new transportation planning and execution.

Delivery-based Integration

In addition to order-based integration, SAP EWM can also be integrated based on DTRs. In contrast to the previous scenario, where the SAP ERP outbound delivery was created as a consequence of OTR-based delivery proposals and transportation planning in SAP TM, this integration is mainly based on outbound deliveries created and scheduled in SAP ERP.

Logistics execution

Delivery processing in SAP ERP involves grouping deliveries to pick, pack, and ship and performing all the functions associated with the delivery process. Based on split criteria or combination indicators, you can group together entire orders or individual items or split orders into partial deliveries.

Process overview

In the delivery-based integration scenario, SAP ERP is the leading system for logistics execution and inventory planning. As shown in Figure 12.23, outbound deliveries are created and scheduled in SAP ERP and are the basis for transportation planning in SAP TM. As a result of the transportation planning in SAP TM, a freight order is created in SAP TM, which results in the creation of a shipment in SAP ERP and a TU in SAP EWM. Similar to the previous scenario, goods are typically picked by waves, directly put into shipping HUs, staged, and finally loaded on a truck. Posting goods issue in SAP EWM immediately adjusts the inventory in SAP ERP, updating the outbound deliveries and the freight order in SAP TM for a later freight cost settlement.

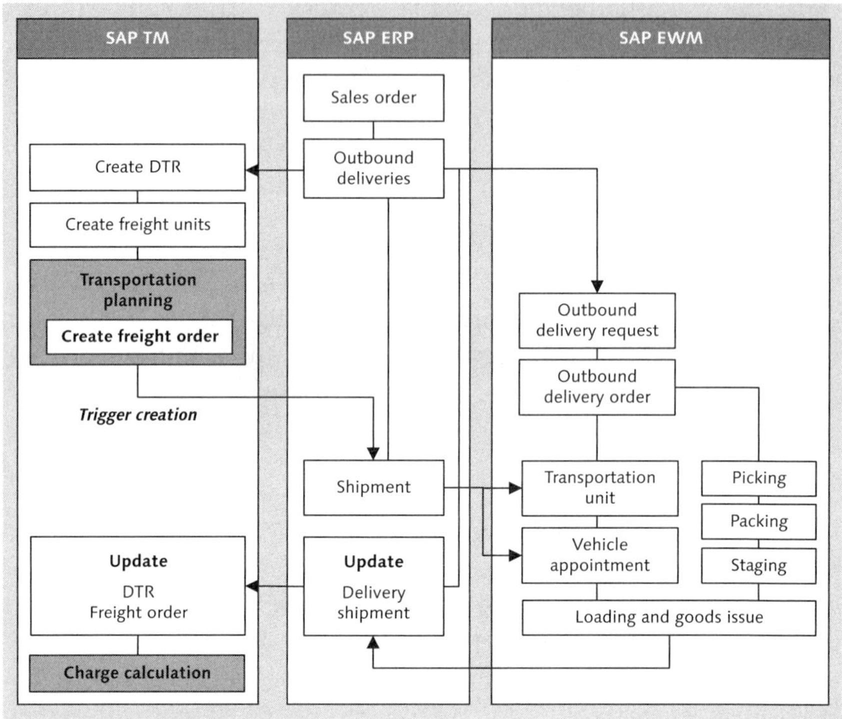

Figure 12.23 Delivery-based SAP EWM Integration Process Flow

Delivery and transportation requirements

The process flow of the delivery-based integration is quite similar to the order-based integration planning, so we'll just focus on the main differences here.

Because SAP ERP inventory planning and logistics execution has priority over the transportation planning process in SAP TM, the main difference is that the deliveries are already created based on existing sales orders in SAP ERP; that's why the delivery-based transportation planning starts with DTRs and freight unit creation in SAP TM. At the same time, SAP ERP deliveries are automatically transferred to SAP EWM. Similar to the previous scenario, SAP EWM creates outbound delivery requests and outbound delivery orders. As long as they have not been assigned to a TU, these outbound delivery orders are locked for execution.

Transportation planning

After planning the transportation requirements, and once the freight order has been created in the transportation system, SAP TM triggers the creation of SAP ERP shipments. SAP ERP assigns the outbound deliveries to the shipments and transfers the shipments to SAP EWM, where the outbound delivery orders are still blocked for processing.

Warehouse execution

Sending the SAP ERP shipments to SAP EWM creates a TU and automatically assigns the (locked) outbound delivery orders. With this assignment, the outbound delivery orders are unlocked and updated with the related carrier information from SAP TM. Unlocked outbound delivery orders, assigned to a TU, are now the basis for warehouse execution in SAP EWM.

Integration with Planning in SAP EWM

In the two previous examples, we explained an integrated warehousing and transportation process to send ordered goods to external customers from a warehouse managed with SAP EWM. SAP TM was used for planning OTRs and DTRs and ultimately triggered the creation of shipments in SAP ERP. SAP ERP was used for delivery execution and shipment processing, while the warehouse execution was based on both planning results from SAP TM that created TU activities in SAP EWM and the warehouse requests that were sent from SAP ERP. In this context, to reflect the SAP TM planning results, the corresponding outbound delivery orders have been automatically assigned to the TUs.

Integration with SAP ERP

For integration with planning in SAP EWM, though, this assignment is done manually in SAP EWM. The process starts with the execution of outbound deliveries created in SAP ERP. Like in the previous scenarios, these SAP ERP deliveries have been transferred to SAP EWM and then created outbound delivery requests as well as locked outbound delivery orders. In contrast to the order- and delivery-related transportation planning, transportation planning is now executed in SAP EWM. As a result of this SAP EWM transportation planning, you create a TU and assign the outbound deliveries to it, as shown in Figure 12.24.

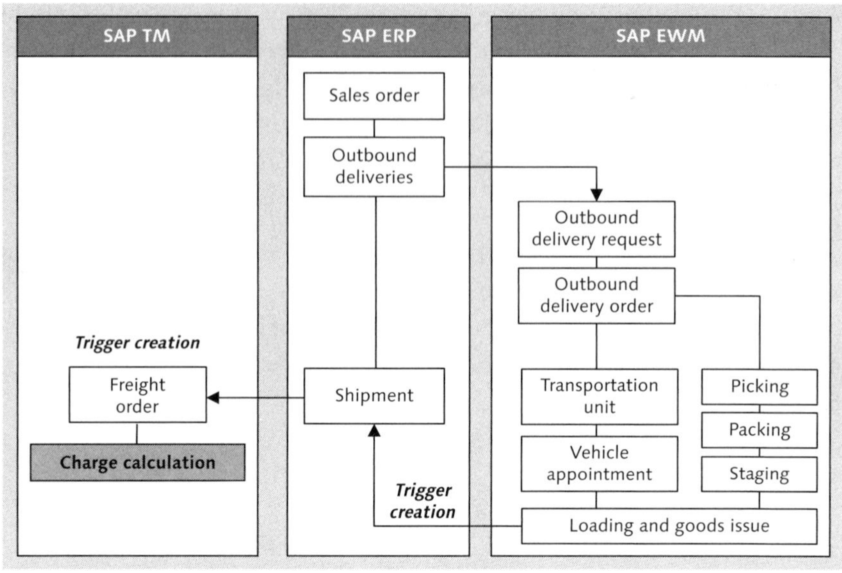

Figure 12.24 Integration with Planning in SAP EWM Process Flow

Creating a TU in SAP EWM

Before outbound delivery orders are supposed to be assigned to a TU, the necessary transportation capacity is ordered for a specific route and confirmed by the shipper. With the confirmation of the capacity for a specific route, the shipping office typically receives an identifier for each transport. Based on that information, the TU is created for the scheduled route and is typically assigned to the corresponding warehouse door. Once the TU has been created and the truck arrival has been scheduled, the outbound delivery orders are assigned. With this assignment, they are unlocked and are now the basis for warehouse execution in SAP EWM.

SAP EWM execution, goods issue

Warehouse execution typically includes picking the goods by waves directly into shipping HUs, staging, and loading them into a truck before posting goods issue. When the goods issue is posted, SAP EWM sends confirmed execution results to SAP ERP.

Creating shipments, freight orders

In SAP ERP, the outbound delivery is updated with weight, volumes, dates, and quantities. The goods movement status is set to completed, and an advanced shipping notification (ASN) can be sent to the customer.

While updating the deliveries, SAP EWM also sends transportation information such as TU data and execution dates, and triggers the creation of an SAP ERP shipment for invoiced deliveries. Once the SAP ERP shipment has been automatically created and completed, a freight order is created in SAP TM.

In the previous sections we talked about integrating systems, supporting cross-application processes, and exchanging data for reporting and execution purposes. SAP TM not only provides system integration capabilities, but also supports collaborative scenarios. In the next section we introduce both the technology and the processes that support collaborative transportation management.

12.3 Collaborative Transportation Management

Complex supply chains, cost pressures, service commoditization, and customer demands—the requirements of modern transport logistics are becoming increasingly complex and require ever more flexibility. With shippers facing extremely thin operating margins, and carriers facing an especially competitive environment that is forcing them to quote unsustainably low rates, flexible communication is key to meeting future requirements and the challenges of modern supply chains.

Engage and stay in contact

To be able to manage this flexibility in the future, today's decisions to invest in a transportation management system are strategic ones. The objective is to introduce software that not only is comprehensive, flexible, and can be scaled up or down, but will also prove to be the best choice for collaborating with internal and external stakeholders. Freight transportation is all about meeting expectations and keeping promises. Thanks to comprehensive planning and execution software, it is no longer a spreadsheet business but a people business. In this context, collaboration is important and covers multiple aspects of communication to share information, stay informed, and find synergies to reduce cost.

Collaboration with SAP TM

For the shipper-carrier collaboration, communication is therefore not only about negotiating rates to the lowest possible level that carriers can bear and still stay in business; it is mainly about forming longer-term relationships with carriers and fostering an operating partnership in which both parties can work profitably. In this section, we provide you with a brief overview of how SAP TM supports carrier communication by leveraging both mobile applications (Section 12.3.1) and portal collaboration (Section 12.3.2).

Topicality of Information

People using mobile solutions want the same ease of use with their enterprise software that they get with their mobile applications. In this context, carrier collaboration and mobile technology both leverage the latest innovations. The mobile applications described in this book reflect the currently available mobile solutions for SAP TM 9.0.

Especially when it comes to user interface technology, SAP renews the most broadly and frequently used business functions with a simple and easy-to-use user interface. This technology is known as SAP Fiori; it currently provides a collection of applications with a simple, easy-to-use, consumer-grade user experience for broadly and frequently used SAP software functions.

These new applications—not yet including mobile applications for SAP TM—allow users to seamlessly experience the applications across all interaction channels. To learn more about SAP Fiori, visit *http://help.sap.com/fiori*.

12.3.1 Mobile Applications

When it comes to consumer interaction, mobile applications are seen as a mandatory commodity to stay in contact with customers and consumers. That means being able to connect and collaborate using virtually any device, providing a seamless, unified mobile customer experience.

From a supply chain perspective, working anytime and anywhere is the new normal. Shippers, carriers, and customers expect instant responses and seamless integration. To engage in and execute on supply chain opportunities faster, employees need to be able to work effectively wherever their work takes them.

Mobility with SAP TM

The interfaces and underlying technology of SAP TM offer multiple ways to communicate with internal and external stakeholders and systems. There are multiple examples where partners or project team members, using SAP technology, have built custom or project-specific mobile applications to access data in SAP TM, SAP ERP, or SAP Event Management. In this section, we focus on carrier collaboration from a transportation tendering and tracking perspective, introducing the mobile applications for SAP TM, provided by SAP:

- SAP Transport Tendering
- SAP Transport Notification and Status

Both of these applications are well documented and can be downloaded and tested from the SAP Store at *https://store.sap.com*.

First, let's briefly explain the mobile applications, as well as the core functionality and principles of the underlying architecture.

Background information

Note

In this book we can only describe the basics and the core functionality of mobile infrastructure, architecture, and device access to SAP TM. Additional information, such as administration guides, installation and operation of mobile applications in SAP Business Suite, and additional technical documentation can be found in the SAP Service Marketplace, in the SAP Help portal, or through the SAP Community Network:

- To learn more about installation, security aspects, and operating these mobile applications for SAP TM, please read the comprehensive *Administrator's Guide*. This document is regularly updated and can be downloaded from the SAP Service Marketplace at *http://service.sap.com/instguides* via follow the menu path SAP COMPONENTS • SAP MOBILE APPLICATIONS • SAP BUSINESS SUITE • SAP TM.
- To learn more about SAP NetWeaver Gateway and its configuration, deployment, and security information, please consult the online application help in the SAP help portal (*http://help.sap.com/nwgateway*).
- To learn more about the Gateway Consumption Tool, we recommend the SAP Community Network (SCN) found at *http://scn.sap.com*. You can use the search functionality and search for "Gateway Consumption Tool".

SAP NetWeaver Gateway

Mobile device access to SAP TM occurs via the SAP NetWeaver Gateway architecture, which is a key enabling technology for the SAP Business Suite and is integrated with the Sybase Unwired Platform (SUP). In this context, SAP NetWeaver Gateway can be considered a framework that provides a simple way to connect mobile devices, environments, or platforms to SAP back-end applications. Making use of REST services and standard OData protocols, it supports database-like access to business data in SAP TM. Consumption of data through the SAP NetWeaver Gateway interface requires nothing more than the ability to make an HTTP(S) request for OData services and does not require the use of any SAP proprietary software or protocols.

Using commonly available development tools such as *.NET*, *Xcode*, or open source languages such as PHP, the SAP NetWeaver Gateway interfaces allow mobile application developers to access and interface SAP data without writing software in ABAP or having a fundamental understanding of the SAP back-end system.

SAP NetWeaver Gateway integration

Communication between the different software components in the mobile system landscape uses different channels and typically proceeds as follows and as shown in Figure 12.25:

- The mobile application is deployed on the mobile device, which is hooked up to the Internet. The mobile application now sends a request for an OData service to the backend system, which operates in a corporate, secured LAN environment.
- The communication involves a *relay server*. In general, this relay server is a web server with a set of two web extensions, supporting HTTP or HTTP(S) communication with both clients and servers. The client extension handles client requests made from applications running on mobile devices. The server extension handles requests made by the outbound enabler on behalf of a back-end server (in our case, SAP TM).
- SAP TM initiates the communication to the relay server from inside the corporate LAN. This provides are more secure deployment because it does not require inbound connections from the DMZ into the corporate LAN.
- Inbound and outbound communication from SAP TM to the mobile device is handled by SAP NetWeaver Gateway and SUP.
- The request channel from the mobile application to the SUP relies on SUP Messaging Channel, while communication between SUP and the

SAP NetWeaver Gateway relies on the HTTP protocol. In this context, user authentication is based on secured HTTP(S) requests.

- The last communication step, which is between the SAP NetWeaver Gateway system to the SAP TM server, uses a trusted RFC connection.

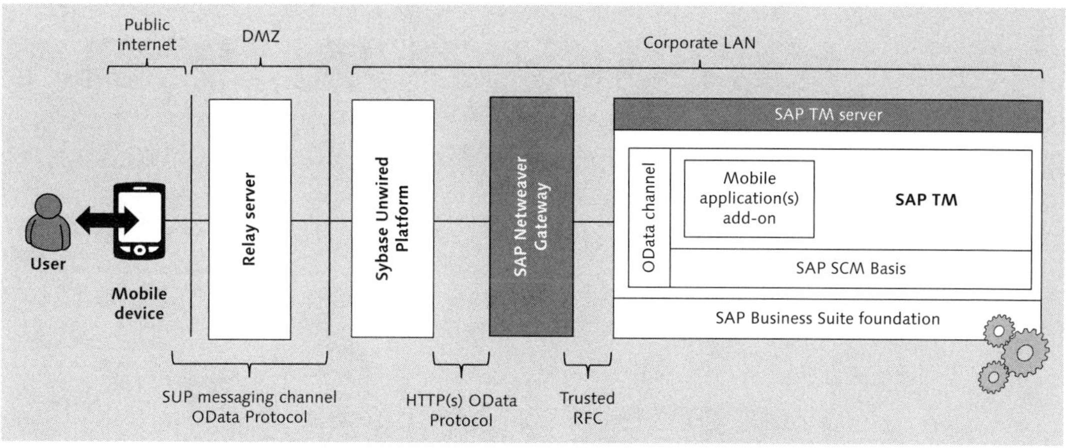

Figure 12.25 SAP NetWeaver Gateway Integration Architecture

To make the OData consumption process even easier, SAP delivers several versions of the Gateway Consumption Tool. This tool generates an OData proxy object in a variety of languages such as C#, Objective C, Java, and PHP. It can also generate a basic application to wrap the generated object.

Gateway Consumption Tool

Transport Tendering

Shippers are increasingly interested in improving collaboration with their carriers. From a carrier's perspective, where equipment, drivers, and capacity are dependent on seasonal fluctuations in demand, forward visibility into future shipping needs helps allocate capacity and better manage supply and demand.

Mobile tendering

SAP TM not only supports shippers in creating, managing, and settling transport requests to third parties, such as carriers, other LSPs, or their own fleet, but also helps them to manage supply and demand using mobile communication for freight tendering (please also refer to Chapter 7, Section 7.3 to learn more about carrier selection and freight tendering). As the tendering process becomes even more competitive and the time in which quotations must be submitted shortens, SAP mobile transport tendering for the iPhone allows carriers to react to these freight requests for quotations (RFQs) immediately (see Figure 12.26).

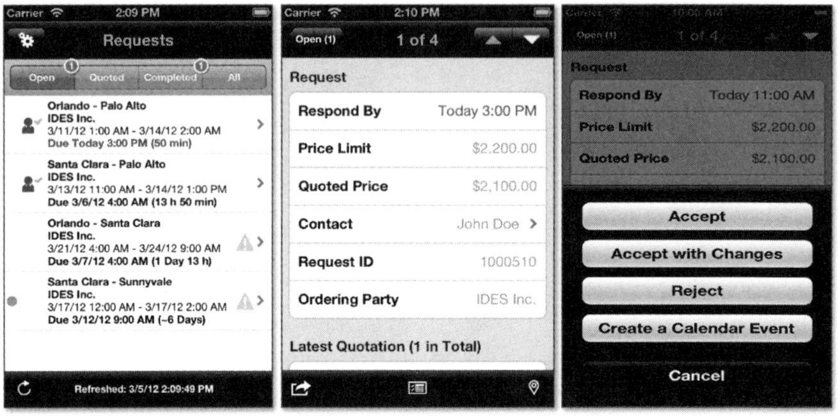

Figure 12.26 Mobile Tendering

From a carrier's perspective, they receive RFQs for a freight delivery from companies such as shippers or LSPs anywhere and anytime. From a shipper or LSP's perspective, the mobile freight tendering application can be used as a channel to send freight RFQs to one or more carriers. The process can be manually triggered or automatically executed, while mobile tendering from a carrier execution perspective supports both direct peer-to-peer tendering and broadcast tendering.

Tendering options

In direct tendering, the tendering manager specifies and selects a carrier and then sends a freight order without creating freight RFQs. With broadcast tendering, RFQs for the freight orders are sent to multiple carriers and include information about the goods to be transported, the dates, and the price limit. In both cases, carriers can accept RFQs directly, reject them with a specified reason, or accept them with changes such as alternative dates or prices. Using core iPhone capabilities, they can check addresses and routes on the map, email or call the specified contact person, or create a calendar entry for tenders while checking other appointments on their mobile device.

Process flow

A tendering process usually starts if the tendering manager or the system determines that there is more than one possible carrier. As shown in Figure 12.27, the tendering process is typically initiated by selecting an existing tendering profile, while the tendering steps, definition of price and time limits, and visibility settings (the option to allow carriers to change prices and dates) can also be automated and stored as a tendering plan.

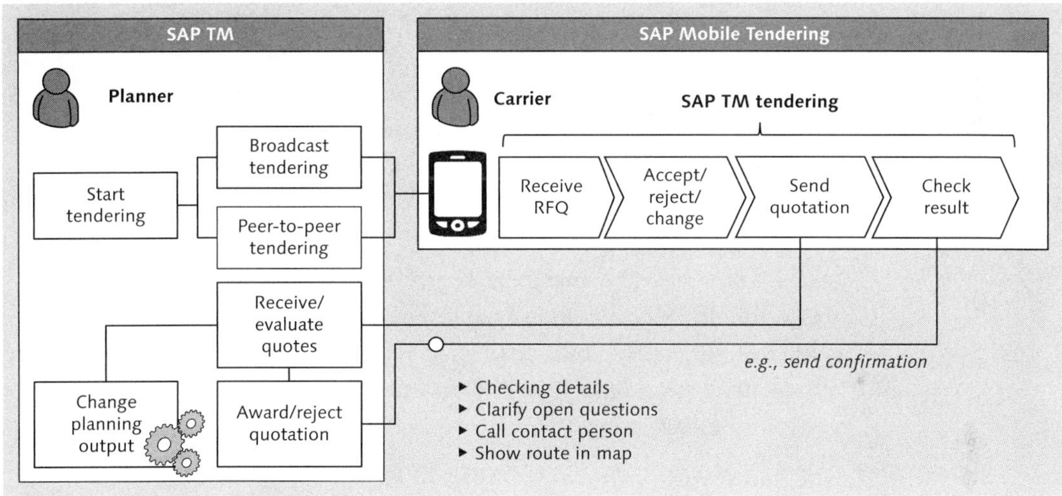

Figure 12.27 Mobile Tendering Process Flow

Carriers receive the RFQ and respond by submitting their quotation. Carriers are supposed to respond within the time limit given in the RFQ. Depending on the visibility settings defined in Customizing, they can not only accept or request the quotation, but also propose different prices and dates. Once the data has been received in SAP TM, their response is evaluated. In this context, SAP TM can automatically accept the quotation and send the freight order back to the awarded carrier for execution.

Transport Notification and Status

SAP Transport Notification and Status is a mobile application for Android devices that lets drivers view freight order details and report delivery statuses, as shown in Figure 12.28.

Figure 12.28 SAP Mobile Transport Notification and Status

Status reporting

We covered the integration of SAP TM with SAP Event Management for tracking and monitoring the execution of freight orders and recording status updates and events in Chapter 8, Section 8.2. To monitor execution, SAP Transport Notification and Status connects to SAP TM and allows drivers to access freight orders that have been assigned to their vehicle resources, get details, display stages, and report the status via predefined events to provide the most accurate processing information to the transportation system. Drivers can show the corresponding routes on a map, or, using other applications on their mobile devices, can use mobile route navigation.

Process flow

The mobile application assists a truck driver in his daily work, displaying all events of a freight order that are relevant for a selected stop, while the actual location and the location of the selected stop are provided to the navigation application. The navigation application then can calculate and display the best route to the next stop of the tour. The main use case typically starts with a driver checking all relevant freight orders and tours. He can display details of the assigned orders during the tour and use his mobile device to report the status via predefined events, as shown in Figure 12.29.

Stops can be confirmed both individually and collectively. Although he receives a planned tour with a predefined schedule, the driver can also capture unexpected events to reflect traffic conditions and capture delays, damages, or other incidents. Status reporting is sent to SAP TM to update the event handler for the corresponding freight order in SAP Event Management.

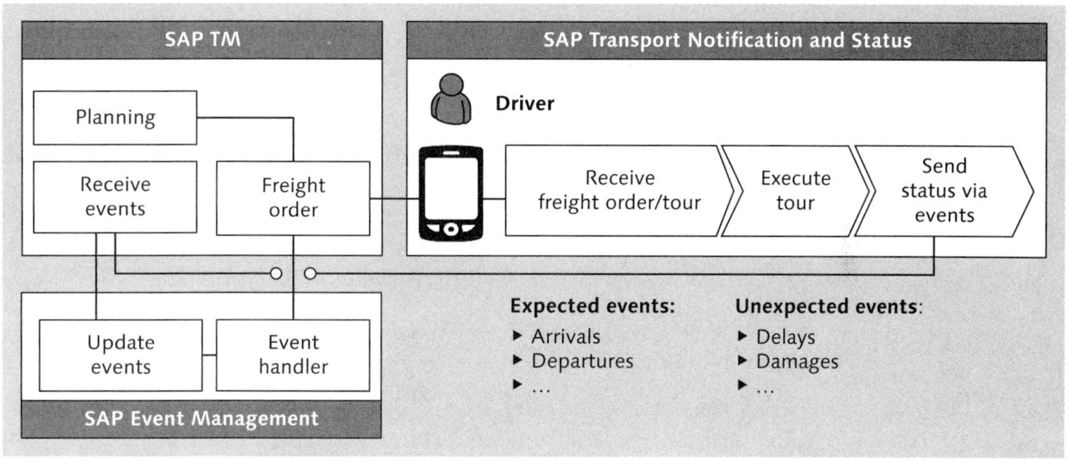

Figure 12.29 Transport Notification and Status Process Flow

Collaboration

Mobile applications are all about wireless communication. Customers communicate with their vendors, shippers communicate with freight-forwarders, and both collaborate with carriers. In most cases, collaboration is pragmatically processed via phone, fax, or email, while administration is often done using Excel or legacy applications. The next section explains the portal technology and functionality SAP TM offers to support carrier collaboration.

12.3.2 Collaboration Portal

Carrier collaboration in SAP TM 9.1

Empowering shippers and carriers to collaborate instantly and directly creates efficiencies and saves money for both parties. In general, by relying on the Internet as a real-time business platform, carriers are making their service offerings increasingly accessible to shippers through the web. This significant technology shift opens opportunities to more easily manage carrier selection, performance, and cost.

In this section we provide you with a sneak preview of the Transportation and Logistics Collaboration portal, which is currently scheduled for

SAP TM 9.1. The purpose of such a portal is to provide web applications for all collaboration work that consumes SAP back-end functionality. To learn more about the final scope and functionality, refer to the Release Notes of SAP TM 9.1. You will find this information in the SAP Service Marketplace at *http://service.sap.com/*.

Tendering application

Transportation and Logistics Collaboration portal functionalities in SAP TM 9.1 will mainly focus on the collaborative processes for freight tendering. Similar to the mobile freight tendering application mentioned in the previous section, it will cover all steps a carrier needs to participate in the tendering process:

- View, select, and filter freight requests for quotation (see Figure 12.30)
- Show details of the tendered freight order such as general information, business partner and contact information, cargo characteristics, dates and time, location, and requested stop sequence (see Figure 12.31)
- Monitor workload and history based on the request status
- Respond to the shipper or freight-forwarder (e.g., rejecting, changing, or accepting the RFQ)
- Provide a real-time status overview to check the quotations that have been awarded, rejected, or still need to be evaluated

Figure 12.30 shows the main screen of the carrier portal where carriers can select and filter the freight RFQs, accept or reject the proposal, or accept with changes.

Technology and integration

The carrier collaboration portal will offer similar functionality to the mobile transport tendering application. With regard to back-end communication, it uses an SAP NetWeaver Gateway infrastructure and services to access data in SAP TM. These services to access the portal functionality are delivered with SAP TM, while the portal application itself is packed and delivered within an additional add-on. This add-on can be installed on the SAP TM backend system or on SAP NetWeaver Gateway.

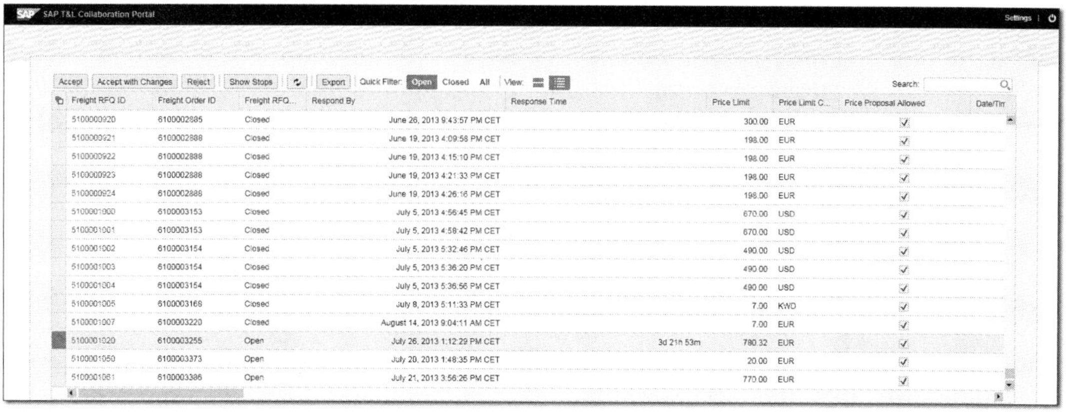

Figure 12.30 Main Screen of Carrier Portal

Figure 12.31 shows the details and stop sequence of the freight RFQ, including freight order details and quotation history. The current release will primarily focus on tendering. Using existing technology, it makes use of sidebars, additional toolbars, and widgets; future releases of the portal will provide more information and make these accessible for the carrier.

Figure 12.31 RFQ Details and Stop Sequence

In this chapter we explained both the prerequisites and content SAP provides to connect SAP TM to SAP NetWeaver BW, as well as the system integration to support cross-application processes with SAP EWM. In addition, we briefly introduced collaborative scenarios supported by SAP TM, how mobile devices can be integrated, and how carriers can access business-critical information via the SAP TM collaboration portal. Integration and collaboration are important to support business decisions and cross-system processes.

In the next chapter, we start with the transportation process for shippers.

Now that we have taken a look at the functionalities of SAP Transportation Management (TM) individually, it's time to bring them together. SAP TM delivers predefined scenarios for different industries and use cases. Most of the scenarios cater to the shipper industry.

13 Transportation Processes for Shippers

Recall from Chapter 1 that the re-launched SAP TM in version 8.0 concentrates on the shipper industry. The *shipper industry* is defined as those companies whose main focus is not the transportation of goods, but the production of goods. As a result, shippers usually want to outsource the transportation of goods to carriers or logistics service providers (LSPs).

However, because transportation costs are a big factor in the company's accounts, shippers tend to focus more on transportation processes, while the transportation execution is still done by external business partners. The shippers take over the planning and routing of their goods themselves and then send transportation orders to the carriers or LSPs.

SAP TM provides several predefined scenarios for shippers that were designed for SAP TM 8.0. These processes were updated throughout the new product releases and are still valid with SAP TM 9.1.

13.1 Overview of Scenarios

Five scenarios have been designed for the shipper industry:

- Outsourced transportation
- International outbound transportation

- International inbound logistics
- Domestic inbound transportation
- Domestic outbound transportation

SAP Solution Manager supplies scenario descriptions and configuration guides to get you started with a template scenario. All these scenarios use a different process flow to demonstrate the different process variants that can be covered with SAP TM. In this chapter, we take a look at each of these scenarios. After we discuss the process flows of these scenarios, we then take a look at how one of these scenarios is configured.

Outbound and Inbound Logistics

In shipper scenarios, we can differentiate between outbound and inbound logistics. In outbound logistics, the shipper sells goods to a customer and processes the transport from the warehouse to his customer.

In inbound logistics, the shipper is on the buying side and organizes the transport from the provider of goods to his own warehouse.

SAP TM supports both transportation directions, which is represented in the five different shipper scenarios.

13.1.1 Outsourced Transportation

Focus on communication shipper-carrier

The focus of the outsourced transportation (OT) scenario is the communication between the shipper and a third party that performs transportation activities. In this scenario, we are dealing with an outbound logistics flow, which means the shipper sells goods to the customer and organizes the transport out of the warehouse to the customer.

Figure 13.1 shows the outsourced transportation scenario process. Let's walk through each step.

The outsourced transportation scenario starts in SAP ERP, where a sales order is created. The sales order is transferred to SAP TM so that SAP TM can take over the process of transportation organization. Because this scenario doesn't focus on the planning of freight, it uses the one-step scenario, meaning that a freight order is created during freight unit building. The freight order therefore represents exactly the sales order that was created in SAP ERP before.

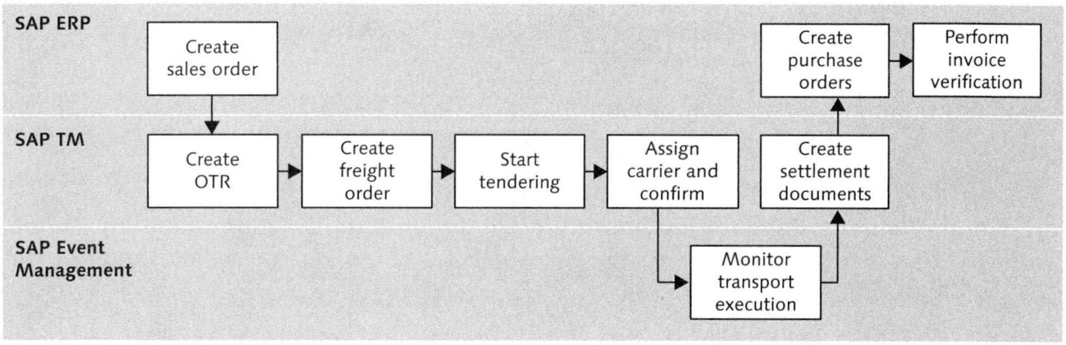

Figure 13.1 Outsourced Transportation Process

Peer-to-peer, broadcast tendering

To find a carrier for the freight order, this scenario uses the tendering functionality. *Peer-to-peer tendering* is automatically started once the freight order is created, and the request for quotation is sent out directly to the first carrier defined in peer-to-peer tendering. An incompatibility between the sales order's customer and a carrier excludes one carrier from the tendering process.

While the tendering is still running, the sales order's quantities are changed in SAP ERP. The updated quantities are communicated to SAP TM and both the order-based transportation requirement (OTR) and freight order are updated. If the quantity update is within a predefined range, the tendering is not stopped but continues. If the change is *not* within a tolerance range, the tendering is stopped and a new tendering is started automatically.

If all carriers contacted during peer-to-peer tendering reject the request for quotation, then a broadcast tendering is automatically started. One of the carriers contacted during the broadcast tendering accepts the request for quotation, and the response is evaluated as OK by the shipper. The carrier is therefore assigned to the freight order, and a confirmation is sent out.

Since the transportation has now become concrete, we can start the delivery creation with the information gathered in SAP TM. The delivery proposal functionality is used to do this.

Transportation execution with SAP Event Management

Now that the planning and carrier assignment has finished, the execution can start. The outbound transportation scenario can also be used for printing documents for the transport.

Outbound Transportation Hits the Road

Although not explicitly stated in the scenario description, the outbound transportation scenario is made for road transportation, which means that print documents configured in this scenario are documents used for road transportation only.

The execution of the transport is monitored by SAP Event Management. You can track the flow of the transport in the SAP Event Management transaction and also see whether exceptional events have occurred. Once the transport has been executed successfully, the transportation service performed by the carrier can be invoiced.

Invoice integration to SAP ERP

Charges have already been calculated automatically, so now freight settlement documents are created and transferred to SAP ERP, where purchase orders are created, so that invoice verification can start.

The rapid deployment solution (RDS) for tendering using SAP TM provides a very similar process flow. The charge calculation and SAP ERP invoice integration are not part of the RDS.

13.1.2 International Outbound Transportation

For the outbound transportation scenario, the transportation planning itself is rather simple: only one transportation stage, one transportation mode, and no customs are involved. But the shippers' transportation scenarios are not always as simple as described in the previous section. In a globalized world, transportation is often an international process with many business partners involved.

The international outbound transportation (IOT) scenario oversees this higher level of complexity by covering an international transport involving a global company's transportation division in the selling country as well as in the receiving country.

Multiple stages and transportation modes

In this scenario, three containers are supposed to be transported from the shipper in the United States to the customer in Germany, as shown in Figure 13.2. The shipper takes over the transport from their plant to the transportation division's warehouse. Therefore the transportation covered in SAP TM starts at the transportation division's warehouse.

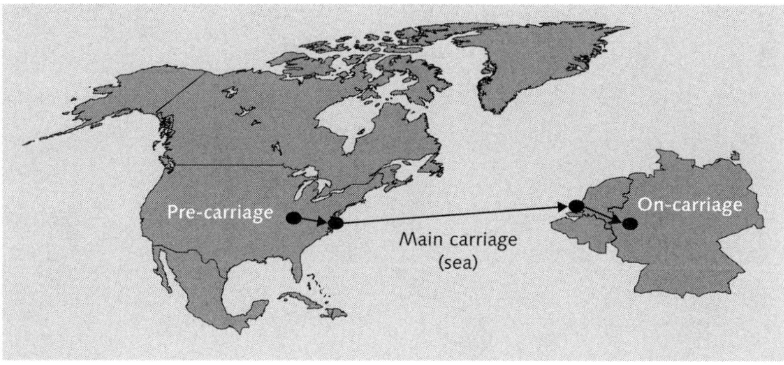

Figure 13.2 Transportation Stages of International Outbound Transportation

As described in Chapter 4, OTRs and DTRs are usually used in shipper scenarios. The international outbound transportation scenario is an exception, because the shipper company is clearly divided into a production division and a transportation division. Therefore, the transportation division acts as an LSP for the production division.

As a result, this shipper scenario starts not in SAP ERP with a Sales and Distribution (SD) or Materials Management (MM) document, but instead in SAP TM with a forwarding order. The forwarding order encompasses the entire transport from the warehouse in the United Stated to the customer in Germany. The containers are recorded as items of the forwarding order.

Due to the movement type and transportation mode used, three stages (the pre-carriage, main carriage, and on-carriage) have been created. After creating freight units, the transportation division enters the ports where the containers are supposed to be loaded and unloaded.

Using freight booking capacities

Meanwhile, the transportation division has also communicated with an ocean carrier and has booked capacity on a vessel based on a schedule. The booking is confirmed with the carrier, so the capacity can be used by the transportation division. Since the freight units' main carriage stage uses the ports for which the booking was made, they can be planned on the existing booking. Shipping instructions can be sent to the ocean carrier now to inform them that actual goods have been assigned to the booking.

After the main carriage is planned, the dates of the booking act as a prerequisite for the pre-carriage and on-carriage stages. In the transportation cockpit, the pre-carriage and the on-carriage stages of the freight units are planned on vehicle resources, and the resulting freight orders are assigned to carriers.

SAP Event Management now monitors the execution of all freight orders and therefore all freight unit stages. Once all expected events of the freight order representing the pre-carriage have been reported, the pre-carriage is completed, which means the containers are ready to be loaded onto the vessel.

Container loading is monitored by SAP Event Management, where the exact time and location of the loading activity are stored. When the containers are loaded, the master bill of lading and the house bill of lading can be issued.

The execution monitoring of the main carriage, and of the on-carriage later, is very similar to the other monitoring we have seen in this and the previous section. Once a freight booking or freight order is completed, the corresponding freight unit stage is also completed, and the next stage is ready for execution.

Settlement of freight documents

Since more than one carrier is involved in the entire transportation process in the international outbound transportation scenario, settling the transportation charges is more complex than in the outbound transportation scenario. Now each freight order and freight booking needs to be settled separately because each document was issued to a different carrier. After the charges are calculated, three freight settlement documents are created and transferred to SAP ERP for invoice verification.

Settlement of forwarding order

This next step differs from the outsourced transportation scenario. Now the transportation division, acting somewhat independently, can charge the production division for the transportation services. Therefore, charges are also calculated on the forwarding order and sent to SAP ERP using a forwarding settlement document. The production division, acting as the shipper of the forwarding order, is then charged by the transportation division.

Figure 13.3 summarizes the international outbound transportation process flow. If you compare this figure with Figure 13.1, you can see that more process steps are needed here; also, some process steps in this scenario occur in parallel, whereas the outsourced transportation process was linear.

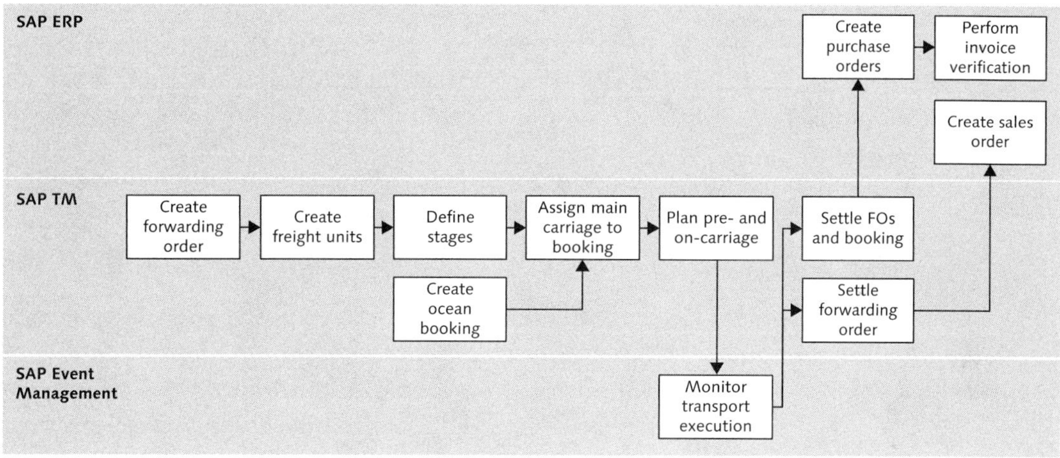

Figure 13.3 International Outbound Transportation Process

13.1.3 International Inbound Logistics

While the previous two sections dealt with outbound logistics, the international inbound logistics (IIL) scenario deals with inbound processes.

In SAP TM, inbound and outbound processes do not differ very much; they can even be combined if necessary. When organizing inbound logistics, the shipper can either outsource the entire transportation to a third-party logistics services provider (3PL) or organize the inbound transportation with their own transportation division. The first option can also be covered by the outsourced transportation scenario. (If you follow the rules, you'll have noticed that the outsourced transportation scenario is designed for outbound logistics but can also be used for inbound logistics without altering the scenario too much.)

In this scenario, the buying shipper orders several goods from a vendor in China and wants to transport these goods to a plant in Germany. To reduce costs, you should consolidate the goods, if possible.

Figure 13.4 summarizes the international inbound logistics scenario.

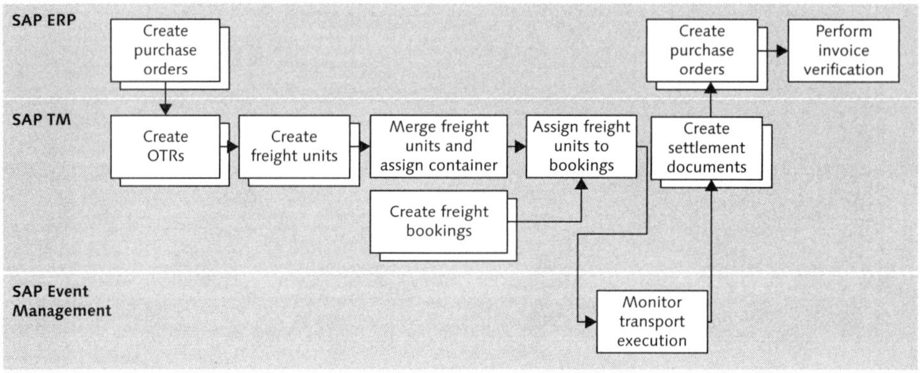

Figure 13.4 International Inbound Logistics Process

Let's walk through this scenario like we did with the previous two.

The scenario starts again in SAP ERP when the shipper creates several purchase orders. The purchase orders are transferred to SAP TM, where OTRs are created. Due to the high number of purchase orders and the freight unit building rule (FUBR) used, several freight units are created.

Some OTRs now have been assigned more than one freight unit because of the FUBR. We can merge these freight units into one container to use consolidation advantages. This scenario focuses on the consolidation possibilities, so several freight units are to be merged. Due to several incompatibilities, the merge in this scenario will only work in certain cases. During the merge, container items are also added to the freight units. Whereas the freight units only contained product items after the creation out of the purchase order, the merge process assigns the product items of both merging freight units to a newly created container item.

Meanwhile, freight bookings have been created that support the entire transportation instead of only the main carriage. The freight units are assigned to the bookings, and the transportation planner can manually reassign freight units from one freight booking to another.

Once the transportation planning is complete, transportation execution monitoring is taken over by SAP Event Management. In this case, even though we have three transportation stages, only one document needs to be tracked; this is because, in contrast to the international outbound transportation scenario, all three stages are planned on the freight booking. SAP Event Management therefore tracks only the freight booking.

Like in the other scenarios, the freight booking is then settled after execution and transferred to SAP ERP for invoice verification.

13.1.4 Domestic Inbound Transportation

The domestic inbound transportation (DIT) scenario is another inbound scenario delivered for shippers with SAP TM 9.1. It is shown in Figure 13.5.

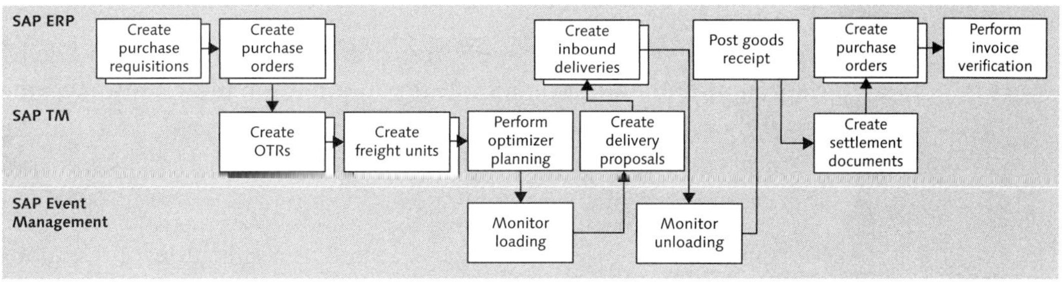

Figure 13.5 Domestic Inbound Transportation Process

The idea behind this scenario is that if the shipper orders a high volume of goods each day from several domestic vendors, freight costs can be lowered if the shipper himself takes over responsibility of the transportation organization. Because the different vendors would transport the goods individually, the shipper has the opportunity to combine the purchases from several vendors and thereby take advantage of lower freight costs due to consolidation effects.

The domestic scenario is based on land-based transportation within Germany. The shipper orders goods and materials from different vendors in different production locations. To do so, the shipper places purchase orders in SAP ERP based on purchase requisitions. The purchase orders are then transferred to SAP TM to create OTRs for which freight units are created automatically.

Multipick freight orders

A large number of freight units are created. In the transportation cockpit, freight unit planning is done with the optimizer. The optimizer planning supports the idea of this scenario by building multi-pick freight orders, which pick up goods at several production locations before going to the shipper's location. Upon creation of the freight orders, the optimizer executes a carrier selection so that carriers are assigned to the new freight orders. After the freight orders are created, the documents are sent to the carrier using the EDI interface.

As in all the scenarios we have discussed so far, SAP Event Management is used to monitor transportation execution. The difference between this scenario and the others is that in this one, the freight orders contain several loading locations; this means that loading events have to be reported several times before the unloading can be documented.

After loading of the freight units has been reported, the inbound deliveries can be created based on the information given in SAP TM using the delivery proposal functionality. Since the freight units are based on SAP ERP purchase orders, it is automatically determined that in this scenario, inbound deliveries have to be created.

Now that the inbound deliveries are created, the unloading of the goods at the shipper's location can be reported. After the events have been posted in SAP Event Management, the inbound deliveries are updated to post the goods receipt.

The scenario is concluded by charge calculation of all freight orders and the creation of freight settlement documents. Like in all other scenarios, the freight settlement documents are transferred to SAP ERP for invoice verification.

13.1.5 Domestic Outbound Transportation

The last of the five shipper scenarios in SAP TM is the domestic outbound transportation (DOT) scenario. If you have read through the other four shipper scenarios, you will have noticed that with each successive scenario, new functionality is used that was not used by any of the previous scenarios. Here, though, the functionalities of SAP TM already used in the other shipper scenarios are put together in a different way, making the domestic outbound transportation scenario to be a completely new scenario.

As the name suggests, this scenario is an outbound scenario. Because it is domestic, only truck-based transportation is used here. Again, consolidation effects are supposed to reduce the costs of transportation.

Typical of a shipper scenario, this process starts in SAP ERP with the creation of sales orders. The sales orders are transferred to SAP TM, where OTRs and freight units are created. Having a high volume of freight units, the shipper uses predefined selection and planning profiles to plan the freight units in the transportation cockpit. The scenario offers several possibilities for planning: automatic planning using the optimizer and manual planning.

After planning, the shipper can review the planning output by displaying the freight orders on a map. If the manual planning or the optimizer planning have had consolidation effects, the shipper can see this consolidation on the map, with the freight order having multiple unloading stops.

The outbound deliveries in SAP ERP are created again using the information SAP TM has gathered during planning. The delivery proposals are sent from SAP TM to SAP ERP, where the deliveries are created accordingly.

Peer-to-peer tendering

The freight orders created by the planning process do not yet have carriers assigned to them. To find the right carrier in this scenario, peer-to-peer tendering is used. Once the carriers' responses have arrived and have been evaluated, the best carrier is assigned and the freight order is sent.

Like in all other shipper scenarios, the execution of the freight orders is tracked with the integration of SAP Event Management. Once the loading of goods has been reported, the documents are printed from the freight order for the driver.

After the documentation is handed over to the driver, the truck leaves the shipper's premises and the transportation execution is recorded by events in SAP Event Management.

After the transportation is executed, the freight orders are settled using charge calculation, freight settlement documents, and the SAP ERP invoice integration. Since this was discussed in detail in the other scenarios, we'll skip it here.

Note that all process steps in the previous scenarios appear in Figure 13.6, but in a different context. This is a reflection of the flexibility of SAP TM's functionalities because they can be used in many different sequences, contexts, and scenarios.

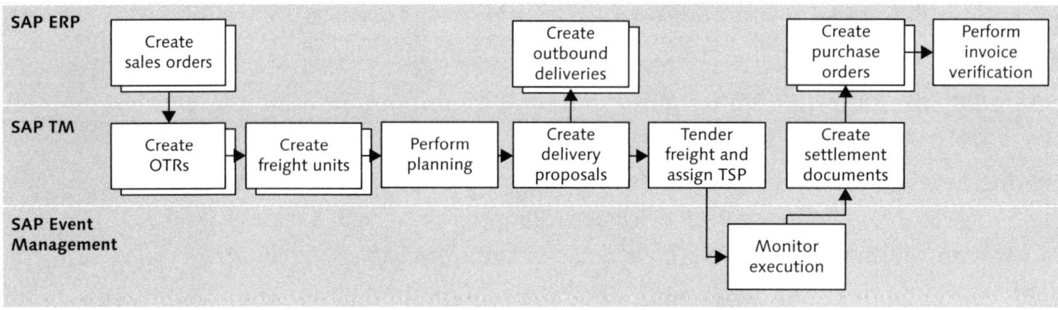

Figure 13.6 Domestic Outbound Transportation Process

Cost distribution As of SAP TM 9.0, the domestic outbound transportation scenario has been enhanced by the cost distribution aspect. While the scenario itself remains as it was in SAP TM 8.0, there is a new process variant to it. In the process variant with cost distribution, the planning has to be done manually, and then the carrier assignment should be done with carrier selection instead of automatic tendering.

After charge calculation, you can check on the freight order's COST DISTRIBUTION tab to see whether the costs of the freight orders have been distributed to the different freight units.

Once the freight settlement documents are transferred to SAP ERP, additional information can be checked in the SAP ERP system. The invoice verification can be monitored, and the freight cost allocation can be seen on the respective documents.

The domestic outbound transportation scenario combines the most common scenarios of the SAP TM RDS for planning and the SAP TM RDS for tendering.

13.1.6 Shipper Scenarios Summary

Before looking into the setup of a shipper scenario, let's briefly summarize and compare the scenarios in terms of what they offer and what they focus on.

The outsourced transportation scenario focuses on the tendering of a freight order that directly represents a sales order in SAP ERP. Planning is not included in this scenario.

International outbound transportation is a more sophisticated scenario in terms of planning. The scenario focuses on the interaction with the carrier and the use of prebooked capacities on ocean liners' vessels.

Consolidation effects come into play in the international inbound logistics scenario. While transportation planning is used only peripherally, the entire transportation is supposed to be executed by a carrier including pre-carriage and on-carriage. The focus is on consolidating ordered goods into containers.

The scenario covering domestic inbound transportation also works with consolidation effects, but on a different level. Because transportation is only done domestically by truck, the consolidation is achieved by multi-pick transports. The optimizer is used to create sensible multi-pick transports.

Domestic outbound transportation encompasses many of the functionalities and process steps previously discussed. The focus is on quick but detailed processes, including delivery creation, planning, carrier selection, and charge settlement. As of SAP TM 9.0, you can also include freight cost allocation on the SAP ERP sales orders in this scenario.

Table 13.1 summarizes the different functionalities that are used with the shipper scenarios.

	OT	IOT	IIL	DIT	DOT
SAP ERP logistics integration	X		X	X	X
Freight unit merging		X			
Delivery proposals	X			X	X
Optimizer planning				X	X
Manual planning		X	X		X
Freight bookings		X	X		
Tendering	X				X
Carrier selection				X	X
Execution monitoring	X	X	X	X	X
Charge calculation	X	X	X	X	X
Cost distribution					X
Freight settlement	X	X	X	X	X
Forwarding settlement		X			
SAP ERP invoice integration	X	X	X	X	X

Table 13.1 Comparison of Shipper Scenarios

So far in this book, we've covered the general possibilities for setting up an SAP TM system. This particular chapter discusses the different scenarios offered by the application. In the next section, to demonstrate the possibilities of the scenario, we blend these two aspects by describing how you must set up such a scenario. We use the domestic outbound scenario as an example of how to set up SAP TM to use the application as a shipper.

13.2 Shipper Scenario Configuration

So let's take a look at how to set up a shipper scenario. The example uses the domestic outbound transportation scenario because it encompasses the most important functionality for shippers, such as order integration, tendering, charge calculation, and invoice integration.

> **Note**
>
> Because the configuration guides for the scenarios can be found in SAP Solution Manager content for SAP TM, we dedicate these pages not just to listing the steps in complete, sequential order, but to explaining why we need to do these configuration steps. To configure the system to get a running shipper scenario, we recommend that you refer to the configuration guides in the SAP Solution Manager content.

13.2.1 Basic Setup in SAP TM

Before configuring the scenario-specific content, we first need to take a look at the SAP TM system in general. Some configuration is used by all scenarios and is therefore considered as basic setup in the SAP TM application. If you have other scenarios already up and running, you might not need to do these steps again. If you have a new SAP TM system in front of you, you need to perform these steps up front because they are prerequisites for the setup of the scenario-specific content.

> **Basic Setup, not Basis Setup**
>
> The configuration guide for the basic setup of SAP TM also contains steps for Basis settings such as activating queues, business functions, and so on.
>
> In this section we concentrate on the settings that we reuse in the scenario setup and that have a business impact. For Basis setup, please refer to the corresponding configuration guide.

Basic setup for charge calculation

As described in Chapter 10, the charge items are calculated based on category codes and subcategory codes that are also used later for mapping in the SAP ERP integration. For the shipper scenarios, only a few of the category codes described in the configuration guide are used. However, to have the basic setup completely done, we recommend that you define all category codes so that you will already have them in the system when you decide to configure another scenario in SAP TM.

Category codes are used to classify the charge items. They become important in SAP ERP integration, where the category codes are mapped to charge elements in SAP ERP. You can define category codes in Customizing via the IMG path SAP TRANSPORTATION MANAGEMENT • TRANSPORTATION MANAGEMENT • BASIC FUNCTIONS • CHARGE CALCULATION • BASIC SETTINGS FOR CHARGE CALCULATION • DEFINE CHARGE CATEGORY CODES.

Now that you've defined category codes, you also need to create subcategory codes. You can find this Customizing activity via the same IMG path, except ending with DEFINE CHARGE SUBCATEGORY CODES.

When defining rate tables, specify the charge types that should be used for the charge line determined by the rate. The charge type encompasses category and subcategory codes and defines further characteristics (e.g., whether it can be negative). The corresponding Customizing activity DEFINE CHARGE TYPES can also be found in that IMG path.

Basic order integration setup

Since the domestic outbound transportation scenario is a typical shipper scenario, we are dealing with the setup of SAP ERP Logistics integration as well. Once the SAP ERP order is transferred to SAP TM, a document type has to be defined. Recall from Chapter 4 that the determination of OTR document types is unique because the condition cannot be assigned to any Customizing or profile. Therefore, there may be only one singleton condition in the system. This is also the reason why this condition is described in the basic setup of SAP TM and not in the scenario-specific setup.

As you can see in Figure 13.7, the condition contains the OTR types for all shipper scenarios.

Find: Next Previous

TR: Original Order C	TR: ERP Order Type	OTR Type
001 (Purchase Order)	DIT1	DIT1 (DIT: Order-based Trsp. Requirement)
001 (Purchase Order)	IIL4	IIL2 (PO Transport Requirements)
114 (Sales Order)	YDOT	DOT1 (Sales Order)
114 (Sales Order)	OTTL	OTTL (Truck Load Order)
114 (Sales Order)	ODOT	DOT1 (Sales Order)

Figure 13.7 OTR Type Determination for Shipper Scenarios

In SAP ERP the scenarios are differentiated by the SAP ERP order type. You can define the condition in SAP NWBC by following the menu path APPLICATION ADMINISTRATION • GENERAL SETTINGS • CONDITIONS • CREATE CONDITION. Then select condition type /SCMTMS/OTR_TYPE.

13.2.2 Basic Setup in SAP ERP

You need to go through some basic setup in SAP ERP as well.

Basic setup: invoice integration

When a settlement document is sent to SAP ERP, the purchase order or sales order needs to be assigned to organizational units. To do this, you can follow the IMG path INTEGRATION WITH OTHER SAP COMPONENTS • TRANSPORTATION MANAGEMENT • INVOICE INTEGRATION • INVOICING • MAPPING OF ORGANIZATIONAL UNITS • ASSIGN ORGANIZATIONAL UNITS FOR PURCHASING. Whereas the SAP TM application only uses a purchasing organization and, optionally, a purchasing group, the mapping in SAP ERP can be more sophisticated. In this Customizing activity you specify which purchasing organization and group the organizational units of SAP TM represent, which company code should be used, and what document type is used for the purchase order to be created for the settlement document.

The charge types, category, and subcategory codes we defined in Section 13.2.1 are now reused and mapped to the SAP ERP service master records so that the charge lines generated in SAP TM will end up on the correct accounts. The IMG menu path INTEGRATION WITH OTHER SAP COMPONENTS • TRANSPORTATION MANAGEMENT • INVOICE INTEGRATION • INVOICING • ASSIGNMENT OF TRANSPORTATION CHARGE TYPES • ASSIGN SERVICE MASTER RECORD AND ACCOUNT ASSIGNMENT CATEGORY leads you to this mapping activity. Make sure you have already defined the necessary service master records before you start the mapping activity here.

Additional setup for cost distribution

If you want to use the new variant of the domestic outbound transportation scenario using cost distribution, you need to perform one more configuration step in SAP ERP. In Customizing you need to map the SAP TM charge types to condition types in SAP ERP. This mapping activity in the SAP ERP Customizing was also newly introduced with SAP TM 9.0. You can find it via the IMG menu path INTEGRATION WITH OTHER SAP

Components • Transportation Management • Invoice Integration • Mapping for Cost Distribution • Assign SAP TM Charge Types to SAP ERP Condition Types.

13.2.3 Setup for Cost Distribution in SAP ERP

As mentioned in Section 13.1.5, the domestic outbound transportation scenario offers a variant that allows cost distribution among the sales orders created in SAP ERP. To enable this process, you need to define some additional settings in SAP ERP. The cost distribution functionality makes use of the agency business, so we have to go into this part of SAP ERP to configure it.

Cost Distribution

This chapter about the shipper scenarios is designed to give you an insight into which shipper scenarios are available and how an exemplary scenario is set up.

Because the cost distribution is only an optional variant in the domestic outbound transportation scenario, we only briefly mention the most important configuration steps of the cost distribution in SAP ERP, without going into as much detail as for the mandatory setup for the general scenario.

For more information about cost distribution, please return to Chapter 11.

The first configuration step in SAP ERP is the creation of a new billing document type and a new payment type. These types represent the documents and items that will be created after integration with SAP TM.

Agency business condition types

The agency business condition types are used to categorize the charges in controlling. We now create condition types like the SAP TM charge types, meaning the creation of basic condition types for freight such as FB00.

After the condition types have been set up, we can define a calculation schema. The calculation schema for the domestic outbound transportation scenario is shown in Figure 13.8. As you can see, the condition types we have just created are reused in the calculation schema.

Reference Step Overview

Step	Co...	CTyp	Description	Fro	To	Ma...	R...	St...	P	SuTot	Reqt	CalTy...	BasT...	Acc...	Accr...
100	0	TMF1	Freight Cost Distrib			☑	☐	☐		9				FR1	
200	0	TM01	Basic amount - TM			☑	☐	☐		9				FR1	
300	0	TMF2	Freight Cost Distrib			☑	☐	☐		9				FR1	
400	0	FB00	Freight Cost FB00			☑	☐	☐		9				FR1	
500	0	FB01	Freight Cost -fb00			☑	☐	☐		9				FR1	
600	0	BAF	Freight Cost -BAF			☑	☐	☐		9				FR1	
700	0	CAF	Freight Cost -CAF			☑	☐	☐		9				FR1	

Figure 13.8 Calculation Schema for Cost Distribution in DOT Scenario

As a final step in SAP ERP, we now set up some general settings concerning the creation of agency documents that are used to execute the cost distribution.

13.2.4 Defining Master Data in SAP ERP

We have now completed all steps described in the configuration guides for the basic setup of SAP ERP and SAP TM. Next up is configuring the domestic outbound transportation scenario. Please note again that we do not go through every single step of the configuration guide but highlight the most important steps.

As we saw in Chapter 4, for shipper scenarios the master data-leading system is SAP ERP. The SAP ERP system provides the system landscape with the most basic master data that is reused in other applications such as SAP TM, SAP Event Management, and SAP Environment, Health and Safety (EHS) Management. The creation of master data in SAP ERP is mandatory if you want to use order integration or invoice integration between SAP ERP and SAP TM.

For the domestic outbound transportation scenario, you need to set up the following master data in SAP ERP, according to the configuration guide:

- **Shipping points**
  Shipping points are used as the source location of the freight units later on because they represent the location where goods are handed over to carriers.

- **Customers**
  In the domestic outbound transportation scenario, customers are the recipients of the goods. The most important information of the customer master data in the domestic outbound transportation scenario is the customers' addresses because they act as the destination location of the freight units.
- **Organizational units**
  Organizational units are important for the creation of sales orders and purchase orders in the context of invoice integration. We recommend that you transfer the organizational units to SAP TM to simplify the mapping described in Section 13.2.2.
- **Material**
  The material represents the items that will be sold to the customer. They appear on the sales order and are transferred to SAP TM. In SAP TM, the material master data is stored as products.
- **Vendors**
  Vendors in SAP ERP can have two meanings for SAP TM. In inbound logistics vendors are the business partner from whom goods are shipped to the shipper. In the domestic outbound transportation scenario, however, vendors represent carriers that SAP TM uses as the executing business partners for the freight orders.
- **Conditions**
  The conditions are used not for freight costing, but for the sales order price itself.

After the master data is created it is transferred to SAP TM using the Core Interface Framework (CIF) and the transaction to transfer the transfer of organizational units.

13.2.5 Defining Master Data in SAP TM

Organizational data

We recommend that you transfer organizational units from SAP ERP to SAP TM in case they need to be mapped for order integration or invoice integration. However, SAP TM also contains planning and execution organizations that are used to allocate responsibility for a transport to a department. The organizational units can be created in Transaction PPOCE.

Transportation network

The transfer of master data from SAP ERP to SAP TM has created locations for both shipping points as well as customers. If you have a Geographical Information Server (GIS) connected to your SAP TM system, then the locations have been geocoded automatically, or you can trigger geocoding with report /SAPAPO/MASS_GEOCODE_LOC. If you do not have a GIS connected, please enhance the location master data in SAP TM with the geo-coordinates because they will be needed to calculate the correct transportation times between the locations.

For the simplified setup of the transportation network—but also to define selection profiles more easily—transportation zones are also used in the domestic outbound transportation scenario. In our case, the shipping point's location is situated in one transportation zone (Midwest), and the customers' locations are situated in different transportation zones, according to their region or postal codes.

The missing setup for the transportation network is the transportation lane, which represents a physical and logical link between two locations or zones. This link is always established based on a means of transport. Therefore, before defining transportation lanes, we need to create the means of transport that we want to use in the transportation lane but also later on as the basis for vehicle resources. In this scenario, we differentiate between full truck load (FTL) and less than truck load (LTL). For both possibilities, a means of transport is created.

Once the means of transport are created, we can now create the transportation lanes. We assign all means of transports that we have created above to all possible connections. As you can see in Figure 13.9, the transportation lanes have been assigned means of transport-specific transportation costs that will be used later in planning as planning costs. In addition, you can see that each means of transport also has a carrier assigned on the lane. This carrier assignment is necessary for both enabling automatic peer-to-peer tendering (notice the P2P tendering checkbox column in the figure) and for carrier selection (notice the priority of carriers column for carrier selection).

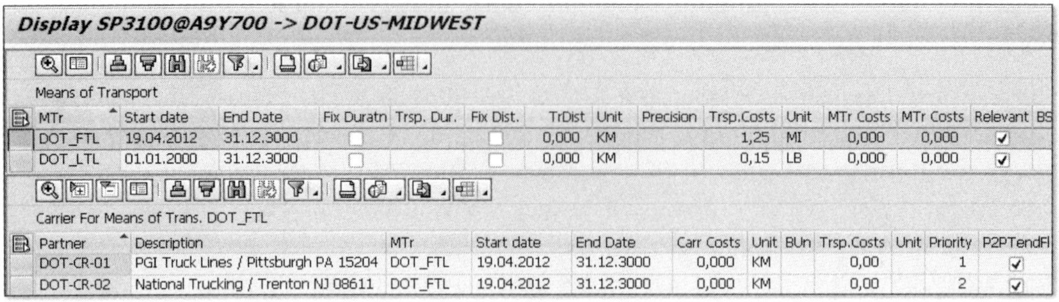

Figure 13.9 Transportation Lane Example for DOT

Vehicle resources

The last master data element missing might be the most important for transportation: the vehicle resources. Because the transportation process is owned by the SAP TM application in the SAP system landscape, SAP TM is also the master data-leading system for vehicle resources. Therefore, the resources were not created in and transferred from SAP ERP.

The vehicle resources need to be specified with the means of transport we have just defined. In addition, we assign capacities to the resources and enable finite scheduling to respect the capacity restrictions in planning.

Having created the vehicle resources, we now have created all the master data necessary for the domestic outbound transportation scenario (except for charge management master data, which we cover later in this chapter). We can now start the Customizing of the scenario.

13.2.6 Order Management Customizing

Setup of SAP ERP Logistics integration

If you want to use CO-PA functionality in SAP ERP with the invoice integration, you need to define more Customizing steps in SAP ERP, such as maintaining allocation structures or settlement profiles. Here, though, we focus on the very basic order management Customizing in SAP ERP to enable the basic flow. For more information on what needs to be configured for CO-PA functionality, please refer to the configuration guide for domestic outbound transportation.

In SAP ERP you need to define a sales order type that will be used to create sales orders that are transferred to SAP TM. As you know from

SAP ERP SD, sales areas need to be assigned to the new sales order type; assign the organizational units that were created before.

Once the sales order type and all necessary configuration have been set up (please refer to the configuration guide for all necessary steps), we can activate the transfer of sales documents as described in Chapter 4, Section 4.1. Since the domestic outbound transportation scenario also works with delivery creation and integration of the deliveries to SAP TM, we also need to activate the transfer of outbound deliveries. For outbound deliveries, the scenario uses the standard document type LF instead of a newly created type.

Setup of TRQ types in SAP TM

In SAP TM we already have defined which OTR type should be used for OTR creation with the singleton condition mentioned above. However, that OTR type does not yet exist. Therefore, we create it now. In SAP TM, follow the IMG menu path SAP Transportation Management • Transportation Management • Integration • ERP Logistics Integration • Order-Based Transportation Requirement • Define Order-Based Transportation Requirement Type and configure OTR type DOT1, as shown at the top of Figure 13.10.

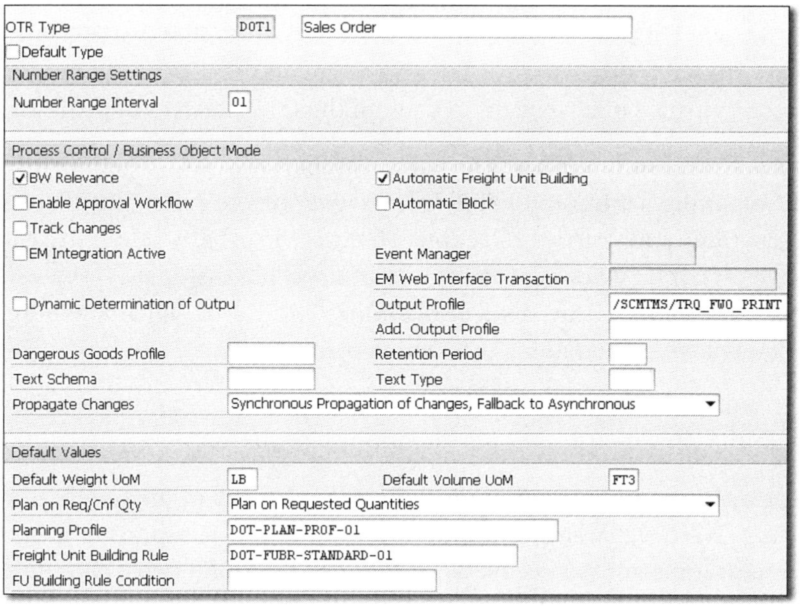

Figure 13.10 OTR Type for Order Integration in DOT Scenario

The most important setting of the OTR type is that freight units are created automatically. Therefore, the corresponding checkbox (AUTOMATIC FREIGHT UNIT BUILDING) is selected, and an FUBR is assigned. If the FUBR has not yet been created in the system, leave both entries empty for now because we create it in Section 13.2.7.

Delivery Integration

Again, we also need to integrate deliveries into SAP TM. Take care when configuring the condition for determining the DTR type. As already mentioned, the domestic outbound transportation scenario does not use its own delivery type, so you need to check whether other scenarios also make use of this delivery type for integration with SAP TM. If they do, consider using a separate delivery type in SAP ERP or maintaining the condition differently from the condition described in the configuration guide. The DTR type Customizing looks exactly like the OTR type Customizing in Figure 13.10. The only difference is the name of the type. Because both OTR and DTR are documents of the TRQ business object, they cannot carry the same type name. Therefore, the DTR type is called DOT2.

13.2.7 Configuration of Planning in SAP TM

Thus far, we have done the configuration in SAP ERP and SAP TM to transfer the sales order to SAP TM and to create the OTR. Now we can start configuring what we want to do with the OTR in terms of planning.

Creating incompatibility for carrier selection

In the domestic outbound transportation scenario, we are dealing with a lot of planning settings and restrictions. One of the restrictions is the incompatibility for carrier selection. The incompatibility defines that a specific carrier may not take transports to a specific customer. However, this can be overruled by manual planning because the incompatibility can be breached when doing interactive planning.

The incompatibility is recorded in new incompatibility settings that only contain the carrier-customer incompatibility for now.

Freight unit building

The next step towards our planning setup is the setup of the freight unit building. We have already created an OTR type that is supposed to create freight units automatically. Now we need to define how the freight unit building should work. A prerequisite for freight unit building is an existing freight unit type, which we don't yet have. The freight unit type for the domestic outbound transportation scenario resembles the default freight unit type. Execution tracking is enabled with the application object type ODT20_FU, which is also the default application object type for freight units. Remember to enter a freight order type in the FREIGHT ORDER DETERMINATION section, but we create this freight order type later in this chapter.

We enter the purchasing organization and purchasing group that we previously created as default organizational units for the freight unit type.

Once the freight unit type is created, we need to set up the FUBR. As shown in Figure 13.11, the FUBR for the domestic outbound transportation scenario is going to create one freight unit per OTR, but it splits the items if they exceed the split quantity of 35,000 lbs.

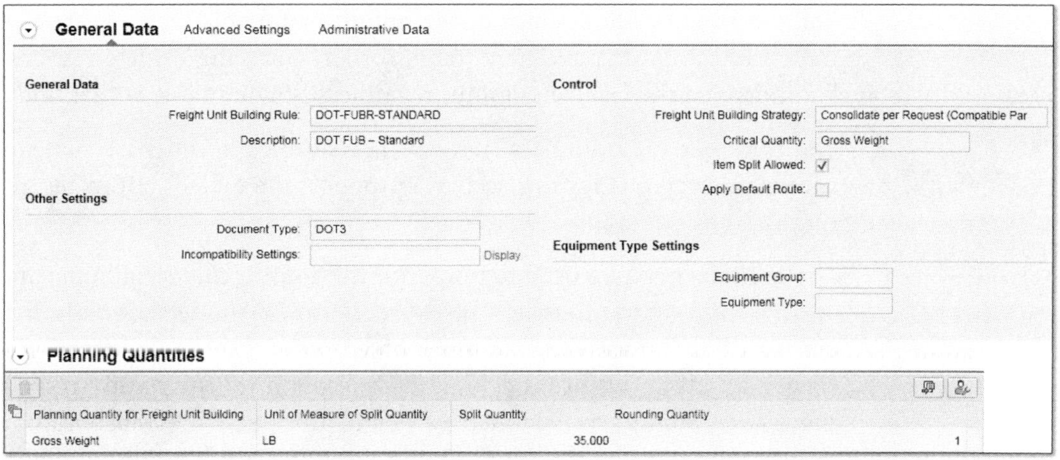

Figure 13.11 Freight Unit Building Rule for DOT Scenario

If you already created the OTR type, remember to go back to the OTR type Customizing to enable automatic freight unit building and define the FUBR we have just created.

The delivery proposal is generated out of a delivery profile that is determined on what information the delivery proposal should be created. In our case, it is a delivery creation per OTR item with no incompatibility settings.

Planning and selection profiles

Now we come to the most important configuration settings for the planning: the planning and selection profiles. We start creating a selection profile. The selection profile for the domestic outbound transportation scenario only contains selection criteria concerning the geographical information of freight unit stages and freight orders. Select all freight unit stages and freight orders that leave the shipper's shipping point and go to any location of the domestic outbound transportation scenario.

The selection profile only selects the transactional data such as freight unit stages and freight orders, but we also need to select resources to be able to plan new transports. Resources are selected using a capacity selection profile, which selects all means of transports used for this scenario. We created these means of transport in Section 13.2.5.

The scenario we are talking about is one of two shipper scenarios that use the optimizer. Therefore, optimizer settings need to be configured. The most important settings of the optimizer for this scenario are that the optimizer will create new freight orders once the vehicle resource has unloaded the last freight unit, regardless of where it is at that time.

You can alter the optimizer performance settings if you wish, but the configuration guide for this scenario proposes the default optimizer settings.

The optimizer can be used not only for the routing and scheduling and creation of freight orders in SAP TM, but also for the selection of carriers. To do this, we need to define some carrier selection settings. The carrier selection settings for domestic outbound transportation respect the incompatibility we defined earlier in this section. Transportation allocations and business shares are not used because we did not define any in Section 13.2.5. The carrier selection differentiates between three strategies: cost-based, priority-based, and priority- and cost-based; this scenario uses the cost-based strategy. Refer back to Figure 13.9, where we assigned possible carriers to the transportation lanes. We did not define any carrier costs, which means we have to take the transportation charges from SAP TM Transportation Charge Management to determine the costs for each carrier. Just in case both carriers have the same price, we have defined priorities on the transportation lane.

If the carrier ranking has been adjusted manually in the freight order, the ranking should not be influenced by the carrier selection again. After the carrier selection has determined a carrier ranking, the tendering should automatically be started. To do so, we need to define a tendering manager, which can be your user ID.

The optimizer only returns the optimal result if we define barriers and preferences. We do this by defining planning costs settings. The planning costs in this scenario should define preferences between the full truck load and the less than truck load, which are represented by two means of transport. The means of transport representing FTL has a fixed cost imposed as well as costs per additional stop. In addition, the variable costs are increased by the costs per distance we defined on the transportation lane (remember Figure 13.9). The LTL variant has neither fixed costs nor variable costs defined.

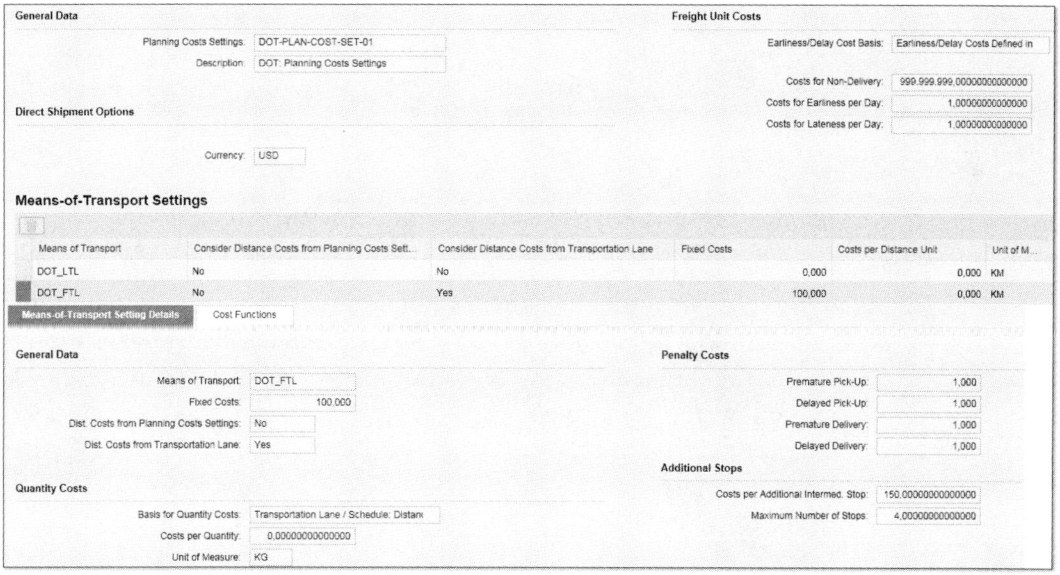

Figure 13.12 Planning Cost Settings for DOT Scenario

Recall from Chapter 6 that the planning profile acts as a frame for different planning-relevant settings and profiles. Now that we have created all these settings and profiles, we can put them together into the planning profile for the domestic outbound scenario. We assign the following settings and profiles:

- Capacity selection profile
- Optimizer settings
- Carrier selection settings
- Planning costs settings

Since the incompatibility settings we created at the beginning of this chapter are only valid for the carrier selection and not for planning, we do not enter any incompatibility settings in our planning profile.

In addition to assigning the settings and profiles, we need to define the output of the planning, meaning which freight document type should be created when performing manual or automated planning. If you haven't yet created the domestic outbound transportation-specific freight order type, remember to come back to the planning profile to assign it later. Due to the assignment of the profiles, we do not need to do anything else about the planning profile. All relevant settings are taken from the assigned profiles.

13.2.8 Freight Order Management Settings in SAP TM

Freight order type

We have talked about the output of the planning, which is a freight order. Like any other document in SAP systems, the freight order has to be of a special type. The freight order type contains many details that are crucial for the domestic outbound transportation scenario. In SAP TM we follow the IMG menu path SAP Transportation Management • Transportation Management • Freight Order Management • Freight Order • Define Freight Order Types and create the new freight order type DOT4.

The freight order type's basic settings define that the freight order is relevant for subcontracting, which is necessary for the tendering and the freight settlement that we want to do in this scenario. Shippers and consignees are moved over from the predecessor documents (in this scenario, the freight units).

Assign a quality tolerance condition in the change controller settings. The condition determines that the change controller should *not* be called if the quantity change is below a certain tolerance because it would terminate a running tendering and start a new carrier selection. We assign the standard strategy CARR_SEL as the creation strategy to the freight order. This creation strategy starts the carrier selection upon creation.

The execution monitoring of the freight orders should be done with SAP Event Management. Therefore, the execution tracking with SAP Event Management needs to be enabled, and an application object type has to be assigned. As for the freight unit type, the standard application object type is used.

We assign the organizational units we created before as default organizational units to the freight order. This is important because the charge calculation will determine a freight agreement based on the combination of the carrier (result of the tendering) and the purchasing organization that is assigned here in the freight order type.

Since we defined earlier that this freight order type is relevant for subcontracting, we have to determine a freight settlement document type that should be used when doing settlement of the freight order. We use the freight settlement document type that we create in Section 13.2.9.

As discussed earlier, we have assigned the creation strategy CARR_SEL, which automatically starts a carrier selection upon creation of the freight order.

However, we need to define what carrier selection settings should be made. To do so, we assign the carrier selection settings we created in Section 13.2.7 as additional settings in the freight order type as you can see in Figure 13.13. The carrier selection settings start a tendering. You might be wondering why we have not yet defined any tendering settings that should be triggered by the carrier selection in the previous section. The tendering settings are also taken from the freight order type we currently define. In the freight order type Customizing, you can see a section for the tendering settings. Define that freight order type-specific tendering settings should be used and enter the process and communication settings that we create in the following section.

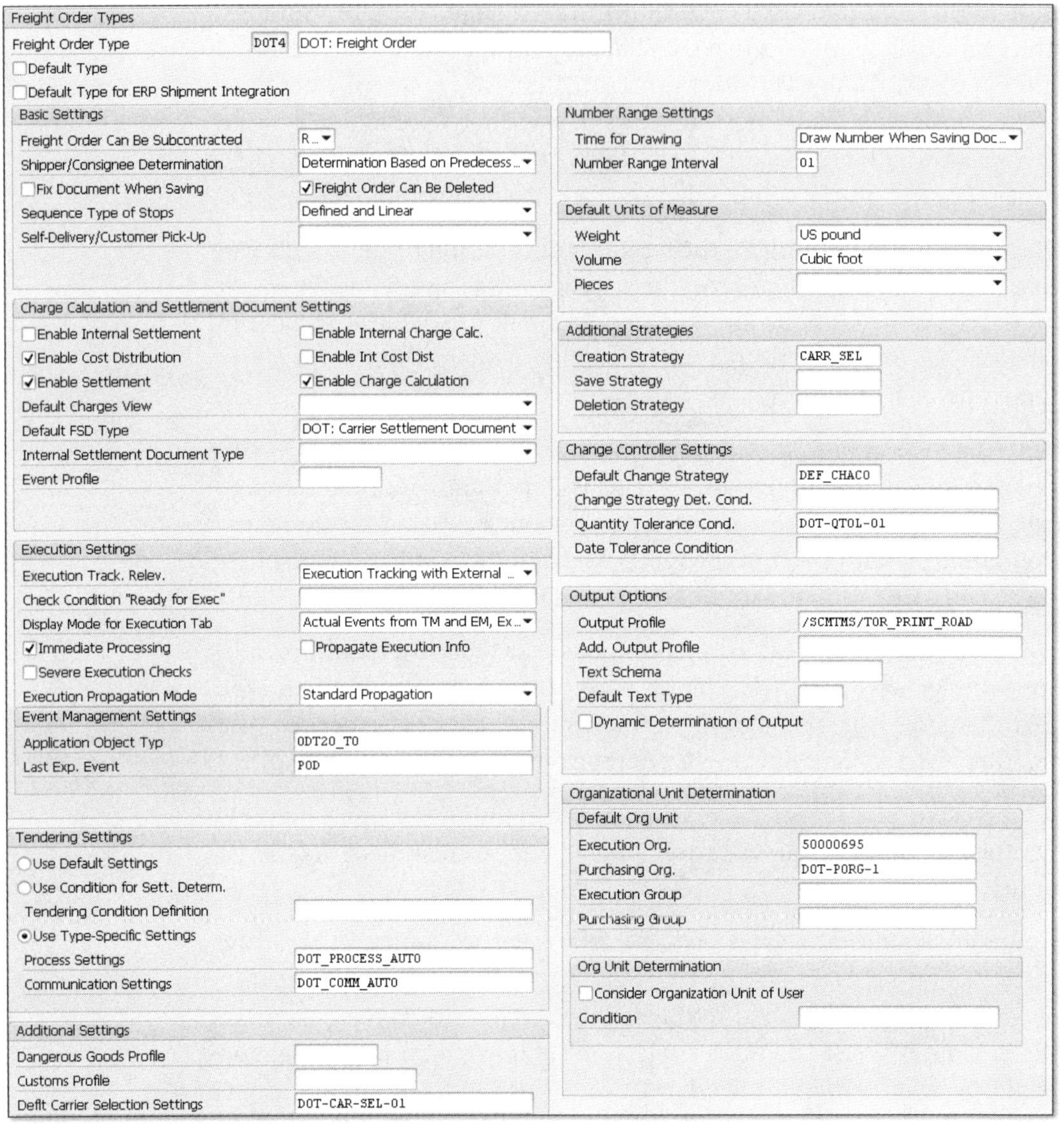

Figure 13.13 Freight Order Type Used in DOT Scenario

General tendering settings

To create a tendering profile, we need to define some general tendering settings in Customizing. These settings can be created via the IMG menu path SAP TRANSPORTATION MANAGEMENT • TRANSPORTATION MANAGEMENT •

Freight Order Management • Tendering • Define General Settings for Tendering.

The first step is to create visibility settings. These are used to determine what the carrier will be able to see and edit when he logs on to his view of the SAP TM system and reviews the RFQs sent to him. In this scenario, the carrier can see the stop dates of the freight order but can't change these in the response. He cannot see the price limit.

The most important general settings for tendering are the process settings that are defined in step 4 of the same Customizing activity. The process settings define how the tendering should work. The visibility settings we just created are assigned to the process settings for the domestic outbound transportation scenario. In addition, we assign the carrier selection settings because the tendering profile will execute peer-to-peer tendering based on a carrier ranking. The carrier ranking is created according to the carrier selection settings. The process settings also contain the tendering profile, which we create now.

Tendering profile

As shown in Figure 13.14, the tendering profile determines the way the tendering is executed. In this scenario, we perform peer-to-peer tendering. To create a tendering profile, follow the SAP NWBC menu path Application Administration • Tendering • Tendering Profiles • Create Tendering Profile. The tendering profile contains one tendering step: being defined as peer-to-peer tendering where a response is required. This means that if no response is received within the response time defined in the process settings, it is considered as a rejection.

Peer-to-peer tendering is done using direct tendering, which means there is no tendering to find a price but only to find a carrier that will execute the transportation. The price is taken from the freight agreements set up in SAP TM Transportation Charge Management. We also assign the visibility settings defined earlier to this tendering step.

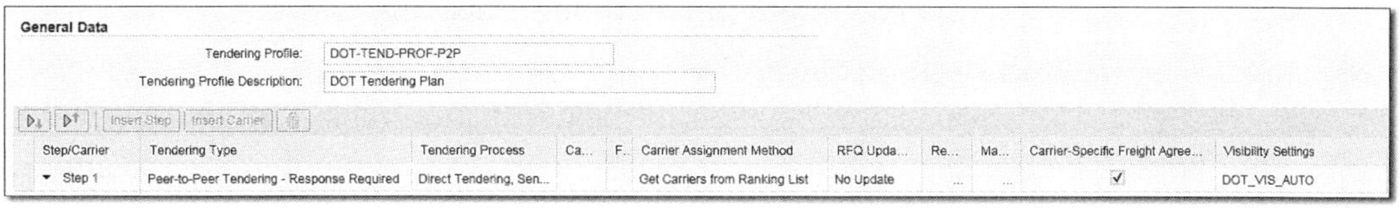

Figure 13.14 Tendering Profile for DOT Scenario

Now that we have defined some general settings and the tendering profile, we can add these settings to the freight order type. In the freight order type, we need to assign the tendering process settings as well as tendering communication settings. The communication settings described in the configuration guide define how the communication between our SAP TM system and the carrier should work.

13.2.9 Configuration of Transportation Charge Management and Freight Settlement in SAP TM

The last process step in SAP TM for the domestic outbound transportation scenario is the charge calculation on the freight order and the settlement in SAP ERP. For the settlement we need to define a freight settlement document that we can do in Customizing via IMG menu path SAP TRANSPORTATION MANAGEMENT • TRANSPORTATION MANAGEMENT • SETTLEMENT • FREIGHT SETTLEMENT • DEFINE FREIGHT SETTLEMENT DOCUMENT TYPES.

Our new freight settlement document type is called DOT6, which is relevant for cost distribution and has some default output profiles assigned to it. Remember to assign the freight settlement document type to the freight order type.

Creating scales

The next step is to set up the master data for the rates. Recall from Chapter 10 that the master data is in a hierarchical form. First, we need to define some scales to reuse in the rate tables. The following scales are created:

- **Source location scale**
  This scale will provide us with the possible source locations of the freight order that we can use in charge calculation
- **Destination region scale**
  The destination region scale will be used to differentiate rates according to the region the freight order is heading to.
- **Destination zone scale**
  This scale is very similar to the destination region scale. However, in this scale the differentiator (the zone) cannot be found in the address

of the destination location but only in the transportation network set up (recall the creation of transportation zones in Section 13.2.5).

- **LTL break-weight scale**
  The break-weight scale is a relative scale, which means it is divided into different break-weight spans (0-500, 500-1,000, 1,000-2,000, etc.).
- **Product group scale**
  The product groups have been created already. The scale allows the use of the product groups that were previously created.
- **Stop-off scale**
  This scale determines how many stops there are in the freight order because the charge calculation will increase the stop-off charge with every stop.

Rate tables

The carrier rate will be based on rate tables. As described in Chapter 10, rate tables consume scales and are embedded later in calculation sheets.

In the domestic outbound transportation scenario, there are four different rate tables for the carrier rates that differentiate the carriers by the offered prices. The carrier rate tables contain the source location scale and the destination region scale.

The rate found in the rate table is multiplied by the actual distance covered in the freight order. Figure 13.15 shows an example rate table for a carrier.

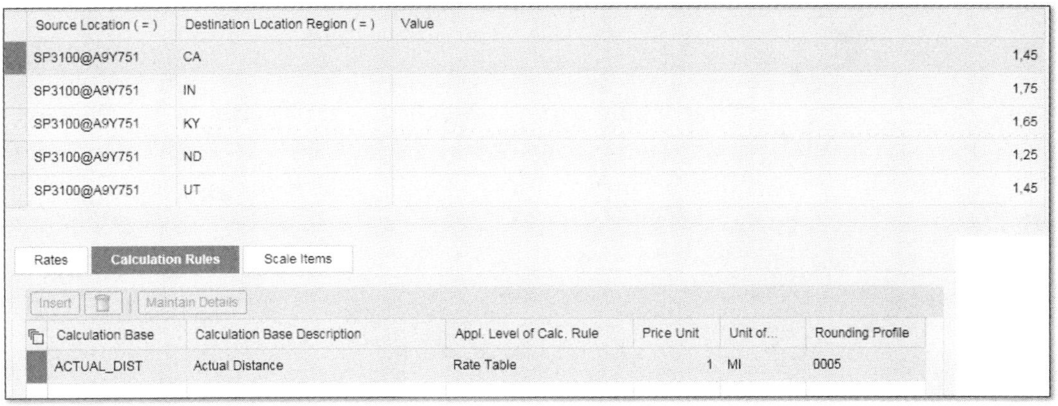

Source Location (=)	Destination Location Region (=)	Value
SP3100@A9Y751	CA	1,45
SP3100@A9Y751	IN	1,75
SP3100@A9Y751	KY	1,65
SP3100@A9Y751	ND	1,25
SP3100@A9Y751	UT	1,45

Rates | Calculation Rules | Scale Items

Insert | Maintain Details

Calculation Base	Calculation Base Description	Appl. Level of Calc. Rule	Price Unit	Unit of...	Rounding Profile
ACTUAL_DIST	Actual Distance	Rate Table	1	MI	0005

Figure 13.15 Carrier Rate Table for Domestic Outbound Scenario

In addition to the carrier rates, two general rate tables are created. The first general rate table is the fuel surcharge. This rate table determines the surcharge used later during the charge calculation. The surcharge is based on the destination location region.

The other general rate table is the rate table for the stop-off charges. This rate table imposes a surcharge based on the number of stops of the freight order.

Defining calculation sheets

These rate tables are embedded in calculation sheets; we now have to create calculation sheets for each carrier. Each calculation sheet contains several items. Each item is either a rate table we created above or a direct amount. Let's look at the calculation sheet for one example carrier.

Figure 13.16 shows the calculation sheet for a carrier. The three rate tables we created now represent the items of the calculation sheet.

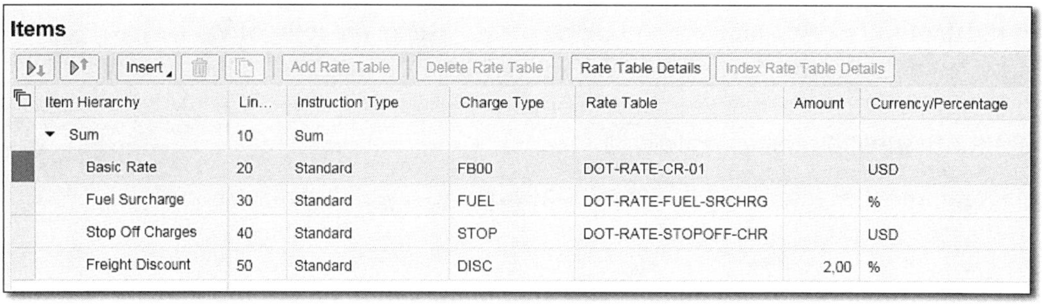
Items

Insert | Add Rate Table | Delete Rate Table | Rate Table Details | Index Rate Table Details

Item Hierarchy	Lin...	Instruction Type	Charge Type	Rate Table	Amount	Currency/Percentage
▾ Sum	10	Sum				
Basic Rate	20	Standard	FB00	DOT-RATE-CR-01		USD
Fuel Surcharge	30	Standard	FUEL	DOT-RATE-FUEL-SRCHRG		%
Stop Off Charges	40	Standard	STOP	DOT-RATE-STOPOFF-CHR		USD
Freight Discount	50	Standard	DISC		2,00	%

Figure 13.16 Calculation Sheet for a Carrier in the DOT Scenario

Each item of the calculation sheet is assigned a charge type that will later be used for mapping during the SAP ERP invoice integration. The carrier rate table represents the basic rate. Currency conversion is not done because the currency maintained in the calculation sheet is the same as that defined in the rate table.

The fuel surcharge is a relative surcharge on the basic rate. Therefore, it references the basic rate. In Figure 13.16 you can see that the stop-off charges are also assigned to this calculation sheet using the general rate table just defined.

This particular carrier offers a fixed discount of 2% on the basic rate. This is represented by another item in the calculation sheet. During carrier selection, all of these items are considered and added up to calculate a price that is then compared to the prices of the other carriers.

Creating a freight agreement

When performing either carrier selection or charge calculation on the freight order, the right calculation sheets for the carriers have to be found. This determination of rates starts with the freight agreement. After creating a freight agreement type, we can create freight agreements for each carrier. The freight agreement represents the contract between a purchasing oranization and a carrier and shows which calculation sheets are involved.

In our scenario, each freight agreement is assigned only one calculation sheet (recall that we have created one calculation sheet per carrier earlier).

Charge calculation profiles

Having defined the master data for the charge calculation, we now have to do some Customizing activities to steer the calculation and settlement of charges for this scenario. The settlement profile for the domestic outbound transportation scenario determines that the settlement document is based on the planned data in the freight order. In addition, the settlement document is created for each freight order instead of each freight order item.

The calculation profile for this scenario is defined such that the charges are calculated based on the actual route of the freight order, no matter the ordered route.

If you want to use the variant of the domestic outbound transportation scenario that includes cost distribution, you also need to define a cost distribution profile. This profile determines for our scenario that the costs should be distributed based on the gross weight of the SAP ERP item.

Cost Distribution

While the creation of the cost distribution profile is a very quick step in SAP TM, please remember that you also need to set up the agency business in SAP ERP as described in Section 13.2.3.

The next step is to define general settings for charge calculation. These general settings are defined for each organizational unit. Since we are only doing charge calculation on the freight order, we need to define general settings for the purchasing organization. The calculation profile, settlement profile, and cost distribution profile we have just created are assigned to these general charge calculation settings for the domestic outbound transportation purchasing organization.

13.2.10 Configuration Summary

In this section, we took a look at how a typical shipper scenario can be set up. First, we set up some basic Customizing in SAP TM (Section 13.2.1) and SAP TM (Section 13.2.2) that concerned not only the domestic outbound transportation scenario, but all shipper scenarios and therefore acts as prerequisite for all shipper scenarios.

After an excursion into the setup of cost distribution in SAP ERP (Section 13.2.3), which is an optional process variant of the domestic outbound transportation scenario, we defined master data first in SAP ERP (Section 13.2.4) and then transferred it to SAP TM. After the transfer of the SAP ERP master data, we then set up further transportation-specific master data in SAP TM (Section 13.2.5), such as a transportation network.

The order management Customizing (Section 13.2.6) was concerned with the setup and activation of the SAP ERP logistics integration, meaning that SAP ERP sales orders and deliveries being transferred to SAP TM where OTRs and DTRs could be created.

Because the domestic outbound transportation scenario was intensely about transportation planning, we have dedicated Section 13.2.7 to the setup of transportation planning in SAP TM. Planning profiles as well as optimizer settings and FUBRs were created here.

As you know from Chapter 6, the result of transportation planning is one or several freight orders or freight bookings. Therefore we had to set up the freight order management, including tendering and carrier selection in Section 13.2.8.

The charge calculation and freight settlement concludes the domestic outbound transportation scenario. In order to enable charge calculation, we set up transportation charge management-specific master data such as calculation sheets and freight agreements and then defined general settings for charge calculation in Customizing. This was described in Section 13.2.9.

In general, this section about the configuration of the domestic outbound transportation scenario is to give you an insight at how an SAP TM system needs to be set up and what considerations need to be undertaken. Consider this section to be a starting point for configuring the SAP TM application according to your use case.

13.3 Retail Scenario

We defined shippers as companies producing goods and transporting these to customers or importing material for production. Other industries can use SAP TM with scenarios similar to the shipper scenarios.

You might be wondering where the difference between "similar" and "different" is now. From an SAP TM perspective, the differentiation between shippers and other industries is that the core business of these companies is not the transportation of goods. Transportation is only a supporting process. Shippers are defined as manufacturing companies that do not have their own fleets, whereas other industries and companies that use transportation also have a supporting activity for their business.

We want to delve into the retail scenario, which is a valid scenario for SAP TM where implementation projects have already shown the challenges and synergies of retail customers when it comes to transportation planning.

Note

The shipper scenarios we have discussed are standard scenarios that SAP delivers with configuration guides, process flow documents, and so on. The retail scenario, however, is not a scenario that is provided by SAP.

So what's the difference? The shipper scenarios can be covered with the standard SAP TM application, performing only Customizing. The retail scenario, on the other hand, needs extra functionality and interfaces to be developed.

However, we want to show you in this section that SAP TM is also aware of the retail scenario.

We do not go into detail where extra functionality is required because we only want to sketch the particularities and challenges of the retail scenario.

The retail industry is defined as an industry that buys goods in order to resell them. The transportation of goods is a central business task for retail companies, but it's not the main differentiator between them and other companies.

Retail customers have a high number of transportation scenarios, both inbound and outbound. Figure 13.17 depicts some of the most common transportation scenarios of retail companies. The highest portion of transports in retail is the distribution logistics, which is shown on the left. In this scenario, the ordered products are delivered to the customer. By doing multi-drop transports, retail companies try to reduce the costs of distribution. However, they need to respect the transportation requirements of different assortments, which often force them to transport goods to the same store several times per day.

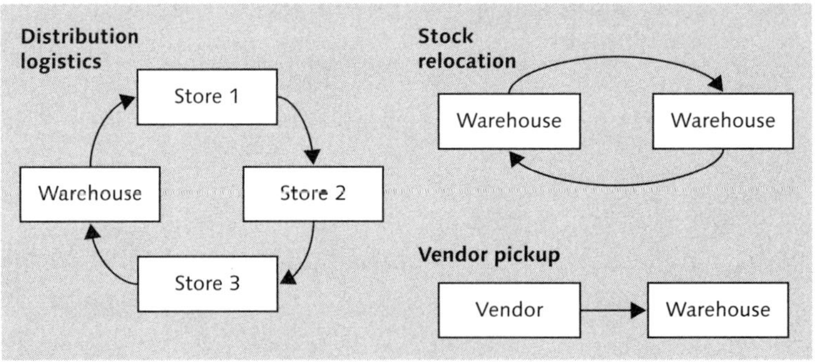

Figure 13.17 Typical Retail Scenarios with SAP TM

Retail transportation challenges

Many stores are located in urban areas, adding another factor to the transportation process: store accessibility. When you transport goods to ports, airports, or warehouses, the locations of loading and unloading are usually in industrial areas where trucks can easily maneuver. In urban areas, store delivery might be restricted to certain truck sizes depending on the time of the day. Standard shipper scenarios take it for granted that locations are easily accessible.

Because transportation is a supporting activity for retail companies, retailers try to reduce the costs of transportation by combining different transportation scenarios. Therefore, stock relocation transports could stop somewhere between the warehouses to deliver goods to stores.

Another challenge of covering retail transportation processes in a transportation management system is the high volume of goods and products that are transported daily. Sophisticated filtering and selection needs to be in place to cope with the high number of freight units.

Strategic freight planning

The high volume of freight units to be transported identifies another challenge for transportation planning. Not only does a high number of goods need to be planned using a low number of vehicle resources, but it needs to be done in a very short time. Stores demand a short time frame between ordering goods and receiving goods. This is because the stock reduction caused by customers buying goods in the stores cannot be predicted with certainty because the consumption depends on things like weather, political circumstances, and so on.

For example, there might be less than 24 hours between order entry and store delivery, which makes it almost impossible for the transportation planner to create a delivery plan from scratch. Therefore, retail companies perform strategic transportation (or freight) planning.

Strategic freight planning is done several weeks before the actual delivery date, which is before any stores have ordered their goods. The warehouses predict the order volume and order composition either using applications that derive predictions out of historical order data (such as SAP Forecast and Replenishment) or based on negotiations with the store itself.

These predictions are then sent to SAP TM, where freight units are created for strategic forecast deliveries. The transportation planner can now plan all strategic freight units on strategic freight orders.

Once the strategic transportation plan is finished, it can be used as reference for upcoming weeks because the ordering pattern of stores is often very similar from week to week.

When the actual store orders are now placed, the freight units created for the goods ordered can now be put on freight orders more quickly and more easily because the strategic freight planning has already provided a template.

Figure 13.18 shows an example process flow of transportation planning for retail companies. As you can see, there's a lot more going on than we have discussed so far. What we haven't discussed yet is the link between the strategic freight planning and the operational freight planning.

Carrier contracting

Another reason for the strategic freight planning is the contracting of carriers. Since transportation isn't the retail companies' focus, they rarely own their own fleets, but it's possible that they require many vehicle resources each day because of their high volume of goods. Furthermore, these vehicle resources often need to fulfill certain standards, such as cooling facilities, loading separation for dangerous goods, and so on.

When retail companies do not own their own fleets, it is crucial for them to proactively reserve a number of vehicle resources at carriers (in most cases, even contracting carriers to perform daily transports). The number and quality of required vehicle resources can be taken from the strategic freight planning. Because this transportation planning derives from an order forecast, it is a reliable source for determining how many vehicles are needed and what special features they might need.

Making long-term contracts in advance is often advantageous for retail companies. This increases the chance of a long-term business relationship, which can lead to potentially lower transportation rates.

As already mentioned, the order volume needs to be forecasted in the best way possible, but it cannot always be predicted accurately due to unforeseen events. Therefore, on the day of delivery, more vehicle resources might be needed than are actually available based on the long-term contracts. The tendering functionality can also be used to quickly find another carrier to execute the transports.

Warehouse management system integration

Figure 13.18 shows another crucial aspect of retail scenarios: the interaction between the warehouse management system (WMS) and SAP TM. While the shipper scenarios usually concentrate on the transportation process only, explicitly excluding the processes in the warehouse, the retail scenarios include communication between the WMS and SAP TM. Once the freight orders are eventually planned, they are communicated to the WMS, where picking can start.

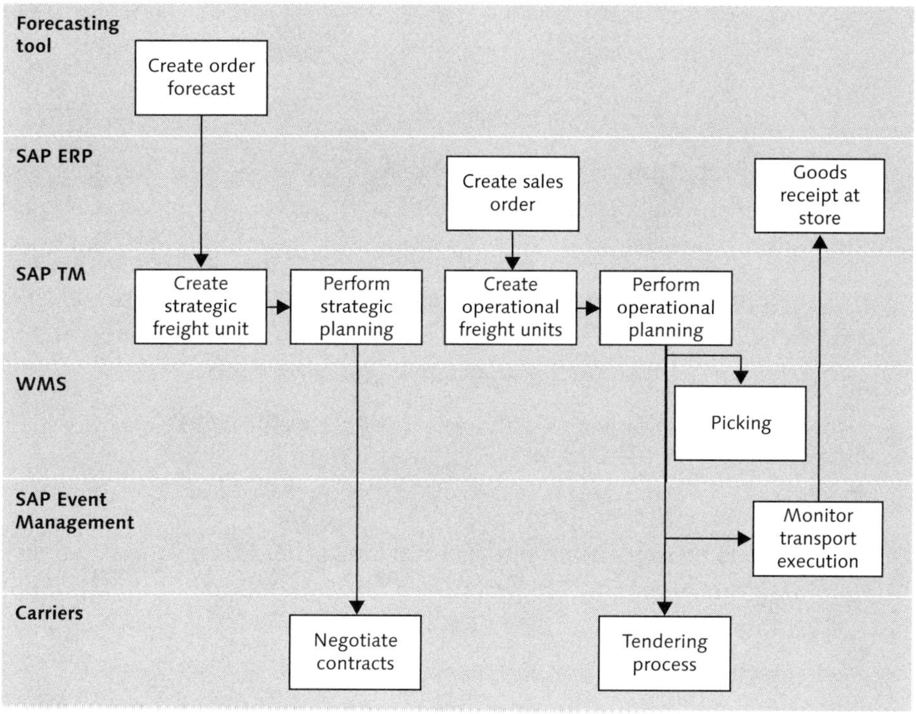

Figure 13.18 Exemplary Process Flow of Retail Scenario

Goods receipt triggering from SAP Event Management

SAP offers an industry-specific ERP system for the retail industry. This can be exploited by the transportation process as well. In Figure 13.18 you can see that the integration of SAP Event Management is also included in the retail scenario. Like in the shipper scenarios, SAP Event Management tracks the current status of the execution of freight orders. However, SAP Event Management is integrated with the mobile devices that the drivers use to monitor loading and unloading of the goods. This integration can also be used to trigger goods receipt in the store's retail store solution. Once the driver registers unloading of the goods at the customer on his mobile device and the store has confirmed the receipt, the goods receipt is created.

Since the retail scenario is not delivered by SAP and cannot be fully accomplished with the standard SAP TM application, it cannot be considered together with the shipper scenarios. However, in many cases, the capabilities of SAP TM support the realization of a retail company's transportation planning processes.

13.4 Conclusion

Having dealt with detailed functionality of SAP TM over the course of this book, this chapter began assembling the different functionality puzzle pieces together.

The shipper scenarios were developed for SAP TM 8.0 and are still valid in the current release. We have also discussed a scenario that would be categorized with shipper scenarios: the retail scenario.

Let's review the different scenarios and what we've covered in this chapter:

- **Outsourced transportation**
  Focuses mainly on the interaction with carriers. Transportation planning is completely excluded in this scenario. We took a detailed look at this scenario in Section 13.1.1.
- **International outbound transportation**
  Designed for large international companies with their own transportation division. The scenario covers the reservation of capacities at an ocean carrier. You can have a look at the process flow of this scenario in Section 13.1.2.
- **International inbound logistics**
  We discussed this scenario in Section 13.1.3. This inbound logistics scenario aims at consolidating different SAP ERP orders in SAP TM to reduce the administrative effort of transportation planning and reduce the costs of transportation as goods from different orders are consolidated into containers.
- **Domestic inbound transportation**
  This scenario puts the optimizer's capabilities into focus. You can see this in Section 13.1.4. The optimizer uses the geographical proximity

of manufacturers to create multi-pick transports and reduce the costs of transportation compared to single transports for each SAP ERP order.

- **Domestic outbound transportation**
  The domestic outbound transportation scenario deals with many of the functionalities contained in the other scenarios. The focus is on carrier selection and charge calculation these two components interact. This scenario has been enhanced in SAP TM 9.0 by the addition of a process variant covering the cost distribution functionality. We looked at the process flow in Section 13.1.5 and took this scenario as an example to demonstrate how a shipper scenario is set up in SAP TM. Section 13.2 was concerned with the setup of this scenario in SAP TM as well as SAP ERP.
- **Retail scenario**
  Not being a standard SAP-delivered scenario, the retail scenario still uses many of the process steps in the shipper scenarios. However, some challenges and particularities such as WMS communication and strategic freight planning differentiate the retail scenario from the standard shipper scenarios.

You have seen that the transportation process can be looked at from many different industry angles. This chapter used the angle of the shipper and retail industry, using transportation as a supporting activity of their business.

We now continue with discussing the transportation process from another angle: the LSP's angle. Because transportation planning and execution is a primary activity in LSPs' business process portfolio, their requirements for a transportation management system are very different than shippers' requirements.

The shipment of goods via ocean freight is of tremendous importance in global supply chains because it is a cheap and efficient method of transportation between countries and continents. This chapter teaches you how to set up and execute the consolidation process for ocean freight to support a freight-forwarder in SAP TM.

14 Transportation Process for Less than Container Load in Ocean Freight

SAP TM has been supporting the shipment of cargo via ocean freight since release 8.0. It offers the capability to manage both less than container load (LCL) and full container load (FCL) cargo for freight-forwarders. With the recent releases of SAP TM, more and more functionalities were provided to support specific scenarios and use cases. We begin this chapter by describing the high-level flow chart of the process example we chose for an end-to-end LCL shipment from Japan to the United States.

The process steps are outlined on a high level from an industry perspective. Section 14.1 provides you with the necessary process flow and introduces to you our imaginary freight-forwarder, Interforwarders Corp. In Section 14.2 of this chapter, we dive into the system configuration and all the necessary settings which are required to enable such a scenario in SAP TM 9.1. We introduce the key settings specific to the scenario in all related areas to this process, such as capacity management, order management, transportation planning, and charge management and settlement. In Section 14.3, we systematically run through the process in the SAP TM system.

In this chapter, we introduce how to manage space bookings with carriers, how to capture customer orders for LCL, and the capabilities in transportation planning for a multi-modal shipment, including trucking and ocean freight services. Moreover, we cover the system capabilities for the tracking and tracing of shipments, documentation handling, and consolidation functionality. We highlight export and import handling, and integrated charge calculation and settlement.

14.1 Ocean Freight Process Overview

Scenario background

We want to construct an end-to-end scenario for an export/import transportation of LCL cargo. Let's start by introducing our freight-forwarder, Interforwarders Corp., and its organizational structure.

Interforwarders has its headquarters in Japan, and acts as an international logistics service provider (LSP) for air and ocean freight services. It offers door-to-door service with an ocean leg from Yokohama, Japan, to Long Beach, USA, with different liners and voyages. Independent customer orders are received from different shippers within Japan. On the importing side in California, the containers are deconsolidated and delivered to the consignees in San Diego and Phoenix.

Organizational structure

Interforwarders has stations around Japan's biggest cities that take orders directly from regional customers. Other stations are spread internationally in the United States, Europe, and Asia to serve local demands. Such stations are called forwarding houses in SAP TM.

Two forwarding houses, in Tokyo and Nagoya, Japan, are involved in our scenario to take orders and arrange for the pre-carriage with local truckers. Next to the very customer-centric forwarding houses, Interforwarders has a container freight station (CFS) in Yokohama, near the container terminal in charge of procurement of space and cargo execution. The CFS procures and manages carrier contracts with rates and capacities for the port of Yokohama. Similarly, on the importing side, Interforwarders has a CFS in Long Beach and destination stations in San Diego and Phoenix that arrange the local deliveries.

Table 14.1 summarizes the involved organizational units and their responsibilities.

Organizational Unit	Responsibility
Origin stations (forwarding houses)	▶ Capture customer orders ▶ Manage customer contracts ▶ Initial route planning ▶ Arrange pickup services ▶ Handle export haulage ▶ Customer billing and settlement ▶ Local truck carrier settlement ▶ Export clearance
Export CFS (purchasing organization, planning, and execution organization)	▶ Request for quotation management with carriers ▶ Long-term contracts for quantity commitments/allocations ▶ Carrier bookings and handling of shipping instructions ▶ Consolidation of shipments ▶ Drayage arrangements ▶ Pre-advice to import CFS ▶ Charge calculation and settlement with ocean carriers
Import CFS (purchasing organization, planning, and execution organization)	▶ Drayage arrangements ▶ Deconsolidation of received cargo ▶ Import operations ▶ Import haulage planning and management
Destination stations (forwarding house)	▶ Arrange delivery services ▶ Customer billing and settlement ▶ Local truck carrier settlement ▶ Import customs clearance

Table 14.1 Responsibilities and Structure of Organizational Units

Export process The process starts by capturing four customer orders from four shippers in the forwarding houses of Tokyo and Nagoya. In each customer order, a door-to-door service is requested, including pickup and delivery and import customs clearance to consignees in the United States.

Table 14.2 shows the multiple orders. The two forwarding houses in Nagoya and Tokyo arrange for pickup services to their local warehouses and plan the transportation for pre-carriage and main carriage. The cargo is measured, weighed, and transported to the CFS in Nagoya.

Shipper	Consignee	Cargo	Gross Weight, Volume	Via Origin Station	Via Destination Station
Kan Xinson, Tokyo, JP	Dosena, Sedona, AZ, USA	Clothing	800 kg, 3 m^3	Tokyo	Phoenix
Wabei, Tokyo, JP	Ten Media, Tucson, AZ, USA	Electronic components	1 400 kg, 3 m^3	Tokyo	San Diego
Japan Cable, Gifu, JP	Southwire, Brawley, CA, USA	Fiber optic cables	2 000 kg, 3.5 m^3	Nagoya	Phoenix
Samurai, Tokyo, JP	Scottsdale Fireworks, Scottsdale, AZ, USA	Furniture	2 200 kg, 4 m^3	Tokyo	Phoenix

Table 14.2 Overview of Ocean Freight Shipments

The Yokohama CFS receives the two haulage truck loads. It unloads the trucks and consolidates the cargo into a D20 container for a voyage with the carrier SOCAL to Long Beach. The CFS handles the entire interaction with the ocean liner, such as issuing the booking confirmation, submitting the shipping instructions, and receiving the (draft) bill of lading. Finally, the container is transported to the terminal of the ocean liner SOCAL with a drayage order. The house bill of ladings for the individual shippers are generated, and the cargo is reported shipped-on-board. Figure 14.1 gives an overview of this exporting process.

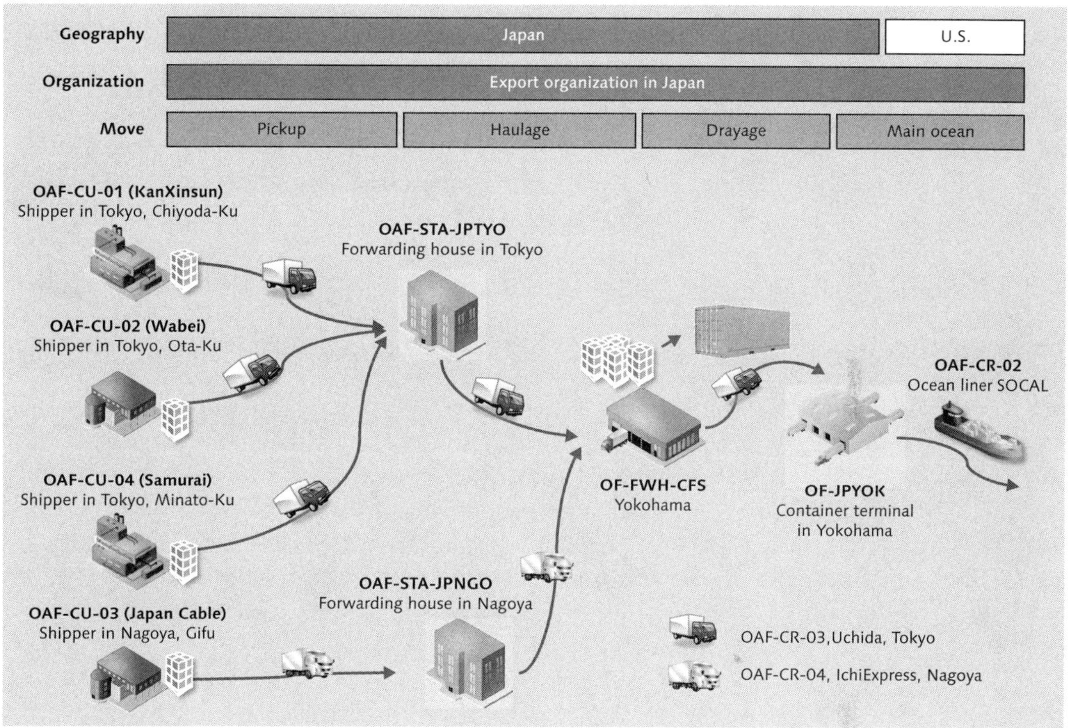

Figure 14.1 Export Scenario Overview

Import process

After the vessel of our carrier SOCAL has arrived safe and sound in the port of destination in Long Beach, the importing CFS starts with the deconsolidation. The cargo is reported as having arrived at destination. The entire import handling is managed by the U.S. branch of our LSP Interforwarders, which is an independent legal company. The CFS arranges the haulage services from Long Beach to the forwarding houses in the United States. The importing forwarding houses not only perform import customs clearance, but also manage the deliveries to the final destinations of the consignees.

Figure 14.2 shows the importing process of our scenario. Once the shipments arrive at their final destinations, the proof of delivery (POD) can be issued.

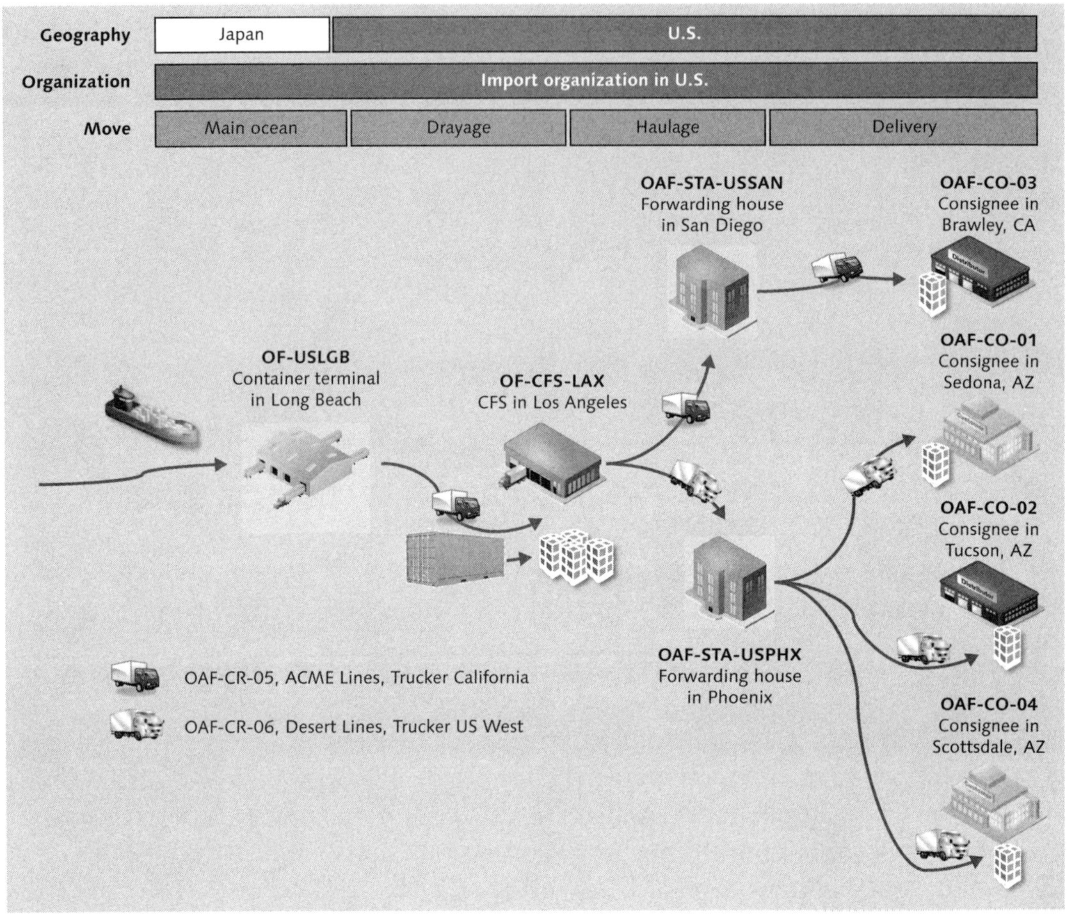

Figure 14.2 Import Scenario Overview

Financial process

The charge calculation and settlement is performed along with the execution of the process. Customer charges are calculated and billed depending on the incoterms. One order has the incoterm free-on-board (FOB), which means the cargo is prepaid by the shipper up to the port of loading. The rest of the charges have to be shouldered by the consignee. All other orders have the incoterm delivery duty paid (DDP), so the shipper pays for the entire transportation.

For subcontracted and procured services, the charges are calculated similarly and settled for invoice verification by the CFS in Yokohama for the main leg and individual forwarding houses for trucking. An internal settlement is performed by the CFS in Yokohama to recover the main leg charges.

In the next section, we walk through the core Customizing and setup required to execute this LCL ocean freight scenario.

14.2 Ocean Freight Process Configuration

Reference information

To set up and configure this LCL ocean freight process, a good amount of customizing, system setup, and master data is required. We want to give you an overview of the core steps that are necessary for the configuration.

> **Additional Information**
>
> If you are an employee, customer, or partner of SAP and you want to read in more depth about all the necessary steps required, you can refer to the LCL Ocean Configuration Guide, provided by the SAP Standard Development. Navigate to *http://help.sap.com/transportationmanagement90* and select Configuration and Deployment Information • SAP Service Marketplace • Scenario Configuration Guides • LCL Ocean Freight.

The first prerequisite is to set up the technical integration of your logical systems between SAP TM and SAP ERP and to complete the basic configuration for SAP TM. You can find the basic configuration guide at *http://help.sap.com/transportationmanagement90* • Configuration and Deployment Information • SAP Service Marketplace • Integration Guides • Basic Settings for SAP TM.

14.2.1 Partner and Organizational Master Data Setup

The first required setup is the master data for business partners. Even in an LSP scenario, both customers and vendors are set up first in SAP ERP and then transferred to SAP TM via the Core Interface Framework (CIF) because the financial processes for billing, invoicing, and accounting require similar master data.

Table 14.3 shows the parties in the process.

Partner Type	Partner ID	Name	Country
Vendor (truck carrier)	OAF-CR-03	Uchida Trucking	Japan
	OAF-CR-04	Ich-Ni Express	Japan
	OAF-CR-05	ACME Lines	U.S.
	OAF-CR-06	Desert Express Lines	U.S.
Vendor (ocean liner)	OAF-CR-01	East Asia Shipping Corp (EASC)	Japan
	OAF-CR-02	Southern Anna Camilla Shipping (SOCAL)	U.S.
Customer (shipper)	OAF-CU-01	Xinsan Clothing	Japan (Tokyo)
	OAF-CU-02	Wabei Electronics	Japan (Tokyo)
	OAF-CU-03	Japan Cable	Japan (Gifu)
	OAF-CU-04	Samurai Fireworks	Japan (Tokyo)
Customer (consignee)	OAF-CO-01	Dosena Clothing	U.S. (Sedona)
	OAF-CO-02	Ten Media	U.S. (Tucson)
	OAF-CO-03	Southwire	U.S. (Brawley)
	OAF-CO-04	Scottsdale Fireworks	U.S. (Scottsdale)

Table 14.3 Customer and Vendor Master in SAP ERP and Business Partner in SAP TM

In SAP ERP, you need to assign all vendors to the company codes in the United States or Japan. In addition, they need to be assigned to purchasing organizations and account groups. Ensure that you enter the complete address information for the shippers and consignees. In the replication of the master data to SAP TM via the CIF, locations in the transportation network are created for all business partners who can be used in SAP TM planning. You also need to enter reconciliation accounts and payment terms for both customers and vendors. All your customers need an assignment to SAP ERP sales organizations and distribution channels and, similarly, to reconciliation accounts.

CIF replication

After you have generated all the required master data in SAP ERP, use the CIF to replicate it with business partners in SAP TM. You need to generate and activate an integration model by using Transaction CFM1 or navigate to the SAP ERP menu path LOGISTICS • CENTRAL FUNCTIONS • SUPPLY CHAIN PLANNING INTERFACE • CORE INTERFACE ADVANCED PLANNER AND OPTIMIZER • INTEGRATION MODEL • CREATE. Do not forget to select the CREATE LOC./BP checkbox before the replication to generate locations automatically in SAP TM.

Organizational setup

In addition to the business partner, you need to generate the organizational structure for the business of our freight-forwarder, Interforwarders Corp. Use Transaction PPOME in SAP TM to generate the organizational structure. You are required to create one corporate unit to represent the head organization of Interforwarders Corp.

In our process, we interact with two country organizations that need to be created as companies for Japan and the United States, so you need to create entities for the forwarding houses in Tokyo and Nagoya, as well as the planning and execution organization/purchasing organization (the CFS) in Nagoya. The same setup is required for the importing side in the United States. In Chapter 15, we include an example of a similar organizational setup for the air freight process directly in SAP TM.

14.2.2 Freight Order Management

Equipment types

For the capacity management, you need to maintain equipment types that can be used for booking and handling purposes, as shown in Figure 14.3. Generate the equipment groups and types in SAP TM Customizing via the menu path IMG • TRANSPORTATION MANAGEMENT • MASTER DATA • RESOURCES • DEFINE EQUIPMENT GROUPS AND EQUIPMENT TYPES. You can maintain all container types for dry and refrigerated cargo. Make sure to assign modes of transport.

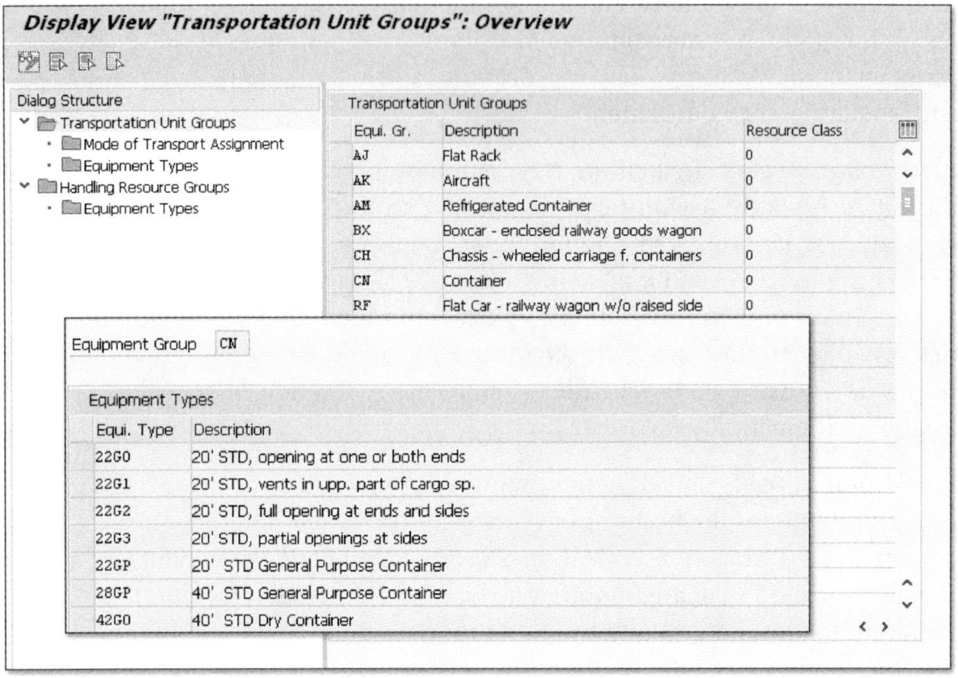

Figure 14.3 Equipment Groups and Types

Freight order type

To be able to generate pickup and delivery orders and drayage or haulage orders, you need to generate freight order types. Navigate to the SAP TM Customizing: IMG • Transportation Management • Freight Order Management • Freight Order • Define Freight Order Types. Generate two different types: one to be used for pickup and delivery, one for the pre-carriage and on-carriage. In Figure 14.4 you can see the freight order type OF05 for the pickup and delivery legs. Make sure to select all the checkboxes for charge calculation, cost distribution, and settlement. You can assign default means of transport and output profiles. If you want to use the SAP TM carrier selection functionality to choose the trucking providers, you can assign a carrier selection profile. In our process, this is not required; we choose the truckers manually.

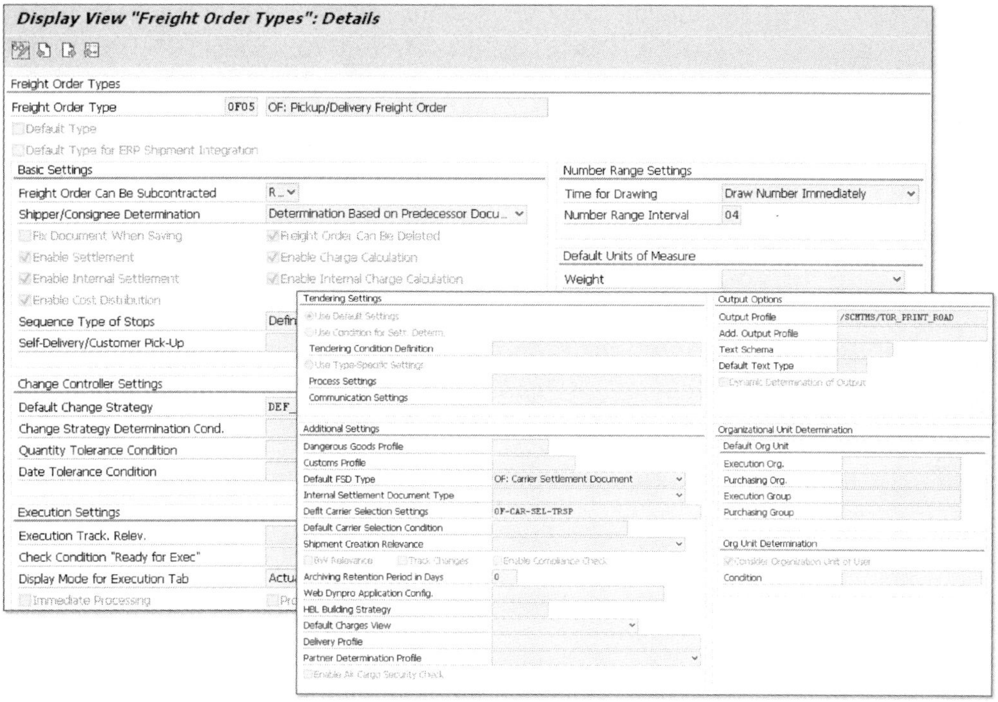

Figure 14.4 Freight Order Type

14.2.3 Schedule and Booking Setup

Schedule types, master data

To offer transportation services for ocean freight, you can work with carrier and sailing schedules that allow you to generate voyages. In trucking, you can maintain haulage schedules. Interforwarders has its own sailing schedule from Yokohama to Long Beach, which needs to be maintained, and different haulage schedules with local truckers. You can maintain three schedule types in Customizing via the menu path IMG • Transportation Management • Master Data • Transportation Network • Schedule • Define Schedule Types.

In addition, create schedules in the SAP TM master data via the menu path Master Data • Transportation Network • Create Schedule:

1. **Carrier schedule and type**
   In the carrier schedule type, use the MoT 03 – Sea, and deselect any checkboxes under Basic Settings. Create an appropriate schedule in

SAP TM master data. The schedule is from the ocean liner SOCAL (OAF-CR-02) and between the ports of Yokohama (OF-JPYOK) and Long Beach (OF-USLGB). The corresponding locations have to be maintained up front. Maintain departure rules with times and dates. Now, choose the validity for and generate the voyages. You can assign a vessel name to each voyage.

2. **Sailing schedule and type**
   In addition to the carrier schedule, you can maintain your own service offerings as a subset of the official carrier schedules. Consequently, you need to maintain your own schedule type. Make sure you select the GATEWAY and REFERENCE checkboxes to maintain the schedule from CFS to CFS and not from terminal to terminal. In SAP TM schedule master data, generate a new schedule with reference to the carrier schedule of SOCAL. You can see the required entries in Figure 14.5. Generate new voyages to represent the service offering of Interforwarders.

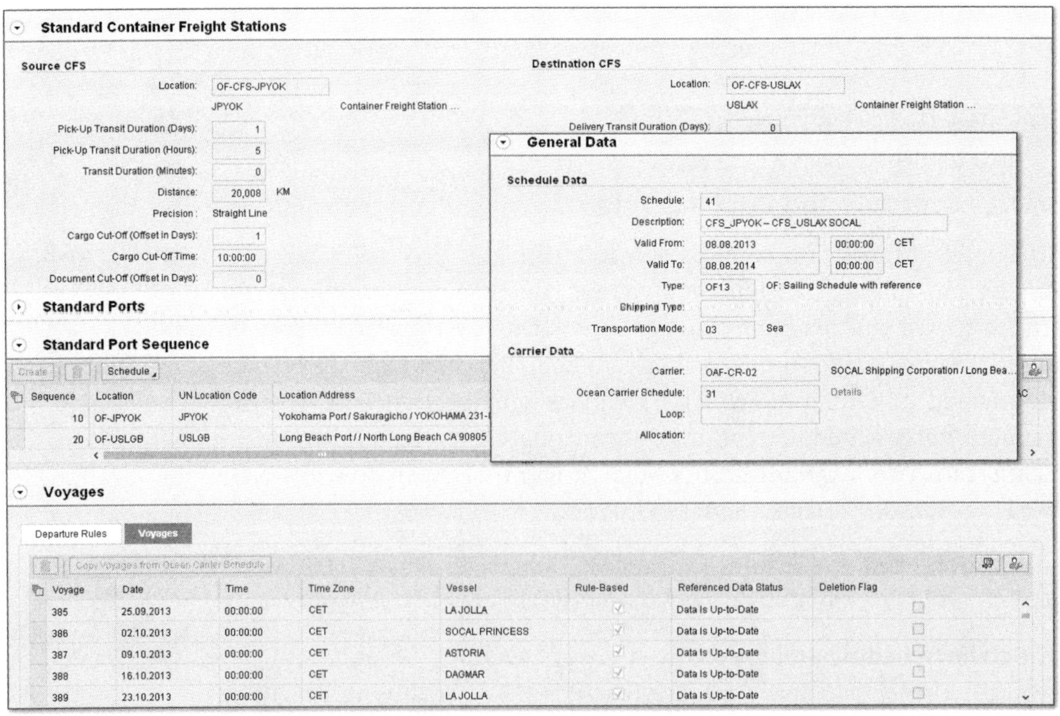

Figure 14.5 SAP TM Sailing Schedule with Reference to Carrier Schedule

3. **Trucker schedules and type**
   Maintain a separate type in Customizing with MoT 01 – ROAD and select the GATEWAY and DIRECT checkboxes. Maintain four schedules to connect each forwarding house with the CFS on the exporting and importing side. Assign one of the local truckers to each schedule. Generate daily departures and cutoff times to offer a haulage run between each location pair for every day of the week.

Freight booking type

To create capacity bookings, shipping instructions, and the bill of lading for the main ocean carriage, you need to create a freight booking type. An international shipment process in SAP TM requires the setup of separate document types for export and import bookings. In our process example, you will see that an import booking is generated from the export document as soon as it is reported shipped on board and serves as the pre-alert for the importing organization. You can generate booking types in SAP TM Customizing via the menu path IMG • TRANSPORTATION MANAGEMENT • FREIGHT ORDER MANAGEMENT • FREIGHT BOOKING • DEFINE FREIGHT BOOKING TYPE. Start by generating the import booking type. Ensure that you make the following settings:

- Choose mode of transport 03 – SEA and identify it as an import booking by setting the traffic direction as IMPORT.
- Select the ENABLE SETTLEMENT, ENABLE CHARGE CALCULATION, and ENABLE COST DISTRIBUTION checkboxes.
- Choose the change strategy, output profile, and execution settings.
- To create drayage and haulage orders directly from a freight booking, you can assign the freight order types created earlier.

Repeat this process by generating an export freight booking type. You can also default certain information, such as the purchasing and execution organization. Figure 14.6 shows the required settings for the export freight booking. Make sure you set the traffic direction as EXPORT. We have boxed the assignment of the import freight booking type to the export freight booking type in Figure 14.7.

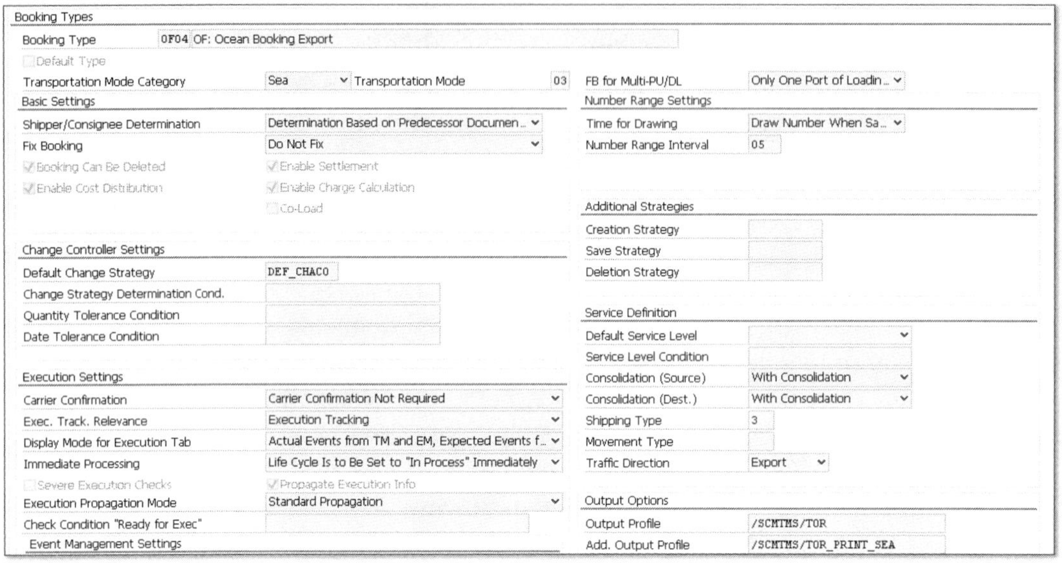

Figure 14.6 Definition of Export Freight Booking Type

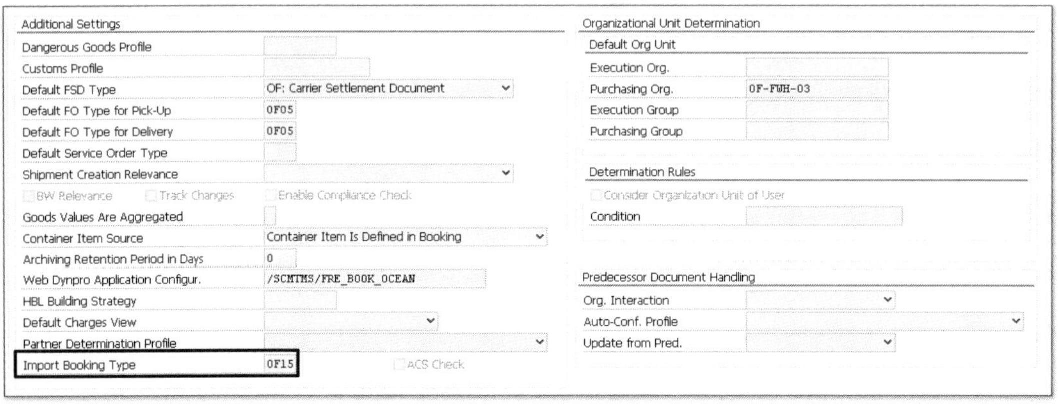

Figure 14.7 Assign Import Freight Booking to Export Freight Booking Type

14.2.4 Order Management Customizing

Forwarding order type

To fully use SAP TM Order Management and the capability to take and process forwarding orders, you need to enable various settings. You can find all the Customizing tables via the menu path IMG • TRANSPORTATION MANAGEMENT • BASIC FUNCTIONS • FORWARDING ORDER MANAGEMENT.

One essential setting is to define the forwarding order type. Choose DEFINE FORWARDING ORDER TYPE in Customizing. Similar to the freight bookings, this is required for generating an export and import forwarding order type. Start by generating the import forwarding order type and later assign it the export forwarding order type, as shown in Figure 14.8. The assignment of import forwarding order types to the export forwarding order type can optionally be managed by conditions. In the export forwarding order type, select the checkboxes for charge and internal calculation, and forwarding and internal settlement. Set the traffic direction to EXPORT. Maintain the transportation mode and a default sales organization.

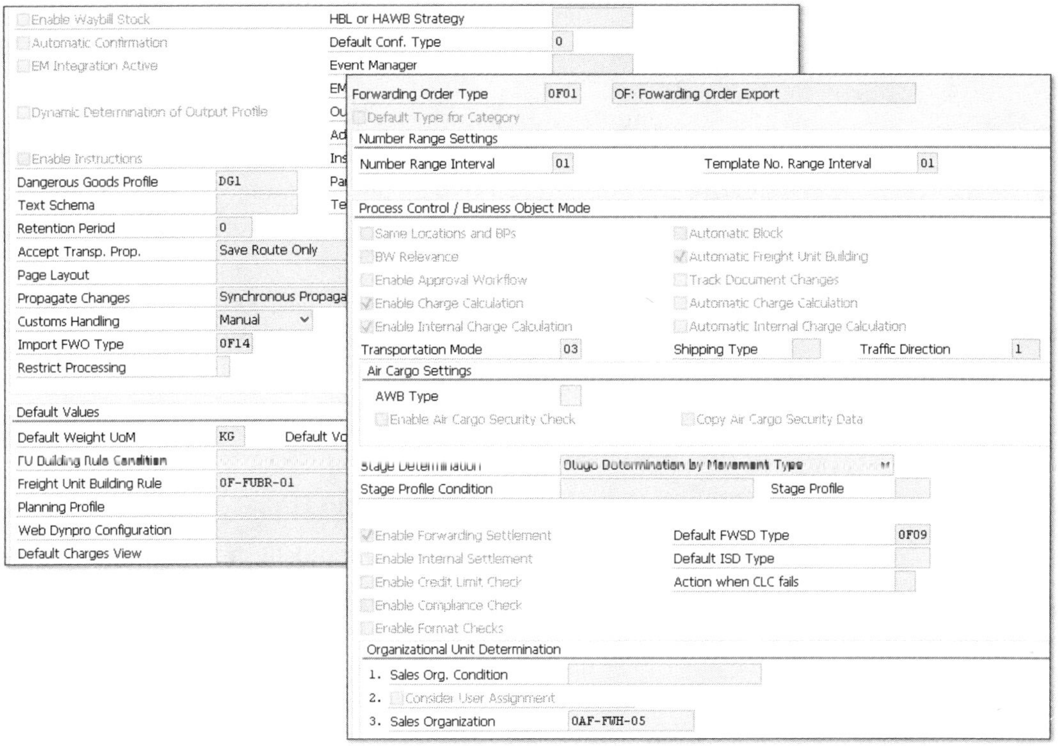

Figure 14.8 Export Forwarding Order Type

Basic setup

1. First, create package type codes (e.g., boxes, cartons, pallets). In LCL example, it is enough to maintain one entry for packages. Assign the

package type to a new item type (OF03) and maintain the item category as PACKAGE under DEFINE ITEM TYPES FOR FORWARDING ORDER MANAGEMENT.

2. Maintain a movement type for the door-to-door scenario we want to execute. Ensure that (in our scenario) at least five stage types are already maintained: for pickup, pre-carriage, main carriage, on-carriage, and delivery. Assign all five stage types to the movement type in the DEFINE STAGE SEQUENCE FOR MOVEMENT TYPE field.
3. An important setting required for the correct use of incoterms is to maintain default agreement party roles. We use the two incoterms free-on-board and delivery duty paid in the process. Consequently, you need to maintain a value for each stage type (pickup, main carriage, etc.) against the incoterm and define the agreement role as either prepaid or collect.
4. In addition, you need to define stage profiles for your import order. Create an entry for each stage type in the stage profile and mark all legs that are performed before the importing stage as NEVER RELEVANT FOR PLANNING because they are executed by the exporting organization.
5. For the various freight orders and freight bookings that are going to be created in this process, you can maintain default types for each stage type. Choose DEFINE DEFAULT FREIGHT DOCUMENT TYPES FOR STAGES and assign your freight order type for each stage type and shipping type.

14.2.5 Network and Planning Configuration

Freight unit type, building rule

Order management Customizing is a prerequisite for planning, but the key objects for planning are the freight units, which are created from a forwarding order. Create freight unit types in SAP TM Customizing by following the menu path IMG • TRANSPORTATION MANAGEMENT • PLANNING • FREIGHT UNIT • DEFINE FREIGHT UNIT TYPES. When you save your forwarding orders, SAP TM generates the freight units based on a freight unit building rule (FUBR), which you need to create by following the menu path SAP NBWC • APPLICATION ADMINISTRATION • PLANNING • GENERAL SETTINGS • CREATE FREIGHT UNIT BUILDING RULES. You can configure any rule, depending on the required planning granularity. For an LCL scenario, we suggest maintaining one building rule per item in a forwarding order or by a certain weight. Assign the freight unit type to the building rule. Eventually you need to assign the FUBR to your forwarding order type.

Vehicle resources, means of transport

Create three means of transport to represent a vessel, a common truck, and a regular haulage service by following the menu path IMG • TRANSPORTATION MANAGEMENT • MASTER DATA • RESOURCES • DEFINE MEANS OF TRANSPORT. Now generate at least one resource in the SAP TM master data for a truck in Japan, a truck in the United States, and an ocean vessel. Mark the resources as MULTI-RESOURCES and assign the means of transport. Resources are required not to represent your actual fleet, but as a virtual capacity that is required for the automated planning functionality.

Transportation locations, lanes, zones

The first transportation locations were generated after you replicated the SAP ERP master data to the SAP TM business partners. In addition to this, you need to maintain locations for the CFSs, the water ports, and your forwarding houses in Japan and the United States. To enable the carrier selection and transportation proposal functionality, you must connect your locations with transportation lanes, so create two transportation zones: one representing the West Coast of the United States and one for Japan. For each zone, create an intrazone transportation lane and assign the corresponding means of transport and local truckers. All related master data entries can be made via the SAP TM menu path NWBC • MASTER DATA • TRANSPORTATION NETWORK.

Default routes and carrier selection

There are many ways to execute and perform planning. Our freight-forwarder works with a combination of manual and semi-manual routing and transportation capabilities. For the initial decision of how to route cargo from a customer location via a forwarding house to an appropriate CFS and water port, a routing guideline is used.

Routing guides are defaulted routes that are followed for regular LCL cargo. In SAP TM you can implement a routing guideline by setting up DEFAULT ROUTES. Figure 14.9 shows the default route from Nagoya to Phoenix. The selection of an appropriate ocean carrier and schedule is supported by the default route, which suggests the voyages that can meet the required cutoff times. The selection of inland truck carriers is done manually by choosing local truckers.

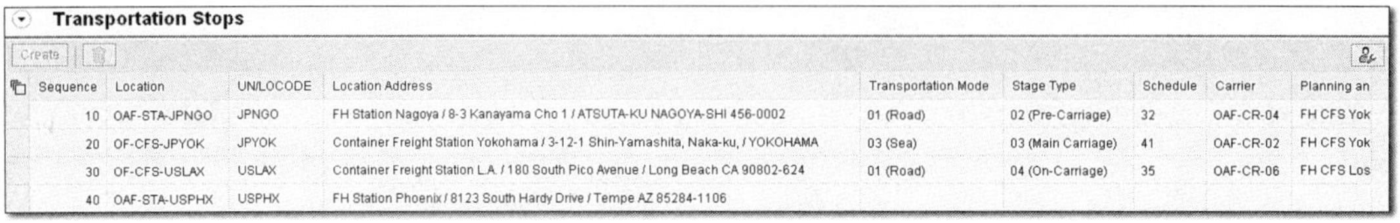

Transportation Stops

Sequence	Location	UN/LOCODE	Location Address	Transportation Mode	Stage Type	Schedule	Carrier	Planning an
10	OAF-STA-JPNGO	JPNGO	FH Station Nagoya / 8-3 Kanayama Cho 1 / ATSUTA-KU NAGOYA-SHI 456-0002	01 (Road)	02 (Pre-Carriage)	32	OAF-CR-04	FH CFS Yok
20	OF-CFS-JPYOK	JPYOK	Container Freight Station Yokohama / 3-12-1 Shin-Yamashita, Naka-ku, / YOKOHAMA	03 (Sea)	03 (Main Carriage)	41	OAF-CR-02	FH CFS Yok
30	OF-CFS-USLAX	USLAX	Container Freight Station L.A. / 180 South Pico Avenue / Long Beach CA 90802-624	01 (Road)	04 (On-Carriage)	35	OAF-CR-06	FH CFS Los
40	OAF-STA-USPHX	USPHX	FH Station Phoenix / 8123 South Hardy Drive / Tempe AZ 85284-1106					

Figure 14.9 Default Route

Planning and selection profiles

You can use the SAP TM transportation cockpit to consolidate cargo in the inland transportation on haulages between the stations and CFSs. To enable the configuration cockpit, the following settings are required which you can find in the following SAP TM menu path (not the TM Customizing): APPLICATION ADMINISTRATION • PLANNING. Because Inter-forwarders has two planning organizations in our process (one for trucking in Japan and one in the United States), you have to maintain the corresponding profiles for each planning organization separately with a different scope.

1. Maintain optimizer settings.
2. Create a capacity selection profile for a planner in the CFS in Japan and a separate profile for a planner in the United States (Long Beach).
3. Maintain two planning profiles and assign the different capacity selection profiles and the same optimizer settings.
4. Generate at least two selection profiles that differentiate the planning demands for Japan and the United States. In addition, you can maintain a third profile the main ocean leg to make it possible to perform planning in the cockpit for the main stage.
5. Finally, navigate to NWBC • PLANNING • TRANSPORTATION COCKPIT and create three new layout sets. You can customize the screen design by selecting your layout set and choosing LAYOUT PERSONALIZATION.

14.2.6 Transportation Charge Management

After you have configured all these settings to execute the physical cargo process, you need to define the Customizing for SAP TM Transportation

Charge Management. In the process with Interforwarders, multiple charge calculations are executed and settlement documents are generated:

- Customer invoicing to shippers in Japan
- Customer invoicing to one consignee (incoterm free-on-board) in the United States
- External settlement between the import and export organization
- Internal settlement between the CFS and the export stations for main leg
- Carrier invoicing to the ocean liner
- Trucker invoicing for pickup, delivery, and haulage services

Charge types

First, you need to set up all the different charge types. Consider a few examples of charge types that you can create on top of the SAP TM standard charge types: international port security charge (IPSD), advanced manifest system-fee (AMS), Spanish banking surcharge (SBS), and piracy risk surcharge (PRS). You can find the Customizing table by following the menu path IMG • TRANSPORTATION MANAGEMENT • BASIC FUNCTIONS • CHARGE CALCULATION • BASIC SETTINGS FOR CHARGE CALCULATION • DEFINE CHARGE TYPES.

Rate tables

Create rate tables for your key customer charges to simulate a realistic charge calculation scenario. You can find the master data via the SAP TM menu path MASTER DATA • CHARGE MANAGEMENT AND SERVICE PRODUCTS • RATE TABLES • CREATE. Set the charge usage as CUSTOMER. Maintain the rates (e.g., per unit of weight or per ton of chargeable weight) and then create rate tables in the structure shown in Table 14.4.

Charge Type	Calculation Base	Scale Items
Pre-carriage trucking charge (PRE)	SOURCELOC DESTLOC CHRG_WEIGHT	List of all customer locations to the CFS in Yokohama
Basic sea freight charge (BSF)	SOURCELOC DESTLOC CHRG_WEIGHT	CFS in Yokohama to the CFS in Long Beach
Terminal handling charge (THC)	DESTLOC CHRG_WEIGHT	Destination Long Beach with weight tiers

Table 14.4 Rate Tables

Continue with the setup of these rate tables for additional charge types: PRS, emergency bunker adjustment factor (EBAF), and on-carriage charge for trucking (ON).

Now you need to maintain the rate tables for your service providers. Set the charge usage in these rate tables as SERVICE PROVIDER, and set up different rates for the following charge types: FB00, BAF, THCO, THCD, and PEAK. Use the calculation bases equipment type (EQUIP_TYPE), origin (SOURCELOC), and destination (DESTLOC).

Last, set up four rate tables for your trucking providers, again with the charge usage of SERVICE PROVIDER. It is sufficient to maintain a simple rate table for each leg, pickup, export haulage, import haulage, and delivery for the charge type of a basic trucking charge (FB00) as full truck load (FTL) charges.

Forwarding agreements, calculation sheets

In our LCL process example, we ship orders from four customers to four consignees. We will maintain only five agreements—one agreement for every shipper in Japan and one generic agreement to cover rates for all four consignees in the United States—by selecting the MULTIPLE PARTIES checkbox in the agreement type Customizing. Once you have defined a forwarding agreement type, navigate to the SAP TM charge calculation master data and create each contract. You can choose to generate different forwarding agreement line items to differentiate between pre-carriage, main leg, and on-carriage charges to your customers. Alternatively you can store all charges for pre leg, main leg, and on-carriage in the same agreement line item, especially when working with service products. Create one transportation charge calculation sheet (TCCS) per agreement item.

In Table 14.5, you can see an example of how to maintain a shipper's agreement with corresponding charge calculation sheets and rate tables. Maintain this structure for each of the four shipper agreements and make sure that you enter the two forwarding houses in Tokyo as the organizational units. For the consignee agreements, maintain only the main carriage and on-carriage items—because due to the incoterms in our scenario, the shippers always pay at least the pre-carriage—but list the two forwarding houses in the United States as the organizational units.

Agreement Item	Calculation Sheet	Charge Type	Rate
Pre-carriage	OAF-TCCS-PRE	▸ PRE ▸ FUEL ▸ TOLL	▸ Rate table ▸ Percentage ▸ Fix amount
Main carriage	OAF-TCCS-MAIN	▸ BSF ▸ BLAD ▸ CAF ▸ BAF ▸ PRS ▸ EBAF ▸ IPSD ▸ THCD ▸ AMS	▸ Rate table ▸ Fixed amount ▸ Percentage ▸ Rate table ▸ Rate table ▸ Rate table ▸ Rate table ▸ Rate table ▸ Fixed amount
On-carriage	OAF-TCCS-ON	▸ ON ▸ FUEL	▸ Rate table ▸ Percentage

Table 14.5 Structure for Forwarding Agreement

Freight agreements, calculation sheets

On the service provider side, you need to create a freight agreement type and set up the freight agreements in master data. Maintain the purchasing organizations for the corresponding local truckers and carriers—that is, the CFS in Yokohama for the agreement with SOCAL or the forwarding house in Tokyo for the pickup truck agreement.

1. Create one agreement between your ocean carrier SOCAL (OAF-CR-02) and the CFS in Yokohama. Define two agreements between the haulage truck providers and the CFS.
2. Maintain one trucking agreement between each station in Tokyo and Nagoya and the two local trucking providers for pickup services.
3. Set up two agreements for the CFS in Long Beach and the two trucking providers for the haulage on the import side.
4. Last, define one more agreement between each of the two import stations in San Diego and Phoenix and their two trucking providers.

Internal agreements

It is sufficient to maintain only one freight agreement item and assign one calculation sheet. In the TCCS, you need to set up all the service provider charge types and rate tables as defined before. Remember that the main ocean charges are FCL charges in the corresponding rate tables, and the trucker charges are maintained as FTL rates. To set up an internal charge calculation, you need to define a separate internal agreement between the exporting and importing organizations with rate tables and calculation sheets to cross-charge for intra- and intercompany payments. Refer to Chapter 10 for more details about setting up the internal charge calculation.

Calculation profile, charge profile

The final steps in setting up the charge management functionality are to define a calculation profile and a charge profile. Maintain the calculation level as STAGES in the calculation profile. Make sure to assign the calculation profile to a charge profile that is to be referenced in each purchasing organization and forwarding house of Interforwarders.

14.2.7 Settlement Configuration

Settings in SAP TM, SAP ERP

The settlement process in SAP TM requires only a few settings for a basic scenario. You need to define a settlement profile that can be used for both freight and forwarding settlement documents. Assign the profile to the charge profile that is already assigned to the organizations. You also need to maintain forwarding and freight settlement document types and at least one internal settlement document type. Assign the different types to your forwarding order and freight order types. To use the SAP TM profitability functionality, maintain a cost distribution profile and assign it to your charge profile. For the invoice integration with SAP ERP billing and MM, you need to configure both the pure SAP ERP functions and the integration to SAP TM, such as SD condition types, pricing procedures, a mapping of SAP TM charge types, or assignment of organizational units for purchasing. Refer to Chapter 11, Section 11.1 to learn more about the detailed setup of the integration.

14.3 Running the Ocean Freight Process

Now that we've configured the LCL ocean freight scenario in SAP TM, let's run through the process from end to end.

14.3.1 Export Process

The export process starts by procuring capacities with the ocean liner SOCAL. We describe all steps from taking the customer orders until the cargo is successfully reported shipped-on-board. To keep a clear picture of the entire process chain and where we are standing at each point in time, we outline the process flow in Figure 14.10. We refer to the numbered steps in the figure while building up the scenario.

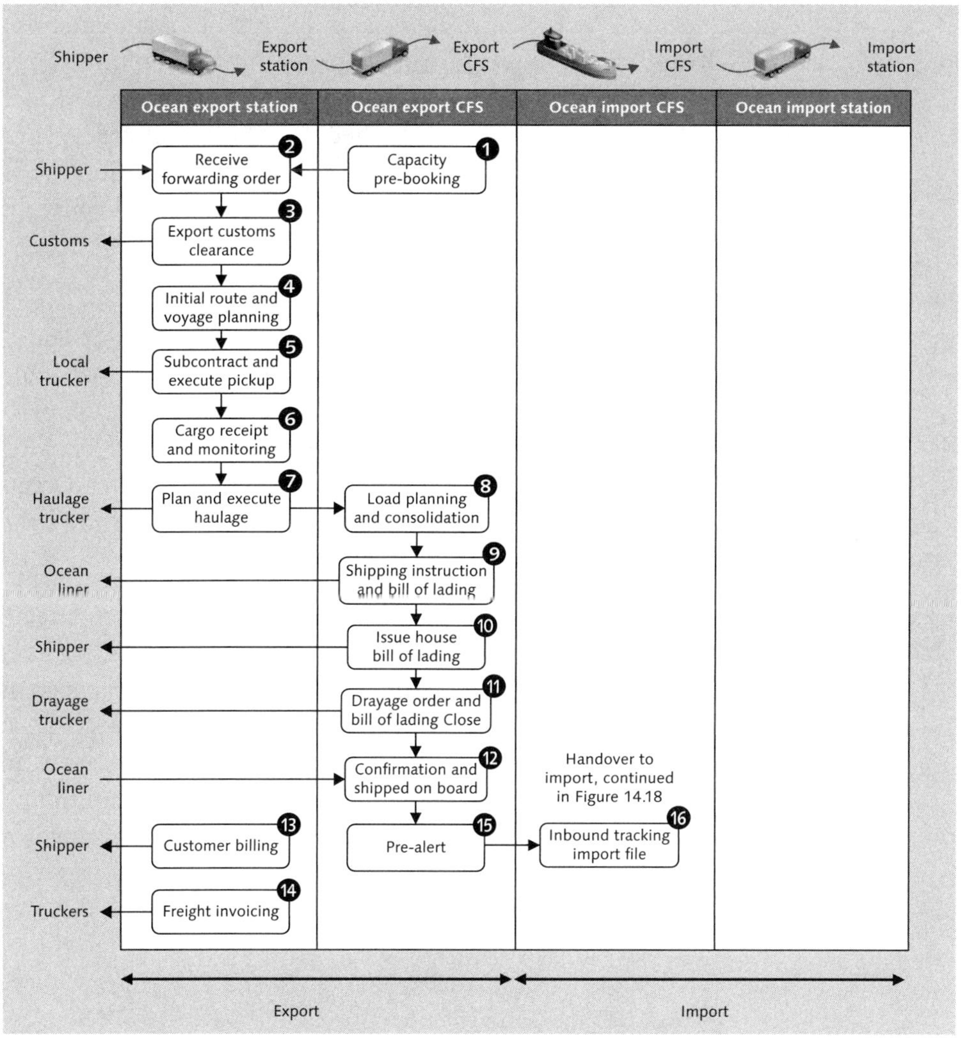

Figure 14.10 Detailed Export Process

Capacity Management and Booking

Allocations and bookings

Our freight-forwarder Interforwarders has a decentralized procurement organization. The different CFSs in each country autonomously procure capacities with ocean liners. Recall from Chapter 5 that SAP TM supports long-term capacity management. As part of a carrier contract, you can generate capacity allocations that represent the quantity commitments between the ocean liner and the freight-forwarder. In our process, we can assume that we do *not* have capacity commitments or allocations with the liner SOCAL but have contractually agreed upon rates for freight of all kinds from the origin Yokohama to the destination in Long Beach with different equipment types. Interforwarders offers a service for this trade lane with its own sailing schedule on different voyages. To meet the internal demand for shipments, the CFS regularly prebooks capacity on a popular loop with SOCAL. This is how we start the process ❶.

Creation of a freight booking

You can create freight bookings directly from SAP NWBC. Navigate to FREIGHT ORDER MANAGEMENT • OCEAN • CREATE OCEAN FREIGHT BOOKING. To create the booking with reference to an existing sailing schedule of Interforwarders from the gateway in Yokohama, choose SCHEDULE • ASSIGN and select the carrier and voyage you want to book. The booking references the relevant dates and times (e.g., cargo cutoff, document cutoff, and sailing date) and vessel information. Figure 14.11 shows the prebooking with all dates and times according to the schedule.

Submission and confirmation of booking

Once you have created the prebooking in your system, based on the sailing schedule, you have to get confirmation from the ocean liner about whether the capacity is available on the requested vessel and voyage. If you have set up a direct interface with your carrier, the booking request submission happens completely automatically, and the ocean liner can respond similarly to confirm the booking. Alternatively, you can submit the request manually via email and change the status to confirmed or rejected in the booking.

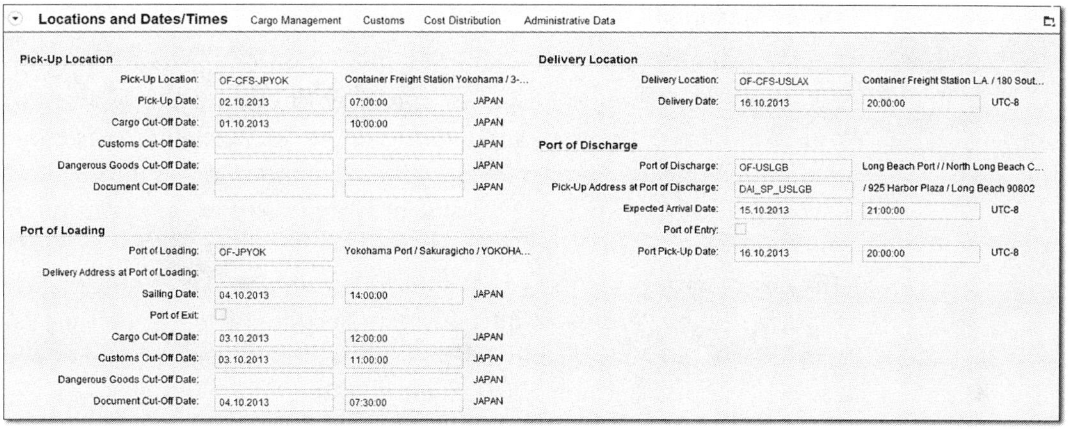

Figure 14.11 Ocean Freight Booking with Reference to Carrier Schedule

Figure 14.12 shows the confirmed booking for a voyage with SOCAL and the liner booking reference number. You can see that the purchasing organization is automatically defaulted to the CFS in Yokohama. The capacity you just booked with the ocean liner will be consumed automatically from an existing allocation for the specific carrier and origin-destination pair.

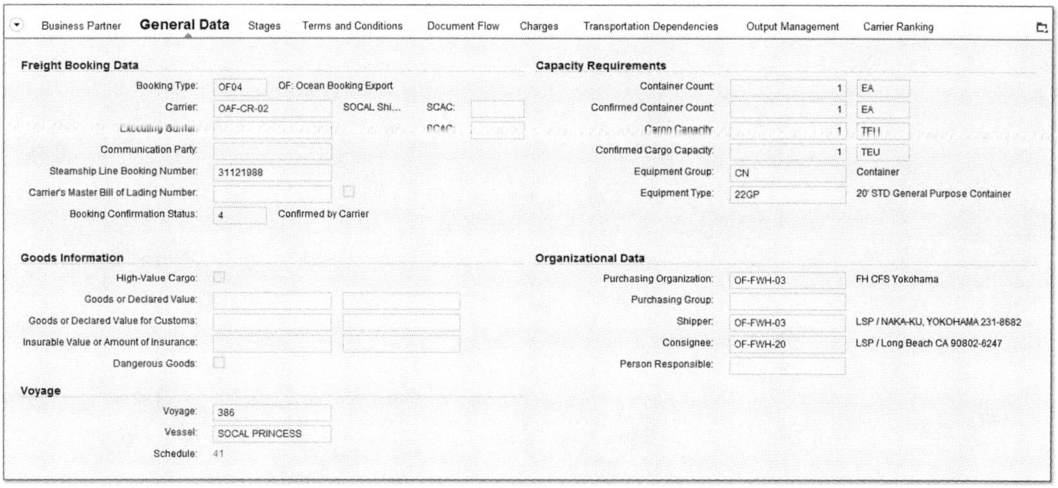

Figure 14.12 Confirmed Ocean Freight Booking

Order Management and Export Customs

Create forwarding order from template

Independent of the procurement activities of our CFS in Yokohama, the two forwarding houses in Tokyo and Nagoya capture the forwarding orders of local customers ❷. You can generate forwarding orders either from the scratch or with reference to a template to reuse information. The benefit of working with templates is that you save a lot of time in the data entry for regular customers who do not have an EDI interface. The customer representatives of Interforwarders in Tokyo and Nagoya capture the four orders as introduced in Section 14.1, by entering origin, destination, pickup, and delivery times, shipper and consignee, and cargo details like amount, weight, and volume. In Figure 14.13, you can see the details that have been captured in our export forwarding order. After you save the forwarding orders, the appropriate freight units are generated automatically based on the FUBR.

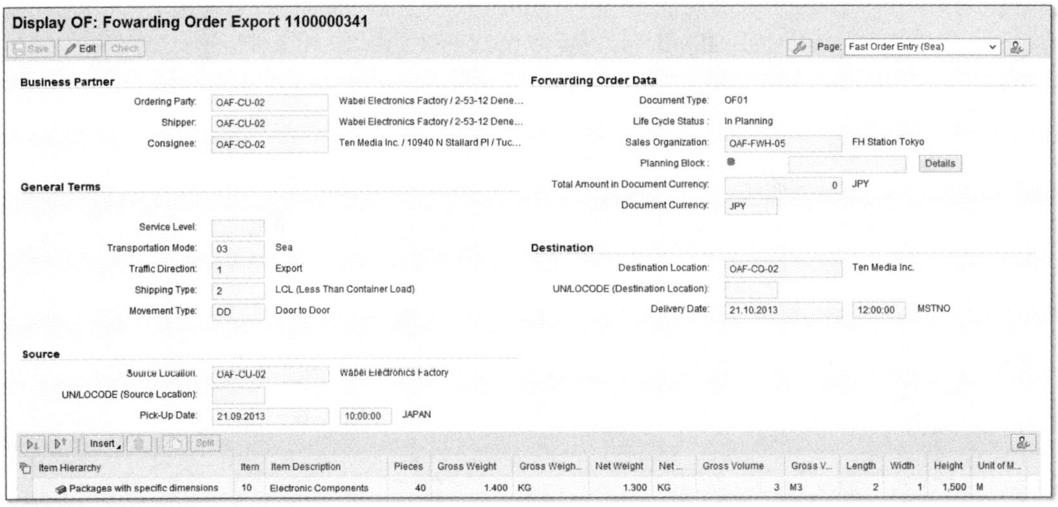

Figure 14.13 Forwarding Order Entry

Export customs clearance

Before you can start planning the transportation, you need to trigger the export customs services to avoid any delays in the process ❸. SAP TM automatically checks whether an export customs clearance is required and blocks the forwarding orders for planning and execution, so you need to enter all customs-relevant data (e.g., commodity codes) for each item in the forwarding order. You can use the interface with SAP GTS to automate the interaction with the authorities as we described in Chapter 9. Alternatively, you perform the customs clearance manually and outside of the SAP system. After a successful procedure, enter the customs reference number in the forwarding order and change the customs status to CUSTOMS CLEARANCE APPROVED, which releases the planning and execution block.

Initial route planning, capacity interaction

Once an order has been created and is not blocked for planning anymore, an initial routing decision has to be made to determine which CFS or port is to be used for the main leg of the transportation ❹. This impacts the pre-carriage drawn out by the stations. SAP TM offers various functionalities to support a route planning, such as the generation of a transportation proposal and a completely automated optimization, as you can read about in Chapter 6. You can also use a routing guide, based on pre-defined default routes. In our example, we select the default SAP TM routes.

The standard process we highlight in this example is that the forwarding houses of Interforwarders are responsible for choosing a sailing schedule directly for each order. This determines the CFS and the port of loading. In addition, the forwarding houses choose an existing freight booking, provided by the CFS, to block the capacity. The station in Tokyo, for example, plans the LCL cargo on the freight booking for voyage 386, which we just created. You can choose a schedule from the STAGES tab in the forwarding order by selecting the main ocean leg and then selecting SCHEDULE • SELECT.

This not only launches the possible schedules but also provides end-to-end proposals for how to ship the cargo based on your pre-defined default routes. In the pop-up screen shown in Figure 14.14, choose an appropriate voyage based on a sailing schedule. The corresponding legs between the CFS and the container terminal are automatically added. Afterwards, you can get an overview of the available capacity and make a reservation for the selected voyage by choosing FREIGHT DOCUMENT • SELECT FREIGHT BOOKING for the main ocean leg in the forwarding order.

The booking you choose is still owned by the respective CFS in Yokohama, which is in charge of the utilization and management of the space. As a result, the CFS as the planning organization still has the power to overrule any preplanning decisions from a forwarding house. This is supported by the organizational interaction. The CFS sees all preplanned freight units on its bookings (such as in the transportation cockpit) and can choose to confirm or reject the preplanning by the forwarding house.

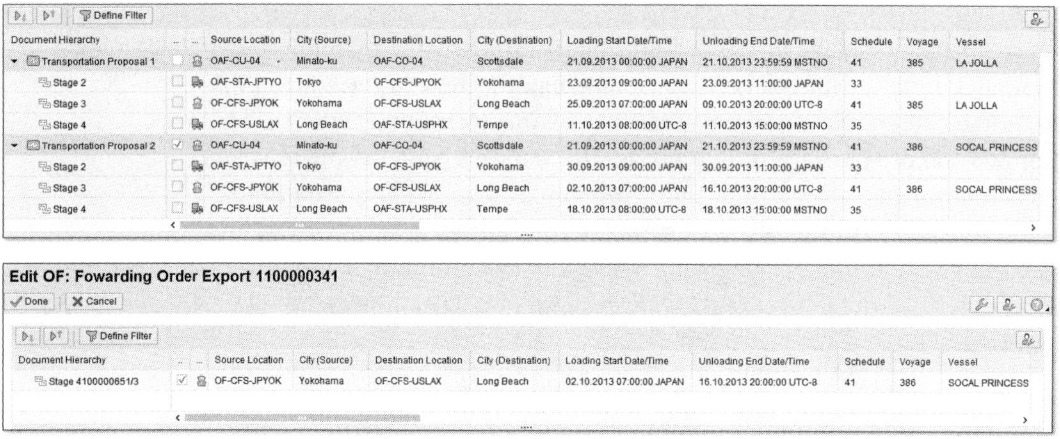

Figure 14.14 Selection of Schedule and Booking for Forwarding Order

After you have confirmed that enough space is available in a voyage and you can meet the requested transit time, you can confirm the order. SAP TM automatically generates an output file to submit the confirmation to the customer via EDI or by email, fax, and so on. You can see the print preview of an output file in your forwarding order.

Planning of Inland Transportation for Export

Arrange pickup

After the route for each forwarding order is preplanned, you need to arrange for the pickup of the cargo to transport it from the customer site to the warehouses of the stations in Tokyo and Nagoya ❺. Again, you have different options to plan the pickups (e.g., by using the transportation cockpit or directly from each forwarding order). In our process, we arrange an individual pickup for each order without any multi-pickup or consolidations for trucking.

You can choose the first leg of the transportation for each forwarding order in the STAGES tab and select FREIGHT DOCUMENT • CREATE. SAP TM automatically generates a pickup order with the dates as requested in the corresponding customer order. You can now select a trucking provider, either manually or by using the carrier selection and tendering functionality (refer to Chapter 7). For each generated freight order, you can manage the interaction with your trucking provider, track the progress of the transportation, calculate charges, and print any required transportation documentation.

Now that the cargo is successfully picked up and on its way to the forwarding houses, you need to plan for the haulage between the stations and the CFS in Yokohama. Because multiple orders and cargo items are waiting to be transported to the CFSs, you can use the transportation cockpit for this planning to consolidate as much cargo as possible into the trucks.

It is a common practice to have fixed haulage schedules that run frequently between busy stations and CFSs. In our process, we plan the three orders that are to be transported from Tokyo to Yokohama on an existing haulage schedule with a local trucker. In the transportation cockpit, you can see all freight units that need to be transported between the two locations, as well as existing schedules and truck order. In Figure 14.15 one freight unit has already been planned via a schedule, and a freight order was generated for the given day.

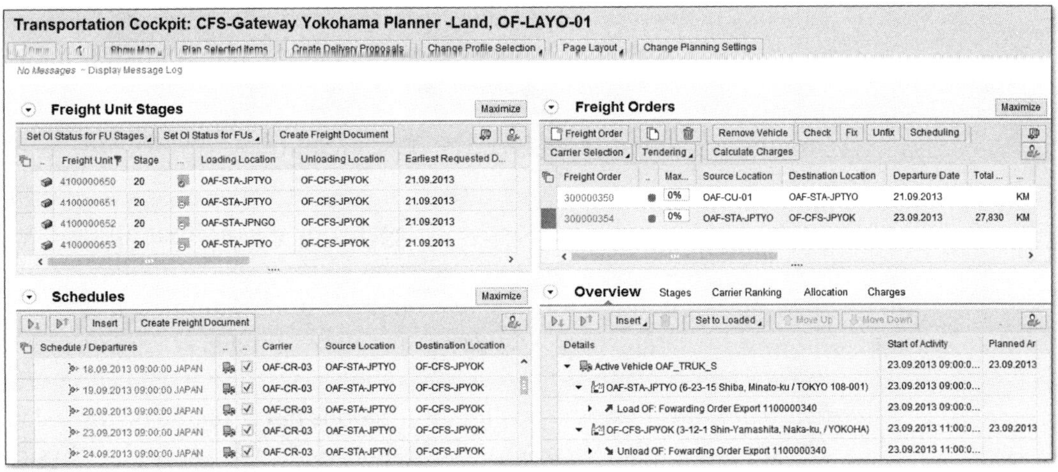

Figure 14.15 Transportation Cockpit for Export Haulage

Execution and Monitoring

Cargo receipt, haulage execution

Before the cargo can be loaded onto the haulage truck and is being sent on its way to the CFS (❻ and ❼), you need to perform a goods receipt in the exporting stations. A warehouse communication is required for cargo weighing and possible discrepancy handling. You also need to issue loading instructions. Once the cargo is packed on the haulage trucks, you continue monitoring the shipments. You know from Chapter 8 that SAP TM is fully integrated with SAP Event Management. To have clear visibility of the status and location of your forwarding orders/freight units and freight orders, you can report events with SAP Event Management.

Monitoring

In every step of the transportation processes, you can report events manually or automatically using SAP Event Management. Once the cargo is loaded onto the pickup truck at the customer site, the trucker can confirm (e.g., via a cell phone application, a simple phone call, or completely automatically via GPS) that the cargo has been loaded. Consequently, whenever a stage has been executed, you can see the current status. In SAP TM you can see the updates of each order and freight document in the EXECUTION tab. The Interforwarders truckers who are responsible for the pickup of cargo at the customers near Nagoya and Tokyo update all relevant statuses via text messages to an execution clerk at the stations in Nagoya and Tokyo, who update all statuses manually in SAP Event Management. Customers can now track the latest status of their orders.

Cargo Stuffing and Transportation Documentation

Cargo consolidation

Once the cargo has finally arrived at the CFS in Yokohama, it is possible to wait for further cargo to arrive. Triggered by the cargo cutoff time, the local warehouse staff can start with the stuffing of the containers and close them for further consolidation with other shipments. SAP TM supports an interaction between a dispatcher who is in charge of the transportation planning and the warehouse for physical loading. The dispatcher can create the loading instructions based on the planned consolidation and send them to the warehouse floor ❽. After the physical loading, the warehouse can respond with discrepancies or confirmation (the loading report).

Impact Organizational Interaction Status

Organizational interaction statuses allow forwarding houses to request and purchasing organizations to confirm or reject a requested capacity, as described earlier. Please note that you will not be able to plan or load the freight units into containers in the freight booking if you have not confirmed the space for each freight unit yet. The system assumes that no answer means rejection, and you need to navigate to the transportation cockpit or to a freight unit to confirm the space.

The CFS warehouse in Yokohama successfully consolidates all four customer orders into a 20-foot general purpose (GP) container that is available on the freight booking with SOCAL. Figure 14.16 shows the Cargo Management tab of the freight booking and shows that all forwarding orders were consolidated. You can see that the utilization of the container allows more cargo to be loaded.

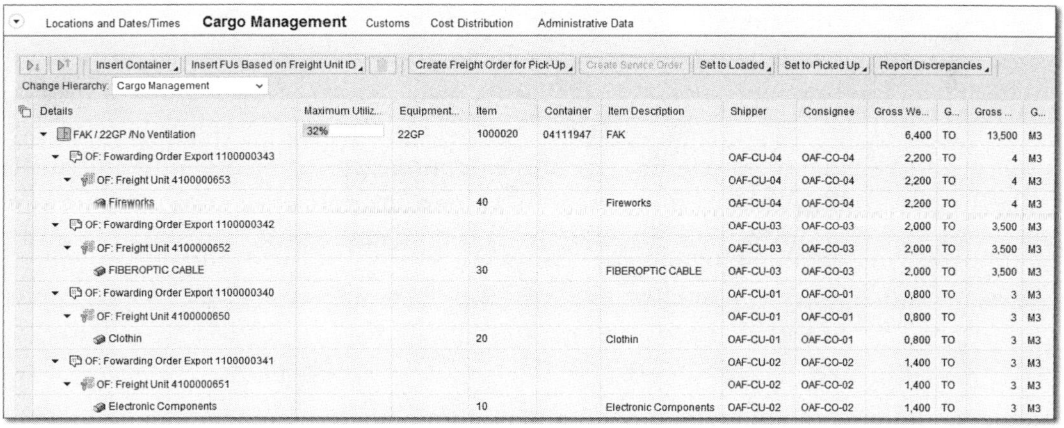

Details	Maximum Utiliz...	Equipment...	Item	Container	Item Description	Shipper	Consignee	Gross We...	G...	Gross ...	G...
FAK / 22GP /No Ventilation	32%	22GP	1000020	04111947	FAK			6,400	TO	13,500	M3
OF: Fowarding Order Export 1100000343						OAF-CU-04	OAF-CO-04	2,200	TO	4	M3
OF: Freight Unit 4100000653						OAF-CU-04	OAF-CO-04	2,200	TO	4	M3
Fireworks			40		Fireworks	OAF-CU-04	OAF-CO-04	2,200	TO	4	M3
OF: Fowarding Order Export 1100000342						OAF-CU-03	OAF-CO-03	2,000	TO	3,500	M3
OF: Freight Unit 4100000652						OAF-CU-03	OAF-CO-03	2,000	TO	3,500	M3
FIBEROPTIC CABLE			30		FIBEROPTIC CABLE	OAF-CU-03	OAF-CO-03	2,000	TO	3,500	M3
OF: Fowarding Order Export 1100000340						OAF-CU-01	OAF-CO-01	0,800	TO	3	M3
OF: Freight Unit 4100000650						OAF-CU-01	OAF-CO-01	0,800	TO	3	M3
Clothin			20		Clothin	OAF-CU-01	OAF-CO-01	0,800	TO	3	M3
OF: Fowarding Order Export 1100000341						OAF-CU-02	OAF-CO-02	1,400	TO	3	M3
OF: Freight Unit 4100000651						OAF-CU-02	OAF-CO-02	1,400	TO	3	M3
Electronic Components			10		Electronic Components	OAF-CU-02	OAF-CO-02	1,400	TO	3	M3

Figure 14.16 Cargo Management in Ocean Freight Booking

Shipping instructions

Once the container has been successfully stuffed, it is time to submit the shipping instructions to the ocean liner, before the deadline of the documentation cutoff date and time runs out ❾. You can submit shipping instructions directly from the ocean freight booking by selecting Subcontracting • Send Shipping Instructions. You can directly submit the shipping instructions via an EDI interface to your carriers. Alternatively, manual transmission (e.g., via email) is supported.

Master bill of lading

Once the carrier accepts the shipping instructions, he issues first a draft master bill of lading and after your confirmation the original master bill of lading. You can store the master bill of lading on the freight booking ❿ and the master bill of lading number is automatically taken over into your freight booking. In addition, you can generate the house bill of ladings for your customers. Select FOLLOW UP • BUILD HOUSE BILL OF LADING, and SAP TM generates them based on the building strategy you chose in the freight booking type (e.g., for each forwarding order).

Handling of Drayage Orders and Cargo Shipped on Board

Drayage order

After you have successfully printed all freight documents, especially the master bill of lading, you need to transport the consolidated container to the actual container terminal. The CFS in Yokohama is still around 20 kilometers away from the water port, so you need to subcontract a trucking company for the execution. Choose the container in the CARGO MANAGEMENT tab of your freight booking and select CREATE FREIGHT ORDER FOR PICKUP to generate a drayage order ⓫. SAP TM automatically generates another freight order with one stage from the CFS to the port of loading. You can select the trucking company separately as well as do the event reporting for this execution step. To represent the successful execution of the drayage leg, report the event ARRIVAL AT DESTINATION from SAP TM or live via SAP Event Management.

Report to shipped on board

Once the container has arrived at the vessel, you can wait for the confirmation of the ocean liner for the cargo to be shipped on board ⓬. You can change the status of the freight booking manually or use SAP Event Management for the reporting. This is the last step in the physical execution of the export process.

Forwarding/Internal Charge Calculation and Settlement for Export

Settlement of prepaid charges

The calculation and settlement of your customer charges depends on the agreed incoterms between the shipper and consignee (refer to Chapter 10 and Chapter 11 for background information). The financial process happens not in sequence with the physical execution, but in parallel whenever a certain payment milestone is reached. In each case, it is the job of a customer service representative to calculate and check the customer charges based on existing agreements.

Each forwarding house in Tokyo and Nagoya calculates their customer charges in this way. One of the shippers in the Nagoya area has agreed with the consignee in San Diego to ship the cargo for free up to the port of loading in Yokohama with the incoterm free-on-board. Consequently, you need to calculate and settle only the prepaid charges with the shipper in Tokyo ⓭. Figure 14.17 shows the results of the successful charge calculation in the export forwarding order from Nagoya.

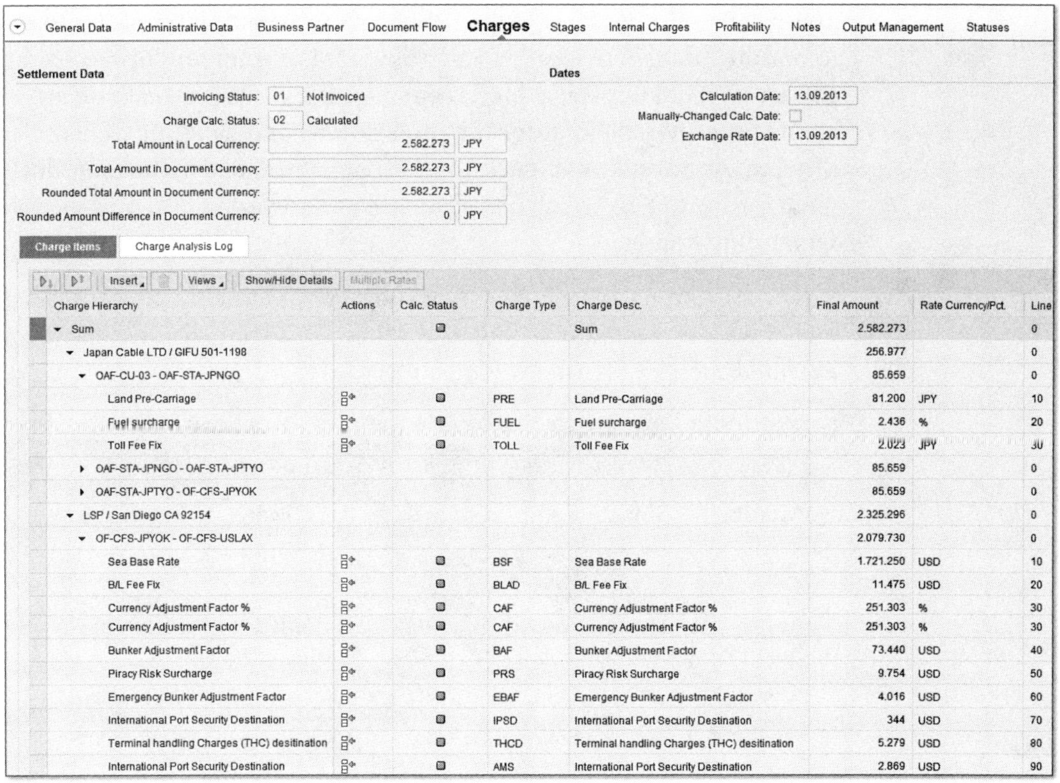

Figure 14.17 Charge Calculation in Export Forwarding Order

Cross-charging, internal settlement

It is important to ensure that you also list the prepaid and collect agreement party in the forwarding order, in addition to shipper and consignee. The charges in the corresponding forwarding order are calculated for each stage that allows a differentiation into prepaid and collect stages. The forwarding house in Nagoya took the forwarding order and hence settles the prepaid charges. The collect charges to the consignee in San Diego are billed by the importing organization in the United States, as they are in close interaction with the consignee on the importing side. To ensure that the exporting organization (the forwarding house in Nagoya) can still recover their cost for the main ocean leg that is only paid by the consignee, a cross-charging takes place between the station in Nagoya and the station in San Diego as the collect agreement party.

You have seen how the CFS in Yokohama provided the capacity to the forwarding houses in Nagoya and Tokyo. It is a common business practice for a CFS to work as a profit center with its own internal rate agreements to recover the cost for the provisioning and handling of main leg services. As a result, you need to generate an internal settlement document for the CFS in Yokohama against each forwarding order of the two forwarding houses.

Billing

Now let's summarize which forwarding and internal settlement documents we can generate at this time, after the export process has been executed successfully (see Table 14.6).

Billing Party	Settlement Party	Internal/ External	Incoterm	Charges
Forwarding house in Tokyo	Shipper Kan Xinson in Chiyoda-Ku	External	Delivery duty paid	All charges
Forwarding house in Tokyo	Shipper Wabei in Chiyoda-Ku	External	Delivery duty paid	All charges
Forwarding house in Tokyo	Shipper Samurai Firework in Minato-Ku	External	Delivery duty paid	All charges

Table 14.6 Export Forwarding and Internal Settlement Documents

Billing Party	Settlement Party	Internal/ External	Incoterm	Charges
Forwarding house in Nagoya	Shipper Japan Cable in Gifu	External	Free-on-board	Pickup pre-carriage
Forwarding house in Nagoya	Forwarding house in San Diego	External	Free-on-board	Proportioned main carriage
CFS in Yokohama	Forwarding house in Nagoya	Internal	–	Proportioned main carriage per forwarding order
CFS in Yokohama	Forwarding house in Tokyo	Internal	–	Proportioned main carriage per forwarding order

Table 14.6 Export Forwarding and Internal Settlement Documents (Cont.)

After the generation of the various settlement documents, you can transfer them to SAP ERP SD for the creation of billing documents. Note also that the settlement between the stations Nagoya and San Diego generates an external billing document. In SAP ERP SD, the corresponding billing documents are generated after a successful transmission.

Freight Settlement for Export

In addition to the settlement of the customer and internal charges, you have to settle the cost with the different trucking companies and the ocean liner. This happens gradually throughout the process, whenever a particular freight order or freight booking has been fulfilled and completed. For each freight order and freight booking, you can calculate the charges. Let's summarize the different freight settlement documents we need to generate on the exporting side:

1. **Settlement for pickup**
   We generated four freight orders with local truckers in Tokyo and Yokohama for the pickup of the cargo items and the transportation to the warehouses of the freight stations. One settlement document needs to be created per freight order.

2. **Settlement for haulage**
   We issued two haulage services from the two forwarding houses to the CFS in Yokohama. Create one settlement document each.
3. **Settlement for drayage**
   You need to generate a freight settlement document for the drayage order that transported the stuffed container from the CFS to the port.
4. **Settlement for main ocean leg**
   Last, you have to settle the charges with the carrier SOCAL for the biggest cost component, the main ocean leg of one 20-GP container from Yokohama to Long Beach with the commodity freight of all kinds.

After the successful generation of the freight settlement documents, you need to transfer them to the SAP ERP MM system to generate subsequent purchase orders and service entry sheets (SESs) for the posting of accruals and invoice verification. A user in the financials area will verify the received invoice from the different truckers and carrier and trigger the payment in SAP ERP ⓮.

Profitability

After the successful charge calculation for all forwarding and freight orders and the main ocean leg, you can see the margin for each export forwarding order in the PROFITABILITY tab. The internal charges that occurred are considered equally to give an accurate break down of the revenue and cost.

14.3.2 Import Process

After successfully handling the export process and the corresponding settlement, it is time to send pre-alerts to the importing side and start the actual import process, as shown in Figure 14.18. Please note that this process is entirely based on the export-import handling with a separate export and import freight booking and forwarding orders that mirror documents for the importing organizations in the United States.

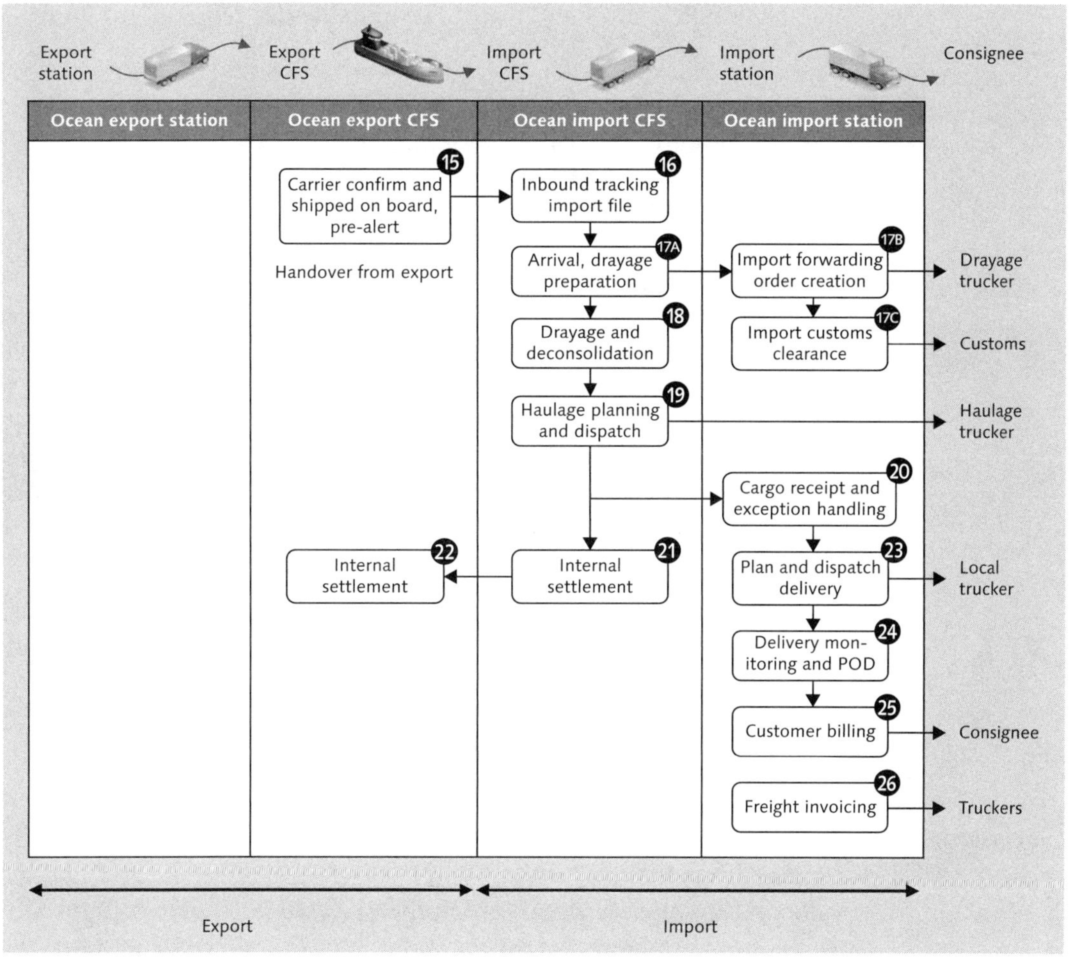

Figure 14.18 Detailed Import Process

Pre-Alert and Import Handling

Pre-alert, import booking

The two key documents for our exporting organizations in Japan are the export booking and the forwarding orders. Most of the workload performed by the export operations has been around these organizations. Now that the SOCAL vessel *Princess* has left the terminal in Yokohama, it is time to notify the importing organization in the United States of the incoming shipments. Keep in mind that so far they have not had any contact with the exporting side from Japan and do not have visibility of the load on the vessel and the different deliveries required for Phoenix and San Diego.

The CFS in Long Beach will be notified first, automatically. By reporting the freight booking as shipped on board, SAP TM automatically created the importing document—the import freight booking that will appear in the worklist of an import-handling agent in the CFS in Long Beach ⓯. You can set up your own specific worklists for pre-alerts to enable tailored visibility of each station and CFS on the importing side.

Import forwarding orders

By changing the status of the automatically created import freight booking from *Draft* to *In Process*, SAP TM generates four import forwarding orders. Similar to the pre-alert for the import operations in the CFS in Long Beach, the involved stations in Phoenix and Dallas will see the import forwarding orders in specific worklists after you have changed them from *Draft* to *In Process* to start the import processing ⓰. The import operations are provided with the same information as the exporting side, such as the final consignee, delivery dates and times, and the cargo load and collect charge information. The two stages that were executed in Japan and the main ocean leg are consistently marked as *Not Relevant* because the execution and planning has been executed. Only the two remaining legs from the CFS in Long Beach to the import forwarding houses remain *Unplanned*.

Import customs clearance

The import forwarding orders that were created are automatically blocked for planning and execution because SAP TM recognized that the import customs clearance is required first (17B). As an import customs broker in the stations in San Diego and Phoenix, you can navigate to the worklist of import forwarding orders and process the customs clearance. You can use the SAP TM import customs clearance functionality (see Chapter 9, Section 9.1.5) to manually submit your customs-relevant documentation to the import authorities (17C). If you have enhanced an additional interface to a customs clearance application (such as SAP GTS) that is not available in the standard versions of SAP TM, you can process the customs clearance in a more automated fashion. Once the import declaration has been retrieved from the authorities, you can set the customs status of each forwarding order to CUSTOMS CLEARANCE APPROVED, which releases the planning and execution block. Then you enter the customs reference number.

Import drayage

To prepare for the drayage of the import shipment from the terminal in Long Beach to the CFS, you need to generate the drayage order from the import booking (17A). Similarly to the exporting side, choose the container in the CARGO MANAGEMENT tab of the import freight booking and select CREATE FREIGHT ORDER FOR DELIVERY. While the CFS in Long Beach prepares for the import, the ship of SOCAL arrives at the container port. Consequently, the execution information is updated by the notification of the ocean liner, and you can set the lifecycle status of the booking to *Completed*.

Import Haulage and Delivery Handling

Deconsolidation

The drayage order you generated earlier has been executed by now and is marked as completed with corresponding reported events (from SAP Event Management or manually in the freight order). With the container available in the CFS in Long Beach, you can now do the consol-split to distribute the cargo items from the container back to individual house bill of ladings. We refer to this as the *cargo deconsolidation* (18). Navigate to the import freight booking and the CARGO MANAGEMENT tab. Select the container and click on SET TO UNLOADED. The unloading event is generated automatically.

Haulage planning and execution

The CFS in Long Beach plans for the haulage to the individual forwarding houses (19). You can use the transportation cockpit for the planning, trucking consolidation, and generation of haulage freight orders. Each planner who uses the transportation cockpit can have a unique layout and selection profile that limits the scope of the planning requirements to their individual responsibility. As the trucking planner in Long Beach, you see a specific planning and selection profile for the import haulage planning. Figure 14.19 shows that only the stages from the CFS to the two import stations are visible for the freight units in the top-left corner. The delivery dates and times are visible, as well as the two haulage schedules to Phoenix and San Diego in the bottom-right corner. We have already planned the freight units, and two new haulage freight orders and their details were generated (visible in the top-right corner).

Remember that working with haulage schedules is just one of many alternatives for managing the inland trucking. You can subcontract local trucking providers on an ad hoc basis rather than having a subcontracted provider via a schedule to transport to stations with less frequent incoming cargo or to procure additional trucking space. You can use the tendering and carrier selection functionality for the trucker selection. SAP TM also supports your fleet if you operate trucks and trailers for haulage and delivery services, which can be managed in the transportation cockpit.

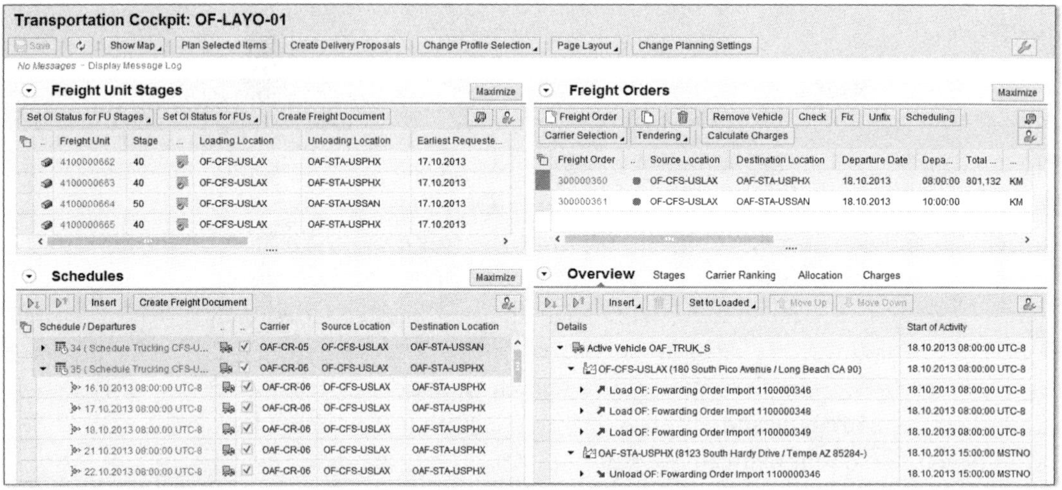

Figure 14.19 Transportation Cockpit for Import Haulage

Haulage execution and delivery

Once the haulage trucks arrive in the stations in San Diego and Phoenix, the local warehouse can report ARRIVED AT DESTINATION and change the status of the two freight orders from *In Execution* to *Completed* ⑳. The last step in the physical execution of the whole transportation process is the final delivery of the cargo to each consignee in the city areas of your two forwarding houses ㉓. A delivery planner of each forwarding house opens the individual import forwarding orders and selects the last stage that remains unplanned in the STAGES tab. The delivery planning can alternatively be executed via the transportation cockpit. The clerk can choose FREIGHT DOCUMENT • CREATE. A new delivery freight order is created. You can choose the desired truck provider or plan the transportation with your own fleet. The communication and documentation handling for the truck move is executed directly from the corresponding freight orders.

For each of the four delivery freight orders, report the corresponding execution events, such as ARRIVAL AT DESTINATION and UNLOADING END with reference to the final destination location of the consignees. The last leg of the forwarding order now automatically changes to *Arrived*. A POD issued by each consignee contains a signature and, optionally, even pictures of the cargo ㉔. This is a very important document for our freight-forwarder to trigger the settlement and billing process upon receiving the scanned POD documents.

Charge Calculation and Settlement for Import

Settlement of collect charges

As part of the export process, we have already settled all prepaid charges of our four customer orders. The three orders with incoterm delivery duty paid were fully billed to the prepaid customers in Japan, and due to the nature of the incoterm, no collect charges apply.

For the order from Nagoya to San Diego with incoterm free-on-board, we still have to invoice our consignee (Southwire Co.) for the main ocean leg and delivery charges. Consequently, one more forwarding settlement document needs to be generated ㉕. Navigate to the import forwarding order and choose the consignee Southwire as the collect agreement party. Ensure that all three stages of the main ocean leg, haulage, and delivery are marked as collect stages. You can now create the forwarding settlement document and choose COLLECT AGREEMENT PARTY in the pop up window for the settlement. The corresponding collect settlement document is created and can be transferred to SAP ERP billing for invoice creation by a colleague in the billing department.

Internal settlement

As described in Figure 14.18, you need to perform internal settlements along the process flow ㉑ and ㉒. Keep in mind that this happens gradually from the importing side after a particular stage has been executed on the importing side, such as haulages and deliveries. Not all internal settlement documents are generated at once, so we describe this process step as part of the general settlement process.

The CFS in Long Beach procured the haulage services based on the schedule to Phoenix and San Diego. To recover the cost for the CFS, additional internal settlement has to happen between the CFS and the station in Dallas for the forwarding order free-on-board, and with the forwarding house in Tokyo for the three forwarding orders with incoterm delivery duty paid. Keep in mind that three of the four forwarding orders did not generate any revenue for the U.S. organization because they were entirely prepaid. This means that you need to cross-charge the cost for the delivery services from Phoenix to the station in Tokyo ㉒.

Table 14.7 gives an overview of all incurred collect and internal settlement documents generated on the importing side.

Billing Party	Settlement Party	Internal/ External	Incoterm	Charges
Forwarding house in San Diego	Consignee Southwire Co. in Brawley	External	Free-on-board	Main carriage delivery
CFS in Long Beach	Forwarding house in San Diego	Internal	Free-on-board	Proportioned import haulage charges to San Diego
CFS in Long Beach	Forwarding house in Tokyo	Internal	Delivery duty paid	Proportioned import haulage charges to Phoenix
Forwarding house in Phoenix	Forwarding house in Tokyo	Internal	Delivery duty paid	Delivery charges for three delivery duty paid forwarding orders

Table 14.7 Import Forwarding and Internal Settlement Documents

Freight settlement for import

The importing organization in the United States generated various transportation orders. During the execution of the process, you can sequentially generate the different freight settlement documents whenever a freight order has been executed ㉖. Send the freight settlement documents to SAP ERP MM for invoice verification and payment by the corresponding finance teams for accounts payables. In total, you need to settle seven freight orders:

- The first freight order was required for the drayage from the terminal to the CFS.
- Two were required for the haulage to San Diego and Phoenix.
- Another four freight orders were generated for the delivery to the final destination.

Final Process States

Now let's have a look at the final results of our LCL process. It is apparent that in order to manage one entire door-to-door LCL ocean shipment, a large variety of tasks are required, internal staff involved, and documents generated. Figure 14.20 shows the end-to-end document flow of one export forwarding order. It represents the comprehensive file containing all relevant transportation-related documents, such as the freight units, the pickup and haulage orders, the export and import freight booking, the import forwarding order, and the import haulage and delivery order. Last, you see the financial documents.

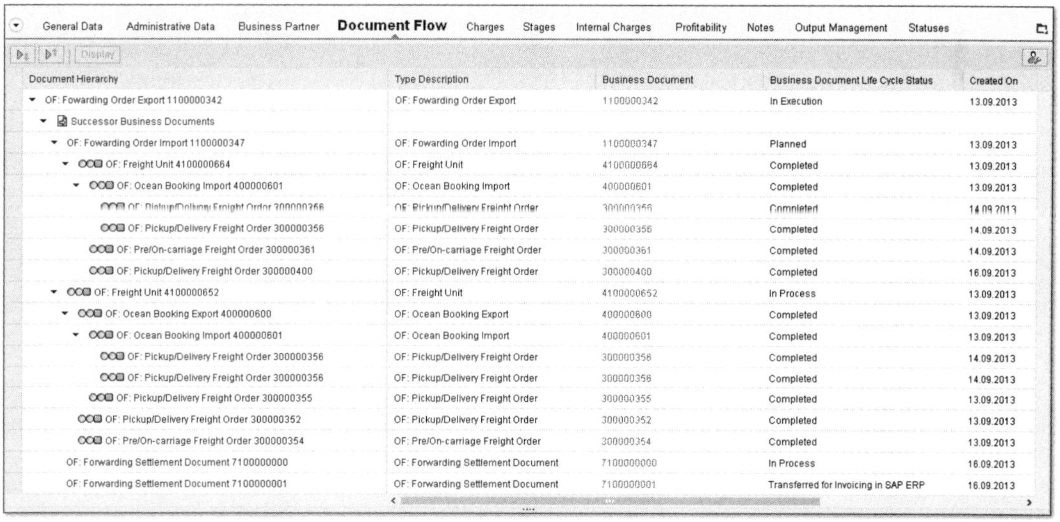

Document Hierarchy	Type Description	Business Document	Business Document Life Cycle Status	Created On
OF: Fowarding Order Export 1100000342	OF: Fowarding Order Export	1100000342	In Execution	13.09.2013
Successor Business Documents				
OF: Fowarding Order Import 1100000347	OF: Fowarding Order Import	1100000347	Planned	13.09.2013
OF: Freight Unit 4100000664	OF: Freight Unit	4100000664	Completed	13.09.2013
OF: Ocean Booking Import 400000601	OF: Ocean Booking Import	400000601	Completed	13.09.2013
OF: Pickup/Delivery Freight Order 300000356	OF: Pickup/Delivery Freight Order	300000356	Completed	14.09.2013
OF: Pickup/Delivery Freight Order 300000356	OF: Pickup/Delivery Freight Order	300000356	Completed	14.09.2013
OF: Pre/On-carriage Freight Order 300000361	OF: Pre/On-carriage Freight Order	300000361	Completed	14.09.2013
OF: Pickup/Delivery Freight Order 300000400	OF: Pickup/Delivery Freight Order	300000400	Completed	16.09.2013
OF: Freight Unit 4100000652	OF: Freight Unit	4100000652	In Process	13.09.2013
OF: Ocean Booking Export 400000600	OF: Ocean Booking Export	400000600	Completed	13.09.2013
OF: Ocean Booking Import 400000601	OF: Ocean Booking Import	400000601	Completed	13.09.2013
OF: Pickup/Delivery Freight Order 300000356	OF: Pickup/Delivery Freight Order	300000356	Completed	14.09.2013
OF: Pickup/Delivery Freight Order 300000356	OF: Pickup/Delivery Freight Order	300000356	Completed	14.09.2013
OF: Pickup/Delivery Freight Order 300000355	OF: Pickup/Delivery Freight Order	300000355	Completed	13.09.2013
OF: Pickup/Delivery Freight Order 300000352	OF: Pickup/Delivery Freight Order	300000352	Completed	13.09.2013
OF: Pre/On-carriage Freight Order 300000354	OF: Pre/On-carriage Freight Order	300000354	Completed	13.09.2013
OF: Forwarding Settlement Document 7100000000	OF: Forwarding Settlement Document	7100000000	In Process	16.09.2013
OF: Forwarding Settlement Document 7100000001	OF: Forwarding Settlement Document	7100000001	Transferred for Invoicing in SAP ERP	16.09.2013

Figure 14.20 Final Document Flow of Export Forwarding Order

Our shipments have successfully arrived at the four consignees and you have settled all vendor bills and invoiced the provided services to the customers. The process has successfully been executed.

In this chapter you learned that this simple and straightforward process example, as introduced in Section 14.1, required multiple Customizing settings. We have lined out the key Customizing settings and how to set up appropriate master data for the process in Section 14.2. Last we went step by step to the entire export and import process in SAP TM in Section 14.3.

We now switch sides from transportation on the water to transportation up in the air with SAP TM for air freight. Chapter 15 gives you an entire run-through of an air freight export and import process with gateway consolidation. You will find that we follow a very similar structure as in this chapter: First we introduce the general process flow, then Customizing and master data setup in SAP TM, and finally each step in the SAP system.

Air freight is an important way to ship cargo on a worldwide scale, especially because it produces very short transit times. In this chapter, we cover how to configure and run a consolidated air freight scenario.

15 Transportation Process for Air Freight-Forwarding

With the release of SAP Transportation Management (TM) 9.0 came comprehensive functionality for handling of air freight processes in the freight-forwarding industry. This allows forwarding companies to sell, organize, and execute air freight shipment services on a worldwide scale. It includes the standard gateway consolidation shipments with pickup and delivery as well as direct IATA shipments and block space operations. SAP TM 9.0 provides features to handle air freight-specific topics such as capacity allocation, master airway bill (MAWB) stocks, air cargo equipment, and IATA settlement. It also adds general functionality to support door-to-door moves via air freight gateways, export and import organizational handling, and internal billing.

In this chapter, we explain the setup (Section 15.2) and operation (Section 15.3) of an air freight scenario run by a freight-forwarding company.

15.1 Air Freight Process Overview

Scenario background

The air freight scenario with gateway consolidation shipments deals with a set of shipments that have to be moved from four shippers in Japan to four consignees in the United States. The main actor is an air freight-forwarder (Interforwarders Corp.) that operates a business unit that offers the requested service between origin and destination countries.

All forwarding orders are defined as door-to-door shipments. Table 15.1 shows a rough overview of the orders communicated by the shippers to the customer service of Interforwarders Corp. in the origin stations.

Shipper	Consignee	Cargo	Gross Weight, Volume	Via Origin Station	Via Destination Station
Xinson in Tokyo, JP	Dosena in Sedona, AZ, U.S.	Clothing	800 kg, 3 m^3	Tokyo	Phoenix
Wabei in Tokyo, JP	Southwire in Brawley, CA, U.S.	Fiberoptic cable	2, 000 kg, 3.5 m^3	Tokyo	San Diego
Japan Cable in Gifu, JP	Ten Media in Tucson, AZ, U.S.	Electronics components	1,400 kg; 6 m^3	Nagoya	Phoenix
Samurai in Tokyo, JP	Scottsdale Fireworks in Scottsdale, AZ, U.S.	Furniture	2,200 kg 4 m^3	Tokyo	Phoenix

Table 15.1 Overview of Shipments in the Air Freight Scenario

Involved organizations and responsibilities

Because there are multiple organizational units or locations of Interforwarders involved in organizing, handling, and moving the shipment, we provide an overview on the operational responsibility of the units. Table 15.2 shows the responsibilities and tasks for the involved organizational units or locations. In Section 15.3, we provide deeper insight into the process steps of each unit.

Unit or Location	Responsibility and Tasks
Origin station	▸ Order taking ▸ Definition of the used route ▸ Pickup subcontracting ▸ Pre-carriage arrangement ▸ Customer billing and settlement ▸ Export clearance ▸ Local truck carrier settlement

Table 15.2 Responsibilities and Tasks of Organizational Units and Locations

Unit or Location	Responsibility and Tasks
Export gateway	▸ Capacity management for air cargo space and master booking ▸ Control of air cargo equipment and paperwork ▸ Consolidation of shipments ▸ Feeder arrangement ▸ Pre-advice to import gateway ▸ Air carrier settlement
Import gateway	▸ Defeeder arrangement ▸ Deconsolidation of received cargo ▸ Gateway operations import ▸ On-carriage arrangement ▸ Local truck carrier settlement
Destination station	▸ Delivery arrangement ▸ Customer billing and settlement ▸ Local truck carrier settlement

Table 15.2 Responsibilities and Tasks of Organizational Units and Locations (Cont.)

Export process

The operational view of the process is shown in Figure 15.1 (export side) and Figure 15.2 (import side). On the export side, the shipments of our four shippers are picked up by local trucking companies and moved to the origin station. The three shipments from the Tokyo area are taken to the Tokyo station; the shipment from Gifu moves to the station in Nagoya.

In the origin stations, shipments are cross-docked and consolidated with other shipments onto long-haul trucks that operate on a scheduled basis between the local stations and the assigned air freight gateways. In the export gateway, the consolidation of the shipments into air cargo equipment (unit load devices [ULDs] or pallets) takes place. Subsequently, the consolidated cargo is fed to the ground handling station of the airline. The decision of which airline to use has already been fixed by the routing decision of the origin station dispatchers, but the export gateway may override that decision for better profitability.

Import process Even before closing the MAWB, the import gateway may get access to information on the shipments planned to be moved by air. After MAWB closure, the complete set of related data is provided for import preparation and handling so the import gateway team can efficiently prepare the receipt of cargo, deconsolidation, and import clearance. In Figure 15.2, the import gateway deconsolidates the cargo, does the paperwork and administrative work for each shipment, and consolidates the shipments onto the long-haul trucks to the destination stations. As in the export case, the import long haul is done based on fixed schedules.

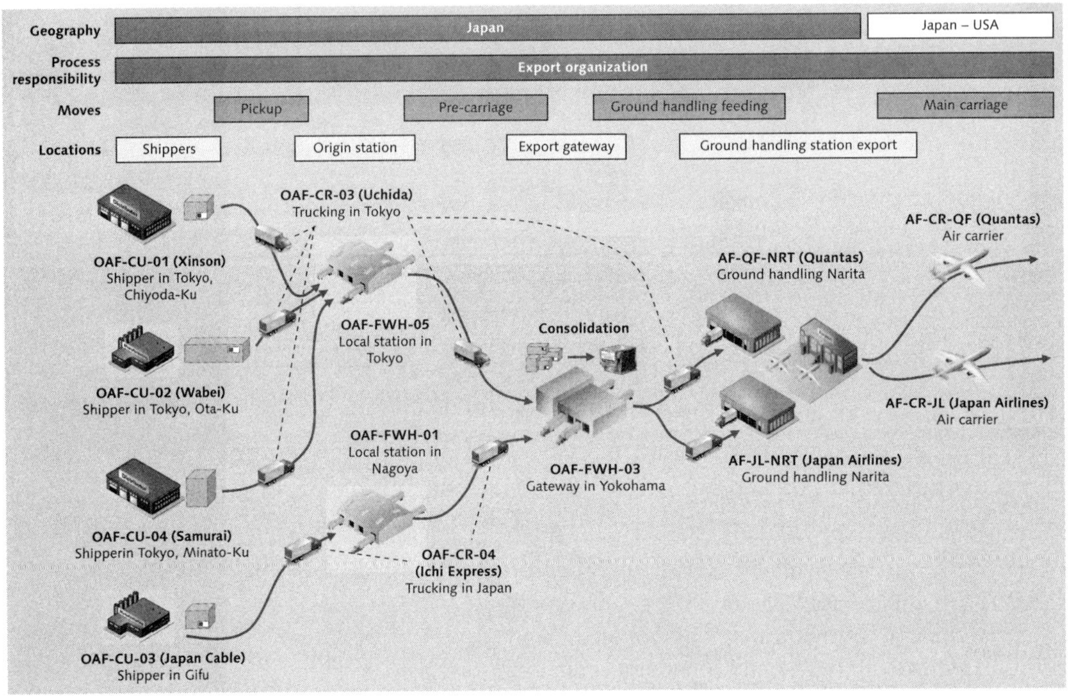

Figure 15.1 Export Air Freight Process

Once the shipments arrive in the destination station, arrangements for local delivery are made with appropriate trucking companies, and the shipments are delivered to the consignee. Finally, proof of delivery (POD) is reported to indicate the end of the operational process.

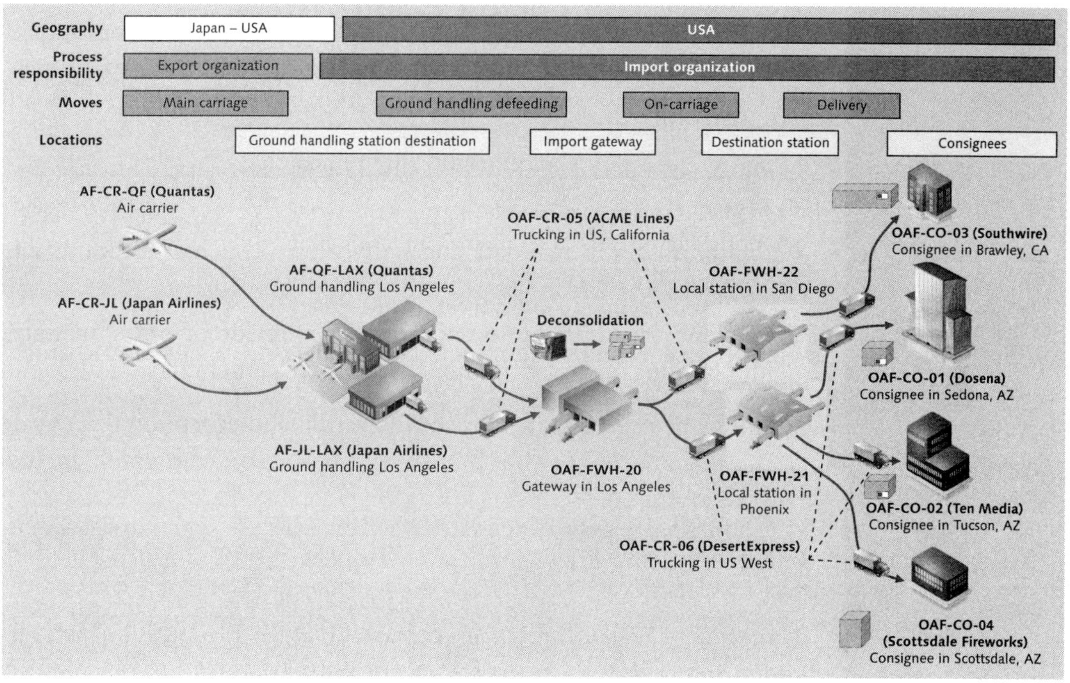

Figure 15.2 Import Air Freight Process

Financial process

However, financial settlement is still open. The origin and destination gateway have to do the settlement with the locally subcontracted trucking companies. The gateways may share the cost for the long-haul trucks with the local stations. The origin station sends an invoice to the shippers. If shipments are done with incoterms other than delivery duty unpaid or delivery duty paid, the destination station also sends a customer invoice to the consignees. For a free-on-board shipment, for example, the destination station may also invoice the air carriage to the corresponding consignee, and an internal settlement for the prorated air freight cost has to be done between the import side (who gets the money from the consignee) and the export side (who has to pay the airline). A second case is settlement of import haulage cost from the import gateway to the import station. Finally, the invoice for the flight has to be settled between the export gateway and the IATA's Cargo Account Settlement System (CASS).

15.2 Air Freight Process Configuration

To enable the air freight process in an SAP TM system environment, you have to configure the various aspects and entities:

- Set up master data for shippers, consignees, and carriers in the SAP ERP system.
- Maintain organizational data and network data such as locations and schedules in SAP TM.
- In SAP TM Customizing, configure the behavior of order management, planning, subcontracting, charge management, and settlement.

There are some prerequisites for configuration and execution that we do not describe in detail here but that have to have been completed, including the following:

- Configuration of logical systems, business system groups, and RFC destinations
- Basic SAP TM configuration described in SAP Solution Manager under SAP TRANSPORTATION MANAGEMENT • CONFIGURATION STRUCTURES • BASIC SETTINGS FOR SAP TM 9.0

In SAP ERP, you have to configure the purchase-to-pay process in MM including purchase order entry, logistics invoice verification, invoice document creation, and transfer to accounting. In FI-CO, the processes supporting purchase-to-pay have to be configured including country-specific legal requirements. In SD, you have to set up customer billing. Your user in SAP ERP needs to have an assignment of the role TM_INVOICE_CLERK.

15.2.1 Partner and Organizational Master Data Setup

Customers, vendors in SAP ERP

Master data for the business partners used in the air freight process needs to be set up in SAP ERP because it is the foundation for financial settlement. For the scope depicted in Figure 15.1 and Figure 15.2, we need to define the vendors and customers as shown in Table 15.3.

Partner Type	Partner ID	Name	Country
Vendor (truck carrier)	OAF-CR-03	Uchida	Japan
	OAF-CR-04	Ichi	Japan
	OAF-CR-05	ACME	U.S.
	OAF-CR-06	Desert Express	U.S.
Vendor (air carrier)	AF-CR-JL	Japan Airlines	Japan
	AF-CR-QF	Qantas	Australia
Vendor (local air carrier business units)	AF-JL-NRT	Japan Airlines stations	Japan, Narita
	AF-JL-LAX	Japan Airlines stations	U.S., Los Angeles
	AF-QF-NRT	Qantas stations	Japan, Narita
	AF-QF-LAX	Qantas stations	U.S., Los Angeles
Customer (shipper)	OAF-CU-01	Xinsan Clothing	Japan, Tokyo
	OAF-CU-02	Wabei Electronics	Japan, Tokyo
	OAF-CU-03	Japan Cable	Japan, Gifu
	OAF-CU-04	Samurai Fireworks	Japan, Tokyo
Customer (consignee)	OAF-CO-01	Dosena Clothing	U.S., Sedona, AZ
	OAF-CO-02	Ten Media	U.S., Tucson, AZ
	OAF-CO-03	Southwire	U.S., Brawley, CA
	OAF-CO-04	Scottsdale Fireworks	U.S., Scottsdale, AZ

Table 15.3 Business Partners (Customers and Vendors) Used in the Air Freight Process

Japanese, Australian, and U.S.-based carriers have to be assigned to appropriate company codes, purchasing organizations, and reconciliation accounts. Customers need an assignment to company code and country-specific sales organizations.

All customers and vendors defined in SAP ERP have to be synchronized to SAP TM via CIF. Figure 15.3 shows one of the carrier business partners after transfer to SAP TM.

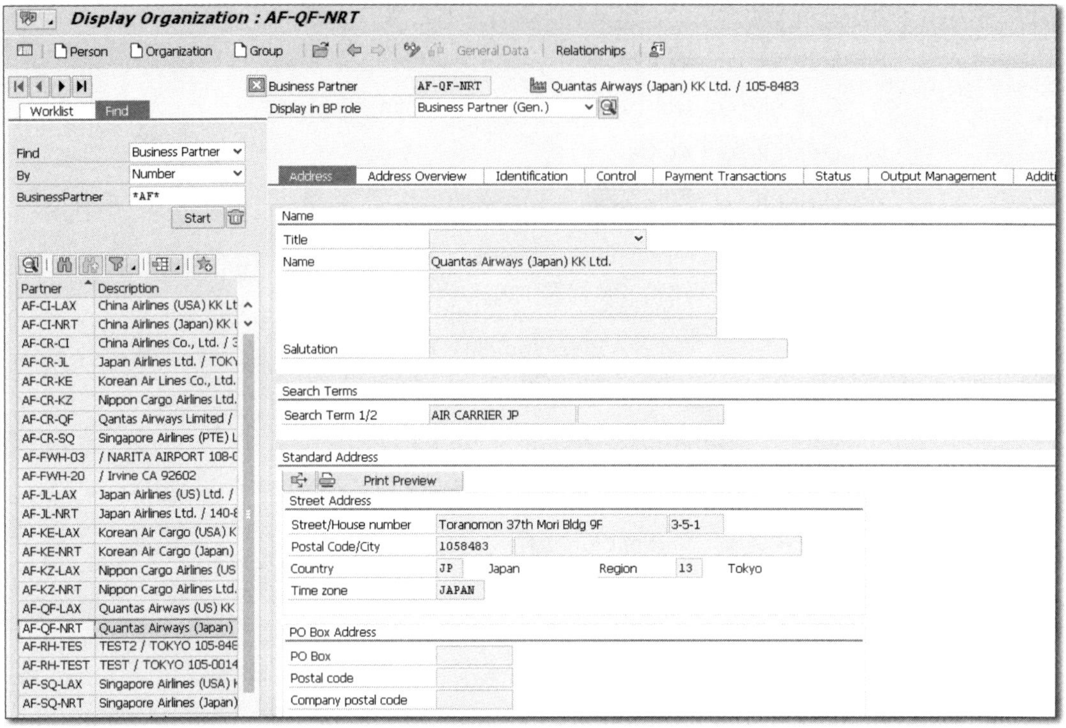

Figure 15.3 Business Partner for an Air Carrier's Local Subsidiary in SAP TM

Business partners in SAP TM

For appropriate air cargo security handling, the created customers need to be assigned the known shipper status. You can do this by updating the customer business partners of the shippers in SAP TM. After creating the vendor role for the business partner, you can maintain the shipper security status, where you have to set the value "KS" (known shipper) and a date range from more than 180 days ago to some future date to comply with legal regulations.

15.2.2 Organizational Setup

The organizational units involved in the air freight process with different tasks need to be defined in SAP TM. Using Transaction PPOME or by following the SAP NWBC menu path Master Data • Organization • Create Organization and Staffing, you can build the hierarchy of the enterprise Interforwarders Corp. with all its subunits such as the companies in Japan and the United States and the forwarding houses that are most actively involved as sales, purchasing, and execution groups. Figure 15.4 shows an overview of the hierarchy and the details for the forwarding house in Nagoya, Japan.

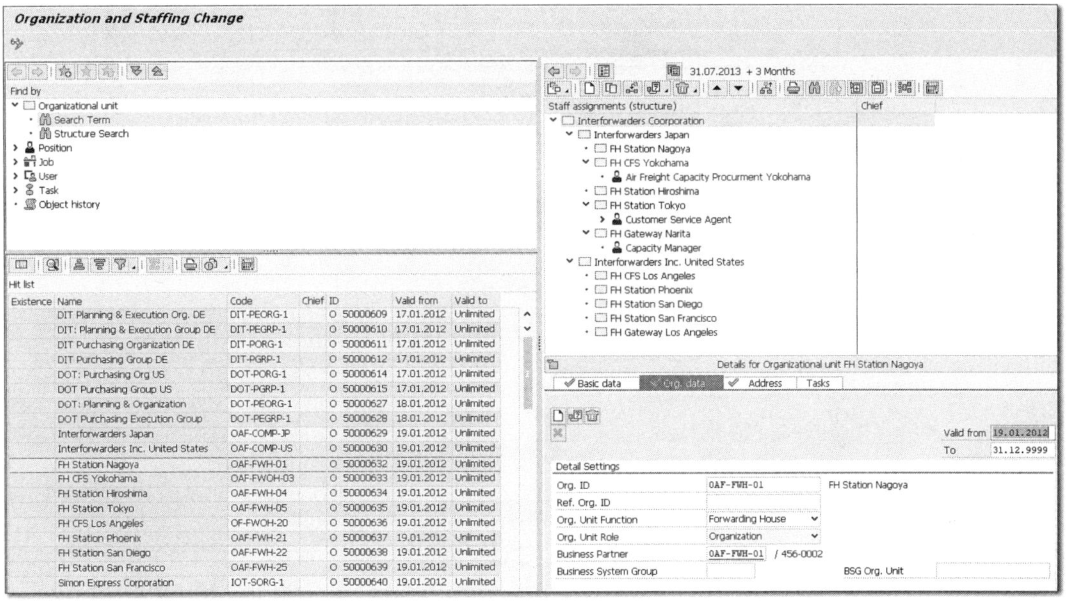

Figure 15.4 Organizational Master Data for the Air Freight Scenario

15.2.3 Basic Customizing and Settings

Equipment and resource information

Because of security reasons, air freight processes are quite dependent on the resources and equipment used. Therefore, it is important to define the different types of equipment and aircraft used to ship the cargo. Even if the forwarder is not responsible for flight operations, he needs to be informed about the aircraft used on particular flights, the type of equipment that can be carried, and the dimension and weight restrictions. In IMG Customizing, you can define the aircraft types used in the air freight scenario (see Figure 15.5) as well as equipment groups and types such as air cargo pallets and ULDs. Figure 15.6 shows different equipment types and the details for an LD3 air cargo container.

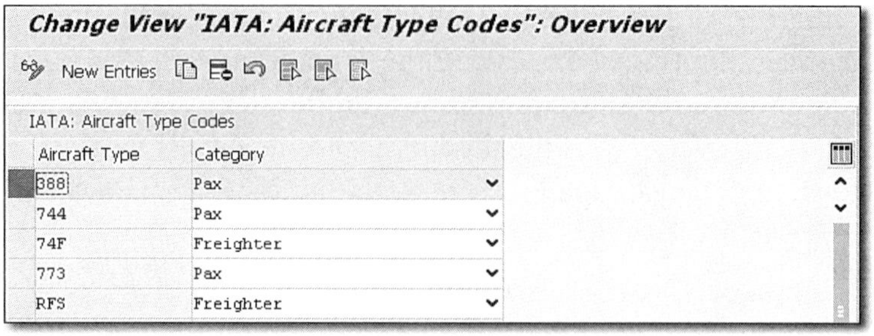

Figure 15.5 Definition of Aircraft Types

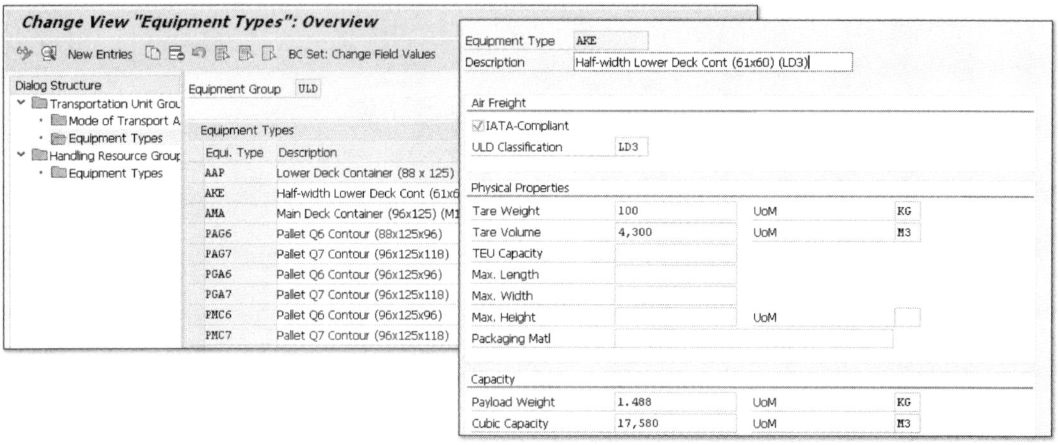

Figure 15.6 Equipment Master Data for Air Cargo Equipment

Air freight-specific settings

You also need to define the means of transport for aircrafts and trucks used for transportation planning. The IATA airline codes and IATA locations are important for building schedules because they are referred to later in the process when freight bookings are built. Handling codes for cargo are important for cargo management, consolidation, and loading in the gateway. Figure 15.7 shows an overview of the relevant Customizing entries.

Definition of MAWB stocks

Setting up MAWB number range stocks is a prerequisite for drawing MAWB numbers for the freight bookings that you will create later in the process based on the flight schedules. MAWB stocks are not Customizing objects but have to be defined via the SAP NWBC menu path Master Data • General • Overview Waybill Stock. You have to create appropriate MAWB stocks for all air carriers used.

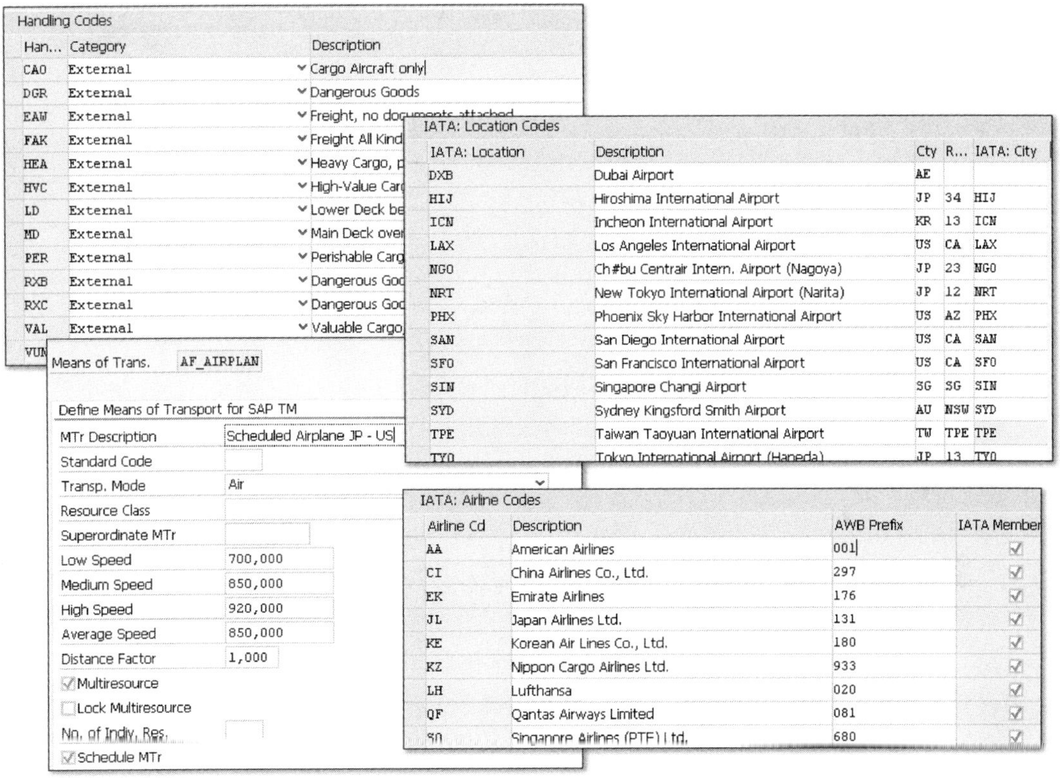

Figure 15.7 Customizing Settings for Aircraft Means of Transport, IATA Airline Codes and Locations, and Cargo Handling Codes

Figure 15.8 shows an MAWB stock defined for Japan Airlines with the airline prefix 131, number range 1,000 to 2,000, validity over the whole year 2013, and activated check-digit and prefix generation. The application also shows the already drawn MAWB numbers and their assignment to freight bookings.

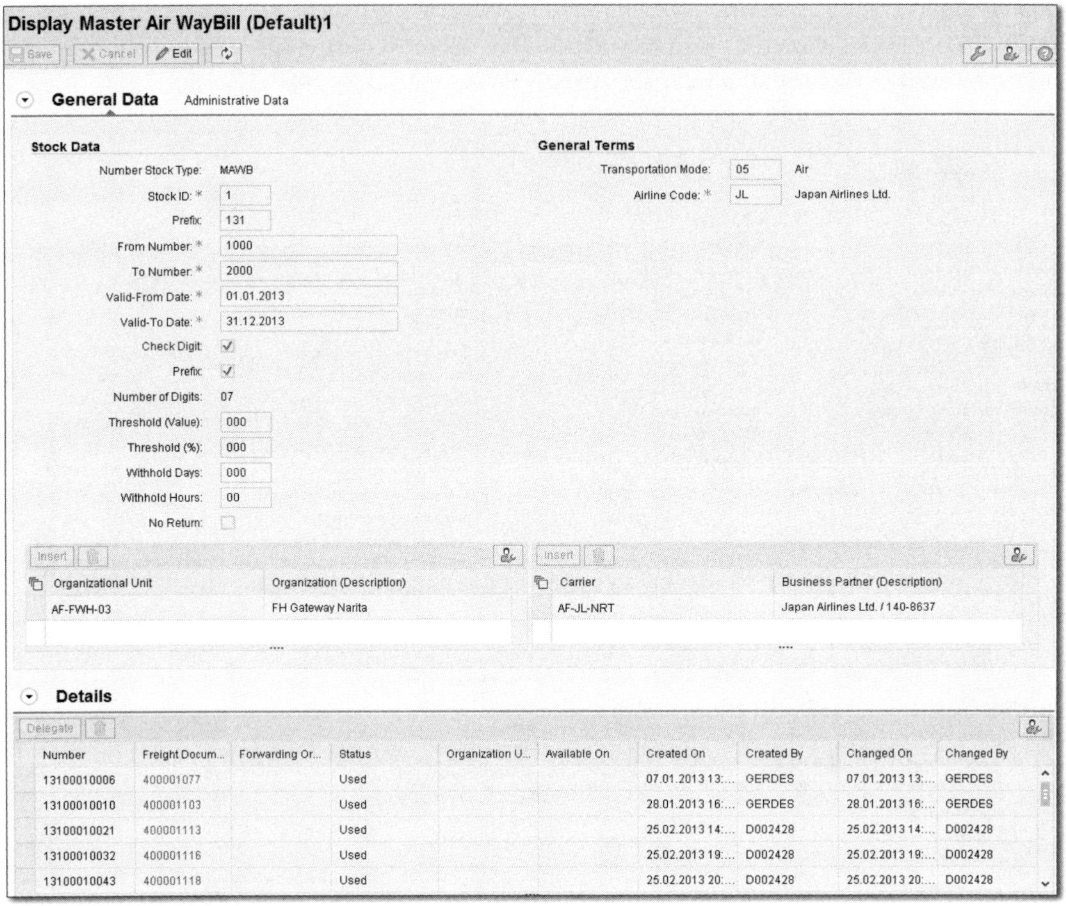

Figure 15.8 Maintenance of MAWB Stocks with Already Drawn Numbers

15.2.4 Flight Plans and Capacity Booking Types

Carrier schedules

When locations such as customer, consignee, gateway, station, and airports have been defined, you can create the interconnections in the form of carrier schedules and master flight schedules. Carrier schedules represent the regularly timed connections of a carrier between different locations. For air freight, the carrier schedule is the flight plan for the freight services on passenger and cargo planes (see Figure 15.9).

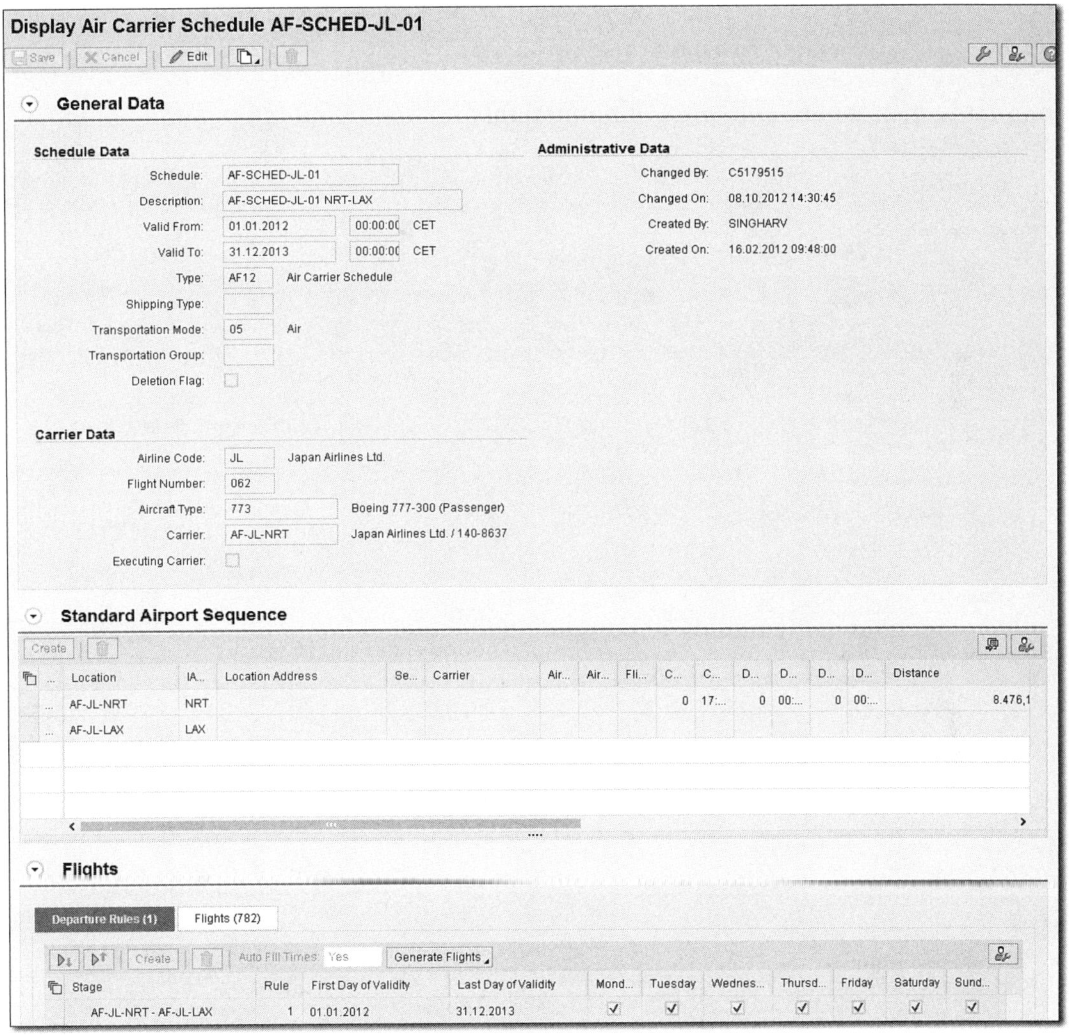

Figure 15.9 Carrier Schedule for a Japan-U.S. Flight

You can download flight plans from the carrier websites or from service providers and define details such as flight number, aircraft type, departure and arrival airport, departure and arrival times, and handover cutoff times for the airline at the ground handling stations. Figure 15.10 shows a flight master plan for our Japan Airlines flight (JL062; see carrier schedule in Figure 15.9) from Tokyo Narita to Los Angeles with a Boeing 777-300 passenger aircraft. The flight plan is valid for the whole year and leaves every day. The individual flights are automatically generated based on this information.

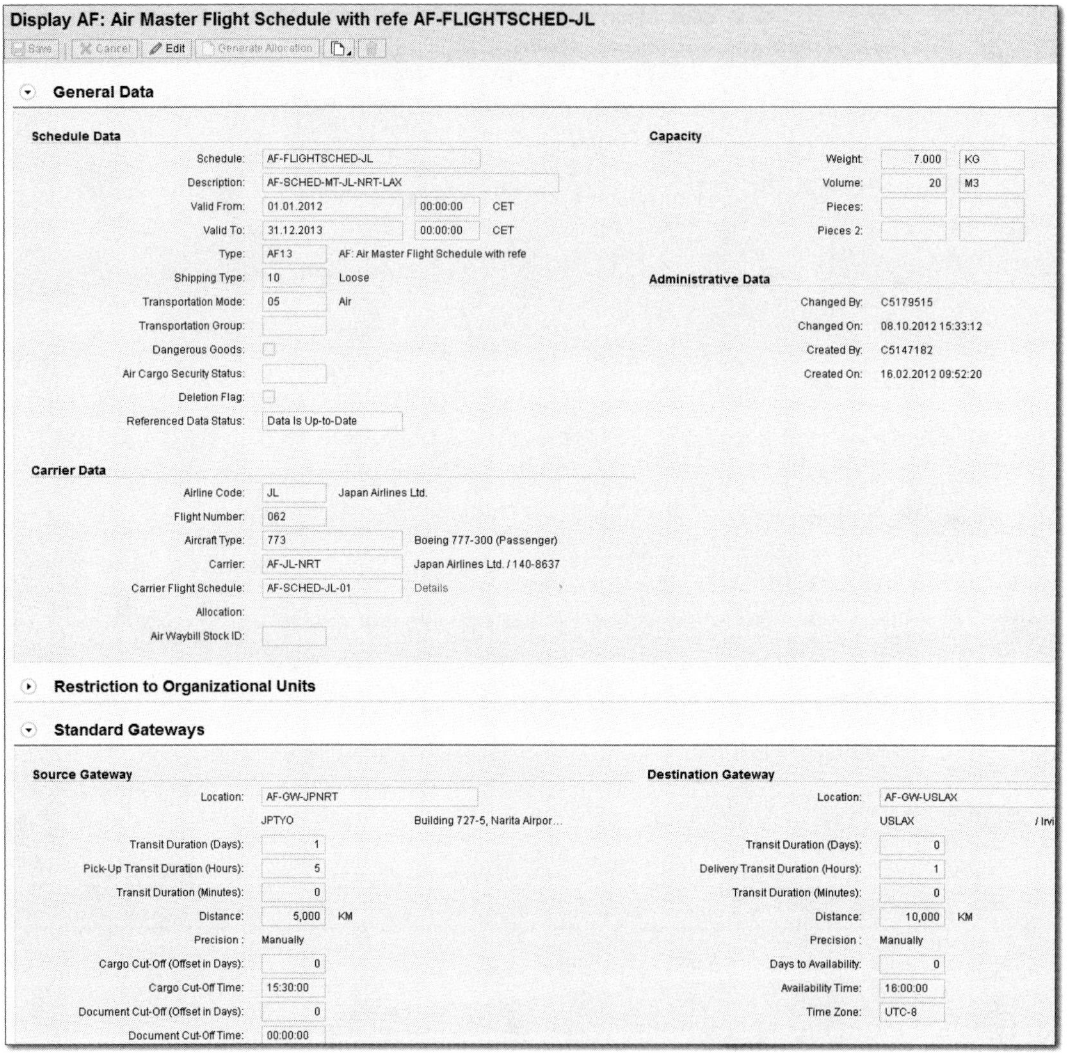

Figure 15.10 Master Flight Plan Covering Connections from Narita Gateway to Los Angeles Gateway

Similar to the carrier flight schedules, you can also create schedules for the long-haul truck legs used in pre-carriage and on-carriage between stations and gateways.

Master flight plan

The master flight plan that our LSP, Interforwarders Corp., uses on its Japan-U.S. relationship is based on these carrier schedule but adds the connections to the forwarder's gateways on the export and import side (see Figure 15.10). Carriers' schedule timing is taken into account and, in addition, you can maintain the cutoff times at the gateway for cargo acceptance, documentation, and dangerous goods acceptance, as well as the availability times on the arrival side. Master flight plans have a direct relation to capacity allocations, which allow you to reserve and maintain cargo capacity on the individual flight departures.

Four-week flight plan

Freight bookings are used to model the four-week (short-term) flight plan. Later in this chapter, when we look to the operational process, we explain how to create the four-week flight plan. At this stage, we show you how it can be configured.

The object type to be created in Customizing is an air freight booking type. You can find the corresponding settings via the menu path SAP Transportation Management • Transportation Management • Freight Order Management • Freight Bookings • Define Freight Booking Types. The main booking type is the export air freight booking shown in Figure 15.11. It is related to consolidation airway bills, which is indicated by the container item source that defines that the equipment for consolidation is part of the booking and not part of the air forwarding order. The consolidation itself is required in the service definition. The related import booking type is defined in the additional settings of the booking type.

Import booking type

The export booking is mainly used as a capacity and execution object on the way from the departure airport to the arrival. For handling in the import country—that is, for all cargo-related operations to be done after the cargo arrives at the import gateway—an import booking is created. The main differences between export and import booking types are different associations with purchasing organizations and different update strategies from the predecessor documents (there is no update of the import booking because it is created in a cut-over process between export and import).

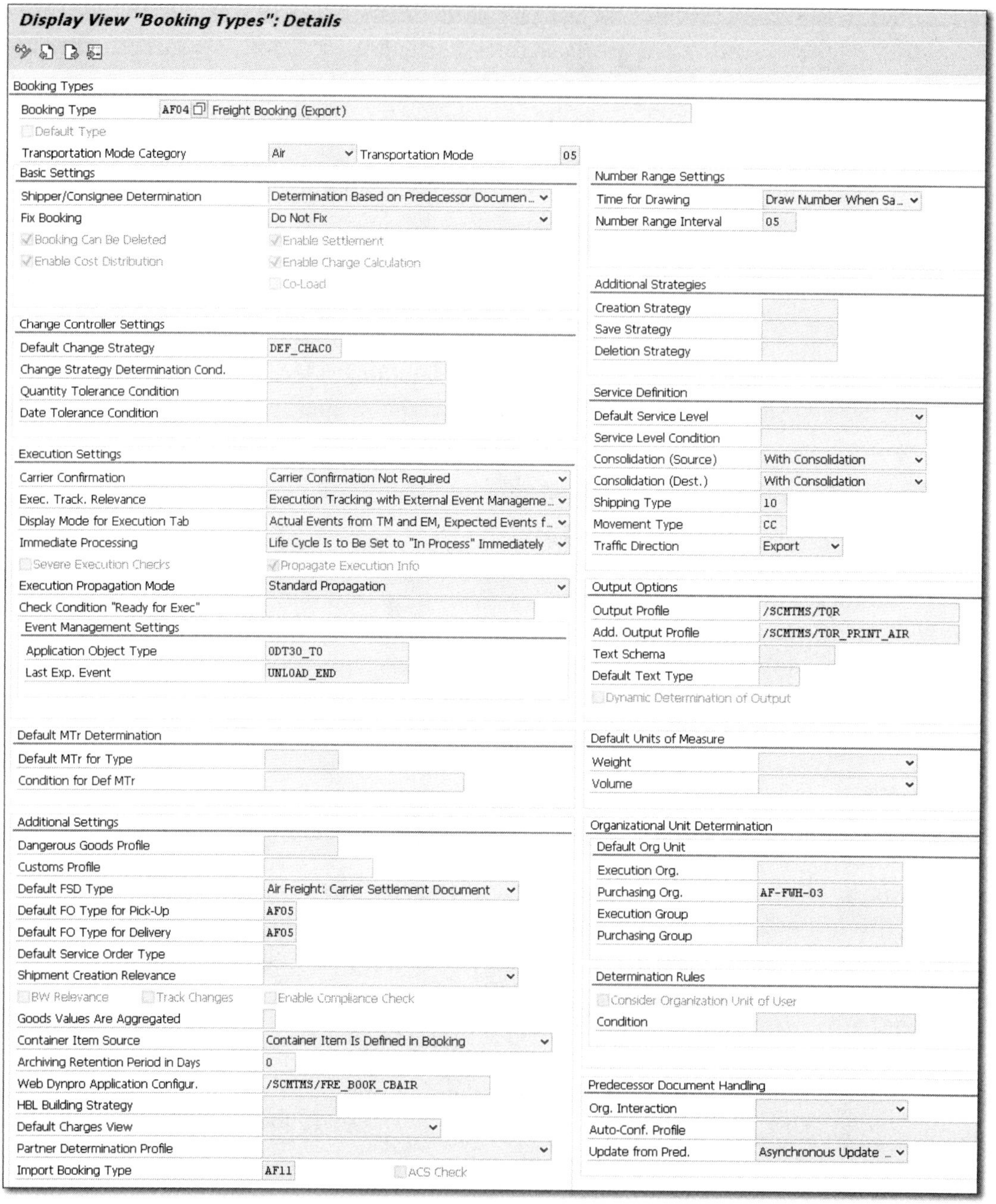

Figure 15.11 Customizing for Air Freight Booking Type (Export)

15.2.5 Order Management Customizing and Configuration

Order management is also done in export and import organizations. Therefore, you have to create an export forwarding order type, which is used to create the customer order (see Figure 15.12), and an import forwarding order type, which allows the import handling for the customer order.

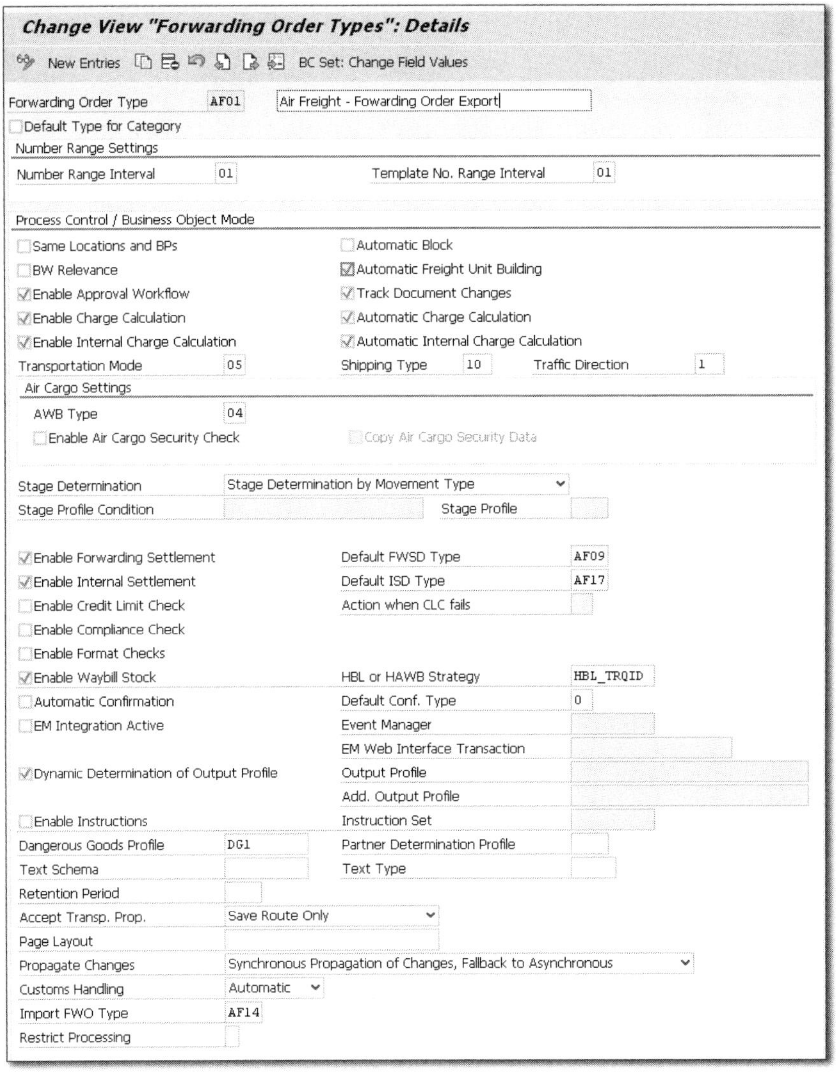

Figure 15.12 Customizing for Air Freight Forwarding Order Type (Export)

Forwarding order types

You can define the forwarding order type in Customizing by following the menu path SAP Transportation Management • Transportation Management • Forwarding Order Management • Forwarding Orders • Define Forwarding Order Types. Similar to the booking order types described in Section 15.2.4, you have to create an export forwarding order type that points to the related import forwarding order type. The stage determination is done based on the movement type given in the corresponding export forwarding order. For a door-to-door shipment (movement type DD), up to five stages are determined. For a station-to-station shipment (movement type CC), pickup and delivery are organized by the shipper or consignee, and therefore only three stages are expected. As result of planning, the route is only saved, but there is no automatic association to freight orders or freight bookings because the routing is to be seen as a pre-arrangement, which can be overridden by the gateway or stations.

Forwarding order item types

For the forwarding orders, you can create specific item types that allow efficient handling of packaged cargo items. This item type should define the default quantities and units of measure, have the item category *package*, and must be assigned to the export and import forwarding order types.

15.2.6 Planning Configuration

Freight unit building rules

Freight unit building in the air freight scenario is configured for handling loose cargo (see Figure 15.13). The freight unit size of an air freight forwarding order should not exceed the maximum allowed cargo net weight and volume of a large air cargo pallet. Therefore, you should set the planning quantities in the freight unit building profile to the corresponding net capacity values of the equipment intended to be used.

Planning procedure configuration

There is a variety of ways to configure the planning procedure for the air freight process. In addition to fully automatic end-to-end planning, which can be done by a central planning group, completely manual or mixed planning procedures are also possible. In our example, a mixed mode is used because it's the best fit for the organization and operation principles of our forwarder.

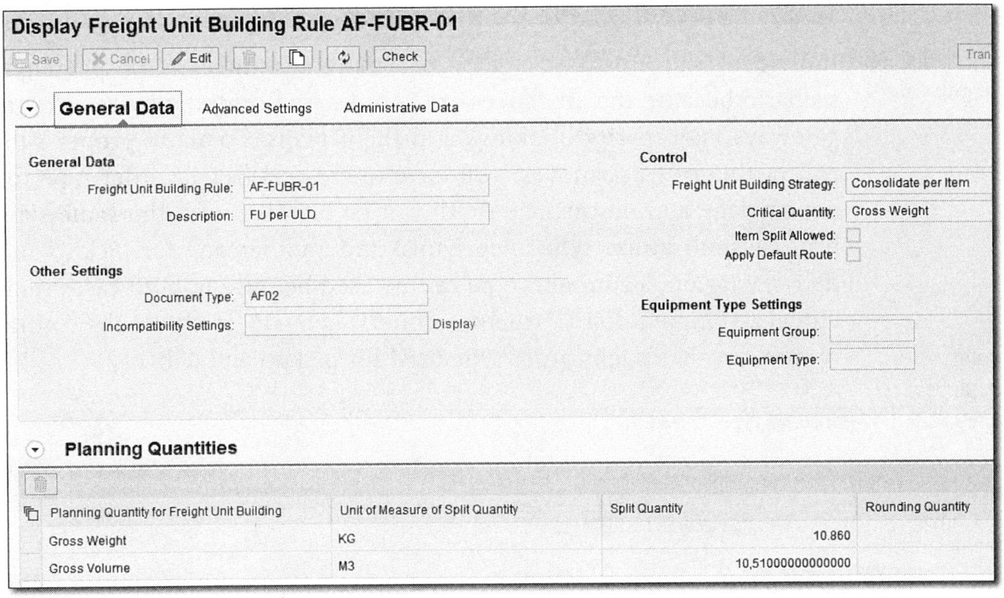

Figure 15.13 Freight Unit Building Rule for Loose Air Cargo

The pickup and delivery legs are manually subcontracted by the origin or destination stations. Therefore, a dedicated planning step is not required, and any use of consolidation potential is up to the decision of the executing truck carrier (although the stations may use an optimizer to preconsolidate pickups and deliveries).

Main leg planning is in the hands of the export gateway. As soon as the gateway is informed about the booking, the corresponding dispatcher can assign it to the appropriate air freight booking to maximize utilization and achieve a good weight:volume ratio. (Ideally, the volume in cubic meters of the consolidated cargo should be six times the weight in tons, resulting in the best price.) Main leg planning is therefore done by assigning the forwarding orders to freight booking via a selection process.

In pre-carriage and on-carriage, optimization is used by the forwarder because he controls the flow of cargo between his stations and gateways. For pre-carriage, a planning profile is created that uses a selection of freight units from the origin stations to the export gateways and allows an optimization using the truck schedules and assigned vehicles for the station-gateway connections.

15.2.7 Subcontracting Configuration

Trucking freight orders

Individual subcontracting is relevant for all truck moves in the air freight scenario because the air moves are coordinated centrally by the export gateways via capacity bookings and flight plans. To allow proper subcontracting of truck moves, you have to create a freight order type for pre-carriage and on-carriage (both can be handled with the same type because both utilize scheduled hauls) and another one for pickup and delivery (again, a common type can be used because in both cases individual assignment to LTL trucks is done). Figure 15.14 shows the configuration for the freight order type used for pickup and delivery.

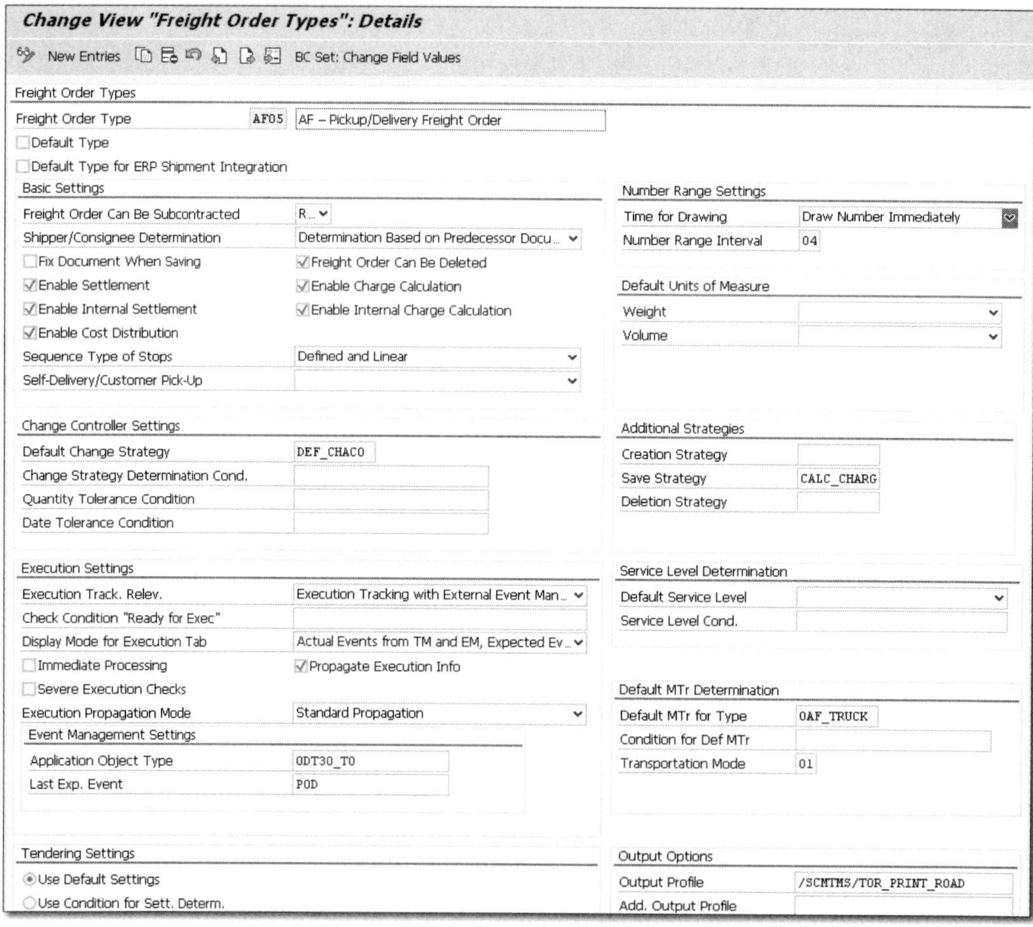

Figure 15.14 Freight Order Type for Pickup and Delivery

15.2.8 SAP TM Transportation Charge Management

In SAP TM Transportation Charge Management, it is important to define relevant charge types, scales, rate tables, calculation sheets, and agreements for the calculation and invoicing cases:

- Customer invoicing to shipper
- Customer invoicing to consignee
- Airline invoicing to export gateway
- Truck carrier invoicing to stations and gateways
- Internal invoicing from import gateway to import station for settlement of haulage cost

Charge elements and scales

A variety of charge elements provided in standard SAP TM 9.1 can be used in the air freight scenario (e.g., FUEL for the fuel surcharge). You only have to define some additional charges that are not available yet (see Table 15.4).

Charge Type	Description
BASE_AF	Basic air freight charge
COO_AF	Certificate of origin per shipment
ECC_AF	Export customs clearance per shipment
GIF_AF	Goods inspection fee per shipment
HD_AF	Handling and documentation fee
PU_AF	Pickup fee
SF_AF	Security fee
TRP_AF	Transportation insurance and CPT value

Table 15.4 Additional Charge Types for the Air Freight Scenario

For air freight rate definition, you have to set up a specific weight scale that reflects the weight breaks used in air transportation and a second one with weight breaks for long-haul trucking. Both scales are base scales with the scale base "weight." Table 15.5 shows the corresponding scale values.

Air Freight Scale AF-NRT-LAX	Air Cargo Scale	Trucking Scale AF-CU-PU
1 kg	1 kg	1 kg
45 kg	45 kg	500 kg
100 kg	100 kg	2,000 kg
2,000 kg	300 kg	5,000 kg
5,000 kg	500 kg	500,000 kg
	1,000 kg	
	2,000 kg	
	5,000 kg	

Table 15.5 Scale Values for Air Freight and Trucking Rate Tables

Rate tables

Multiple rate tables must be created due to the different calculation and invoicing use cases described at the beginning of this section. Most of the rate tables have a quite simple structure. The following list gives an overview of the typical structures:

- **Customer charges for pickup**
  Calculation is done on a pure gross weight basis. A trucking scale is used with a decreasing price for increasing weight starting at 2,100 Yen/kg and dropping to 1,250 Yen/kg.
- **Customer charges for air freight**
  Calculation is done on a chargeable weight basis. An air freight scale is used with minimum price of 8,000 Yen and a decreasing price for increasing weight. Rates start at 750 Yen/kg and drop to 185 Yen/kg. A differentiation of rate table may be done by carrier used. In this case, a rate table condition has to be used to find the appropriate rates for the carrier. Additionally, rates may vary by customer, which is later reflected by integration of different tables into calculation sheets of agreements.
- **Truck carrier charges for pre-carriage and on-carriage**
  Calculation is done on a gross weight basis with a fixed price per kilogram for haulage between station and gateway.

- **Air carrier charges for main carriage**
  Calculation is done on a chargeable weight basis. An air cargo scale is used with an additional same scale for the executing carrier. The price per kilogram decreases with increasing weight. A minimum charge applies.
- **Internal charges between import gateway and import station**
  For simplification reasons, the on-carriage trucking rates are used in the same way as for the customer when charges are calculated between the import gateway and station.

Figure 15.15 shows an example of a rate table (air freight rates from Narita to Los Angeles for customer charging).

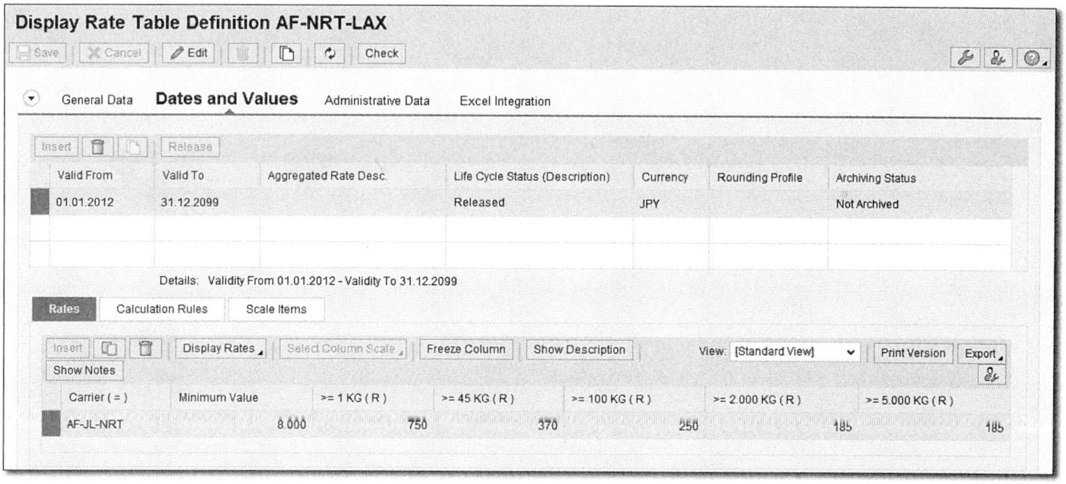

Figure 15.15 Air Freight Rate Table for Customer Charging

Calculation sheets and agreements

To properly calculate the charges for these cases, we need a set of calculation sheets:

- Carrier calculation for pickup
- Carrier calculation for pre-carriage
- Carrier calculation for main carriage
- Carrier calculation for on-carriage
- Carrier calculation for delivery
- Customer/shipper calculation for pickup

- Customer/shipper calculation for main carriage
- Customer/consignee calculation for main carriage
- Customer/consignee calculation for delivery
- Internal calculation import

The calculation sheets are embedded into similarly structured freight and forwarding agreements that are the contractual basis for calculating rates for customers and carriers. Figure 15.16 shows the items of the calculation sheet for the main carriage (air freight) that is used for customer charge calculation. In addition to basic air freight charges and pickup charges, there is a list of applicable surcharges (e.g., for handling, documentation, customs, security, and fuel).

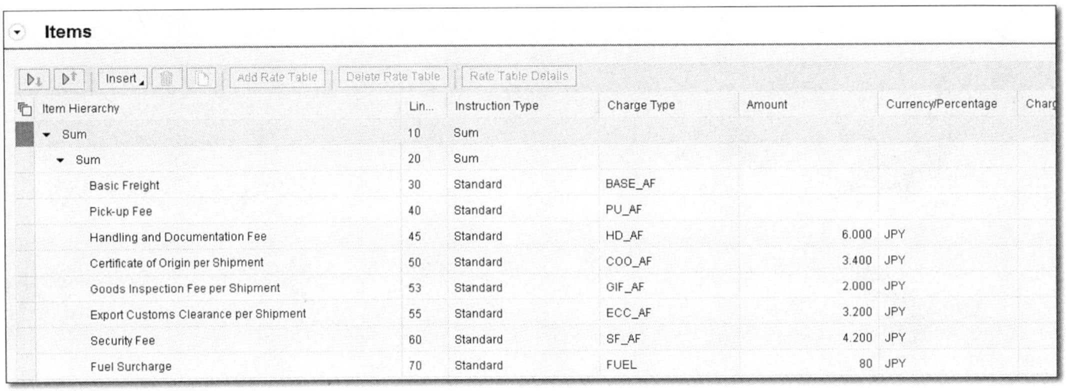

Item Hierarchy	Lin...	Instruction Type	Charge Type	Amount	Currency/Percentage	Charg
Sum	10	Sum				
Sum	20	Sum				
Basic Freight	30	Standard	BASE_AF			
Pick-up Fee	40	Standard	PU_AF			
Handling and Documentation Fee	45	Standard	HD_AF	6.000	JPY	
Certificate of Origin per Shipment	50	Standard	COO_AF	3.400	JPY	
Goods Inspection Fee per Shipment	53	Standard	GIF_AF	2.000	JPY	
Export Customs Clearance per Shipment	55	Standard	ECC_AF	3.200	JPY	
Security Fee	60	Standard	SF_AF	4.200	JPY	
Fuel Surcharge	70	Standard	FUEL	80	JPY	

Figure 15.16 Charge Items of the Main Carriage Calculation Sheet for Customers

15.2.9 Settlement Configuration

Settlement configuration includes the following settings, which are mainly comparable to the standard settings delivered with SAP TM 9.1:

- Creation of settlement profiles that control invoice creation and correct invoicing date selection
- Creation of forwarding and freight settlement document types
- Creation of a cost distribution profile based on order item gross weight, which allows the cost distribution for profitability calculations of the forwarding orders

15.3 Running the Air Freight Process

This section describes the air freight process from an operational and transactional point of view. As you might have gathered from Section 15.1, the air freight process requires a lot of interaction between the participating organizational units. To make this interaction smoother and the responsibilities a little bit more transparent, we explain the big-picture process in Figure 15.17 and Figure 15.18 to show the export and the import view of the overall process.

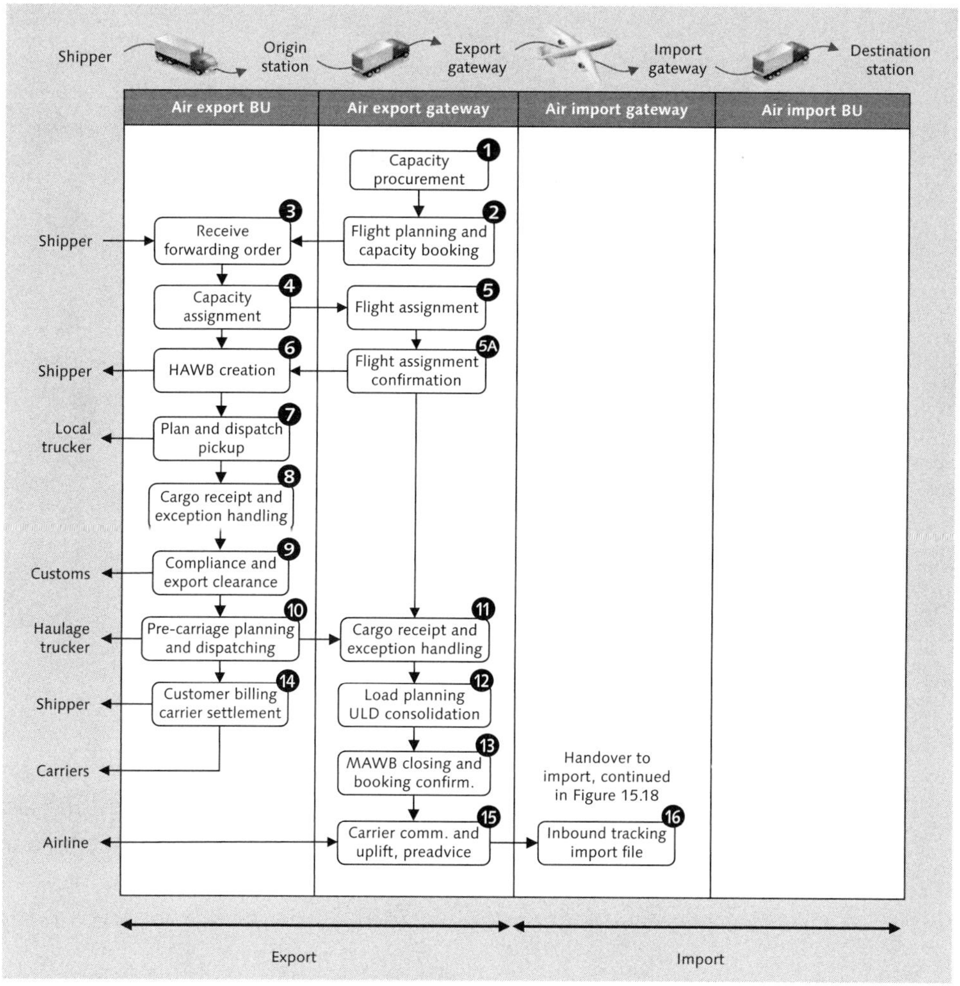

Figure 15.17 Air Freight Export Process Steps

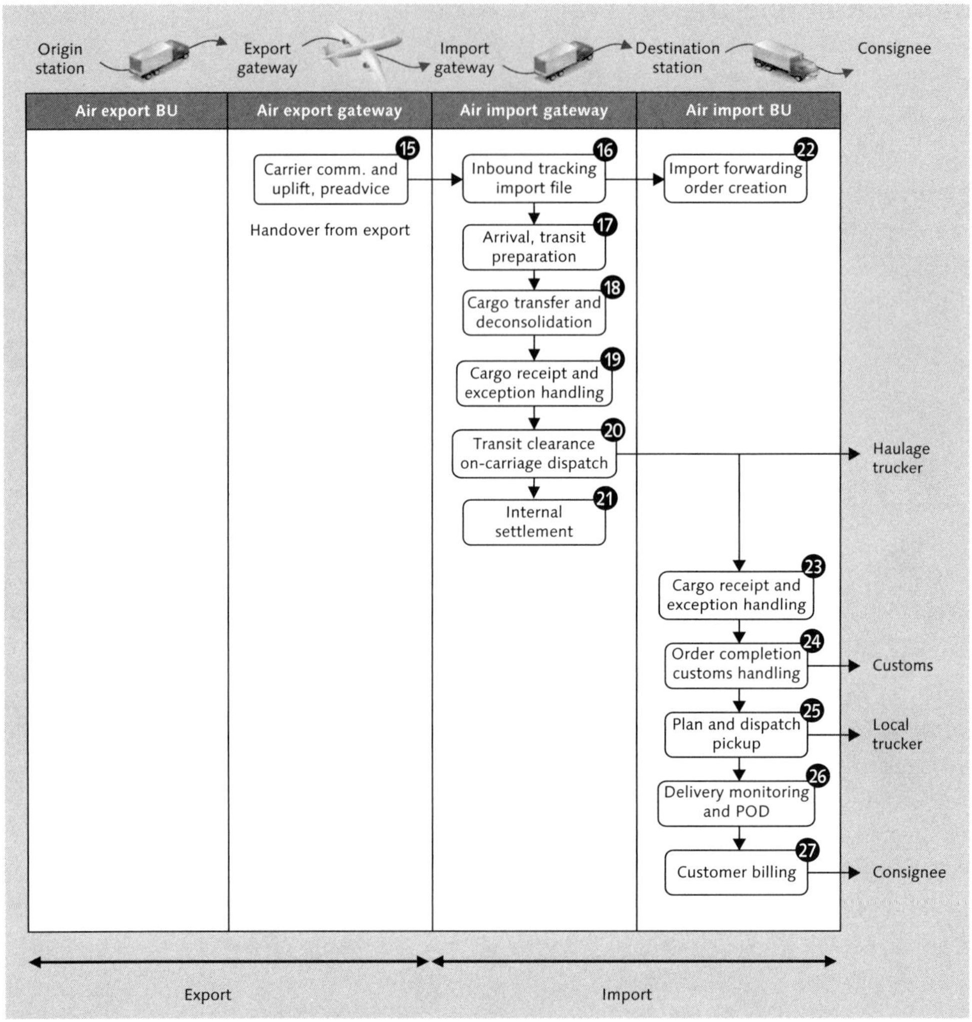

Figure 15.18 Air Freight Import Process Steps

On the top of each diagram, we use columns to describe the acting organizational units. The process moves from top to bottom, where the last step of the first diagram is integrated into the first step of the second diagram. The square boxes in the "swim lanes" are tasks to be carried out by the corresponding organization. In the description, we refer to the number listed on the top-right corner of the box. External parties (customers, carriers) who interact with the organizational units are displayed on the left or right edge of the chart in ovals. For the sake of clarity, the carrier settlement tasks are not displayed.

15.3.1 Creation of Capacity Bookings and Four-Week Flight Plan (Steps 1 and 2)

Flight plans and capacity

The master flight plans described in Section 15.2.4 are usually created several months in advance, where the allocations allow provision of a mid-term capacity planning capability for the flight legs ❶. For short-term operational use, you can use a report that processes the data of the master flight plan and generates multiple freight bookings for a selected period and the flights relevant during this period. The capacity manager of the export gateway analyzes the capacity demand and creates a four-week flight plan from the master flight plan ❶.

Flight plan capacity bookings

You can run the report for the capacity booking creation from the master flight plan worklist in SAP NWBC ❷. By selecting the master flight plan AF-FLIGHTSCHED-JL, for example, and setting the requested planning time frame to next four weeks, SAP TM automatically creates 28 freight bookings with the corresponding booking type (one for each departure), incorporating flight and carrier data as well as allocation capacities. The freight bookings now serve as planning objects for any forwarding orders and freight units to be shipped by air freight on the corresponding origin-destination relation. Figure 15.19 shows an overview of the header data of an air freight capacity booking that shows flight data, MAWB data, MAWB number, and capacity requirements.

Booking confirmation

After finalizing the bookings and completing the required data, the booking can be sent to the airline for capacity confirmation. Via the function SUBCONTRACTING • SEND BOOKING, the booking is sent out and the status is set to *No Confirmation Yet*. Once the carrier confirms, the status is automatically switched to *Confirmed by Carrier*.

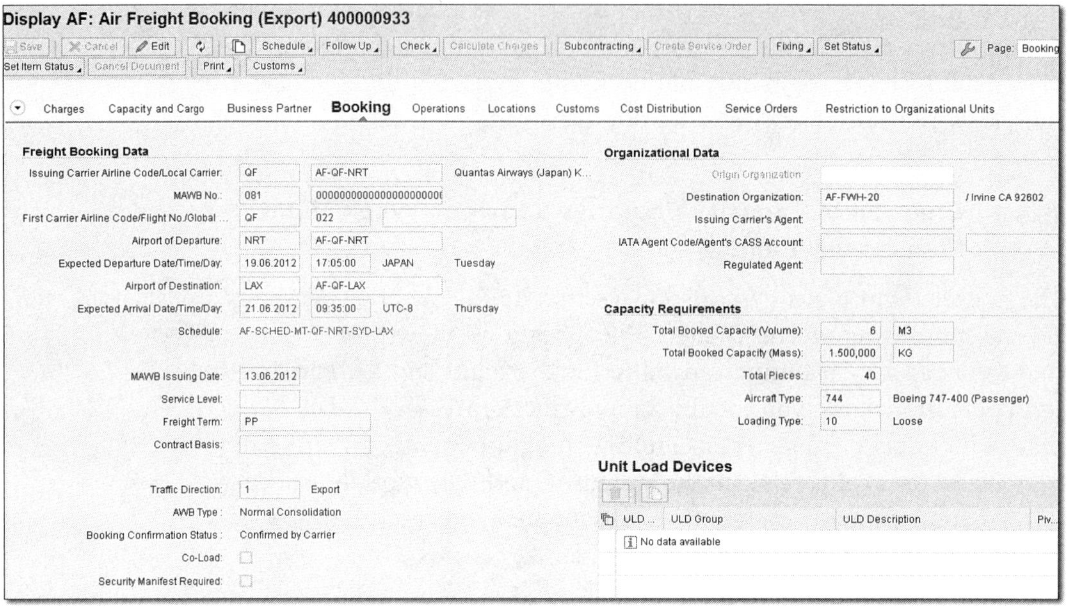

Figure 15.19 Freight Booking for Air Freight (Header Data and Capacity)

15.3.2 Creation of Air Freight-Forwarding Orders and HAWBs (Steps 3–6)

Air forwarding orders can be communicated to our freight-forwarder in various forms (e.g., phone, fax, mail, or via booking services and portals). In our example, the forwarding orders are created manually by a customer service agent assigned to the origin station, who creates a new air forwarding order from the SAP NWBC menu.

Create air forwarding order

The required forwarding order data is entered into the screen field, and relevant checks are carried out ❸. Data to be captured includes shipper information, consignee information, and cargo information such as cargo type, commodities, packaging, weight, volume, and piece count. In addition, shipping and requested arrival dates have to be maintained. To allow the service agent to make a proper shipping decision, information from the customer about the requested service level and movement type is useful. Figure 15.20 and the upper part of Figure 15.21 give an example of an air forwarding order and the corresponding Items tab.

The order is door-to-door for loose cargo and is sold by the sales organization in Tokyo.

The cargo is 40 boxes of electronic components with an overall weight of 1,400 kg. The customer does not give specific information on a requested routing. Therefore, the ordered route consists of just one stage from shipper to consignee.

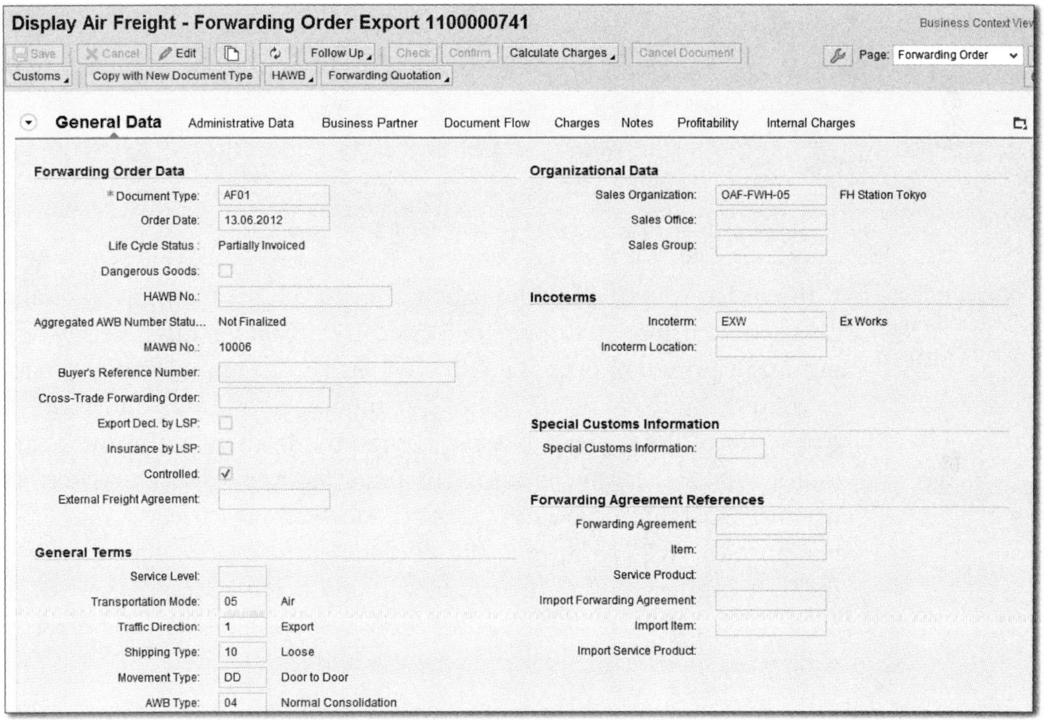

Figure 15.20 Forwarding Order for Air Freight Consolidation

After the order is captured, required checks can be carried out. These checks may include the known shipper verification, blacklist screening, or credit limit checks.

Templates to simplify order entry

If customers frequently order similar services, you can create forwarding order templates to simplify data entry. The templates can, for example, be prefilled with shipper, consignee, origin, destination, and commodity information so that only quantities and shipping dates have to be added to complete order entry.

Items

Insert Split

Item Hierarchy	Item Description	Pieces	Piece...	Gross We...	Gros...	Gross Vol...	Gros...	Net Weight	Net ...
Packages with specific dimensions	Electronic Components	40	EA	1.400	KG	3	M3	1.300	KG

Stage Description	Stage Type	Transport...	Source Location	IATA...	City	Pick-Up ...	Pick-Up T...	Destination Location	IATA Loc.
Ordered Route							00:00:00		
Complete Carriage		05 (Air)	CUOAF-CU-02@A9X700		Ota-ku		00:00:00	CUOAF-CO-02@A9X700	
Item 10		05 (Air)	CUOAF-CU-02@A9X700		Ota-ku		00:00:00	CUOAF-CO-02@A9X700	
Actual Route							00:00:00		
Complete Carriage			CUOAF-CU-02@A9X700		Ota-ku	18.06.2012	08:20:00	CUOAF-CO-02@A9X700	
Stage 1	01 (Pick-Up)	01 (Road)	CUOAF-CU-02@A9X700		Ota-ku	18.06.2012	08:20:00	OAF-STA-JPTYO	TYO
Stage 2	02 (Pre-C...	01 (Road)	OAF-STA-JPTYO	TYO	Tokyo	18.06.2012	10:13:13	AF-GW-JPNRT	NRT
Stage 3	03 (Main C...	05 (Air)	AF-GW-JPNRT	NRT	Narita Airport	19.06.2012	10:40:00	AF-GW-USLAX	LAX
Stage 4	04 (On-Ca...	01 (Road)	AF-GW-USLAX	LAX	Irvine	22.06.2012	08:00:00	OAF-STA-USPHX	PHX
Stage 5	05 (Delivery)	01 (Road)	OAF-STA-USPHX	PHX	Tempe	24.06.2012	00:00:00	CUOAF-CO-02@A9X700	

Figure 15.21 Items and Stages of an Air Forwarding Order

Order routing, confirming capacity

If the order is verified without being set to a hold status, routing and capacity confirmation can be done ❹. You can handle this either by calling a transportation proposal (which gives you an end-to-end sequence of routes), by deciding the routing manual, or with a default routing. The result of the routing decision is captured in the actual route stages (see Figure 15.21). By selecting the main leg and selecting ASSIGN TO FREIGHT DOCUMENT, you can assign the forwarding order to a capacity booking (MAWB) to reserve the necessary capacity on the requested flight. A confirmation of the export gateway may be required in special cases, such as if there are dangerous goods to be shipped or the capacity is tight on specific flights parentheses ❺ and 5A.

Drawing HAWB numbers

Order processing is continued by drawing a house airway bill (HAWB) number for the forwarding order ❻. If there is no requirement to create multiple HAWBs for a single order, this can be easily done directly for the order. Together with creation of the HAWB, you can also confirm the order to the customer by sending a forwarding order confirmation.

If you have multiple HAWBs for one order or a joint HAWB for multiple orders, this can be executed from the MAWB, where a detailed assignment is possible.

15.3.3 Pickup, Cargo Receipt, and Compliance (Steps 7–9)

Pickup planning and execution

Because the order was posted with a door-to-door service level, the origin station has to arrange the pickup and ensure that the cargo is moved from the shipper to the station, where it is received and inspected.

The planning of the pickup ❼ is triggered by the export station dispatcher, who marks the pickup stage in the actual route and creates the freight order for pickup by selecting FREIGHT ORDER • CREATE. From the forwarding order, the dispatcher can navigate to the newly created freight order and assign a truck carrier as long as it has not been done by the system. The freight order can then be communicated to the trucker for execution.

Export cargo receipt

The arrival of the cargo at the export station ❽ can either be communicated via SAP Event Management or be entered manually in the forwarding order's EXECUTION INFORMATION tab by setting the ARRIVAL AT DESTINATION event. This sets the cargo status for the pickup stage to *Arrived*.

As part of the compliance check ❾, the known shipper and blacklist status were verified after the order data was entered. Further activities might be required concerning legal and trade compliance, which are supported by SAP GTS. After receiving and inspecting the shipment, the compliance of the physical goods characteristics and their packaging with the documented data in the forwarding order is confirmed. For compliance handling and documentation issuing, the forwarding order can be transferred to SAP GTS, where compliance rules are applied and export clearance is confirmed.

15.3.4 Pre-Carriage from Export Station to Gateway (Steps 10 and 11)

Pre-carriage planning

At regular intervals, the export station dispatcher starts pre-carriage planning to arrange the haulage from the export station to the export gateway ❿. Because this task is consolidation-relevant and the capacity of the utilized scheduled trucks needs to be managed well, the dispatcher uses the transportation cockpit. These tasks also are repeated quite frequently, so a profile and layout set holding all pre-carriage planning settings is used. The profile set contains the selection profile for freight units and freight orders as well as an appropriate planning profile and capacity selection profile referencing the required schedules and trucks.

Based on the settings, the dispatcher can efficiently build the LTL truck from station to gateway and create the freight orders to subcontract the moves to the truck carrier. Figure 15.22 shows an overview on the cockpit layout with freight units on the left and planned or open freight orders on the right.

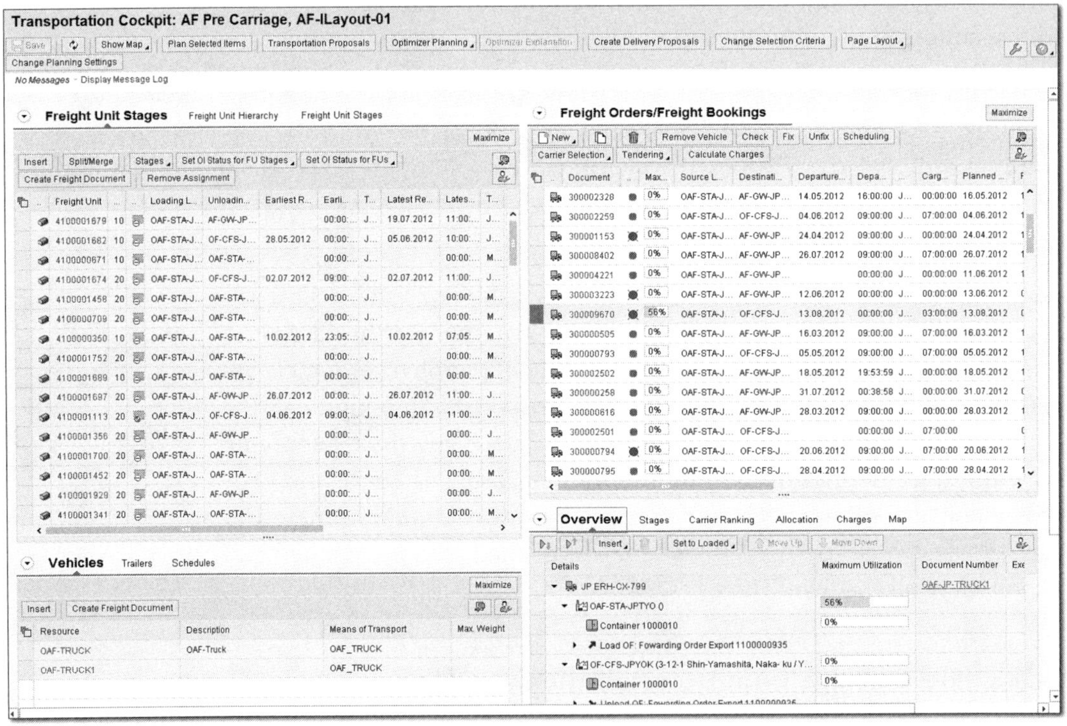

Figure 15.22 Transportation Cockpit for Pre-Carriage Planning

Pre-carriage and gateway receipt

The freight order created by the station is used for execution of the long haul. With a worklist, the user in the gateway can get information about the arriving shipments ⓫. Execution may be tracked using SAP Event Management. Event messages for departure and arrival allow the setting of the pre-carriage status. The arrival event sets the status to *Arrive*.

The physical receipt of cargo can be done as part of freight order handling. In the cargo management view of the freight order, discrepancies of cargo amounts and characteristics can be maintained by the gateway staff.

15.3.5 Load Planning, MAWB Close, and Uplift (Steps 12, 13, and 15)

Load planning for the flight

When the gateway dispatcher starts load planning ⓬ for a flight, the corresponding freight booking is switched to the status *Closed* to prevent further assignments of forwarding orders that might mess up existing plans. From a load guide worklist, the gateway dispatcher can get an overview of the relevant forwarding orders arriving at the gateway; that is, he can see which of the shipments expected to go on his flight have already arrived and are free of discrepancies, and which ones are still in transit to the gateway or are expected to be delayed. The forwarding orders with a critical transit status may have to be taken from the freight booking and assigned to a different flight for proper shipment. In the cargo management view of the freight booking, the load planning activities and status can be maintained (see Figure 15.23).

Execution of load consolidation

Load consolidation activities can be documented directly in SAP TM screens by creating ULDs and assigning freight units via drag-and-drop. Alternatively, the information can be exchanged electronically with a hub management system that receives the freight booking, freight unit, and forwarding order information and sends back the consolidated MAWB data to update the freight booking with ULD information, packing hierarchies, actual weight and volumes, and detailed cargo item status. Using the cargo management view, you can also react to overbooking situations that are indicated by the actual utilization for weight and volume (see Figure 15.23).

MAWB document and status handling

During and after loading and stowing of cargo, variety of documents and labels must be printed and sent with the cargo ⓭. These documents can be created from the freight booking on the OUTPUT MANAGEMENT tab. Figure 15.24 shows the preview of an MAWB label, which is attached to the cargo.

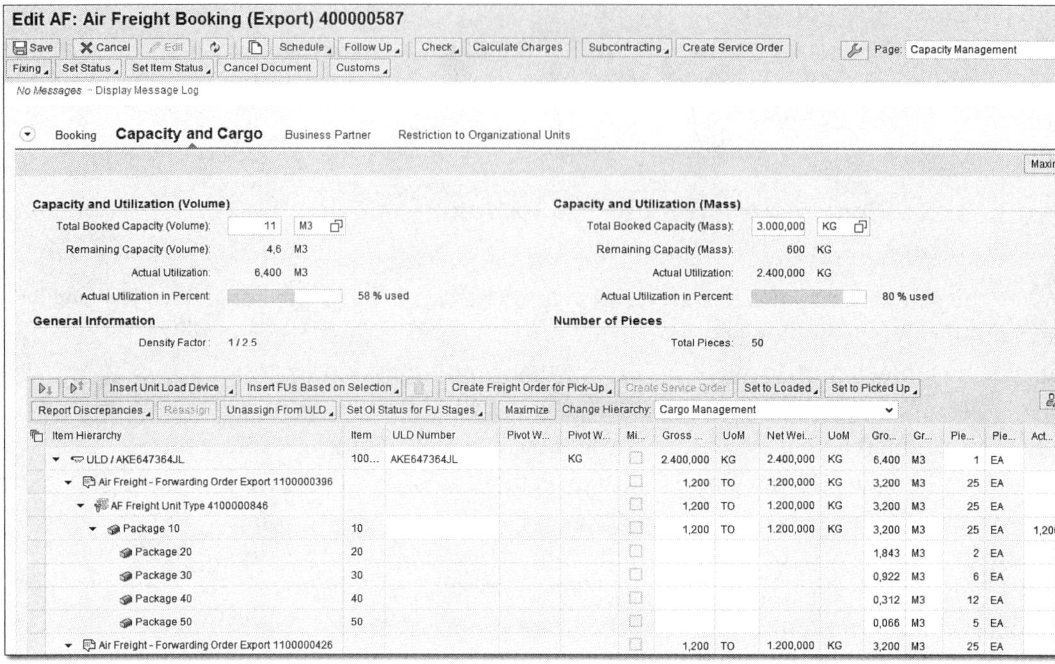

Figure 15.23 Cargo Management for a Capacity Booking

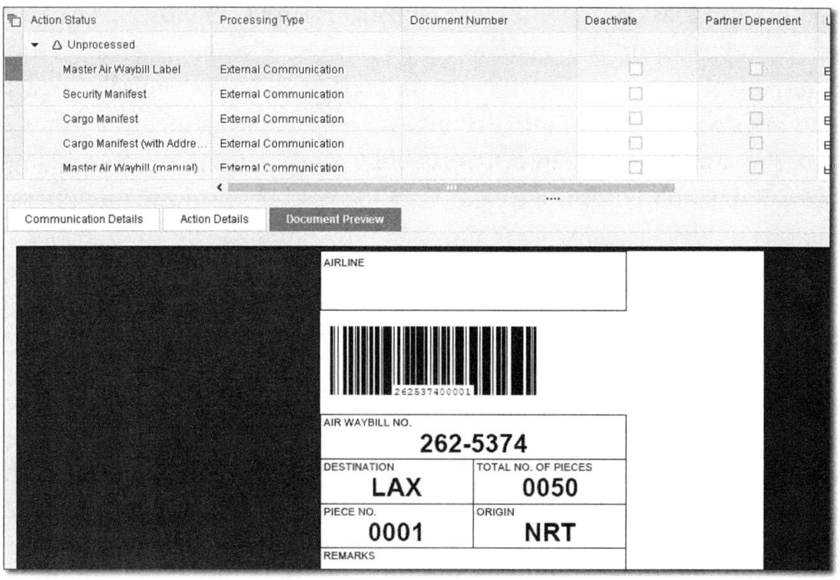

Figure 15.24 Preview and Printing of the MAWB Label

Closing and confirming the MAWB

Once the loading process is completed and the cargo to be shipped is stowed on the ULDs, the MAWB can be closed and final booking confirmation and shipping instructions can be sent out from the freight booking either as an electronic message or in an alternative output format. Confirmation status in the freight booking can be either set by receiving an electronic carrier confirmation or by setting the status manually in the freight booking.

Carrier feeding, uplift confirmation

The overall status of the freight order is reflected in the POWL, which you can use to analyze the freight booking readiness for cargo transfer to the airline's ground handling facility ⓯. Freight bookings that are ready for feeding allow you to create a corresponding freight order and a transfer manifest with all relevant flight and shipment details. The airline receives the cargo and replies with a freight status message (*Cargo IMP FSU*). This message is sent again after uplift of the cargo. The gateway dispatcher then manually sets the uplift status in the SAP TM freight booking. Alternatively, an interface to feed the cargo IMP information into SAP TM or SAP Event Management can be created as a project solution.

15.3.6 Financial Processing in the Export Organization (Step 14)

Settlement with carriers (export)

Financial processing on the export side ⓮ includes four tasks that may have to be carried out depending on organizational and financial criteria. The sequence and timing of these tasks is not fixed and has only limited relation to the timing of the operational process.

- **Carrier and truck vendor settlement**
  Settlement with the truck vendors is initiated by the export station (pickup and pre-carriage) or by the export gateway (airline feeding). In each case, the responsible employee calculates and checks the charges in the freight order related to the pickup or pre-carriage stage of the forwarding order or on the freight order created from the freight booking in the case of an airline feed. Subsequently, freight settlement documents are created that are transferred to SAP ERP for

accrual and at a later stage for payment. There may also be a consolidation of multiple freight order charges into a joint settlement document. These process steps can be seen as decoupled from the actual operation because whether each invoice is to be paid separately or a weekly summary invoice is to be settled depends on the payment regulations.

The air cargo charges are calculated in the freight booking. This is done by an export gateway staff member who calculates and checks the charges and creates a freight settlement document from the freight booking. Air cargo settlement is usually done via CASS.

- **Internal settlement**
  The export gateway has to carry the cost for airline feeding, which they need to allocate to the stations because these earn the revenue for the shipment. To distribute the cost, the gateway employee opens the freight booking, navigates to the related forwarding orders, and selects CREATE INTERNAL SETTLEMENT DOCUMENT. This can be done individually or collectively if there are multiple forwarding orders from the same station.

  Assuming that orders are shipped with incoterms such as free-on-board, it may be necessary to invoice apportioned main leg costs (air carriage) to the import organization because the revenue for air freight is achieved by the import organization, and the cost is still with the contracting export side.

- **Customer billing**
  Customer billing is triggered by the export station. After uplift has been confirmed, the customer service agent in the export station opens the forwarding order and selects CALCULATE CHARGES. Then the follow-up function to create a forwarding settlement document is called. After transferring the settlement document to SAP ERP, billing is triggered from SD.

- **Profitability calculation**
  The customer service agent in the export station can now calculate the profitability. In the forwarding order, he selects CALCULATE CHARGES • CALCULATE PROFITABILITY to collect cost and revenue information relevant for the shipment and summarize it in a profitability figure.

15.3.7 Import Handover and Transactional Processing (Step 16)

Handover to import organization

The import gateway can use a POWL to get information about incoming flights with freight bookings relevant for the gateway ⑯. By viewing the status, the gateway dispatcher can judge whether a flight is already airborne or still on the ground. To take over the process from the export organization, the import gateway dispatcher selects arriving freight bookings from export and creates corresponding import freight bookings with the related import forwarding orders ㉒. As soon as individual handover needs to be triggered, the status of each freight booking can be changed from *Draft* to *In Process* by the import gateway staff. The same activity is done for the import forwarding orders because these also are explicitly set into operation upon cargo arrival in the import country or as soon as transactional handling (e.g., import customs processing) needs to be done.

15.3.8 Import Cargo Handling and On-Carriage (Steps 17–20)

Arrival handling, import planning

The arrival status of the import freight booking is monitored ⑰. Once the arrival and airline pickup date is confirmed, the import gateway arranges the defeeding of the consolidation from the airline to the import gateway by creating a freight order based on the import freight booking to subcontract the move from the ground handling station to the LSP's hub. This is done by selecting CREATE FREIGHT ORDER FOR DELIVERY in the CAPACITY AND CARGO view of the freight booking. The resulting freight order can subsequently be subcontracted to the local defeeding carrier, and execution is tracked via SAP Event Management.

Deconsolidation

Once the cargo arrives at the import gateway, it is checked, counted, and inspected. The gateway team uses the freight booking to do the inbound cargo handling and documentation on receipt status ⑱, deconsolidation, quantities, piece count, damages, and discrepancies ⑲.

Reconsolidation and on-carriage

As soon as import and customs requirements for the shipments are clear, the planning of the import moves can be executed. Some of the shipments may be going into a consolidated transit to the import station to have customs declaration and handling done by the import station. Other cargo is cleared by the gateway and can then go into the on-carriage to the import station. All shipments going to the same station are reconsolidated on the on-carriage truck ⓴. On-carriage planning is done by the import gateway dispatcher using the transportation cockpit. Similar to the pre-carriage planning (see Section 15.3.4), a profile and layout set targeted to gateway station moves based on schedules is used to efficiently plan the on-carriage. The freight orders created from the planning cockpit are again subcontracted to a local truck carrier for long-haul execution. Status of on-carriage–related events and cargo flow can be maintained based on the long-haul freight order either in the EXECUTION tab or via SAP Event Management.

15.3.9 Cargo Delivery and Import Billing (Steps 21–27)

Import cargo receipt and delivery

The cargo arrives at the import station at the end of the on-carriage leg. Cargo receipt is set in the freight order by the station staff. The corresponding cargo items automatically get an assignment of *Arrived* status ㉓. Again, cargo inspection and discrepancy reporting can be done for the freight order ㉔.

Delivery planning is done by the import station dispatcher either in a manual mode or via the transportation cockpit and automatic planning ㉕. Manual mode works like the pickup described in Section 15.3.3, where the dispatcher individually created freight orders for the delivery stage of each import forwarding order. For automatic planning, you have to create planning profiles with selection profiles for the delivery from the import station and capacity profiles for the available delivery trucks. An optimization run then creates consolidated tours to deliver the individual shipments to the consignees. Execution and delivery statuses can be set via the EXECUTION tab and in cargo management of the created freight orders ㉖. Via SAP Event Management, the final proof of delivery that ends the operational process can be confirmed.

Charge calculation and settlement

Financial processing on the import side (27) may continue after execution and includes three activities, whose sequence and timing is quite independent of the operational process:

- **Carrier and truck vendor settlement**
  Settlement with the truck vendors is required for airline defeeding, on-carriage, and delivery. For defeeding and on-carriage, it is in the hands of the import gateway. Delivery settlement has to be done by the import station. Like in export, the responsible employee in the import gateway or station calculates and checks the charges in the freight orders with subsequent freight settlement document creation and SAP ERP transfer for accrual and payment.
- **Internal settlement in import**
  Similar to the export side, the import gateway carries costs for airline defeeding that the station as the final consumer needs to pay.
- **Customer billing**
  All collect charges that are dependent on the incoterm are calculated and settled by the import station. Handling is similar to the customer billing in the export process, but in this case is based on the import forwarding order from where a forwarding settlement document is created. After transferring the settlement document to SAP ERP, billing is again triggered from SD.

15.3.10 Final Process State

The organization and operation of an integrated air freight door-to-door process with several organizational units involved is not trivial. There is a plethora of steps to be done by a variety of people. All activities and documents involved in the process are reflected in the document flow for the forwarding order. SAP TM shows the complete chain of documents created and allows drill-down into all export and import processes. Figure 15.25 shows the final state of the document for a forwarding order after all steps are completed.

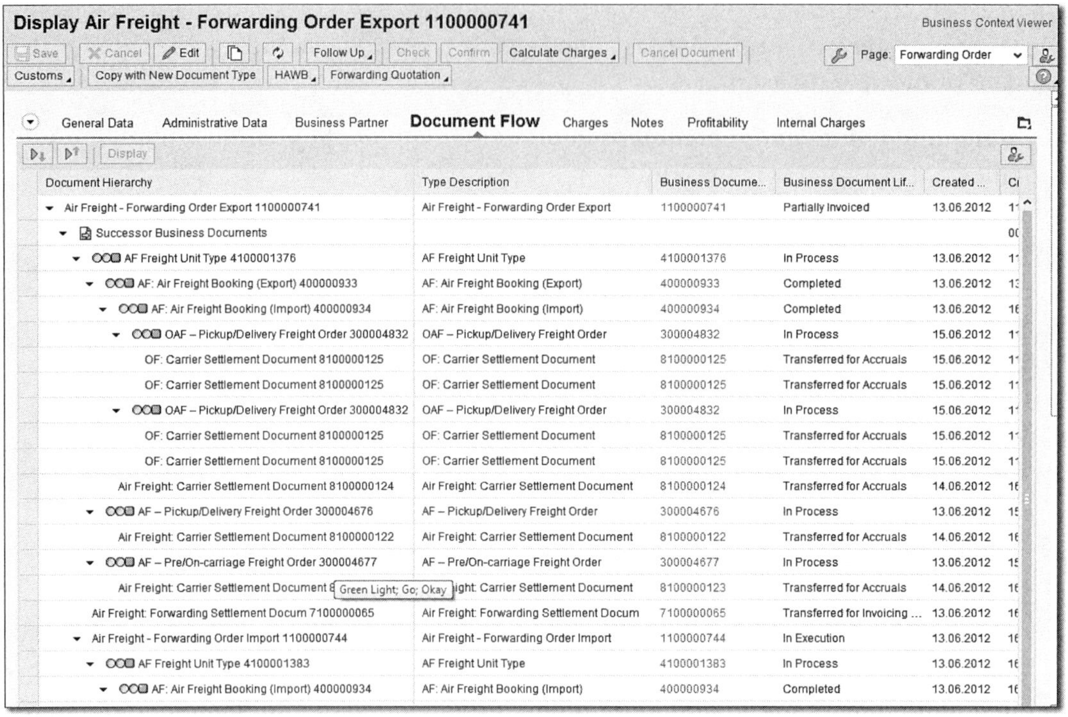

Figure 15.25 Forwarding Order after Air Freight Process Finalization Detailing All Document Dependencies

This chapter explained the setup and execution of an end-to-end air freight process. In the next chapter, we will give tips on what to consider when implementing SAP TM and insight into a specialized industry solution built on top of SAP TM.

For an integrated software suite like SAP TM, it is always helpful to get hints on where to start, what to do next, and what to consider in the overall implementation. Furthermore, this chapter gives insight on process extensions using a liner shipping example.

16 Implementation Best Practices

As you've seen throughout this book, SAP Transportation Management (TM) is a very comprehensive software suite. Its breadth and coverage are comparable with an SAP ERP system because it covers sales, procurement, planning, execution, and costing processes and configurations. From this perspective, it is important to know where to start with implementation, which steps are critical, and where you can find dependencies or characteristics that need to be taken into account to successfully get transportation processes up and running.

Hints on process implementations

In Section 16.1, we give an overview of the challenges involved with implementing SAP TM in large installations. We offer SAP TM configuration hints that fall into the following categories:

- Cross-functional influence of specific settings (i.e., settings that may influence various parts of the SAP TM functionality)
- General Customizing know-how
- Setup sequence
- System improvement dos and don'ts

We also provide an overview of the central control elements of the system to give insight on certain dependencies. Several kinds of pre-configurations are available that can help you in prototyping or scenario implementations.

A second aspect is how to use SAP TM as a foundation for transportation business processes that are not manageable with standard SAP components alone.

As an example, we walk through a container shipping line template that uses the SAP Supply Chain Management (SCM) suite in conjunction with enhancements and add-on modules to provide end-to-end liner coverage.

Liner shipping enhancements

Section 16.2 gives insight into typical transportation functionality required for liner shipping, which is the core business of container shipping lines. Liner shipping has particular challenges because it is a very asset-intensive business where the carrier needs full control of equipment and vessel capacity.

In the quote-to-order process, a liner carrier has to struggle with extra-large requests for quotation (RFQs) with several thousand line items, which have to be answered with a bid within a relatively short time frame.

In order management, where customer booking requests have to be processed, the complexity lies in complex capacity and compliance checks for booking confirmation and receiving of several versions of shipping instructions that are a refinement of the information sent with the booking request.

The customer invoicing is done not on the order, but on a bill of lading level that is created based on confirmed shipping instructions.

Managing liner shipping requires certain software extensions on the basis of SAP TM that have been provided by SAP Custom Development as part of a liner strategy. We give you an overview of the liner-specific processes and how to integrate them with standard SAP TM.

16.1 Best Practices for Implementing SAP TM

Implementation challenges

Organizational setup, process coverage, and the complexity of the process steps and details can make implementing a transportation management system (TMS) quite a challenging process. This is not unusual, and is basically valid for all TMS setups that are used to handle core business of a company.

16.1.1 Large Transportation Management System Installations

First, we want to take a look at typical situations in large or extra-large organizations dealing with transportation management, using the example of a logistics service provider (LSP). We find these challenges at country-wide trucking companies as well as in worldwide freight-forwarding or carrier organizations. It is also sometimes applicable for large manufacturing companies that have a worldwide network, especially if they take care of distribution on their own.

Challenges of large LSPs

First let's pick apart an LSP's company and network setup to better understand its challenges:

- Each LSP might be divided into many global, regional, and local organizations with different responsibilities and thousands of employees interacting in different roles with each other and external parties. Let's look at a fictitious profile of a large, worldwide LSP:
  - 1,500+ stations in 150+ countries
  - 10,000+ employees/users
  - Multilevel organizational hierarchy
  - 10,000+ customers with customer hierarchies (enterprise, companies, then subsidiaries) assigned to one or multiple organizational units
- Its global, regional, or local customers and subcontractors might have individually created contracts with corresponding pricing and service-level agreements (SLAs). It is very important that each of the employees in the logistics chain is, within his responsibility area, aware of the characteristics, commitments, and restrictions of the sold services.
- Support for many modes of transport might be creatively mixed. In regions and countries, the type of vehicles used can vary widely, from a motorboat in Venice, to a 15,000-ton train in the United States, to a motorbike in Africa.

- LSPs need to identify, capture, and respond to many, often very individual customer preferences.
- LSPs must adhere to legal, cargo, and country-related obligations and regulations to avoid penalties.
- It's often important to be able to change the organizational or network structure setup when the business foundation changes, and to introduce new services, products, or geographical regions into the business.

Figure 16.1 shows an example of this type of LSP organization.

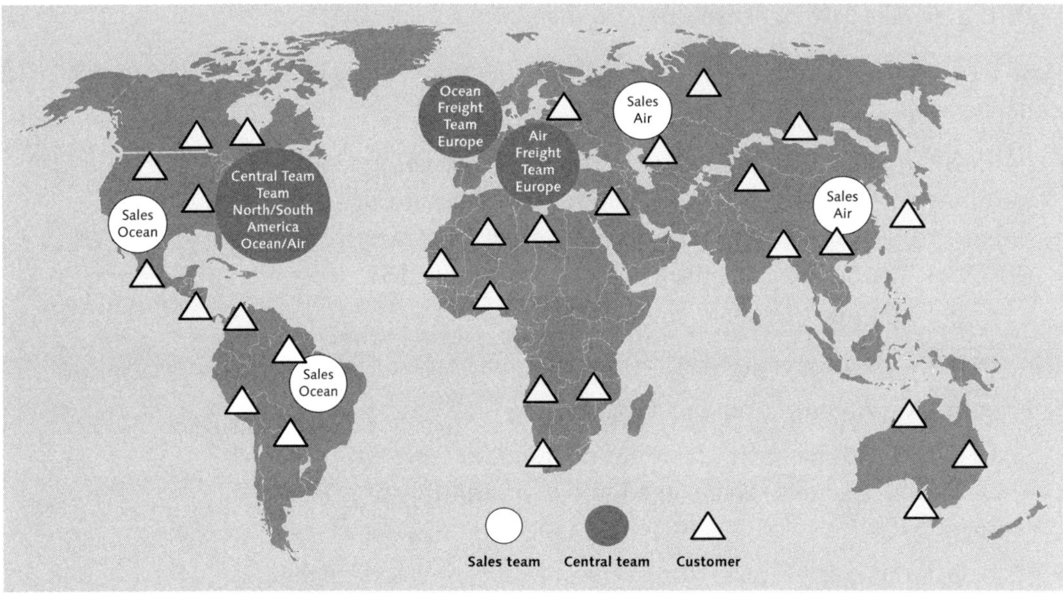

Figure 16.1 Example of a Worldwide LSP Organization and Network

In addition to the distribution of work tasks and work force, individual teams have to deal with many local legal settings in terms of handling cargo, local customs requirements, and local obligations to print or communicate. This needs to be reflected in the authorization and role settings of the users, the menus, and the POWL setup of the individual assigned roles.

Because LSPs are often still organized in a very regional way (where different regions have profit center responsibility, but have to work together on customer orders), there is often the requirement to hide specific data from the eyes of the "competitive" internal organization. This means, for example, that the export organization shouldn't see what price is invoiced by the import organization for a single order. This requires very detailed authorization settings, which often cause difficulties for IT departments in terms of data maintenance. Some LSPs already identified this as a disadvantage and restructured their organization and financial and business processes to simplify the internal collaboration and to achieve increased transparency.

16.1.2 Using Preconfigured Sample Scenarios

SAP delivers descriptions of preconfigured scenarios with the content for the SAP Solution Manager. These scenarios are implemented in terms of providing standardized test cases for typical business scenarios to guarantee the quality of the released software. Customers implementing SAP TM can use these test cases as guidance for their own implementations, as test scenarios for the correct operation of the system environment, or as the basis for in-house employee training.

SAP standard scenarios

Table 16.1 shows scenarios that have been configured and delivered with SAP TM as SAP Solution Manager content.

Scenario Name	Release	Description
Domestic inbound transportation (shipper)	SAP TM 8.0	Inbound shipment processing for purchased goods from domestic suppliers with integrated shipper transportation planning and LSP execution.
International inbound logistics (shipper)	SAP TM 8.0	International, intermodal shipment management for purchased goods from overseas vendors with integrated purchase order and transportation and invoice management.
Outsourced transportation (shipper)	SAP TM 8.0	Domestic outbound transportation management with routing optimization based on transportation requirement mix and vehicle capacity.

Table 16.1 Scenarios Defined and Delivered with SAP TM 9.1

Scenario Name	Release	Description
Domestic outbound freight management (shipper)	SAP TM 8.0	Domestic outbound freight transportation management supporting multiple planning/optimization variants, carrier selection, tendering, dispatching, costing, and invoicing.
Full container load (shipper, forwarder)	SAP TM 8.0	Multimodal full container transport, from forwarding order processing, ocean carrier booking, freight planning, execution and tracking, to order settlement.
LCL ocean freight (forwarder)	SAP TM 8.1	Less than container load (LCL) ocean freight operations of the freight-forwarding company doing door-to-door moves with consolidation from Japan to the United States.
Air freight (gateway consolidation shipment)	SAP TM 9.0	Air freight shipment via a gateway consolidation with capacity allocation in the master airway bill, door-to-door moves, export and import organizational handling and billing.
Rail shipment (carrier)	SAP TM 9.1	Railcar shipments within a North American network handled by railway carriers with switching.
Parcel	SAP TM 9.1	Parcel shipment from a shipper's perspective.

Table 16.1 Scenarios Defined and Delivered with SAP TM 9.1 (Cont.)

Let's shift our attention away from preconfigured support to the implementation support that comes with rapid deployment solutions.

16.1.3 Using Rapid Deployment Solutions

SAP introduced rapid deployment solutions (RDSs) as an integrated delivery approach targeted at defining a concrete implementation scope, containing costs, and streamlining schedules for implementation. Using

RDSs, customers are enabled to quickly and affordably deploy SAP solutions for specific business requirements.

Rapid deployment solutions

An RDS is a package containing Customizing settings, standardized master data, analytics content, and a template set up for a range of configuration settings required to implement a process. In some cases, the RDS can be accompanied by consulting notes that deliver additional functional content. Several available RDSs focus on providing data and process content for transportation management; you can find an overview in the SAP Service Marketplace at *http://service.sap.com/rds*.

Let's look at some now.

SAP Freight Tendering for Shippers

The SAP Freight Tendering RDS supports the communication and decision lifecycle for a shipper running outsourced transportation. It is based on best practices for the trucking industry. Shippers can communicate to carriers during the tendering process with the goal of coming to an agreement with a carrier to execute the transport. Shippers can connect to carriers via a web communication channel or through the SAP Information Interchange On-Demand solution, which offers complete integration to a number of carriers.

SAP Ocean Carrier Booking for Shippers

SAP TM offers a wide variety of ways to model the ocean shipment process. The SAP RDS Ocean Carrier Booking for Shippers supports various booking scenarios, including combinations of different "flavors" of ocean transportation for shippers such as the following:

- Inbound or outbound transportation direction (based on a sales or purchase order)
- Full container load
- Door-to-door transportation, port-to-port transportation including pre-carriage and on-carriage, and port-to-door transportation
- Shipments with or without pre-booking or capacity reservation
- Execution monitoring with SAP Event Management

- Connectivity to ocean carrier directly or via a freight-forwarder via SAP Information Interchange On Demand
- Use of allocations as forecasts and as a basis for carrier selection

SAP TM Transportation Charge Management for Shippers

The SAP TM Transportation Charge Management for Shippers RDS allows companies to easily set up freight charge calculation, cost distribution, and carrier invoicing for their outsourced transportation. The RDS can support end-to-end transportation processes in which freight charges are incorporated into the financial accounting and costing functionality of the SAP ERP application. The complete logistics process from sales order entry to the shipment execution in SAP ERP for truck transports (less than truckload [LTL] and full truck load [FTL]) and sea transports (less than container load [LCL] and full container load [FCL]) is supported, followed by the SAP TM Transportation Charge Management functionality for freight charge calculation in SAP TM.

SAP Transportation Planning for Shippers

This solution is designed for companies that ship goods using their own transportation and network management. It supports order consolidation, routing, scheduling, and carrier selection based on customer requirements and transportation network constraints. The process setup is designed to automate transportation planning operations while taking into account predefined business rules, transportation costs, and user-defined business constraints such as delivery windows, resource capacities, and incompatibilities. By automating the planning process, users can focus on managing exceptions via their own personalized transportation planning cockpit and worklists.

16.1.4 Onboarding Customers in SAP TM Installations

Onboarding

Customer onboarding is the process of setting up the requirements of a new customer in an LSP's existing system installation (e.g., SAP TM). In simple cases, onboarding would just mean a bit of master data maintenance. In more complex situations—for example, a new worldwide customer with custom-tailored logistics service requirements has to be supported—configuration settings for the processes defined for a new customer have to be fitted into the existing system. Setup has to be done in a way that other already running customer processes are not affected and the LSP can manage the new customer with all his requirements. The following list gives you an overview of the various steps you have to consider:

Handling customer onboarding

- In the contracting phase, it is important to properly specify the customer's requirements, including the SLAs.

  Take the following steps to document and close the contract:

  - Document customer-specific requirements
  - Issue contract
  - Approve contract internally
  - Close contract
  - Initiate setup

- During setup, you need to translate requirements into straightforward system settings. This chapter provides valuable hints about what to take into account. Because setup sometimes increases complexity, keep an eye on the impacts of new settings on existing processes (e.g., if you are implementing a completely new air freight service that requires, for example, new means of transport).

  Take the following steps to set up contract-specific settings:

  - Translate contract to system-specific settings
  - Implement settings
  - Implement coding
  - Document settings

- During the test phase, existing processes should be checked for impact and be tested.

  - Test settings with real customer data
  - Test influence of customer settings on other customers or regional or global settings (regression test)
  - Release settings

- In the run phase, run customer orders with the following steps:
  - Receive and process orders
  - Analyze SLAs
  - Execute contract reviews
- Once the customer process is up and running, maintenance has to be done because customer business changes, contract renewals, and requirements adjustments might affect system settings.
- To update settings when required, follow these steps:
  - Update, extend, or close new contracts
  - Identify existing settings
  - Retire unnecessary settings
  - Update or create other settings
  - Retest new and existing scenarios

16.1.5 Using Central Control Elements in SAP TM Setup

As you found out in the previous chapters, SAP TM has quite a lot of settings and configuration capabilities; you can influence these when creating scenarios and processes for transportation management. Most of these settings are straightforward in their setup and have a local or direct influence on the behavior and design of the system (e.g., limited to order management or invoicing). However, some very central control elements need special attention (please note that the list is not complete):

Control elements requiring attention

- Elements that can affect the behavior of multiple process steps (e.g., if you change a setting for a specific process purpose, then that change might unintentionally affect the behavior of quite different process steps)
  - Means of transport
  - Equipment types and groups
  - Freight unit building rules (FUBRs) and freight unit types
- Elements that are critical in terms of well-designed data structuring and long-term maintenance
  - Transportation charge calculation sheets and tariffs
  - Rate tables

- Elements that must be set up in a balanced way
  - Transportation planning cost
  - Forwarding and freight order types
  - Resource types

Figure 16.2 presents an overview of a variety and hierarchy of selected Customizing and configuration settings that control the process. The control elements mentioned are marked in dark with white text.

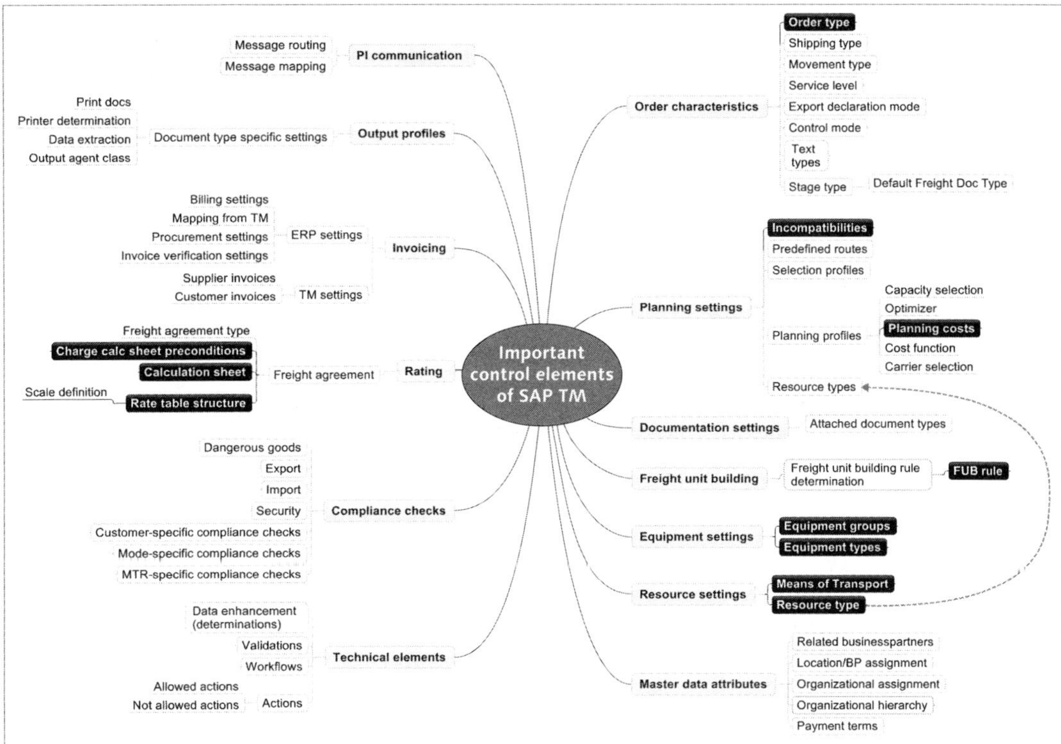

Figure 16.2 Control Elements of SAP TM

Let's consider some of these central control elements further.

Define Means of Transport

As a control element, means of transport is used many ways. Because it is central to transportation planning, cost calculation, and decision making by conditions, it's very important to think hard about the required complexity of means of transport for planning, costing, and conditioning and to select a model that fits best into all three aspects.

The means of transport indicates a categorization of groups of resources that can be coarse or very fine, which makes it a bit tricky. Depending on the regional, country, or organization-specific requirement, you can define the means of transport according to different strategies, for example:

- **Truck, tractor, trailer (to build semi-trailer truck)**
  This is a very coarse definition just to decide between a self-moving vehicle and a passive vehicle. For example, it might be sufficient for planning in *some* overall contexts, but not granular enough for local decisions or costing purposes.
- **Large truck, small truck, tractor, large trailer, small trailer, chassis**
  This is a definition of different self-moving vehicles as well as passive vehicles with different characteristics and loading capabilities, but still valid on a general basis.
- **12-ton truck (Germany), 30-ton trailer (Germany), standard tractor (United States), 80,000-pound trailer (United States), and many more**
  This is the usual case for large LSPs. If the fleet becomes very diverse due to national, regional, local, or subcontractor-specific regulations and capabilities, the definition of means of transport can become manifold, and you may end up with hundreds or thousands of means of transport.

Control means of transport complexity

If the definition of the means of transport is simply for controlling the planning, then the variety would not have a major impact. Otherwise, in many implementations, means of transport definitions are also used as input parameters for calculation of freight costs and prices or for other condition evaluations such as determination of loading durations. In this case, the complexity is linked directly to the number of definitions in planning (e.g., if you define many means of transport because it is so practical for planning, you may create unnecessary complexity in cost calculation).

Structure Equipment Types and Groups

The situation described for means of transport is also valid for equipment types and groups: The more complex the definition of cargo aspects, the larger the impact on pricing definition and condition complexity.

Equipment type and group have an additional challenge. The equipment type is usually directly based on definitions of individual kinds of equipment, such as 20-foot standard container, 20-foot-high cube container, and 20-foot refrigerated container (reefer), which are assigned to the individual equipment master records for each unit (e.g., according to ISO 6346). The grouping can be used to determine which equipment types logically belong together. A grouping defined for cargo handling and movement purposes focused on use of space and compatibility of stowing does not need to be in line with a grouping focused on pricing. Table 16.2 shows some examples of cases in which the grouping criteria may lead to conflicts.

Grouping for Cargo Handling	Grouping for Pricing
Dry containers, high-cube containers, reefer containers, flatbed containers	20-footer dry (standard, ventilated, high-cube), 40-footer dry, and so on
By category (e.g., open railcars, closed railcars, bulk railcars)	By railcar size/length and maximum weight

Table 16.2 Equipment Grouping from Cargo Handling and Pricing Perspective

For equipment grouping, you should therefore also create a well-defined set of equipment types and groups that meet all requirements upfront.

Balance Freight Unit Complexity

Freight units are created using FUBRs. For the creation of a freight unit and the setup of the corresponding rule, you need to find a good balance between the flexibility of freight unit assignment and system performance, and choose the right freight unit granularity.

Best freight unit granularity

If, for example, you want to ship three orders with general cargo (20 forwarding order items with 577 loose packages), you can use freight unit building (FUB) to create various results. Table 16.3 gives an overview of three of these options. If space utilization of vehicles up to the limit is not the first goal, you may want to create fewer freight units because this creates the lowest workload for the employees and the system. You also have the option to split freight unit as required and optimize space utilization selectively.

Strategy of FUB	Result of FUB	Consequence
Freight units per item	20 large freight units	▸ Low flexibility in distributing freight units over trucks and utilizing available smaller remaining capacity ▸ Simple dispatching and cargo management
Freight unit with maximum of 10 packages	58 medium size freight units	▸ Compromise between increased flexibility to use small loading spaces and handling effort in cargo management ▸ No direct control of which package goes into which freight unit
Freight unit per package	577 small size freight units	▸ Best flexibility in using remaining capacity of truck, leading to highest utilization results ▸ High effort in cargo management; many objects on dispatching screens ▸ Higher performance requirements due to many freight units

Table 16.3 Example Results of Freight Unit Building and Consequences

Structure Charging Rules and Rates

Standardization of pricing rules

When working with tariffs, charge calculation sheets, and rate tables, it is advisable to create templates for most of the pricing entities. This allows you to give guidance to the users who have to set up forwarding and freight charge calculation. These templates should align with the company's pricing strategy. A standardized, company-wide strategy simplifies the price definition, negotiation, quoting, and contracting process; it also allows you to do price maintenance (e.g., uplift, rate increase) in a more straightforward way without giving up flexibility of individual pricing.

Standardization means defining a number of rate table structures with sensible scales, a number of charge elements, and a number of template charge calculation sheets based on service products (e.g., air freight transatlantic, ocean freight standard). These are all reused to create customer-specific contracts that then may have individual prices but have a structure that follows the template scheme.

Other Planning Configuration Settings

You may want to think twice about how simple or diverse you set up some other configuration elements. These settings should be well-balanced between easy maintenance, process flexibility, and practical value, and you should answer some typical questions to help you to find the right solution:

Avoid unnecessary complexity

- **Transportation planning costs in optimization**
  What factors drive decisions in transportation planning and optimization? Do you need to include virtual penalty costs for late or non-deliveries to get the best result, or is it sufficient to plan just by cost-driving factors such as miles driven, space utilization, or vehicles used? The answers to these questions drive the complexity of cost models in optimization.
- **Forwarding and freight order types**
  How many order types do you need for your processes? Is an express order a separate order type, or is it simply modeled by characteristics such as the shipment type or a service product? We generally recommend that you reduce the number of order types to those that have a clear-cut focus (e.g., air freight order, ocean freight order, and perhaps a distinction between LCL and FCL).
- **Resource types**
  Do you need separate resource instances for each vehicle that is used to transport cargo? In many cases, it may be sufficient to model resources just as resource pool (e.g., 20 instances of a generic 40-ton truck resource instead of each truck resource as separate instance).

Distinct resources are required if you also need to maintain the resources. Mixed models are often suitable; with these, you might have instance-based models in your core regions or core business and pool-based models for regions where you do not have direct control.

Structure and Name Your Data Well

Make settings identifiable

Before starting configuration, think about naming schemes for your configuration data. Once the employees are used to these schemes, they can more easily find their way around a complex implementation project. Naming schemes like the following should be applied for all settings:

- **Naming of forwarding order types**
  You can name the forwarding order type in a way to indicate mode of transport and service (e.g., "AFST" for air freight standard and "OFLC" for ocean freight LCL).
- **Rate tables**
  You can include the scope, service, and customer significance in the naming. BSF_ASUS_C1234 could be a rate table applicable for basic sea freight (BSF) on a route from Asia to the United States (ASUS) with specific prices for customer 1234.

If you run multiple scenarios in a single client, it's helpful to segregate the settings by selecting consistent identification names, such as starting all air freight settings and purely air freight-related master data with "AF" and starting those for ocean freight with "OF." This can be beneficial in setting up consolidation settings like planning profiles, where you need to select other entities by referencing. By consistently segregating the name spaces for unit-specific settings, you can reduce the maintenance effort considerably.

16.1.6 Optimizing the Sequence of SAP TM Configuration

You already understand the importance of having well-structured configuration data before going into a larger process implementation project. This is especially important if the configuration is done in a system where processes are already running and may be affected.

Basic setup of SAP TM

The following system setup sequence should give you a rough guideline (though you should note that it's not a complete list of steps):

- **Set up basic systems settings.**
  If you start in a new system, there are several settings to be made as prerequisites.
  - Set up logical system connections (SAP NetWeaver)
  - Create an active version and model for SAP SCM master data; otherwise, no master data can be maintained (SAP TM)
  - Set up geo-location services (SAP NetWeaver)
  - Set up archiving, attachments, and PPF basics (cross-application)
- **Set up basic SAP TM settings.**
  Basic SAP TM settings are all settings that should be done in the beginning because they can be referenced in a later phase of configuration.
  - Set up code lists (e.g., air craft type codes)
  - Set up fundamental types and categories (e.g., means of transport types, forwarding agreement item types, charge element types)

SAP TM business process setup

- **Set up business object configuration.**
  Business object configuration includes setting up the behavior of different object types used. If you use the order integration with SAP ERP, the general layout of the process should have already been done before starting SAP TM configuration (e.g., SAP ERP sales order types and their use in SAP TM).
  - Define customer order types and order types used for SAP ERP integration (e.g., forwarding order types, forwarding quotation types, OTR and DTR types)
  - Define planning and capacity object types (e.g., freight unit types, freight booking types)
  - Define subcontracting order types (e.g., freight order types)
  - Define agreement types (e.g., freight agreement types, forwarding agreement types)
- **Set up processes and their configuration.**
  Process configuration defines the behavior of the system when it works with data and functions that spans multiple business objects

and creates new information or updates existing information. Because some of the setup required may be heavily dependent on master data, you may need to make some refinements later or define some master data before.

- Set up charge calculation rules (e.g., forwarding agreement determination, charge calculation sheet structures that reuse the charge element types, rate tables)
- Set up planning rules (e.g., FUBRs, planning profiles, incompatibilities)

Master data and integration

- **Define master data.**
  Master data can be set up in parallel with the process configuration.
  - Business partner master data
  - Organizational model
  - Location and network master data (e.g., locations, schedules, resources)
  - Rate and tariff data
- **Set up integration with other systems such as SAP ERP, SAP Event Management, and SAP GTS.**
  - Set up system integration (RFC, CIF, enterprise services, etc.)
  - Transmit master data
- **Set up ancillary services.**
  Ancillary services can be set up relatively late in the implementation process.
  - Set up output or printing rules and printing forms
  - Integrate with analytics system
  - Configure archiving integration and document management within the process (archiving itself has already been set up)

User interface setup

- **Set up UI configuration.**
  User interface setup is done in SAP NetWeaver Business Client.
  - Set up POWL layouts based on the standard layout so that they meet their purpose and show valuable information
  - Set up transportation cockpit layouts
  - Adjust screen layouts with the Floorplan Manager

- Set up menu structure and roles
- Give menu topic names and adjust to country-, industry-, or customer-specific naming

It's generally wise to run small tests during the configuration so you can estimate how far there are design flaws in the whole setup. In some cases you have to go back to a previous step and complete or adjust the configuration (e.g., add a default output profile to an existing forwarding order type).

16.1.7 Understanding SAP TM's Strengths and Limitations

In this section we offer some additional hints that can be helpful to understanding both capabilities and current limitations of SAP TM and its ecosystem.

Limit the Network and Plan What Is Necessary

Because SAP TM has many means of defining a transportation network and using it for planning and routing, you might be tempted to set it up in a very detailed way—perhaps to define customer locations down to the dock door as part of the transportation network.

Instead, we recommend that you keep the network as simple as possible so that the planning performance is good and the definition of planning rules does not become overly complicated.

Exercise Caution when Using SAP TM Web Dynpro Screens on the Web

You can use SAP TM's Web Dynpro screens on the Internet. Some screens are prepared for this (e.g., the tendering portal where a carrier can place bids on tenders), but many of the SAP TM screens are not to be exposed to external access in this way. Of course, intracompany use on the Web via a virtual private network is no problem, but do not try to expose a forwarding order entry screen as a customer portal screen for online ordering because the data security is not high enough for external use. For example, by pressing F4 while the cursor is in the business partner field, an external user may see a large list of customers that he is not authorized to see.

If SAP TM screens are reused for external web usage, you should create a new screen (also as a copy of an existing one) and adjust the layout, security, and authorization so that it meets the required standard.

Understand Configuration versus Personalization

Many tasks that are done during SAP TM setup are classified as configuration; they control the system for all users with the appropriate authorization (the other users simply cannot use a configured feature). In UI setup, some tasks can be classified as *personalization*, which means that the setup is made only for the user who is currently doing it. This is applicable to all non-administrative activities in the UI adjustment (such as hiding fields) or to POWL setup, where you change the column layout and sequence or selection criteria or the queries. If you need to provide changes to a larger group of users, do it with the administrative tools.

Create New POWL Applications

POWL applications cannot be reused directly and adjusted to a different context. If, for example, you want to create a menu structure with different branches for order management (such as industrial customer orders and end customer orders) and reuse the forwarding order POWL in both menu branches, then adjustment of the POWL in one branch will automatically lead to the same changes in the other branch. If you need to reuse a POWL, you therefore have to do the following to create a new POWL application (do this in the SAP GUI, not SAP NWBC):

1. Call Transaction FPB_MAINTAIN_HIER. Create a new POWL application as a copy of the one you want to reuse (e.g., the order management POWL SCMTMS_POWL_OM).
2. Go to Transaction POWL_TYPER and set the role assignment for the new POWL application.
3. Use Transaction POWL_QUERYR to define the role assignment details.
4. Finally, run the report POWL_D01 using Transaction SE38 to reset the POWL caches.

Use Cross-System Processes in Marketing, Sales, and Execution

The process integration between SAP CRM and SAP TM is based on the objective of using as much SAP TM master data as possible in SAP CRM sales processes. Therefore, the main sales objects such as quotes, contracts, and orders are not used as SAP CRM objects but are used as their corresponding SAP TM instances. This has the advantage of using a single source of information for all pricing and rule management activities in sales. Marketing-related documents and early sales document (e.g., opportunities) are used as CRM objects. The handover to SAP TM is done from opportunities.

16.1.8 Enhancing SAP TM via Custom Development

Even though SAP TM is a very powerful tool, it is not always complete in terms of covering everything customers envision or require. Therefore, the SAP Custom Development organization provides solutions for single customers or small groups of customers requiring deeper functionality in specific areas. Several *custom development projects* (CDP) have been created for SAP TM, including the following:

- *Business rules framework* can be used to easily set up validation rules in order processing that gives users hints on how to check or complete an order.
- The *product configurator* allows the use of the Internet pricing and configuration (IPC) capabilities in SAP TM contract creation and order management.
- *Resource planning Gantt-chart* is a scheduling board for vehicle and driver resources in the trucking business. It comes with a very practical "yard in-out view," allowing you to clearly see when certain resources are expected to come into a yard and which shipments still need assignment of resources to go out on time.
- *Incident Management Integration* between SAP Event Management, SAP CRM, and SAP TM allows the handling of incidents by customer service in a very structured and auditable way.

Several other CDPs have been developed. By decision of SAP, it may be possible that some of the CDPs will be retrofitted into a future SAP TM release.

In this section, we looked at best practices for setting up an SAP TM environment. Let's shift gears and take a look at best practices for enhancing SAP TM for a specific industry using the example of a liner shipping solution based on SAP TM.

16.2 Liner Shipping with SAP TM

Liner solution for SAP TM

SAP TM has been growing from a pure shipper/consignee solution in the very early releases to a comprehensive logistics platform. Today, SAP TM supports shippers, LSPs, and, with standard SAP TM 9.1, also carriers (mainly rail and truck) and parcel businesses.

But it is important to mention that with the given architecture and the configurable design of SAP TM, other logistics business can be supported if industry-specific enhancements are utilized. This flexible architecture of SAP TM is used by SAP Custom Development and SAP partners to, for example, enable an SAP TM-based solution with add-ons for commercial and operational container shipping line processes. We will refer to it as "the liner solution"; it is shown in Figure 16.3. Neither the add-ons nor the liner solution as a whole are part of SAP's standard software delivery.

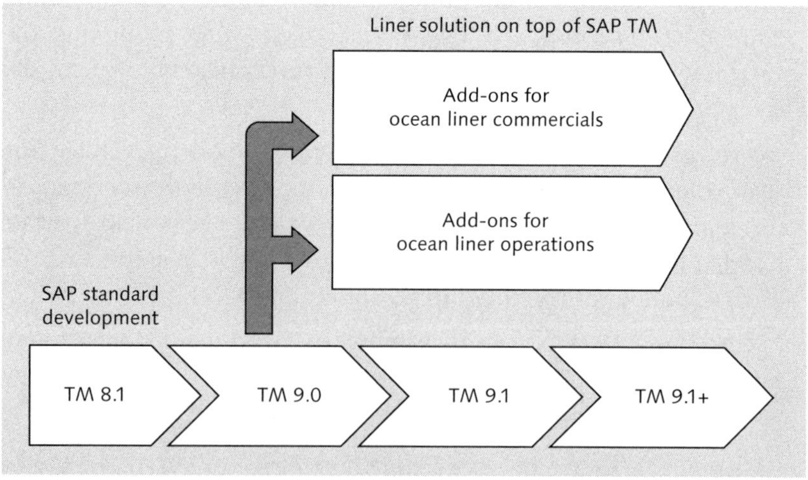

Figure 16.3 Exemplary Componentization for a Liner Solution on Top of SAP TM

The liner add-ons use SAP TM as the central platform and functional framework. The core functional areas of SAP TM are both enhanced and integrated with dedicated liner-specific add-ons. Future releases of SAP TM by SAP Custom Development will continue to provide new features and innovations that may be embedded into the functional scope of the liner solution.

Scope and Design of the Liner Solution

It is important to emphasize that the detailed scope in certain areas of the liner solution is still in the design phase. Dedicated solution extensions are being developed for several SAP customers. The functionality and processes described in this chapter are a common denominator of main processes of leading ocean carriers.

We describe the generic functional components to cover the core business requirements of an ocean liner. As a result, this chapter represents an overview of the functional content of a liner solution, which can be individually adjusted to customer requirements. Depending on the final scope of a customer installation, the liner add-ons are individually scoped, contracted, and priced. They are *not* part of standard SAP TM software delivery.

Extensions for other industry verticals (e.g., accessorial management for railways) are provided in a similar way: enhancement of SAP TM functionality by dedicated add-ons delivered by SAP Custom Development.

In addition to SAP TM as the foundation, other applications are used, including SAP Billing and Revenue Management (BRIM), SAP Event Management, SAP CRM, and SAP Global Trade Services (GTS). So in this section, we give you an overview on architecture and functional aspects of a liner solution on top of SAP TM. We look into various liner-specific enhancements and add-ons and provide an overview of a general liner process that is going to be supported.

16.2.1 Architecture and Functional Coverage of the Liner Solution

So let's take a look at the overall process coverage of the liner solution. The liner solution supports core processes such as customer, proposal, and pricing management, a comprehensive contract and rating repository, booking and documentation handling, route planning, capacity management, cargo execution, and customer invoicing. In Figure 16.4 you can see the different process areas that are part of the upcoming liner solution.

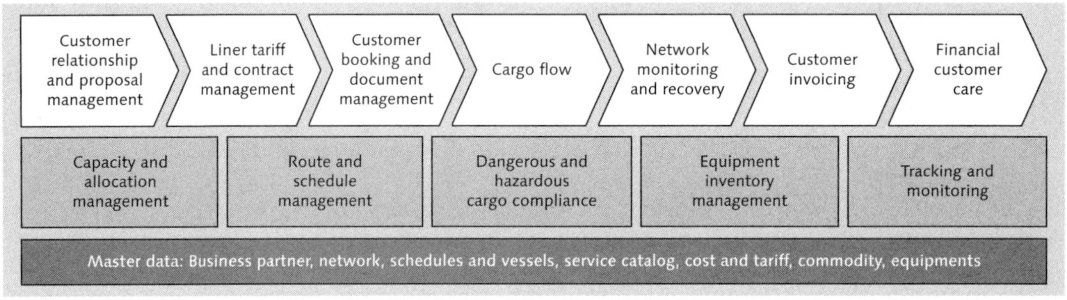

Figure 16.4 Process Coverage of Liner Solution

Functional Coverage

Functional overview

Figure 16.5 gives insight into the functional areas which are handled in the liner solution. The central platform is based on standard SAP applications and capabilities, such as SAP TM Transportation Charge Management, SAP Event Management, and SAP BRIM Convergent Invoicing. These are displayed in the middle of the figure. In addition, the figure shows the specific liner extensions that are to be deployed.

There are two kinds of extensions in Figure 16.5. The functional enhancements perched above and below tailor existing capabilities to the liner industry by providing specific user interfaces, additional fields, logic, and features. The functional add-ons enable entirely new capabilities and are optional.

RFP and pricing management

As an example, let's take a look at the Quotation Manager, which offers a central cockpit to analyze, price, and respond to customer requests for a proposal.

The first functional area we want to highlight is the management of requests for proposals (RFPs) and pricing. In the ocean liner business, it is very common—and in some countries legally required—to publish the ocean liner rate tariffs. Such tariff rates are normally the high end of a price scale and are rarely booked by customers. As a result, it is a business practice to offer more individual rates, perhaps dedicated to a specific industry segment or market. Such segmented tariffs are common, especially if customers are reluctant to agree to long-term contracts.

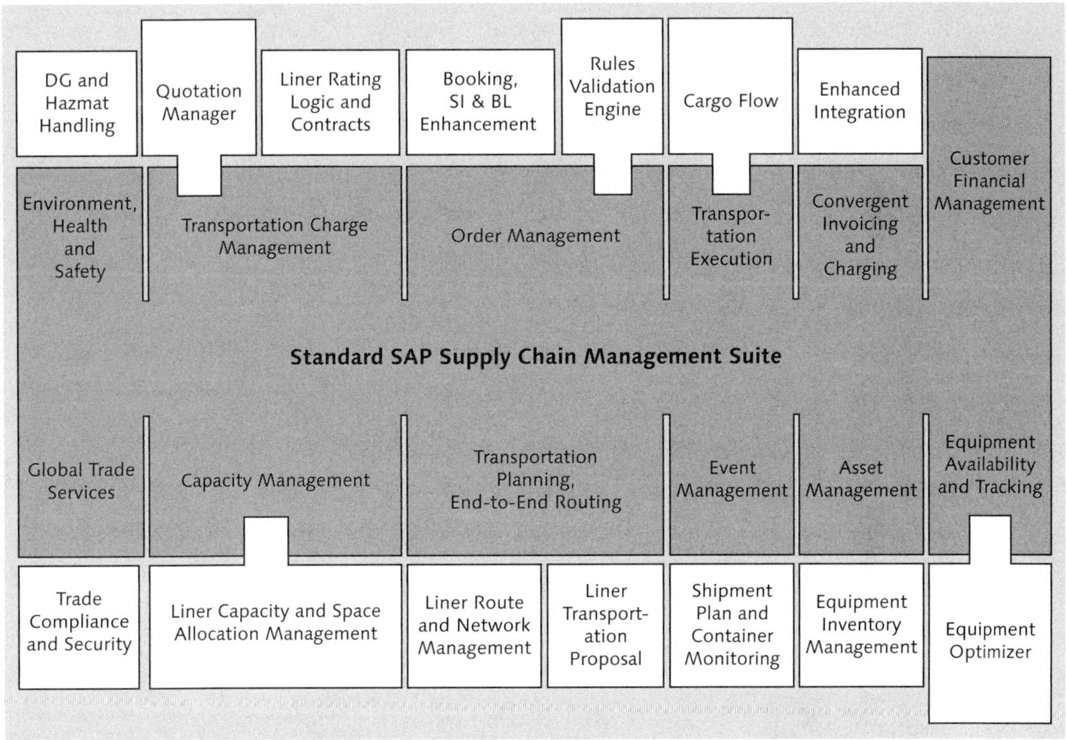

Figure 16.5 Functional Overview of Liner Solution on Top of SAP TM

The most specific form of committing to rates is to offer unique prices to a customer either on an ad hoc basis or, alternatively, in the form of yearly contracts. The risk for an ocean carrier of agreeing on long-term contracts is that it can't flexibly react to the volatile market conditions in demand and supply of container space. In times of over-capacity, the liner will be forced to grant cheaper spot-rates to get their vessels utilized. In times of space scarcity, customers must pay more expensive rates to buy space on vessels, with or without a contract, which is a disadvantage.

One state-of-the-art way of dealing with this issue is to agree on index-linked container contracts. Such agreements adjust the rates throughout the duration of a contract based on an external index, which is influenced by the demand/supply situation and other external factors. An ocean liner receives hundreds and even thousands of RFPs from different shippers and freight-forwarders every year, asking for discounted rates on specific services. A single RFP for a yearly contract can become highly complex and can contain up to ten or even hundreds of thousands of port pairs, equipment types, and commodities. Such RFPs often include volume commitments that result in space allocations, specific requirements in rating structures and exclusivity rules, complex contract clauses, and additional service requirements, such as extra free days and accessorials. The RFP and pricing management supports the handling of large-volume requests and simple spot quotations. Return to Section 16.2.2 for more details on the Quotation Manager add-on.

Customer contracts and rating segmentation

After a successful proposal process, the individual customer contracts or segmented liner tariffs need to be managed and executed in operations to ensure a consistent rating of bookings and bills of lading, adherence to space allocations and contract terms, and efficient service delivery. Recall from Chapter 10 that SAP TM has strong capabilities in contract and rating management. Consequently, the SAP TM Transportation Charge Management component is used to manage both customer contracts and a liner tariff. An enhanced charge calculation enables a liner-specific auto-rating in bookings and bills of ladings. A service product catalog is used in the RFP process and in contracts to standardize service offerings and enable cross-selling opportunities for specific services offered on top of basic ocean freight shipments (e.g., expedited services, fumigation, or specific cold-chain services). Allocations are handled as part of the capacity management.

Network and schedule management

The core service of any liner is to provide ocean transportation services on different trade lanes. Based on different loops/vessel services, with deviating port sequences and schedules, the offering of voyages and transit times can be quite heterogeneous between liners. Because with the liner solution you have the capability to perform end-to-end multimodal routing for bookings, you need to maintain your transportation network with locations, such as container yards, ports, and terminals.

You need to define vessel services, loops, a global schedule master, and regular voyages for the liner routing engine. A strategic or tactical network optimization is currently not part of the liner solution.

Capacity and allocation management

Every ocean liner faces capacity limitations based on schedules, vessels, and voyages. Consequently, it is an absolute core business requirement to manage available capacities in the most profitable way, prioritizing the most profitable bookings. This needs to be in line with capacity allocations for customers and agents. The liner solution offers capabilities to manage capacities and allocations (refer to Section 16.2.2 if you want to learn more about the liner capacity management):

- Vessel capacity master data
- Assignment of vessel-specific capacity profiles to voyages
- Generation of port allocations for equipment groups and services
- Generation of capacity allocations for customer commitments, alliance partners, and sales agents or for empty containers
- Automated consumption of voyage- and port-specific capacities based on space allocations; utilization of space allocations by incoming bookings
- Voyage capacity management cockpit with live allocation consumption

Customer booking and route planning

SAP TM Forwarding Order Management is used for the creation of customer bookings. Liner-specific user interfaces and enhancements allow an efficient booking process. To ensure compliance with numerous business rules, prohibitions, and restrictions, you can use a specific add-on to do any booking validations. The Rules Validation Engine is based on the generic BRF+ repository and brings business rules into a liner or logistics context. It enables validations and defaulting of information in bookings. With established contracts, standard tariffs, and segmented tariff rates, a fully automated rating is supported based on SAP TM Transportation Charge Management with liner enhancements.

The liner solution offers end-to-end routing capability for bookings via different enhancements in the area of transportation planning. A key enhancement is the liner transportation proposal based on the existing SAP TM transportation proposal. It concentrates key information for a user to make a routing decision, such as possible voyages, and existing rates, as provided in the TM 9.1 standard release and enrich this data with available capacities and the yield for each option.

Note

Refer to Section 16.2.2 if you want to learn more about the liner transportation proposal. Other core liner requirements in customer order management are covered by shipping instructions, a manifest plan, and the bill of lading object.

Tracking and monitoring cargo flow

Once a booking has been accepted, a manifest plan helps you to manage the complex documentation process for a booking. After you assign a booking to a voyage and the booking is confirmed, the shipment needs to be tracked. You can generate a shipment plan and report any incoming events based on SAP Event Management for each booking, bill of lading, and container. This enables end-to-end visibility of both the cargo and equipment flow. In addition, you can use delivery orders based on SAP TM freight orders to steer the release of cargo to the customer and manage inland cargo movements.

Liner-specific master data

Enabling new functionality requires additional master data. The liner solution uses several existing standard master data entities, such as schedules, locations, and the service product catalog. The existing functionality is enhanced in certain areas. Additional master data is enabled, such as a vessel master, capacity profiles, and an enhanced commodity hierarchy. Another core master data element is cost information. The liner solution can store various cost components and use them to manage yield. The standard cost information needs to be maintained manually to represent route-dependent, equipment-related, cargo-specific, and service-dependent costs.

Network monitoring and recovery

The actual execution of the cargo transportation for an ocean liner is significantly impacted by external events such as weather conditions on the oceans, port conditions/capacity bottlenecks, damage and defects on vessels or containers, and other influences. As a result, it is important to react to any issues along the transportation chain immediately and efficiently. The voyage recovery cockpit of the liner solution provides visibility and allows a quick reaction to incidents. A user is informed of any anomalies in the execution and can replan the execution of impacted bookings. Figure 16.6 shows the design of the voyage recovery cockpit.

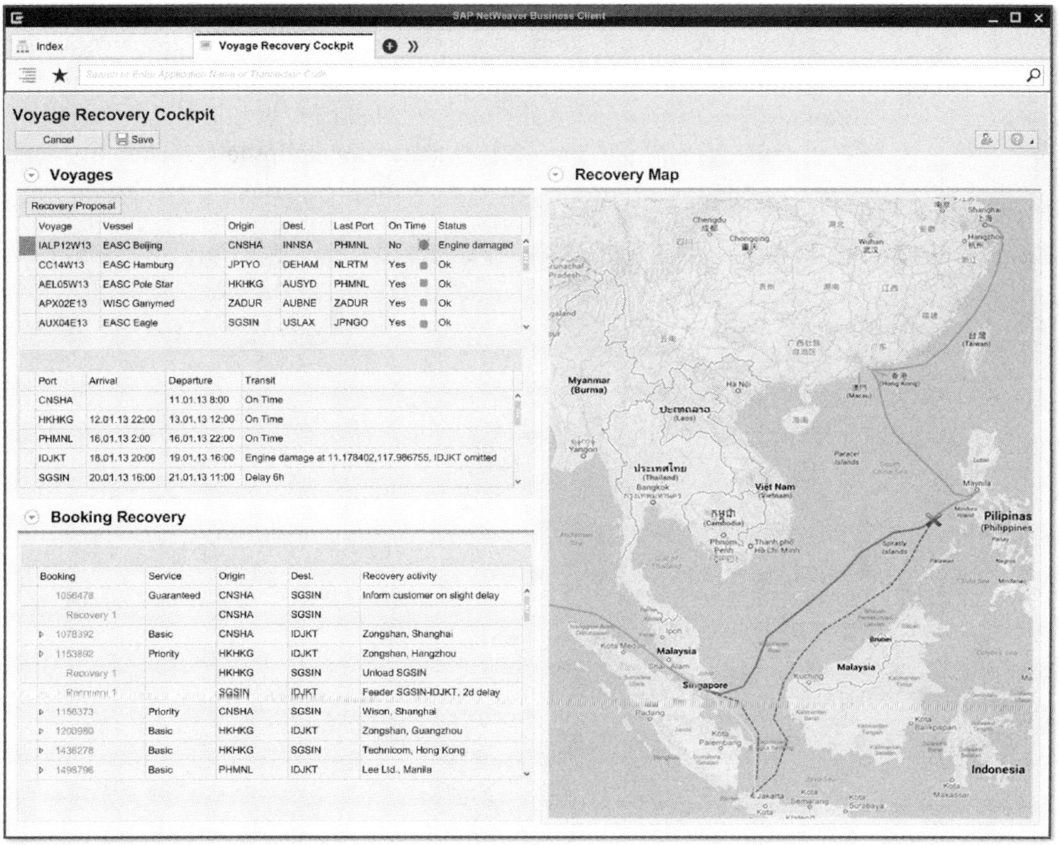

Figure 16.6 Voyage Recovery Cockpit

The liner shipping industry faces geographically diversified demands for transportation. The strong production and manufacturing sector in Asia and the availability of natural resources and agricultural products leads to a tremendous demand for exporting cargo from Asia. Trade lanes originating in Asia and exporting goods to Europe or the United States generally face much higher trade volumes than the opposite directions.

Equipment inventory and optimization

This leads to an over-capacity of empty containers in the western hemisphere and consequently a shortage of empty containers in Asia. Freight rates are cheaper to Asia because voyages are not fully utilized, and empty containers need to be provided to Asian countries.

The liner solution addresses equipment inventory management and optimization capability by using existing standard functionalities, such as SAP Event Management and a new add-on. Inventory management is utilized for equipment availability check during booking, and short-term equipment planning. Figure 16.7 shows an equipment availability check with a regional equipment overview during a booking process.

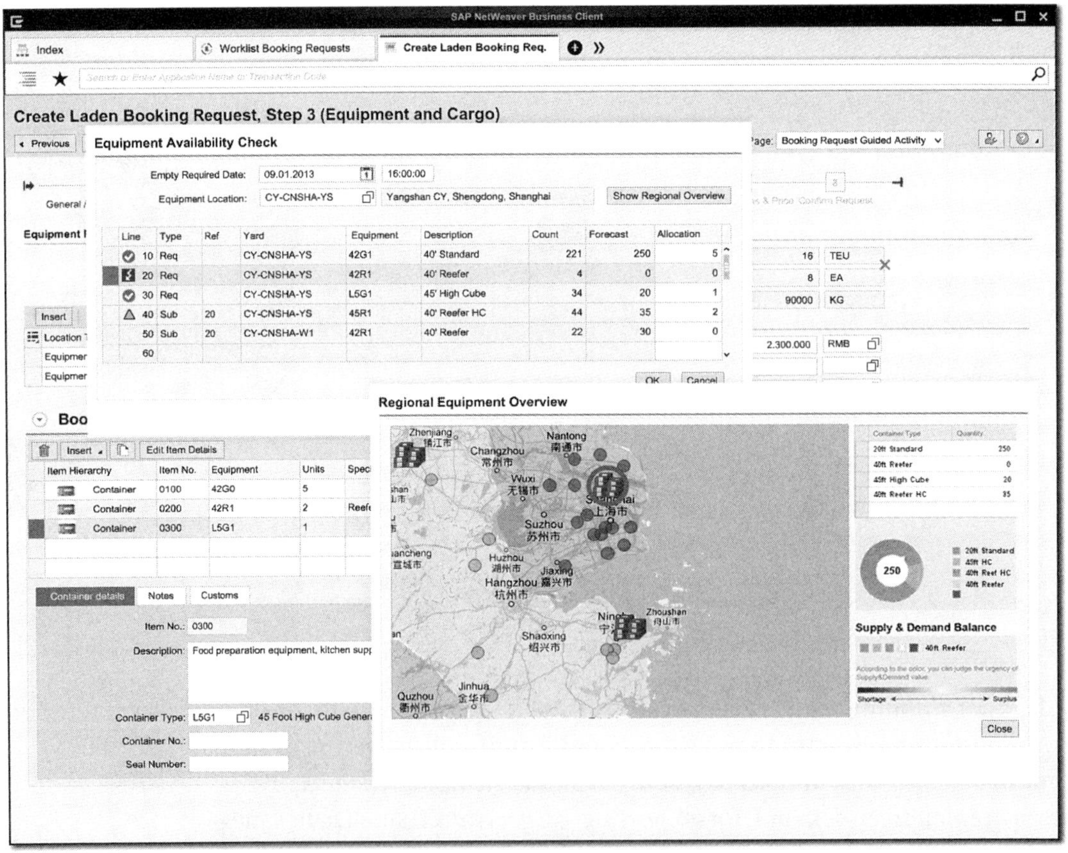

Figure 16.7 Equipment Availability and Regional Overview during Booking

Customer invoicing and financials

Customers need to be billed for services provided. The individual billing requirements of specific customers can be demanding and challenging to implement. Billing items need to be generated flexibly enough to allow customer-dependent billing rules. Like any other industrial business, a liner needs to ensure the collection of payments and efficient dispute handling for any issues to avoid delays or omission of payments. To prevent such issues, efficient credit management is required to grant credit terms. With the amounts and volumes shipped by large companies, this can have a huge impact on the cash flow of an ocean liner. SAP BRIM and the customer financial care applications support this functionality and are fully embedded as part of the liner solution (see Chapter 11, Section 11.3).

Application Architecture

SAP SCM suite

The applications of the SAP SCM suite are used to enable the functional areas we have described before. Figure 16.8 shows the SAP components used as the foundation of the liner solution.

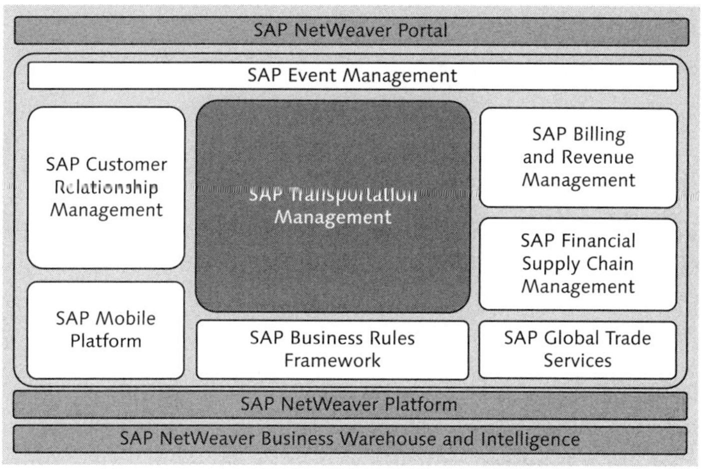

Figure 16.8 SAP Components Used for the Liner Solution

- **SAP Customer Relationship Management**
  SAP CRM is mainly used for the lead to opportunity process, supporting sales and account management and campaign management before customer RFPs are received and processed.

- **SAP Transportation Management**
  SAP TM is the central application used for the core processes of RFP management, liner tariff, segmentation and contract management, booking and documentation handling, capacity management, transportation planning and routing, cargo execution, equipment management, and yield optimization.
- **Business Rules Framework Plus**
  BRF+ the central repository for any specific business rules, including prohibitions and restrictions, pricing and ocean tariff rules, and validations and check rules.
- **SAP Billing and Revenue Management**
  The SAP BRIM solution is integrated with SAP TM mainly for the flexible billing capabilities of SAP Convergent Invoicing (CI). This allows a very sophisticated invoicing design. In addition, dispute and collections capabilities are available.
- **SAP Financial Supply Chain Management**
  To manage and track credit limits with customers, the SAP FSCM Credit application is integrated with the booking management.
- **SAP Mobile Platform**
  Mobile applications for a liner can include sales support via customer fact sheets, event tracking and tracing, and even quotation management.
- **SAP Global Trade Services**
  As the key application for export-import customs management, SAP GTS is used for document handling and interactions with authorities.
- **SAP Portal**
  Use SAP Portal to allow registered customers and users to submit e-bookings, enquiries for rates, and requests for quotations.
- **SAP Event Management**
  Use the integration between SAP TM and SAP Event Management to leverage all requirements of tracking and tracing. This enables a comprehensive container tracking solution and integration with geographical information systems (GISs). Generate and manage shipment

plans and manifest documentation plans based on routing results for end-to-end visibility.

- **SAP NetWeaver Business Warehouse and SAP BusinessObjects BI**
  You can use SAP NetWeaver BW to store all your historical transactional data. With SAP BusinessObjects BI you can run retrospective reports, for example, on revenues or volumes. With SAP TM 9.1 it is also possible to forecast future demands and volumes, as you can read about in Chapter 10, Section 10.4.
- **SAP NetWeaver platform**
  The technical foundation of the entire application architecture is the SAP NetWeaver platform. SAP NetWeaver brings SAP NetWeaver PI to enable interfaces, standard output management, and the core technical framework, standardized for any SAP application.

16.2.2 Specific Liner Enhancements and Add-ons

Having examined the general scope and design of the liner solution on top of SAP TM, let's consider four specific enhancements. All four areas are essential in an ocean liner business and have high capabilities to enable cost savings, improvement of revenue, efficiency, and visibility in the liner solution.

Quotation Manager

First, we want to highlight the Quotation Manager. We have looked at the business background; we now look more from a process perspective and highlight steps to be taken in an add-on. Any LSP, rail carrier, trucking company or ocean carrier that is selling transportation services can use the Quotation Manager to process and handle large RFPs from customers. It offers a comprehensive bid management tool with various functionalities.

> **Note**
>
> One missing puzzle piece is the actual data upload and download of customer RFP files, such as from Microsoft Excel. This capability is currently not planned for the Quotation Manager but can be enabled based on an implementation project (e.g., with tools like Opentext Capture Center and other partner solutions).

The following are supported functionalities:

1. **Customer interaction**
   The Quotation Manager contains a request tool that allows a customer to submit both ad hoc and long-term contract RFPs. It contains enquiry functionality for existing rates in standard tariffs, agreements and contracts, which can be enabled as a customer self-service.
2. **RFP scope management**
   You can analyze requested bid items from small to highly complex RFPs regarding the availability of services, routes, and business rules and validate them against prohibitions and restrictions.
3. **Pricing**
   It is possible to define complex rating structures with included and exempted surcharges. You can use a comprehensive toolset of pricing procedures and analysis functionalities for the core pricing. A live visibility of yield and profit margins are provided based on pricing decisions. Last, you can use the automated pricing capability for mass pricing updates in RFPs. A detailed pricing logic and rating rules can be developed by SAP Custom Development in a customer-specific project only.
4. **Proposal and contract generation**
   After successful RFP processing, you can automatically generate proposal documents and quotations to your customers. Based on the acceptance by the customer you can assign an agreement to a predefined rating segment, together with numerous other customers or generate a unique contract. The created rates are consistently used in bookings and bills of lading, as well as in future proposal processes for an automated charge calculation.

Figure 16.9 shows the main pricing table of the Quotation Manager, which is where you execute some core functionalities. You also have live visibility of the yield, based on changes in your pricing.

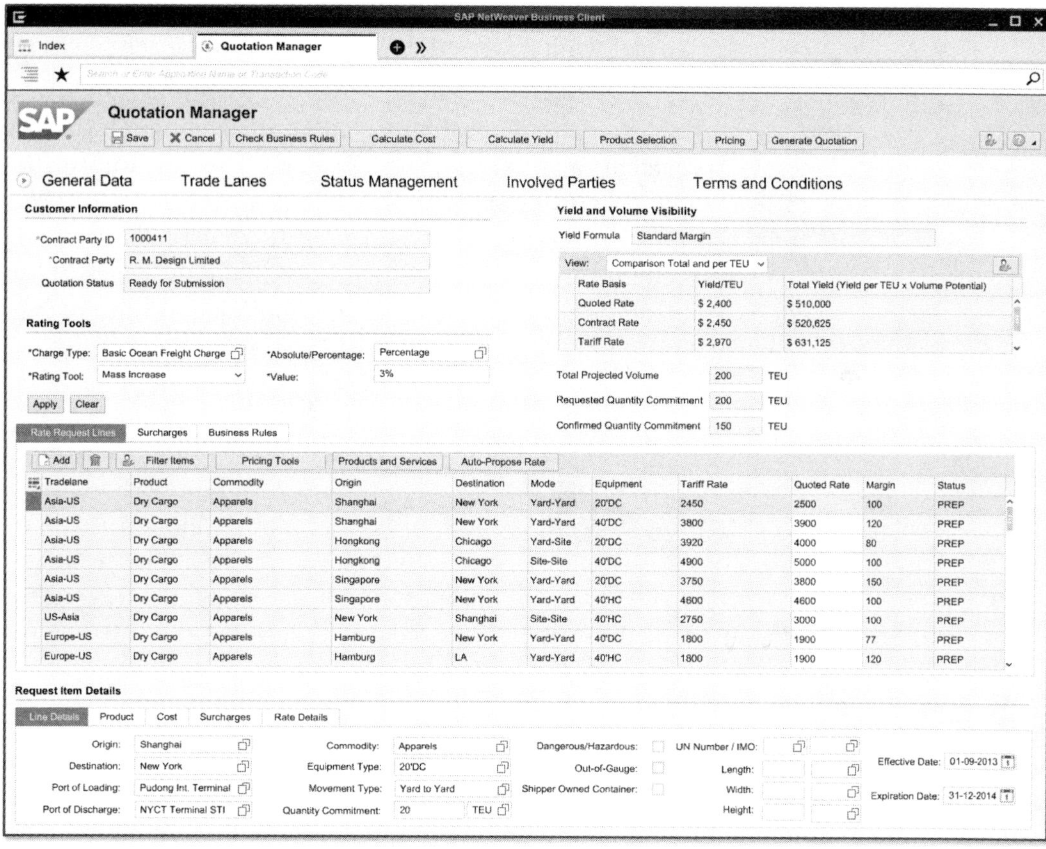

Figure 16.9 Liner Solution Quotation Manager

The liner transportation proposal is the heart of the order management. It combines key information to give a user comprehensive insight regarding the feasibility of a booking request, any capacity constraints, and the yield per proposed voyage, even before accepting a booking. The proposal provides the following information and capabilities:

Liner transportation proposal

- The booking contains the transportation demand concerning container count and equipment types, as well as the requested port pairs.
- The end-to-end routing engine suggests all voyages, combined with multi-modal inland legs, and proposes how a booking can be executed.

- The capacity and allocation management allows you to enrich each proposed voyage with allocated, booked, and free capacity between each origin-destination port combination.
- A rate for the booking request is available from the liner tariff, customer agreement, and contract management. The cost information for each voyage is stored in cost tables, such as the standard ocean freight cost, repositioning cost, and inland cost. Yield can be calculated and displayed for each option. Another capability of the proposal is to check equipment availability based on the equipment inventory management.

This comprehensive cockpit is used for each laden booking. Figure 16.10 shows the layout of the liner transportation proposal.

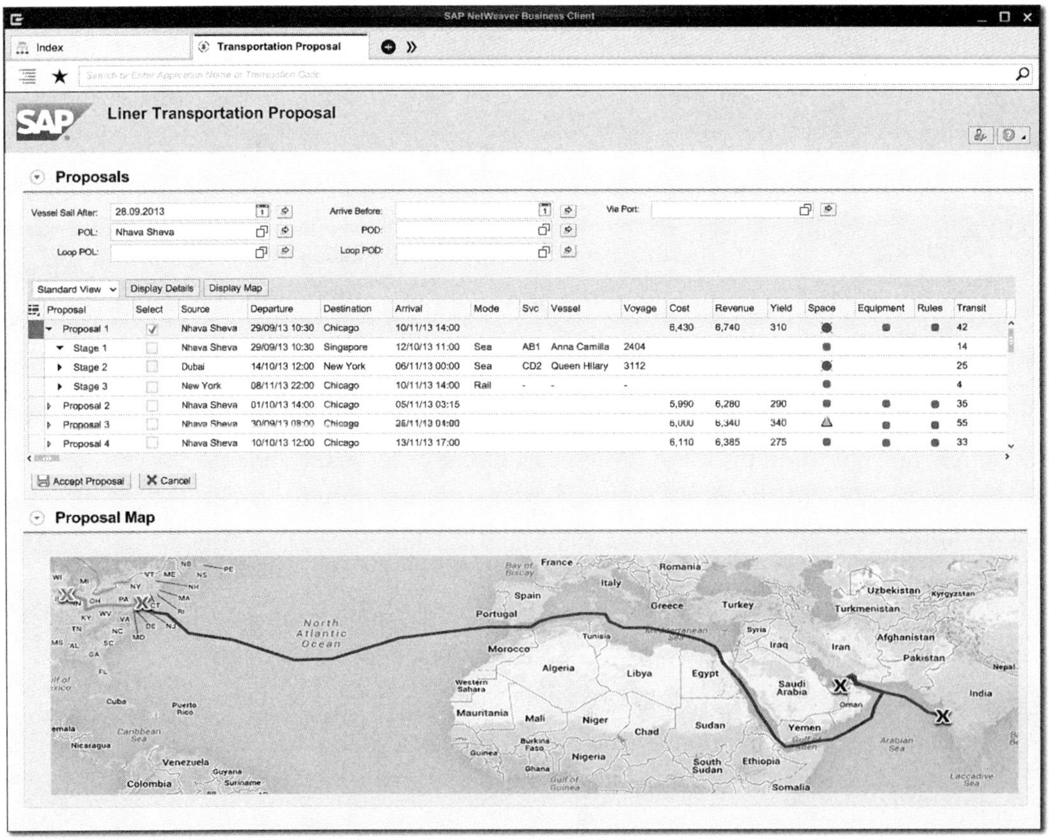

Figure 16.10 Liner Transportation Proposal

Capacity management

In liner operations, capacity management is very important to ensure that the most profitable bookings are given priority by respecting capacity allocations and commitments. The liner solution capacity management offers following functionality:

1. You can generate vessel-specific capacity profiles and decide how much space per equipment group (e.g., dry, reefer) and size (20, 40High Cube) you want to allocate based on the overall capacity of a vessel.
2. You maintain your vessel service master with schedules and voyages, which are required for end-to-end route planning.
3. By assigning a vessel to a voyage, the predefined capacity profile is used for the voyage. You can then create sub-allocations between the different port pairs.
4. Finally you generate capacity allocations for specific voyages and port pairs. Allocations can be for customer commitments, booking agents, alliance partners, and even for empty container repositioning.

The defined allocations consume the space of the vessel-voyage capacity profile. Incoming bookings utilize the allocations. As a result, you have a comprehensive overview of the available space for each voyage, allocation space, and booked space.

Figure 16.11 shows a design layout for a voyage capacity overview. Here a user can gain insights regarding the capacity situation per vessel. This information is dynamically adjusted with changes in allocations or bookings.

Yield optimization

The final enhancement we want to highlight is the capability of yield optimization. The liner solution contains the essential elements to calculate the yield: revenues from contracts and tariffs and costs for the transportation and accessorials. It is important to note that the actual cost data is not suggested or proposed. The yield definition workbench is the cornerstone that is being enhanced to support yield optimization. This allows the flexibility to define mathematical expressions based on SAP TM charge types (refer to Chapter 10). This formula can be plugged into the different areas where the yield needs to be analyzed, such as the Quotation Manager or the voyage proposal cockpit.

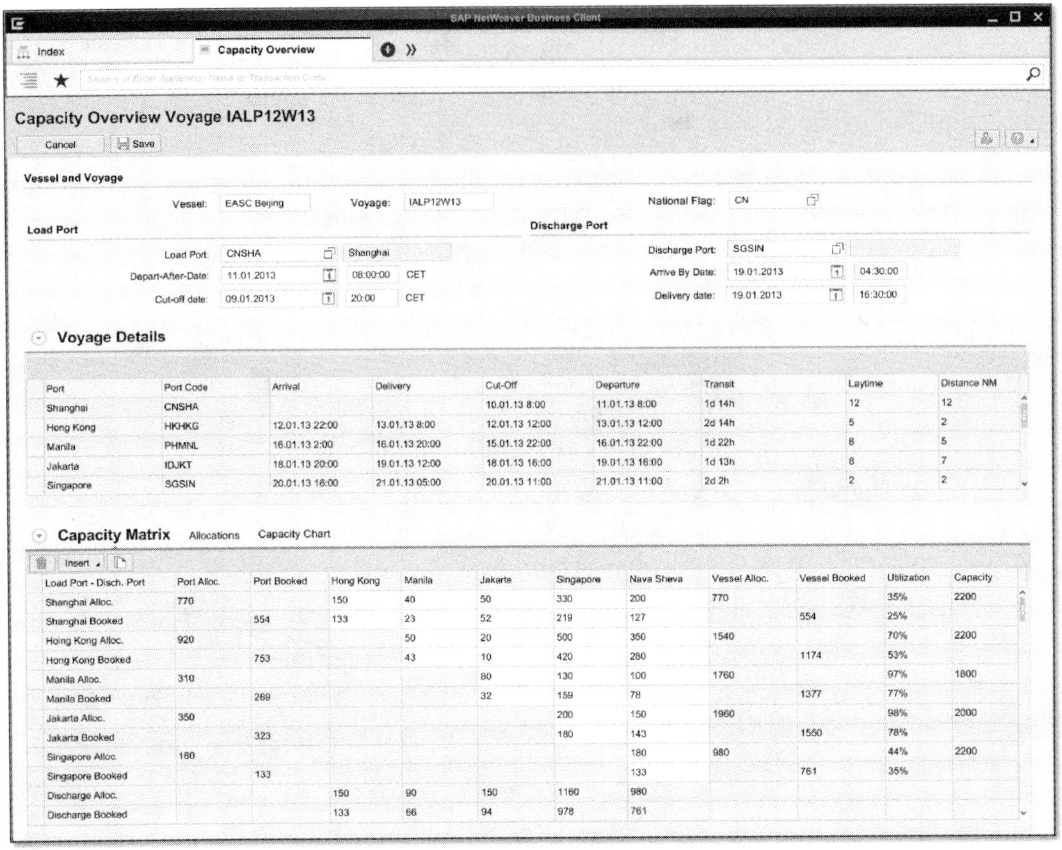

Figure 16.11 Voyage Capacity Overview

16.2.3 Enabling the End-to-End Liner Process

Now that you have learned about the overall solution coverage and specific add-ons and enhancements for the ocean liner solution, let's describe an end-to-end liner process. Figure 16.12 shows a high-level process of some core steps in a liner process from a customer RFP to a paid invoice. Let's assume that the key master data is already available, such as business partner, commodity master, network and schedules, and equipment inventories.

Let's walk through the process now.

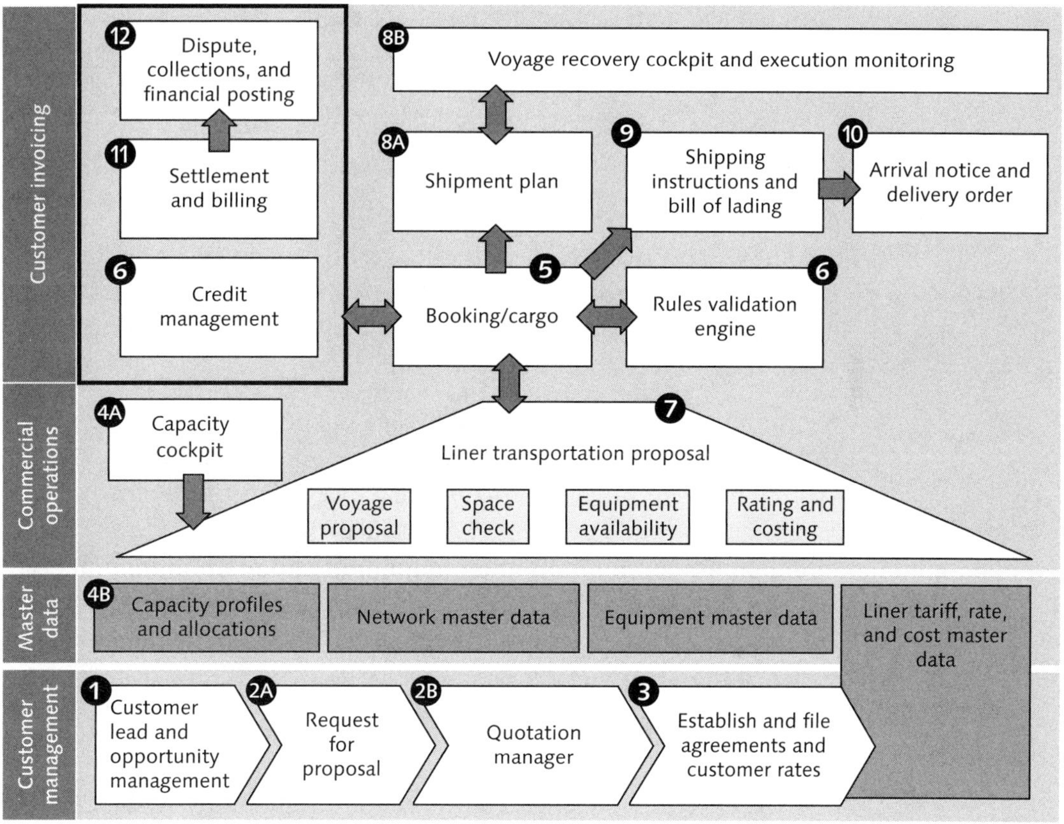

Figure 16.12 High-Level Liner Process Chart

Customer management

The process starts with a new business opportunity that is managed by a sales representative ❶. After a successful sales approach, the customer includes our ocean liner in a request for a proposal. The customer has opened new distribution centers in the United States and requires shipments from its manufacturers in Asia and Europe. The trade lane manager, sales representative, and team of pricing experts use the Quotation Manager to evaluate, analyze, and price the RFP. Based on the volume, market conditions and potential future business with the customer, the pricing and sales team decides to offer customer individual rates instead of applying industry segmented tariff rates.

A new quotation is generated and submitted to the customer ❷. After an evaluation period and further negotiation rounds, the customer agrees to sign a new contract with rates and capacity allocations. Specific service agreements are part of the contract, such as additional free times for detention and demurrage charges or inland services ❸. Based on the new contract, the capacity experts of the ocean liner generate space allocations for the new VIP customer on upcoming voyages ❹.

Commercial operations

After a few weeks, the first bookings are received from the new customer ❺. To evaluate any conflicts with business rules or credit terms, the booking is checked with the Rules Validation Engine, and certain business rules are applied. Neither the credit terms nor any rules prevent the continuation of the process ❻.

A customer agent who is processing the booking uses the liner transportation proposal to check for voyage, space, equipment availability, and yield. A highly profitable departure that meets the critical transit time is found, but no unblocked capacity is available. Due to the customer allocation, an allotment is still reserved for our customer, and consequently, the proposal can be accepted. The equipment is substituted with a different type, free of charge to the customer, to achieve a container re-positioning ❼.

Based on the routing decision, a shipment plan is generated containing the different milestones in the cargo execution and documentation process. The network cockpit is used in order to monitor any incidents in the execution of the transportation and to react with immediate recoveries ❽. Based on a manifest plan, all documents are processed, the shipping instructions are received, a draft and final bill of lading is issued, and the customs clearance is processed ❾. With the cargo arriving at the port of discharge, an arrival notice is issued to the customer. A delivery order is generated but put on hold until the payment of all collect charges has been reported ❿.

Customer invoicing

Based on the booking, a new settlement document is created. In addition to the main transportation service, accessorial services are executed, such as a container cleaning and under-deck stowage, and a vessel is chosen that sails under the American flag. Consequently, different billable items are generated and grouped for invoicing based on the customer's discretion. The corresponding customer invoices are generated ⓫. Last, the collections management is triggered to ensure timely payment. In the event of any disputes, follow-up and clarification is supported. Upon payment confirmation, the delivery order and cargo are released and the customer acknowledges the receipt of the cargo ⓬.

Mirroring the final journey of our container in the simulated liner process, you have also concluded the journey through this book together with us and successfully achieved an overview of SAP Transportation Management 9.1. You have learned about the various functional areas and components of SAP TM, examined the technical foundation, integrated surrounding SAP applications, and explored industry-specific scenarios. In our last chapter, you learned useful best practices for implementing SAP TM and innovative methods for marking use of SAP's extendibility to implement a liner solution.

We hope you have achieved what you desired with this book and found valuable answers to nerve-stretching questions. But most important, we hope we have sparked or encouraged your passion for transportation and logistics with SAP Transportation Management.

The greatest challenge facing a number of managers in recent years has been to regain lost market share and secure new competitive advantages. The impetus for this trend continues to be the ubiquitous tendency toward globalization and the ensuing intensification of international competition.

17 Summary and Outlook

Customer orientation, lean management, and reengineering are the buzzwords that characterize efforts to regain lost market share and secure new competitive advantages. Many companies need to reorganize their value-added processes, paying special attention to the interfaces between the sales and procurement markets, which are increasing in importance. Within this context, hardly a corporate function has grown in significance in recent years as much as logistics.

More and more, logistics is being functionally mapped in standard business software. Accordingly, there is great demand for transportation functionality in connection with the know-how surrounding the implementation of transportation management in complex supply chain systems.

Treated until just a few years ago as an operational aid and an object of isolated rationalization efforts—especially in the age of supply chain management—logistics is now considered an essential element of strategic corporate leadership.

Rationalization potential of logistics

The operational significance of logistics for many companies still lies in its rationalization potential. In general, a reduction of logistics costs should improve corporate success by achieving a competitive advantage. Surveys of businesses have demonstrated that, for the coming years, companies are still counting on a considerable cost-reduction potential of 10-15% of total costs.

Cost component of logistics

This statement does not contradict the fact that the share of logistics costs of many companies was more likely to increase in the past because, for example, it is directly related to whichever operational processes are included in the logistics process. Thus, the scope of logistics in recent years has continually expanded; for example, it now might include production planning or quality control systems. In addition, significant investments are made in IT technology in areas such as supply chain management planning. In the near future, this will lead to a decrease in administrative logistics costs (e.g., through shipment tracking, transport organization, or Internet-based ordering).

Outsourcing logistic functions

In highly competitive supply chain environments, outsourcing traditional core logistic processes, for example, is still a stable trend and a decisive means of cost cutting and competitive differentiation, especially among manufacturers and retailers. In this context, companies expect further savings by subcontracting logistics services (logistics outsourcing) and outsourcing these services to LSPs. In particular, operative logistics tasks such as transport, storage, commissioning, and packaging have already been outsourced to a high degree to external LSPs. Modern logistic solutions will support all aspects of corporate and outsourced logistics: warehousing and transportation, as well as a seamless integration with an organization or its customers' backend systems.

System integration

In today's fast-moving, modern society and the corresponding "I want it now" mentality and expectations of many customers, it is vital for suppliers to have a fully integrated solution. From the perspectives of both the shipper and the LSP, several logistics processes either include interfaces with customers or have effects on the customer. That is why logistics processes must be oriented toward customer needs and performed in a service-friendly manner. In an era when logistics demands are becoming ever more exacting and, by the time the consumer is reached, ever more customized, transportation systems that master these processes to the advantage of their customers will experience a competitive edge that, at least in the short term, cannot be bridged by the competition.

Logistic performance

When the quality of competitors' products continues to become more comparable—and in most industries there is hardly room to lower prices any more—competition takes place on the level of service performance. Within these services, transportation ranks high. Delivery dependability, rapid returns processing, and high-quality of customer service are characteristics with which a company can set itself apart from its competitors.

Conclusions

The commercial world is shrinking as modern communication channels give everyone the ability to share information. Companies are beginning to think in terms of their global logistics operations rather than individual hub operations. They want technology providers to deliver an integrated logistics and supply chain system that is capable of seamlessly delivering goods from one end of the globe to the other. In this context, transportation and logistics is a major growth and priority investment area today. That's no surprise that, given that logistics costs represent around 10% of a country's gross domestic product, in addition they have a major impact on a company's profitability, customer service level, and sustainability balance sheet. Supply chain, transportation, and warehouse managers are now under pressure to automate, transform, and optimize their business processes.

The future of SAP TM

Without sophisticated transportation systems, logistics could not bring its advantages into full play. Besides a good transport system in logistics activities could provide better logistics efficiency, reduce operation costs, and promote service quality. The improvement of transportation systems needs the effort from both customers as well as software vendors.

In this context, SAP's vision is to help companies achieve their supply chain execution excellence, increase transportation and logistics efficiency and productivity, reduce supply chain execution costs, and improve sustainability and compliance.

Following SAP's strategic mission to deliver the world's best supply chain execution platform, SAP TM began a new era in the market for transportation management software and is already a best-in-class solution that supports integrated and connected supply chain execution processes, with a verticalized offering for shippers, freight-forwarders, and carriers.

This book is intended to provide you with a comprehensive reference work and workbook for transportation management with SAP TM. The authors' experience as SAP TM architects, developers, and product managers offers you comprehensive insight into application-oriented process handling, operation, and the technical background of transportation management with SAP. SAP will extend and improve SAP TM with additional versions in the years ahead. Some parts of this book have already pointed to functions of SAP TM 9.1 that enable a major step forward in terms of practical relevance, functionality, and user-friendliness.

Appendices

A Additional References

Bibliography

3PL (2013). "2013 Third-Party Logistics Study: The State of Logistic Outsourcing." *http://www. www.3plstudy.com.*

Chopra and Meindl, *Supply Chain Management: Strategy, Planning, and Operation* (New York: Prentice Hall, 2012).

Ihde, *Transport, Verkehr, Logistik: Gesamtwirtschaftliche Aspekte und einzelwirtschaftliche Handhabung* (Munich: Vahlen, 1991).

Kappauf, Lauterbach, and Koch, *Logistic Core Operations with SAP: Inventory Management, Warehousing, Transportation, and Compliance* (New York: Springer, 2012).

Kappauf, Lauterbach, and Koch, *Logistic Core Operations with SAP: Procurement, Production, and Distribution Logistics* (New York: Springer, 2011).

Recommended Resources

Council of Supply Chain Management: *http://cscmp.org/*

International Chamber of Commerce (ICC): *http://www.iccwbo.org*

International Air Transport Association (IATA): *http://www.iata.org*

SAP Community Network: *http://scn.sap.com*

SAP Help Portal: *http://help.sap.com*

SAP Service Marketplace: *http://service.sap.com*

SAP Store: *https://store.sap.com*

B Abbreviations

A2A	Application to application
ABAP	Advanced Business Application Programming
ABD	Agency business document
ALE	Application Link Enablement
AMS	Automated Manifest System
APO	SAP Advanced Planner and Optimizer
ASEAN	Association of Southeast Asian Nations
B2B	Business to business
B2C	Business to customer
BAdI	Business Add-in
BI	SAP BusinessObjects Business Intelligence
BO	Business object
BOBJ	SAP BusinessObjects
BOPF	Business Objects Processing Framework
BPM	Business process management
BRIM	SAP Billing and Revenue Innovation Management
CAP	Common Agricultural Policy
CASS	Cargo Account Settlement System
CCAD	Charge Correction Advice Document
CDP	Custom development project
CFR	Code of Federal Regulations
CIF	Core Interface Framework
CO-PA	SAP ERP Controlling & Profitability Analysis
CRM	SAP Customer Relationship Management
CRUD	Create, Read, Update, Delete

CSL	Container shipping line
CSR	Cargo sales report
CSV	Comma-separated value (file type)
DCM	Debit/credit memo
DD, D2D	Door to door
DDIC	SAP Data Dictionary
DIT	Domestic inbound transportation scenario
DOT	U.S. Department of Transportation
DOT	Domestic outbound transportation scenario
DP	SAP APO Demand Planning
DSD	Direct store delivery
DSO	DataStore Object
DTR	Delivery-based transportation requirement
ECC	SAP ERP Core Component
EDI	Electronic Data Interchange
EFTA	European Free Trade Association
EhP	Enhancement pack
EHS	SAP Environment, Health, and Safety Management
EP	SAP Enterprise Portal
ERP	SAP Enterprise Resource Planning
ESOA	Enterprise Service-Oriented Architecture
ESR	Enterprise Services Repository
EWM	SAP Extended Warehouse Management
FA	Freight agreement
FB	Freight booking
FBI	Floorplan Manager BOPF Integration
FCL	Full container load

FI-CO	SAP ERP Finance and Controlling
FO	Freight order
FPM	Floorplan Manager
FSD	Freight settlement document
FTL	Full truck load
FU	Freight unit
FUBR	Freight unit building rule
FWA	Forwarding agreement
FWO	Forwarding order
FWQ	Forwarding quotation
FWSD	Forwarding settlement document
GCR	General cargo rates
GIS	Geographical Information Service
GRC	SAP Governance, Risk, and Compliance
GTS	SAP Global Trade Services
GUI	Graphical user interface
GUIBB	Generic user interface building block
HAWB	House airway bill
HBL	House bill of lading
IATA	International Air Transport Association
ICAO	International Civil Aviation Organization
ID	Identifier
IGS	Internal Graphic Server
II	SAP Information Interchange
IIL	International inbound logistics scenario
IMG	Implementation Guide
IMO	International Maritime Organization

IOT	International outbound transportation scenario
IPC	SAP Internet Pricing and Configurator
ISF	Importer Security Filing
LCL	Less than container load
LE	SAP ERP Logistics Execution
LO	SAP ERP Logistic General
LSP	Logistics service provider
LTL	Less than truck load
LVD	Long-term vendor declaration
MAWB	Master airway bill
MDM	SAP NetWeaver Master Data Management
MM	SAP ERP Material Management
MRN	Movement reference number
MTR	Means of transport
NAFTA	North American Free Trade Agreement
NCTS	New Computerized Transit System
NWBC	SAP NetWeaver Business Client
OT	Outsourced transportation scenario
OTR	Order-based transportation requirement
P2P	Peer-to-peer tendering
PDF	Portable document format
PI	SAP NetWeaver Process Integration
PM	SAP ERP Plant Maintenance
POWL	Personal Object Worklist
PP, P2P	Port to port
PPF	Post-Processing Framework
RCS	Repeatable custom solution (SAP custom development)

RDS	Rapid deployment solution
RFC	Remote function call
RFQ	Request for quotation
SCAC	Standard Carrier Alpha Code
SCM	SAP Supply Chain Management
SCR	Specialized commodity rates
SD	SAP ERP Sales and Distribution
SES	Service entry sheet
SICF	Service Interface Component Framework
SLA	Service-level agreement
SLD	System landscape directory
SNP	SAP APO Supply Network Planning
SOA	Service-Oriented Architecture
TAL	Transportation allocations
TCCS	Transportation charge calculation sheet
TD	SAP Transportation and Distribution (Oil & Gas)
TM	SAP Transportation Management
TMS	Transportation management system
TO	Transportation order
TPVS	SAP APO Transportation Planning and Vehicle Scheduling
TRQ	Transportation requirement (business object)
TSW	SAP Traders and Schedulers Workbench (Oil & Gas)
UI	User interface
UIBB	User interface building block
ULD	Unit load device
URL	Uniform resource locator
VB	SAP Visual Business

VSR	Vehicle scheduling and routing
WD	Web Dynpro
WMS	Warehouse management system
XML	Extensible markup language

C The Authors

Dr. Bernd Lauterbach studied and received a PhD in electrical engineering. After spending several years doing Assembler, C /C++, and Occam programming for image processing, GIS, and auto routing systems, he joined SAP in 1995 at the SAP headquarters in Walldorf, Germany as part of the SAP ERP LE-TRA development team. From 2000 to 2007 he was the lead architect for SAP Event Management and SAP Auto ID Infrastructure, and subsequently led development of SAP Transportation Management as a program/development manager.

A second focus of his work is supporting transportation projects of strategic SAP customers from shipper and LSP industries. Since 2008, Bernd has worked as a senior business solution architect in the IBU Travel & Transportation area of SAP responsible for cargo and logistics.

Dominik Mark Metzger studied business management at the Cooperative State University in Mannheim, Germany. After joining SAP AG in 2007, he gained a deep insight into SAP's processes while working in business development, consulting, and SAP Solution Marketing in New York.

Dominik joined the SAP Business Transformation Services in 2010, specializing in the transportation and logistics industry as an SAP TM consultant. After helping win one of the biggest customer projects with SAP TM in Europe, he successfully delivered this demanding implementation project while driving go-to-market and business development activities. Since 2012 he has been managing the first project with SAP TM for an ocean liner in both the proposal and entire implementation as a senior enterprise consultant for SAP in Singapore.

Dominik is highly specialized in the ocean liner, freight forwarding, trucking, and contract logistics industry. He offers both a very strong functional SAP background in and a deep knowledge of many end-to-end business processes. He manages complex implementation projects as a solution architect while supporting market development activities.

Stefan Sauer has studied international business information technology at the Cooperative State University in Mannheim, Germany. After joining SAP in 2007, Stefan gathered experience in development and consulting teams in Germany as well as with SAP Research in Australia.

With the ramp-up of SAP Transportation Management 8.0 in 2010, he accompanied the success of the software product as an SAP TM consultant, being involved in implementation projects in retail and ocean freight industries and participating in customer engagements in other industries such as fashion, rail, air freight, and manufacturing.

Stefan supports projects both from a business process perspective by creating blueprints and implementation concepts, and from a technical perspective as the link between project and custom development. Furthermore, he is a trainer for SAP TM for both functional and technical training courses.

Jens Kappauf is a senior business solution architect for the Travel and Transportation SAP AG Industry Business Unit. In this role, he is responsible for the strategic direction of SAP's third-party and contract logistics solutions, while also focusing on technical integration, deployment, and implementation.

With over 16 years of industry and international consulting experience, he is a hands-on expert on SAP solutions, templates, and global roll-outs, with a special focus on warehouse operations and spare parts logistics, in the area of after-market sales and service operations and supply chain planning, fulfillment, and execution.

Jens holds a degree in business administration, is the author and co-author of several logistics-related books and publications, and lives with his family near Heidelberg, Germany.

Dr. Jens Gottlieb received a PhD in computer science and has worked in software development at SAP AG in Walldorf, Germany since 2000. He has written three books and more than 25 scientific publications in the areas of transportation management and heuristic optimization algorithms, and has edited six books.

Jens was the development architect for the vehicle scheduling and routing optimizer and the carrier selection optimizer, which are used in SAP TM, APO-TPVS and MRS. He headed development projects both in SAP TM and APO-TPVS, served as development manager for SAP TM, and has been a product owner for SAP TM responsible for transportation planning and network.

Dr. Christopher Sürie has worked for more than ten years as an expert consultant in supply chain optimization and transportation management at SAP Deutschland AG & Co KG in Walldorf, Germany. In this position he has been involved in numerous international customer projects, implementing both SAP Advanced Planner and Optimizer and SAP Transportation Management and focusing on their planning capabilities and optimization engines in production and transportation. He worked as a solution architect in several TM projects and taught numerous TM customer trainings.

Prior to joining SAP he worked in the department of production and supply chain management at Darmstadt University of Technology and completed his PhD thesis in the area of production planning for process industries, which was awarded several prizes.

Index

D

G

Q

R

S

T

W

Y

Z